Hybrid
Microelectronics
Handbook

Electronic Packaging and Interconnection Series

Charles M. Harper, Series Advisor

ALVINO • *Plastics for Electronics*

CLASSON • *Surface Mount Technology for Concurrent Engineering and Manufacturing*

GINSBERG AND SCHNOOR • *Multichip Module and Related Technologies*

HARPER • *Electronic Packaging and Interconnection Handbook*

HARPER AND MILLER • *Electronic Packaging, Microelectronics, and Interconnection Dictionary*

HARPER AND SAMPSON • *Electronic Materials and Processes Handbook, 2/e*

LAU • *Ball Grid Array Technology*

LICARI • *Multichip Module Design, Fabrication, and Testing*

Related Books of Interest

BOSWELL • *Subcontracting Electronics*

BOSWELL AND WICKAM • *Surface Mount Guidelines for Process Control, Quality, and Reliability*

BYERS • *Printed Circuit Board Design with Microcomputers*

CAPILLO • *Surface Mount Technology*

CHEN • *Computer Engineering Handbook*

COOMBS • *Electronic Instrument Handbook, 2/e*

COOMBS • *Printed Circuits Handbook, 4/e*

DI GIACOMO • *Digital Bus Handbook*

DI GIACOMO • *VLSI Handbook*

FINK AND CHRISTIANSEN • *Electronics Engineers' Handbook, 3/e*

GINSBERG • *Printed Circuits Design*

JURAN AND GRYNA • *Juran's Quality Control Handbook*

JURGEN • *Automotive Electronics Handbook*

MANKO • *Solders and Soldering, 3/e*

RAO • *Multilevel Interconnect Technology*

SZE • *VLSI Technology*

VAN ZANT • *Microchip Fabrication*

To order or receive additional information on these or any other McGraw-Hill titles, please call 1-800-822-8158 in the United States. In other countries, contact your local McGraw-Hill representative.

BC15XXA

Hybrid Microelectronics Handbook

Jerry E. Sergent

Charles A. Harper

Second Edition

McGraw-Hill, Inc.

New York   San Francisco   Washington, D.C.   Auckland   Bogotá
Caracas   Lisbon   London   Madrid   Mexico City   Milan
Montreal   New Delhi   San Juan   Singapore
Sydney   Tokyo   Toronto

Library of Congress Cataloging-in-Publication Data

Hybrid microelectronics handbook / [edited by] Jerry E. Sergent.
  Charles A. Harper. — 2nd ed.
      p.   cm. — (Electronic packaging and interconnection series)
  2nd ed. of Handbook of thick film hybrid microelectronics /
Charles A. Harper.
  Includes bibliographical references and index.
  ISBN 0-07-026691-3 (acid-free paper)
  1. Microelectronics—Handbooks, manuals, etc. 2. Hybrid
integrated circuits—Design and construction—Handbooks, manuals,
etc. I. Sergent, Jerry E. II. Harper, Charles A. III. Harper,
Charles A. Handbook of thick film hybrid microelectronics.
  IV. Series.
  TK7874.H912 1995
  621.381—dc20   95-10061
                                                       CIP

The first edition was published under the title *Handbook of Thick Film Hybrid
Microelectronics.*

1 2 3 4 5 6 7 8 9 0 AGM/AGM 9 0 0 9 8 7 6 5

ISBN 0-07-026691-3

*The sponsoring editor for this book was Steve Chapman, the editing supervisor
was Frank Kotowski, Jr., and the production supervisor was Pamela A. Pelton. It
was set in Century Schoolbook by Don Feldman of McGraw-Hill's Professional
Publishing composition unit.*
Printed and bound by Quebecor/Martinsburg.

McGraw-Hill books are available at special quantity discounts to use as premi-
ums and sales promotions, or for use in corporate training programs. For more
information, please write to the Director of Special Sales, McGraw-Hill, Inc.,
11 West 19th Street, New York, NY 10011. Or contact your local bookstore.

 This book is printed on recycled, acid-free paper containing 10%
postconsumer waste.

To my family: No author ever had more love and support.   *[J.S.]*

Contents

List of Contributors   xi
Preface   xiii

Chapter 1. Introduction to Hybrid Microelectronics   1-1

    1.1 Interconnection Substrates   1-2
    1.2 Thick Film Technology   1-2
    1.3 Thin Film Technology   1-7
    1.4 Cofired Ceramic and Glass–Geramic Technology   1-11
    1.5 Printed Wiring Board Technology   1-12
    1.6 Laser Trimming   1-15
    1.7 Assembly   1-16
    1.8 Packaging   1-20
    1.9 Advantages of Hybrid Microelectronics   1-20
    1.10 Applications and Examples   1-20
    1.11 Summary   1-28

Chapter 2. Ceramic Substrates for Hybrid Applications   2-1

    2.1 Introduction   2-1
    2.2 Basic Properties of Hybrid Substrate Materials   2-2
    2.3 Ceramic Compositions for Thick Film Circuits   2-17
    2.4 Ceramic Substrate Manufacturing Methods   2-20
    2.5 Substrate Cleaning   2-25
    2.6 Postfire Processing   2-26
    2.7 Substrate Sources and Costs   2-33
    References   2-34

Chapter 3. Thick Film Materials and Processes   3-1

    3.1 Introduction   3-1
    3.2 Thick Film Conductors   3-1
    3.3 Thick Film Dielectrics   3-38
    3.4 Resistor Materials and Processing   3-69
    3.5 Rheology and the Screen Printing Process   3-103
    3.6 Quality Control and Manufacturing Processes   3-137

    3.7 Nonhybrid Applications   3-141
    References   3-143

Chapter 4. Thin Film Materials and Processing   4-1

    4.1 Introduction   4-1
    4.2 Thin Film Deposition Techniques   4-1
    4.3 Substrate Materials   4-8
    4.4 Conductor Materials   4-10
    4.5 Resistor Materials   4-11
    4.6 Capacitor Materials   4-15
    4.7 Inductor Materials   4-18
    4.8 Emerging Thin Film Materials   4-18
    4.9 Thin Film Design Guidelines   4-23
    4.10 Fabrication Sequence for Thin Film Resistor-Conductor Circuits 4-31
    4.11 Characterization of Thin Films   4-34
    4.12 Emerging Technology: Multichip Module Deposited (MCM-D)   4-37
    4.13 Conclusion   4-41
    Appendix A Deposition Techniques   4-42
    Appendix B Metal Etchants for Thin Film   4-53
    References   4-53

Chapter 5. Pure Copper Metallization Technologies   5-1

    5.1 Introduction   5-1
    5.2 Substrates   5-4
    5.3 Direct Bond Copper Technology   5-7
    5.4 Plated-Copper Technology   5-17
    5.5 Active Metal Brazing Copper Technology   5-21
    5.6 Comparisons   5-27
    5.7 Assembly   5-30
    5.8 Thermal Performance   5-36
    5.9 Electrical Performance   5-39
    5.10 Reliability   5-41
    5.11 Applications   5-48
    References   5-52

Chapter 6. Assembly of Hybrid Microcircuits   6-1

    6.1 Introduction   6-1
    6.2 The Decision Support System   6-5
    6.3 The Solder Joint   6-10
    6.4 Hybrid Assembly Using the Surface Mount Technology   6-17
    6.5 Cleaning   6-29
    6.6 Repair and Rework   6-41
    6.7 Component Attachment   6-46
    6.8 Intermetallic Formation   6-52
    6.9 Wire Bonding   6-62
    6.10 TAB and Flip-Chip Bonding   6-73
    6.11 Statistical Process Control   6-80
    6.12 Factory Automation   6-90
    Bibliography   6-99

Chapter 7. Electronic Packaging of Hybrid Circuits   7-1

7.1 Introduction   7-1
7.2 Requirements of Electronic Packaging   7-2
7.3 Types of Electronic Packages   7-7
7.4 Material Considerations for Electronic Packaging   7-15
7.5 Electronic Package Fabrication   7-19
7.6 Package Sealing   7-23
7.7 Evaluation of Electronic Packages   7-26
7.8 Applications   7-42
7.9 Conclusion   7-43
References   7-44

Chapter 8. Discrete Passive Components for Hybrid Circuits   8-1

8.1 Introduction   8-1
8.2 Capacitors   8-1
8.3 Inductors and Transformers   8-25
8.4 Resistors   8-32

Chapter 9. Cleanrooms for Hybrid Manufacturing   9-1

9.1 Introduction   9-1
9.2 Cleanroom Design   9-4
9.3 Layout Considerations   9-27
9.4 Cleanroom Management   9-44
References   9-60

Chapter 10. Failure Analysis of Hybrid Microelectronics   10-1

10.1 Introduction   10-1
10.2 Who Should Perform a Failure Analysis?   10-3
10.3 What You Should Expect to Get Out of a Failure Analysis   10-5
10.4 Why Hybrid Circuit Failure Analysis is Different from Other Microelectronic Failure Analysis   10-5
10.5 Analytical Approach to Failure Analysis   10-10
10.6 Failure Analysis Documentation and Reports   10-17
10.7 Failure Analysis Tools   10-18
10.8 Typical Hybrid Failure Modes   10-35
10.9 Typical Failure Analysis Examples   10-36
10.10 Failure Analysis in the Future   10-43
10.11 Summary   10-45
Appendix 10.1 Sample: Basic Failure Analysis Report   10-46
Appendix 10.2 Sources of Additional Reading on Failure Analysis   10-53

Chapter 11. Design Methods for Hybrid Circuits   11-1

11.1 Introduction   11-1
11.2 The Design Process   11-2
11.3 The Design Sequence   11-3
11.4 Design of Thick Film Resistors   11-11
11.5 Design of Thin Film Resistors   11-20
11.6 Electrical Considerations in Hybrid Design   11-25

x   Contents

11.7 Themal Considerations   11-32
11.8 Determination of Conductor Width   11-59
11.9 Determination of Wirebond Size   11-63
11.10 Selection of Substrate Technology   11-64
11.11 Design Guidelines   11-67
11.12 Assembly Materials and Process Selection   11-72
11.13 Reliability Considerations   11-74
11.14 Summary   11-103
References   11-103

Index   1

Contributors

Fred D. Barlow   *Bradley Department of Electrical Engineering, Virginia Polytechnic University, Blacksburg, Virginia (CHAP. 4)*

Imran A. Bhutta   *Bradley Department of Electrical Engineering, Virginia Polytechnic University, Blacksburg, Virginia (CHAP. 7)*

John Buono   *Analytical Answers, Inc., Woburn, Massachusetts (CHAP. 10)*

Roger Cadenhead   *Communication Associates, Anniston, Alabama (CHAP. 6)*

Aicha Elshabini-Riad   *Bradley Department of Electrical Engineering, Virginia Polytechnic University, Blacksburg, Virginia (CHAPS. 4, 7)*

Thomas C. Evans   *Thomas C. Evans Consulting, LaPorte, Colorado (CHAP. 5)*

Dana Hankey   *Electronics Materials Division, Ferro Corporation, Santa Barbara, California (CHAP. 3)*

R. S. ("Sig") Jensen   *Applied Laser Technology, Beaverton, Oregon (CHAP. 2)*

R. Wayne Johnson   *Alabama Microelectronics, Science, and Technology Center, Auburn University, Alabama (CHAP. 1)*

Chandra S. Khadilkar   *Electronic Materials Division, Ferro Corporation, Santa Barbara, California (CHAP. 3)*

Jerry Sergent   *BBS PowerMod, Victor, New York (CHAPS. 7, 8, 11)*

Asiz S. Shaikh   *Electronic Materials Division, Ferro Corporation, Santa Barbara, California (CHAP. 3)*

Roy Trowbridge   *Xerox, Inc. (CHAP. 9)*

S. Vasudevan   *Electronic Materials Division, Ferro Corporation, Santa Barbara, California (CHAP. 3)*

Preface

The first edition of this book, the *Handbook of Thick Film Hybrid Micro-electronics*, was published in 1974 when the hybrid microelectronics industry was relatively young, and represented, at the time, the most comprehensive reference available. Today, 21 years later, the industry has changed drama-tically, and the second edition, titled the *Hybrid Microelectronics Handbook*, is intended to fill the same role. Even the term "hybrid microcircuit" has expanded in scope. Previously limited to only thick and thin film hybrid microcircuits, the hybrid technology encompasses new metallization technolo-gies, such as direct bond copper (DBC), and new substrate materials, such as aluminum nitride.

Certain aspects of the technology that merited entire chapters in the first edition have matured to the point that they may be addressed as subtopics in more comprehensive chapters, allowing more space to cover new technologies and materials.

This edition consists of all new material by authors who are recognized experts in their respective fields, and is intended to be a practical reference for those who are active in all aspects of the hybrid technology. For the conve-nience of the reader, the authors attempted to write each chapter as an inde-pendent, "stand-alone" document, with all the pertinent data included for that topic. The result was a small degree of redundancy, with some material appearing in two different chapters.

Chapter 1 is an introduction to the hybrid microelectronics technology and its appplications, and provides a broad overview of the remainder of the book. Chapter 2 describes certain of the ceramic substrate materials and provides a valuable insight into laser machining and how this approach can be used to the designer's advantage. Chapters 3 through 5 consider methods of metalliz-ing substrates. Chapter 3 is an in-depth treatise of how thick film materials are manufactured and processed, and deals extensively with applications beyond the traditional conductor, resistor, and dielectric materials. Chapter 4 considers the thin film technology, which was not covered in the first edition. Methods of depositing and etching a wide variety of materials are covered, including superconducting materials. Chapter 5, which describes pure copper metallization processes, is also a new chapter. This technology has led to many new applications for hybrid circuits in the power and high frequency markets. Chapter 6 discusses the assembly processes used to attach components to metallized substrates. The topic of cleaning in light of the Montreal Protocol is extensively considered. Chapter 7 describes the methods by which hybrid microcircuits are packaged, and provides data on the electrical and thermal characteristics of common package types. Chapter 8 deals with the passive components used in hybrid circuits, and is a convenient source for the designer when selecting a component for a specific application. Chapter 9 describes cleanroom construction and practices. Rarely considered in this detail in other books, this topic is one of the most important aspects of reliability. Chapter 10, on failure analysis, provides a description of modern methods of failure analysis of hybrid circuits. By their nature, hybrid circuits are small and rely on material interactions to perform. A well-performed failure analysis can quickly isolate problems and eliminate guesswork, saving both time and money. Chapter 11, on the topic of design, provides a systematic approach to the overall design process. Design guidelines for manufacturability and reliability of both thick and thin film hybrid circuits are presented, along with methods and tables for calculating wire bond and conductor sizes. A method for calculating the thermal resistance of various structures, considering both conduction and convection, which allows a quick determination of the approximate device temperature is also provided. This method is presented in a simple, easy-to-follow, step-by-step manner which allows the designer to determine whether or not a more exact method is necessary.

It is the hope of the editors that this book will provide a handy reference for the majority of questions which arise during the course of the hybrid design process, and will also provide additional sources of information which are too extensive to cover in a volume of this size.

Jerry E. Sergent
Charles A. Harper

Hybrid
Microelectronics
Handbook

1

Introduction to Hybrid Microelectronics

R. Wayne Johnson

Hybrid microelectronics is a packaging and interconnection technology for combining two or more semiconductor devices on a common interconnect substrate, typically to create a specific electrical function. The semiconductor devices may be in the form of bare, unpackaged die or in miniature packages which mount on the interconnect substrate surface. The interconnection pattern may include deposited resistors, capacitors, and inductors, or these passive components may be mounted in chip form on the surface of the substrate. Figure 1.1 illustrates this interconnection concept. The basic interconnect substrate and assembly techniques used in hybrid microelectronics are listed in Table 1.1. The assembled hybrid microcircuit may be packaged in a metal,

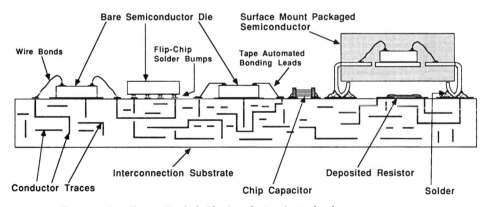

Figure 1.1   Cross section illustrating hybrid microelectronics technology.

TABLE 1.1   Hybrid Microelectronics Substrate and Assembly Techniques

Interconnection substrate	Assembly
Thick film	Chip and wire
Thin film	Flip-chip solder attachment
Cofired ceramic or glass–ceramic packaged devices	Solder reflow of surface mount
Printed wiring board	Tape automated bonding

ceramic, or plastic package coated with a protective coating or may require no additional packaging depending on the construction and the application.

1.1   Interconnection Substrates

The substrate provides (1) conductive traces for electrical interconnection between the various active and passive elements, (2) mechanical support for the components, and (3) a path for heat removal from the devices. The substrate contains alternating conductor and dielectric layers. Layer-to-layer conductor connections are created using vias in the dielectric. In addition, resistive layers may be included to form resistors. Typical materials and processes for the various interconnection substrate technologies are presented in Table 1.2, and typical properties are given in Table 1.3.

Thick and thin film hybrids utilize a base substrate on which the conductor, dielectric, and resistor layers are deposited. Typical base substrate properties are given in Table 1.4. Alumina is the most widely used base substrate material. Substrates for thick film hybrids are 96% Al_2O_3 and thin film hybrid substrates are 99.5% Al_2O_3. Alumina combines a reasonable thermal conductivity and coefficient of thermal expansion (CTE) with high temperature process compatibility, high strength, and low cost. Aluminum nitride and beryllia are used in applications requiring high thermal conductivity. The CTE of aluminum nitride nearly matches that of silicon, while the CTE of alumina nearly matches gallium arsenide.

1.2   Thick Film Technology

In the thick film process, the individual layer is deposited by screen printing as illustrated in Fig. 1.2. The thick film material to be printed is referred to as an ink or paste. The ink contains three components: a functional phase which defines the electrical properties of the fired film, a binder which provides adhesion between the fired film and the substrate, and the vehicle which establishes the printing characteristics.

For conductors the functional phase may be gold, silver, copper, palladium–silver, platinum–silver, palladium–gold, or platinum–gold. Other conductor metallurgies have also been used. The metallurgy chosen depends on a number of engineering factors including wire bondability, solderability, environmental requirements, electrical conductivity, and cost. A comparison of conductors is provided in Table 1.5.

TABLE 1.2 Summary of Typical Materials and Processes Used to Fabricate Interconnection Substrates

Materials	Thick film	Thin film	Cofired glass–ceramic	Cofired ceramic	Printed wiring board
Base substrates	Al_2O_3, AlN, BeO	Al_2O_3, AlN, BeO, Si, Cu, glass/ceramic	N/A	N/A	N/A
Conductors	Au, PdAu, PtAu, Ag, PdAg, PtAg, PtPdAg, Cu	Au, Al, Cu	Au, Ag, PdAgCu	W, Mo	Cu
Dielectrics	Glass–ceramics, recrystallizing glasses	SiO_2, polyimide, benzocyclobutene	Glass–ceramic tape	Ceramic (Al_2O_3) tape	Epoxy–glass, polyimide–Kevlar, BT-epoxy glass, cyanate ester–glass laminate sheets
Resistors	RuO_2 doped glass	NiCr, TaN	RuO_2 doped glass	N/A	N/A (carbon-doped polymers)
Processes	Sequentially print, dry, and fire conductor, dielectric, and resistor pastes	Sequentially vacuum deposit, spin coat, and/or plate conductors, dielectrics, and resistors; photolithography; etch	Punch vias, print and dry conductors on tape, collate layers, laminate, cofire	Punch vias, print and dry conductors on tape, collate layers, laminate, cofire	Photolithography, etch, collate sheets, laminate, drill vias, plate

TABLE 1.3   Summary of Typical Interconnection Substrate Properties

	Thick film	Thin film	Cofired ceramic	Cofired glass–ceramic	Printed wiring board
Linewidth, μm min	125	10	100	100	75
Via diameter, μm min	250	15	125	175	200
Number of metal layers (range)	1–6	1–8	1–100	1–100	1–50
Conductor resistance, mΩ/sq	2–100	3–35	8–12	3–20	0.15–1
Relative permeability:					
Substrate	7–10	4–12	N/A	N/A	N/A
Dielectric	5–9	2.8–4	9–10	5–8	4–5
Resistor values	0.1–1 GΩ	0.1–100 kΩ	N/A	0.1–1 MΩ	N/A
CTE, ppm	4–7.5	3–7.5	6	3–8	4–16
Thermal conductivity:					
Substrate	25–260	25–260	N/A	N/A	N/A
Dielectric, W/(m·°C)	2	0.15–1	16–20	1.2–2.5	0.15–0.35
Relative cost:					
Low volume	Medium	High	High	High	Medium
High volume	Medium	Medium	Medium	Medium	Low
Tooling and setup costs	Low	Medium	High	High	Low
Capital outlay	Low	Medium	High	Medium	Low

TABLE 1.4   Material Properties for Common Base Substrate Materials Used in Thick and Thin Film Hybrid Circuits

	Si	Al_2O_3 (99.5%)	Al_2O_3 (96%)	BeO	SiC	AlN	Glass–ceramic
Electrical:							
Volume resistivity, Ω cm	10^4–10^6	$>10^{14}$	$>10^{14}$	$>10^{15}$	$>10^{11}$	$>10^{14}$	$>10^{14}$
Dielectric constant at 1 MHz	11.9	9.9	8.9–10.2	6.5	40	8.8	5–8
Dielectric loss at 1 MHz	N/A	<0.0004	0.001	0.004	0.05	<0.001	0.002
Dielectric strength, kV/mm	N/A	25	23	10	1.0	14	1.5
Thermal:							
Thermal conductivity, W/m·K	150	25	20	250	70–270	140–220	1–4
CTE from 25–400°C, ppm	3.5	6.5	7.1	9.0	3.8	4.1	3–8
Specific heat, J/g·K			0.88	0.88	1.0	0.8	0.7
Mechanical:							
Density, g/cm³	2.33	3.89	3.75	2.85	3.2	3.26	2.9
Grain size, μm	Single crystal	1–5	5	9–16	2–5	5–10	1–5
Compressive strength, kpsi		375	340	225	560	300	
Tensile strength, psi		30	25	22	26	28	10–15
Bonding strength, kg/mm²	10–50	25–35	25–35		45	40–50	10–20
Young's modulus, Mpsi		23.5	40	50	59	40	15–20
Poisson ratio	0.2782	0.22	0.22	0.26	0.15	0.25	0.26
Sintering temp., °C	1412 (MP)	1300	1300		2250	1900	850
Max. use temp. (nonoxidizing atmosphere), °C	1400	1500	1500	1500	1900	1800	500

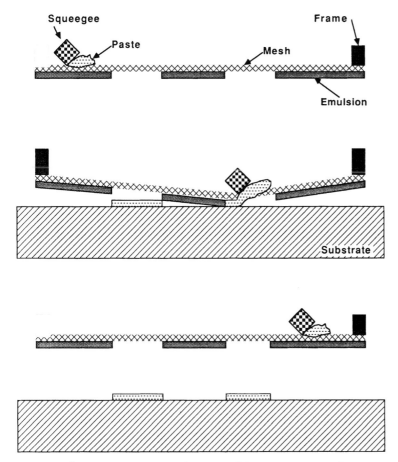

Figure 1.2 Screen printing process for material deposition onto a substrate.

TABLE 1.5   Thick Film Conductor Properties

| | Die bond methods | | | Wire bondability | | | | |
Material	Eutectic	Solder	Organic	Gold	Aluminum	Solderability	Solder leach resistance	Corrosion resistance
Au	Good	Poor	Excellent	Excellent	Good	Poor	Poor	Excellent
PtAu	NG	Good	Excellent	Fair–poor	Fair	Excellent	Good	Excellent
PdAu	NG	Fair	Excellent	Fair–poor	Fair	Good	Good	Excellent
Ag	NG	Good	Excellent	Good	NG	Good	Fair	Poor
PtAg	NG	Good	Excellent	Good	NG	Good	Good	Good
PdAg	NG	Good	Excellent	Good	Good	Good	Good	Good
PdPtAg	NG	Good	Excellent	Good	Good	Good	Good	Good
Cu	NG	Good	Excellent	Fair	Fair	Excellent	Excellent	Poor
Ni	NG	NG	Excellent	NG	NG	NG	NG	Excellent
Polymer	NG	NG	Excellent	NG	NG	Fair	Poor	Good

Resistor systems are formulated with ruthenium (RuO_2, $Bi_2Ru_2O_7$, etc.) doped glasses. Thick film resistor pastes provide a wide range of resistor values by varying the doping level of the glass. Key resistor properties are value, temperature coefficient of resistance, and stability. Resistor formulations exist for general purpose, high voltage, and potentiometer applications. Although thick film resistors have been used for decades, the exact conduction mechanism for thick film resistors has never been fully explained.

Dielectrics serve three primary roles in thick film technology: insulation between conductor layers, formation of capacitors, and encapsulation of the hybrid substrate. Dielectrics for insulation are typically glass–ceramic compositions formulated for a low dielectric constant, low dissipation factor, high voltage breakdown strength, high insulation resistance, and a CTE matched to the substrate to minimize bowing of large area substrates with many layers. Screen-printed parallel plate capacitors are not widely used. High permittivity dielectric materials are difficult to achieve; therefore, large value capacitors will require excessive substrate area. Barium titanate is commonly used as the functional phase to achieve higher dielectric constant inks ($k = 150$). It is also difficult to maintain tight tolerances of the capacitance value. Thick film dielectric encapsulants are formulated with low melting point glasses to allow the encapsulant to be fired at 500°C. This minimizes the change in resistor values due to multiple firings of the resistor at high temperatures (850°C). The encapsulant should be pin hole–free and compatible with laser trimming of resistors under it.

The glass in resistors and dielectrics also serves as the binder which adheres the film to the substrate through substrate wetting during the firing process. Conductors, however, require the addition of a binder to the ink. Glass, oxide, and mixed (glass and oxide) bonded conductors are used. Some glass bonded systems use a significant quantity of glass which, depending on processing conditions, can float to the top during the firing process, impairing soldering and wire bonding. Oxide bonded conductors may contain CuO or other metal oxides which chemically react with the substrate during firing. These inks generally require firing temperatures of 900°C or higher for the chemical reaction. Mixed bonded conductors use a small quantity of glass and the metal oxide. The firing temperature for mixed bonded conductors is 850°C, which has evolved as the industry standard firing temperature.

The final component of the ink, the vehicle, is a solution of polymer resin dissolved in a volatile solvent. Wetting agents are also included so the organic vehicle will wet the inorganic metal and glass powders. The vehicle, in large part, determines the printing and drying characteristics of the ink.

After printing, the inks are typically dried at 150°C for 15 min to remove the volatile solvent component of the vehicle. Next, the ink is fired through a furnace with a temperature profile which includes 10 min at a peak temperature of 850°C and an overall firing cycle time of 30 to 60 min. During the early portion of the profile (300 to 500°C) the nonvolatile resin portion of the vehicle is pyrolyzed. In the temperature range 600 to 850°C, the glass flows, sintering of particles occurs, and chemical reactions take place to form the final film and to provide adhesion to the substrate. Precious metal conductors are fired in air while copper conductors require firing in a nitrogen furnace atmosphere. The print, dry, and fire steps are repeated to fabricate the interconnection structure. The thick film process can be automated for cost effective, high volume production.

1.3 Thin Film Technology

Thin film materials are deposited by vacuum deposition (evaporation, sputtering, chemical vapor deposition), spin coating, and plating. The patterns are formed by photolithography and etching. The process is shown in Fig. 1.3. The resulting fine-line patterns are particularly suited to high density interconnections and high frequency applications. In deposition by evaporation, the material to be deposited is heated under vacuum to create a vapor of the material which is deposited on all surfaces in the vacuum chamber—including the substrate. The heating may occur by passing current through a filament coated with the source material, passing current through a refractory metal boat containing the source material, or directing an electron beam into

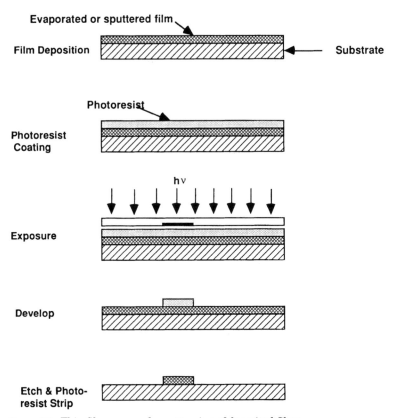

Figure 1.3 Thin film process for patterning of deposited films.

a crucible containing the source material. Alloys such as NiCr are generally not deposited by evaporation owing to the different vapor pressures of the constituents at a given temperature.

Sputtering is the most widely used deposition technique in thin film processing. It is a plasma process in which argon ions are created and accelerated toward a target of the material to be deposited. Upon impact, target atoms are ejected or sputtered from the target surface. These sputtered atoms then coat the substrate. DC magnetron sputtering is widely used for metal deposition. Magnets increase the plasma density, increasing the deposition rate. Dielectrics are deposited by rf or rf magnetron sputtering to prevent charge buildup on the insulating target. Reactive sputtering can be used to deposit compounds such as TaN, which is deposited by sputtering a Ta target while maintaining a nitrogen partial pressure in the vacuum chamber.

In chemical vapor deposition (CVD), a source vapor (for example, SiH_4 and NH_3) is decomposed through thermal and/or plasma energy to deposit the desired material (Si_3N_4). CVD systems are commonly used in the semiconductor industry and can be used to deposit thin film dielectrics.

Liquid organic precursors and solutions of polymers may be dispensed onto substrates, which are then spun at 1000 to 5000 rpm to produce a uniform coating. The coating is subsequently thermally treated to remove the solvent and/or to initiate the chemical reaction in the case of an organic precursor. Both organic and inorganic dielectrics can be deposited in this manner.

Electroplating is often used to increase conductor thickness. Electroplating is the electrodeposition of an adherent metallic coating upon an electrode (substrate conductor) by electrolysis—the production of chemical changes by the passage of current. Electroless plating, the deposition of a metallic coating by a controlled chemical reduction that is catalyzed by the metal or alloy being deposited, is also used in thin film processing.

With the possible exception of electroplating, in which the conductor may be plated in the desired pattern, the deposited layer must be patterned. For conductors, resistors, and some dielectrics (silicon dioxide and certain polymers) this is generally accomplished by photolithography and wet chemical etching. The process sequence is: spin coat a layer of photoresist on the substrate, soft bake to remove the solvent from the photoresist, expose the photoresist to a uv light source through the appropriate photomask, develop the photoresist, hard bake to improve the chemical resistance of the photoresist, then wet chemically etch the material in the appropriate etchant, as illustrated in Fig. 1.3. Some polymer dielectrics are photosensitive and can be patterned like photoresist. Many polymer dielectrics are not wet chemically etchable and require plasma or reactive ion etching.

Typical thin film conductor properties are given in Table 1.6. Aluminum is widely used in the semiconductor industry for interconnection on the integrated circuit. Because of its higher resistivity, aluminum is of limited use in thin film hybrids. Gold and copper are commonly used. However, they do not adhere to substrate and dielectric materials and require an adhesion layer of chrome or titanium. The resistivity of the adhesion layer must be considered

TABLE 1.6   Properties of Common Thin Film Metallizations

	Material		
Property	Aluminum	Copper	Gold
Resistivity	2.66 mΩ cm	1.67 mΩ cm	2.35 mΩ cm
Adhesion	Good	Ti or Cr adhesion layer used	Cr, TaN, or NiCr adhesion layer used
Deposition	Sputtering	Sputtering and plating	Sputtering and plating
Corrosion	Corrodes (H_2O and Cl)	Corrodes without barrier layers	Does not corrode
Special notes	Easiest to process	Reacts with polyamic acid— requires barrier layer (Ni or Cr)	

in design and processing of microwave hybrids because of the skin effect at high frequencies.

Multilayer thin film hybrids, also known as deposited multichip modules (MCM-D), are finding widespread acceptance as a means of interconnecting high density digital devices. Currently, polyimide is the most widely used dielectric for these multilayer structures. Typical dielectric properties are given in Table 1.7. Dielectric selection involves a number of trade-offs including dielectric constant, high temperature stability, as-deposited film stress, moisture absorption, and processibility.

Thin film resistor properties are shown in Table 1.8. NiCr resistors offer the lowest temperature coefficient of resistance but must be encapsulated to

TABLE 1.8   Thin Film Resistor Properties

	NiCr	TaN
Sheet resistance	25–300 Ω/sq	20–150 Ω/sq
Sheet resistance tolerance	± 10% of nominal	± 10% of nominal
TCR, ppm/°C	0 ± 50 ppm/°C 0 ± 50 ppm/°C with special anneal	−75 ± 50 ppm/°C 0 ± 25 ppm/°C with vacuum anneal
TCR tracking (−55 to +125°C)	2 ppm	<2 ppm
Resistor drift (1000 h at 150°C)	<2000 ppm <1000 ppm with special anneal <200 ppm, sputtered films with 350°C anneal	<1000 ppm
Ratio tracking	5 ppm	5 ppm
Resistance tolerance after anneal and laser trimming	± 0.1%	± 0.1% standard ± 0.03% bridge trim
Noise (100 Hz to 1 MHz)	<−35 dB	<−40 dB

TABLE 1.7 Summary of Thin Film Interlayer Dielectric Properties

Property	Standard polyimide	Fluorinated polyimide	Silicone polyimide	Acetylene terminated polyimide	Low stress polyimide	Benzocyclo-butene	Polyphenyl-quinoxaline	Silicon dioxide
Dielectric constant at 1 kHz	3.4–3.8	2.7–3.0	3.0–3.5	2.8–3.2	2.9 (z)—3.4 (xy)	2.7	2.7	3.75
Dissipation factor at 1 kHz	0.002	0.002	0.002	0.002	0.002	0.0006	0.0005	0.0001
Decomposition temp., °C	520–550	>470	450	500–520	620–650	430	500	>1000
Glass transition temp., °C	300–320	<300	<300	215–230	>400	350–360	365	N/A
CTE, ppm	20–40			38	3–6	45–70	55	0.5
Moisture absorption, % wt	1.1–3.5	0.7	0.9	0.8–1.3	0.4–0.5	0.3–0.5	0.9	<0.1
Young's modulus, kpsi	400–1000		50–250	300–450	1280	340		10,000
Stress, dyn/cm × 10^8	3	5.3	3	4	0.4–0.5	3.7		± 20
Degree of planarization, %	24–34	25–30	22	91–93	18–28	91		0–5
Percent solids	14–17	15–19	26	35	10–13.5	35–62	35–47	N/A
Viscosity, poise	11–19	15–25	12	15	20–110	0.15–1.5	11–15	N/A
Deposition methods	Spin coat Spray	Spin coat Spray	Spin coat Spray	Spin coat Spray	Spin coat Spray	Spin coat Spray	Spin coat	CVD
Curing temp., °C	350–400	350–400	300–400	300–350	350–400	230–250	400–450	300–400

prevent corrosion. A Ta_2O_5 passivation layer can be formed on the surface of TaN resistors by heat treating in air to provide environmental protection of the resistor. Thin film resistors are very stable and thin film hybrids are often used in high precision analog applications.

1.4 Cofired Ceramic and Glass–Ceramic Technology

Cofired ceramic substrates evolved from multilayer ceramic capacitor and ceramic single chip package manufacturing technologies. The process is similar to thick film processing with two exceptions: no base substrate is used and an unfired ceramic tape is used for the inner layer dielectric instead of a printed dielectric paste. The process flow is shown in Fig. 1.4. Vias in the tape are formed by punching and are filled with conductive paste by screen printing to provide electrical connections in the z axis. Conductor patterns are screen printed onto the unfired, ceramic green tape—one conductor layer

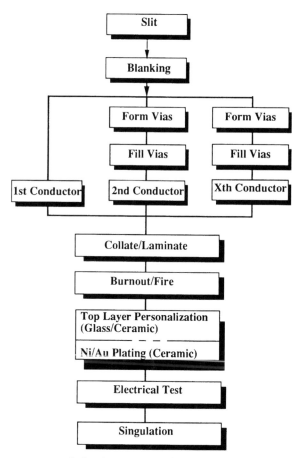

Figure 1.4 Cofired ceramic and cofired glass–ceramic process.

per tape. The printed layers (tape) are collated, laminated, and fired to produce a sintered, monolithic substrate. On firing, the substrate shrinks; this dimensional change must be considered and accounted for in the design. The parallel processing (and inspection) of the individual layers prior to lamination and firing allows substrates to be fabricated with many layers at acceptable yields. The resulting substrates provide high interconnection densities.

The fabrication process for cofired glass–ceramic substrates is similar to that for cofired ceramics except that the firing temperatures (850 to 900°C vs. 1500 to 1600°C) are lower. This allows higher conductivity conductor metallurgies to be used. In addition, engineering of the glass–ceramic composition permits tailoring of specific properties, for example, dielectric constant or coefficient of thermal expansion. Conductor and dielectric tape properties are given in Table 1.9. Thick film resistors can be incorporated with cofired glass–ceramic substrates, either buried in the substrate or deposited on the surface.

Cofired ceramic substrates are compatible with subsequent thin film processing, while additional layers can be added to a cofired glass–ceramic substrate by either thin film or thick film processing. This allows the customization of technologies for specific applications. Cofired ceramic and glass–ceramic substrates are also referred to as ceramic multichip modules (MCM-C) in the microelectronics industry.

1.5   Printed Wiring Board Technology

Printed wiring boards are constructed with reinforced organic dielectric sheets. Typical dielectric properties are given in Table 1.10. The conductor

TABLE 1.9   Typical Cofired Ceramic and Glass–Ceramic Properties

	Cofired cerami (Al_2O_3)	Cofired glass–ceramic
Conductor resistance, mΩ/sq	8–12	3–20
Relative permeability at 1 MHz	9–10	5–8
Dissipation factor at 1 MHz	$5–15 \times 10^{-4}$	$15–30 \times 10^{-4}$
Insulation resistance, Ω cm	$>10^{14}$	$10^{12}–10^{15}$
Breakdown voltage	550 V/25 mm	800 V/25 mm
Resistor values (range)	N/A	0.1 Ω–1 MΩ
Coefficient of thermal expansion, ppm/°C	6.5	3–8
Thermal conductivity, W/m•°C	15–20	2–6
Camber, mils/in	1–4	Conforms to setter
Surface roughness, min	10–25	8–10
Flexural strength, MPa	275–400	150–250
Firing temp., °C	1500–1600	850–1050
Firing shrinkage, %		
x,y	12–18	12.0 ± 0.2
z	12–18	17.0 ± 0.5
Repeatability, %		
x,y	0.3–1.0	0.3–1.0
z	0.2	0.5

TABLE 1.10 Typical Properties of Dielectric Laminate Materials

Property	Material					
	Difunctional epoxy	Multifunctional epoxy	High performance epoxy	Bismaleimide–triazine–epoxy	Polyimide	Cyanate ester
Electrical:						
Dielectric constant of neat resin				2.9	3.5–3.7	2.7–2.8
Dielectric constant at 1 MHz (8-ply 7628 E-Glass)	4.6	4.5	4.4	3.9–4.2	4.6	3.8
Dissipation factor at 1 MHz	0.020	0.019	0.012	0.015	0.010	0.004
Volume resistivity, Ω cm	1.0×10^{14}	3.8×10^{14}	4.9×10^{14}	4.0×10^{12}	2.1×10^{14}	1.0×10^{13}
Surface resistivity, Ω cm	1.1×10^{12}	2.7×10^{12}	2.4×10^{15}	2.0×10^{12}	3.7×10^{14}	1.0×10^{13}
Arc resistance, s	128	128	123	105	180	130
Dielectric breakdown, kV	>55	>55	50		>55	
Electrical strength, V/mil	1200	1400	1300	900	1600	1100
Physical:						
Water absorption, %	0.06–0.15	0.05–0.13	0.1–0.13	0.1–0.25	0.3–0.4	0.3–0.6
Specific gravity	1.85	1.85			1.7	
Chemical resistance (solvents)	Good	Excellent	Good	Excellent	Excellent	Excellent
Mechanical:						
T_g by DMA, °C	125	130–160	165–190	175–200	220–280	180–260
CTE	25–125°C	25–135°C	25–180°C	25–220°C	25–220°C	25–245°C
z axis, in/in/°C	5.5×10^{-5}	4.4×10^{-5}	4.4×10^{-5}	7.0×10^{-5}	3.5×10^{-5}	8.1×10^{-5}
CTE	25–125°C	25–135°C	25–180°C	25–220°C	25–245°C	
xy axis, in/in/°C	1.7×10^{-5}	1.5×10^{-5}	1.5×10^{-5}	1.4×10^{-5}	1.5×10^{-5}	
Dimensional stability after elev. temp., in/in	0.0004	0.00035	0.0004	0.0005	0.0005	0.0003
Flexural strength over 20 mils, psi:						
Lengthwise	68,000	70,000	150,000		80,000	
Crosswise	57,000	60,000	82,000		65,000	

metallization is copper. A number of options exist for the fabrication of printed wiring boards. A basic four-layer process is illustrated in Fig. 1.5. In the first step, a copper foil laminated to a dielectric sheet is photolithographically patterned and etched. The surface of the copper is then treated to produce an adherent surface oxide to improve inner layer adhesion. The copper-dielectric layers are then stacked with intervening layers of prepreg (partially cured dielectric sheets) and laminated under heat and pressure, completing the cure of the prepreg to produce a monolithic substrate. Holes are drilled in the laminated substrate and the sidewalls are treated (desmeared) to remove polymer which may have been smeared onto the ends of the exposed copper traces during the drilling operation. The holes are catalyzed, typically with PdCl; then the board is electroless copper plated to produce a thin copper

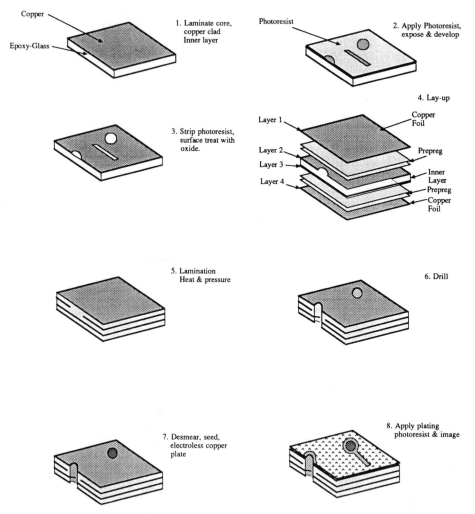

Figure 1.5 Four-layer printed wiring board process.

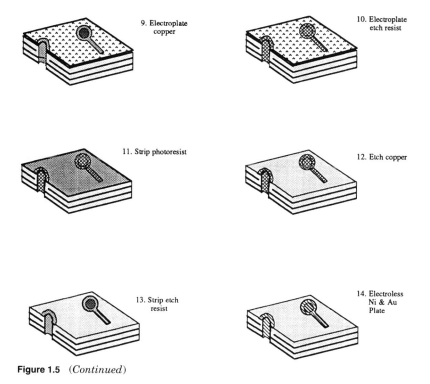

Figure 1.5 *(Continued)*

layer in the holes to allow electroplating. After photoresist application and patterning, the board is electroplated to increase the thickness of the copper in the holes and the surface conductor traces. The copper plated into the holes contacts the ends of the inner copper traces exposed during the drilling operation to provide z axis connection between those x and y axis traces which intersect the drilled hole. The copper electroplating is followed by the electroplating of an etch mask such as tin. The photoresist is then stripped and the exposed copper is etched, leaving behind the plated copper-tin pattern. The tin etch mask is stripped and an organic solder mask is applied. Finally, a nickel and gold surface finish is plated to provide a wire bondable surface. Printed wiring boards are a cost effective substrate option. Hybrids which use printed wiring board substrates are often referred to as chip-on-board (COB) assemblies or laminated multichip modules (MCM-L).

1.6 Laser Trimming

As-deposited resistors exhibit a distribution of values. To produce resistors with tolerances closer than ±10% of the desired value, the resistors are designed 10–20% below this value. After fabrication, the resistors are laser trimmed. A laser beam is used to vaporize resistor material, decreasing the width of the resistor and increasing the resistance value. In passive laser trimming, the resistor value is continuously monitored while the laser beam is directed across the resistor. When the desired value is achieved, the laser is switched off. Typical thick film resistor tolerances after trimming are 1% with 0.1% possible. Thin film resistors can be trimmed in special cases to <0.05% tolerance. In active trimming, the hybrid is fully assembled and power is supplied to the circuit. The desired circuit parameter such as gain or cutoff frequency is monitored as the appropriate resistor is trimmed. The trimming operation is terminated when the parameter is within specification. In this case, laser trimming is used as a one-time adjust potentiometer and is a widely used feature in analog hybrids.

Each of the substrate types will be discussed in greater detail in the following chapters.

1.7 Assembly

Hybrid assemblies use both unpackaged and surface mount packaged semiconductor die. Passive components are typically in surface mount chip form. The use of unpackaged die minimizes size and weight, and reduces electrical parasitics and chip-to-chip propagation delays. However, packaged devices are easier to pretest, easier to assemble, and do not require additional packaging of the completed hybrid circuit.

Attachment of the die to the substrate with epoxy adhesives and wire bonding is the most common method of bare die assembly. This approach does not require any special processing of the semiconductor die, increasing the number of devices available for hybrid assembly. However, the ability to fully pretest the die prior to assembly is limited, which adversely affects hybrid yield.

Unfilled die attach epoxies are thermally and electrically insulating. By adding ceramic fillers such as Al_2O_3 or AlN to the epoxy, the thermal conductivity can be increased. If electrical conductivity is also required, Ag or Au particles are added instead of the ceramic fillers. Flexible epoxies can be formulated to minimize the stress associated with the assembly of large Si die on an alumina substrate where the mismatch in CTE is approximately 2:1. If the subsequent assembly operations involve high temperatures (for example, solder package sealing at 350°C), epoxies may not be suitable and Au-Si eutectic bonding may be used. Gold and silicon exhibit a eutectic melting temperature of 370°C at a composition of 6 wt % Si. For Au-Si eutectic assembly, the substrate metallization must be gold and the back side of the semiconductor die may be either silicon or gold metallized. With the die and substrate die bond pad heated to 420°C, the die is brought into contact with the gold die bond pad. A slight scrubbing action is used to break up any surface contamination, and the Au and Si interdiffuse. A eutectic liquid layer forms and grows as additional Au and Si dissolve. When the temperature is lowered below 370°C, the eutectic solidifies, forming a strong metallurgical bond. Au–Si eutectic die attach requires high temperature processing and does not provide stress relief for large area Si die.

In attaching high power dissipating semiconductor die to the substrate, the die attach material often produces the highest thermal resistance in the packaging structure. High lead content solders (90%Pb/10%Sn, 95%Pb/5%Sn) are often used for the attachment of large power transistors and other high power dissipating devices. Solders have a higher thermal conductivity than filled epoxies. However, solders can fatigue and crack under the cyclic strains due to thermal or power cycling.

The electrical connection between the aluminum pads on the top of the semiconductor die and the substrate metallization are made by wire bonding. In the wire bonding process, the wire is brought into contact with the metallized pad and the wire is deformed, producing a shearing action at the interface. The shearing action removes contaminants from the wire and the bond pad at the atomic level. When sufficient cleaning has occurred, the exposed surface atoms from the wire and the bond pad metallurgically bond. Thermosonic bonding of gold wire combines heat, pressure, and ultrasonic energy to produce the required deformation, while ultrasonic bonding of aluminum wire requires only ultrasonic energy and pressure. Figure 1.6 shows thermosonic gold wire bonding in a thin film hybrid circuit.

A second method for assembly of bare die is by flip-chip solder attachment. In this method solder bumps are formed on the die input/output (I/O) pads as the final step in wafer fabrication. The bumping process is illustrated in Fig. 1.7. To assemble the hybrid, the die are "flipped" face down with the solder bumps in contact with corresponding metal pads on the substrate as shown in Fig. 1.1. The substrate and die are then heated to simultaneously reflow all of

Figure 1.6 Thermosonic gold wire bonding to a thin film hybrid circuit. (*Courtesy of Kulicke & Soffa.*)

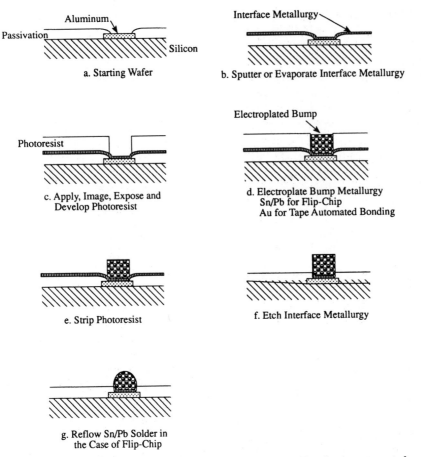

Figure 1.7 Wafer bumping process for flip-chip solder assembly of tape automated bonding.

the solder bumps, creating the mechanical and electrical connection. Flip-chip assembly requires the smallest footprint and provides the lowest inductance (shortest) interconnection. The solder bumps can be placed as an array over the surface of the semiconductor die. This allows a significantly higher number of I/O than can be achieved with perimeter I/O pads necessary for wire bonded interconnections. Since special wafer processing is required, the availability of die for flip-chip assembly is limited.

Tape automated bonding involves first bonding the die to a tape containing a copper lead pattern as shown in Fig. 1.8. The die are bonded to the metal leads through bumps on the die I/O pads. The bumps are fabricated in a process similar to that shown in Fig. 1.7 for solder bumps. A number of different bonding methods and metallurgies are used including gold bump–to–gold plated lead with thermosonic bonding and gold bump–to–tin plated lead by laser solder reflow. The metal leads fan out the I/O to test pads to allow pretesting of the device prior to assembly on the substrate. The die can be

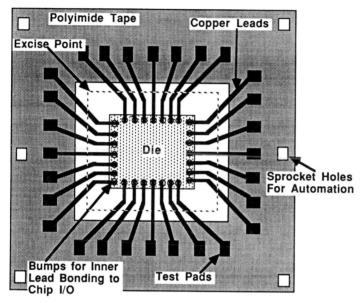

Figure 1.8   Illustration of die interconnection for tape automated bonding.

assembled to the substrate in either a face-up or face-down configuration. For assembly the die and segments of the metal lead are excised from the tape. If the die is mounted face-up, the leads are formed as shown in Fig. 1.1 and typically soldered to the substrate metallization. Thermocompression outer lead bonding is used for fine pitch assembly. In face-down bonding, the leads protrude horizontally beyond the edge of the chip and are bonded to the substrate metallization. Tape automated bonding requires bumping of the chip as a last step in the wafer processing and custom tape for each chip I/O pattern. These factors have limited the availability of die in tape format.

Other techniques including adhesive bonded microbumps and z axis elastomeric connectors have been used to attach and interconnect bare die.

Surface mount components come in a variety of package styles including small outline transistor (SOT), small outline IC (SOIC), plastic leaded chip carrier (PLCC), quad flatpacks, and leadless ceramic chip carriers (LLCCC). These components are typically reflow soldered to the substrate metallization. In this process, a solder paste containing solder powder, flux, and a solvent is screen or stencil printed onto the hybrid substrate. The components are then placed with their leads in the solder paste and the substrate is heated to melt or reflow the solder. The typical solder metallurgy is 62Sn/36Pb/2Ag.

Electrically conductive epoxy is often used to attach surface mount passive components to substrates assembled with bare semiconductor die.

Hybrid microcircuits often combine multiple assembly techniques on a single substrate to meet electrical, mechanical, thermal, and component availability constraints. This adds to the complexity of the assembly process.

1.8   Packaging

A package provides mechanical and environmental protection for the circuit, provides I/Os to interface with the next level of system interconnections, and acts as a heat spreader. In determining hybrid packaging requirements, a number of factors must be considered including cost, reliability, electrical performance, end-use environment, and thermal management. A number of package styles are shown in Sec. 1.10.

Hybrids assembled with unpackaged die typically require some type of packaging or protection of the die. In high reliability applications such as military and medical, the assembled substrate is sealed in a hermetic ceramic or metal package. The internal cavity of the package provides a benign environment (typically nitrogen or vacuum) for the hybrid circuit.

In other applications, plastic packaging is used. The plastic is typically either molded around or coated and cured on the circuit. In some cases, liquid resins are applied directly over the bare die and cured. In others, the assembled substrate is mounted in a plastic cavity package and silicone is poured over the substrate and cured. A number of variations are possible. Plastic packaging does not provide a hermetic environment—moisture permeates through the plastic. However, the adhesion between the plastic and the hybrid surfaces prevents condensation of water, which minimizes the potential for corrosion.

Hybrid circuits assembled with packaged surface mount components typically do not require additional packaging.

1.9   Advantages of Hybrid Microelectronics

Hybrid microelectronics encompasses substrate, assembly, and packaging options which can be combined to optimize the critical parameters for a given application. Advantages of hybrid microelectronics are summarized in Table 1.11. Electrical performance, size, weight, thermal characteristics, reliability, environmental requirements, design cycle time, and cost are all variables which factor into the decision to use hybrid microelectronics.

1.10   Applications and Examples

Early hybrid microcircuits combined discrete transistors, diodes with capacitors, and resistors to produce functional analog building blocks such as amplifiers, voltage regulators, comparators, and filters. With the development of the integrated circuit, predictions of the demise of the hybrid industry were frequently heard. To the contrary, the hybrid industry has grown with the IC industry, combining ICs to create even more complex electrical functions. Today hybrid microcircuits are found in all electronic market segments including military, medical, computer, automotive, instrumentation, industrial, and consumer. Examples of various hybrid constructions and applications are highlighted in the following paragraphs.

TABLE 1.11   Hybrid Microelectronics Advantages

Over discrete circuits:

1. Smaller size and less weight

2. Higher performance due to shorter circuit paths, closer component spacing (tighter thermal coupling), better control of parasitics, excellent component tracking

3. Simpler system design and reduced system cost due to simplified assembly, and functional trimming capability

4. Higher reliability due to fewer connections, fewer intermetallic interfaces, higher immunity to shock and vibration

5. Easier system test and troubleshooting due to hybrid circuit pretested functional blocks

6. Higher power handling capability due to high thermal conductivity substrates

7. Suitable for use in harsh environments because of material and construction options

Over custom monolithic ICs:

1. Lower nonrecurring design and tooling costs for low to medium volume production

2. Readily adaptable to design modifications

3. Fast turnaround for prototypes and early production

4. Higher performance subcomponents available (both substrate and add-on components), for example, $\pm\ 0.1\%$ resistors, $\pm\ 1\%$ capacitors, and low TCR Zener diodes

5. Ability to intermix device types of many different technologies (bipolar, CMOS, power, analog, digital, silicon, GaAs), leading to increased design flexibility

6. Ability to rework allows complex circuits to be produced at reasonable yields, and allows a certain amount of repair

7. Reduced parasitics, better isolation

SOURCE: Adapted from Licari and Enlow.

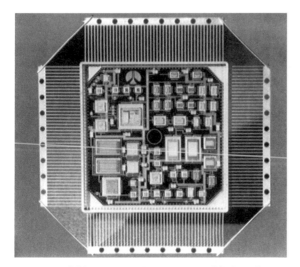

Figure 1.9 Military avionics processor fabricated on a multilayer thin film substrate. (*Courtesy of Rockwell.*)

The military avionics processor module shown in Fig. 1.9 contains 53 ICs and 40 discrete devices on a 2.2 × 2.2 ft silicon substrate with thin film, aluminum–polyimide interconnections. The module is a flight computer system with 64 kbytes of static data memory, 64 kbytes of nonvolatile program storage, system timer, multiple interrupt inputs, discrete output ports, and an advanced 16-bit microprocessor operating at 30 MHz. The module is packaged in a Kovar package (2.4 × 2.4 in) with 180 I/Os.

A second military application is the velocity and position deception module (VPDM) shown in Fig. 1.10, which is used in various fighter aircraft electronic warfare countermeasure platforms. The VPDM is a sophisticated rf input network and receiver housed in an aluminum module that integrates a variety of rf–microwave hybrid microstrip subassemblies including seven custom multicavity circuits. As enemy radars are detected, the VPDM processes and transmits amplitude and frequency shifted pulses across a broad spectrum up to 18 GHz.

A heart pacemaker circuit is shown in Fig. 1.11. The circuit uses an eight metal layer, cofired glass–ceramic substrate. For implantable medical electronics, circuit size, weight, reliability, and power consumption are critical. The circuit contains 4 LSI chips and over 75 additional discrete components, including resistors, capacitors, diodes, and telemetry devices. To achieve maximum circuit density, components are assembled on both sides of the substrate.

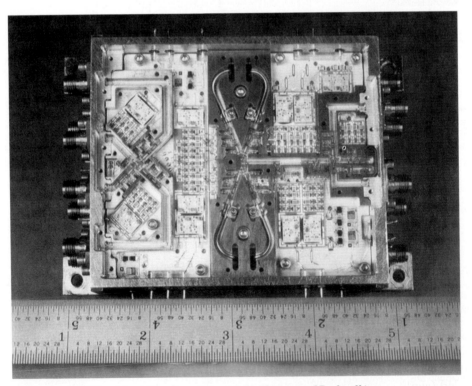

Figure 1.10 Velocity and position deception module. (*Courtesy of Rockwell.*)

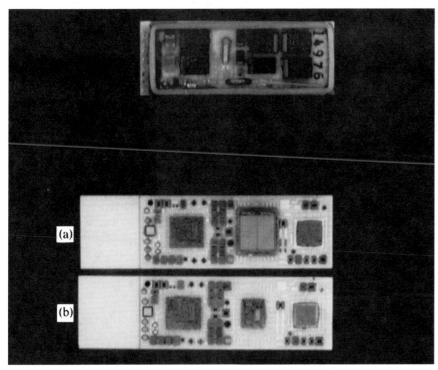

Figure 1.11 Heart pacemaker circuit fabricated on a glass–ceramic substrate. Components are assembled on both sides of the substrate to increase density. In addition, a ceramic lid is bonded over the smaller IC in the center of the substrate (*a*) and a larger IC is then bonded to the lid (*b*). (*Courtesy of Pacesetter Systems, Inc.*)

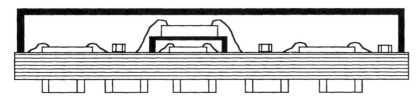

Figure 1.12 Illustration of pacemaker circuit construction.

To further increase the density, a ceramic cap was bonded over the center IC in the lower circuit (*a*) and a second larger IC was then bonded to the top of the ceramic cap as seen in (*b*). The concept is illustrated in Fig. 1.12.

A second pacemaker circuit is shown in Fig. 1.13. The cofired ceramic substrate was fabricated using 175 mm thick ceramic tape and contains six conductor layers and approximately 400 filled vias. Components are mounted on both sides of the substrate. Some components remain exposed after the lid is sealed (center of photograph). This allows the functional conversion between

Figure 1.13 Heart pacemaker circuit fabricated on a cofired ceramic substrate. (*Courtesy of Pacesetter Systems, Inc.*)

a single and dual channel pacemaker by changing the exposed components. Dual channel pacemakers function in two chambers of the heart, while single channel units operate on only one chamber.

The cross section shown in Fig. 1.14 is of a 63 copper metal layer, cofired glass–ceramic substrate with two thin film copper–polyimide layers on top. The semiconductor chips (bipolar ECL) are assembled on the substrate using flip-chip solder bumps. The circuit is used in the system 390/ES9000 family of IBM mainframe computers. A maximum of 121 ICs can be assembled on the 12.75×12.75 cm substrate. Heat (the circuit can dissipate up to 2 kW) is removed from the exposed IC back sides by pistons which contact the chips. A total of 2772 signal I/O and power and ground pins are brazed to the bottom of the substrate.

Figure 1.14 Cross section of a 63 metal layer glass–ceramic substrate with two thin film metal layers on top. The ICs are attached by flip-chip soldering. The circuit is used in the system 390/ES 9000 mainframe computers. (*Courtesy of IBM.*)

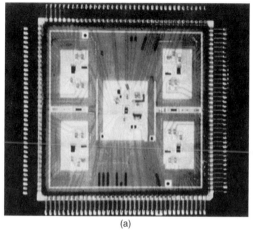
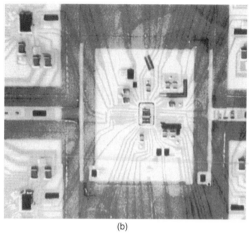

(a)   (b)

Figure 1.15 (*a*) Thick film circuit for a digitizing oscilloscope. (*b*) A close-up view of the circuit. (*Courtesy of Hewlett-Packard.*)

The thick film circuit in Fig. 1.15 is a track/hold and A/D converter for a digitizing oscilloscope. The hybrid dissipates approximately 17 W to the heat sink by use of both the alumina substrate and an integral metal heat spreader. The two GaAs chips are mounted directly on the metal heat spreader.

The circuit shown in Fig. 1.16 combines surface mount components and chip-and-wire assembly on a printed wiring board substrate. The wire bonded chips are encapsulated for mechanical and environmental protection. The circuit is used in the display assembly for a palm-top computer. The five chip-and-wire ICs drive the LCD rows and columns.

The thick film hybrid in Fig. 1.17 is an automotive pressure sensor. The circuit monitors atmospheric and engine manifold pressure. This information is

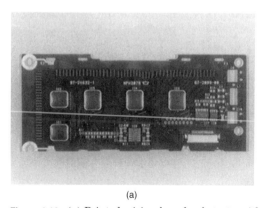
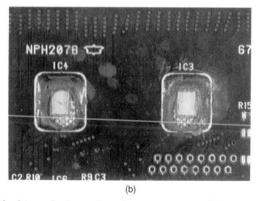

(a)   (b)

Figure 1.16 (*a*) Printed wiring board substrate with both chip-and-wire and surface mount assemblies. The chips (IC 1–5) are coated with epoxy. The circuit is part of a display assembly for palm-top computers. (*b*) A close-up view of bare die with the epoxy partially removed. (*Courtesy of Hewlett-Packard.*)

Figure 1.17   Thick film automotive pressure sensor. (*Courtesy of Delco Electronics.*)

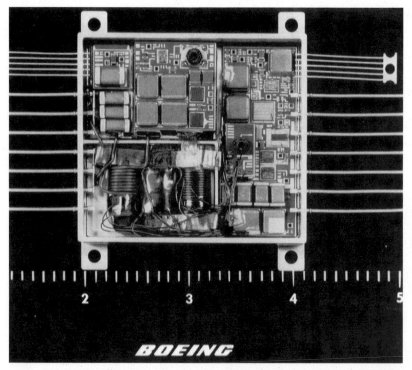

Figure 1.18   Thick film 50 W, triple output DC-DC converter. (*Courtesy of Boeing Electronics.*)

sent to the engine computer, which electronically regulates engine function for better fuel economy and lower emissions. The automotive environment is harsh, exposing the hybrid circuits to high temperatures, rapid temperature changes, repeated thermal cycles, vibration, corrosive gases, solvents, salt spray, and high humidity.

The 50-W, triple output DC-DC converter in Fig. 1.18 processes 28 V input to supply two isolated +15-V outputs and a +5-V output. Output currents are rated at 7 A for the 5-V output and 500 mA for each 15-V output. The power supply is designed for military and aerospace applications. The circuit includes a number of magnetic components and capacitors common to power circuits.

The trigger speed control for cordless hand tools shown in Fig. 1.19 contains two thick film hybrids mounted on opposite sides of a metal heat sink. The lower hybrid uses surface mount components. The resistor stripe along the lower right-hand edge of the substrate provides a variable resistor (potentiometer) as wiper fingers which slide along the resistor surface when the trigger is squeezed. The power components are mounted as bare die on the second (upper) substrate for improved heat dissipation.

The three conductor layer thick film hybrid, shown in Fig. 1.20, is used for the sensing circuitry in a prototype biomedical system designed to sense and

Figure 1.19 Thick film trigger speed control for cordless hand tools. (*Courtesy of Milwaukee Electric.*)

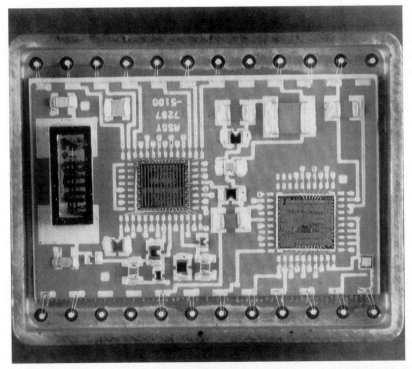

Figure 1.20 Multilayer thick film hybrid circuit with chip-and-wire assembled design to sense and inhibit self-injurious behavior. (*Courtesy of the Applied Physics Laboratory, Johns Hopkins University.*)

inhibit self-injurious behavior. The hybrid detects the output from an accelerometer switch (≥2g's acceleration), processes the signal, and sends an encoded transmission pulse on an alternating magnetic field to a similar receiver hybrid located some distance away. The circuit contains two LSI chips, a surface mounted crystal oscillator, one diode, nine chip capacitors, and nine chip resistors.

The thin film circuit in Fig. 1.21 is used to test the operation of a photodetection diode. The circuit contains circuitry for detecting and amplifying high frequency signals from the photodetection diode. The hybrid incorporates microstrip signal and clock lines capable of 0.5 to 3.5 GHz operation. The corners are mitered on the microstrip lines to minimize reflections at the bends.

The double-sided thick film hybrid in Fig. 1.22 includes over 50 printed and laser trimmed resistors on each side. Complex active trimming routines are used for this signal conditioning circuit which includes high performance differential amplification with programmable gain, offset, and low pass filtering. The circuit is assembled with surface mount components.

1.11   Summary

Hybrid microelectronics is not characterized by any single technology. As can be seen from the examples presented, the strength of hybrid microelectronics

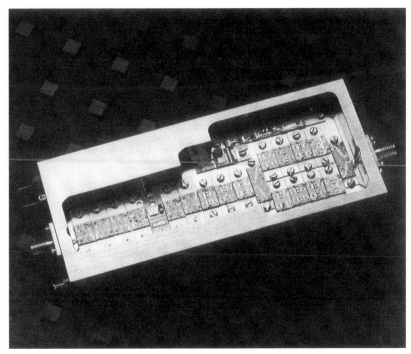

Figure 1.21   Microwave thin film amplifier. (*Courtesy of Avantek, Inc.*)

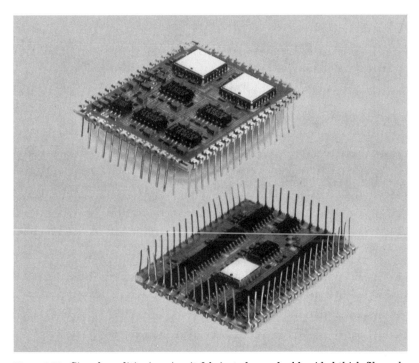

Figure 1.22   Signal conditioning circuit fabricated on a double-sided thick film substrate with surface mount components. (*Courtesy of MIni-Systems, Inc.*)

is the ability to choose and mix different technologies to best meet the requirements of the application. The following chapters detail the various technologies available to the hybrid microelectronics engineer. While all of the possible combinations cannot be addressed, the challenge of creating new combinations will continue to be the future for the hybrid microelectronics industry.

2

Ceramic Substrates
for Hybrid Applications

R. S. "Sig" Jensen

2.1 Introduction

Hybrid substrate processing has evolved extensively in the last 25 years and substrate properties remain a common denominator of circuit design. It is incumbent on the reader, however, to consider this chapter as a design nutrient. Greater insight into substrate properties, processing, and costs structure will assist the circuit designer in obtaining project acceptance by the hybrid processor and end user. Better designs and more robust and less expensive circuits are the inheritance of the fully enlightened designer. This chapter was written to provide guidance in comparing and selecting these base materials, and is directed toward the user of these materials rather than the manufacturer. The most common types of materials, specifically aluminum oxide (alumina, Al_2O_3), beryllium oxide (beryllia, BeO), and aluminum nitride (AlN), are comprehensively discussed to provide a more complete understanding of critical hybrid design criteria.

The subject of this chapter is intended to show the relationship of material properties of thick film substrates and the associated means of processing them. Electronic devices are requiring smaller, more dense circuitry to manage their electrical energy thirst and provide a manageable size package sufficient to house the ever-increasing sophisticated and complicated apparatus.

It is hoped that this materials information will allow the designer or engineer involved with producing these "high-tech" devices the latitude of choosing the most economical and effective materials and material processing methods for use as substrates. Later chapters will show the importance of these choices.

A broad examination of material properties and material processing basics and hints follows. The format includes various ceramics and their mechanical, electrical, thermal, and other properties. Further, the processing methods include those in use today which lend themselves to efficiency and effectiveness with available mechanization.

2.2   Basic Properties of Hybrid Substrate Materials

Ceramic substrates are primarily metal oxides and nitrides mixed with glasses and fired at an elevated temperature. The resultant structure is a hard, brittle composition with many advantages in hybrid circuit applications. Compared to organic circuit board materials, for example, ceramics have a much higher thermal conductivity, a higher tolerance to temperature extremes, and a lower temperature coefficient of expansion. These properties and others make ceramics the most desirable materials for the foundation of hybrid circuits.

General consideration needs to be given to ceramic materials used in thick film processing not only to avoid pitfalls associated with electrical and mechanical properties and their composition, but to take full advantage of their desirable properties. Ceramic materials are in widespread use in thick and thin film circuits because of their inherent mechanical strength, electrical properties, and thermal attributes. Chemical reactivity (or the converse, chemical inertness) is a process consideration because of the expanding methods of manufacture and availability of new chemical processes and materials with their associated manufacturing compatability.Greater circuit component density requirements have limited the range of materials available to use in the shrinking spaces demanded by the customer. Multichip modules and their associated high density interconnection systems are a solution to the increasing density requirements where the monolithic chip application specific integrated circuit (ASIC) is impractical, both thermally and economically.

The use of ceramic materials as substrates, capacitors, resistors, sensors, and actuators has paralleled the growth in integrated circuits, computers, telecommunications, and military, industrial, and consumer electronic applications. The properties and capabilities of these materials depend on the ratio of glass to the nonadditive laden pure ceramic in the fired structure.

Substrates for hybrid applications are characterized by the properties listed in Table 2.1.

2.2.1   Mechanical properties

The mechanical properties of ceramic substrates are important in applications where the hybrid circuit must undergo severe thermal or mechanical stresses during operation or storage. In these applications, the mechanical strength and thermal expansion characteristics may be the determining factors in material selection.

TABLE 2.1   **Parameters Which Characterize Ceramic Substrates**

Mechanical:
  Compression strength
  Tensile strength
  Young's modulus (modulus of elasticity)
  Dimensional stability
  Flexural strength
  Thermal coefficient of expansion
Physical:
  Camber
  Surface finish
  Specific gravity
  Water absorption
Chemical:
  Compatibility with thick and thin film materials
  Chemical reactivity
Electrical:
  Volume resistivity
  Dielectric constant
  Dissipation factor
  Dielectric strength
  Permeability
Thermal:
  Thermal conductivity
  Specific heat

Ceramic materials are crystalline in nature and are quite brittle. The brittle properties of ceramic substrate materials precipitate the breakage problems normally associated with handling, and are intensified by stresses in the edge zone on the substrate perimeter. The edge zone is an area that retains residual stresses from the original firing process. Unless these stresses are relaxed by annealing or other means of stress relief, they can cause brittle fracture and microcracking adjacent to, in, and around the edge zone during and after further processing.

2.2.1.1   Young's modulus (modulus of elasticity in tension).   Plastic deformation is at a minimum even under very large physical loads because of the high modulus of elasticity in alumina ceramics. Young's modulus for alumina ranges from 33 million psi for 85% purity to 52 million psi for 99.5% purity, even under high loading at room temperatures. Alumina exhibits only a very slight elastic deformation under high loads and functions as an extremely rigid material. These values change when the temperatures rise above standard room temperature. For example, the annealing temperatures for substrates cause plastic deformation during loading. Refractory furniture and setter plates must therefore be utilized in annealing to avoid permanent deformation.

Hooke's law [Eq. (2.1)][15] is followed by ceramic substrates during mechanical loading to the point of fracture, which is significantly different from that of metals because of the lack of plastic deformation in the ceramic.

$$\sigma = E\varepsilon \qquad\qquad (2.1)$$

where σ = stress, psi (or N/m^2)
E = Young's modulus, psi (or N/m^2)
ε = strain, in/in (or m/m), the net elongation or contraction per unit length

Ceramic strength properties are generally measured using modulus of rupture and transverse bending tests rather than tensile tests because of this brittle nature. In this test, the substrate is placed upon two supports while load is placed on the center of the span until fracture of the substrate occurs. Test standardization is necessary to ensure specimen uniformity and inhibit use of specimens which are quite sensitive to defects such as microcracks and voids. Test methods used to evaluate hybrid substrate materials are listed in Table 2.2.

2.2.1.2 Compression strength. Compression strength is measured by placing a sample of the material under compression and measuring the strain. By convention, the strain under compression is considered negative. Compared to metals, ceramics have generally greater compression yield strengths. This high compression strength offers design engineers a very strong material that is much less expensive than exotic materials. For example, the compression strength of alumina at 85% purity is about 290 ksi (290,000 psi) at room temperature, increasing to 380 ksi at 99% purity. By contrast, tempered and hardened high strength alloy steels have room temperature compressive strengths ranging only from 275 to 318 ksi, as shown in Table 2.3. At higher

TABLE 2.2 Tests and Test Methods for Evaluating Ceramic Substrates

Property	Test
Specific gravity	ASTM C20-46
Hardness, Rockwell 45N	ASTM E1867
Surface finish, arithmetic average	Profilometer
Crystal size	
Water absorption	ASTM C373-56
Gas permeability	
Color	
Compressive strength	ASTM C528-63T
Flexural strength	ASTM C369-56
Tensile strength	Brazil Test
Modulus of elasticity	Sonic method
Shear modulus	Sonic method
Bulk modulus	Sonic method
Sonic velocity	Sonic method
Poisson's ratio	Sonic method
Thermal coefficient of expansion	ASTM C372-56
Maximum use temperature	

SOURCE: Coors Corporation.

TABLE 2.3   Compression Strength of Selected Materials

Material	Compression strength, ksi
Alumina	290–380
Ultrahigh-strength steel	284
Stainless steel	165
Phosphor bronze	75

temperatures, alumina maintains a much higher compressive strength than do metals. Alumina ceramics will retain their shape well above the melting point of most metals. The compression strength rises by a small amount as the temperature decreases toward that of liquid nitrogen (77 K).

2.2.1.3   Tensile strength.   The tensile strength of ceramics is only about 10% of the compression strength and is less than that of many metals. For example, the tensile strength of alumina ranges from 17 to 35 ksi. In applications where this parameter is significant, it is preferable to utilize a thicker substrate or to design in such a manner that the substrate is in compression. Compared to other crystalline materials, ceramics have a tensile strength approximately two to five times that of electrical porcelains, and are much stronger that glass and steatite.

2.2.1.4   Thermal coefficient of expansion (TCE).   The thermal coefficient of (linear) expansion is the ratio of the change in length of a material per °C to the original length at 0°C. Over the range of interest, the TCE of most materials is linear and may be expressed by a single number. Figure 2.1 illustrates the percentage change which can be expected in certain materials over temperature. The TCE of a given material is the slope of the thermal expansion versus temperature curve. The differential magnitudes of thermal expansion between one material and another are often the most significant characteristics or considerations in the design of hybrid circuits. If the TCE differential is too large, sufficient stresses may be created in the structure to cause failure. The problem is most acute when ceramics are attached to metal or plastic. The TCE of ceramics is substantially less than that of either of these materials, and the bond must be capable of absorbing the resultant stresses.

2.2.1.5   Dimensional stability.   One reason for use of ceramics as a base for circuitry is the extremely high dimensional stability. Highly dimensionally stable ceramics have properties of high hardness, minimum thermal expansion, and great rigidity in order to provide material that does not warp or distort as metals do as a result of the heat generated during the machining process. Aging, pressure loading, and heating causes plastics to flow and change dimensions, and they do not maintain their shape during thermomechanical processing. Ceramic, on the other hand, is almost perfectly elastic at these same temperatures and pressures which deform many other materials.

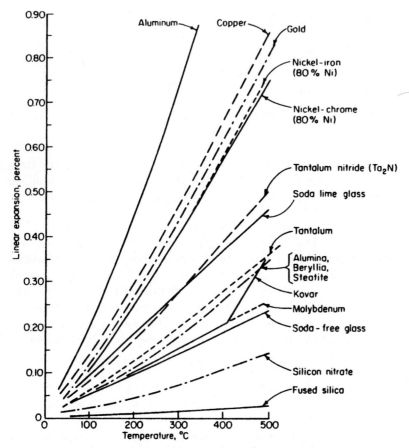

Figure 2.1 TCE vs. temperature for selected materials.[7]

2.2.1.6 Thermal shock failure mechanisms. Ceramic substrates withstand large temperature excursions and thermal shock conditions extremely well, partially because of the low temperature coefficient of expansion and the high modulus of elasticity. The density, specific heat, and thermal conductivity of ceramic are other conditions in a complex relationship governing this thermal shock survival. It can be empirically shown that some compositions of alumina ceramics maintain better shock resistance through some thermal cycles than others, especially during high temperature annealing cycles (see annealing in Sec. 2.6). Accordingly, the thermal shock resistance is an empirical determinant of acceptable ceramic substrate materials.

An empirical approach to estimating the survivability of a ceramic during thermal shock conditions can be described by a factor called the coefficient of thermal endurance (F) developed by Winklemann and Schott,[11] which is expressed by

$$F = \frac{P}{\alpha E} \sqrt{\frac{k}{\rho c}}$$

where P = tensile strength
α = linear coefficient of thermal expansion
E = Young's modulus
ρ = density
c = specific heat
k = thermal conductivity

This equation has been found to give qualitative agreement with the empirical observations, and is valuable in comparing differing substrate materials. The coefficient of thermal endurance of various substrate materials is shown in Table 2.4. Alumina exhibits a thermal endurance factor similar to that of beryllia but greater than that of glass and much less than that of silica. A comparison of this factor with the TCE for temperatures under 300°C suggests that the high thermal endurance of silica is due to several factors: the low TCE of the material and because the TCE exerts the greatest amount of influence over the ability of the material to withstand shocks. Paradoxically, different glasses in the table are found to have the lowest shock resistance while fused silica is at the top of the list.

Nonuniform stresses and the ability of the substrate to resist them are the main differences between resistance to thermal stress versus shock resistance. Material adjacent to a localized heated area may not expand at the same rate, setting up a condition of thermal stress which may be produced anisotropically. Polycrystalline anisotropic substrate materials and two-phase materials, i.e., glazed metals, ceramics, and glass ceramics, are susceptible to the effects of thermal stresses. Homogeneous bodies are also subject to thermal stresses when temperature gradients are present during heating and cooling. Substrate surfaces respond quickly to the change in temperature as compared to the interior of the substrate. Therefore, temperature gradients can be induced perpendicular to the surface. These temperature gradients can be transverse when a thick film component is electrically loaded, heating the surface. These components may cover only part of the surface.

TABLE 2.4 Thermal Endurance Factor for Selected Materials[7]

Material	F	TCE, ppm/°C
Silica	13.0	0.56
Alumina	3.7	6.0
Beryllia	3.0	6.1
Glass	0.9	9.0

Mechanical and thermal shock and thermal stress may be the primary reason for substrate cracking and chipping and excessive stress failures.

2.2.2 Physical properties

2.2.2.1 The substrate surface. The ceramic surface finish is a function of the microgranular structure and density of the ceramic–glass composite, which may be semiporous in nature. The smoother substrates are formed using small grains in as-fired high density ceramic structures, whereas the larger grained substrates form a relatively coarse surface. The microscopic rippled structure of small vias is formed as a function of the drilling process and post-drilling processes such as annealing. The larger grained surface exposed to metallization creates anchoring sites for adhesion of the cured paste. The degree of surface roughness required depends on the selected metallization process.

Thick film deposition requires a rougher, more porous surface which can be attained by the normal firing process of the substrate without subsequent processing. Fundamentally, a rougher surface has a higher equivalent surface area per linear square inch, providing a larger interface area between the film and the substrate. By contrast, thin film deposition requires a smoother surface. A substrate suitable for thick film applications will have a surface in which the distances between the peaks and valleys are much greater than the thickness of a thin film. A thin film circuit deposited on such a substrate will have unpredictable characteristics with a wide variation in properties.

General thick film surface requirements are in the centerline average (CLA) range of 15 to 40 μin. Those substrates with surface finishes under 5 μin are used primarily for thin film purposes. As-fired surfaces are preferred for thin film applications as opposed to glazing or polishing.[16] Glazing involves the application of a thin layer of vitreous glass which is fired at approximately 550°C to provide a smooth surface. The glass degrades both the electrical and thermal properties of the substrate, and also has a larger TCE.

Polishing ceramics is a difficult and expensive operation, because most ceramics are extremely hard and other considerations. Alumina, for example, is the third hardest substance known, behind diamond and carborundum. Polishing therefore requires that the surface be ground with a diamond or carbide surface or lapped in a slurry made from the powder of one of these substances. The polishing process creates a surface which contains small microcracks which may propagate and create particles during processing or use.

2.2.2.2 Measurement of surface characteristics. A profilometer is utilized primarily for measuring thick film substrate surface characteristics of roughness where resolutions greater than 1 μin are not required.[7] The fine profilometer stylus is attached to a tracer arm and drawn across the surface. The tracer arm magnifies the amplitude of the trace and provides output to a plotter or computer. The resolution of the profilometer is limited to the stylus tip configuration and size.

One method of characterizing the surface roughness is by means of the centerline average, or CLA. This attribute is derived by averaging traces of roughness of the surface; that is, a centerline is placed midway between equal areas under each of the roughness and valley curve traces. A mathematical definition of CLA or centerline average is given in Eq. (2.2) and pictured in Fig. 2.2.[7, 13]

$$\text{CLA} = \frac{a + b + c + d + \cdots}{ML} \qquad (2.2)$$

where a, b, c, d,... = area under peaks and valleys of the trace
$\qquad\qquad L$ = assigned length of stylus travel
$\qquad\qquad M$ = vertical magnification used

Alternatively, the root mean square roughness, or roughness, may be substituted for the CLA measurement. While roughness is a continuous type of surface condition, a waviness or flatness may be a periodic distortion of the surface. Peaks and valleys of this periodic display can be divided into segments of height and averaged under a square root form, yielding a roughness value as defined in Eq. (2.3).

$$\text{RMS} = \sqrt{\frac{y_1^2 + y_2^2 + y_3^2 + \cdots + y_n^2}{n}} \qquad (2.3)$$

where $y_1, y_2,..., y_n$ = heights of traces as shown in Fig. 2.1

The roughness is useful for evaluating the periodicity of waviness and flatness in these substrates as depicted in Fig. 2.3. Surface roughness does not describe all aspects of the surface condition, and a combination of roughness and CLA is required for this determination. A comparison of CLA values to roughness values indicates the CLA value to be lower by some 10 to 30%, and indicates the difference in their respective mathematical definitions. While the CLA value can be measured by means of a planimeter or derived by use of an electronic integrating instrument, the roughness values are primarily statistical averages and do not completely represent a full description of the surface texture. The mechanism of scanning the substrate obfuscates some of these hidden digs and scratches that might otherwise be caught by a profilometer check. These hidden deep scratches or defects can cause discontinu-

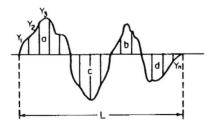

Figure 2.2 Calculation of center-line average.[7]

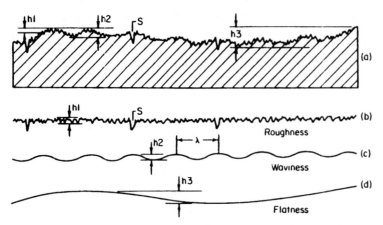

Figure 2.3   Roughness, flatness, and waviness in a ceramic substrate (a)–(d).[7]

ities in thin film structures while not being critical to thick film architecture. Flatness and waviness of substrates are not always reflected in the roughness or CLA values. However, these properties are very important in substrate processing. Identical average values can be achieved when using both methods of surface analysis, and this problem causes some ambiguity in their use. The average values can be identical when both methods have the same amplitude surface profile but different periodicities.

The tip radius of the profilometer also determines the type of data to be gained from the profilometer during surface analysis. The tip radius (typically 0.5 to 1.0 mil) limits the resolution in the direction of travel.[7] Different magnifications in the horizontal and vertical axes of the profilometer's trace can give a clearer picture or an indication of mechanical surface working in a given direction of tip travel. A large tip radius can also hide the surface striations.

2.2.2.3 Surface flatness of substrates.   The amount of camber or warpage and the degree of flatness are two of the greatest concerns associated with thick film substrates processing. Camber is defined in Fig. 2.3 as the overall deviation from one side of the substrate to the other as measured along the diagonal, or hypotenuse, of the surface, while waviness is more periodic in nature and may appear on only one surface. The units of camber are in/in, and it may be measured by using micrometer-adjustable parallel plates set at the substrate thickness plus camber and tolerance. Substrates which pass through the parallel plates are deemed acceptable. It should be noted, however, that error is associated with this method, which may amount to as much as 10 percent of the total amount of camber. Substrates at the lower end of the thickness tolerance may exhibit a higher degree of camber than thicker substrates and still be of acceptable quality for processing.

Printing of electronic circuit thick film pastes on a cambered substrate causes instability and variability of their values. For example, deposited resistors are printed to a stated thickness and must be stable in order to be trimmed effectively. Substrates are held in place on a fixture by a vacuum during most processes, and warped substrates can cause interruptions in the screen printing process by lifting from the fixture during printing. It is also difficult to transfer heat from a warped surface, which complicates soldering processes and thermal management design. While the substrate can be flattened during printing by added vacuum, this causes additional problems in the stability of electrical values for screen printed resistors and capacitors when the vacuum is released. The effect is worse when large components are screen printed.

For thinner substrates, camber plays more of a part in the processing of the substrate through screen printing and multiple firings. Substrates 25 mil thick with 2- to 3-mil/in camber are now available as standard, with 1-mil/in camber available at a premium. For a 2×2 in substrate, a substrate with a 3-mil camber will have a thickness deviation of 8.4 mils, or a third of the dimension of a 25-mil-thick substrate. Thicker substrates from 40 mils or greater are sometimes used to avoid camber problems in screen printing. A thicker substrate, however, results in added cost and in certain applications may add to the thermal resistance of the circuit.

2.2.2.4 Specific gravity and water absorption. The specific gravity is the ratio of the density to that of water, while water absorption is a measure of the amount of water which may intrude into the substrate by capillary action or other means. These two parameters are discussed together since they are both a measure of the degree of porosity of the substrate. A substrate of a given composition should have a mass capable of structurally supporting circuit board processing and use. The robust and tough nature of the structure is predictable based on material properties. If the mass is lower than predicted, some internal porosity may be present. Water absorption is more a measure of surface porosity, since water may not penetrate into areas not exposed to the air.

2.2.3 Chemical properties

2.2.3.1 Compatibility with thick and thin film materials. Thick film pastes have been developed to bond to oxide-based ceramics, such as alumina and beryllia. They have been optimized for use with alumina and do not adhere as well to beryllia unless special additives are used. Thick film materials for use with aluminum nitride are of an entirely different composition and may require extensive preoxidation of the substrate surface prior to use.

The adhesion mechanism of a thin film is an oxide of the deposited metal which forms at the interface between the film and the substrate. Both alumina and beryllia are very compatible with thin film processes compared with other ceramics because surface oxides enable the deposited metal atoms to attach to the substrate. As with the thick film process, the growth of a surface oxide is often necessary to permit adhesion on aluminum nitride, because most metals do not form nitrides as readily as oxides.

2.2.3.2 Chemical reactivity. Ceramic oxide-based substrates are generally impervious to chemical reactions with the exception of fluorine-based acids, which may attack the glasses used in the bonding agents. Acids of this type may be used to remove surface glass from laser-drilled holes to improve adhesion or to slightly increase the surface roughness to improve adhesion.

2.2.4 Electrical properties

Substrates for hybrid applications must have good electrical properties to prevent interference with circuit performance. The specific properties defined as critical depend on the circuit application. For example, the loss factor may be significant for high-frequency applications but may not apply to low frequency circuits.

2.2.4.1 Volume resistivity. Electrical resistivity is primarily a measure of the resistance that a material offers to current flow in a DC field. Materials used to fabricate substrates are composed of electronic grade ceramics with a high electrical resistance and are used in the electronics industry in abundance. It can be seen that all the substrate materials resistivity values drop sharply as a function of temperature as shown in Fig. 2.4. Notice that this effect is nearly the opposite of the mechanical property linear thermal expansion versus temperature (Fig. 2.1). Most insulating materials have these same temperature characteristics with respect to temperature. By contrast, the metal resistance of metals increases with temperature.

2.2.4.2 Dielectric constant. The dielectric constant is the measure of ability of the material to store charge relative to vacuum. This is a key property of electrically insulating materials. The factors of greatest influence in determining the dielectric constant are compound purity and composition. Figure 2.5 shows the change in the dielectric constant for various materials is shown with temperature. Table 2.5 shows the change in dielectric constant with frequency.

2.2.4.3 Dissipation factor. In AC applications, the current and voltage across an ideal capacitor are 90°C out of phase, with the current leading the voltage. In practical cases, the resistive component of the capacitor causes the current to lead the voltage by an angle less than 90°. The dissipation factor (DF) is a measure of the real or resistive component of the capacitor and determines the energy loss from the material per cycle in the form of heat. The DF is defined by Eq. (2.4)[13] for the parallel equivalent circuit

$$DF = \tan(90 - \Theta) = \frac{I_R}{I_C} = \frac{\text{Capacitive reactance}}{\text{Parallel resistance}} \tag{2.4}$$

which is usually used in low-frequency applications, or by

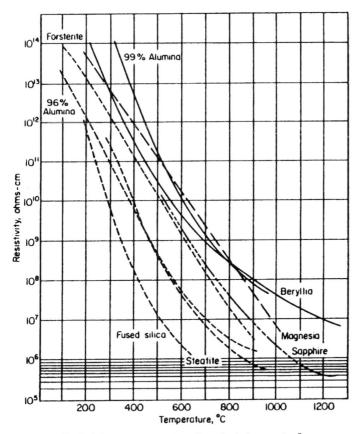

Figure 2.4   Resistivity vs. temperature for selected ceramics.[7]

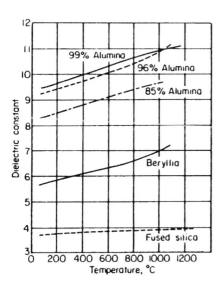

Figure 2.5   Dielectric constant vs. temperature for selected materials.[7]

TABLE 2.5   Dielectric Constant vs. Temperature and Frequency for Selected Materials[7]

Ceramic	1 MHz 25°C	1 MHz 500°C	1 GHz 25°C	1 GHz 500°C	10 GHz 25°C	10 GHz 500°C	25 GHz 25°C	25 GHz 500°C
Fused silica	3.78	3.78	3.78	3.78	3.78	3.78	3.78	3.78
Steatite	5.7	6.7	5.5	6.5	5.2	6.0	5.2	
Forsterite	6.2		5.9		5.8	6.3	5.8	
Beryllia, 99%	6.4	6.9			6.1	6.3	6.0	6.3
Alumina, 96%	9.0	10.8		9.5	8.9	9.4	8.7	9.0
Alumina, 99%	9.2	11.1	9.1	9.88	9.0	9.86	8.9	9.85

$$DF = R_S \omega C_S = \frac{\text{Series resistance}}{\text{Capacitive reactance}}$$

where $\omega = 2\pi f$
$\quad f$ = frequency

and where most dielectics are low loss materials and are used ostensibly in series circuits where $\tan \delta$ (DF) is <0.032 in high-frequency applications.[3]

The dissipation factor of selected materials is presented in Table 2.6. High frequencies such as microwave frequencies are an area of importance for dissipation factor which increases as a function of temperature and frequency.

2.2.4.4 Dielectric strength.   The dielectric strength is a measure of the voltage required to cause breakdown for a given configuration. The dielectric strength is actually a field effect and the term V/unit length is constant for a given dielectric material. Temperature variations cause a proportional shift in the dielectric strength of most insulating materials. Other material properties also affect the dielectric strength of the ceramic, including frequency, material thickness, density, porosity, and purity. Temperature, thickness, and frequency are the most important of the parameters based upon their significance in applications. Material dielectric strength decreases largely with increases in frequency, material thickness, and temperature. Figure 2.6 compares the dielectric strength of several materials, and Table 2.7 provides guidelines for alumina in high voltage applications.

2.2.4.5 Permeability.   Most ceramics do not exhibit magnetic properties. However, certain oxides (ferrites) can be magnetically polarized and are widely used to make magnetic cores for inductors and transformers. These materi-

TABLE 2.6   Loss Tangent vs. Temperature and Frequency for Alumina and Beryllia[7]

Ceramic	1 MHz 25°C	1 MHz 300°C	1 MHz 500°C	1 GHz 25°C	1 GHz 300°C	1 GHz 500°C	10 GHz 25°C	10 GHz 300°C	10 GHz 500°C	25 GHz 25°C	25 GHz 300°C	25 GHz 500°C
Alumina (85%)	0.0004	0.002	0.009	0.001	0.002	0.004	0.0015	0.002	0.003			
Alumina (96%)	0.0003	0.003	0.013	0.0003	0.0007	0.0015	0.0006	0.001	0.002	0.0007	0.0009	0.002
Alumina (99%)	0.0002	0.0006	0.002	0.0002	0.0003	0.0015	0.0001	0.0001	0.0002	0.0003	0.0003	0.0003
Beryllia (99%)	0.0001	0.0001	0.0004	0.0002	0.0003	0.0006	0.0001	0.0001	0.0001	0.004	0.004	0.004

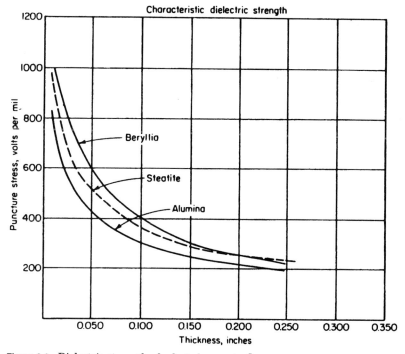

Figure 2.6 Dielectric strength of selected ceramics.[7]

TABLE 2.7   Dielectric Strength of Alumina 0.055 in Thick at Room Temperature[7]

Frequency	Dielectric strength rms V/mil
60 Hz	298
1 kHz	298
38 kHz	253
180 kHz	253
2 MHz	178
18 MHz	112
100 MHz	69

als have not been widely exploited for use as substrates to this point, but considerable experimentation is taking place to investigate their use in applications requiring buried magnetics.

2.2.5   Thermal properties

2.2.5.1   Thermal conductivity. The rapid growth of power hybrid applications requires substrates to be thermally conductive to maintain circuit functionality. Consequently, the thermal conductivity of the substrate has become one of the most characterized design criteria. One of the most significant parameters is the thermal conductivity as a function of temperature. Figure 2.7

depicts this parameter for a number of materials, and Fig. 2.8 compares alumina, beryllia, and aluminum nitride directly.

The thermal conductivity of the glass used as a binder is slightly lower than that of the ceramic material, which results in a somewhat lower thermal conductivity than for the pure material. Tables 2.8 and 2.9 list the thermal conductivity of alumina and beryllia, respectively, as a function of purity.

The glazing process can have an extremely adverse effect on the thermal conductivity. As an example, a 1 mil thickness of 2.5% sodium oxide glass commonly used for glazing is equivalent to about 30 mils of alumina, which is further equivalent to 190 mils of beryllia.

2.2.5.2 Specific heat.

The specific heat of a material is a measure of the temperature rise as a function of the amount of stored heat and is measured in $W \cdot s/g \cdot °C$ or $cal/g \cdot °C$. The specific heat of alumina, beryllia, and aluminum nitride as a function of temperature is presented in Fig. 2.9. It can be seen that for the same amount of heat introduced, beryllia and alumina will remain cooler to the touch than molybdenum or nickel.

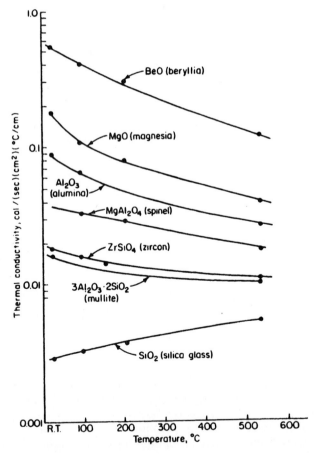

Figure 2.7 Thermal conductivity vs. temperature for selected materials.[7]

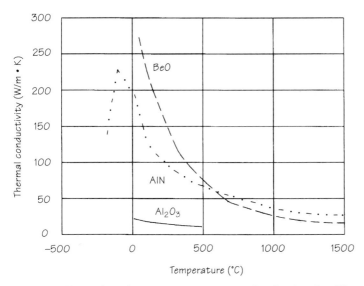

Figure 2.8 Thermal conductivity vs. temperature for alumina, beryllia, and aluminum nitride. (*Courtesy of Tegmer–Brush Wellman Corp.*)

TABLE 2.8 Thermal Conductivity of Alumina vs. Alumina Content[7]

Alumina, %	Thermal cal/s•cm•°C	Change in thermal conductivity, %[7]
99	0.070	
98	0.061	−13
96	0.043	−39
85	0.035	−50

TABLE 2.9   Thermal Conductivity of Beryllia vs. Purity[7]

Percent of thermal conductivity of 100% pure, 100% dense BeO

BeO, %	BeO + porosity	BeO + alumina	BeO + silica
100	100	100	100
99	98.8	92.1	85.2
98	96.8	85.2	78.8
97	94.8	78.8	62.0
96	93.2	74.8	58.6
95	90.6	66.0	48.0

2.3 Ceramic Compositions for Thick Film Circuits

Three primary ceramic materials are used in hybrid substrate applications: alumina, beryllia, and aluminum nitride. Silicon carbide and boron nitride are used primarily in packaging applications and will not be discussed in detail in this chapter.

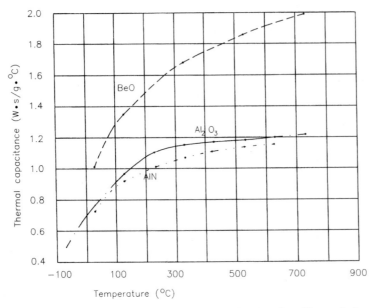

Figure 2.9 Specific heat vs. temperature for alumina, beryllia, and aluminum nitride. (*Courtesy of Tegmen–Brush Wellman Corp.*)

2.3.1 Alumina

The continuing requirement for relatively inexpensive and electrically and mechanically appropriate substrate material leads directly to aluminum oxide, Al_2O_3, the most widely used hybrid circuit substrate material. It has excellent dielectric insulating properties, mechanical strength, a relatively high degree of chemical inertness, and good surface texture characteristics, and it is easy to fabricate. This composition is found in purities ranging from 90–92% for the so-called opaque material, to 94–96% for the white materials, with 96% being the most commonly used, to the 99% materials (99.5–99.8%), which are primarily used in thin film applications. The glass binder, which has a lower thermal conductivity than alumina, limits the thermal conductivity of the lower purity materials.

Certain additives (or dopants) to the main ceramic mixture improve the thermal, electrical, and mechanical properties. Substrate strength and thermoelectric characteristics are also proportional to grain size, which is highly influenced by the degree of sintering which takes place during firing. Heat and pressure are applied in proprietary proportions at various points along the manufacturing process to improve the sintering process. The resultant congealed and taut mass forms the basis for substrate dielectric circuit board material.

2.3.2 Aluminum nitride

The primary advantages of aluminum nitride are the high thermal conductivity and the close TCE match with silicon. Aluminum nitride can draw heat away from a silicon device better than alumina, and the resultant mechanical stresses are substantially reduced.

2.3.2.1 Composition. Aluminum nitride is a difficult substrate to manufacture. The density and high thermal conductivity result from both the basic composition and the sintering process. Rare earths or calcium are added to promote sintering. Yttrium, when added in small amounts, produces a tan or gray substrate appearance, while erbium gives the substrate a pink tinge. By contrast, substrates using calcium appear white.

Single crystals of pure aluminum nitride have a thermal conductivity of 319 W/(m•K).[1] This value is close to 80% of that of pure copper. The distribution and content of oxygen and yttrium (in secondary phases) largely determine the thermal conductivity of aluminum nitride substrates in the fired state. Depending on the additives and the processing, the thermal conductivity of aluminum nitride substrates ranges from 170 to 200 W/(m•K). Higher values are available but are more expensive.

2.3.2.2 Thermal expansion coefficient. The TCE of AlN is typically 4 ppm/°C in the temperature range of 25 to 200°C. The TCE of silicon at the same temperatures is approximately 5 ppm/°C. This proximity makes a relatively good match for these two materials in high power applications. Die attach performance in withstanding thermal shock is significantly improved relative to alumina or beryllia and thermal cycling.

Typical properties of AlN substrates are given in Table 2.10.

2.3.3 Beryllia

Beryllia has long been in use in hybrid substrate applications. The high thermal conductivity and the low dielectric constant make it attractive for high power and high frequency applications.

The wider use of BeO has been limited by two factors, the relatively high cost compared to alumina and the toxicity of BeO dust. The cost has dropped dramatically in recent years, owing primarily to the increasing use of BeO in automotive applications. The economies of scale which can be realized in the automotive industry have an impact on the remainder of the electronics industry, which is enjoying the benefits.

The toxicity of BeO dust has been an issue for some time. It is important to note that this is a consideration only when BeO is being sawed, ground, or otherwise machined. The inhalation of dust particles from BeO is considered to be the only toxicity concern. During ordinary processing, such as screening and firing (so long as no toxic levels of BeO are expelled from the oven), there is no danger from beryllia, and it may be handled with complete safety, the same as alumina.

TABLE 2.10   Typical Properties of AlN Substrates[2]

Thermal conductivity at 25°C	190 W/(m•K)*
Thermal expansion from 25 to 200°C	4×10^{-6}/°C
Surface roughness (Ra)	<0.6 μm (<25 μin)
Camber	<0.003 mm/mm
Length and width tolerances	±1%†
Thickness tolerance	±10%‡
Squareness	<0.006 mm/mm
Color	Tan
Density	3.3 g/cm^3
Average flexural strength	>300 MPa
Volume resistivity	>10^{13} Ω cm
Dielectric strength	>15 kV/mm
Dielectric constant	8.9
Dielectric loss at 1 kHz at 25°C	<0.05
Dielectric loss at 10 kHz at 25°C	<0.01
Dielectric loss at 1 MHz at 25°C	<0.001
Dielectric loss at 10 MHz at 25°C	<0.001

*Other thermal conductivities are available.
†But not less than ± 0.1 mm (0.004 in).
‡But not less than ± 0.03 mm (0.001 in).

The properties of beryllia and other materials are summed up in Table 2.11.

2.4   Ceramic Substrate Manufacturing Methods

There are several methods by which a substrate may be manufactured: roll compaction, tape casting, powder pressing, isostatic powder pressing, and extrusion. The characteristics of substrates made by these processes differ in properties such as camber, surface finish, and porosity. The basic process is depicted in Fig. 2.10.

Roll compaction.   Roll compacting is a process which uses a set of very large rolls to press, closely pack, and flatten the mixture of binders, ceramic pow-

TABLE 2.11   Properties of Substrate Materials[5]

Property	Material			
	AlN	Al$_2$O$_3$	BeO	SiC
Density, g/cm^3	3.28	3.8	2.9	3.2
Thermal conductivity, W/(m•K)	70–250	20	240	70
TCE, ppm/°C	4.6	7.3	7.5	3.7
Dielectric strength, kV/mm	14	18	26	0.07
Volume resistivity, Ω cm	>10^{14}	>10^{14}	>10^{14}	>10^{13}
Dielectric constant at 1 MHz	8.8	9	6.7	40
Dissipation factor at 1 MHz	0.005–0.001	0.0001	0.0012	0.05
Bending strength, ksi	43	43	28	45

der, and other additives into a sheet. This sheet is then heat treated to sinter the mass.

Initially, the mixture is prepared by cleaning and spray drying high purity powders which were previously ball-mill processed with other dispersant materials and sized to provide a particle size consistent with the required material specification. Organic binders and plasticizers are added to maintain the necessary cohesiveness and rheology of the material flow for further processing. The mixture is spray dried from the slurry to form a sheet of feedstock with flow characteristics similar to those of bread dough. The powdered feedstock is presented to the large rolls by a reservoir fed hopper system and is congealed by the rolls and compacted to form tape. Roll compaction requires sufficient binders and placticizers in the proper proportions to create and maintain a continuous sheet of tape which remains flexible enough to be reeled and hold the original shape.

Tape casting. In this process, depicted in Fig. 2.11, sheets of ceramic are manufactured through a low pressure process involving a slurry flowing under a knife-edge to form a sheet of uniform thickness. This is a low pressure process compared to the roll compaction method. The ceramic oxides are combined with plasticizers, binders, solvents, and other additives to form a

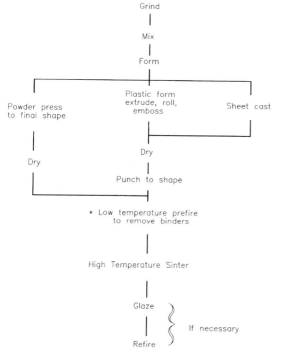

Figure 2.10 Process flow for substrate manufacturing.[7]

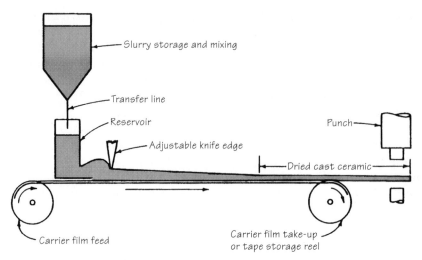

Figure 2.11   Tape casting of ceramic substrates.

slurry which is deposited on a Mylar or cellulose acetate film "carrier." The spread is moved under a knife-edge positioned a calculated distance above the carrier. A thin sheet of wet ceramic forms as the slurry-laden film moves under the knife-edge. The "green tape" thickness is achieved by careful adjustment of the knife-edge distance above the carrier. Air drying the slurry removes the solvents and allows the tape to be formed. Typically the tape is punched 30 to 50% larger than required to create alignment edges. The tape will air cure over time and therefore should not be overexposed to air.

Powder pressing.   This process requires a dry or slightly damp powder to fill a hard die cavity. The die cavity is made with abrasion resistant materials to maintain tolerances over the large number of cycles required of the project. The powder packed in this die is subjected to high pressures (8000 to 20,000 psi) over a time period sufficient to rheologically distribute the ceramic particles for semiequilibrial stress management. A dense ceramic body is formed which can be reproduced automatically to acceptable tolerances. The sintering process further stabilizes the body. Shrinkage is controlled or minimized during the sintering process by preloading or compressing the package.

The limitations of this process preclude it from producing parts larger than 6 in. square. Unavoidable pressure variations and the accompanying excessive part warpage result from process inconsistencies like uneven filling of complex dies which may be long or have pins close to the cavity edge (this causes holes to be formed close to the part edge).

Isostatic powder pressing.   Dry powders are used to fill a flexible die and surrounded with an even pressure produced by the addition of water or glycerin at a high (5000 to 10,000 psi) pressure. The even application of pressure causes a uniform compaction of the part. A relatively large aspect ratio (length to width) is processible. The pressed part must be machined to refine the rougher finish of the isostatic die interface. Some types of high volume products cannot support the added cost of this postpress machining and finishing operation. Isostatic pressing is therefore uneconomical for the production of many types of substrates.

Extrusion. Ceramic powders can be formed into uniform cross section parts by a combination of powders with organic materials. The material is forced through a die in a plastic, very fluid form. It is an economical, fast process which can produce fine detail extrusions. Tight tolerances of the final dimensions are hard to obtain owing to the amount of shrinkage involved during the drying and firing phases. This process replicates the roll compaction process when extruding thin sheets up to 0.1 in. The addition of glass to the slurry mix allows the extruded material to avoid problems associated with high temperature sintering and radial cracking around vias. Substrates and extrusions of pure alumina require sintering temperatures of approximately 1900°C, which makes the firing of the furnaces uneconomical in the manufacturing operation.

The sintering process. The oxide powder formed from a slurry is often prefired at 300 to 900°C to remove the organic binders. Firing at higher temperatures, or sintering, removes the plasticizers, organic binders, and lubricants used to maintain rheological consistency during the forming process. The sintering process contracts, shrinks, and densifies the small particle aggregate. Recrystallization of the particles or grains by heating up to the liquid phase during firing promotes grain growth, which greatly strengthens the substrate mechanically. Smaller grains produce a smoother surface, and conversely, the growth into larger grains during firing yields a surface roughness referred to as the "as-fired" surface. Grain growth is inhibited by the addition of certain additives which provide not only stability in the particle packing density but also grain uniformity. The scaling laws help predict the particle size.[14] However, the particle packing density and uniformity are a major determinant, a trait shared by all powder systems. Inside the grain structure the matrix increases in compressive strength as the crystals come to lie against each other and interlock. This process increases with organic removal.

Organic removal includes thermal decomposition and evaporation, which precede mass transfer of the volatiles out of the porous ceramic body. Oxidation of the remaining binder then takes place, further removing any adsorbed materials from the surface. The low temperature cofired substrate often used in the multichip module technology requires an even greater attention to the organic composition of the binders, lubricants, plasticizers, and solvents. The lower temperature ceramics sinter only slightly above the temperature of the organic burnout. Removal of 99.95% of the organics before sintering is a prerequisite to successful firing. The binder burnout characteristics are determined by the chemical kinetics, binder distribution in the substrate, and the heat and mass transport. The amount of heat available determines the organic volatile removal rate. The thickness of the ceramic influences the binder distribution, while the open porosity determines the resistance to mass transfer. Diffusion variations are responsible for the differential rates of removal from the surface as compared to the interior. The amount of heat available determines the rate of diffusion and the volume of liquid. Liquids are thought to be transported from large pores in the interior by capillary action to small pores on the surface where diffusion takes place.[3] The surface organics are removed first because of their shorter diffusion distance. Liquefied solids fill the smaller pores rather than homogeneously distributing throughout the body.

Ceramic binder systems produce free radicals and other materials from decomposition and burnout. During this thermal process several types of reactions are concurrently volatilizing the organic, losing hydrogen atoms and forming free radicals. Polymer scission, or splitting, to produce monomers or cross-linked products can occur at the same time. The principal importance of these reactions lies in the extent of decomposition and the remaining surface and intergranular impurities which can and do inhibit thick film processing.

Large grains primarily govern the shrinkage rates during sintering while size distribution also affects the sintering kinetics. Particle packing plays a significant part in the material densification. Matrix grains are constrained by the larger grains and prevent a uniform shrinkage. Considerable shrinkage is evident in the sintering process, approximately 50% by volume and 18–25% in the planar dimensions. The shrinkage affects all internal hole and slot locations which may be punched in the "green" material prior to firing.

The various methods of substrate fabrication are summarized in Table 2.12.

Substrate fabrication methods for aluminum nitride substrates.   Substrates of AlN are formed using the same types of processes as other ceramic substrates. Thin sheets of AlN (less than 1 mm thick) are typically tape cast, with thicker sheets and parts formed by pressing. The majority substrates are below 4.5 in square. Larger substrates in excess of 6 in square are available but at a greater cost per unit area.

The manufacturing process for AlN is more difficult than for alumina or beryllia. The sintering atmosphere is nitrogen, which complicates the organic removal process. The binder materials cannot be burned away owing to the lack of oxygen and must be evaporated and carried away by the nitrogen stream. Temperatures in excess of 1800°C are used in the heating processes, which require elaborate fixturing (often manufactured from boron nitride) to support and contain the substrates during sintering. Consistent substrate properties are maintained by the precise control of sintering conditions. Color consistency of the AlN substrate is a good indication of control in the sintering process.

TABLE 2.12   Methods of Substrate Fabrication[7]

Forming process	Material	Firing shrinkage, %	Commercial tolerances
Powder pressing	Alumina, 94–99%	16	±1%, but not less than ±0.005 in in any direction
	Steatite	10	
	Beryllia, 94–98%	16	
	Barium titanate	14	
Extrusion	Same as for powder processing	13–14	±0.5% but not less than ±0.003 in in any dimension
Isostatic processing	Alumina, 94–99%	16–18	Same as for powder processing
	Steatite	10–12	
	Beryllia, 94–98%	16–18	
Tape casting	Alumina, 94–99%	18–22	Length and width: ±0.5%
	Steatite	18–22	Thickness: up to 0.040 in, ±10%
	Beryllia, 94–98%	18–22	
	Barium titanate	18–22	

2.5   Substrate Cleaning

Cleaning processes are critical to good thick film adhesion and microstructure integrity. Alumina and beryllia are quite chemically resistant, allowing mild or strong aqueous organic solvents and mild aqueous solutions to be used for standard cleaning. Acid base–alkali alternation processes can be effective with unreactive substrates like alumina. Fluorine-based acids are not recommended because of their potential reactivity with glass binders in the intergranular substrate regions. Other water-based soils may be removed with one or a combination of deionized water, isopropyl (not methyl) alcohol, or various surfactant-based detergents or cleaning agents. If the specific soil is known, the cleaning process can often be simplified.

Most organic materials can be removed by simply prefiring the substrates in a thick film furnace prior to subsequent processing. This heating process should be performed only on substrates with high temperature stability, such as low stressed alumina or other unglazed nonreactive ceramics. Some stress-sensitive substrates should not be subjected to heated cleaning processes which may further aggravate, crack, or shatter the substrate.

Water-soluble coatings are often used as a slag deterrent and are also effective in preventing oil-based coatings from depositing on the surface during laser processing. These coatings may either be dip coated, spray coated, or brush coated.

Ultrasonic degreasing with trichloroethylene or Freon® solvent compounds is no longer environmentally acceptable and is being replaced by degreasing with oil-based solvents. Any solvent-based cleaning process must be maintained and frequently cleaned to avoid a buildup of undesirable stain-causing residues. If a solvent-carried residue causes a stain, the firing process may remove it. If solvents are not effective, plasma cleaning may be used despite the added cost. Oxygen ions in the plasma will decouple otherwise stubbornly adherent stains, and the process is comparable in many ways to firing. The effectiveness of cleaning and the oxidation process are two of these similarities.

Strong cleaning agents should be avoided with AlN (aluminum nitride). Alkaline solutions in particular rapidly attack the microstructure of AlN, etching it severely. Contaminant removal by prefiring prior to subsequent processing is often used with all types of substrate materials, including AlN.

2.6   Postfire Processing

While the surface of ceramic substrates is ready for processing as soon as sintering is completed, a degree of postfire processing is necessary. The outer edges of the substrates are generally unsuitable for the deposition of thick or thin films and must be removed or scribed for later removal. Holes required for connection between the top and bottom surface must be formed and annealed to promote adhesion of the deposited film or coating. The substrate must be scribed for later separation if multiples of the circuit configuration are to be fabricated.

It is informative to define the terms which will be used in the description of postfiring processing.

Annealing: The process of heat treating a ceramic structure to achieve the required measure of micro- and macrostructure modification. This fundamental surface change results from the thermochemical oxidative phase change that the substrate undergoes. The heat-affected zone is modified to provide improved metal adhesion to newly formed "tooth" sites on the previously glassy laser drilled hole walls.

As-Fired Edge: The substrate edge formed by firing the previously mechanical punched unfired (green) ceramic tape. This edge is not expected to be linear as on mechanically or thermally laser formed edges.

Ball Milling: A rotary method of grinding using a rotary drum filled with abrasive ball media used for removing sharp or uneven substrate edges and excess slag deposits.

Bead Blasting: A cleaning process employing high pressure glass bead spraying to remove slag. This process also enjoys a second-order peening–stress relieving effect.

Cell: The circuit board single entity.

Chamfered Corner: A purposely machined or scribed and snapped corner which functions as a quick reference for the working side and the datum(s).

Datum: The axis lines from which all hole coordinates, scribe line dimensions, and other substrate features are referenced.

Drilling: The machining of round holes 100% through the thickness of the substrate using laser or diamond tipped drill bits.

Fiducial: An alignment or substrate orientation reference mark usually scribed into the substrate on a specified corner.

Flat: A laser machined indentation on the substrate edge used as a pin guided precision plate alignment feature.

Heat-Affected Zone (HAZ):   A material region or zone adjacent to the laser thermally affected machined, scribed, or drilled edge, occasionally exhibiting a changed phase state.

Hole Shoulder: The edge of the hole as seen from the drilled side where the laser induced melt causes slump or rounding from the substrate surface to the hole wall.

Laser: The acronym for light amplification by stimulated emission of radiation.

Laser Tick: A small scribe mark intended for proper side orientation.

Locating Hole: Either a hole acting as a fiducial or a customer designated hole acting as a reference point.

Machining: Laser cutting a profile or contour at 100% depth.

Multiple-Up: The array or substrate master containing multiple cells which enables efficient print, fire, and assembly processes.

Pulse Depth: The average pulse penetration into the substrate; often measured by dye penetrant and fracture line testing.

Pulse Spacing or Period: The distance between adjacent scribe pulses as measured from center to center of the pulses.

Scribing: Pulse semiperforation of the substrate on a linear path. The substrate may then be separated into individual parts by snapping.

Slag: Redistribution of resolidified ceramic material adsorbed on the substrate surface during laser scribing and machining.

Snapping the Substrate: The process of singulating into individual parts using either mechanical or manual dexterity.

Tab: An unmachined section of the substrate profile used to retain the individual part(s) in the master substrate or array.

Taper: The angle of slope that the hole forms during drilling and machining.

2.6.1 Laser processing

Now material processing will be covered using "high-tech" tools new to the industry since the early 1960s. The ablation of materials by a laser takes place by thermal and photochemical as well as by other means. The type of laser and the wavelength determine the ablation method. The CO_2 laser is the primary laser type for drilling, machining, and scribing ceramic substrates.

The primary CO_2 wavelength used in industrial lasers for scribing, drilling, and machining is 10.6 μm. Other wavelengths are used for these processes other than the medium infrared wavelength. The Nd:YAG laser is a solid-state laser using a neodymium "doped" crystal resonating rod made from yttrium-aluminum-garnet. The lower power "Q-switched" Nd:YAG lasers produce a very controllable beam which can ablate materials sufficiently to be marginally effective in scribing and drilling thin ceramic (0.010 and 0.005-in thick). The pulsed Nd:YAG laser is much more effective but still not as efficient as the Q-switched Nd:YAG laser. Both lasers work in the 1.064-μm range, which is considered near infrared.

The other industrial laser used in processing materials is the excimer laser. This word (excimer) is a combination of "excited" and "dimer." The dimer can be (among others) XeCl, XeF, or KrF. These gases lase at different wavelengths in the ultraviolet band. An advantage of the excimer laser is in the size hole that can be resolved due to its short wavelength. 1-mil-diameter holes can be drilled in thin layers of uncured ceramic tape. Low-fire tape is an example of one material.

The process of crystal-produced frequency multiplying is allowing the Nd:YAG to process materials with a shortened wavelength, again in the ultraviolet band or the visible band. The principle involved is one of high energy ablation. Short wavelengths produce more energy per photon than larger wavelengths. This is demonstrated by the formula $E = h\nu$, where E is photon energy (possibly in Joules), h is Planck's constant, and ν is frequency. One can readily see the relationship increasing energy dramatically with the increase in frequency.

Ceramics and other materials are little affected until a threshold energy quantity is absorbed by the material surface. Ceramic binders and glass are readily ablated by the medium energy beams. The amount of percussive force impacting the ceramic is perhaps the most surprising aspect of medium to high energy beams in use, for example, with the excimer laser. This force enhances its capability in drilling fired ceramics in limited requirements.

The efficiency of the process and the corresponding absorption of the impacting wavelength by the ceramic are some of the reasons why in general the laser process is used. The edge quality and consistency of the process are a result of this absorption. The thermal expansion, thermal conductivity, and semisolid liquid granular structure of the material affect multiphase consistency of the material as it moves within itself during the thermal process of cutting, drilling, or scribing. The small breadth of the beam and the sharp peak of energy over time guarantee that a relatively small heat affected zone (HAZ) will result from this ablation. Lower stresses and therefore smaller strains result from the correct application of laser parameters to the material. Some of the parameters considered in the ablation process are:

Beam current

Pulse width

Pulse diameter

Beam mode

Focus (spot size)

Depth of focus

Polarization

Assist gas type

Gas pressure

The CO_2 laser is capable of retaining 86% of the beam energy within the central maximum cross-sectional profile of the beam because of diffraction limiting optics and the laser resonating structure.

A useful measure of beam quality can be found in the formula[8]

$$M^2 = \frac{\theta_{act}}{\theta_r}$$

where M is measurable at beam diameter locations, and $\theta_{act} \equiv$ the actual beam divergence of the laser in question, and $\theta_r \equiv$ the divergence of a Gaussian beam of the same size as the laser beam.

The mode of the beam and the related M^2 profile is a beam divergence determinant and contributes to the consistency of the ablation process. Proper and timely maintenance of the laser system is a primary contributing factor to this consistency, as the beam of the so-called slow flow laser design requires periodic adjustments and optical train cleaning and alignment to give a low fluctuating beam profile. In ordinary operation, however, the laser beam will always fluctuate as a normal response to the dynamic flow of plasma in the resonating chamber. The slow flow type of CO_2 laser has provided the smallest deviation in this fluctuation and the most consistent beam mode as compared to fast flow type and TEA type lasers based upon the CO_2 media. TEA is an acronym for transversely excited atmospheric and is merely another resonance format strongly resembling that of the excimer laser.

The "fast flow" laser type utilizes a dynamic gas flow mechanism to present freshly ionized gas media in plasma form to the resonating cavity. This flow circulates transversely across the beam axis. Multimode beam profiles of high power are possible from this type of laser.

"Slow flow" lasers are of an older design and are physically larger systems (Fig. 2.13) for the same power output. Axial flow gases are excited and flow several meters in this long resonating cavity where they are eventually evacuated by a vacuum pump.

The single-beam profile stabile mode is achievable with the slow flow CO_2 laser for uniform processing results. The higher peak power of the slow flow CO_2 laser improves the state of vaporization and reduces glassing at the HAZ in ceramic materials.

The laser system is composed of a laser power supply coupled to a workstation using an optical train of lenses, bending mirrors focusing mechanisms, and possibly boring head and beam splitting devices referred to as the beam delivery system. These components are used in conjunction with a motion system and numerical (NC) controller to provide the precise placement of smaller holes or vias, larger holes, perimeter cutting coordinates, and scribing pulses. The NC (controller) is a device to cause the motion system tables to move or transport the beam to precise coordinates.

2.6.1.1 Laser drilling with the boring head. Trepanning or boring holes in substrate materials using circular motion was accomplished by numerical (NC) controlled motion systems until the advent of optical trepanning. The boring head forms holes by rotating optics. This produces more uniform and economical holes than the table method of trepan contouring where the numerical (NC) controlled motion system moves to allow the laser beam to complete the perimeter of a hole. The two types of boring heads are rotating lens and counterrotating wedges.

Rotation of the focusing lens provides a manual method of hole trepanning. The lens traverses under the beam near the lens edge in a circumferential path. The beam refracts by offsetting the lens under the beam from the lens major axis by adjusting the lens holder slide. This method takes time to readjust for different hole sizes.

Optical wedges can produce round holes very economically when used as part of a numerical (NC) controlled laser system. Wedges with a predetermined angle (nominally 15' to 20') are offset electrically and out of phase with each other, and counterrotated. These wedges must be matched to each other so that a round hole results. The laser beam is refracted first through one wedge and then through the directly opposed circular mating wedge to produce an offset beam directed by a programmable controller. This controller is switched by the main X-Y motion system controller. Multiple hole sizes can be entered into the hole drilling program and then quickly executed.

2.6.1.2 Ceramic substrate drilling, machining, and scribing. The CO_2 laser can perform a multitude of process operations depending upon the fixture clamping or holding method, optical beam delivery system, process gas mixtures, and process program parameter settings. The type of process determines the optimum settings for each particular setup. No one setup is completely optimum for all processes, however. Trade-offs are to be considered for a mixture of processes required to drill, machine, and/or scribe the thick film substrate.

The standard optical setup for ceramic substrate drilling is similar to that for machining. A very small (0.004 in) diameter beam is required from the laser output objective at the focal point. The ideal beam profile or mode is TEM00 having a single maxim profile when considering the pulse and beam cross section. The appearance of the spot is gaussian in energy distribution shown by cross section or profile. This cross section provides the laser operator with the ideal spot size and condition substrate for thermal ablation.

A small spot size and proper beam profile or mode result in a small HAZ. The smaller HAZ will describe less localized heating and less resultant localized stress. The magnitude of slag buildup is also related partially to the focal spot condition.

A degree of slag around the exit side of a CO_2 laser drilled hole or via is to be expected. The amount of slag deposited depends upon the degree of beam alignment, mode, polarization, air jet injection, and cover gas used. Substrate coatings which dissolve in water are available for use on the substrate surface. These coatings are sacrificial as they provide a removable surface for the inevitable slag buildup, while it is primarily the exit side of the drilled hole which collects slag and it is the entrance side of the scribe pulse where surface slag develops. Within the scribing process molten slag is blown usually across the substrate surface by a horizontal air jet. Machining and drilling are not usually performed until after the scribing process because of the potential for slag to attach to the machined or drilled sidewall. Slag is easily removed from the surface, however, by light scraping with another piece of ceramic.

Laser machining utilizes the optical setup of the drilling operation to effectively cut the material with 100% depth. By varying assist gases and removing or adding electronic pulsing, the laser beam can be made to scribe or drill materials. Electronic pulsing is added to the normally continuous beam for scribing. Pulse durations from 0.1 to 1 ms are achieved. Circular beam polarization is another parameter which becomes necessary for symmetrical cutting and a quality edge condition.

Laser scribing can pose a stress generation problem. The reason scribing works is that the pulse string shape, consistency, frequency, and depth combine to create a stress riser (a point of material solidification where localized stress is much greater than surrounding stress) at the tip of each pulse (see Fig. 2.14). There may be microcracking at the very tip of that pulse, and the added strain at the pulse string causes the crack initiators (points of increased stress) to accelerate to fracture. This procedure further develops in a process of manually or mechanically "snapping the substrate along the scribe line." Adding scribe lines in the presence of already stressful areas can cause the crack initiators to proceed with crack acceleration, mostly along the scribe lines but also along the path of least resistance. Indeed, scribe lines will weaken a substrate more than a high density hole formation because of the crack initiator's integration with the scribe pulses. Annealing the substrates can further cause movement of the material around the stress area. Substrate composition and annealing temperatures can play a large part in the ultimate degree of cracking.

A standard tolerance for a scribed cell dimension edge to an internal feature is ±0.003 in. It is possible for a cell edge to crowd a cavity dimension causing further frailty and instability to the already brittle structure. This can be avoided, and emphasis must be added for the need of advanced and thorough circuit layout planning. Further, great amounts of practice and care must be given to the priority and technique of the "snapping" method because the snapped scribe line does not always snap cleanly without leaving a hook or other edge protrusion. Machined dimensions are more expensive but are often necessary to avoid the potential dimensional crowding experienced by the scribed and snapped perimeter. The snapped edge dimension can also be exacerbated by the annealing process.

Annealing is a thermal heat treating process designed to thermally modify the grain structure of the ceramic. The process is necessarily a low temperature annealing because the glass binding the grains melts and is diminished at higher temperatures (greater than 1450°C).

Annealing can provide a "tooth" mechanism for thick film adhesion to the laser drilled hole wall. Temperatures above 1350°C can cause glass to redistribute within the intergranular structure. Some compositions of ceramic have been responsible for low temperature meniscus loading of the pulses near the tips.

General substrate processing hints. Processing through the various preprint operations requires an understanding of physical and economic relationships concerning substrates.

Smaller substrate sizes cost less due to reductions in manufacturing yields from induced stress. The method of substrate manufacturing entails processes where stresses are induced into flat, thin plates. Subsequent hardening freezes these stresses in place. This causes unhomogeneous distributions of the stresses in predictable areas of the substrate plate. Smaller plate sizes have a smaller number of these stress concentration locations.

Stress concentration causes nonplanar movement or strain called cambering or warping. Excessive camber in substrate plates produces defective and unusable materials governed by print requirements for flatness (as cosmetically defined in ASTM F109 and ISHM SP009).

As-fired edges should be removed prior to printing to relieve tension at the perimter of the plate.

Smaller substrate sizes cost less owing to the manufacturer's higher yield. The designer should always plan for a substantial (up to 0.25 in) waste edge (nominally 0.150 in) around the cell array. This may not cost much more but can save a great amount in yield loss through further processing.

Aluminum nitride must be not only sintered in a dry, reducing atmosphere but also oxide inhibited during other parts of the preprint manufacturing process. The amount of laser beam energy must be carefully monitored during laser trimming, because the laser wavelength can reduce the underlying aluminum nitride to aluminum metal, which would effectively short circuit the cut. The problem can be overcome with the proper setup of the laser trimmer coupled with flooding the area with an oxidizing gas.

The efficiency of laser processing on ceramic can be compared to UIGs. The ultrasonic methods form the basis for low-stress adiabatic material processing. This process cannot be favorably compared costwise to laser processing.

2.6.1.3 Ultrasonic impact grinding. The ultrasonic impact grinding (milling) system (UIGS) can provide machined holes, cavities, and slots. This process is not as efficient as laser machining of substrate perimeters because of the large amount of time used to vibrate and eject ceramic material, although many holes can be ultrasonically drilled at one time. The UIGS is composed of an ultrasonic power supply, toolholder, consumable tool, and a numerical

(NC) controlled motion system. A fine abrasive is used to direct the ultrasonic energy to the part.

Ultrasonic impact grinding is accomplished by coupling a 20 kHz (nominal) frequency to the part and grinding away the brittle material. This is done by using a stainless steel tool with a slurry of boron carbide flowing at 6 gal/min. The tool continually impacts the surface of the brittle material and breaks off minute granules which are flushed away by the slurry. The stroke of the tool amplified grinding sequence can be as small as 0.0005 in to a maximum of 0.0025 in. Tools are necessarily fabricated to the precise dimensions of the required hole or cavity pattern of the part.

The advantages of hole drilling and cavity or slot grinding are derived from the nonthermal and chemical or electrical process of forming these configurations. Stresses which may be imparted to the workpiece by the above processes are not generated by the ultrasonic machining method.

2.7   Substrate Sources and Costs

2.7.1   Sources

Substrate producers in the alumina substrate market include Coors Ceramics Company and TechCeram in North America, Kyocera and NTK in Asia, and Hoechst Ceramtec in Europe. The Aluminum Nitride Producers include Sherritt Inc. in North America and Maruwa in Japan. Some other producers have chosen to concentrate their aluminum nitride efforts on packages, and also sell substrates on a commercial basis, however. These include Carborundum in North America; Toshiba, Tokuyama Soda, Sumitomo, and Kyocera in Japan; and Hoechst in Europe. The most prolific substrate suppliers of aluminum nitride thick film substrates in North America are Sherritt, Toshiba (via Toshiba America), and Tokuyama Soda (via General Ceramics). A few companies specializing in substrate machining also stock aluminum nitride substrates in North America, most notably Stellar Industries and Applied Laser Technology. Producers of note for beryllia are Brush Wellman in North America and Consolidated Beryllia Limited (CBL) in Europe. Both sell product in the United States.

2.7.2   Costs

Current (1994) pricing for 96% alumina approximates $0.06 per square inch, whereas aluminum nitride is about $1 per square inch and beryllia is ~$4 per square inch for $4.0 \times 4.0 \times 0.025$ in thick substrate in volume. The price for AlN is expected to fall by a factor of about 4 as the aluminum nitride market grows and production volumes increase. Alumina pricing has increased by 5–10% per year over the past 5 years. Beryllia is experiencing a price-technology competition with AlN and has increased only slightly over the past 5 years.

References

1. Richard L. Prober, *Ceramic Technology for Electronics,* 2d ed., International Society for Hybrid Microelectronics, 1984.
2. Brad Palmer, Aluminum Nitride Substrates for Thick Film Circuits, Applications Paper, Sherritt Gordon Limited, 1993.
3. Relva C. Buchanan, *"Ceramic Materials for Electronics Processing—Properties and Applications,"* 2d ed., Marcel Dekker, Inc., 1991.
4. Brush Wellman, "BeO Engineered Materials," *Technical Bulletin* GI-8510,1985.
5. John B. Snook, "Aluminum Nitride—General Characteristics & Specifications," *Stellar Industries Corporation Technical Bulletin,* 1988.
6. "Thick Film Substrates Technical Specifications," *Bulletin* 37-2-0893, Coors Ceramics Company, Electronics Division, 1993.
7. *Substrates for Thick Film Circuits, Basic Ceramic and Thick Film Substrate Materials,* McGraw-Hill, 1969.
8. William M. Steen, *Laser Material Processing,* Springer-Verlag, 1991.
9. S. L. Ream (ed.), "Focus on Laser Materials Processing," *Proceedings of the 6th International Congress on Applications of Lasers and Electro-optics,* IFS Publications, Springer-Verlag, 1987.
10. Special thanks to Lyle Gibson, Kyocera Industrial Ceramics, for the use of his library.
11. A. Winklemann and O. Schott, *Ann. Phys. Chem.,* vol. 51, p. 730, 1984.
12. This volume, chapter xx.
13. R. D. Jones, *Hybrid Circuit Design and Manufacture,* Marcel Dekker, New York, 1982.
14. For a definition of scaling law, see S. P. Parker (ed.), *McGraw-Hill Dictionary of Scientific and Technical Terms,* 4th ed., McGraw-Hill, New York, 1989, p. 1758.
15. R.L. Armstrong and J.D. King, *Mechanics, Waves, and Thermal Physics,* Prentice-Hall, Inc., Englewood Cliffs, N.J., 1970.
16. Rao R. Tummala, "Ceramic Packaging," in *Microelectronics Packaging Handbook,* edited by Rao R. Tummala, Eugene J. Rymaszewski, and Alan G. Klopenstein, Van Nostrand Reinhold, New York, 1989.

Thick Film Materials and Processes

Dana L. Hankey, Aziz S. Shaikh, S. Vasudevan,
and Chandra S. Khadilkar

3.1 Introduction

During the past several decades, the progressive development of complex integrated circuits has resulted in demanding requirements such as increased device densities, higher signal speeds, higher operating frequencies, miniaturization, increased reliability, and reduced costs. This has presented a great challenge to materials scientists who must accelerate development of materials and associated processes to keep pace with the rapidly changing environment. These materials include both active and passive components such as conductors, resistors, dielectrics (capacitors and insulators), varistors, filters, substrates, and sensors.[1–3] This chapter will focus on the basic technology for thick film materials with an emphasis on composition, design, processing, properties, test measurement methods, and structure/property relationships. The primary topics include conductors, resistors, and dielectrics.

3.2 Thick Film Conductors

3.2.1 Background

Materials classified as thick film conductors play a major role in hybrid microelectronics, electronic packaging, components, displays, and photovoltaic applications. Generally speaking, the functionality of thick film conductors is analogous to wiring in predecessor technologies. However, the current

material requirements are much more demanding. These challenges, as identified in the introduction, are further intensified when one considers the
broad number of processing conditions to which conductors may be subjected
as well as the compositional variations of other circuit elements that require
compatible interfacing. The net result requires optimization of conductor
properties and related processing parameters for specific applications.

The primary function of a thick film conductor is to provide an electrically
conducting path from one location on the circuit to another. Examples in
basic hybrid circuitry include signal traces, crossover connections, and interconnection of one level to another or to an external device. However, conductor materials perform a wide variety of additional functions in microcircuitry,
packaging, and component applications. These include the following:

1. Terminations for thick film resistors, capacitors, and inductors

2. Electrodes/terminations for multilayer and disk capacitors as well as
   other fine ceramics such as varistors, thermistors, sensors, filters, and
   inductors

3. Low value resistors, typically less than 10 Ω/cm^2 (sheet resistivity) for
   applications such as surge arrestor circuitry

4. Wire bond attachment

5. Die bond attachment

6. Discrete component attachment

7. Base for electroplating

8. Shielding materials

9. Metallizations for photovoltaic materials (solar cells)

10. Electrodes for flat panel displays

The above list is not all-inclusive but is a reasonable representation of most
high volume applications.

Thick film conductors consist of three basic phases: (1) a metallic conductive or functional phase (i.e., gold silver, copper); (2) a permanent inorganic
binder phase (i.e., high temperature glass and/or oxide phase); and (3) an
organic vehicle for dispersion and application to the substrate.

The properties of thick film conductors that are typically specified in
order to determine formulation selection for a given application include the
following:

1. Resistivity

2. Solderability and solder leach resistance

3. Line resolution

4. Physicochemical compatibility with other components such as resistors,
   dielectrics, and, most important, the substrate

5. Long-term stability in hostile environments such as heat, humidity, and
   thermal cycling (e.g., resistance to adhesion degradation)

6. Migration resistance

7. Cost

8. Wire and die bondability

Although many properties of conductor formulations are dependent on the bulk properties of the specific metallurgy and inorganic binder compositions, numerous techniques exist for property modification. These include formation of binary and ternary metal alloys, process modifications, and alternative material selections for such tasks as soldered-lead/device attachments and wire bonding. The physicochemical compatibility of thick film conductors with other circuit elements and various substrate types is one of the most critical parameters in formulation development. As a result of emerging technologies and more stringent packaging requirements, alternative substrates such as aluminum nitride (AlN), porcelainized steel, low temperature cofired ceramics (LTCC) of various compositions, low temperature glass (displays), and silicon (photovoltaics) present challenges for material formulators. The ensuing sections will address the numerous topics presented in this introductory section.

3.2.2 Basic constituents

The basic constituents of a thick film conductor can be divided into three distinct phases:

1. A metallic, conductive phase (or functional phase) that typically consists of finely divided metal powders or alloys, such as gold (Au), silver (Ag), copper (Cu), silver–palladium (Ag–Pd), aluminum (Al), and nickel (Ni)

2. A permanent binder phase which is inorganic in nature and typically consists of a mixture of glass powders and/or oxides

3. An organic medium, or vehicle, which acts as the carrier agent for the inorganic constituents and provides the rheology appropriate for the deposition technique utilized in application of the pastes on respective substrates

The constituents of polymeric thick film conductors exhibit variations to the above description. These type conductors typically consist of a silver, carbon (C), or Ag and C mixture functional phase dispersed in a thermoplastic or thermosetting polymer and solvent.[4] In the case of polymeric thick film conductors, the polymer serves as an insulative binder and adhesive for the functional phase.

Metallurgy. The conductive (or functional) phase in thick film conductors consists of elements as described in Table 3.1[5-7] which are categorized into noble and base metals. These metals are utilized in the form of powders which can have different morphological characteristics. The particle sizes range from submicrometer to several micrometers, and the selection of size, distribution, surface chemistry, and shape depends on the application and interfacing materials. Most metal powders are prepared by precipitation from various solutions. Powder morphology and agglomeration state are determined by the processing conditions. These powder characteristics determine packing efficiency and subsequent sintering kinetics which determine fired film densities. These sintering mechanisms will be discussed later.

As can be seen from Table 3.1, there is a broad spectrum of basic properties in the various metals. Those of utmost importance in material selection for a specific application include resistivity, melting point, and relative cost. Metals exhibiting high electrical conductivity, or low resistivity, come from Group IB [gold (Au), Ag, Cu] in the periodic table and have a valence of 1. Metals with valences of 2 or 3 (Groups II and III in the periodic table) cannot attain electrical conductivity values of Group I metals. However, these metals which include Pd, platinum (Pt), Ni, tungsten (W), and Al (not in Table 3.1) are also good conductors and are widely utilized in conductor formulations. Sheet resistivity, a parameter that is typically specified, is a material property that is independent of film geometry. The derivation of sheet resistivity as specified for a given thickness is presented later in the section on resistors. The values in Table 3.1 are theoretical sheet resistivities which are not fully attained in practice because of impurities and inability to achieve 100% dense structures. Polymeric materials, not included in Table 3.1, have sheet resistivities in the range of 0.01 to 0.08 Ω/sq. The melting points of several of the metals in Table 3.1 are high relative to standard thick film processing temperatures, which range from 500 to 1000°C, depending on the substrate material and application. The noble metals as well as Cu and Ni can be utilized for standard thick film processing, while W and molybdenum (Mo) are utilized in high temperature processes (1100 to 1800°C) such as cofired processing of alumina (Al_2O_3) and AlN packages.

Metals with strong bonding characteristics exhibit high melting temperatures, as shown in Figure 3.1.

The thermal coefficient of expansion (CTE) varies inversely with melting temperature of metals.[8,9] Except for W and Mo, the materials exhibit thermal

TABLE 3.1   Elements Used in Thick Film Conductors

Element	Density, g/cm^3	Electrical resistance, $\mu\Omega \cdot$cm at 25°C	Sheet resistivity, mΩ/sq/25 μm	Melting point, °C	Linear thermal expansion coefficient, ppm/°C	Thermal conductivity, W/m$\cdot$°K	Relative cost
			Noble Elements				
Ag	10.5	1.6	0.64	961	19.7	429	1
Au	19.3	2.3	0.92	1063	14.2	317	70
Pt	21.4	10.5	4.2	1769	9.0	72	100
Pd	12	10.8	4.3	1552	11.7	72	30
			Base Elements				
Cu	8.96	1.7	0.68	1083	16.5	401	0.02
Mo	10.2	5.2	2.04	2610	5.1	138	0.07
W	19.3	5.6	2.24	3410	4.6	174	0.06
Ni	8.9	6.8	2.72	1453	13.3	91	0.03

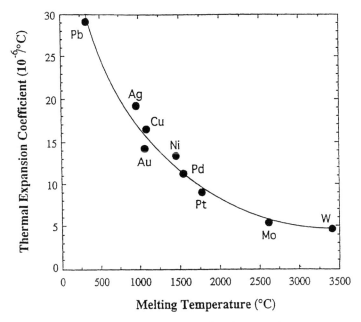

Figure 3.1 TCE vs. melting temperature of various metals.

expansion coefficients that greatly exceed the expansion coefficients of substrate materials on which they are deposited. This will create thermal stresses which develop during the cooling segment of the firing profile. Brittle materials such as ceramics would experience severe tensile forces which can lead to brittle fracture. However, in more ductile materials such as metals, the thermal stresses can be eliminated or relieved by plastic deformation. Resultant interfacial shear stresses can lead to metallization delaminations and even cracking in applications where there is repetitive temperature and/or mechanical cycling.[9,10] The wide variations in thermal conductivity of the metals could become important in applications where significant heat is generated, such as power hybrid circuits. This property can also be utilized for selection of heat dissipation in applications where thermal vias are employed.

The final general characteristic to be discussed from Table 3.1 is cost. In many small volume, high reliability applications, cost is not an issue and Au-based materials may be selected. However, high volume consumer and commercial applications will generally utilize Ag and Ag–Pd based materials as well as certain base metal conductors. One aspect of cost not factored into Table 3.1 is the associated processing costs of base metals. Refractory base metals such as W and Mo not only require specific controlled atmospheres, but also high temperatures and, therefore, significantly more power consumption. Base metals such as Cu can be processed at more conventional temperatures, but the costs associated with controlled atmospheres, such as nitrogen, for processing and storage are still significant.

TABLE 3.2 Solubility of
Precious Metals in Molten Tin
at 250°C

Metal	Solubility, wt %
Au	15.0
Ag	6.0
Pd	~0.5
Pt	1.0

Table 3.2[11] compares the solubility of various precious metals prior to solid phase precipitation in molten tin at 250°C. Au and Ag exhibit the highest solubilities, while Pd and Pt have very limited solubility. These inherent characteristics affect solderability, leach resistance, and film degradation.

Conductor compositions. Materials selection is a complex issue that is driven by the basic properties of various metals (as noted above), processes, substrate materials, the specific application, and required performance characteristics. The following discussion will review the various metallurgies and applications in more detail. Table 3.3[5] compares attributes of the various conductor system metallurgies.

Similarly, Table 3.4 compares sheet resistivity, leach resistance, and adhesion of various conductor types based on in-house experience. The adhesion values are stated in general terms because there are numerous adhesion tests and soldering methods which can produce a variety of results and discrepancies. Tables 3.1 through 3.4 can be referred to throughout the following sections.

3.2.2.1 Noble metal conductors

Gold and gold alloys. Pure Au conductor formulations exhibit high conductivity, excellent resistance to corrosion and migration, and are generally nonre-

TABLE 3.3 Attributes of Various Conductor System Metallurgies (Thick Film)

Material	Die bondability			Wire bondability		Solderability	Corrosion resistance
	Eutectic	Solder	Organic	Gold	Aluminum		
Au	Good	Poor	Excellent	Excellent	Good	Poor/good*	Excellent
Pt–Au	NG	Good	Excellent	Fair to poor	Fair	Excellent	Excellent
Pd–Au	NG	Fair	Excellent	Fair to poor	Good	Good	Excellent
Ag	NG	Good	Excellent	Good	NG	Good	NG
Pt–Ag		Good	Excellent	Good	NG	Good	Good
Pd–Ag		Good	Excellent	Good	Good	Good	Good
Pd–Pt–Ag		Good	Excellent	Good	Good	Good	Good
Cu	NG	Good	Excellent	NG	Fair	Excellent	Poor
Ni	NG	NG	Excellent	NG	NG	NG	Excellent
Polymeric	NG	NG	Excellent	NG	NG	NG	Good
W or Mo–Mn	NG	NG	Excellent	NG	NG	NG	Good

NG: Not good
*Depending on solder type.

TABLE 3.4 Typical Properties of Various Thick Film Conductors

Conductor type	Resistivity*	Leach resistance†	Initial adhesion‡	Aged adhesion
Gold	3–5	Use In solder	Fair	Fair
Gold/platinum	30–50	Excellent	Good	Fair
Gold/palladium	5–7	Use In solder	Fair	Fair
Gold/via fill	20–40	Poor	Poor	Poor
Silver	3–5	Use 62 Sn–36 Pb–2 Ag	Good	Good
Silver–platinum	4–6	Fair	Excellent	Good
Silver–palladium:				
25 Ag–1 Pd	4–7	Poor	Good	Good
6 Ag–1 Pd	12–16	Fair	Good	Good
4 Ag–1 Pd	15–20	Good	Good	Good
Copper	2–4	Good	Good	Fair
Nickel (air-fired)	40–70	Not solderable	Fair	Fair

*mΩ/sq based on fired thickness of 0.5 mils (12.7 μm).

†Based on results with 60 Sn–40 Pb solder at 230°C. If another solder type is recommended, it means that results in 60 Sn–40 Pb are generally poor.

‡Al$_2$O$_3$ substrate.

active when interfaced with other materials. Gold also exhibits good to excellent wire bondability with Al and Au, while also demonstrating the ability to be die-bonded with both eutectic and organic materials. Therefore, Au is suitable for applications in military and telecommunication circuits requiring high reliability interconnections, resistor terminations, complex multilayer configurations, high frequency stripline configurations, and pads for both wire and die attachments. However, factors such as cost and the inability to withstand commonly used tin–lead (Sn–Pb) solders[2,12] limits the applications of pure Au. Special solders such as Au–Sn alloys or indium (In) compounds must be used when soldering pure Au.[13] Since Pt is even more costly than Au and has an electrical resistivity almost 5 times that of Au, it is typically only used in its pure form for high temperature applications such as electrodes for sensors and high-fire multilayer ceramic capacitors (MLC). However, additions of Pt to Au conductor formulations can result in materials with excellent solderability and leach resistance at the expense of conductivity.[14] Typical Au–Pt resistivity values are 30 to 50 mΩ/sq at 12.7 μm fired thickness. Gold–platinum conductors are utilized widely in high reliability applications where discrete components are attached with solder or where circuit complexity requires frequent component replacement. Although initial bond strengths of Au–Pt compositions tend to be slightly lower than those of Ag–Pd alloys, they tend to have greater resistance to solder bond strength degradation.[15] However, the cost issue severely limits applications of Pt-containing golds to those requiring high reliability and solderability.

Addition of Pd to Au compositions results in properties similar to those of Pt additions while reducing the cost. These compositions provide some economical solutions to situations where military specifications exclude the use of Ag because of migration issues. The solder leach resistance and aging of these alloys are somewhat inferior to those of Au/Pt compositions. Although Al wire bonds can be made to Au conductors, bond strengths degrade after aging at

150°C or higher.[12] The degradation is due to the formation of intermetallic compounds such as Al_2Au_5 and $AlAu_4$. This formation leads to Kirkendall void formation within the Au because of diffusion. The addition of a few percent of Pd to Au reduces the susceptibility to wire bond strength degradation.

This is the result of a reduced concentration of Au and a high concentration of Pd at the leading edge of Al. Instead of void formations, a stable Al–Pd–Au intermetallic phase is formed which has high electromechanical integrity. Therefore, Au–Pd alloys are typically employed for Al wire bonding applications and can be die-bonded under suitable conditions (organic).

In addition to binary Au alloys, ternary alloys are sometimes utilized to optimize properties of interest. For example, Au–Pt conductors do not typically fire to high densities and tend to crack, or fissure, in high temperature processing. Small amounts of Pd to form a Au–Pt–Pd alloy usually increase the density of the fired film, reduce the tendency to fissure, and improve solder leach resistance.

Silver and silver alloys. Of the noble metals, Ag has the lowest resistivity (Tables 3.1 and 3.4) and is the most economical. Although Ag is readily wetted by molten Sn–Pb solders, its leach resistance is quite poor. Modifications to the solder composition, such as adding 2% Ag, improve the leach resistance significantly. Pure Ag is the material of choice for numerous applications including conductors for plasma displays, metallizations for solar cells, terminations and/or electrodes for multilayer and disk capacitors or other fine ceramics, and high volume hybrid applications where environmental conditions are not corrosive. However, the primary concern with Ag is its inherent tendency to migrate in the presence of an electrolyte, such as molten glass or water, and an applied electric field. The migration can occur very rapidly, as one can witness in a simple water drop test in which two Ag electrodes (250 μm apart) are subjected to droplets of distilled water and 5 V (Fig. 3.2). Electromigration results in Ag dendrite growth between the two electrodes,

SILVER MIGRATION TEST

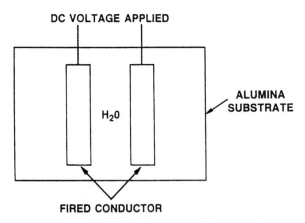

Figure 3.2 Electromigration test configuration.

causing a short circuit. Research has also shown that Ag can readily migrate through thick film dielectrics during the firing cycle because of its high solubility in many molten glass compositions.[16] Numerous molten glass compositions can dissolve up to 2 wt % of Ag at peak firing temperatures of 850°C by dissolution, migration in the molten glass electrolyte, and absorption/reprecipitation at the surface of the second electrode. This not only produces dendritic growth, but it also produces the release of gases (such as oxygen) and resultant bubble/blister formation.

Proper design of thick film dielectric materials, such as crystallizing formulations with optimized remnant glass subsequent to firing, can greatly reduce the tendency for Ag migration in multilayer configurations.[17–19] The referenced work presents data for both thick film multilayer and low temperature cofired tape formulations based on crystallizing glass formulations. Cross-sectional analyses using energy dispersive spectroscopy (EDS) for elemental Ag detection coupled with 2000-h bias humidity tests (85°C, 85% relative humidity, 30-V bias) suggest little or no Ag migration with specific dielectric compositions. Similarly, capacitance and insulation resistance measurements remained stable, supporting the idea that Ag migration is absent. Work by Alexander and Shaikh[18] suggests that buried Ag can be interfaced with surface Au compositions to eliminate surface migration problems and enhance wire bondability. When material selection is limited, Ag migration can also be reduced by proper circuit design and the use of special encapsulants. Significant technical advances are in progress to reliably enhance the use of Ag to take advantage of its cost and conductivity.

Binary Ag alloys are used extensively in the manufacture of thick film hybrids and electronic components. The most commonly used alloy is Ag–Pd. As illustrated in Fig. 3.3,[6] Ag and Pd exhibit complete solid solubility. These two metals can be prealloyed prior to preparation of the formulation or can be added as two distinct metallic components which alloy during the ensuing fir-

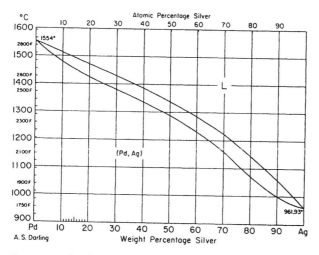

Figure 3.3 Ag–Pd phase diagram. (From *Metallography, Structures and Phase Diagrams*; courtesy of A. S. Darling.)

ing cycle. X-ray diffraction analysis[20] can be utilized to determine whether one has a true alloy prior to firing. Similarly, this technique can be utilized to determine if complete solid solutions have been formed during the firing cycle.

The addition of Pd to pure Ag improves solder leach resistance and reduces the tendency of Ag to migrate because of solid solution formation. These formulations are typically used for terminating resistors in network applications. Also, Ag–Pd conductors with low Pd content generally have good gold and aluminum wire bonding capability[21] (Table 3.3), although they are not suitable for eutectic die bonding. However, the addition of Pd introduces several undesirable attributes. The first significant issue is illustrated in Fig. 3.4[5] and Table 3.4, which show the substantial electrical resistivity increase with increasing Pd amounts. The maximum resistivity of approximately 40 $\mu\Omega\cdot$cm occurs near the composition of 40 wt % Ag–60 wt % Pd. This coincides with the temperature coefficient of resistance approaching zero which is useful in the design of low value resistors (less than 10 Ω/sq) for numerous applications, particularly in telecommunication and power hybrid circuits. The conductivity of Ag–Pd compositions can be further diluted if additional inorganic binder is necessary in order to achieve full densification as Pd content increases[5] (Fig. 3.5).

The second consideration surrounding Pd additions to Ag is the oxidation potential and kinetics of Pd, which is uncommon for noble metals. The reaction for palladium oxidation can be written as follows[22]:

$$\text{Pd (s)} + \tfrac{1}{2}\text{O}_2\,(\text{g}) \Leftrightarrow \text{PdO (s)} \qquad (3.1)$$

where s indicates a solid and g indicates a gas. The oxygen (O) partial pressure (P_{O_2}) can be expressed by the following equation[22,23]:

$$P_{O_2}^{-\tfrac{1}{2}} = \exp\left(-\Delta G_f^0 / RT\right) \qquad (3.2)$$

where ΔG_f^0 is the standard free energy of formation of one mole of PdO from Eq. (3.1), T is the absolute temperature in kelvin, and R is the gas constant. Standard thermodynamic data compilations[23] illustrate that if $P_{O_2} = 0.21$

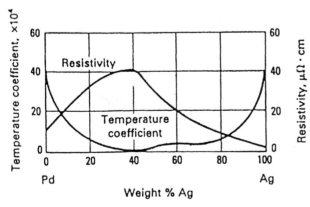

Figure 3.4 Electrical resistivity and temperature coefficient of resistance of Ag–Pd alloys. (From *Electronic Materials Handbook.*)

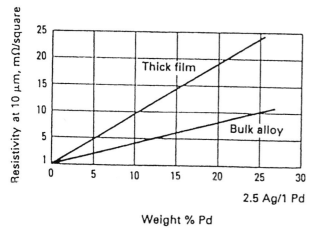

Figure 3.5   Effect of binder content on Ag–Pd alloys. (From *Electronic Materials Handbook.*)

atm (the value for air), the calculated temperature is 802°C. Therefore, although Pd metal is the stable phase above 802°C in air, palladium oxide is the stable phase below 802°C, which means that PdO will form on the surface of the conductor during cooling from a typical firing cycle. This can result in reduced conductivity and solderability.

The above analysis assumes unit activity for pure Pd. In a Ag–Pd alloy, the activity of Pd will be reduced and the temperature for Pd oxidation will depend on Pd content. Rapid cooling of the composition during firing can reduce the amount of PdO formation by reducing the oxidation kinetics. As discussed earlier, Ag–Pd formulations can utilize either prealloyed powders or individual constituents that can alloy in situ. Palladium oxidation can inhibit the solid solution process as illustrated by Vest,[22] where 80% Ag–20% Pd (by weight) did not completely form a solid solution even after 24 h at 850°C. In this study, an equivalent degree of solid solution was attained after 20 min at 450°C when an atmosphere of nitrogen (N_2) containing less than 1 ppm O_2 was utilized. Although one apparent solution would be to use pre-alloyed compositions, finely dispersed alloy powders are not always readily available or suitable for specific applications.

Although silver to palladium ratios of 3:1, 4:1, and 6:1 are commonly used in hybrid applications, thrifted versions utilizing Pd contents less than 5% have been shown to have good leach resistance in 62% Sn–36% Pb–2% Ag solders.[24,25] Table 3.4 compares properties of various Ag:Pd ratios. These materials meet the requirements of high conductivity materials needed for circuits with higher speeds and densities. Similar performance can be met by small additions of platinum to silver (1:100).[26] The leach resistance of a 1% Pt addition is shown to be equivalent to a 10% Pd addition with minimum effect on conductivity. However, Ag migration such as dendritic growth and diffusion into resistors or capacitors when used as a termination material is still an issue with only minor Pd or Pt additions.

The soldered adhesion of silver conductors, particularly Ag–Pd compositions, appears to undergo more degradation on thermal aging than their Au counterparts. Numerous articles have been written on the subject of intermetallic formation leading to mechanical failure during thermal or power cycling[27–31] after application of tin–lead solders. The last article,[31] by Chiou et al., focuses on intermetallic formation at Ag–Pd conductor interfaces. The research concluded that the formation of intermetallic compounds such as Pd_3Sn_2, Pd_2Sn, Pd_3Sn, $PdSn$, Ag_5Sn, and Ag_3Sn results in adhesion degradation above 130°C for times exceeding 40 h. Tin diffusion also causes conductor swelling due to volumetric changes of intermetallic phases.

3.2.2.2 Base metal conductors.

Copper conductors have gained some acceptance over the past decade because of their relatively low metal cost, low resistivity, good adhesion on Al_2O_3 substrates, excellent solder leach resistance, and low migration tendency. Advances in compatible thick film dielectric formulations have resulted in significant use of Cu in multilayer interconnect boards, primarily for military applications. Also, uses of Cu conductor materials have included power hybrid and microwave-related applications. Their applications in more complex systems and networks have been limited by the availability of state-of-the-art nitrogen-firable resistor systems. However, there are additional factors that complicate the widespread usage of this versatile material.

Because of copper's strong oxidation potential, this material must be processed in nitrogen atmospheres containing very low partial pressures of oxygen (typically less than 10 ppm O_2). The cost of processing and storage in nitrogen can be considerable because of the high gas flow rates necessary to prevent oxidation. Most important, however, there are a number of deleterious effects associated with processing in a controlled atmosphere that must be addressed. These include complex phase equilibria scenarios at elevated temperatures for processing Cu (900 to 1000°C) and the difficulty of organic removal in a nonoxidizing atmosphere. The thermodynamics of processing Cu in various atmospheres was studied in detail by Vest[7] as well as by Palanisamy and Sarma.[32] Figure 3.6 is a phase stability diagram[7] for various elements critical to thick film technology as derived from theory and thermodynamic data from Turkdogan[23] and Gaskell.[33] Since the free energy of formation becomes more negative as the oxygen partial pressure decreases [Eq. (3.2)], the more stable oxides at elevated temperatures include alkaline earth materials, while oxides of bismuth (Bi) and lead (Pb) are significantly less stable. Referring to a peak firing temperature of 900°C and O_2 partial pressure of 1 to 10 ppm O_2, it would appear that C would oxidize in this region and Cu metal would be the stable phase. In actuality, cuprous oxide (Cu_2O) is stable in this region.[7,32] Bright metallic films are produced in this region, however, because of the extremely slow copper oxidation rate at 900°C compared to the rapid diffusion rate of O_2 into Cu. The presence of copper oxide on copper particles during the initial firing stage actually promotes adhesion. It would also appear that lead oxide (PbO) and bismuth trioxide (Bi_2O_3) would remain stable at the 900°C firing range in 1 to 10 ppm O_2. However,

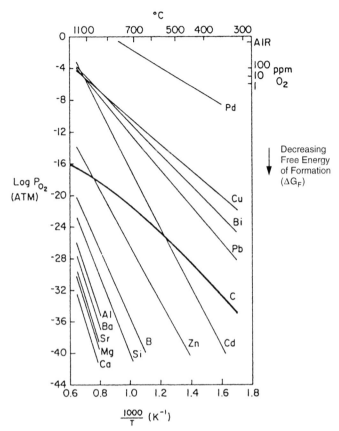

Figure 3.6 Phase stability diagram. (From *Ceramics Materials for Electronics.*)

lack of complete organic removal leads to localized reducing atmospheres (C line in Fig. 3.6) and subsequent reduction to the metallic phase.

Research by Pitkanen et al.[34,35] concluded that the presence of Cu_2O in copper conductor and dielectric films produced blistering phenomena. Lund[36] proposed that Cu_2O can be reduced to a copper precipitate that forms protuberances that eventually lead to a shorting phenomenon. Additional research has also suggested that reduction of lead and bismuth oxides leads to vaporization/condensation of these metals in furnaces and onto substrates creating cosmetic as well as functional problems. Reduction of Bi_2O_3 can also lead to grain boundary penetration into Cu conductors and subsequent formation of brittle alloys.[37] From a design standpoint, the ideal situation is to eliminate lead- and bismuth-containing oxides or glasses from all formulations to be processed in nitrogen. However, they have numerous benefits and cannot always be eliminated.

Although Cu exhibits excellent leach resistance and good solderability, the most severe limitation is the deterioration of adhesion on Al_2O_3 substrates when it is soldered with Sn–Pb alloys and subjected to thermal aging and/or temperature cycling. The formation of various copper/tin intermetallic compounds[38] such as Cu_3Sn and Cu_6Sn produces a thermal expansion mismatch at the conductor-substrate interface. The degradation is accelerated by voids in the solder and conductor porosity. This has prevented Cu from being accepted into the automotive industry where corrosive environments abound and acceptable thermal shock and power cycling behavior is imperative.

In addition to Cu, other base metals not requiring high firing temperatures include Ni and Al. As seen in Table 3.4, Ni compositions are typically modified for firing in air and exhibit sheet resistivities in the 40 to 70 mΩ/sq range (12.7 μm fired thickness). These compositions are generally not solderable and have fair adhesion values on Al_2O_3 substrates. However, Ni and/or Ni-containing compounds such as nickel borides and silicides are utilized in many nonhybrid applications such as end terminations for capacitors and electrodes for gas discharge display devices.[39] Similarly, Al compounds have found nonhybrid applications as back metallizations for photovoltaic devices. The requirements are quite different than those for traditional thick film conductors. Nickel and aluminum requirements in these applications will be discussed in more detail in a later section.

Refractory metals. Refractory metals, such as W, Mo, and molybdenum–manganese (Mo–Mn), have been used for many years in high temperature packaging applications. Although Mo–Mn compositions can be used as postfired metallizations on particular substrate types such as Al_2O_3, AlN and beryllium oxide (BeO), the majority of the W and Mo paste consumption is in high temperature multilayer cofired applications. These conductors, as seen in Table 3.1, exhibit high resistivity values compared to Au, Ag, or Cu. As a result of their high melting points, high temperature processing is necessary for production of dense fired films. These materials are also nonsolderable and must be plated with a solderable metal or alloy such as Ni prior to soldering. They require processing in moist, reducing atmospheres. Refractory metallizations are typically utilized in high volume commercial areas such as automotive applications (i.e., ignition module, voltage regulators).

Polymeric conductive formulations. The conductive systems used in polymeric formulations include C, Ag, Ni, Cu, and various mixtures. Table 3.5 illustrates typical sheet resistivities for various metal powders and flakes. Silver flakes produce the most conductive system, since silver oxide remains conductive and exhibits slow oxidation kinetics, resulting in wide processing latitude

TABLE 3.5 Typical Sheet Resistivity of Conductive Fillers

Conductor	Resistivity for 25-μm film, Ω/sq
Silver	0.01–0.1
Silver mixture	0.1–1.0
Nickel–copper	1.0–10.0
Carbon	10.0–100+

with respect to curing cycles. Base metals oxidize in air and, therefore, require very limited curing profiles. As seen in Table 3.3, polymeric conductors have reasonable corrosion resistance and can be used for organic die attachment applications. However, wire bonding and solderability are typically poor for these systems.

3.2.3 Inorganic binders

Thus far the discussions have focused on thick film conductor formulations from a metallurgical standpoint. This section addresses inorganic binder compositions and their importance in property development of thick film conductors, particularly adhesion. These inorganic binder systems are typically composed of glass compositions and oxide additives.

Glass compositions are primarily utilized in thick film formulations for two reasons: (1) to promote and develop adhesion to the substrate and (2) to affect the sintering kinetics and mechanisms of the composite. General requirements for glass compositions include the following properties[7]: good dielectric and mechanical strength; low dielectric loss; corrosion resistance, particularly with respect to moisture and acids; high electrical resistivity; thermal shock resistance; abrasion resistance; and low dielectric constants except in the case of capacitor dielectrics. In addition to the above general properties, additional physical properties of the glass that will be relevant to the particular application and substrate type include thermal expansion coefficient, viscosity as a function of temperature, glass transition temperature, surface-tension/temperature relationship, and compatibility with each material with which the glass will interface.

Typical thick film glasses include a variety of borosilicate compositions ranging from Pb and Bi borosilicate compositions to alkaline earth borosilicate systems including barium (Ba), calcium (Ca), and magnesium (Mg). Although alkali ions serve as excellent fluxes in glass, their content should remain low in thick film glass compositions in order to minimize electrical conductivity and dielectric loss. During the firing cycle, the glass must exhibit an optimum change in viscosity with temperature in order to control the sintering kinetics of the metal powders. This will lead to optimum densification of the fired film without bubble entrapment, blister formation, capillary movement of excessive amounts of glass to the fired film surface, and exaggerated grain growth of the conductive phase. The glass must also wet the substrate in order to penetrate the ceramic to a certain extent to provide mechanical and chemical bonding to enhance adhesion. This is accomplished by mechanical interlocking with substrate surface grains augmented by interdiffusion of ions. Typical weight percentages of glass additions range from 1 to 10%. The glass should penetrate the functional phase or metallic network as well as the substrate. Processing conditions and glass viscosity/temperature relationships must be optimized in order to prevent a continuous glass film formation between metal and substrate. This can result in brittle fracture and crack propagation at the interface under conditions of thermal and power cycling or relatively low mechanical stresses. The chemical composition of the glass is not only critical to the above requirements, but determines the chemical compatibility with the substrate in question.

The majority of thick film research in the past has been conducted on alumina substrates. However, rapidly changing technology has introduced new substrate materials which require utilization of alternative glass compositions. For example, typical lead/bismuth borosilicate compositions compatible with alumina are rapidly reduced when processed on aluminum nitride substrates, as seen in Fig. 3.7.[40] Therefore, alternate compositions such as alkaline earth borosilicates with lower thermal expansion coefficients must be developed to optimize compatibility with the substrate.[41-43]

A scanning electron micrograph illustrates a cross section of an AlN-compatible glass in Fig. 3.8.[43] A topographic scanning electron micrograph (Fig. 3.9) illustrates the fracture surface of a conductor containing the compatible glass on AlN.[43] This micrograph shows the excellent mechanical interlocking of the referenced glass composition by the presence of fractured glass structures still adhering to the substrate after a pull test. Figure 3.10[43] shows a cross section of an optimized Ag–Pd thick film formulation on AlN.

Similar research has been necessary to optimize formulations for applications on alternative substrate types such as BeO, low temperature tape, and nonhybrid substrates such as glass substrates for displays, silicon (Si) for photovoltaic applications, and various ceramic bodies for component applications such as titanates for MLCs, zinc oxide varistors, and zirconia sensors. In each application, not only will the chemistry vary, but the particle size, surface area, and volume percentage of the respective glass compositions will also vary.

As previously mentioned, the glass composition is also critical to compatibility of thick film conductors with other thick film materials such as resistors and dielectrics. This requires additional considerations when the formulation is designed. Figure 3.11[44] shows conductor/resistor incompatibilities at

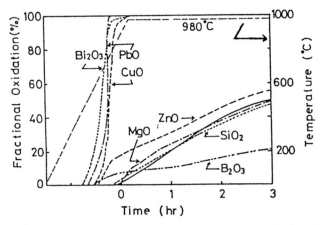

Figure 3.7 Effect of 5 mole % oxide additions on the oxidation of AlN (solid line: oxidation of pure AlN). (From *Proceedings of IMC.*)

Figure 3.8 Cross section of compatible glass film on AlN substrate.

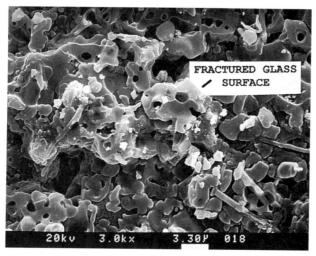

Figure 3.9 Fracture surface underneath conductor film illustrating bonding with AlN substrate.

the interface resulting in cosmetic as well as performance-related issues. Glass chemistry selection is also very dependent on the firing atmosphere, as discussed earlier in the section on Cu conductors.

The second class of inorganic binders includes oxide additives that perform various functions. Numerous oxides such as copper oxide and cadmium oxide promote adhesion on alumina substrates by formation of spinel compounds[23] such as copper aluminate ($CuAl_2O_4$). This type of bonding mechanism is a chemical or reactive bond. Reactive oxides are typically present in concentrations of less than 2 wt %. The benefit of this type of adhesion promoter is the

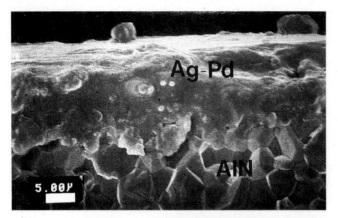

Figure 3.10   Cross section of Ag–Pd conductor on AlN.

Figure 3.11   SEM photomicrograph of incompatible conductor–resistor interface and conductive phase thinning.

absence of deleterious effects on wire bonding and solderability. Other oxide additions, such as Bi_2O_3, have been found to flux many glass systems and aid in wetting the substrate and enhancing adhesion. This is called *flux bonding* and typically utilizes additions of less than 5 wt % of the respective oxides. M. Hrovat and D. Kolar[45] investigated the Bi_2O_3–Al_2O_3–CuO system and concluded that, although no ternary compounds exist in this system, numerous binary compounds exist that melt incongruently at various temperatures and assist in adhesion promotion. Thick film conductor compositions are, therefore, commonly regarded as having three different types of bonding mechanisms to the substrate:

1. Frit bonding, which is totally dependent on the glass-substrate interaction

2. Reactive bonding, which depends on the chemical reaction between oxide addition and the substrate

3. Flux bonding, which is a result of oxide additions to flux mixed-bonded conductors typically containing all of the above components

Reactive-bonded conductors offer the best bond between noble metals and the substrate, which is an oxygen-metal bond. Frit-bonded systems offer surprisingly strong bonds via mechanical interlocking mechanisms, but remain dependent on appropriate firing conditions. Excessive interactions can significantly reduce the strength of mechanical bonds. Mixed-bonded systems offer unique solutions for good adhesion by combining the above mechanisms.

Organic vehicles. Organic vehicles, which act primarily as the transfer medium in thick film compositions, consist of one or more high molecular weight polymers, such as ethyl cellulose, and low vapor pressure solvents, such as terpineol or glycol ethers. The vehicle also usually contains rheology modifiers such as surfactants and thixotropes. The composition of an organic vehicle for a given formulation depends on the following: (1) the concentration, densities, and morphological characteristics of the metallic and inorganic binder phases; (2) the method of paste manufacturing; (3) the deposition method (screen printing, dipping, spraying); and (4) the subsequent processing environment such as drying and firing profiles and atmospheres. Rheology is a very complex issue and is discussed later in the chapter as a separate topic.

3.2.4 Microstructure development

Microstructural development of the thick film compact during the firing process is critical since the resultant microstructure will largely determine the respective properties of this material. The relationship between processing, microstructure, and properties is extremely important in the production of all materials, especially in complex systems such as thick film composites which involve metals, ceramics, and polymers. The physicochemical properties of the raw materials will affect green packing densities, sintering kinetics and mechanisms, and the resultant material properties.

Initially, the particles in a given system, as printed and dried, are in intimate point contact prior to the specific firing cycle. As the temperature is raised during the firing cycle, however, several forces and resultant processes arise that lead to densification of the compact. This process is referred to as *sintering*.

The principal driving forces that are responsible for material transport and sintering are capillary-type forces that are a result of surface and interfacial tensions. Subsequently, these tensions or forces produce stresses that are related to the curvatures of the respective surfaces and interfacial areas. The pressure difference ΔP produced across curved surfaces of two different particles or grains can be represented as[46]

$$\Delta P = \gamma \left(\frac{1}{r_1} + \frac{1}{r_2} \right) \tag{3.3}$$

where γ is the surface tension and r_1 and r_2 are the radii of curvature of the surfaces. This pressure difference, for example, causes liquids to rise in capillaries. This equation is usually presented with stress σ replacing ΔP and is called the *LaPlace equation*.[47]

Similarly, there is a free energy difference ΔG across a curved grain boundary represented by[46]

$$\Delta G = \gamma \, \overline{V} \left(\frac{1}{r_1} + \frac{1}{r_2} \right) \tag{3.4}$$

where γ is the boundary energy, $\overline{V}$ is the molar volume, and r_1 and r_2 are the principal radii of curvature as above.

This free energy difference applied to material on two sides of a grain boundary is the driving force that moves a boundary toward the center of curvature and leads to densification and grain growth. Therefore, material transfer, on a microscopic scale, is related to both the pressure differences [Eq. (3.3)] and changes in free energy [Eq. (3.4)] across a curved surface. The free energy change that gives rise to densification processes in materials is the lowering of the surface free energy and respective surface areas. This results in elimination of higher energy solid-vapor interfaces by the formation of lower energy solid-solid interfaces.

Although sintering is typically divided into three stages, the major focus is on the initial stage of sintering where vapor pressure differences between the neck area and particle surface results in transfer of material, resulting in neck formation[46] (Fig. 3.12). In the case of solid-state sintering, a general equation for isothermal neck growth between two spheres can be expressed

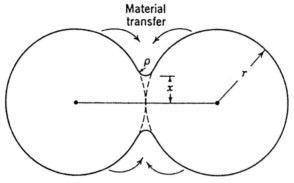

Figure 3.12 Neck formation during initial stages of sintering. (From *Introduction to Ceramics.*)

as[48]

$$\left(\frac{x}{r}\right)^{N} = Bt \tag{3.5}$$

where x = neck radius
    r = particle radius
    N = constant dependent on the sintering mechanism
    B = constant
    t = time

The small negative radius of curvature ρ at the neck produces a vapor pressure that is significantly lower than that of the particle. This results in the driving force for material transport.

Material transport during the initial stages of solid-state sintering can occur by a number of mechanisms, as displayed in Table 3.6.[46] If the material is transferred from the surface to the neck by surface, lattice, or vapor transport, there is no resultant shrinkage of the compacted material or decrease in porosity. However, if the matter source is in the bulk material—grain boundaries or dislocations, for example, then shrinkage and pore elimination occur. Figure 3.13 illustrates some of these mechanisms.

Thick film formulations exhibit different types of dominant sintering mechanisms, depending on their composition. For example, thick film conductors can sinter like solid-state metal particles or spheres, while sintering of resistors and dielectrics is more complex and largely dependent on glass sintering behavior.

Nordstrom and Yost[48] studied the sintering behavior of thick film reactively bonded Au films and concluded that surface diffusion is the controlling mechanism in these systems, as in prior findings about sintering of high purity Au spheres[49] and pure Au wires.[50] Intermediate and final stage sintering was controlled by volume or grain boundary diffusion mechanisms. However, most thick film conductor formulations include low temperature glasses and fluxes to aid in the densification process at relatively modest time/temperature profiles. This can result in additional sintering mechanisms, referred to as *vitrification* and *reactive liquid sintering*. Vitrification is densification with the aid of a viscous liquid phase which results in microrearrangement by viscous flow of material into the pore regions. Kingery[46] defined the initial neck growth rate by this type of mechanism by the following equation:

TABLE 3.6 Alternative Paths for Matter Transport during the Initial Stages of Sintering

Transport path	Source of matter	Sink of matter
Surface diffusion	Surface	Neck
Lattice diffusion	Surface	Neck
Vapor transport	Surface	Neck
Boundary diffusion	Grain boundary	Neck
Lattice diffusion	Grain boundary	Neck
Lattice diffusion	Dislocations, neck	

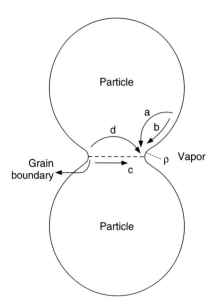

Figure 3.13 Material transport mechanisms during initial stage of solid-state sintering: (*a*) volume diffusion, (*b*) surface diffusion, (*c*) grain-boundary diffusion, and (*d*) volume diffusion from grain boundary.

$$\frac{x}{r} = \left(\frac{3\gamma}{2\eta l}\right)^{1/2} t^{1/2} \tag{3.6}$$

where η is the viscosity of the viscous component.

Sintering with a reactive liquid refers to a process in which the solid phase has limited solubility in the liquid at the sintering temperature. During this sintering process, the solid phase dissolves in the glass at the high energy region and is transported to the low energy neck region where reprecipitation of the solid phase results in increased grain size and densification. Densification rates depend on the amount of liquid phase, the solid phase solubility in the liquid, and the solid phase wettability.

Cole[51] reported significant increases in sintering rates of Ag/Pd compositions attributable to the solution/reprecipation process described above. As it is important to control the viscosity of the respective glasses for optimum densification in the above processes, control of heating rates, peak firing temperature, and glass compositions are critical parameters for attainment of optimum densification. Sintering mechanisms for thick film resistors and dielectrics which have high concentrations of glass constituents are addressed by Vest.[52]

The sintering process is undoubtedly very complex and various mechanisms compete and occur simultaneously for a given system. Sintering of thick film conductors typically involves all of the mechanisms described above. Variables such as composition, particle size, surface area, packing efficiency (green density), glass viscosity, surface tension, wettability, impurity levels, atmosphere, time, and temperature all affect the sintering rates and mechanisms. Figure 3.14[5] is an example of the type of data that can be

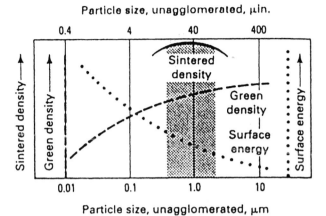

Figure 3.14 Effect of particle size on green and sintered densities. (From *Electronic Materials Handbook.*)

generated for a specific metal system to understand processing latitude of selected variables. In this figure, the preferred particle size can be determined in order to achieve optimum green density and maximum sintered density. Generally speaking, particle size increases result in decreases in surface area and associated surface energy. Green density is determined by these factors as well as interparticle forces that can control the degree of agglomeration. The final level of densification is controlled by many factors, including the green density and related particle/pore structure as well as sintering profiles.

Wu and Vest[53] conducted research on Ag conductor systems with varying amounts of glass frit. For compositions containing 2% glass frit, initial porosities were approximately 50%, which is comparable to theoretical values (48%) for simple cubic packing of spheres. As glass frit content increased, initial porosities increased because of inefficient packing. With respect to particle size of Ag, relative shrinkages of conductors with particle sizes less than 1 μm were much higher (as expected) than conductors with Ag particle sizes greater than 8 μm. These studies illustrate the significance of optimization of all sintering parameters for achieving maximum densification. An example of complete densification can be seen in Fig. 3.10, noted earlier. The relationship between microstructure of fired thick films and various properties will be addressed in later sections for conductors, resistors, and dielectrics.

3.2.5 Conductor properties and test methods

The combination of several factors including bulk material properties, conductor composition (i.e., metallurgy and glass content), substrate type, and the various processing steps (i.e., deposition process and firing profile) will determine the properties of the fired conductor. In most cases, the testing configuration depends on the conductor type and the specific customer requirements, which usually depend on the application. In this section, the basic tests and various factors that can affect test results will be addressed.

Test pattern. The basic conductor properties can be measured using a single test pattern, as illustrated in Fig. 3.15. These include resistivity, print definition and film thickness, film density, solder leach resistance, wettability, adhesion, and wire bondability. Each property will be discussed individually with reference to Fig. 3.15. Many applications require functional use tests which usually require specific test patterns and even multilayer construction processes. Similarly, numerous applications require standard conductor tests on thick film dielectrics instead of the bare substrate.

Fired film thicknesses can be measured by a number of instruments such as light-section microscopes and surface profilometers. Thickness uniformity is an important parameter, since variations can affect properties such as resistivity, adhesion, and solder leach resistance. Print definition is usually measured with surface profilometers and provides a measure of print quality, since it relates to both printing conditions and paste rheology. Undesirable conditions such as excessive line spreading and splining (commonly referred to as *drying lines*) can be detected. Figure 3.16 illustrates a data printout from a surface profilometer. The data were collected on the conductor test pattern in Fig. 3.15 by profiling a line perpendicular to the serpentine pattern (*b*). The trace illustrates the print definition as well as the actual thickness and uniformity of the print. A similar profile could be utilized on pattern (*c*) to study line definition. In this particular pattern, line widths and spaces down to 125 μm are utilized to determine print resolution. Microscopy techniques can also be used to study print resolution.

Conductivity. The conductivity for various metallurgy types varies significantly, as discussed in earlier sections. This property is especially important for applications requiring rapid signal speed. Electrical conductivity σ can be defined by the following relationship[46]:

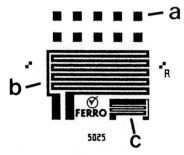

Figure 3.15 Conductor test pattern serving various functions: (*a*) pads for adhesion, wettability, solder leach resistance and wire bondability; (*b*) serpentine pattern for measuring print thickness and resistivity; and (*c*) pattern for print resolution.

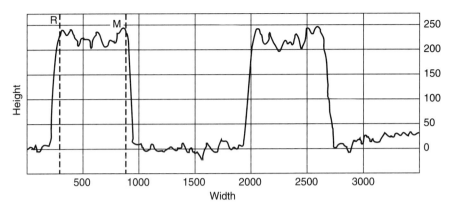

Figure 3.16 Surface profilometer tracing of thick film conductor. Dashed lines represent width of line at peak height.

$$\sigma = \frac{j}{E} \tag{3.7}$$

where j is the electric current density (charge transported through a unit area in unit time) and E is the electric field strength.

In practice, the figure of merit for conductors is resistivity, which is the inverse of conductivity. Resistivity is the ability of a material to resist the passage of current (units are ohm-centimeters for bulk resistivity, as illustrated for metals in Table 3.1). Surface resistance or line resistance is expressed in ohms. The common expression for thick film conductors is sheet resistivity (expressed in milliohms per square), which is the electrical resistance measured across opposing sides of a square for a normalized thickness.

Resistivity is measured by utilizing a four-point probe ohmmeter and very low currents. As illustrated in Fig. 3.15*b*, a serpentine pattern of 200 to 400 squares in length is usually employed for this measurement. The pattern in Fig. 3.15*b* is 360 squares. Resistivity depends on the metallurgy, composite formulation, fired density, and thickness.

Wettability. Wettability is defined as the extent to which a metal surface is clean and free of oxides so that metal contact can be made between solder and the metal. This term is used interchangeably in hybrid microelectronics with *solderability*. Various conductor systems exhibit different degrees of solder acceptance, as illustrated in Table 3.3. The tests for wettability in the past have been largely subjective and qualitative and have been based on visual observations. Most solderability tests involve dipping a fired metallized substrate into a particular solder system with a specified flux, time, and temperature. After the substrate is cleaned with solvent, the solderability is qualitatively determined by the degree of coverage and evidence of dewetting and/or voiding. Moderate differences can be readily visualized, but subtle differences cannot be determined from this qualitative approach. This visual test can be done on the pads on Fig. 3.15*a*.

Solderability or wettability depends on the wetting characteristics of a solid surface by a liquid or molten solder. The angle θ between a solid surface and the tangent to the liquid surface at the contact point specifies the conditions for minimum energy according to the following equation[46]:

$$\gamma_{LV} \cos \theta = \gamma_{SV} - \gamma_{SL} \qquad (3.8)$$

where γ_{SV}, γ_{SL}, and γ_{LV} are the interfacial energies between the phases present in a particular system. The boundary condition between wetting and nonwetting is θ = 90°. If we review Fig. 3.17,[46] we can see the various conditions for wetting, nonwetting, and complete spreading of a liquid on a solid.

The above principles were utilized to develop a more quantitative solder-wetting test within the North American Rockwell Electronics Group.[11] Engineers developed a formula for calculating a spreading factor based on the weight and density of the solder as well as the height of the solder after spreading on a fired substrate surface. Cooper and Monahan[54] also utilized these principles to characterize solder wetting forces using a meniscograph.

More recently, precision wetting balances have been developed to provide qualitative wetting force values that arise from interfacial energies developed by solders on fired metallized substrates. The apparatus consists of a precision microbalance and electromechanical dipping device that executes preset substrate immersion and emersion rates. The test substrate should be fully metallized on both sides and should be standardized with respect to size. During the test, the metallized substrate is suspended from the microbalance and is stationary while the pot containing molten solder is raised and subsequently lowered to produce constant immersion and emersion rates. Figure 3.18 illustrates a typical wetting balance curve for a solderable material.[55] The zero wetting force line must first be calculated to make corrections for buoyancy forces produced during the immersion phase. This force depends on

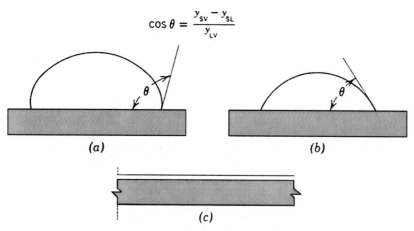

$$\cos \theta = \frac{\gamma_{SV} - \gamma_{SL}}{\gamma_{LV}}$$

(a)   (b)

(c)

Figure 3.17 Illustration of (a) nonwetting (θ >90°C), (b) wetting (θ<90°C), and (c) spreading (θ = 0) of liquid on a solid.

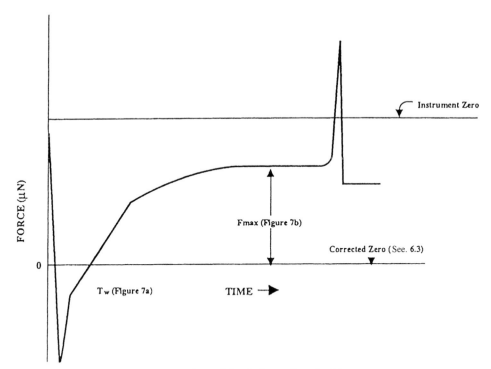

Figure 3.18 Wetting balance curve. (From *Joint Industry Standard.*)

solder density and coupon volume. The time to commence wetting is designated by T_W on Fig. 3.18, while F_{max} is the maximum wetting force of the solder on the substrate. For materials that are nonsolderable, such as bare ceramic, the net wetting forces are rarely significant enough to exceed the buoyancy force.

Pinch and Sjostak[56] utilized this type of equipment to characterize solderability of various Cu pastes on ceramic circuit boards. This type of test is particularly useful in distinguishing minor performance variations in materials that typically exhibit good solderability. There are numerous suppliers for this type of wetting balance test apparatus.[55]

Adhesion. Adhesion measurements provide a figure of merit for the degree of conductor bonding to the substrate. The various mechanisms for bonding to the substrate are composition-dependent and were discussed in Sec. 3.2.3 under "Inorganic Binders". This property is one of the most difficult to measure in terms of reproducibility and correlation. This is a result of the many factors that affect adhesion test results. These include, but are not limited to, the following: specific conductor formulation, substrate, test pattern, test preparation, wire attachment method, solder type, and adhesion test method.

Although adhesion mechanisms have already been discussed, it should be noted that the relative concentrations and combinations of adhesion promoters in the formulation will have a significant effect on relative adhesion values.

Substrate variables that affect adhesion include bulk and surface chemistry, surface roughness, thickness, size, and shape. For example, increased surface roughness results in an increase in the respective surface area. The net result is increased conductor adhesion due to the additional area for mechanical and chemical bonding as well as the fact that adhesion test loading will not always be perpendicular to the potential fracture surface.[57] In this situation the delamination forces produced by the adhesion test may result in nonplanar interfacial stresses that are less than those in the planar direction because of stress relaxation arising from plastic flow behavior. The fracture mechanics of this type of behavior is discussed by Anderson et al.[58] In addition to substrate type (i.e., Al_2O_3 versus AlN) and the resultant effect on reactivity with various bonding agents, the intergranular strength and chemistry of the substrate can affect the adhesion strength.

When specifying the results of a particular adhesion test, it is important to also specify the details of the test pattern being utilized. These parameters should include the size, shape, quantity, and arrangement of the test pads as well as the recommended and actual thickness of the fired conductor. An example is the test pattern illustrated in Fig. 3.15, which utilizes 80×80-mil square test pads in a double-row configuration. Individual paste suppliers will provide the recommended fired thickness targets.

The wire attachment process involves a number of variables that must also be considered. Wire considerations include metallurgy, shape, size (diameter), and finish. Preparation processes prior to wire attachment could include a pad burnishing operation to remove oxides from the surface, a pretinning step, and cleaning or fluxing operations that employ various flux compositions. The attachment process can utilize various solder compositions that require different solder times and temperatures. Methods of wire attachment include hand soldering, solder reflow such as belt reflow and vapor phase reflow, and solder dipping. Each of these methods involves specific processing variables with respect to time, temperature, and solder type. These variables have a profound effect on the solder leaching characteristics of the metallization and, therefore, have an impact on adhesion values. Williams[59] concluded that the solder dip test is an easy and rapid test for generating comparative adhesion values. However, this study also concluded that total pad lift should be attained to fully understand true adhesion values and mechanisms. Hand soldering methods were determined to be the most appropriate for this approach. Undoubtedly, aging of the soldered test coupon or assembly prior to testing also has an effect on the final value.

As noted earlier and readily seen by the complexity of variables, adhesion testing is a very inexact science. Many tests have been devised in order to characterize the adhesion property. These include both qualitative and quantitative test methods. Examples of qualitative tests are the Scotch tape test and razor blade tests which certify that the adhesion exceeds some arbitrary threshold strength. These tests are not recommended and are not discussed further here. The quantitative tests typically used are not ideal either in terms of generating complete adhesion information or their dependence on operator experience, technique, etc. However, they do provide a quantitative comparison for a particular material type, substrate, and process that can be utilized to determine differences or changes in the adhesion performance. These tests can also indicate the robustness of a particular material and process with respect to adhesion performance.

Some of the quantitative adhesion test configurations are illustrated in Fig. 3.19.[15] These include shear, tensile, and various wire peel tests. The shear test (Fig. 3.19a) involves pulling a wire in a direction that is parallel to the

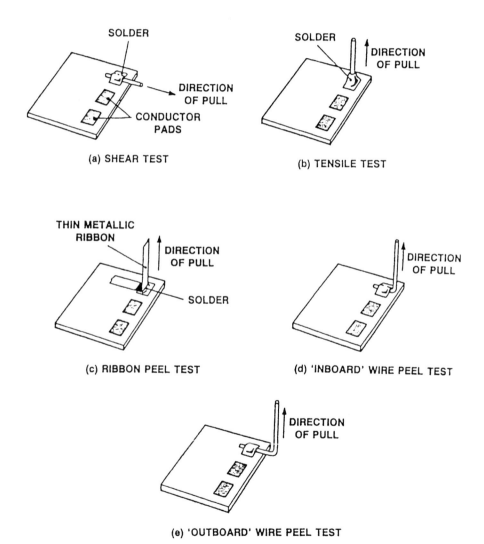

Figure 3.19 Quantitative adhesion test configurations. (a)–(e) (From *Handbook of Thick Film Technology.*)

substrate. The primary drawback of this test is that the failure mode is typically within the substrate, which prevents testing the true strength of the conductor-substrate interface. Reported adhesion values of this test are very high and are reported as load values in kilograms. The tensile test (Fig. 3.19b) employs a wire or nail with a small head which is attached to the center of a conductor pad. The nail head should extend to the edges of the pad in order to render the test insensitive to the amount of solder used to produce the joint. These values are often inaccurate if tensile failure occurs only in the center of the pad. Adhesion values of tensile tests are typically lower than those from shear tests.

Peel tests are a combination of tensile and shear forces and more accurately simulate the types of forces that are encountered during actual circuit assembly and operation.

The ribbon peel test, shown in Fig. 3.19c, is not a popular test. It requires soldering a thin metal ribbon (e.g., Cu) to a conductor pad, followed by application of a force perpendicular to the pad. Ensuring solder uniformity is difficult and test results can therefore be variable. The most popular test employed is the wire peel test (Fig. 3.19d and e). This test involves attaching a wire (which may be prebent to form a 90° angle) across one or more conductor pads in a row and pulling the wire in a direction perpendicular to the substrate surface. This test is sensitive to the pull rate, solder method, position of the wire bend, and pad configuration. Peeling or twisting the lead against the soldered conductor produces failures at much lower forces than normal tension or shear tests because of the enhanced stresses at the rather brittle conductor-substrate interface.

A detailed description of one form of this test can be found in Chap. 5 of Ref. 11. Figure 3.20[11,57] illustrates various detailed aspects of the peel test. Figure 3.20a shows bending of the wire lead, positioning of the bend with respect to the substrate, and the ideal solder fillet geometry over the conductor pad. Figure 3.20b shows a popular pad configuration which utilizes three 80 × 80-mil pads in a linear geometry. Figure 3.20c illustrates a popular lead attachment method to the back end of the substrate, known as the *shepherd's crook* method. Although test configurations must be standardized, many variations are acceptable. Figure 3.15a shows a test pad configuration that utilizes the shepherd's crook on two rows of five pads.

Adhesion values for the peel test method are generally lower compared to tensile and shear test methods. As noted earlier, Williams[59] concluded that the hand solder test combined with the wire peel test for a specified pull rate produces the most significant data for an adhesion test, since it generally results in total pad lift from the substrate. After adhesion values have been measured, close observation can reveal the true failure mode and provide accurate conclusions regarding the limiting factors. As many of these tests require room temperature aging of the soldered fixture at some specified condition prior to the peel test, adhesion values versus time should not be overlooked. There is usually an increase in value over a given time, resulting from an annealing process in the solder which renders it more ductile. The net result is a more uniform load distribution over the conductor pad. Although soldered peel tests

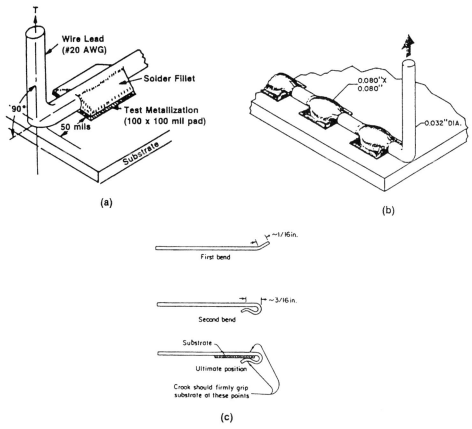

Figure 3.20   Detailed aspects of wire peel test. (*a*) Wire bend and solder fillet configuration. (*b*) Specific pad geometry configuration. (*c*) Shepherd's crook lead attachment method.

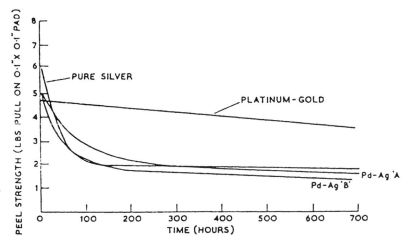

Figure 3.21   Aging of thick film conductors at 125°C. (From *Handbook of Thick Film Technology.*)

are not usually done on Au conductors due to their solubility in Sn–Pb solders, work has been done using indium-containing alloys.[60]

Although not usually thought of as an adhesion test method, the wire bond test can measure the adherence of a film. Thermocompression bonding of even 1-mil wires can result in visible film separation of conductors with poor adhesion. The test severity can be increased by increasing the diameter of the wires utilized in the bonding test. However, the failure mode must be analyzed because the concentrated load often results in breakdown of the cohesive forces of the fired film. The appearance of this type of failure indicates tearing of the intergranular particles or layers of the film. This type of failure is most often noted in Au films. The wire and die bondability as a function of conductor metallurgy was discussed earlier in this chapter. Additional discussions will occur in Sec. 3.2.6, "Structure-property relations."

Aging characteristics. The most severe tests that ultimately limit the number of applications for thick film conductors are aging tests, including storage at elevated temperatures as well as thermal and power cycle tests. Specific conditions are usually determined by the application. Typical storage temperatures exceed 100°C, with 150°C being the most common. Storage times can extend to 1000 or 2000 h, although most quality control tests specify 48 h. Figure 3.21[11] illustrates adhesion aging of several thick film conductors. Since property degradation is basically exponential with time, most of the change is often seen within the first 48 h. However, this can be misleading, since it is possible for the degradation to continue to completion (zero adhesion). Thermal cycle tests that monitor adhesion as a function of time measure the thermomechanical stability of the conductor-substrate interface. The test parameters vary with respect to the end point temperatures and dwell and excursion times. A typical profile could include end point temperatures of −55°C and 150°C coupled with excursion and dwell times on the order of 10 min. This type of test is imperative for the automotive industry because of the variable environmental conditions both external to the car and under the hood.

Adhesion degradation under these conditions results from various phenomena. The diffusion of tin (Sn) from the solder into the interfacial regions and subsequent formation of intermetallic compounds is the primary failure mechanism. This was discussed earlier for Ag, Ag–Pd, and Cu formulations. Formation of these compounds creates interfacial stresses as a result of thermal expansion mismatches of the various phases. The result is mechanical failure during thermal and power cycling. Another mechanism proposed by Crossland and Hailes[61] suggests that Sn diffuses more rapidly into the conductor than Ag diffuses outward during thermal aging. The net effect is growth of the composite and fracture of the glass network resulting in adhesion loss. In general, Au-containing conductors exhibit less adhesion loss during aging than Ag-bearing conductors. The use of low Sn solders (i.e., 10 Sn–90 Pb) can reduce this degradation.

Solder leach resistance. Solder leach resistance can be defined as the ability of a specific conductor to withstand the attack of molten solders and resulting dissolution of the fired film. The solder leach characteristics of various precious metals and conductor metallurgy systems were discussed earlier. The solder leach test typically involves immersion of a test coupon with conductor pads and/or lines into a molten solder for a specified time. Visual and/or microscopic observations are utilized to determine the extent of solder leaching. The variables include the solder composition, solder temperature, immersion rate, time of immersion, and number of immersions. The test was originally conducted by single immersions for long time periods. However, the solder adjacent to the immersed metal area would attain saturation and reduce the severity and sensitivity of the test. Therefore, this test was replaced by tests involving multiple dips for relatively short times. For example, one form of the test requires that 80% of a 20-mil line or 80-mil-square pad remains after 20 successive 1-s dips into a specified solder. The integrity of the conductor line or pad should be observed after each immersion. The observer should also look for breaks in the line or discontinuities in the pad. Backlighting the substrate can be useful in this type of observation. Mild conditions are utilized for Ag conductors such as 62 Sn–36 Pb–2 Ag solders at 215°C with a mildly active flux, while more durable compositions such as Au–Pt conductors may utilize an active flux with a 63 Sn–37 Pb solder at 250°C. The severity of this test is usually determined by the specific application.

Microstructure. As the microstructure of the fired conductor dictates many of the performance characteristics, various techniques can be employed to assess different features. Macroscopic features such as blistering, leaching, and line spreading can often be seen by visual observations. Metallurgical microscopes can be utilized to see finer features not discernible with the naked eye. Backlighting the substrate provides information regarding the density of the film and continuity of fine lines. Toplighting of the substrate can provide information regarding the surface of the film and the print resolution, as well as more detail on visual observations described above. Compatibility of the conductor with the substrate, resistors, dielectrics, etc., can be initially assessed with microscopic examination. Advanced microscopic techniques such as dark field or bright field imaging can provide more detailed information such as elemental diffusion, phase separation, and internal porosity. The photomicrographs in Fig. 3.22 were prepared using dark field imaging techniques and compare two Cu multilayer systems (cross sections), with Cu diffusion or "blushing" being visible in Fig. 3.22b. The use of polarization techniques can provide additional microstructural data.

More complex characterization tools can be employed in research or troubleshooting. Scanning electron microscopy (SEM) provides detailed microstructural information by topographic and cross-section examination. Topographic features are best characterized by secondary electron imaging (SEI), while cross-section features can be studied using SEI or backscattered electron imaging (BEI), which provides atomic number contrast for elemental or phase distinction. Figures 3.9 and 3.11 provide examples of SEI and BEI, respectively. Energy dispersive spectroscopy is utilized to determine elemental diffusion across critical interfaces such as the substrate–conductor, resistor–conductor, or dielectric–conductor interfaces. Surface analytical tech-

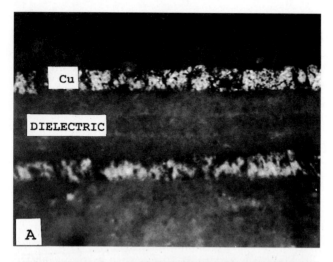

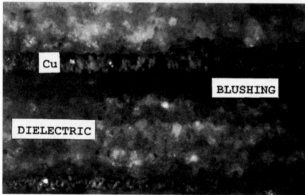

Figure 3.22 Photomicrographs of Cu multilayer cross sections using dark field imaging microscopy. (*A*) No diffusion of Cu into dielectric; (*B*) excessive Cu diffusion into dielectric, termed *blushing*.

niques can be utilized to determine the elemental composition of the first several atomic layers of an exposed surface. These techniques include Auger electron spectroscopy (AES) and elemental spectroscopy for chemical analysis (ESCA). The latter technique can produce actual phase information, while AES is limited to elemental information. Finally, x-ray diffraction (XRD) analysis provides phase information such as the extent of solid solution formation of an Ag–Pd composition (described earlier) or the crystallizing phase in a particular glass composition. Numerous examples of the above techniques will be discussed in the section on structure–property relations.

3.2.6   Structure–property relations

Although some examples of structure–property relations have been alluded to in the text, this section will focus on more specific examples.

In the study by Nordstrom and Yost,[48] the sintering behavior of thick film Au conductors was monitored by electrical resistance measurements and scanning electron microscopy. Initial-stage sintering kinetic equations coupled with resistance/neck area correlations were used to develop a relationship between resistance R and sintering time T:

$$\frac{1}{R} = Ct^{2/n} \tag{3.9}$$

where C is a constant and n depends on the sintering mechanism. Resistance changes were correlated with structural changes during the initial and intermediate stages of sintering. This leads to a good understanding of the relationship between resistivity and densification. The sintering or firing profile can be optimized to maximize the fired density and resultant performance with respect to resistivity, solderability, leach resistance, and bondability.

Pinch and Sjostak[56] studied the effects of surface structure and processing parameters of Cu conductors on solderability. They utilized AES to study the differences in the distribution of oxide components on the surface as they relate to solderability. The formation of very thin layers of surface oxides due to local fluctuations in the oxygen content of the furnace atmosphere adversely affected the solder acceptance of the fired Cu conductor. A similar study by Webb[62] utilized AES and SEM to study the effects of firing conditions on solderability and adhesion of Cu conductors. Adhesion of Cu films was increased when nitrogen flow rates or the partial pressure of the oxygen in the nitrogen gas was increased. This is due to the greater copper oxide concentration at the conductor–substrate interface under these conditions, resulting in spinel compound formation. Adhesion was also increased when multiple firings were employed because of the reduction of residual carbon at the interface. However, soldering to the conductor became increasingly more difficult under these conditions as thin oxide layers formed on the surface. Soldering to Cu becomes difficult when the oxygen concentration in the nitrogen gas exceeds 15 ppm. Adhesion degradation mechanisms for Cu- and Ag-bearing conductors due to intermetallic formation were also observed.

The surface of the fired conductor is critical to the success of bonding wire and die. The surface should be clean of oxides and metal-rich, smooth, and dense. Overfiring of a conductor or the use of glasses in the formulation that are not appropriate for a specific firing profile can result in capillary forces that cause migration of the molten glass to the conductor surface. The firing atmosphere can also lead to oxide formation on the surface of the fired film. Similarly, mishandling of processed parts can produce the same result. Figure 3.23 illustrates an oxide-containing surface as verified by AES. The black region is glass that has penetrated the surface by capillary action. Wire bonds to a fired film with glassy regions and/or oxides on the surface are not acceptable, since the metal bonding process cannot be completed, resulting in little or zero adhesion of the wires. A brief acid etch of the surface results in acceptable bond characteristics by removing the oxide layer.

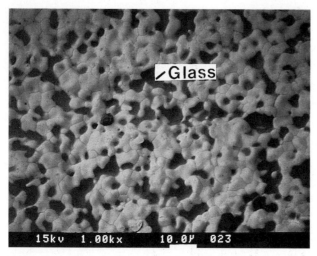

Figure 3.23 SEM photomicrograph of Au conductor surface illustrating presence of oxides in form of glass.

The design of thick film components must take into account the compatibility of other materials being considered. As discussed earlier, Ag migration is a major concern in the presence of an electrolyte and an applied voltage. Many thick film glasses act as electrolytes in their molten state, with up to 2% solubility of Ag. These glass formulations cannot be utilized in dielectric compositions that interface with Ag. With proper design, however, the dielectric can accommodate the use of Ag in multilayer configurations. Figure 3.24 illustrates the energy dispersive spectra (elemental mapping of Ag) of multilayer cross sections of Ag electrodes and dielectric before and after 1000 h of

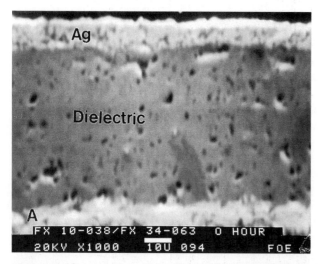

Figure 3.24 Cross sections of Ag–Pd multilayer structure. (*A*) SEM photomicrograph; (*B*) Energy dispersive spectroscopy (EDS) Ag elemental map of as-fired structure; (*C*) EDS Ag elemental map after 1000-h humidity bias test.

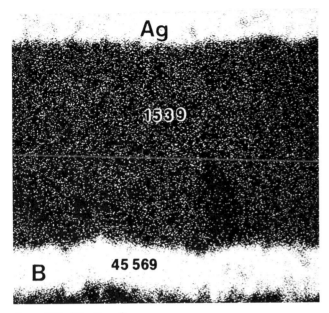

Figure 3.24 *(Continued)*

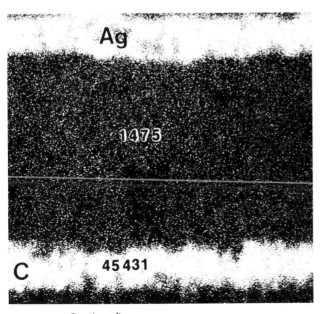

Figure 3.24 *(Continued)*

bias humidity testing.[19] The ratio of Ag concentration at the center of the Ag conductor to the concentration at the center of the Ag dielectric remains essentially constant after the stringent test for Ag migration. The statistical count at the center of the dielectric corresponds with the background noise in this configuration. Capacitance measurements for this simple capacitor structure also suggest that there is no migration of the Ag, as there is little change in capacitance or insulation resistance values following the bias humidity test up to 2000 h. Proper design of the materials system can eliminate or control phenomena that can typically lead to failure during circuit testing or operation. In this situation, the design of a crystallizing glass with very low Ag solubility provided a stable environment for the use of Ag and its beneficial properties. The utilization of formulations and processing knowledge to engineer microstructures and the resultant properties is critical for meeting the performance challenges in today's electronics industry.

3.3   Thick Film Dielectrics

3.3.1   Introduction

Thick film dielectrics are used in a variety of hybrid microelectronics applications. The most common functions of thick film dielectrics are to:

1. Provide electrical insulation
2. Provide mechanical and environmental protection
3. Store electric charge as a capacitor

Dielectrics used to satisfy the first and the second requirements are electric insulators with relatively low dielectric constants (less than 15). Capacitor dielectrics with dielectric constants greater than 20 are used as charge storage devices. Thus, thick film dielectric formulations can be broadly categorized as either insulator or capacitor dielectrics. Insulator dielectric materials can be further categorized as sealing glasses, crossover dielectrics, and multilayer dielectrics.

Sealing glasses are used for environmental, chemical, and mechanical protection. Key properties for sealing glasses are:

1. Coefficient of thermal expansion
2. Sealing temperature
3. Hermeticity
4. Chemical resistance
5. Compatibility with other materials
6. Stability at high temperature and humidity
7. Electrical insulation resistance

The substrate material, environmental conditions, and firing temperature limitations determine the sealing glass selection for a particular application.

Crossover dielectrics are used as electrical insulation layers between two conductor traces. The requirements for sealing glasses also apply to crossover dielectrics. Additional requirements include dielectric strength and interactions with conductor materials. Unlike sealing glasses, crossover dielectrics are typically fired at the same temperature as the conductor material. Interactions between the dielectric and conductor materials influence their respective properties.

Multilayer dielectric materials are used to build three-dimensional interconnect structures, whereby vias provide connectivity in the direction perpendicular to the substrate between several layers of signal conductor traces, voltage planes, and ground planes. The requirements listed for crossover dielectrics are applicable to multilayer dielectrics. Additional requirements complicate design of multilayer dielectrics. During multilayer construction, dielectric and conductor materials are subjected to multiple firings. The chemical interaction between dielectric and conductor materials in a multilayer configuration should be minimized. The CTE of the dielectric material should be closely matched with that of the substrate and should be stable on refiring to prevent warpage of the substrate.[63,64] Since multilayer dielectric layers contain vias, the printability of multilayer dielectric materials is critical. The dielectric paste, when printed, should retain via holes, which are as small as 8 mils in diameter, and should level quickly to fill in the pinholes created by the printing process. These two characteristics require opposing rheological properties and require an optimum rheology.

For high speed multilayer circuitry, low dielectric constant (K) materials are required. Furthermore, the impedance along the signal traces needs to be tightly controlled, which requires uniform thickness of the dielectric layers over the entire printed area. The compatibility of multilayer dielectrics with conductor materials is also a critical parameter. Gold, Ag, and Cu are the most commonly used conductor materials. Gold is the least reactive of the three materials. Since Cu must be processed in an inert atmosphere, the dielectric materials used for Cu-based multilayer interconnect boards (MIBs) must also be stable in an inert atmosphere. The temperature-dependent sintering kinetics should be conducive to the removal of organics from these materials during the firing process.

The key properties of thick film capacitor materials are

1. Dielectric constant

2. Temperature coefficient of dielectric constant

3. Insulation resistance

4. Dielectric strength

5. Dielectric loss

6. Stability of the dielectric properties in humid environments

7. Laser trimmability

8. CTE

Thick film capacitor dielectric materials provide a cost effective and high performance alternative to externally mounted discrete monolithic chip capacitors.

3.3.2 Dielectric compositions

3.3.2.1 Insulator dielectrics. An insulator glass is the major component of any insulator dielectric formulation. The use of glass by itself does not provide adequate mechanical strength to prevent submerging of the conductor film at high temperatures encountered in the firing process. To achieve this strength, refractory oxides such as Al_2O_3, zirconia (ZrO_2), silica (SiO_2), and titania (TiO_2) are added to the glass. These thick film dielectric formulations are referred to as *glass filled with ceramic* thick film dielectrics. Crystallization of the glass after full densification can also provide high temperature strength. These formulations are referred to as *crystallizing dielectrics*.

3.3.2.2 Ceramic-filled glasses. These dielectric compositions consist of glass and one or more refractory oxides. During firing of these materials, the glass melts and the ceramic oxide partially dissolves in the glass. The glasses used in these formulations have low alkali content, high insulation resistance, high dielectric strength, and excellent resistance to property changes in humid environments. The glass in the dielectric forms a hermetic structure. The ceramic oxide provides high temperature strength and also acts as a CTE modifier. Table 3.7 lists the properties of some of the commercially available dielectrics of this type.

The chemistry of glasses used in thick film dielectrics covers a broad range of formulations. Barium aluminosilicate or lead borosilicate glasses have reportedly been used in these formulations.[65,66] Figure 3.25 is a differential thermal analysis (DTA) trace of a noncrystallizing glass used in a thick film dielectric. The glass transition temperature is about 670°C and no exothermic

TABLE 3.7 Typical Properties of Ceramic-Filled Glass Dielectrics

Properties	Typical values
Dielectric constant	7–11
Dissipation factor, %	<0.2
Insulation resistance, Ω	>10^{12}
Breakdown voltage, V/mil	>500
Hermeticity, $\mu A/cm^2$, as measured by leakage current test	<1

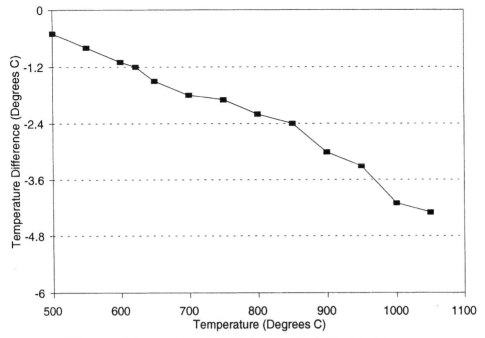

Figure 3.25 DTA trace of glass used in a glass-filled ceramic thick film dielectric.

peak associated with crystallization of the glass can be seen. The endothermic step at about 750°C is due to the glass fusion. The oxide fillers partially dissolve in the fluid glass and increase the viscosity of the glass, thereby providing mechanical support to the film. Glasses filled with refractory ceramics are normally recommended to be used with Au-based conductors. Silver is chemically more reactive than Au and typically dissolves in glasses to a certain extent, which leads to poor dielectric properties. Also, interaction of glasses in the dielectric with Ag-based conductors may affect the solderability of these conductors[66] as well as basic properties of interfacing resistors. Crystallizing dielectrics offer solutions to some of these problems.

3.3.2.3 Crystallizing dielectrics. The major constituents of crystallizing dielectrics are a crystallizable glass and crystallization agents. The crystallization agents are added to control the crystallization behavior of the base glass. Generally, fine powder of the phase being crystallized is used as a seed crystal. The amount of crystallization and type of phase affects the densification and CTE of the dielectric. Figure 3.26 is a DTA trace of a typical crystallizing dielectric. The glass transition temperature of the dielectric glass is about 660°C, and the onset of crystallization occurs at 802°C. The crystallization is completed by 850°C. Glass densification occurs mostly between the glass transition temperature and the onset of the crystallization. Crystallization of the dielectric minimizes the interaction of the dielectric with other components (e.g., conductor, resistors, etc.) at standard thick film

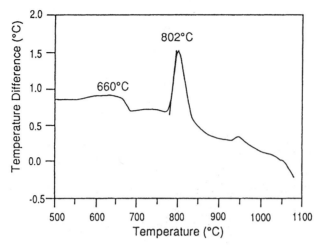

Figure 3.26 DTA trace of a crystallizing dielectric.

peak firing temperatures of 850°C. Table 3.8 lists a range of properties of some commercially available crystallizing dielectrics.

3.3.2.4 Specialty dielectrics

Low *K* dielectrics. For high speed digital circuitry, insulator dielectrics with low dielectric constants are needed. In general, there are three types of low *K* dielectrics: polymeric dielectric, porous dielectric, and glass ceramic dielectric materials. Table 3.9 lists dielectric constants and CTEs of various ceramic materials.

Fused silica has the lowest *K* among ceramic materials. The dielectric constants of polymeric materials are in the range of 2.7–3.7. Thick film dielectric materials based on fused silica filled with quartz, cordierite, sol-gel silica, and other proprietary glass ceramic formulations have been developed.[68,69] Table 3.10 summarizes properties of low *K* dielectric materials reported in the literature.

Attempts have been made to lower the dielectric constant of thick film dielectrics below 4 by introducing controlled porosity. Porous thick film dielectrics with *K* values as low as 2.5 have been reported.[70–72] Thus far, use of these low *K* dielectric materials has not been very successful in the production of complex hybrids.[73] These dielectric materials are not stable during refiring. For example, the materials based on fused silica crystallize a quartz phase during refiring, leading to significant changes in thermal expansion and subsequent cracking. Materials based on controlled entrapped porosity tend to exhibit growth of porosity during multiple firings.

Nitrogen-firable dielectric materials. Copper has been considered an alternative conductor material to precious metals for more than a decade. Conductivity, solderability, solder leach resistance, and microwave properties of Cu make it a very desirable material for building MIBs and radio frequency (rf) cir-

TABLE 3.8 Typical Properties of Crystallizing Dielectrics

Properties	Typical values
Dielectric constant	7–10
Dissipation factor, %	<0.2
Insulation resistance, Ω	$>10^{12}$
Breakdown voltage, V/mil	>500
Hermeticity, $\mu A/cm^2$	<20

TABLE 3.9 CTE and *K* Values for Various Ceramic Materials

Material	CTE, ppm/°C	K
Quartz (crystalline)	10	4.5
Fused silica	0.55	3.78
Spinel ($MgAl_2O_4$)	6.6	7.5
Mullite (Al_2SiO_5)	4.3–5	6.2–6.38
Steatite ($MgSiO_3$)	7.8–10.4	5.9–6.1
Forsterite (Mg_2SiO_4)	10.6	5.8–6.7
Cordierite ($Mg_2Al_4Si_5O_{18}$)	2.3	4.1–5.4
Wollastonite ($CaSiO_3$)	7.0	6.6

TABLE 3.10 Typical Properties of Low *K* Thick Film Dielectrics

Properties	Typical values
Dielectric constant	3.9–5.2
Dissipation factor, %	<0.3
Insulation resistance, Ω	$>10^{11}$
Breakdown voltage, V/mil	>1000
Hermeticity, $\mu A/cm^2$	<1

cuitry. In addition to performance advantages, Cu is less expensive than precious metals. However, Cu conductors need to be processed in a nitrogen atmosphere. Special dielectric materials need to be used for nitrogen processing. Glasses used in these dielectric materials must be stable in reducing atmospheres. Sintering kinetics of these dielectrics should be rigorously controlled so that a hermetically dense film is formed that still permits removal of organics under the reducing conditions. Furthermore, special glass compositions must be carefully selected to minimize reactivity of Cu with the dielectric material. Incomplete pyrolysis of organics, copper–dielectric interactions, and porosity in the dielectric film are the leading causes of failures of Cu-based MIBs. These problems have been the topic of numerous technical articles.[34–37,74–79] These problems can be addressed by selecting a crystallizing nitrogen-firable dielectric with a high temperature glass softening point. Figure 3.27 is a DTA trace of a crystallizing dielectric.

Glass fusion begins above 830°C, which allows complete pyrolysis and organic removal as well as total densification during a standard 900°C N_2 firing profile. Glass densification occurs between 830 and 880°C. The glass crystallizes at the recommended peak firing temperature of 900°C. Glass crystallization reduces the interaction with copper conductors during firing. The crystallization kinetics are carefully controlled to obtain a dense microstructure (Fig. 3.28).

Table 3.11 lists typical properties of commercially available nitrogen-firable dielectrics.

Thick film dielectrics for other applications. So far the information presented in this section has dealt with thick film dielectric materials for Al_2O_3 substrates. Other substrate materials used in the hybrid industry include beryllia (BeO), AlN, and porcelainized steel substrates. The display industry uses a variety of glass substrates. These other substrate materials all require the design and development of specific formulations because of their chemical and mechanical differences.[41–43,80,81]

The CTE of BeO (8 ppm/°C) is close to that of alumina (7.3 ppm/°C).[67] Thus, dielectric materials for alumina substrates have been used on BeO sub-

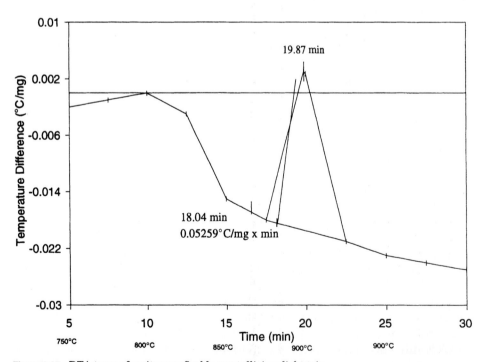

Figure 3.27 DTA trace of a nitrogen-firable crystallizing dielectric.

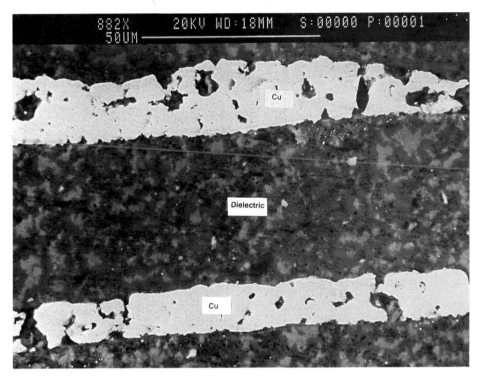

Figure 3.28 Cross section of a hermetic dielectric and copper conductors in a multilayer configuration.

TABLE 3.11 Typical Properties of N$_2$-Firable Dielectrics

Properties	Typical values
Dielectric constant	6–9
Dissipation factor, %	<0.5
Insulation resistance, Ω	>10^{12}
Breakdown voltage, V/mil	>500
Hermeticity, μA/cm^2	1–1000

strates, although chemical compatibility and adhesion issues often surface as problems.

Aluminum nitride is a nonoxide ceramic material. The CTE of AlN (4.5 ppm/°C) is substantially lower than that of alumina. Early attempts to use thick film dielectric materials designed for alumina substrates on AlN were unsuccessful.[80] Many of the glasses used in thick film formulations designed for alumina substrates are chemically incompatible with AlN. Furthermore, the CTE differences between dielectrics designed for alumina and AlN substrates lead to cracking. Special thick film dielectric formulations have been

TABLE 3.12   Typical Properties of an AlN-
Compatible Dielectric

Properties	Typical values
Dielectric constant	<7
Dissipation factor, %	<0.4
Insulation resistance, Ω	>10^{12}
Breakdown voltage, V/mil	>750
Hermeticity, $\mu A/cm^2$	<1
CTE, ppm/°C	4.4

developed for use on AlN substrates.[23] Table 3.12 shows properties of a typical thick film dielectric on AlN substrates.[41–43]

Displays are produced with soda lime silica or other types of glasses. These glasses have softening points of less than 650°C. Thus, dielectric formulations with a CTE matched to display glass panels and firable at temperatures less than 600°C have been developed.

3.3.3   Thick film capacitor dielectrics

Figure 3.29[67] shows working voltage-capacitance areas for the principle types of capacitors. Polymer dielectrics have low dielectric constants and a limited capacitance range. Electrolytic capacitors have high dissipation factors at high frequencies due to high series resistance of electrode materials, which limit their use in applications requiring low dielectric loss. Ceramic dielectrics offer high capacitance per unit volume, low losses, and a controlled temperature coefficient of capacitance. There are several challenging issues in the development of thick film capacitor materials firable at commonly used temperatures for thick film processing. Ferroelectric materials such as barium titanate ($BaTiO_3$) and lead titanate ($PbTiO_3$) possess high dielectric constants. The temperature dependence of K can be altered by numerous additives.[82–84] These materials sinter at temperatures of 1350 to 1400°C. The sintering temperature can be lowered by sintering aid additions which lead to a dramatic decrease in the dielectric constant. Other materials known as *relaxor ferroelectric materials* sinter at relatively low temperatures, but they are extremely process and composition sensitive. Resultant compositional modifications to match TCE also lower K and create additional undesirable effects. Although there are ceramic materials with dielectric constants in excess of 15,000, only capacitor dielectric thick film materials with dielectric constants of less than 1500 are commercially available.[85–87]

Thick film capacitor materials with K values of 500 to 1000 have been produced by mixing ferroelectric phases such as barium titanate, lead titanate, strontium titanate and their solid solutions with sintering aids. Most commonly used sintering aids are cadmium borosilicate glass, bismuth oxide, or lead-based glasses.[88–91] Thick film capacitor formulations with K values of

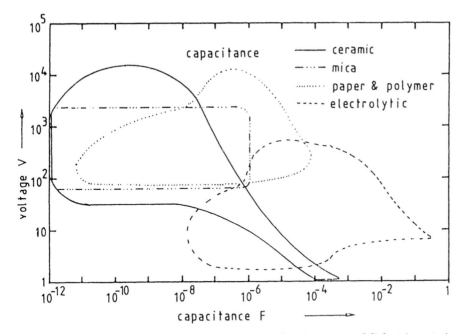

Figure 3.29 Working voltage and capacitance ranges of various types of dielectric materials. (From *Ceramic Dielectrics and Capacitors.*)[67]

400 to 800 which crystallize high dielectric constant phases such as barium titanate, lead bismuth titanate, and barium bismuth titanate have also been developed.[92–94] More recently, thick film capacitor dielectrics based on relaxor dielectric materials such as lead iron tungstate/niobate were synthesized. Dielectric constant values of 1600 to 4300 have been achieved.[95–97] Table 3.13 shows typical properties of commercially available high K thick film dielectrics.[98,99]

A typical thick film capacitor dielectric material with K of approximately 25 has a dissipation factor less than 1%, insulation resistance greater than 10^{12}, and breakdown voltage greater than 500 V/mil. In contrast, thick film capacitor materials with K of 2000 have dissipation factors as high as 5%, insulation resistance greater than 10^9, and low breakdown voltages (about 200 V/mil).

TABLE 3.13 Typical Properties of High K Thick Film Dielectrics

Properties	Typical values
Dielectric constant	25–2000
Dissipation factor, %	<5
Insulation resistance, Ω	>10^9
Breakdown voltage, V/mil	>200

3.3.4 Microstructure development

Microstructural features of dielectric materials include the chemical composition and structure of the various phases as well as porosity and its distribution. These microstructural features, coupled with macroscopic defects such as pinholes, mesh marks, foreign matter, and interaction with conductor materials, determine dielectric properties. Very little information is published about composition of thick film dielectrics, which are complex materials systems. The following section is a general discussion of dielectric microstructure development.

Insulator dielectrics. Microstructure development of insulator thick film dielectrics may be divided into three stages: (1) organic removal, (2) sintering, and (3) crystallization or ceramic dissolution.

Dielectric paste is a mixture of organic vehicle, glass powders, and oxide powders. Most commonly used ethyl cellulose–based resins are completely removed in air before 550°C. Figure 3.30 is a thermogravimetric analysis of a dielectric paste. The organic removal in the dielectric paste appears to be a two-step process. The weight loss prior to 200°C is associated with solvent removal and that between 250 and 550°C is related to binder removal.

It is important that the organic removal be completed prior to any appreciable sintering of the dielectric. Since the glass transition temperature of typical glasses used in air-firable dielectrics is greater than 600°C, organic entrapment in these dielectrics is not a major concern. However, low temperature sealing glasses or overglazes often use glasses with glass transition temperatures in the range of 350 to 400°C. Copper-compatible dielectrics must be processed in a nitrogen atmosphere in which ethyl cellulose–based binders are not completely removed until 800°C. This problem can be solved by using an acrylic-based binder system. However, these binders do not exhibit the good screen printing characteristics of ethyl cellulose–based binder systems. Thus, for nitrogen-firable dielectrics, glasses with high glass transition temperatures are desired. Incomplete removal of organics prior to

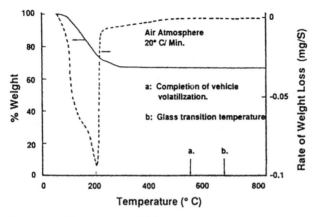

Figure 3.30 Thermogram of dielectric paste.

sintering may lead to organic entrapment resulting in porosity and blistering (Fig. 3.31). Some of these issues have also been discussed in Sec. 3.2, "Thick Film Conductors."

Sintering of dielectrics is dominated by the glass phase. Glasses undergo sintering by viscous flow. Initial sintering kinetics as measured by volumetric shrinkage are given by the following equation[46]:

$$\frac{\Delta V}{V} = \frac{9\gamma t}{4\eta r}$$

(3.10)

where $\Delta V/V$ is the volume shrinkage, γ is surface energy of glass, η is the viscosity of the glass, r is the radius of glass particles, and t is time. The sintering kinetics of glass-ceramic materials are governed by viscosity and surface energy as well as their temperature dependencies. As temperature increases, the glass viscosity decreases and the amount of densification increases. In the final stages of sintering, the glass contains isolated pores. The final densification of glass occurs as a result of surface tension forces and is given by the following expression[46]:

$$\frac{d\rho}{dt} = \frac{3\gamma}{2r_0\eta}(1 - \rho)$$

(3.11)

where ρ is the relative density of the glass (the bulk density divided by the true density) and r_0 is the initial radius of glass particles.

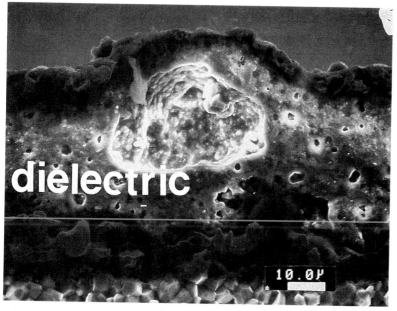

Figure 3.31 Blistering of dielectric related to organic entrapment. Solid curve = % weight. Broken curve = rate of weight loss. (a) T (°C) for complete vehicle volatization. (b) T (°C) at which glass transition occurs.

Crystallization of glasses has been well researched for a variety of glass compositions used for houseware. The basic research on crystallization of thick film dielectrics is not well-published. However, the nucleation and growth theories regarding crystallization of other glass ceramics apply to crystallizing thick film dielectrics. A book by Z. Strand[100] describes the crystallization of glass ceramic materials in great detail. Crystallization processes are governed by the free energy of the liquid-to-solid transformation, the interfacial energy of the solid-liquid interface, and other kinetics-related parameters. For diffusion-controlled crystallization processes, the rate of crystallization of glasses is inversely proportional to the viscosity of glasses. Sintering and crystallization processes compete with each other during the microstructure development of thick film crystallizing dielectrics. As viscosity of the glass decreases, the extent of sintering and crystallization may increase. Crystallization of the glass reduces the amount of liquid phase and retards the sintering process. Crystallization of the dielectric prior to densification can lead to an interconnected porosity in the fired structure.

Crystallization of dielectric materials also reduces the interaction with conductor materials and diffusion of conductor materials into the dielectric. Some of these interactions have been discussed in Sec. 3.2, "Thick Film Conductors."

The crystallization temperature must be carefully controlled to reduce interactions with the conductor material and still attain full densification of the dielectric. Particle size of the glass powders and the amount and type of nucleation agents in the dielectric formulation affect the crystallization temperature and the structure of the crystallized phases.

In the case of glass-filled ceramics, the rate of dissolution of the filler oxide increases as the viscosity of the glass decreases. Dissolution of refractory oxides in the glass increases the viscosity, thereby reducing reactivity with conductor materials.

Capacitor dielectrics. Capacitor dielectrics contain barium titanate and its solid solutions, which possess high dielectric constants. The melting point of these materials is over 1600°C. Sintering additives are often used to lower the sintering temperature to 850 to 1000°C. The liquid phase formed by reaction between sintering aids and refractory dielectrics controls the sintering kinetics. For rapid and complete densification the following are necessary[46]:

1. An appreciable amount of liquid phase

2. Appreciable solubility of the solid in the liquid

3. Wetting of the solid by the liquid

The driving force for densification is the capillary pressure of the liquid phase located between the solid particles.

The liquid phase, on solidification, has a much lower dielectric constant than the ferroelectric phase. For good densification, complete wetting of the ferroelectric phase by the liquid phase is desired. On the contrary, to achieve high dielectric constants, the amount of liquid phase should be minimized and the liquid phase should not completely surround the ferroelectric phase. The composition of capacitor dielectrics is arrived at by a compromise between sintered density and the respective dielectric constant.

3.3.5 Dielectric properties

This section describes the important properties of dielectric materials, structure–property relationships, and property measurement techniques.

Dielectric constant. The dielectric constant K of a material is the ratio of dielectric permittivity ε of the material to the permittivity ε_0 of a vacuum:

$$K = \frac{\varepsilon}{\varepsilon_0} \quad \text{or} \quad \varepsilon = K\varepsilon_0 \tag{3.12}$$

Most often, the K of a material is obtained from capacitance measurements of a parallel plate capacitor made from thick film dielectric and conductor materials. The electrical capacitance C of a parallel plate capacitor with electrode area A and dielectric thickness d may be expressed as follows:

$$C = \frac{\varepsilon A}{d} \tag{3.13}$$

Substituting Eq. (3.13) into Eq. (3.12) gives

$$C = \frac{K\varepsilon_0 A}{d} \tag{3.14}$$

where $\varepsilon_0 = 8.85$ pF/m. Rearranging Eq. (3.14) to solve for K gives

$$K = \frac{Cd}{\varepsilon_0 A} \tag{3.15}$$

Equation (3.15) does not take into account the effect of fringing fields shown in Fig. 3.32. The effect of a fringing field is significant under the following conditions:

1. The dielectric constant K_m of the material is close to that of the medium (generally air) in which the measurement is being made.

2. The separation d of electrodes is comparable to the area A of electrodes.

3. Electrodes do not extend to the edge of the dielectric.

The effect of the fringing field can be taken into account as an increase in the electrode area, δA, as expressed by the following equation:

$$\delta A = XUd \tag{3.16}$$

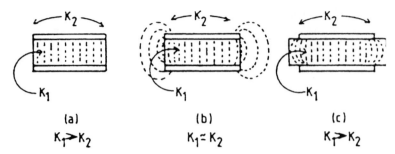

Figure 3.32 Fringing fields across various capacitors.[67]

where X is a factor dependent on the ratio K_m/K and capacitor structure and U is the perimeter of the electrode material. The factor X has a maximum value of 0.3 when $K = K_m$ and the electrode covers the entire dielectric area. However, $X = 0$ when $K >> K_m$ and the electrode covers the entire area.

For a square electrode geometry with sides w, the ratio of capacitance due to fringing, C_p to that due to area between the electrode, C_e, is given by

$$\frac{C_f}{C_e} = \frac{\delta A}{w^2} = \frac{4\,dX}{w} \tag{3.17}$$

Normally $d/w \ll 0.1$. In this case, the correction due to fringing fields is insignificant.

Effects of material structure on dielectric constant.   Electrical capacitance is the ability of a material to store an electric charge. When an electric field E is applied to a material, charges within the material are realigned in such a way as to neutralize part of the applied electric field (Fig. 3.33). The realignment of charges in the dielectric is referred to as *polarization*. Equation (3.18) shows the relationship of polarization P of the dielectric, the dielectric constant K, and the electric field E:

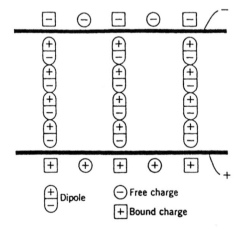

Dipole

Free charge

Bound charge

Figure 3.33 Schematic of rearrangement of charges within a capacitor body.

$$P = \varepsilon_0(K - 1)E \qquad (3.18)$$

Equation (3.18) is valid for linear dielectric materials (i.e., polarization is directly proportional to electric field). Ferroelectric materials used in high K capacitor dielectric materials are nonlinear. Figure 3.34 illustrates the relationships between polarization, electric field, and dielectric constant for a ferroelectric material as given by the following equation:

$$\frac{\Delta P}{\Delta E} = \varepsilon_0(K - 1) \qquad (3.19)$$

In the case of ferroelectric materials which have not been subjected to a polarizing dc field, K is the slope of the P vs. E curve at the origin, as shown in Fig. 3.34. The K of ferroelectric materials is dependent on the prior electric field history, the frequency of the electrical signal used for the measurements, the DC bias voltage applied across the sample, and temperature. Figure 3.35 shows polarization mechanisms contributing to K. Figure 3.36 shows frequency ranges in which these mechanisms operate. Electric polarization is a result of the shift of the center of gravity of the negatively charged electronic cloud in relation to the positively charged nucleus in an atom. Electronic polarization is common to all dielectric materials. Movement of anions and cations under an electric field leads to ionic polarization. Some materials are classified as polar materials. The charge distribution in polar materials is nonuniform, which gives rise to a permanent electric dipole even in the absence of an electric field. When an electric field is applied to such materials, the dipoles are aligned in the direction of the electric field, resulting in orientation polarization. Water is a polar material. In a macroscopic

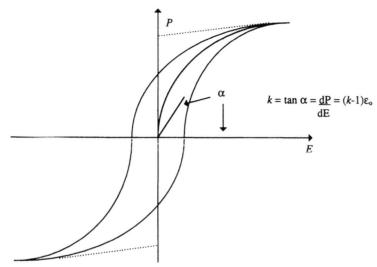

$$k = \tan \alpha = \frac{dP}{dE} = (k-1)\varepsilon_0$$

Figure 3.34 Polarization vs. electric field behavior of a nonlinear dielectric material.

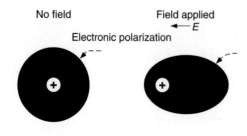

No field   Field applied
           ← E

Electronic polarization

Atomic polarization

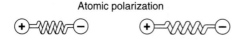

Orientation polarization

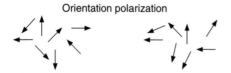

Space charge polarization

Figure 3.35 Schematic of polarization mechanisms. (From *Introduction to Ceramics.*)

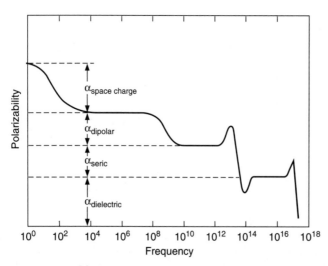

Figure 3.36 Polarization mechanisms for the full spectrum of frequencies. (From *Introduction to Ceramics.*)

sense, ferroelectric materials are considered to contain permanent electric dipoles. However, the most fundamental theories of ferroelectricity are based on lattice vibration modes (phonons). The total polarization is a sum of contributions from various mechanisms.

Insulator dielectrics. Insulator thick film dielectrics are multiphase materials. The electronic, ionic, and interfacial polarization mechanisms all contribute to the dielectric constant of glass–ceramic materials. Electronic polarization is directly proportional to the density of electrons in the glass–ceramic. Thus, dielectrics based on glasses containing oxides of high atomic number elements (e.g., lead) or high density exhibit high dielectric constants. The electronic contribution to dielectric constant, K_e, can be calculated from the refractive index n as shown in the following equation:

$$K_e = n^2 \tag{3.20}$$

The ionic contribution to the dielectric constant also can be calculated using the physics of lattice vibrations. Reference 46 shows a method of calculating ionic contribution to the dielectric constant. Since low K glass ceramics do not contain permanent dipoles, the contribution of orientation polarization is not significant in these thick film dielectrics. However, if glasses contain mobile alkali ions, these mobile ions can jump between the two sites in an open network structure of glasses. This leads to an increase in dielectric constant at low frequencies. The dielectric constant of a multiphase glass ceramic material may also increase at low frequencies due to accumulation of space charges at the interface between phases with different electrical conductivity.

Capacitor dielectrics. For capacitor dielectrics, dielectric constant and its temperature dependence are critical properties. Capacitor dielectric materials contain ferroelectric materials with high dielectric constants and sintering additives which have low dielectric constants. Barium titanate and its solid solutions are the most commonly used ferroelectric materials in capacitor dielectrics. At the Curie temperature, barium titanate also undergoes a cubic to tetragonal structural transformation. The cubic form of barium titanate is paraelectric, whereas the tetragonal form of barium titanate is ferroelectric. Tetragonal barium titanate changes to the orthorhombic phase at a lower temperature. Both structural transformations lead to an increase in the dielectric constant. Figure 3.37 shows the temperature dependence of K for barium titanate and barium strontium titanates. The addition of strontium titanate to barium titanate decreases the Curie temperature, thereby increasing the room temperature dielectric constant. A similar effect is produced by addition of barium stannate, barium zirconate, or rare earth oxides to barium titanate.[67] Table 3.14 shows the effect of various additives on the phase transition temperature of barium titanate.

Other materials that possess high dielectric constants are relaxor materials such as lead magnesium niobates. The structure of relaxor materials is extremely sensitive to stoichiometry and processing. Small changes in processing conditions or stoichiometry lead to formation of undesirable paraelec-

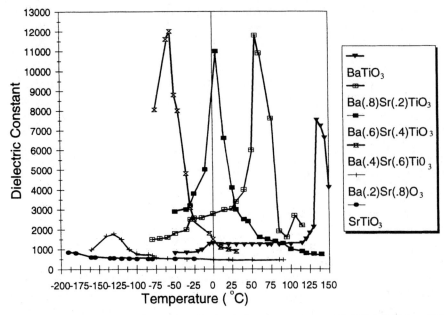

Figure 3.37 Temperature dependence of the dielectric constant of $BaTiO_3$–$SrTiO_3$ solid solutions.

TABLE 3.14   Effect of Additives on the Phase Transition Temperature of Barium Titanate

Class	Modifier ions	Substitution site Ba	Substitution site Ti	Amount of substitution Partial	Amount of substitution Complete
IA	Pb^{2+}, Sr^{2+}	X			X
IB	Zr^{2+}, HF^{4+}, Sn^{4+}		X		X
IIA	MG^{2+}, Ca^{2+}, Cd^{2+}, Zn^{2+}	X		X	
IIB	Th^{4+}, Ce^{4+}, Si^{4+}, Ge^{4+}		X	X	
IIIA	La^{3+}, Ce^{3+}, Gd^{3+}, Tb^{4+}, Sc^{3+}, Y^{3+}, Pr^{3+}, Nd^{3+}, Sm^{3+}, Eu^{3+}, Dy^{3+}, Ho^{3+}, Er^{3+}, Bi^{3+}	X	X	X	
IIIB	Nb^{5+}, Ta^{5+}, Sb^{5+}, W^{6+}, Mo^{6+}		X	X	
IVA	Na^+, K^+, Rb^+, Cs^+		X	X	
IVB	Fe^{3+}, Co^{3+}, Ni^{3+}, Mn^{3+}, Cr^{3+}, Rh^{3+}, Tm^{3+}, Yb^{3+}, Lu^{3+}		X	X	

tric phases and a resultant low dielectric constant. The extreme sensitivity of these materials has limited their use in capacitor dielectrics.

The fired microstructure of a thick film dielectric contains ferroelectric phases and other phases produced by sintering additives or modifiers. The dielectric constant of this complex mixture of phases can be estimated by using various mixture rules.[46] For example, one of the mixture rules used in estimating the dielectric constant of a mixture is

$$K = \frac{V_m k_m \left(\dfrac{2}{3} + \dfrac{k_d}{3k_m} \right) + V_d k_d}{V_m \left(\dfrac{2}{3} + \dfrac{k_d}{3k_m} \right) + V_d} \qquad (3.21)$$

where V_m is the volume fraction of the matrix phase, k_m is the dielectric constant of the matrix, V_d is the volume fraction of the dispersed phase, and k_d is the dielectric constant of the dispersed phase.

Dielectric loss. Ideal dielectric materials store electrical energy without dissipating energy in the form of heat. In reality, however, part of the electrical energy supplied to the dielectric material is lost in the form of heat. The dielectric loss factor is an indicator of the amount of this lost energy.

To understand dielectric loss, the current flowing through a capacitor and the voltage applied across a capacitor should be treated as vectors. Figure 3.38 is an electric phase diagram, a graphical representation of current and voltage across a capacitor.

The component of the current which is in phase with the voltage that causes the resistive heating is called the loss current I_l. The component of the current which leads the voltage by 90° is called the charging current I_c. The details of derivations of equations for charging current and loss current can

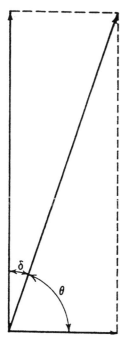

Figure 3.38 Electric phase diagram for a lossy capacitor.

be found in textbooks.[46] The following equation is commonly used to measure dielectric loss:

$$\text{Loss tangent (tan } \delta) \text{ or dissipation factor} = \frac{I_L}{I_C} = \frac{\varepsilon''}{\varepsilon'} \qquad (3.22)$$

where ε'' is absolute dielectric loss and ε' is dielectric permittivity. For high frequency applications, the quality factor is used as a figure of merit:

$$\text{Quality factor} = \frac{1}{\tan \delta} \qquad (3.23)$$

Figure 3.39 shows various loss mechanisms over a spectrum of frequencies.

Dielectric losses at lower frequencies are associated with movement of ions, which generates heat. Mobile alkali ions or free charge carriers in glass–ceramic materials can drift when a DC electric field is applied. The movement of ions under an AC electric field ceases above a certain frequency. Thus, the contribution to dielectric losses from the conduction of ions is negligible at frequencies greater than 1 kHz.

Glasses possess an open structure in which ions such as silicon and boron form a network and monovalent or bivalent ions are randomly distributed in the network. Under an electric field, the randomly distributed bivalent or monovalent ions can jump from site to site in the open network. These ionic jumps also contribute to dielectric loss at frequencies up to 1 MHz. The ions are unable to follow the electric field at frequencies greater than 1 MHz. Ionic jumps are also possible in crystalline ceramic materials. Crystalline ceramics possess lattice defects which act as sites for ionic jumps.

The presence of porosity in ceramic materials can also lead to an increase in the dielectric loss. The surfaces of pores contain large numbers of crystalline defects. These surface defects can provide sites for the hopping of ions. Further absorption of moisture in the pore and subsequent leaching of the

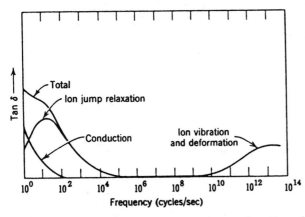

Figure 3.39 Dielectric loss mechanisms as a function of frequency.

ions in the ceramic by adsorbed moisture can lead to increases in dielectric loss.

In capacitor dielectrics containing ferroelectric materials, movement of domain walls near and below the Curie temperature leads to additional dielectric losses. At frequencies greater than 10^{10} Hz, ion deformation or movement of the electronic cloud surrounding atoms leads to absorption of energy, which results in dielectric losses.

Resistance of the electrode material and stray impedance of leads attached to the capacitors also cause the dielectric loss to increase at high frequencies. The part of the dissipation factor attributable to resistance of the electrode is given by[67]

$$(\tan \delta)_s = 4\pi f C \left(\frac{\rho_e}{t_e} \right) \left(\frac{L}{w} \right) \tag{3.24}$$

where $(\tan \delta)_s$ = contribution to the dissipation factor due to series resistance
f = frequency
C = capacitance
ρ_e = resistivity of the electrode
t_e = thickness of the electrode
L = overlapping length of the electrode
w = width of the electrode

Insulation resistance. In ceramic materials, electrical charge may be carried by electrons, electron holes, or ions. The total electrical conductivity is the sum of the contributions of the conductivities due to various charge carriers:

$$\sigma = \sum \mu_i (n_i z_i e) \tag{3.25}$$

where n_i is the concentration of charge particle i, μ_i is the mobility of the particle, and $z_i e$ is the charge on the particle. Electrical resistivity is the inverse of electrical conductivity. Alkali metal ions or free electrons or holes show high electronic mobility. Glasses containing high amounts of alkali metal oxides or semiconducting oxides show low insulation resistance or high conductivity. Figure 3.40 shows the effect of various oxides used in typical thick film dielectric glasses on electrical resistivity. Glasses containing CaO, BaO, PbO, and MgO exhibit higher resistivity than those containing alkali metal oxides. Therefore, insulator dielectrics are predominantly free of alkali metal oxides and semiconducting oxides.

Capacitor dielectric formulations contain mixtures of ferroelectric and nonferroelectric phases. The ferroelectric phases are complex oxides with crystal defects. The crystalline defects in ferroelectric phases may create highly mobile electrons or holes for charge balance. Furthermore, some of the crystalline defects are charged and quite mobile. Thus, insulation resistance of capacitor dielectric materials is lower than insulator thick film dielectrics.

As mentioned in the discussion of dielectric loss, the presence of porosity in dielectric materials can lower the insulation resistance because of absorption of moisture and crystalline defects at the surfaces of pores. The migration of

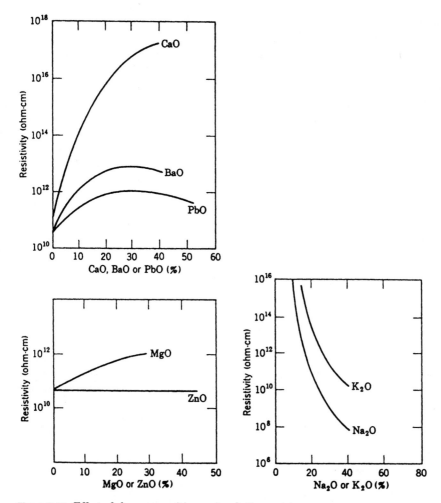

Figure 3.40   Effect of glass composition on insulation resistance.

electrode material into the dielectric during firing can lower the insulation resistance. Migration of metals or other conducting ions through interconnected porosity in a humid atmosphere can also cause loss of insulation resistance.

Dielectric strength.   The intrinsic dielectric strength of insulator dielectric materials is very high. However, observed dielectric strength of thick film dielectric materials is 10 to 60 V/μm. Dielectric breakdown is generally caused by localized defects which increase the intensity of the local electric field to a level as high as the intrinsic breakdown strength. The largest flaw determines the breakdown strength of the material. As the electrode area or dielectric thickness increases, the probability of finding a critical flaw in the material also increases and the resultant breakdown strength decreases. Mesh marks, pinholes related to printing, porosity, and nonuniformity of the electrode increase electric field concentration and promote reduction of the breakdown strength. Intrinsic dielectric breakdown occurs when the electric field levels are sufficient to cause a field emission of electrons into the dielectric. These electrons experience additional acceleration due to the electric field. Accelerated electrons possess enough energy to generate additional electrons on collision with the material and initiate electron avalanches.

Dielectric materials may also undergo thermal breakdown. Application of an electric field for an extended period of time may cause localized heating. Dielectric materials with high dissipation factors may increase the local temperature even further. This increase in temperature generates additional charge carriers which increase the dissipation factor and accelerate the temperature rise, ultimately leading to localized melting of the ceramic.

Coefficient of thermal expansion. The coefficient of thermal expansion is one of the most important thermomechanical properties of the dielectric. The CTE is determined by the CTEs of the individual phases present in the dielectric. An empirical relationship for the CTE of a ceramic composite containing multiple phases is[46]

$$\alpha_c = \frac{\displaystyle\sum \frac{\alpha_i K_i F_i}{P_i}}{\displaystyle\sum \frac{K_i F_i}{P_i}} \tag{3.26}$$

where α_c = CTE of the composite
α_i = CTE of the individual phases
F_i = weight percent of individual phases
K_i = bulk modulus of individual phases
P_i = density of individual phases

The mechanical stresses created by CTE differences between the substrate and the dielectric cause bowing of the substrate and, in extreme cases, cracking of the dielectric. Strobeck et al.[185] proposed a model for substrate bowing using the theory of plates. Figure 3.41 illustrates bowing of the substrate under various conditions. The degree of bowing w is given by:

$$w = \frac{3}{16}(a^2 + b^2)\left(\frac{D}{C}\right)(\Delta T) \tag{3.27}$$

where ΔT is $T_0 - 25°C$ and T_0 is the temperature below which thermal stresses in the dielectric are not relieved.

$$D = -E_1\alpha_1(h_1 + z_0)^2 + E_2\alpha_2(h_2 - z_0)^2 + (E_1\alpha_1 - E_2\alpha_2)z_0^2 \tag{3.28}$$

and

$$C = E_1(h_1 + z_0)^3 + E_2(h_2 - z_0)^3 + (E_2 - E_1)z_0^2 \tag{3.29}$$

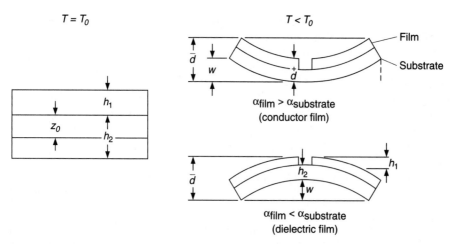

Figure 3.41 Warpage of substrate due to thermal stress created by CTE mismatch.[185]

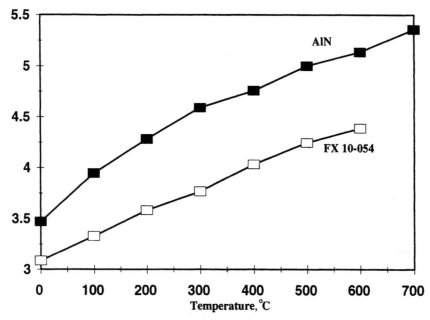

Figure 3.42 Comparison of CTE of a thick film dielectric and AlN substrate.

z_0 is the position of a generalized neutral axis and is given by

$$z_0 = \frac{h_1}{2}\left(\frac{E_2 h_2^2}{E_1 h_1^2} - 1\right)\bigg/\left(\frac{E_2 h_2}{E_1 h_1} + 1\right) \tag{3.30}$$

where E_1 = Young's modulus of the dielectric
E_2 = Young's modulus of the substrate
α_1 = CTE of the dielectric

α_2 = CTE of the substrate
h_1 = thickness of the dielectric film
h_2 = thickness of the substrate

If the CTE of the dielectric is less than that of the substrate, the dielectric film is under compressive stress and the substrate is under tensile stress after cooling during the firing process. Since ceramics are typically brittle, it is desired to place them under compression. Excessive tensile stresses in the substrate can produce unacceptable warpage. Since the thickness of the substrate is much larger than the thickness of the dielectric film, tensile stresses created in the substrate due to thermal expansion mismatch seldom exceed the fracture strength of the substrate. If the CTE of the dielectric is greater than that of the substrate, the dielectric material is under tension. These tensile stresses often exceed the fracture strength of the dielectric, resulting in dielectric cracking. Ideally, the CTE of the dielectric should be slightly less than that of the substrate and the CTE should be stable on refiring. Figure 3.42 shows a comparison of the CTE of a dielectric designed for an AlN substrate and that of the substrate. The resultant warpage is shown in Fig. 3.43.

Hermeticity and stability in humid environments. Hermeticity of a dielectric depends on its ability to seal and form a dense glass–ceramic body. Reliability of dielectrics in a humid environment is a function of hermeticity and inherent stability of the glasses. Interconnected porosity in the fired structure can provide paths for ion migration. The migration of ions under an electric field ultimately leads to the formation of a metallic film in the interconnected

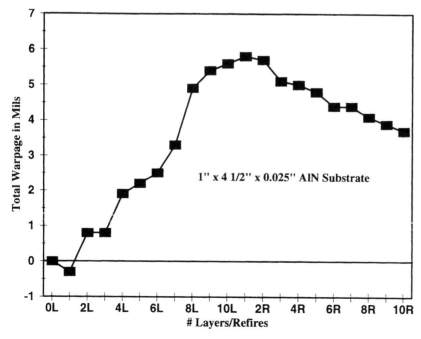

Figure 3.43 Warpage of an AlN substrate caused by CTE mismatch.

pores and results in an electrical failure. The correlation between hermeticity and failures in the humidity bias test is proposed by C. R. S. Needes.[76] For a copper–dielectric system, a leakage current value in an electrolytic hermeticity test of less than 20 μA/cm^2 is considered acceptable for prevention of failures in a humidity bias test. The electrolytic leakage current test does not take into account the stability of glasses in humid environments. Certain glasses are soluble in water. In general, glasses containing high amounts of boric oxide and strontium oxide are susceptible to failure in moist environments even though they may pass a hermeticity test.

3.3.6 Property measurements

Dielectric constant and dielectric loss. Impedance measurement bridges are the most commonly used instruments for dielectric constant and dielectric loss measurements for frequencies up to 10 MHz. A capacitor structure is built as a bottom electrode, top electrode, and two or more separation layers of dielectric material. Figure 3.44 shows a typical electrode pattern. The outside guard ring on the top electrode pattern prevents measurement errors due to fringe effects. Dielectric constant and dielectric loss measurements at frequencies higher than 10 MHz are much more difficult and require special equipment and measurement techniques.

Figure 3.45 shows various dielectric property measurement techniques over a broad frequency range. Details of these measurement techniques are available in various references.[102,103]

Dielectric strength. Dielectric strength is measured by applying an AC or DC electric field across a capacitor structure. The voltage across the capacitor is increased incrementally until dielectric breakdown occurs. Dielectric strength under a DC electric field is lower than that under an AC electric field. Sample preparation, the rate of increase of electric field, and the environment surrounding the sample all affect dielectric strength measurements.

Warpage. Warpage of the substrate is determined using several methods which involve measurement of distortion of the substrate due to deposition, firing, and refiring of dielectric layers. Generally, a flat granite block and spherometer or dial gauge with a stylus is used for warpage measurements. The substrate is kept on two supporting pieces on the granite block and the position of the center of the substrate with respect to the flat granite block is measured with the dial gauge (Fig. 3.46). More than 95% of the substrate area is covered with dielectric except for a small area at the center of the substrate, which is left open for the dial gauge measurements (Fig. 3.47).

Hermeticity. Hermeticity of thick film dielectrics is evaluated by an electrolytic leakage current test. This test measures leakage current through the fired dielectric printed on an electrode pad when 10 V is applied between a Pt foil anode and an electrode. A saline solution of 1 normal concentration is used as the electrolyte. Details of the leakage current test can be found in Refs. 101 and 104. Figure 3.48a illustrates a test coupon used for this test.

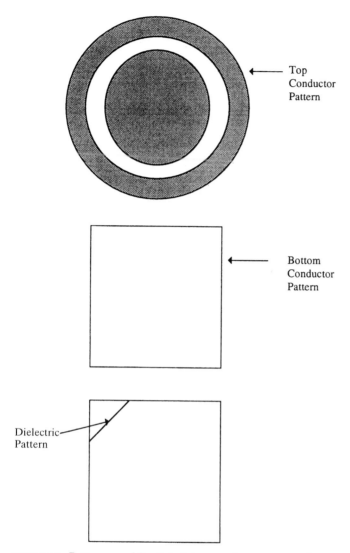

Top
Conductor
Pattern

Bottom
Conductor
Pattern

Dielectric
Pattern

Figure 3.44   Patterns used for dielectric property measurements.

Figure 3.48*b* illustrates the test configuration. Since a single print defect can lead to very high leakage current, the leakage current test results exhibit very high degrees of scatter. Sample preparation and interaction between the dielectric and the conductor can strongly influence leakage current values.

Life tests. The high humidity bias test is the most commonly used test to determine reliability of a conductor/dielectric system. The most commonly used test conditions are 85% relative humidity, 85°C, and 10- or 30-V bias voltage. Typically, the test is run for 1000 h. This test simulates more than 20 years' life at average ambient conditions.

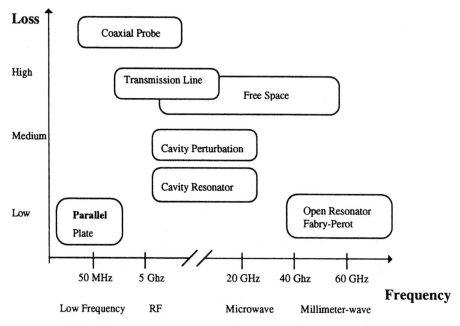

Figure 3.45 Dielectric property measurement techniques used for various frequency ranges.

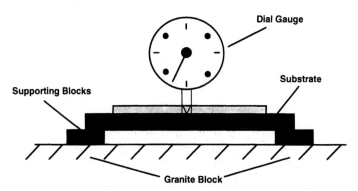

Figure 3.46 Warpage test configuration.

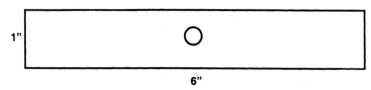

Figure 3.47 Warpage test pattern.

Conductor Test Pattern

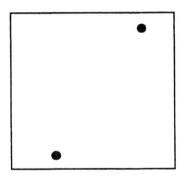

Figure 3.48a Test patterns used for the leakage current test. Conductor test pattern; dielectric test pattern.

Dielectric test Pattern

For a quick evaluation of the stability of materials in humid environments, an accelerated humidity bias test is conducted in a pressure cooker. The test coupons shown in Fig. 3.49 are subjected to atmospheric pressure of water vapor and 10- or 30-V bias voltage.

3.3.7 Effects of processing conditions on dielectric properties

Printing. The screen printing process for dielectric materials should be carefully controlled to eliminate print defects. Print defects such as pinholes, mesh marks, and foreign matter can reduce the breakdown strength of the dielectric and, in extreme cases, lead to shorts. As discussed in Sec. 3.5, "Rheology and the Screen Printing Process," the rheology of the inks and printing parameters affect print quality. In an automated production line, printing defects can lead to substantial yield loss if the process is not properly controlled.

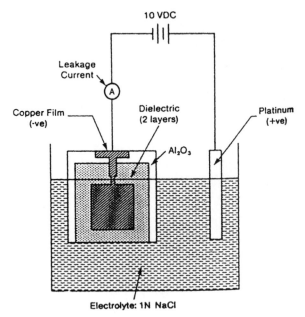

Figure 3.48*b* Leakage current test configuration.

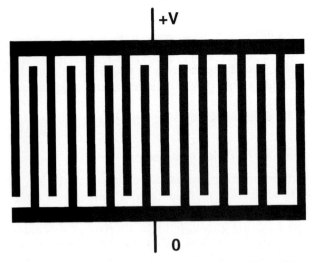

Figure 3.49 A test pattern for humidity bias testing.

Drying. Drying parameters are not known to have a significant effect on dielectric properties. However, in the case of copper-conductor-based multilayers, drying can affect the oxidation of copper conductors. The oxidized copper conductor may interact with the dielectric. In certain instances these interactions can result in dielectric blistering.

Firing. Firing is the most critical process. It can be broken into three segments: heating, isothermal hold, and cooling. The organic removal process is completed during the early part of the heating cycle. The heating rate and airflow rates influence organic removal. Very high heating rates and insufficient airflows can lead to organic entrapment, which can cause blistering and porous microstructure development.

The heating rate between the glass transition temperature and peak firing temperature and the isothermal hold temperature govern sintering and crystallization of the glasses. Thus, CTE and hermeticity of dielectrics are greatly affected by this region of the firing cycle. In return, warpage and leakage current are greatly affected by the firing cycle. The isothermal hold temperature and time also affect the conductor–dielectric interactions. In general, higher temperatures lead to increased blistering or dissolution of the conductor into the dielectric. These interactions can lead to reductions in insulation resistance and breakdown strength as well as increases in the dissipation factor.

3.4   Resistor Materials and Processing

3.4.1   Introduction

The developments and technological advancements in the area of thick film resistors have played a major role in establishing thick film as a viable technology. The term *thick film resistor* is a generic name applied to a wide variety of resistive materials presently used in hybrid microcircuits and other electronic applications.[15] A thick film resistor is a complex nonequilibrium system made up of at least one conductive phase in an insulating glass matrix. Resistor inks generally contain, as a minimum, an insulating glass frit, an electrically conducting powder, and an organic screening agent. In addition, commercial formulations invariably contain small amounts of additives in order to improve certain electrical properties such as temperature coefficient of resistance (TCR), stability, and voltage handling characteristics. A thick film resistor is processed by screen printing the resistor ink or paste onto a ceramic substrate. The firing process removes organic constituents and fuses the glass frit in order to bond the resistor to the substrate and produce the required microstructure. The two most important features that make thick film resistor technology unique are its ability to vary sheet resistance by adjusting the volume fraction of the conductive phase and the low TCRs it produces. The sheet resistance of a thick film resistor can be varied by at least 6 orders of magnitude (from 10 to 10^7 Ω/sq) by adjusting the volume fraction of the conductive phase. State-of-the-art thick film resistors will exhibit TCR values of less than ± 100 ppm/°C. However, there are several resistor systems now commercially available which exhibit very low TCR values of ± 50 ppm/°C for limited sheet resistance ranges. Advances in thick film resistor materials have paralleled thick film technology progress to date. Thick film resistors enjoy applications in every phase of the electronics industry ranging from consumer electronics to space exploration.

3.4.2 Background

In the 1950s, work with palladium conductors led to the development of resistors formulated with mixtures of Ag, Pd, and glass powder.[105] Electrical properties were controlled by adjusting the ratios of Ag–Pd solid solutions, palladium oxide (PdO), and glass.[106] These materials were successfully used in SLT (solid logic technology) modules by IBM in the late 1960s for the model 360 computer and by other hybrid circuit manufacturers. However, the materials required stringent process controls by the end user. Silver–palladium resistors were designed to fire at 760°C, which is below the palladium oxide reduction temperature (800°C).[107] Palladium oxide is thermodynamically stable below 800°C and reduces to palladium metal[108] above this temperature.

The kinetics influencing the amount of PdO in the fired film and the resulting electrical properties are controlled by the amount of Ag. This system is chemically dynamic and reproducibility of sheet resistance from part to part is very difficult to achieve. Processing limits of time and temperature are very narrow. The shortcomings of this technology resulted in wide variations in sheet resistance due to slight changes in furnace conditions, limited sheet resistance ranges (10 to 500 kΩ/sq), and high TCR values (between 250 and 500 ppm/°C). Because of the deficiencies of Ag–Pd technologies, researchers actively pursued alternative conductive phases to be used in resistors.[109]

Since the late 1960s and early 1970s, several conductive phases, such as bismuth and lead ruthenates ($Bi_2Ru_2O_7$, $Pb_2Ru_2O_7$), ruthenium dioxide (RuO_2), and iridium dioxide (IrO_2), have been successfully utilized in thick film resistor formulations.[109] The sheet resistance of these resistor systems can be varied from 10 to 10^7 Ω/sq by adjusting the volume fraction of the conductive phase. Although conductive phase developments created significant advancements in resistor technology, the importance of glass chemistry and its influence on resistor properties cannot be overlooked.

Studies of glass chemistry and its influence on microstructure,[52] interaction of the glass and conductive phases, and interdiffusion of substrates and resistor glasses have provided understanding of these complex systems. This led to the development of resistor systems with excellent performance and properties. Similarly, studies of thermal expansion coefficients of resistor composites relative to substrate materials has led to development of resistors with excellent laser trim stability.[110] A high performance resistor can be formulated by appropriate control of particle size of the conductive phase and the design of glass chemistry. Currently, component and hybrid manufacturers use RuO_2 as the conductive phase for low sheet resistance values (10 Ω/sq to

100 kΩ/sq) and pyrochlore-type ruthenates ($Bi_2Ru_2O_7$, etc.) for values exceeding 100 kΩ/sq. This optimizes the inherent properties of each resistor system. Recently, silver–palladium resistor inks, consisting of Ag–Pd powder and glass powder, have been developed for surge protection applications.[111] They are typically formulated for lower sheet resistance ranges (0.1 to 10 Ω/sq) and utilize long serpentine resistor designs to achieve the desired resistance. Through appropriate control of particle size of silver and palladium powders and Ag/Pd ratios, TCR values of less than ±100 ppm/°C can be readily achieved.

3.4.3 General requirements

Thick film resistors are extensively used in hybrid microcircuits because they can meet design requirements such as high sheet resistance, good power dissipation, high voltage handling capabilities, and desirable response at high frequencies. These design functions are difficult or impossible to achieve with silicon monolithic integrated circuits. The most important electrical property of a thick film resistor is the attainable sheet resistance range. Figure 3.50 shows a typical variation of sheet resistance with volume fraction of RuO_2 conductive phase.

Electrical resistance R of materials is expressed as

$$R = \rho \frac{L}{A} \tag{3.31}$$

where ρ is resistivity, L is the current carrying length, and A is the current carrying cross-sectional area. Sheet resistivity of thick film resistors is generally expressed in terms of ohms per square, as shown in the following equations:

$$R_s = \frac{R}{n} = \frac{R}{L/W} \tag{3.32}$$

where R_s is the sheet resistance and n is the number of squares, which is the aspect ratio (length/width) of the resistor. Substituting Eq. (3.31) into Eq. (3.32) results in the following:

$$R_s = \frac{\rho(L/A)}{L/W} \tag{3.33}$$

The current carrying cross-sectional area A is equal to W times thickness t. Substituting for A in Eq. (3.33) gives

$$R_s = \frac{\rho(L/Wt)}{L/W} \tag{3.34}$$

and

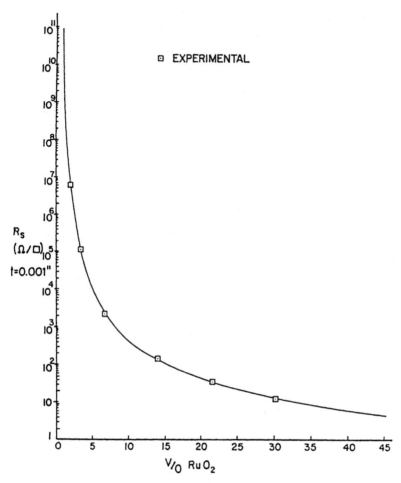

Figure 3.50 Variation of sheet resistance with volume fraction of conductive phase.

$$R_s = \frac{\rho}{t} \tag{3.35}$$

Therefore, sheet resistivity at a desired thickness can be calculated using Eq. (3.35). For example, the sheet resistivity at 25 μm is given by:

$$R_s(\text{at 25 } \mu) = \frac{R_s(\text{measured}) \times t}{25} = \frac{\rho}{25} \tag{3.36}$$

Resistance of thick film resistors of any geometry can then be predicted by the following equation:

$$R = R_s n \tag{3.37}$$

The above analysis assumes that the thickness of the resistors is uniform and a linear relationship exists between the resistance and the thickness of a resistor. This assumption is not always true.

From the above equations, the sheet resistivity of the fired paste can be calculated for a given resistor size. In the same way, the ratio of length to width, or aspect ratio, of a resistor for a required resistance value can theoretically be determined for a paste of known resistance.

Another important parameter which describes the electrical characteristics of thick film resistors is the temperature coefficient of resistance. The TCR can be defined most rigorously as

$$\text{TCR} = \frac{1}{R}\frac{dR}{dT} \tag{3.38}$$

where T is the temperature and dR and dT are incremental changes in resistance and temperature, respectively.

Taking into consideration dimensional changes in the resistor due to temperature changes, one can write:

$$\frac{1}{R} = \frac{dR}{dT} = \frac{1}{\rho}\frac{d\rho}{dT} + \frac{1}{L}\frac{dL}{dT} - \frac{1}{t}\frac{dt}{dT} - \frac{1}{W}\frac{dW}{dT} \tag{3.39}$$

Over a limited temperature range where ρ equals resistivity between T_1 and T_2, TCR values can also be approximated as follows:

$$\text{TCR}_{(T_1-T_2)} = \frac{R_{T_2} - R_{T_1}}{R_{T_1}(T_1 - T_2)} \tag{3.40}$$

where R_T is the resistance of the resistor at temperature T. Since TCR is very small in magnitude, the right side of the above equation is generally multiplied by 10^6 and the result is expressed in parts per million per degree Celsius, ppm/°C.

The temperature coefficient of resistance is usually defined as

$$\text{TCR} = \frac{R_T - R_{25}}{R_{25}(T - 25)} \times 10^6 \quad \text{ppm/°C} \tag{3.41}$$

where R_T is the resistance measured at some temperature T (°C) and R_{25} is resistance at room temperature. In industrial practice, two standard TCR values are typically specified: (1) hot TCR (HTCR), where $T_2 = 125°C$ and $T_1 = 25°C$; and (2) cold TCR (CTCR), where $T_2 = -55°C$ and $T_1 = 25°C$. A standard thick film resistor should exhibit both hot and cold TCR values of less than ±100 ppm/°C and preferably less than ±50 ppm/°C. It is also quite common for specific applications to require some predetermined TCR value over expected circuit operating temperature ranges (i.e., ±50 ppm/°C between

0 and 70°C). Later discussions will focus on formulation changes to control TCR values.

Thick film resistor performance is also measured by the voltage coefficient of resistance (VCR), which can be defined as follows:

$$\text{VCR} = \frac{1}{R_{V_1}} \frac{R_{V_2} - R_{V_1}}{V_2 - V_1} \times 10^6 \qquad \text{ppm/°C} \qquad (3.42)$$

where V is the applied voltage and R_{V_2} and R_{V_1} are the resistances measured at the respective voltages of V_2 and V_1. These voltages are typically 50 V (V_2) and 5 V (V_1). Permanent changes in resistance with application of voltage are characterized by the short-term overload voltage (V_{STOL}). The short-term overload voltage is the voltage required at 5 s duration to produce a 0.1% permanent change in resistance at 25°C. The short-term overload voltage of a thick film resistor should be maximized. Resistor stability under applications of high voltage transforms to good power handling characteristics of the resistor, as power is directly proportional to voltage squared. Resistor power rating can be defined by the following equation:

$$P_R = \frac{(V_{\text{SW}})^2}{R} \qquad (3.43)$$

where P_R is the maximum power rating and V_{SW} is the standard working voltage. V_{SW} is defined as 0.4 times V_{STOL}. Another resistor property of importance is current noise I_n, which should be minimized. The current noise is measured in terms of noise index. It is the decibel equivalent of the root mean square noise voltage to DC bias voltage ratio in a decade of frequency.

$$\text{Noise index} = 20 \log \left(\sqrt{\overline{V^2}}/V_{\text{DC}} \right) \qquad \text{dB} \qquad (3.44)$$

where current noise $\sqrt{\overline{V^2}}$ is the rms current noise voltage measured in microvolts and V_{DC} is the DC applied voltage across the resistor measured in volts. By this definition, 1 μV of current noise at 1-V DC bias would give a noise index of zero decibels.

Resistor stability is a key parameter for fired resistor performance criteria. In addition to tests already described, changes in resistance are typically monitored as a result of hot column testing, thermal aging, thermal cycling, humidity and bias voltage aging, as well as stability following laser trimming. These will be discussed later.

3.4.4 Resistor compositions

Cermet resistors. The name *cermet* is generally applied to materials which result in a fused structure of conductive phase in a vitreous, nonconductive binder. Resistor inks contain an electrically conducting powder, an insulating glass powder, and an organic screening agent. In addition, commercial resistor inks contain small amounts of various additives added to adjust specific electrical properties. The conducting phase used in thick film resistor formulations should satisfy the following requirements: (1) stability at the peak firing temperature, (2) small but finite solubility in the glass at the firing temperature, and (3) ability to be readily wet by the glass. These requirements limit the choice of the conductive phase for air-firable resistor systems to a few binary metallic oxides such as RuO_2 and IrO_2. Both RuO_2 and IrO_2 are thermodynamically stable in air at the typical thick film processing temperature of 850°C. More complex conductive materials commonly used in commercial thick film resistors are $Bi_2Ru_2O_7$, $Pb_2Ru_2O_6$, and Ag-Pd-PdO. These conductive materials exhibit a positive temperature coefficient of resistance and have resistivities somewhat higher than those of metals. Examples of resistivities for various conductive phases are shown in Table 3.15.

In the case of nitrogen firing, typical conductive phases are tin oxide (SnO_2), indium oxide (In_2O_3), strontium ruthenate ($SrRuO_3$), lanthanum hexaboride (LaB_6), titanium disilicide ($TiSi_2$), and tantalum nitride (TaN).

Insulating glass frits generally utilized are lead borosilicate glasses. Glass chemistry plays a major role in the resistor ink formulation. Electrical performance such as voltage handling and properties such as laser trim stability and environmental stability are mainly dependent on the glass chemistry.

TABLE 3.15   Resistivity Values of Various Oxides

Oxide	ρ at 300 K, $\Omega \cdot cm$
Rutile	
RuO_2	3.5×10^{-5}
IrO_2	4.9×10^{-5}
Rh_2O_3	$<10^{-4}$
Pyrochlore	
$BiRu_2O_7$	2.3×10^{-2}
$Ri_2Rh_2O_{6.8}$	3.2×20^{-3}
$Bi_2Ir_2O_7$	1.5×10^{-3}
$PbRu_2O_6$	2.0×10^{-2}
$Pb_2Ru_2O_{6.5}$	5.0×10^{-4}
$Pb_2Rh_2O_7$	6.0×10^{-1}
$Pb_2IrO_{6.5}$	1.5×10^{-4}
$Pb_2Os_2O_7$	4.0×10^{-4}
$Ti_2Ru_2O_7$	1.5×10^{-2}
$TiIr_2O_7$	1.5×10^{-3}
$Ti_2Rh_2O_7$	6.0×10^{-4}
$Ti_2Os_2O_7$	1.8×10^{-4}
Perovskite	
$LaRuO_3$	4.5×10^{-3}
$La_{0.5}Sr_{0.5}RuO_3$	5.6×10^{-3}
$CaRuO_3$	3.7×10^{-3}
$SrRuO_3$	2.0×10^{-3}
$BaRuO_3$	1.8×10^{-3}

Flow properties of the glass such as the glass softening point and viscosity of the glass at the firing temperature are critical to the final resistor properties. In this chapter the discussions are limited to air-firable resistor inks.

Low ohm resistor inks. Resistor inks having sheet resistances lower than 10 Ω/sq are generally classified as *low ohm thick film resistors*. These resistors generally consist of Ag-Pd solid solutions and an insulating glass matrix. Low ohm resistors are mostly used in the telecommunications industry for surge protectors. These are typically thick film resistor networks used to protect electronic devices against lightning strikes and power crosses. Recent advances in telecommunications for lightning strike and voltage protection requires resistors having low sheet resistance and excellent stability under an applied electric surge.[112] These applications also demand resistors with low TCR (less than ± 100 ppm/°C) and sheet resistance as low as 100 mΩ/sq. The TCR values of ± 100 ppm/°C can typically be controlled by optimizing the Ag/Pd ratio and the particle sizes of the glass and metal powders. Further requirements of these resistors include: good stability on life testing such as humidity aging and thermal cycling, good laser trimmability, and stability during multiple refirings. There are several publications[112,113] which describe the use of low ohm thick film resistors for surge protection in telecommunications.

The typical composition of a low ohm resistor paste is given in Table 3.16.

Thick film inks are generally screen-printed onto a 96% alumina substrate and fired through a furnace at a peak firing temperature of 850°C for 10 min. The typical profile used to fire low ohm resistors is given in Fig. 3.51.

Glass softening point and viscosity of the glass at the peak firing temperature affect the final resistor properties.[114] The proper selection of glass and metal powders can aid in complete sintering of the resistor with a dense, pore-free resistor microstructure. A representative scanning electron micrograph of a dense, as-fired resistor cross section is shown in Fig. 3.52.

Figure 3.53 shows the effect of metal content on the sheet resistance of low ohm resistors for various cooling rates. As metal content increases, the sheet resistance of the ink decreases for a given cooling rate.

Typical properties of high performance, low ohm resistors are presented in Table 3.17.

TABLE 3.16   Typical Composition of a Low Ohm Resistor Paste

Material chemistry	Function	Sample
Metal powders	Conductive phase	Ag and Pd powders
Glass powders	Binder	Borosilicate powder
Resin	Rheology (binder to improve green strength)	Ethyl cellulose
Solvent	Carrier for the ingredients	Terpinoel, butyl, carbital, etc.
Surfactant	Dispersion	Proprietary

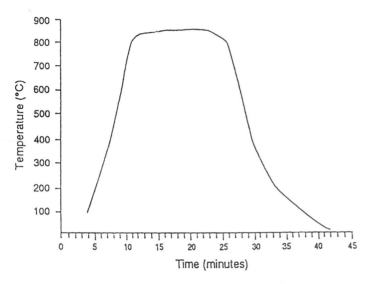

Figure 3.51 Typical firing profile used to process thick film resistors.

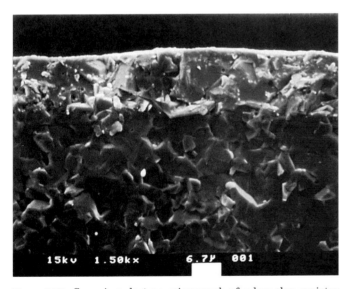

Figure 3.52 Scanning electron micrograph of a low ohm resistor cross section.

Mid/high range resistors. Midrange to high range resistors include those with sheet resistance values exceeding 10 Ω/sq. The inks for these resistors contain an electrically conducting oxide such as RuO_2, Bi_2RuO_7, or Pb_2RuO_7, insulating glass powder, and an organic screening agent. They contain dopants in order to alter one or more of the electrical characteristics of the ink. The most common dopants are TCR modifiers and stabilizers. TCR values are typically ± 100 ppm/°C and are controlled by varying the type and

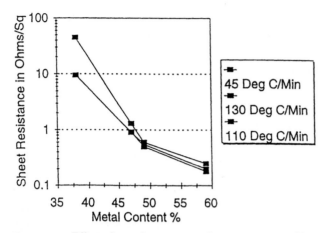

Figure 3.53 Effect of metal content on sheet resistance of low ohm resistors for different cooling rates. (From The International Society for Hybrid Microelectronics, Dallas, TX.)

TABLE 3.17   Typical Properties of High Performance Low Ohm Resistors

Properties	Typical values
Sheet resistance, Ω/sq	0.2–10.0
Tolerance, %	± 10
TCR, ppm/°C:	
Hot TCR	± 100
Cold TCR	± 100
Drift after electrostatic discharge (ESD) at 2000 V, %	<0.1
Drift after thermal aging (150°C for 1000 h), %	<0.1
Drift after thermal cycling (5 cycles), %	<0.5
Drift after humidity aging (85°C/85% for 1000 h), %	<0.5
Drift after overglazing (low temperature overglaze), %	<0.1
Drift under lightning surge (10 × 1000 μs, 200 pulses at 970 V), %	<0.1

amount of TCR modifiers in the ink. Semiconducting oxides such as manganese dioxide (MnO_2), cobalt oxide (Co_2O_3), titanium dioxide (TiO_2), niobium pentoxide (Nb_2O_5), ferric oxide (Fe_2O_3), rhodium sesquioxide (Rh_2O_3), and vanadium pentoxide (V_2O_5) are added to shift the TCR values in a negative direction. These are all called *negative TCR modifiers*. Metal oxides such as cuprous oxide (Cu_2O) and silver oxide (Ag_2O) are added to the inks to shift TCR values in a positive direction. These are called *positive TCR modifiers*. Oxides such as Al_2O_3, SiO_2, and ZrO_2 are added to improve resistor stability and reduce sensitivity to variations in processing conditions. These are called *stabilizers*. Ruthenium dioxide (RuO_2) based thick film resistors are one of the most commonly used commercial thick film resistors. Properties of RuO_2-based thick film resistors will be discussed in detail.

Both sheet resistance and TCR are a function of particle[115–117] size of the respective conductive and glass powder phases. Several studies have been done to understand the effect of conductive phase particle size on electrical properties of resistors. Ruthenium dioxide powder is commercially available in a range of particle sizes from 5 to 100 nm. Since RuO_2 particles are very small, the powders tend to agglomerate. The particle size is a function of the surface area of the powder, which is the key property typically specified for resistor conductive phases. The glass powder generally used is lead borosilicate. A typical composition and properties of a lead borosilicate glass are listed in Tables 3.18 and 3.19, respectively. Physical properties of lead borosilicate glasses have been well-studied by Vest et al.[52] In addition to particle size, glass chemistry plays a significant role in controlling the electrical properties such as *R*, TCR, ESD shift, voltage handling, laser trim properties, and stability of the resistor.

A typical microstructure of a RuO_2-based thick film resistor is shown in Fig. 3.54. The microstructure consists of networks of conductive phase in an insulating glass matrix. Electrical properties of thick film resistors are a function of these conductive chains in the glass. Formation of the conductive chains and their distribution is a function of particle sizes of both conductive

TABLE 3.18 Chemical Composition of Typical Lead Borosilicate Glass

Oxide	Weight %
PbO	55.5
B_2O_3	22.0
SiO_2T	10.5
Al_2O_3	12.0

TABLE 3.19 Physical Properties of Typical Lead Borosilicate Glass

Density	4.31 g/cm^3
Coefficient of linear thermal expansion (25–200°C)	6.03 ppm/°C
Strain point	464°C
Annealing point	469°C
Deformation point	490°C

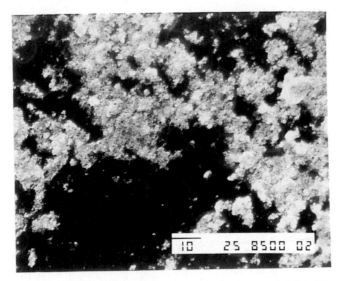

Figure 3.54 Typical microstructure of an RuO_2-based thick film resistor.

TABLE 3.20 Typical Properties of Thick Film Resistors

Property	Resistor value						
	10 Ω	100 Ω	1 kΩ	10 kΩ	100 kΩ	1 MΩ	10 MΩ
Sheet resistance, Ω/sq	10	100	1 k	10 k	100 k	1 M	10 M
Tolerance, %	±10	±10	±10	±10	±10	±10	±20
TCR hot/cold, ppm/°C	±100	±100	±100	±100	±100	±100	±200
Drift on ESD (2000 V), %	<0.1	<0.1	<0.1	<0.1	<0.1	<0.1	<0.1
Drift on thermal aging, %	<0.1	<0.1	<0.1	<0.1	<0.1	<0.5	<0.5
Drift on thermal cycling, %	<0.5	<0.5	<0.5	<0.5	<0.5	<0.5	<0.5
Drift on thermal shock, %	<0.5	<0.5	<0.5	<0.5	<0.5	<0.5	<0.5
Drift on humidity aging, %	<0.1	<0.1	<0.1	<0.1	<0.1	<0.5	<0.5
Short-term overload voltage, V/mm	8	25	75	250	380	370	
Power dissipation, mW/mm²	900	600	800	1000	230	22	
Tech noise, dB	<−30	<−30	<−20	<−15	<−10		

phase and glass, the physicochemical properties of the glass, and thermophysical conditions during firing. Typical properties of resistors ranging from 10 Ω/sq to 10 MΩ/sq are listed in Table 3.20.

Thermistors. Thick film thermistors are generally used for temperature compensation in various electronic circuits. They are a special class of thick film resistors, designed to have a very large and controlled change in resistance with temperature. These materials can be classified into two systems: (1) sensor materials made of glass doped with metal oxides and (2) thermistor materials made of spinel type oxides and conductive oxides embedded in a glassy matrix. Sensor materials can exhibit a TCR which is low and fairly constant over the entire operating temperature range of $-55°C$ to $125°C$. The response curve, resistance vs. temperature, can be described by the following relationship:

$$R(T) = R_0(1 + \alpha \Delta T) \tag{3.45}$$

where R_0 is sheet resistance at $25°C$ and α is the temperature coefficient of resistance. Typical measured α values are $+1300$ ppm/$°C$ for positive type materials and -3000 ppm/$°C$ for negative type materials. The resistance vs. temperature behavior of these materials is shown in Fig. 3.55[118]

Spinel-based inks exhibit a different response curve, more frequently associated with thermistors:

$$R(T) = R_0 \exp\left(\frac{ß}{T}\right) \tag{3.46}$$

where T is expressed in degrees Kelvin, R_0 is the sheet resistance at $25°C$, and ß is the thermistor constant. Temperature dependence is controlled by the parameter ß, and the magnitude of resistance is controlled by the preexponential term R_0.

Negative temperature coefficient (NTC) thermistors are resistors that have an inverse relationship between their electrical resistance and their body temperature. As the body temperature increases, the electrical resistance decreases, and vice versa. This resistance–temperature (R–T) characteristic is predictable and reproducible. Electrical resistance change per degree Celsius is typically -4 to -6%, which provides a significant signal response to changes in temperature. Because of the large signal change with temperature, long lead lengths have virtually no effect on the accuracy of NTC thermistors. The R–T behavior of NTC thermistors is shown in Fig. 3.56.[118]

The materials used in thermistor inks are typically a mixture of doped transition metal oxides, RuO_2, and glass.[119] The metal oxides include Mn, cobalt (Co), Ni, iron (Fe), and Cu. Two-glass systems, each glass having different softening points, combined with compositions in the Cu–Mn–O and Co/Mn/O systems, were found to have good control of ß to ±50 over the range of 1000 to 250.[120] It is a common practice to add a conductive phase such as RuO_2 to lower R_0. A wide range of R_0 can be achieved (from 500 to 10,000 Ω/sq) by varying the RuO_2 content. Thick film resistors with a high positive TCR are also useful as thermistors or temperature sensors. The spinel type thermistors exhibit a 25% tolerance in sheet resistance values, while sensor type materials exhibit a tolerance of around 10%.[118] Sensor type materials are quite stable during aging at $120°C$ in air. Spinel type oxides exhibit a large initial increase in resistance at relatively short times followed by a more stable response.[118]

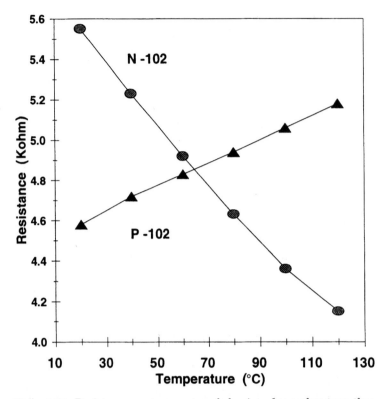

Figure 3.55 Resistance vs. temperature behavior of sensohm type thermistors.

Strain gauge resistors. Thick film resistors experience a resistance value change with the application of stress. This piezoelectric property of thick film resistors has been utilized in several pressure and strain gauge sensor applications. Thick film resistors utilized for strain gauge applications must show a reversible relative change in resistance $(\Delta R/R)$ with strain ε.[121] Strain is defined as fractional change in length $(\Delta L/L)$. The relative resistance change is directly proportional to the applied strain ε and, therefore, the gauge factor GF is defined as

$$\mathrm{GF} = \frac{\Delta R/R}{\varepsilon} \tag{3.47}$$

An ideal strain gauge should exhibit a high piezoresistive effect combined with a negligible thermoresistive effect. More precisely, a high gauge factor (greater than 10) is associated with a low TCR and TCGF, where TCGF is ΔGF/GF and is called the *temperature coefficient of gauge factor*[122] and ΔT is the temperature range where ΔR and ΔGF are considered. The resistance and gauge factor should remain stable with time under operating conditions. Figure 3.57 shows the relative change of resistance under compressive and tensile stress of highly strain sensitive thick film resistors.[123]

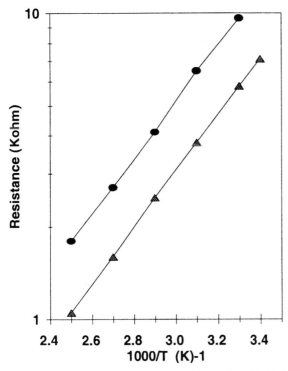

Figure 3.56 Resistance vs. temperature behavior of spinel type NTC thermistors.

Commercially available strain gauges use semiconductor or metal film resistors. The former are characterized by high GF, but also high TCR and TCGF. Thick film resistors used in strain applications must be extremely stable. A minimum in the temperature vs. resistance curve near room temperature is a typical property of high GF, stable thick film resistors. There is no scientific basis to explain this phenomenon. However, this behavior is used to identify a new thick film resistor with improved gauge factors. The low TCR values of thick film resistors make them more attractive as thick film pressure sensors than semiconductor sensors. The semiconductor sensors have very high gauge factors, but their TCR values are very high, typically, ± 1500 ppm/°C. Generally, the large gauge factor in thick film resistors is due to the large particle size of glass and RuO_2 powders. Modifiers do not typically improve the gauge factor. Figure 3.58 shows that resistors with larger RuO_2 particle sizes have higher gauge factors.[124]

Resinates. Precious metal resinates are organometallic compounds in which the metal atom is attached to a sulfur or oxygen atom linked to a carbon atom.[125] When these compounds are thermally decomposed, a film is deposited on the substrate. The thermal decomposition process takes place between 250 and 500°C. Under these conditions, noble metal compounds deposit metallic films while other precursor materials deposit metal oxide films.

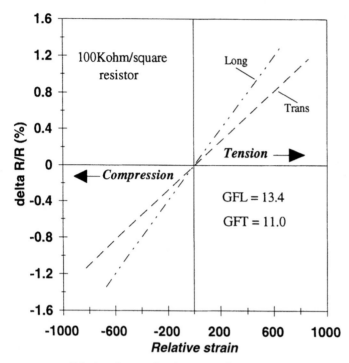

Figure 3.57 Relative change of resistance under compressive and tensile stress of a highly strain sensitive thick film resistor.

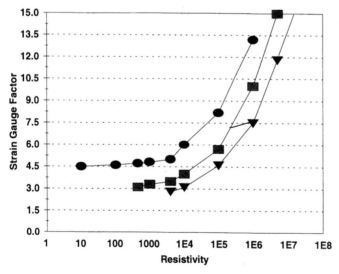

Figure 3.58 Effect of RuO_2 particle size on gauge factor for various resistivities.

Ruthenium resinates can be used to formulate thick film resistors. Ruthenium resinates decompose during firing and form an RuO_2 conductive phase. Thick and thin film ruthenium resinate-based resistors are extensively used in thermal printhead applications.[126,127]

3.4.5 Resistor properties

Microstructure development. Thick film resistors are prepared by formulating pastes consisting of submicrometer size conductive phase particles and micrometer size glass particles. Since conductive particles are of submicrometer size, complete mixing of conductive phase particles and glass powder is difficult to achieve because the smaller particles tend to agglomerate to a certain extent. During screen printing and drying, conductive particles/agglomerates can segregate at the interstices of glass powder because of the density differences. Also, the smaller conductive particles can fill interstices of the glass particles. A segregated microstructure is formed even prior to firing thick film resistors because of these factors. Although some of the conductive phase agglomerates coat the surface of the glass particles, a higher concentration of the small conductive phase agglomerates can be seen[128] in the interstices of the large glass particle matrix. During firing, the conductive phase particles rearrange themselves to form a continuous network of electrically conducting chains. These preexisting chains of conductive phase can undergo further development or degradation, depending on the thermophysical conditions during firing.

Networks of conductive phases have frequently been observed by many investigators.[128–130] Electrical conduction occurs through the networks of electrically conducting chains embedded in the glass matrix. The formation and connectivity of these conducting networks determine the microstructural development and resultant electrical properties. Thick film resistors are very complex nonequilibrium systems whose microstructure is a nontrivial function of various materials' properties and processing parameters.

Extensive investigations have been conducted in order to develop a comprehensive model for microstructural development of RuO_2-containing thick film resistors.[52,128,129] The various stages of microstructure development and the respective sintering kinetics were studied. The important processes were determined to be glass sintering, glass wetting and infiltration, conductive phase rearrangement, and conductive phase network formation.

Resistor noise. The random motion of current carriers causes spontaneous fluctuations in electric current, termed *electrical noise.* Current noise in thick film resistors is called *excess noise,* as it is considered the excess noise over thermal noise. Thermal noise is predictable and is a function of resistance and temperature, whereas current noise has no definite relationship. Noise in thick film resistors depends on the sheet resistance value, thickness of the resistor print, and geometry of the resistor. Figure 3.59[131] shows a typical plot of measured noise vs. resistor area for a 200-kΩ/sq resistor. Current noise decreases as resistor area increases. Resistor noise also depends on processing conditions. Generally, noise decreases with an increase in peak firing temperature. The typical noise value for commercial thick film resistors is shown in Table 3.20.

Sheet resistance. The sheet resistance can be adjusted by altering the conductive phase concentration. Figure 3.60[132] illustrates the variation of sheet resistivity as a function of conductive phase concentration for various systems. The relationship between volume fraction and electrical properties of a heterogeneous system consisting of a conducting phase and an insulating phase was studied by several people, including Avogadro, Faraday, Lorents Clausias, Lorentz, and Maxwell. Various theoretical models such as percolation theory, effective medium theory, and perturbation theory have attempted to explain the variation of sheet resistivity of thick film resistors in terms of conductive phase concentration. The theoretical models developed by Pike,[133] Smith and Anderson,[134] and Ewen and Robertson[135] all assume a fixed idealized microstructure for thick film resistors. Pike developed a bond percolation model to explain the connectivity in thick film resistors, commencing with a small volume fraction of the conductive phase. According to his model, the larger size glass particles are closely packed and the smaller size conductive particles occupy the interstices. The conductive particles form partial bonds in a lattice of connecting interstices. Smith and Anderson[134] developed a model based on effective medium theory. In their model, they considered the size difference between conductive phase and glass particles. They developed an equation for resistivity in terms of volume fraction, parti-

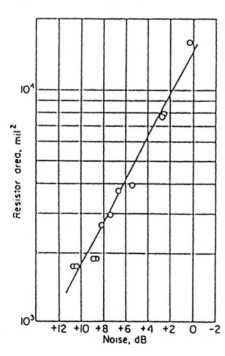

Figure 3.59 Measured noise vs. resistor area for a 200-kΩ/sq resistor.

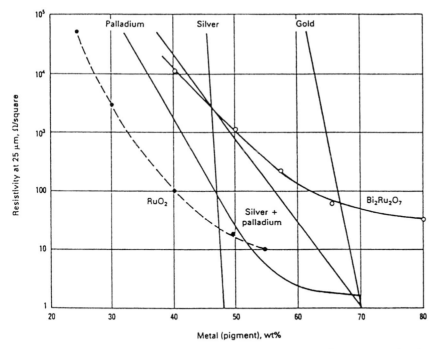

Figure 3.60 Variation of sheet resistivity as a function of conductive concentration for different conductive phases.

cle size ratio, resistivity of a representative unit in the network, and a probability parameter related to the distribution of conductive phase particles.

Ewen and Robertson[135] developed a site percolation model to explain the variation of resistance as a function of conductive phase concentration. In their model, they used a packing fraction for glass and conductive phase particles and an empirical parameter which explains how effectively the conductive phase particles fill the random lattice.

Statistical loading curve models have been proposed that take into account the segregated nature of thick film resistors. A systematic microstructure development study has been conducted by Vest[52] for RuO_2 model thick film resistors. Vest[136] proposed a statistical loading curve to explain the sheet resistivity variation as a function of volume fraction of the conductive phase and particle sizes of both the conductive phase and glass. Several theoretical models have been developed to explain the variation of resistivity as a function of volume fraction of the conductive phase. A comprehensive model to explain the sheet resistivity variation in thick film resistors in terms of composition and physicochemical properties of powders over the entire resistance range has not been developed to date.

Temperature coefficient of resistance. The TCR of a thick film resistor is typically less than ±100 ppm/°C. The TCR of the conducting oxide used in the resistor formulation is typically several thousand ppm/°C (the TCR of RuO_2 is 5760 ppm/°C).

Lead borosilicate glass has an extremely large negative TCR, with its resistivity showing an exponential temperature dependence. The typical temperature dependence of resistance behavior for a thick film resistor is shown in Fig. 3.61. This behavior has led to the development of large numbers of diverse explanations and resultant models to explain this behavior. The R vs. T curve typically shows a minimum in sheet resistance near room temperature. The TCR of a thick film resistor is not a constant because the resistance is not a linear function of temperature. The more commonly observed behavior is a shallow minimum in resistance at some temperature as shown in Fig. 3.61. As a result of this behavior, TCR values of thick film resistors are low. The variation of resistance with temperature is determined by the microstructure and charge transport mechanisms of thick film resistors. The temperature dependence of resistance is a function of the arrangement of conducting chains in the insulating matrix and the charge transport through these chains. Variables such as material parameters and processing condi-

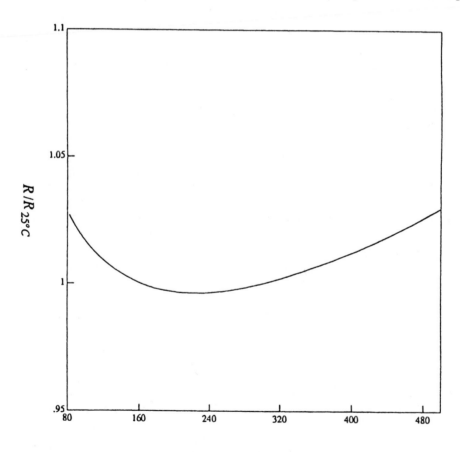

TEMPERATURE IN K

Figure 3.61 Typical temperature dependence of resistance behavior for an RuO_2 thick film resistor.

tions can, therefore, influence the microstructure development and resultant temperature dependence of resistance and TCR values of thick film resistors.

Charge transport through the conductive particle is metallic conduction. Resistance to charge transport through the conducting chains is due to particle and grain boundary resistance. However, charge transport through the insulating glass is very different and has to be explained by charge transport mechanisms that occur in insulating films.

Voltage stability. Thick film resistors are required to be stable under an applied voltage. Voltage stability of resistors depends on the microstructure of thick film resistors, glass chemistry, and conductive phase types used to formulate the resistors. Recent trends in hybrid microelectronics require resistors with higher voltage stability. The power handling characteristics of commercial thick film resistors are shown in Table 3.20.

Surge properties. Fractional ohm resistors are commonly used in telecommunications for surge protection applications. The resistors used in this application require excellent stability under electric surge with low sheet resistance and TCR. Voltage surges due to lightning and switching are very short and typically subside within microseconds. Such pulses can be extremely destructive, since surge currents can produce intense heat. Any thick film resistor used in a protective circuit should withstand instantaneous current and effectively dissipate the heat generated. During this process, the resistor should remain stable. Generally, these resistors are tested for stability against repetitive pulses.

During lightning strikes, very high voltages can be transmitted through telephone lines. It is very hard to predict the magnitude of the voltage and type of voltage pulse that may be produced during lightning strikes. There are several standards, such as those of the U.S. Federal Communications Commission (FCC), Underwriters Laboratories (UL), the Institute of Electrical and Electronics Engineers (IEEE), and Bellcore, which specify the exact shape of a pulse or its wave form and the number of pulses to be used for testing.

Wave shapes that are typically used to test the surge resistor are shown in Fig. 3.62. The surge handling capability of resistors is considered excellent when the drift due to repetitive voltage pulsing of different shapes is low as shown in Table 3.17. These low drift values are typical of state-of-the-art thick film resistors for surge applications.

Thermal stability. After laser trimming and overglazing, resistors are generally tested for shift in resistance in thermal aging, thermal shock, and hot column tests. Typically, commercial thick film resistors change less than 0.5% in resistance after these tests. Thermal aging is done by storing parts at 150°C for 1000 h. Typical change in resistance due to thermal aging for stable resistors from 10 Ω/sq to 1 MΩ/sq is less than 0.1%. (The relative change in resistance as function of storage time at 125°C is shown in Fig. 3.63 for 1 kΩ/sq resistors).

Hot column stability testing is done by placing the test substrate with resistors on a hot stage at 400°C for 5 min, with a subsequent quench to

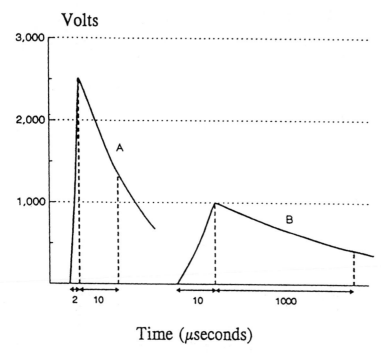

Figure 3.62 Various surge waveforms used to test resistors.

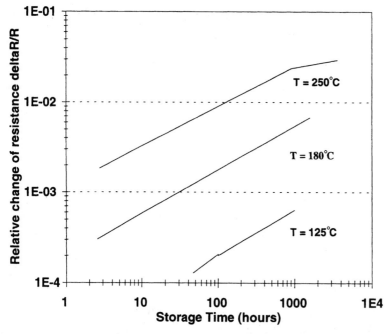

Figure 3.63 Relative change in resistance as a function of storage time for 1-kΩ/sq resistors.

room temperature. The shift in resistance values is then recorded. Stable resistors experience minimum change from this test. Stability of resistors after thermal shock is shown in Fig. 3.64. The stability of resistors can also be tested by subjecting them to a thermal cycle test which consists of 5 cycles of 5 min at −65°C, transfer within 10 s to +150°C, and a dwell of 5 min before transfer back to −65°C. The stability of commercial thick film resistors is considered acceptable if changes in resistance of less than ±0.2% result from this test.

3.4.6 Effects of variables on electrical properties

Peak firing temperature. The effect of peak firing temperature on resistance of different decade value resistors is shown in Fig. 3.65. Sheet resistance values decrease with an increase in peak firing temperature for high value resistors. As firing temperature increases, the glass viscosity is reduced. The primary way in which glass affects the electrical properties is by controlling the formation of three-dimensional conductive phase networks. As temperature increases, the conductive particles rearrange to form an increased number of conductive chains, which can result in decreased sheet resistances. The effect of peak firing temperature on TCR values of resistors is shown in Fig. 3.66. TCR values typically become more positive for high value resistors as temperature increases. However, for low value resistors, the increase in firing temperature results in an increase in the resistance and more negative TCR values. The peak firing temperature has a different effect on the microstructure formation of low value resistors because of the low glass concentrations; glass

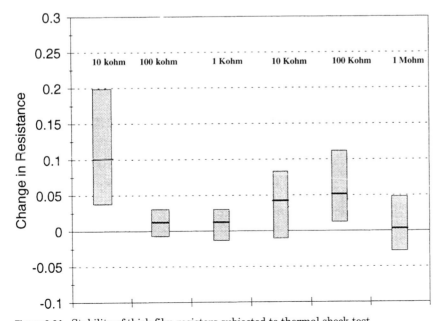

Figure 3.64 Stability of thick film resistors subjected to thermal shock test.

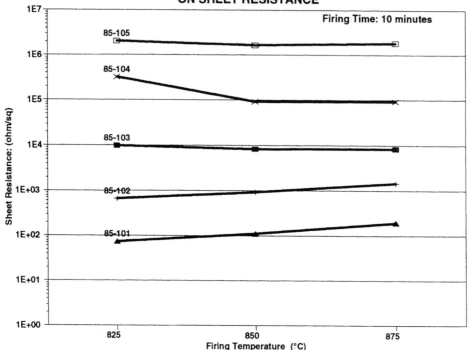

Figure 3.65 Effect of peak firing temperature on sheet resistance for decade value resistors.

may flow better at higher temperatures and form thin barriers around conductive phase particles.

Firing time at peak firing temperature. An increase in firing time at peak temperature produces conductive phase rearrangement, which alters the number of three-dimensional conducting chains. If this rearrangement can increase the number of conducting chains, the resistance will decrease and TCR values will become more positive. The effect of firing time on R and TCR at peak firing temperature is shown in Fig. 3.67. As in most kinetic processes, temperature has a greater impact on properties than time and must be more tightly controlled.

Particle size and electrical properties. The effect of glass particle size on sheet resistance of a model thick film resistor is shown in Fig. 3.68[137] for various conductive phase concentrations. The effects of glass particle size are more pronounced at low conductive phase concentrations. The effect of conductive phase particle size on electrical properties of thick film resistors was studied by several researchers.[115–117,137] For a given volume fraction of conductive phase, smaller particle sizes will decrease the sheet resistance and render the TCR more positive.

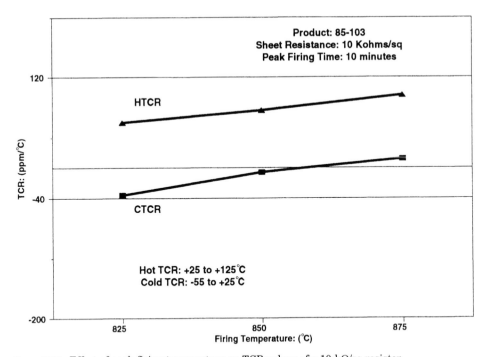

Figure 3.66 Effect of peak firing temperature on TCR values of a 10-kΩ/sq resistor.

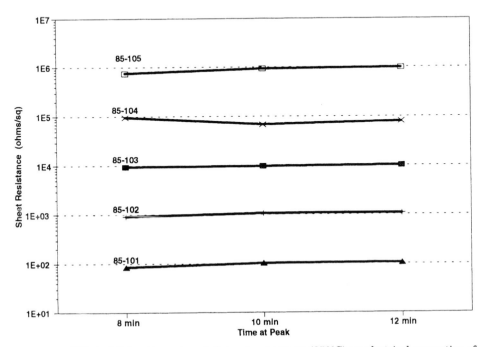

Figure 3.67 Effect of firing time at peak firing temperature (850°C) on electrical properties of (10-kΩ/square) resistors.

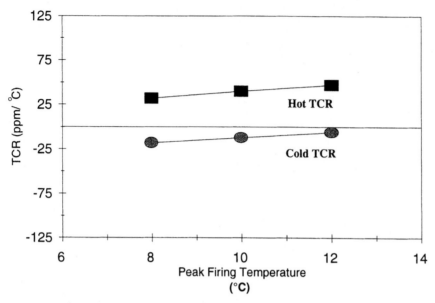

Figure 3.67 *(Continued)* Effect of firing time at peak firing temperature (850°C) on electrical properties of (10-kΩ/sq) resistors.

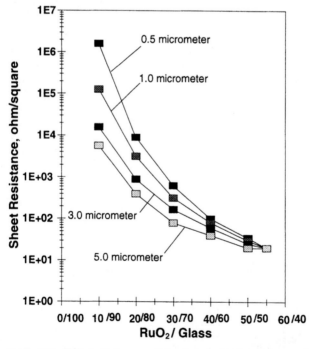

Figure 3.68 Effect of glass particle size on sheet resistance of model thick film resistors.

Termination effects. Sheet resistance can be affected by the type of conductor material used to terminate the resistor. Several techniques have been developed to demonstrate the interaction between conductor and resistor at the termination interface. Figure 3.69[138] illustrates the observed resistance as a percent of total resistance when the profile point is moved from one termination interface to another. This technique shows that the resistance value and interaction between conductor and resistor can be studied by observing the nonlinearities in the resistance vs. length curve. The region very close to the conductor shows a nonlinear relationship between length and resistance because of conductor–resistor interfacial interactions.

The ideal geometry of a terminated resistor is shown in Fig. 3.70a. The resistor overlap area will always be thicker since the resistor is typically printed on top of the fired conductor. If the current flux is uniform throughout the resistor and conductor–resistor interface, there are no geometry effects. However, if there is conductor bleed-out during resistor firing, then the current flux will not be uniform and current crowding can lead to hot spots at the edge of the conductor, as shown in Fig. 3.70b. A change in resistance due to a local increase in temperature will occur at the hot spots. This will result in a nonuniform resistance at the conductor–resistor interface.

The observed variation in resistance along resistor length is due to chemical interactions at the conductor–resistor interface. The ideal voltage distribution along the length of a resistor is shown in Fig. 3.71 for low and high resistance values. The diffusion of conductor material into the resistor during resistor firing will result in low resistance. This is most commonly observed in conductors which contain silver. This correlates well with the diffusion rates of metals. The chemical interactions between resistor and conductor materials can change the microstructure of the resistor near the interface

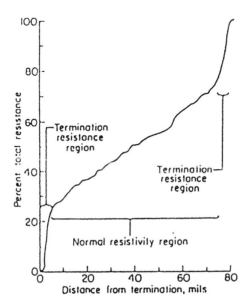

Figure 3.69 Effect of resistance as a function of length of a thick film resistor.

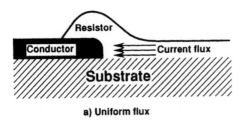

a) Uniform flux

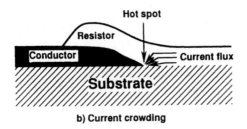

b) Current crowding

Figure 3.70 Current flux at the resistor–conductor interface.

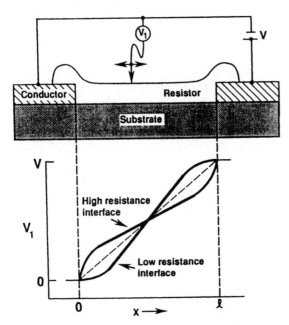

Figure 3.71 Ideal voltage distribution along the length of a resistor.

and will result in higher resistance in that region. This can also increase current noise in thick film resistors. The effect of various types of conductor terminations on sheet resistance and TCR is shown in Fig. 3.72.

Another technique used to demonstrate conductor–resistor interactions is the determination of aspect ratio. The aspect ratio variations for both an ideal resistor and a resistor experiencing interfacial interactions are shown in

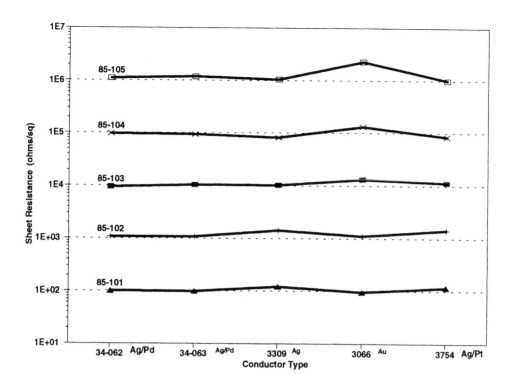

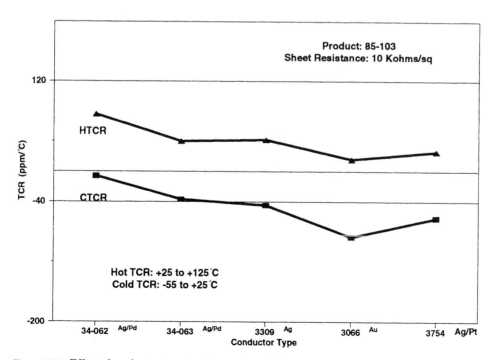

Figure 3.72 Effect of conductor termination on electrical properties of resistors.

Fig. 3.73. The data presented in these figures are very useful for the design engineer, who must account for all geometry effects as well as termination effects for thick film resistor design. The effect of resistor length and geometry on resistance of thick film resistors is shown in Fig. 3.74A and 374B.

3.4.7 Conduction mechanisms

A considerable amount of research has been done in attempts to understand conduction mechanisms in thick film resistors. However, there is no common agreement among researchers in explaining these mechanisms. Pike and Seager[133] proposed a tunneling barrier model to explain the small activation energy associated with the shallow minimum in resistance vs. temperature behavior of thick film resistors. In explaining the conduction mechanism, they assumed a simple metal–insulator–metal (MIM) unit as the typical microstructure of thick film resistors, and concluded that the conduction mechanism through the insulating glass is thermally activated tunneling. According to their model, the small activation energy is associated with electrostatic charging of the fine conductive particles. The model also assumed that the presence of impurities within the tunnel barrier can act as resonant centers to increase the barrier transmission coefficient. Hill[139] proposed an alternative model with the assumption that doping impurities in the glass interface form narrow bands. Smith and Anderson[140] developed another model by combining the tunneling barrier and narrow band models and concluded that both mechanisms occur together. Halder[141] proposed that the electron transport in thick film resistors is due to both tunneling and hopping mechanisms. Prudenziati et al.[142] proposed a model with a temperature-dependent tunneling barrier height and width without taking into account the resistance of the conductive phase. Vest[143] proposed a multipath model for thick film resistors in which the dominant carrying paths were viewed as a series of parallel resistors, each having a parallel path consisting of resis-

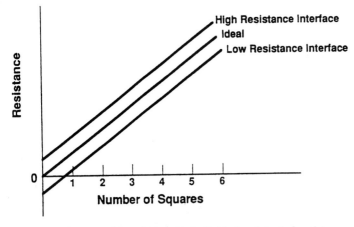

Figure 3.73 Aspect ratio variations for both ideal and typical resistors.

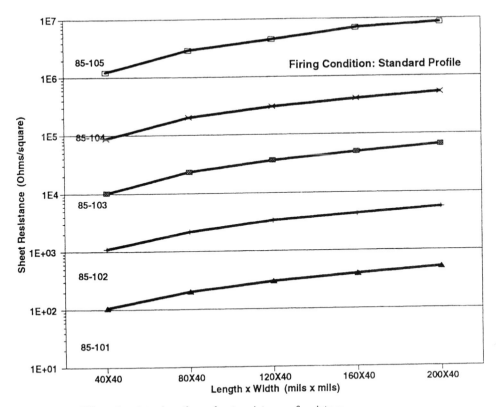

Figure 3.74A   Effect of resistor length on sheet resistance of resistors

tors with different numbers of sintered and nonsintered contacts. The equivalent resistance of these paths and their temperature dependence is a function of the number of sintered and nonsintered contacts, the geometry of the contacts, and the respective conduction mechanism.

It has been demonstrated that, by considering simple resistor microstructures and utilizing the multipath model, conduction mechanisms with very high activation energies (0.5 eV) can be used to explain the temperature dependence of resistors which exhibit very low activation energies.[144] However, no single quantitative model developed thus far completely explains the conduction mechanism of thick film resistors.

3.4.8   Laser trimming

Hybrid thick film resistors are required to meet tight tolerances of $\pm 0.1\%$. Because of the complex nature of thick film resistors, they cannot be consistently fired to a predetermined resistance value with the required tolerances. Therefore, the resistors are laser-trimmed to the target value by removing part of the resistor material with a laser. The material removal process in thick film resistors is called *laser trimming*. Generally, design engineers will

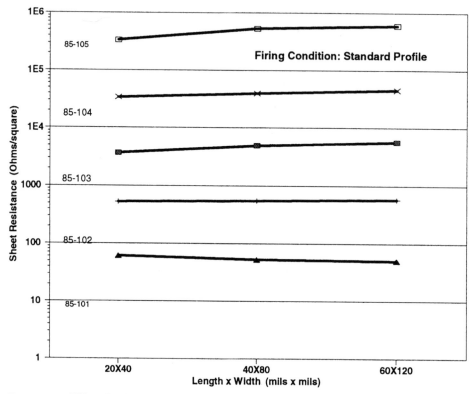

Figure 3.74*B*   Effect of resistor geometry on sheet resistance of resistors.

choose resistance values for a given circuit such that the as-fired distribution will remain below the final target value. The resistors can then be trimmed to the target value by decreasing the effective width of the resistor, which in turn increases the resistance value.

·In laser trimming, materials are actually removed by short duration, high intensity, coherent light pulses. The material absorbs the light energy, which causes it to heat rapidly and vaporize. The typical volume of thick film resistor removed by a single laser pulse is on the order of several cubic micrometers (μm^3). A laser cut is produced by a succession of overlapping laser pulses as shown in Fig. 3.75.[145]

For thick film resistor trimming, a neodymium-doped yttrium aluminum garnet (Nd:YAG) laser operating at 1.06-μm wavelength is typically utilized. There are several important parameters in laser trimming. Q rate is defined as the number of laser pulses issued per second. Byte size is the amount of incremental material affected by each laser pulse. Kerf width is the outer width of the cut. Trim speed is the rate of materials removed in inches per second. The laser operation is typically in the Q-switch mode and the laser beam pulses are focused by galvanometer mirrors. The resistance is continuously monitored and as soon as the desired resistance values are obtained, a closed-loop control system shuts off the laser power. Because of the complexi-

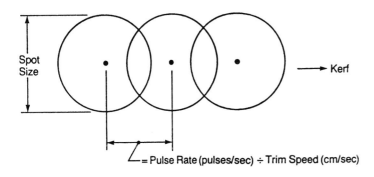

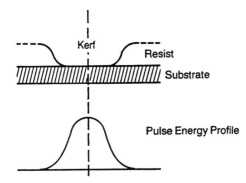

Figure 3.75 Overlapping of successive laser pulses which produce a laser cut.

ties and variations of laser cut parameters, the combination of byte size, Q-rate, and laser power which can produce an optimum laser cut for a particular material must be experimentally determined. High throughput, trim accuracy, and stability are goals of these experiments for ultimate trim parameter recommendation.

Types of cuts. Laser trims made on thick film resistors can be categorized into three major groups: plunge cuts, L cuts, and double cuts.

Plunge cuts. The fastest cut that can be used to trim a resistor is the plunge cut. It also provides the lowest amount of laser exposure to the resistor. The disadvantage of this type of cut is that resistors with high accuracy cannot be achieved because of rapid change in the resistance value as the cut approaches the other side of the resistor. The increase in resistance with laser cut length is shown in Fig. 3.76. In general, if the initial value of the resistor prior to trimming is close to the target value, then reasonable control and accuracy can be achieved.

L cuts. L cuts are used to avoid the problem of rapid resistance change which is inherent in the plunge cut. First, the resistor is trimmed linearly toward the opposite side of the resistor until the resistance begins to change rapidly. The laser trim direction is subsequently changed to run along the length of the resistor for more accurate control. The laser cut, however, exposes the resistor to a greater amount of laser damage. Studies suggest that the L cut produces a narrow distribution of resistance about the target and maintains a low current noise after trimming. The lower noise index in thick film resistors is achieved by removing resistor material parallel to the current flow, which retains uniform flux lines. In the case of the L cut, the length of the laser cut and trim times are longer. Also, a larger amount of resistor material remains. If the L cut extends too far into the other side of the resistor, microcracks will form and propagate and cause large variations in resistance after completion of the laser trim.

Double plunge cuts.   Double plunge cuts are also used to avoid rapid resistance changes during trimming. During the first cut the resistor is trimmed to a specified resistance value (e.g., 80% of total trim target), which permits the second cut to trim the resistor at a slower rate. The second cut works in the "shadow" of the first laser cut. In order to make this method effective, the second cut should be shorter than the first cut. Laser damage due to double cuts is less than that of the L cut because microcracks perpendicular to the kerf do not cause the resistor to open. Double cuts also have less laser trim length than L cuts.

Serpentine cuts.   The serpentine cut is very similar to a double cut except that it is used to increase the resistance by a large amount. It consists of a series of short plunge cuts on opposite sides of the resistor. This is done by increasing the length of the current path. The increase in the length of current path is a function of the depth of cuts and their locations.

Drift due to trimming.   Laser trimming produces a very servere thermal shock in thick film resistors and can lead to microcracking in the neighborhood of

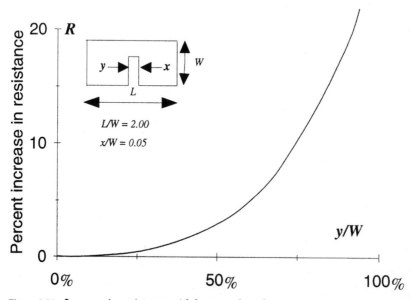

Figure 3.76 Increase in resistance with laser cut length.

the kerf. These microcracks can cause resistance drift during storage at elevated temperatures.

Thick film resistors with tight tolerance values can be predetermined by calculating the drift due to trimming and subsequent processes prior to setting target values. This is possible only if the resistor drift is both predictable and repeatable.

3.4.9 Overglazing of resistors

Thick film resistors are often overcoated with a glaze material to protect against the environment, chemical attack, and mechanical abrasion. The typical material used is lead borosilicate glass, which usually fires at or below 550°C. When thick film resistors are refired, there is a drift in sheet resistance associated with this process. Overglazes are typically designed to fire at lower temperatures in order to minimize this drift. The color of overglazes is generally green to facilitate laser trimming of resistors; the laser energy absorption of green material is excellent because of the wavelength matching with Nd:YAG laser light. The thermal expansions of the glazes are matched to that of the substrate to avoid any cracks due to thermal stresses.

A typical firing profile for processing an overglaze is shown in Fig. 3.77. Low temperature firing of overglazes can also serve as an annealing treatment for both resistors and glazes to avoid any stress buildup in the resistor.

3.5 Rheology and the Screen Printing Process

3.5.1 Introduction

Rheology is the study of flow and deformation of materials. It is sometimes confused with the viscosity of a material. The viscosity of a material is its resistance to flow. Viscosity is but one aspect of rheology. The study of rheology includes viscous as well as viscoelastic response of the material, the ability

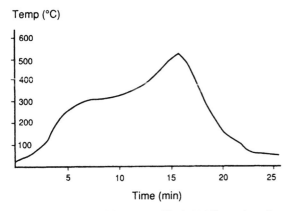

Figure 3.77 Typical firing profile (505°C) used to fire overglaze.

to thin and thicken with increasing stress, and the concept of yield point, i.e., the point at which material begins to flow.

Thick film preparation and applications involve flow phenomena. The control of rheological behavior of thick film inks is important for the screen printing operation. Print thickness and print resolution are directly affected by paste rheology. Paste viscosity is used as a quality control tool by circuit manufacturers. For the paste manufacturer, understanding the correlation between the paste formulation and rheology is important for reproducible product manufacturing and new product development.

A number of excellent books and review articles are available on rheology.[146–153] In this section, rheology concepts and examples related to the manufacture of thick film inks and paste flow behavior during and after screen printing will be examined.

3.5.2   Viscosity

The viscosity of a liquid is the property that defines resistance of the liquid to flow. The viscous nature of the liquid is due to the molecular attraction that offers resistance to flow. To understand the concept of viscosity, consider the model illustrated in Fig. 3.78. Two parallel plates are separated by a distance x and the space between the plates is filled with a viscous liquid. The bottom plate is held stationary. A force F is applied to the top plate, of area A, in a tangential direction so that the top plate moves with a constant velocity V in the y direction, parallel to the bottom plate. A thin layer of liquid adjacent to the plate will move with the same velocity as the plate. This is the *no-slip* assumption and holds true for most liquids. Thus, the liquid molecules near the top plate surface will move with a specific velocity V, while velocity at the bottom plate will be zero. Molecules in the liquid layers between these two plates will move at an intermediate velocity.

The tangential force acting per unit area is called the shear stress σ and is given by the following equation:

$$\sigma = \frac{F}{A} \tag{3.48}$$

The units of shear stress are N/m² or dyn/cm².

The shear rate $\dot{\gamma}$ is related to the velocity gradient (dV/dx) across the gap. Shear rate is defined by the following equation:

$$\dot{\gamma} = \frac{dV}{dx} \tag{3.49}$$

The units of shear rate are (m/s)/m, i.e., 1/s or s^{-1}.

Viscosity η of the liquid can be defined in terms of two measurable quantities, i.e., shear stress and shear rate, as follows:

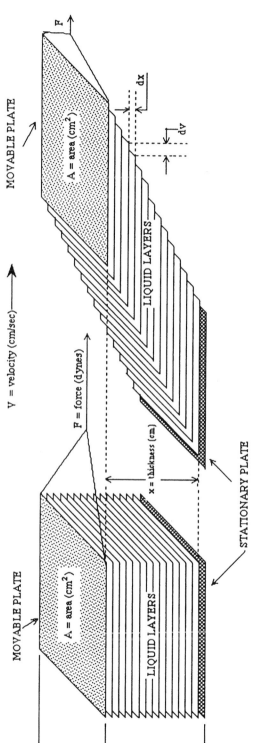

Figure 3.78   Schematic diagram illustrating the flow behavior of a newtonian liquid in a parallel plate arrangement.

$$\eta = \frac{\sigma}{\dot{\gamma}} \qquad (3.50)$$

The International System of Units (SI) unit of viscosity is the pascal-second or $N/m^2 \cdot s$. A commonly used centimeter-gram-second (cgs) system unit of viscosity in the thick film industry is poise ($dyn/cm^2 \cdot s$). The following relation exists between various units:

$$1 \ Pa \cdot s \ (Pascal \cdot second) = 10 \ poise \ (P) = 1000 \ centipoise \ (cP)$$
$$= 1000 \ milli\text{-}pascal\text{-}second \ (mPa \cdot s)$$

For example, the viscosity of water at 20.2°C is 1 cP or 1 mPa•s.

3.5.3 Classification of rheological behavior

Thick film pastes are two-phase mixtures of solid particles dispersed in a continuous organic phase. If the particle size of the solid phase is smaller than 1 μm, the two-phase mixture is called a *colloidal dispersion*. For suspensions, the solid particle size is greater than 1μm. The rheological properties of the dispersion are dependent on the nature of the components and also on the interactions between these two phases. Dispersions can be classified into two broad categories based on rheological behavior: (1) newtonian and (2) non-newtonian.

Newtonian flow. Newtonian flow behavior is characterized by a linear relation between the shear stress and shear rate. In this case, the viscosity is independent of the applied shear rate or stress and is a material constant. Materials exhibiting shear-independent viscosity are called *newtonian* fluids. Newtonian flow behavior is generally exhibited by solvents, dilute polymer solutions, and dispersions. Thick film pastes rarely show newtonian flow behavior, although the concept is sometimes used in the theoretical analysis of screen printing behavior.

Nonnewtonian flow. Materials which do not strictly follow newtonian flow behavior are called *nonnewtonian fluids*. For nonnewtonian materials, it is convenient to define an apparent viscosity, η_a,

$$\eta_a = \frac{\sigma}{\dot{\gamma}} \qquad (3.51)$$

where η_a is a function of $\dot{\gamma}$. Thus, it is not sufficient to characterize materials by measuring viscosity at a single shear rate. The material is characterized by describing the relationship between the shear stress and shear rate. This type of plot is referred as a *flow diagram*. Nonnewtonian materials also show time-dependent flow behavior, whereby the viscosity is a function of shear stress or rate and time.

Dispersions can behave either as pure viscous substances or viscoelastic materials. In the case of viscous substances, the deformation does not recover after the stress has been removed. Part of the deformation can be recovered after the removal of shear stress for viscoelastic materials. Thick film pastes may exhibit both time-dependent and viscoelastic flow behavior. The classification of flow behavior is shown in Table 3.21. Various types of flow behavior are shown in Fig. 3.79. This figure shows the plots of shear stress versus shear rate and corresponding plots of viscosity versus shear rate.

Shear thinning flow. Dispersions showing a decrease in viscosity with shear rate (or shear stress) are described as *shear thinning* or *pseudoplastic.* Shear thinning behavior is generally produced by the reversible breakdown of suspension structures or alignment of anisotropic particles due to shear.

Shear thickening. Dispersions showing an increase in viscosity with increasing shear rate are called *shear thickening* or *dilatant* materials. This type of flow behavior is usually exhibited by dispersions of rigid particles at high concentrations. Thick film pastes rarely show this type of flow behavior.

Bingham plastic. Bingham plastic dispersions exhibit a yield stress σ_y, the minimum shear stress that must be exceeded to initiate flow. Below the yield stress it is generally assumed that the material behaves as an elastic solid (i.e., it can deform but it does not flow). The presence of a yield stress indicates strong interactions between the particles and the presence of a suspension structure. Once the yield stress is reached, flow starts and the Bingham plastic dispersion behaves as a newtonian dispersion; i.e., the excess shear stress is proportional to the shear rate. The slope of this straight line is called the plastic viscosity η_p.

After the yield point, dispersion may show yield shear thinning or yield shear thickening flow behavior. Thick film pastes often show a yield point. The importance of yield point in paste flow behavior will be discussed later.

Thixotropy. For the dispersions discussed above, it was assumed that for a given shear stress there is only one associated shear rate (or viscosity). Thus, it was assumed that the measured viscosity is independent of other factors such as shear history and time.

TABLE 3.21   Classification of Fluid Flow Characteristics

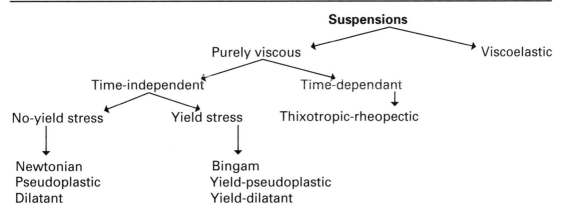

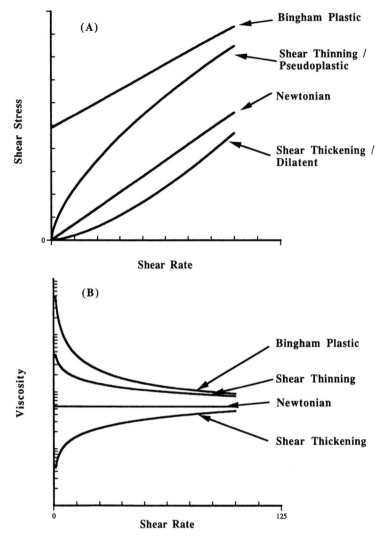

Figure 3.79 Plots of (A) shear stress vs. shear rate and (B) viscosity vs. shear rate for various types of flow behavior.

Dispersions showing time-dependent flow behaviors are called *thixotropic* dispersions. There is a gradual decrease in viscosity with time when a constant shear stress (or shear rate) is applied. After removal of the stress, the viscosity gradually increases. Thixotropic materials generally show shear thinning behavior. The origin of shear thinning and thixotropic behavior is in the breakdown of the suspension structure. The suspension structures arise due to attractive interactions between the particles. In the case of shear thinning materials, the rate of structure breakdown is the same as structure rebuilding at a given shear rate. For a thixotropic material the suspension structure gradually breaks down when the stress is applied.

The presence of thixotropy can be detected either by measuring shear stress at a constant shear rate as a function of time or by studying ascending and descending shear stress–shear rate curves under certain programmed conditions as shown in Fig. 3.80. A thixotropic dispersion under such conditions will produce a hysteresis loop in the shear stress vs. shear rate plot. If the program conditions are kept constant, the hysteresis can be related to the degree of thixotropy.[154]

Thick film pastes often show thixotropic flow behavior. The shear history must be taken into account during paste viscosity measurements in order to explain the paste rheology and printing correlation.

3.5.4 Measurement instruments

To measure viscosity, a sample is placed in a specified measuring geometry to produce a steady simple shear. Most of the viscometers depend on rotational

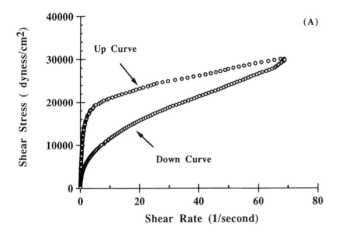

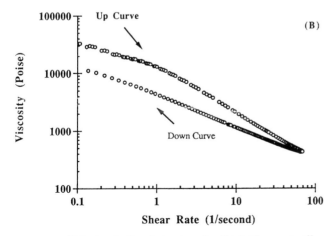

Figure 3.80 Thixotropic flow behavior of a thick film paste illustrating the thixotropic loop.

motion to induce a simple shearing flow. Three types of geometries are commonly used: concentric cylinder, cone and plate, and parallel plate. The shear rate is calculated from the applied or induced angular speed and the measuring geometry. The shear stress is determined by the applied or induced torque and the measuring geometry utilized. From the known value of shear stress or shear rate, the viscosity can be calculated. The dependence of shear stress, shear rate, and viscosity on measuring geometry is shown in Table 3.22.

Instruments used to measure the flow behavior of pastes can be broadly classified into two categories: (1) stress-controlled and (2) strain-controlled viscometers. In the strain-controlled machines, the shear rate is varied with a suitable drive motor and the induced shear stress is measured. The strain-controlled equipment can be further divided into two categories: (1) instruments where the applied shear rate and the induced shear stress are measured on the same member of the measuring geometry (Brookfield, HAAKE-RV, and Ferranti Shirley viscometers are examples of this type) and (2) instruments where the drive is applied on one member of the measuring geometry, e.g., outer cylinder, and the resulting torque is measured on the other member, e.g., inner cylinder (Weissenberg and Rheometrics rheogoniometers and HAAKE CV-100 systems are examples of this type).

In the stress-controlled viscometer, the torque is applied to the material and the resulting deformation is measured. The torque and resulting rate of rotation are measured on the same member of the measuring system. Carri-Med, Physica, Rheometrics, DSR, and Bohlin are stress-controlled viscometers. Stress-controlled viscometers are useful in characterizing stress-controlled phenomena encountered in coating applications such as sagging and

TABLE 3.22   Dependence of Viscosity Parameters on Measuring Geometry

Geometry	Measured quantity	Maximum shear stress	Maximum shear rate	Newtonian flow equation
Concentric cylinder	Torque M, angular velocity ω	$\sigma = \dfrac{M}{2\pi a^2 L}$	$\dot\gamma = \dfrac{\omega_a^2 a^2 b^2}{a^2(b^2 - a^2)}$	$\eta = \dfrac{M}{4\pi L \omega a}\left(\dfrac{1}{a^2} - \dfrac{1}{b^2}\right)$
				a = radius of inner circle
				b = radius of outer circle
				L = length of cylinder
Cone, plate	Torque, angular velocity	$\sigma = \dfrac{3M}{2\pi a^3}$	$\dot\gamma = \dfrac{\omega a}{\alpha}$	$\eta = \dfrac{3\alpha M}{2\pi \omega a^3}$
				a = radius of core
				α = angle between cone, plate ($\approx 1°$)
Parallel plate	Torque, angular velocity	$\sigma = \dfrac{3M}{2\pi a^3}$	$\dot\gamma = \dfrac{\omega a}{h}$	$h = \dfrac{3Mh}{2\pi a^4 \omega}$
				h = gap between parallel plates

leveling of coatings and the sedimentation of particles. The "true" yield point of the paste can be measured using stress-controlled viscometers. The viscosity shear rate behavior over several decades of shear rates can also be measured.

The type of viscometer used to characterize a thick film paste depends on the information required. To measure viscosity accurately, viscometers should be routinely calibrated with a newtonian viscosity standard. The viscosity of the standard should be similar to that of the sample being measured. Viscometers should be properly leveled and the sample temperature should be carefully controlled. The shear and temperature history during storage and shipping of a sample should be closely monitored to obtain consistent results for thixotropic samples. Enclosed measuring systems are preferred to avoid solvent loss during measurements. Brookfield viscometers with the cone and plate geometry are commonly used for checking the lot-to-lot reproducibility of a paste. A single point viscosity measurement is generally not sufficient to correlate viscosity to printing characteristics of a paste. To characterize paste flow behavior over a broader shear rate range, research type viscometers are required.

3.5.5 Measurement methods

Flow curves. In this type of testing, shear stress or shear rate is changed as a function of time and resulting shear rate or shear stress is measured. This type of test is generally used to survey flow behavior over one to two orders of magnitude of shear rate range. Thixotropic behavior of pastes can be determined from the hysteresis in shear stress–shear rate plots. There are several phenomenological models available to curve-fit the experimental data.[150] Flow curves for various thick film pastes are shown in Fig. 3.81.

Creep test. In a creep test, a constant shear stress is applied to a sample for a certain period of time and the resulting deformation is monitored. This type of test is used to characterize viscoelastic properties of the paste. Creep tests can also be used to measure paste viscosity at very low shear rates (approximately 10^{-3} to 10^{-6} s^{-1}). In a strain test, a step shear rate is applied to the sample and the resulting stress relaxation is measured. Examples of creep tests for thick film pastes are shown in Fig. 3.82.

Oscillation test. During an oscillation test, a cyclic shear strain is applied to the sample and resulting shear rate response and time lag of response are measured. From these measurements, viscous and elastic properties of the material can be separated. The elastic modulus G' measures the amount of energy stored, whereas the loss modulus G'' measures the energy loss due to viscous dissipation. Figure 3.83 is an illustration of G', G'', and η as a function of frequency for a Ag paste. The viscoelastic properties of materials are further characterized by measuring the frequency, amplitude, time, and temperature dependence of the elastic and loss modulus.

The rheological response of a material during either measurement or use depends on the ratio of the characteristic time of the material τ, to the char-

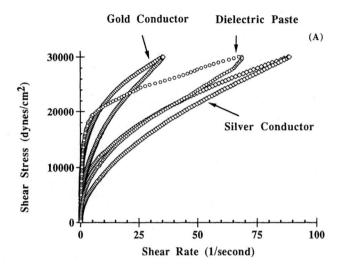

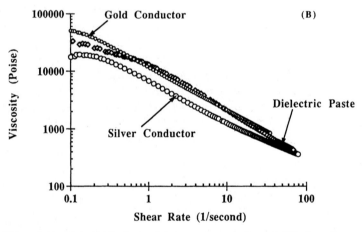

Figure 3.81   Plot of (A) shear stress vs. shear rate and (B) viscosity vs. shear rate for various thick film pastes.

acteristic time T of the deformation process being observed. This ratio is called a *Deborah number*, D_e[146], where

$$D_e = \frac{\tau}{T} \tag{3.52}$$

The time τ is infinite for an elastic solid and zero for a newtonian viscous liquid. Low Deborah numbers indicate liquidlike behavior and high Deborah numbers indicate solidlike behavior. Thus, the normally viscous material will behave as a viscoelatic material if the characteristic time of the process is very small (e.g., ink entry into the rollers).

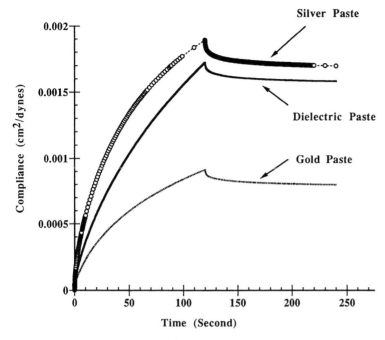

Figure 3.82 Creep flow behavior of various thick film pastes.

3.5.6 Interparticle forces

Thick film pastes are mixtures of solid particulates (metal powders, frit, etc.) suspended in polymeric solutions. The rheological properties of the paste are directly influenced by the relative amounts of each ingredient and interactions between the solid, polymer, and solvent. Additives such as dispersants/surfactants and rheology modifiers are also commonly added to the paste to improve paste performance. The average particle size of the powders is in the range of approximately 0.1 to 10 μm for most thick film pastes. In this size range, the specific surface area is large (several m^2/g to 100 m^2/g) and the interparticle interactions are dominated by the solid–liquid interface characteristics or surface chemistry of the system.

The main factors that affect the rheological behavior of a suspension are brownian motion, hydrodynamic interactions, and the interparticle forces of attraction and repulsion. In the colloidal particle size range (1 nm to 1 μm), gravitational forces on the particle are not very important and particles are moving randomly in a dispersion medium due to thermal energy. Particle encounters due to brownian motion can lead to formation of either doublets or particle aggregates, or particles can remain as individual units depending on the particle interactions. In the absence of any repulsive interactions, these random collisions can lead to aggregation of particles to reduce the free energy of the system.

The origin of attractive interactions between particles is the van der Waals attraction between atoms of the colloidal particles. This attraction energy is

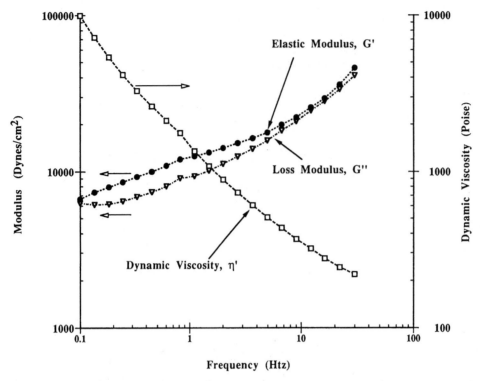

Figure 3.83 Loss modulus G'', elastic modulus G', and dynamic viscosity η' as a function of frequency for a silver paste.

short range, since it is inversely proportional to the sixth power of distance of separation between atoms, but the total interaction energy between colloidal particles (i.e., for a collection of a large number of atoms), is quite large and of long-range order. The energy of attraction V_a between two colloidal particles can be quantified by the following relations:

$$V_a = -\Lambda(A)H(G) \qquad (3.53)$$

where $\Lambda(A)$ is the Hamaker constant of the material and $H(G)$ is determined by the geometry of the system. The Hamaker constant for the medium (i.e., solvent used) and absorption of the polymer at the solid-liquid interface will modify the attraction between the particles.[155] Two mechanisms are available to overcome the attraction between particles: (1) electrostatic interactions and (2) interactions of absorbed polymer.

If the collided particles can develop an electric charge and if all particles have the same sign of charge, then particles will repel one another during approach. Surface charge can be developed in the case of simple oxides by surface group dissociation or by preferential absorption of ionic dispersants. The charge on the solid surface contributes to the distribution of counterions in the solution. This model of the charged interface is often called the *electrical double layer*. The total interaction energy between two particles described by the DLVO theory (for Deryagin, London, Verwey, and Overbeek) is calculated by summation of van der Waals attraction and electrostatic repulsion as shown in Fig. 3.84.[156] The repulsive interaction energy as a function of the distance of separation of particles between two spheres is controlled by the ionic strength of the solution, surface potential (related to charge density on the surface), and the radii of the particles. When the electrostatic repulsive potential barrier is sufficiently large (10 kT), a stable dispersion can be obtained. If the potential barrier is low, the particles will be flocculated due to the van der Waals attraction. Electrostatic interactions are important, for example, during milling of glass flakes in water.

The second method to modify interparticle interactions is by absorption of polymer at solid–liquid interfaces.[157] The absorbed polymer increases the stability by increasing the electrostatic repulsion between particles and/or by decreasing the van der Waals attraction. The absorbed polymer can also impart stability due to an additional steric component of repulsion. In this case, the overlap of absorbed polymer layers gives rise to repulsive interactions. Figure 3.85 illustrates the potential energy of interaction between two particles with an absorbed polymer layer. This protective action due to the absorbed polymer is often called *steric stabilization*. The phenomenon of steric stabilization has a thermodynamic basis. The absorbed polymer should completely cover the particle surface to prevent bridging flocculation. The thickness of the absorbed polymer layer should be sufficient to overcome the van der Waals attraction. The solvent used should be such that the overlap of polymer segments leads to repulsion during particle encounters.

If ionic dispersants are used in paste preparation, then steric as well as electrostatic interactions can occur during particle encounters. There is competition for the particle surface between solvent, binder, dispersant, rheology

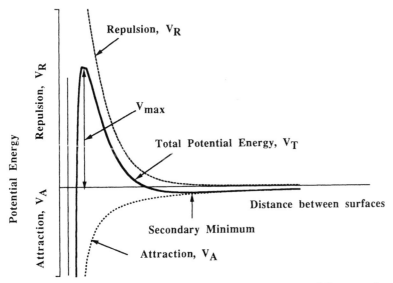

Figure 3.84 The potential energy of interaction as a function of distance of separation for electrostatic interactions.

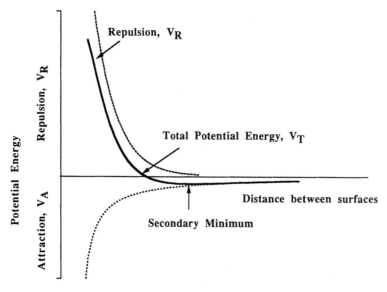

Figure 3.85 Potential energy of interaction as a function of distance of separation for steric stabilization.

additives, etc., and complete quantitative understanding regarding dispersion stability of pastes is not possible.

The third type of force important in paste rheology is hydrodynamic in nature. Viscous force arises from the relative motion of particles with respect to the suspending medium (i.e., localized velocity difference). Since viscosity is a measure of resistance to flow, or the localized energy dissipation rate, presence of particles leads to disturbance in the fluid flow lines and an increased rate of energy dissipation. Hence, the viscosity of the suspension is larger than the suspending fluid. The effect of the suspending liquid is normalized by dividing the suspension viscosity with solution viscosity to get the relative viscosity $\eta_{relative}$ of a suspension:

$$\eta_{relative} = \frac{\eta_{suspension}}{\eta_{solution}} \tag{3.54}$$

From this equation, it is evident that control of solution or vehicle viscosity is important to control the paste flow behavior. The solution viscosity can be controlled by adjusting polymer molecular weight, solvent selection, and polymer concentration.[158]

3.5.7 Suspension structure

The above-described interparticle interactions lead to formation of suspension structures at rest. The type of suspension structure formed depends on whether the interparticle forces are attractive or repulsive in nature. With strong repulsive interactions, solid crystalline structures can be formed.[159]

The attractive interaction appears to be more common with paste materials. The flow behavior of the suspension is strongly affected by the nature of the suspension structure. The extreme cases are the formation of chainlike structures or formation of spherically shaped clusters of particles. The two shapes are the extreme simplifications of the real structures and are often used as structural models. The type of suspension structure developed depends on interparticle interactions, the shape and size of solid particles, solid surface characteristics, particle concentration, mixing conditions, shear history, etc. The basic flow units, called *flocs,* are formed by random packing of primary particles. At low shear or at rest, the flocs group into clusters of flocs called *aggregates,* as shown in Fig. 3.86. These aggregates may form a network which can fill the entire volume of the dispersion and control its plastic or structural properties.[160] Another method to obtain the suspension structure is through the vehicle phase. In paste formulations, rheology additives (generally organic in nature) can form a gel type structure at low shear rates. The observed yield point and viscoelastic response of the materials are indicative of a suspension structure above a critical volume fraction of the solid phase.[161]

When a shear stress is applied, the suspension structure is broken down into smaller units. The vehicle entrapped within the floc aggregates is released. This leads to a decrease in viscosity with increasing shear rate. If the rate of structure breakdown due to shear is equal to the rate of structure buildup due to brownian motion, shear thinning or pseudoplastic flow behavior is observed. The equilibrium floc is generally spherical in shape and the size distribution is narrow.[162]

If the rate of structure buildup is slower than the rate of breakdown due to shear, thixotropic flow behavior is observed. Breakdown of the suspension structure can be reversible and the suspension structure is slowly recovered. The rate of structural breakdown and rebuilding is important to obtain optimum print quality during screen printing of thick film pastes.

INCREASING SHEAR RATE

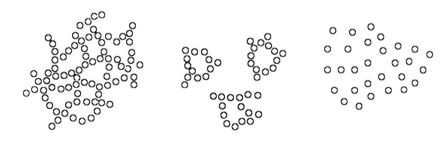

FLOC AGGREGATES   FLOCS

Figure 3.86 Schematic of paste structure.

There are several phenomenological models available to correlate the rate dependence of suspension viscosity.[146-150] The Cross model described below is a general model that requires four parameters to describe the dependence of viscosity on the shear rate of a suspension:

$$\frac{\eta - \eta_0}{\eta_0 - \eta_\infty} = \frac{1}{1 + K\bar{\gamma}^m}$$ (3.55)

where η_0 and η_∞ are asymptotic values of viscosity at very low and very high shear rates and K and m are constants. These parameters are shown in Fig. 3.87. With the Cross model, the shear rate dependence of paste viscosity can be described over a wide shear rate range. The Cross model can be reduced to power law, Sisko, and Bingham model equations with certain approximations[146]:

Power law model: $$\eta = K_2\dot{\gamma}^{m-1}$$ (3.56)

where m is called the *power law index* and K_2 is called the *consistency*.

Sisko model: $$\eta = \eta_\infty + K_2\dot{\gamma}^{m-1}$$ (3.57)

Bingham model: $$\sigma = \sigma_y + \eta_p\bar{\gamma}$$ (3.58)

(A)

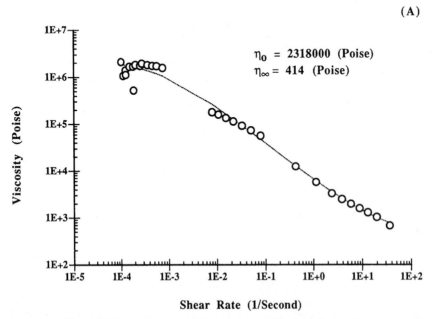

$\eta_0 = 2318000$ (Poise)
$\eta_\infty = 414$ (Poise)

Figure 3.87 Plots of (A) viscosity versus shear rate and (B) viscosity vs. shear stress for a dielectric paste.

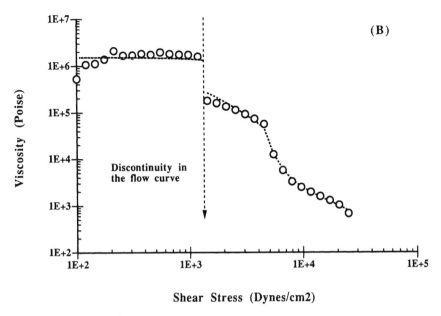

(B)

Discontinuity in the flow curve

Figure 3.87 *(Continued)*

where σ_y is the yield stress and η_p is the plastic viscosity. Another model used in the paint and coatings industry is called the Casson equation:

$$\sqrt{\sigma} = \sqrt{\eta_\infty}\,\sqrt{\dot{\gamma}} + \sqrt{\sigma_y} \tag{3.59}$$

The yield point described by the above models is a calculated value. The true yield point of the material is difficult to measure and depends on the sample shear history and measurement technique.[163] It is sometimes more useful to describe low shear paste behavior by the zero shear viscosity using the Cross model. Still, yield point is of practical importance in describing certain engineering applications.

3.5.8 Effects of paste formulation on paste rheology

The rheological behavior of thick film pastes depends on its composition and the chemical nature of the ingredients. In this section, we will illustrate the effect of some of the compositional variables on paste rheology. It should be noted that each component of the paste (e.g., solid, solvent, dispersant, etc.), has an effect on paste rheology, while interactions between various components can also exhibit a strong effect on paste rheology. The effect of various components is summarized in Table 3.23.[149,153]

The solid constituents of a paste are selected according to the functional requirements of the final products (e.g., metal powders for conductor inks, etc.). Frit is generally added to improve the densification of the film during firing. Most paste formulations generally involve more than one type of solid powder with different densities and particle characteristics such as average

TABLE 3.23   Effect of Paste Components on Paste Rheology

Thick film paste component	Effect on rheological properties
Solid Phase	
Volume concentration of solid, ϕ	Influences hydrodynamic interactions between particles, flocculation and aggregate characteristics.
Particle size, particle size distribution, and particle shape	Influences rheological properties over entire range of shear rates and viscoelastic properties of paste. The particle size etc., affects the particle concentration per unit volume and the rate of flocculation.
Chemical compositions	Influences the interaction force between particles.
Continuous Phase	
Viscosity η_0	Directly proportional to the paste viscosity.
Chemical constitution, polarity	Influences viscosity through the effect on potential energy of interaction between particles, solubility of resin, and viscosity.
Surface Active Agent	
Chemical constitution	Solubility in continuous phase, absorption behavior at solid–liquid interface.
Concentration	Coverage of the surface, viscosity of the continuous phase, η_0
Absorbed layer at interface	Thickness influences the effective particle dimensions and viscosity; it also influences interparticle interactions.
Additives	
Thixotropes, wetting agents, etc.	Modify the interparticle interactions through effect on surfaces of particles, gel formation in vehicle, etc.; can also modify hydrodynamic interactions.

particle size, size distribution, and specific surface area. The rheological properties of the paste are influenced by the fraction of the paste volume occupied by the suspending phase. The volume fraction of the phase is defined by the following:

$$\phi = \frac{\text{volume of the phase}}{\text{total volume of the suspension}} \tag{3.60}$$

As we have seen, hydrodynamic forces influence paste rheology. Hydrodynamic forces act on the surface of particles or aggregates of particles and are generally independent of particle density. The weight percent solids in the paste is sometimes reported by the paste manufacturer. From the known density and weight percent of each ingredient, it is straightforward to convert this information to a volume fraction basis. It is important to understand that the rheological behavior is most influenced by the volume fraction of a particular phase. Addition of similar amounts of diluent thinner can have different effects on a high density conductor paste vs. a low density dielectric paste.

Effects of solid concentration. Einstein showed that for a dilute dispersion of solid particulates the viscosity of the suspension increases linearly with the solid phase volume fraction:

$$\eta_{rel} = \frac{\eta_0}{\eta_{solution}} = 1 + 2.5\phi \qquad (3.61)$$

where η is the suspension viscosity, η_0 is the viscosity of suspending media, and ϕ is the volume fraction of particles. The above equation is valid for rigid, uncharged, spherical particles at very low particle concentrations where hydrodynamic interactions between particles can be ignored.

The dependence of viscosity on volume fraction solids is shown in Fig. 3.88. At high particle concentrations, viscosity of the suspension increases more rapidly than predicted by the above equation due to interparticle interactions. Several empirical equations are available to relate viscosity to the solid concentration behavior of suspensions.[149] As the volume fraction of solids is increased further, a stage will be reached where the particles will be interlocked and no flow will occur (i.e., viscosity approaches infinity). The volume fraction of solids at which this occurs is called the *maximum packing fraction* ϕ_m, and its value will depend on the particle packing, size distribution, shape, etc.

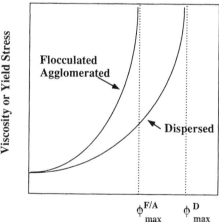

(A)   **Volume Fraction Solids**

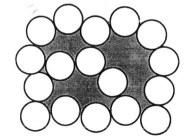

(B)

Figure 3.88 (A) Schematic of viscosity vs. volume fraction of solids in the suspension. (B) schematic of a flocculated suspension

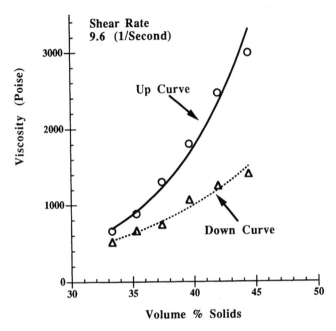

Figure 3.89   Viscosity vs. vol. % solids for a dielectric paste.

The effect of solids concentration on dielectric paste viscosity at a shear rate of 9.6 s^{-1} is shown in Fig. 3.89. These pastes were prepared using the same vehicle with varying solids concentrations from 66 to 76 wt %. A flow test was used in which the shear stress was increased to a maximum value in 2 min and then decreased to zero in 2 min. The viscosity was measured at 25°C using a cone and plate geometry. As shown in Fig. 3.89, a sharp increase in the up curve viscosity with solids loading was observed. The difference in the up curve and down curve viscosity also increased, indicating a higher degree of thixotropy.

The measured yield point of the paste also increased with increasing solids loading, as shown in Fig. 3.90. The yield measurements were conducted by increasing the shear stress from zero to 5000 dyn/cm^2 in 2 min. The reported yield stress indicates the stress at which flow was detected. The measurement of yield stress is sometimes difficult and there is some controversy regarding the existence of yield point.[146,163] Figure 3.91 shows the effect of solids concentration on zero shear viscosity. The low shear flow behavior was determined by applying a constant shear stress to the sample until a constant or "equilibrium" strain rate was obtained. From the known shear stress and shear rate, an equilibrium viscosity at a given shear rate was calculated. The above procedure was repeated again on the same sample at higher shear stresses.

From Fig. 3.91, it is clear that these pastes show a constant (newtonian) viscosity at low shear rates. This viscosity is called *zero-shear viscosity*. The zero-shear viscosity was then calculated from viscosity shear rate data using the Cross equation. The sharp decrease in viscosity with increasing shear

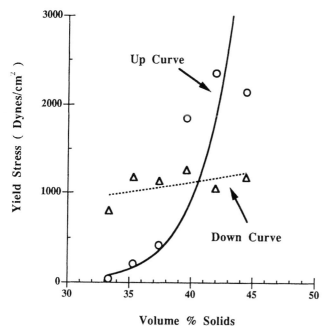

Figure 3.90 Yield stress vs. vol % solids for a dielectric paste.

stress above a critical shear stress indicates the presence of strongly flocculated suspension structures. As shown in Fig. 3.92, the zero shear viscosity increases exponentially with increasing solids concentrations. An increase in elastic modulus G' and the loss modulus G'' was also observed for these pastes with increasing solids concentration.[164]

Effects of vehicle viscosity. The paste viscosity is directly proportional to the vehicle viscosity as defined by Eq. (3.60). Vehicle viscosity is controlled by the chemical nature, average molecular weight, and molecular weight distribution of the resin, and concentration of the resin. The resin concentration and type are generally optimized for paste rheology as well as the green strength of the dried film.[158,165] Solvent viscosity and solvent resin interactions also influence vehicle viscosity. The effect of resin concentration on a crystallizing dielectric paste viscosity at a shear rate of 9.6 s^{-1} is shown in Fig. 3.93.

In the concentration range investigated, paste viscosity essentially increased exponentially (note the log scale for the viscosity) with increasing resin concentration. The increase in viscosity can also be due to absorption of polymer at the solid–liquid interface and an increase in hydrodynamic volume of each particle.[166]

Effects of dispersants and other additives. As described earlier, the rheological behavior of suspensions is strongly influenced by interparticle interactions which are either electrostatic or steric in nature. These interactions are predominantly determined by the solid–liquid interface properties and absorption behavior of the surface active agent at the interface.

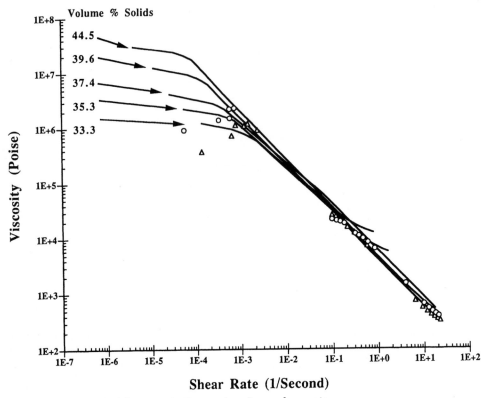

Figure 3.91 Effect of solids concentration on viscosity vs. shear rate.

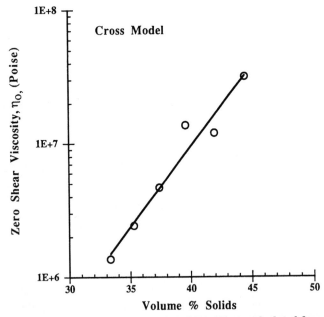

Figure 3.92 Zero shear viscosity vs. vol % solids as calculated from the Cross model.

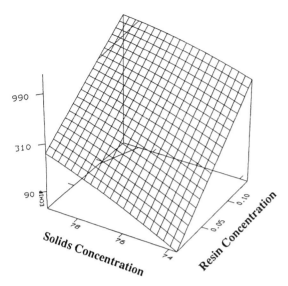

Figure 3.93   Effect of solids concentration and resin concentration on dielectric paste viscosity.

Dispersants are added to thick film pastes to improve the dispersion of particulates in the vehicle. The type and amount of dispersant are generally optimized to obtain a stable paste with dense green and fired microstructures. Dispersants are classified into three categories:

1. Anionic, where the dispersant molecule has a negative charge

2. Cationic, where the dispersant molecule has a positive charge

3. Nonionic, where the dispersant molecule is electrically neutral

Surface active agents also decrease surface tension of the paste and can help wet the screen mesh during printing. The reduced surface tension of the ink will help to wet the substrate. Dispersants also tend to interact with rheological additives and reduce paste viscosity. Excessive amounts of dispersant should be avoided.

Figure 3.94 illustrates the effect of dispersant concentration on paste viscosity at a shear rate of 9.6 s^{-1}. As expected, viscosity decreases with increasing dispersant amounts.

Other additives (i.e., rheology modifiers) are often added to thick film formulations to adjust paste rheology. The interaction between various ingredients is important in understanding paste rheological behavior. Since numerous ingredients are commonly used, statistical techniques must be employed to understand these interactions.

3.5.9   Rheology and thick film screen printing correlation

The properties of thick film printed circuits, such as print thickness, fine line definition, via resolution, and surface smoothness, depend on the screen

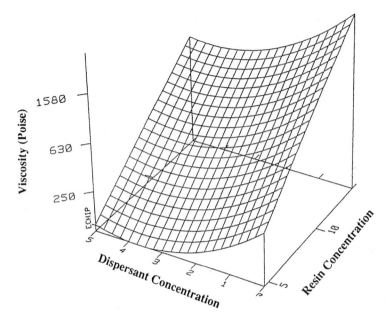

Figure 3.94 Effect of dispersant and resin concentration on dielectric paste viscosity.

printer setup and ink rheology. A large number of variables can affect the screen printing process.[167,168] Analysis of shear stresses or shear rates during the screen printing process is important to establish a correlation between the print properties and rheology. The machine setup parameters such as squeegee hardness, squeegee speed, squeegee angle, screen mesh properties, snap-off distance, and the force that presses the squeegee against the screen and substrate affect the shear rates experienced by the ink. Hydrodynamic analysis of the screen printing process during ink transfer from the screen to the substrate is available.[169–174]

During the screen printing process, a roll of ink is formed in front of the squeegee. Movement of the ink caused by the squeegee generates hydrostatic pressure within the ink. The ink is pumped through the screen opening by the pressure difference and attaches to the substrate by wetting and cohesion between the ink and substrate. The screen emulsion should provide a sealed gasket to avoid spreading of ink due to hydraulic pressure in order to obtain a high definition print.[171] The hydrostatic pressure developed is maximum in the vicinity of the squeegee edge in contact with the screen. Theoretical analysis shows that the hydrostatic pressure generated on the screen and on the squeegee increases with increasing squeegee speed and ink viscosity. This means that the force required to move the ink in front of the squeegee increases with increasing ink viscosity and squeegee speed. The maximum pressure is generated at an angle equal to $\alpha/3$, where α is the squeegee angle. The squeegee angle can decrease during screen printing, depending on the squeegee hardness.[170] The hydrostatic force acting on the squeegee surface can lift the squeegee off the screen and hydroplaning can occur with high viscosity inks at high printing speeds.

As described above, the ink undergoes a complex shear history during the screen printing operation. The estimates of shear rates are shown in Table 3.24.

First we will look at the shear rates involved during the screen printing process. When the ink is stirred and transferred to the screen the shear rate is approximately 10^{-2} to 1 s^{-1}. The shear rates near the vicinity of the squeegee edge are difficult to estimate. These estimates of shear rate are 1 to 10^4 s^{-1}.[169,170,175] In practice, it is difficult to measure the high shear rate viscosity of thick film inks (greater than 500 s^{-1}) because of cohesive failure of the paste.

Shear rate for the flow of ink into the screen mesh can be estimated by the Hagen-Poiseville relation:

$$Q = \frac{\Delta p \pi R^4}{8 \eta L}$$

(3.62)

where Q = volume of the liquid that flows across the cross section in unit time (i.e., flow rate)
R = radius of the tube
L = length of the tube
Δp = pressure gradient
η = viscosity of a newtonian fluid.

The maximum shear rate $\dot{\gamma}_{max}$ occurs near the tube wall and is given by the following equation:

$$\dot{\gamma}_{max} = \frac{\Delta p R}{2 \eta L}$$

(3.63)

The relationship between ink viscosity and squeegee speed can be analyzed using Eq. (3.61).

The volume of ink filled in the wire mesh is related to the fabric thickness L and the diameter of the mesh opening, D (approximately $2R$)

TABLE 3.24 Approximate Shear Rate Ranges Experienced by Thick Film Inks

Operation	Shear rate range, s^{-1}
Paste manufacture, three-roll mill	10^3–10^4
Paste storage	10^{-4}–1
Stirring and transfer to stencil screen	10^{-2}–1
Rolling of ink in front of squeegee	1–10^4
Ink flow through screen	10^3–10^4
Ink leveling after deposition	10^{-2}–10^{-1}

$$\text{Volume of ink in the mesh} \approx D^2 L \qquad (3.64)$$

where D = $1/(M - d_w)$
M = mesh count per unit length
d_w = wire diameter of the mesh
L = thickness of the fabric $\approx 2d_w$

The time t to fill the ink into the mesh is related to the squeegee speed V by

$$t = \frac{D}{V} \qquad (3.65)$$

Substitution of Eqs. (3.64) and (3.65) into Eq. (3.66) yields the relationship between the screen printer setup and the Newtonian ink viscosity to fill the screen mesh:

$$V\eta \le \frac{\Delta p D^3}{128 L^2} \qquad (3.66)$$

Thus, for a given screen (D^3/L^2) and the squeegee angle (related to Δp), the product of squeegee speed and ink viscosity is constant. Smaller screen openings and thicker emulsions need lower viscosity inks and slower printing speeds. A similar equation has been used[169] to explain the emptying of a mesh where Δp is atmospheric pressure. The subtle difference is whether poor printing is due to improper filling of the mesh or the release of ink from the screen mesh. The above equations are modified if the flow behavior is nonnewtonian. Thick film pastes often show yield stress (or very high viscosity at very low shear rate). If a yield stress is present, then no flow will occur into the mesh if

$$\Delta p < \frac{4 L \sigma_y}{D} \qquad (3.67)$$

where σ_y is the static yield stress of the paste. The above equation can help to explain why the ink needs to be worked on before the printing operation is started. The applied shear breaks down the paste structure and the yield stress is decreased. The pressure developed by the squeegee is then sufficient to cause flow into the mesh. The initial few strokes also help to wet the screen with ink.

During the flow of ink through the wire mesh, separation of vehicle and solid can occur. The particles are deflected toward the center of the mesh, and vehicle-rich region is formed on the surface of the mesh. This was experimentally confirmed by measuring the solids concentration of ink left on a screen.[169]

The next event that occurs is adhesion of the ink column to the substrate. At this stage, it is important to have a sealed gasket formed by emulsions so that ink does not flow out to degrade the print quality. The squeegee movement above the screen wipes off the excess ink. After depositing the ink, the

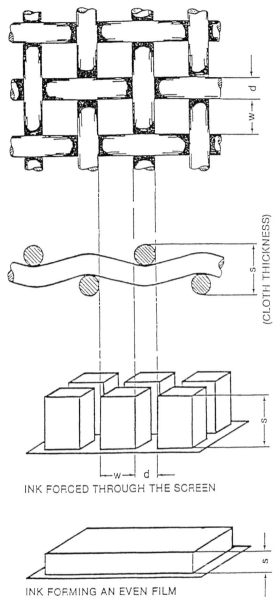

INK FORCED THROUGH THE SCREEN

INK FORMING AN EVEN FILM

Figure 3.95 Schematic of paste columns deposited during screen printing.[168]

screen must separate from the substrate immediately behind the squeegee to deposit ink columns, as shown in Fig. 3.95.

The screen wires are under tension and the pulling force F is determined by the tension in the wire. The pulling force should overcome the viscous resistance of the ink. The situation can be visualized as moving a wire through a viscous fluid. From the definition of viscosity, the following relation can be written:

$$\frac{F}{A} = \sigma = \eta\dot{\gamma} \tag{3.68}$$

The shear rate can be approximated by the velocity gradient in the mesh:

$$\dot{\gamma} \approx \frac{V_a}{D/2} \tag{3.69}$$

where V_a is the snap-off velocity and D is the mesh opening. Substituting Eq. (3.69) into Eq. (3.68) yields

$$\frac{F}{A} \approx 2\eta\frac{V_a}{D} \tag{3.70}$$

If the tension in the screen is not sufficient to overcome viscous drag, the screen is slowly released from the substrate. This leads to formation of a *cling* zone which will degrade the print resolution. The snap-off speed V_a is related to the squeegee speed and snap-off distance. The screen tension needs to be increased for faster printing of a high viscosity paste. Smaller mesh openings will also need higher screen tension. The tension in the screen should be less than the elastic limit of the wires to prevent permanent elongation and distortion of the screen. Near the end of the squeegee stroke, the lifting force decreases because of geometric factors. This can lead to an increase in the width of the cling zone and the degradation of print quality.

It is generally observed that all of the ink inside the wire mesh is not transferred to the substrate.[171] Near the end of screen release from the deposited ink, the upward movement of wires in the paste applies tensile stresses to the paste. The applied tensile stress can lead to cavitation within the ink. The tensile stress required to cavitate the paste can be low because of the presence of particles and dissolved bubbles which can act as nuclei.[176] If the rate of separation of wires from the ink is sufficiently high, the paste can develop high internal stresses that exceed the cohesive strength of the paste and the paste can fracture like a brittle solid.

3.5.10   Wetting and screen print resolution

The release of ink from the mesh and wetting of the substrate depend on the proper surface energy relationships of ink, screen, and substrate.[177,178] Wetting behavior was described earlier in Sec. 3.2, "Thick Film Conductors."

A high surface energy substrate is most easily wet with a low surface tension liquid. In order to obtain good print resolution, the ink should not excessively wet the screen substrate and emulsion. The screen material can be treated to modify the surface energy so that the ink does not wet the screen.[177] If the ink wets the screen but the ink cohesive forces are sufficiently strong to overcome ink–screen adhesion, well-defined ink columns are deposited. Ceramic substrates used in thick film applications have sufficiently high surface energy and the ink wets the substrate. An excessive amount of surfactant can lower the surface tension of the paste and promote excessive wetting and loss of print resolution. Generally, the surface energy of the paste is approximately 30 to 40 dyn/cm. Improved print resolution was observed for an ink printed on low surface energy Teflon compared to that on an Al_2O_3 substrate.

After the ink columns have been deposited on the substrate, low shear rate processes come into effect. The viscosity of the paste begins to increase immediately. The forces acting on the ink are due to gravity, surface tension, and temperature.

3.5.11 Leveling of the printed part

The time to achieve good leveling is given by the following equation:

$$t = K\left(\frac{\eta}{\dot{\gamma}}\right)\frac{(\lambda^4)}{(x^3)}\log\left(\frac{A_0}{A_t}\right) \tag{3.71}$$

where λ = wavelength of the mesh marks
$\quad A_0$ = depth of the mesh mark at time $t = 0$
$\quad A_t$ = depth of the mesh mark at time t
$\quad K$ = numerical constant
$\quad x$ = print thickness

The terms are defined in Fig. 3.96.

The suspension structure of a thixotropic paste is broken down during screen printing. It is important to achieve leveling before the structure can rebuild. To achieve a smooth surface finish, A_0 should be on the order of 1 μm. From the above equation, it is clear that leveling behavior is strongly affected by the wavelength of the mesh marks, which is related to the mesh count M. The leveling time is significantly reduced by using smaller mesh openings, as shown in Table 3.25.

If the paste develops yield stress during leveling, then the leveling stops when the stress due to surface tension becomes equal to the yield stress[180]:

$$A_{ty} = \frac{\sigma_y \lambda^3}{4\pi^3 \dot{\gamma} x} \tag{3.72}$$

The mesh mark depth due to the presence of yield is also shown in Table 3.24. It should be noted that mesh marks will also be affected by drying conditions. During drying, the viscosity of paste can decrease due to a decrease in the vehicle viscosity. The viscosity will increase due to solvent evaporation. Eventually, the solvent loss will overcome the effect of high temperature on vehicle viscosity and the leveling will be completed. If the sample is left at

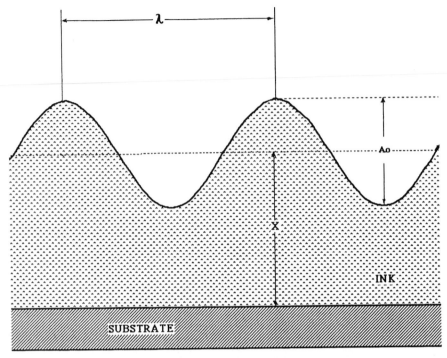

Figure 3.96 Schematic of mesh marks.

room temperature for drying, then the evaporative cooling and solvent loss will increase the paste viscosity and the yield stress. During firing of the ink, some of the mesh marks are further eliminated.

3.5.12 Via retention and line resolution

The rheological behavior required to keep vias open during printing and leveling can be analyzed using a viscous sintering analogy.[181,182] The driving force is the surface tension acting on the surface of a via. The pressure gradient Δp on a via of radius r can be given by the following Laplace equation:

TABLE 3.25   Parameters Affecting Leveling Time

Mesh count, M/in	Mesh count, M/cm	λ, μm	Initial roughness $A_0 = \frac{1}{2} d_0$, μm	Leveling time, s	Min. roughness A_{ty} with yield stress, μm
400	160	63	13	61	0.55
325	130	77	14	38	0.98
250	100	100	18	118	2.15
200	80	125	20	298	4.19

Assumptions: γ is 30 dyn/cm, A_t is 1 μm, x is 25-μm print thickness, η is 10^6 P, d_0 is wire diameter in μm, and yield stress is 2000 dyn/cm^2.

$$\Delta p = \frac{\gamma}{r} \tag{3.73}$$

where γ is the surface tension of the paste.

The pressure gradient will produce shear stresses in the material, and, subsequently, the vias will close. If the material behaves as a Bingham plastic body, then the yield stress will prevent via closure.

For a via, the condition to prevent via closure is given by the following equation[181]:

$$\sigma_c > \frac{\gamma}{2r_1 \ln(r_2/r_1)} \tag{3.74}$$

where r_2 is related to the number of vias per unit area. The effect of via diameter and density on the necessary yield stress to prevent via closure is shown in Fig. 3.97. From this figure it is clear that a higher yield stress is required with decreasing via diameter and increasing via density. The paste yield stress needs to be optimized with respect to leveling and via closure. An excessively high yield stress will result in mesh marks on the printed part. If

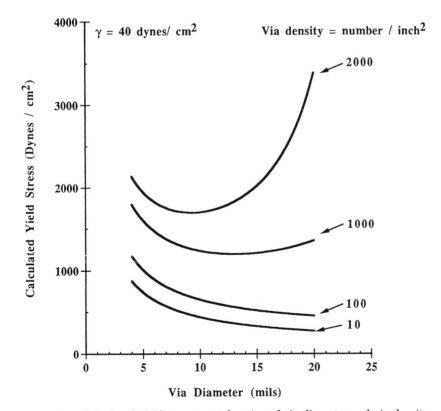

Figure 3.97 Calculated yield stress as a function of via diameter and via density using viscous sintering model.[181]

the paste does flow below the yield point and exhibits creep flow as observed in many thick film pastes, the zero or low shear rate viscosity should be utilized to predict via closure behavior. The rate of via closure will be slower with higher zero shear rate viscosity pastes. The concept of zero shear rate viscosity or yield stress is also important to predict long-term storage stability of pastes. The weak network structure can retard sedimentation and hard packing of particles during storage.

3.5.13   Examples of paste rheology printing performance

In this section, examples of correlation between ink rheology and printing characteristics will be illustrated. The effect of ink viscosity on the wet print thickness is shown in Fig. 3.98, a plot of print thickness vs. paste viscosity. Twenty dielectric pastes were prepared from the same dielectric powder. The relative amounts of solids, vehicle, dispersant, and rheology additive were varied. The paste viscosity was modified from approximately 70 to 7500 P at a shear rate of 9.6 s^{-1} with formulation changes. All parts were printed with a similar printer setting. All pastes were screen printable except one paste sample with a viscosity of approximately 7500 P and a yield stress of approximately 4800 dyn/cm^2. This paste sample did not wet the screen and the paste roll could not be formed. The paste with an up curve viscosity of approximately 3600 P and yield stress of 2900 dyn/cm^2 was screen printable. From Fig. 3.98 it is clear that the viscosity requirements for ink transfer from the screen to the substrate are quite broad. No systematic relation between the wet and fired thickness and the ink viscosity at moderate shear rates was observed. Statistically, significant correlation between the fired print thickness and the solids and dispersant concentration was observed, as shown in Fig. 3.99. The fired thickness increased with increasing solids concentration

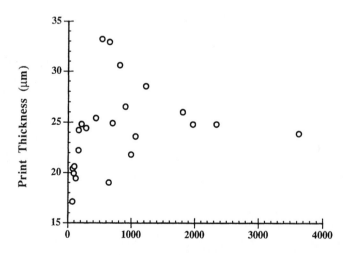

Viscosity (Poise) at Shear Rate 9.6 (1/Second)

Figure 3.98   Dry print thickness vs. dielectric paste viscosity.

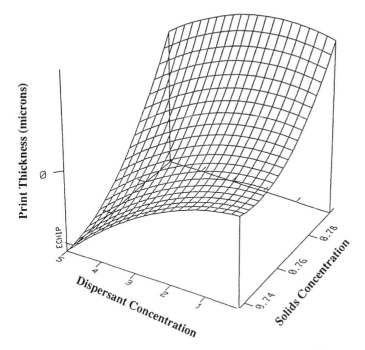

Figure 3.99 Print thickness as a function of paste solids and dispersant concentration for dielectric pastes.

in the paste. This is expected since the higher solids concentrations or particle packing density results in reduced shrinkage on drying and firing. The effect of dispersant on the fired thickness was more significant at lower solids concentrations. At lower particle concentrations, the better packing of particles can lead to increased densification and more shrinkage.[183]

The effect of ink viscosity on the fired microstructure is shown in Fig. 3.100. Pastes with low viscosities formed a dense sintered microstructure with a low leakage current. Samples which yielded thicker fired layers had significant internal porosity and high leakage current. The zero shear viscosity of the high leakage current sample was approximately 1×10^8 P whereas the dense sample had a zero shear viscosity of approximately 1×10^4 P. Very high viscosities at low shear rates prevent the dense packing of particles. A lower green density retards densification.[183] From these photomicrographics, it is clear that the paste rheology not only affects printing performance but also the properties of the fired films.

The correlation between via retention and ink rheology is shown in Fig. 3.101. The paste samples were prepared using the same powder lot. These pastes had a different inorganic composition from the pastes described above. Figure 3.101 shows a plot of paste yield stress versus paste viscosity at a moderate shear rate of 9.6 s^{-1}. At a given viscosity level, increasing yield stress improved via retention consistent with the previous equations. With a further increase in viscosity but constant yield stress, the paste produced vias with a noncircular cross section (i.e., poor acutance). Similar correlation

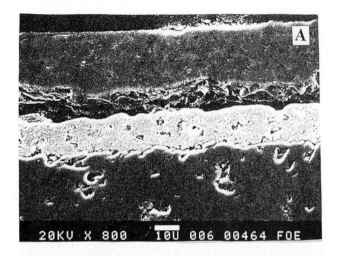

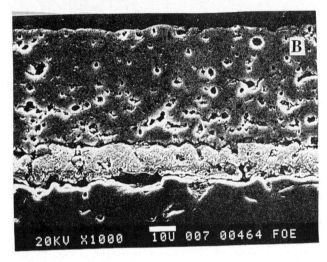

Figure 3.100 The effect of paste rheology on fired microstructure.

between the yield stress and improved resolution was observed for conductor patterns. The yield stress measured in these experiments was a static yield stress. In actual screen printing, yield stress development with time after high shear mixing will be more appropriate. Nevertheless, high yield stress indicates a higher degree of suspension structure which helps to prevent excessive flow and via closure.

Figure 3.102 shows the importance of characterizing paste rheological behavior over various shear rates. This figure shows viscosity vs. shear rate plots for two pastes prepared using the same powder. The viscosity/shear rate behavior was nearly identical at shear rates of 10^{-1} to 10 s^{-1} which are commonly measured in the quality control of pastes. The poor via resolution could

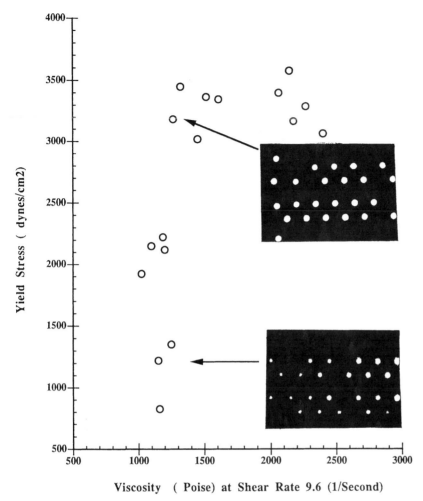

Figure 3.101 Measured yield stress vs. viscosity for various dielectric paste samples.

not be explained with high shear viscosity data. The low shear paste viscosity was approximately 10 times more for a paste with good via retention than the paste with poor resolution. Thus the rheological tests which simulate shearing conditions during ink use will have a better chance of establishing correlation between the paste rheology and print resolution.

3.6   Quality Control and Manufacturing Processes

3.6.1   Raw material and paste characterization

Control of the raw material constituents is critical to the reproducibility and ultimate performance of thick film products. Traceability of each of the com-

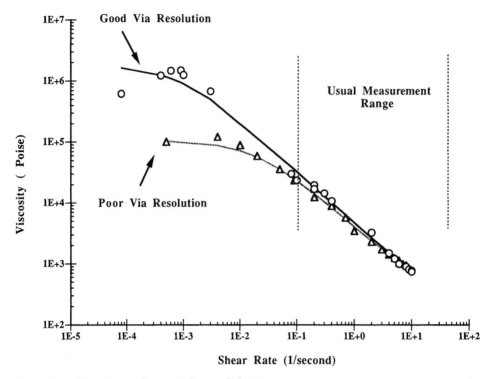

Figure 3.102   Viscosity vs. shear rate for two dielectric pastes.

ponents and the intermediates is imperative for the quality control function in order to comply with the high standards of ISO-9000 and continuous improvement programs. Routing equipment calibration programs, vendor audits, and utilization of statistical process control techniques are also necessary to ensure maintenance of high standards of quality. The raw material constituents can be either produced in house or purchased. In either case, specifications must be determined and satisfied for each raw material in order to reproducibly sustain the relationships between processing, microstructure, and the resultant fired properties.

The properties typically specified for metal or conductive oxide powders may include particle size and size distributions, surface area, tap and/or bulk densities, critical impurity levels, loss on ignition, and morphological characteristics. Metal powders can be utilized in various shapes, including spheres, flakes, and combinations of both. Their particle sizes can range from submicrometer to 20 μm. In the case of conductive oxides such as ruthenium oxide used in resistor formulations, the surface areas can approach 100 m^2/g, which translates to a particle size of less than 10 μm. The state of agglomeration becomes particularly important for these types of powders.

Glass powders can be more difficult to characterize because of their criticality to the densification process. Properties typically specified for glass powders include particle size and size distribution, surface area, purity, TCE, softening point, glass transition temperature, and onset of crystallization. Glass powders are made by high temperature melting of the appropriate mixture of oxides, carbonates, or hydroxides, followed by subsequent quenching of the melt from the peak temperature into a liquid medium such as water or onto cooled rolls. The glass melting conditions must be controlled in order to ensure reproducibility. These parameters include the peak temperature, time, crucible composition, and quench rates. Since the glass must be subjected to particle reduction processes such as ball milling, these parameters must also be controlled, including mill and media composition, milling medium, time, speed, and impurity introductions from the process. The melting process should occur in high purity crucibles such as those made of Pt to produce crystallizing formulations that are very sensitive to minor impurity levels.

The organic constituents are also very important and must be tested to ensure purity, molecular weight, viscosity, and loss-on-ignition characteristics. Utilization of electronic grade products where feasible will improve the tolerance and reproducibility of molecular weight distributions of the critical polymers. The thick film paste manufacturer must determine the critical characteristics for each raw material in the respective formulations and consistently monitor these for each lot of raw material.

3.6.2 Paste production and characterization

The raw material components are weighed out individually and blended prior to three-roll milling by selected blending processes such as planetary mixing or shaking using conventional paint shakers. After the initial blending, or *wetting-out,* phase, the mixture is typically submitted to a three-roll milling process during which the paste is homogenized and coarse particles and agglomerates are reduced in size. Other milling processes can be employed for paste production, although these tend to be proprietary in nature. In the case of three-roll milling, the pastes are subjected to a specific number of passes at specified pressures or milled to a fineness-of-grind specification. This is evaluated by the use of a fineness-of-grind gauge that involves drawing a paste over calibrated grooves that continuously get shallower. The appearance of specks or streaks in the paste defines the upper limit of the paste with respect to particle size. The milling process is continued until the specification is met (e.g., smaller than 10 μm).

The pastes are also tested for conformance to a predetermined viscosity specification. Test methods and the relationship between viscosity (rheology) and screen printing have been addressed in detail in Sec. 3.5, "Rheology and the Screen Printing Process." The percent solids in a paste is typically specified as well, particularly for conductor formulations. The primary purpose of this test is detection of gross weighing errors. This test involves a standard loss-on-ignition test and is described in detail elsewhere.[11] After confirming that the paste meets the preliminary specifications, the materials are subjected to screen printing, drying, firing, and characterization processes as described in Secs. 3.2, 3.3, and 3.4.

3.6.3 Screen Printing

The most commonly used process for thick film deposition is screen printing, although stenciling, dipping, spraying, spin coating, etc., can also be utilized. The specific deposition process, materials performance requirements, and circuit design will determine the necessary rheological characteristics of the paste. Section 3.2, "Rheology and the Screen Printing Process," addressed the paste rheological requirements and also provided a detailed discussion of rheology and screen printing correlations. The important process variables in screen printing were also identified; they include screen printer setup parameters, squeegee parameters, and screen details. Examples of some of these variables include snap-off distance, squeegee hardness, speed, pressure, wire diameter, mesh opening size, emulsion chemistry and thickness, and screen tension.

Vest,[7] Harper,[11] and Holmes and Loasby[15] all provide additional information on the basics of the printing process which supplements our detailed discussions on rheology in this text.

3.6.4 Drying

Thick film pastes are normally dried prior to firing. The drying process evaporates the solvents from the printed films. Most commonly used solvents in thick film inks have boiling points in the range of 180 to 250°C. Due to the high ratio of surface area to volume of deposited films, drying at 80 to 160°C for a period of 10 to 30 min is adequate to remove most of the solvents from wet films. Box furnaces utilizing convection type heating or belt type infrared furnaces are commonly used as dryers. Box type space dryers are suitable for batch processes and are used for small to medium production volumes.

Belt furnaces provide ease of process automation. Some automated production lines utilize a dryer and firing furnace with a common belt. Airflow rates inside the dryer, drying temperature, and time influence the drying process.

3.6.5 Firing

Figure 3.51 shows a typical furnace profile used for thick film materials firing. The heating rate is typically determined by organic removal requirements. The peak firing temperature or isothermal hold temperature is determined by the composition of the thick film materials. The cooling rate influences thermal shock stresses in the substrate and is determined by the substrate material, its dimensions, and the production rate.

Belt furnaces are commonly used for firing thick films. Belt furnaces have several zones through which a belt travels at a constant speed. The zone temperatures and speed of the belt can be controlled independently. By adjusting the zone temperature and the belt speed, a variety of time vs. temperature profiles can be achieved. Typically, belt furnaces utilize resistive heating elements. In muffleless belt furnaces, the heating elements directly provide the heat for the firing process. In belt furnaces with a muffle, the heating elements are wrapped around a metallic muffle which transfers the heat to the substrates being fired. Firing processes in which a controlled firing atmosphere is required utilize belt furnaces with a muffle. Since the muffle increases the thermal mass of the furnace, the response time of this type furnace to the changes in the temperature set points of the furnace zones is sluggish. Prototype facilities in which furnace profiles are often changed employ belt furnaces without muffles. During air firing, the atmosphere inside belt furnaces is typically controlled by adjusting the flow rates of the respective gases. Special furnace designs are utilized to control the furnace atmosphere during the firing of copper conductors and compatible materials.[184]

Thick film metallizations for solar cells are processed in infrared-heated furnaces. These furnaces utilize infrared lamps as the source of heat. With infrared furnaces, it is possible to maintain very short dwell times at peak firing temperatures.

3.7 Nonhybrid Applications

In addition to the numerous traditional hybrid applications mentioned throughout the text, there are several other applications of thick film materials that are worth discussing. These include thick film materials for solar cells, fine ceramics, and the display industry.

An interesting niche application which has emerged for thick film metallizations is in the mass production of single and polycrystalline silicon solar cells. These devices convert incoming solar radiation directly into electrical current with typical conversion efficiencies of 10% to 18%. The typical uses of commercially produced photovoltaics are in cost sensitive markets where electricity is required in locations that are isolated from the conventional power grid. The commercial solar cell is a device similar to that pictured in Fig. 3.103. The most widely used technique for metallizing the n-doped layer that faces the sun involves the screen printing of a grid type pattern of thick film Ag ink to make contact with the silicon. This metallization must be highly conductive at the Ag–silicon interface. Various dopants are added to this formulation to enhance the efficiency of the cell. The requirements of this metallization also include low contact resistance at the interface, good solderability, and fine line resolution. The back contact or metallization of the solar cell can be configured in a variety of geometries. Because the back side of the cell is usually coated with an n-doped layer, back contacts are usually Ag–Al mixtures. The back contact may be in a grid pattern or may completely coat the entire surface. The Al ensures electrical penetration through the n-doped layer. Some back contacts are pure or fritted Al which reflects stray charge carriers back toward the n-p junction. The technical feasibility of thick film phosphorus diffusion sources and screen-printable antireflective coatings based on oxides of tantalum and titanium has been demonstrated, although they are not used in large volume production.

One of the fastest growing categories of electronic devices is flat panel displays. Although there are a number of competing technologies, there is poten-

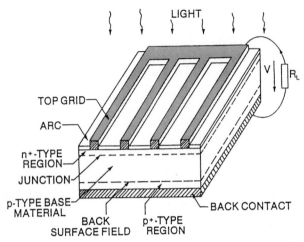

Figure 3.103 Schematic representation of a typical silicon solar cell.

tial for the use of high volumes of thick film materials in the display market. Plasma displays, as depicted in Fig. 3.104, utilize thick film components including air-fired Ni for electrodes, dielectric pastes as barrier ribs (partitioning pixel elements), insulating dielectrics for separation of electrodes, and sealing glass pastes for joining the glass panels. In the liquid crystal display area, thick film opportunities include glass sealing pastes and Ag or Au inks for the driver circuitry and chip mounting. Other display technologies also utilize thick films for sealing panels or driver circuitry. As this technology matures, it is likely that other opportunities for these versatile materials will emerge.

The final applications discussed here are electrodes and metallizations for fine ceramics. This includes ohmic contact pastes for varistors and thermistors as well as materials for complex multilayer components. Figure 3.105 illustrates a cross-sectional view of a ceramic chip capacitor. The internal electrode can be Pt, Pd, or Ag–Pd depending on the type of ceramic formulation and its respective firing temperature. The end termination may be Ag or Ag–Pd and is typically required to be platable. The technical requirements of these materials are extremely challenging.

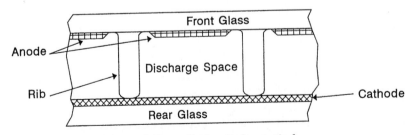

Figure 3.104 Cross section of plasma display discharge pixel.

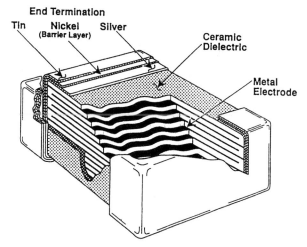

End Termination
Tin   Nickel   Silver
       (Barrier Layer)
Ceramic Dielectric
Metal Electrode

Figure 3.105 Cross section of monolithic ceramic chip capacitor.

References

1. R. R. Tummala, "Ceramic and Glass-Ceramic Packaging in the 1990's," *J. Am. Cer. Soc.,* vol. 74, no. 5, pp. 895–908, 1991.
2. R. W. Vest, "Materials Science of Thick Film Technology," *Am. Cer. Soc. Bull.,* vol. 64, no. 4, pp. 631–636, 1986.
3. R. R. Tummala and E. Rymaszewski, *Microelectronics Packaging Handbook,* Van Nostrand Reinhold, New York, 1989.
4. R. Keeler, "Polymer Thick Film Multilayers: Poised for Takeoff," *Elect. Pack. and Prod.,* pp. 35–38, August 1987.
5. *Electronic Materials Handbook,* vol. 1, *Packaging,* ASM International, Materials Park, Ohio, pp. 106–109, pp. 339–353, 1989.
6. *Metals Handbook,* vol. 8, *Metallography, Structures and Phase Diagrams,* ASM International, Materials Park, Ohio, 1973.
7. R. C. Buchanan (ed.), *Ceramic Materials for Electronics,* 2d ed., Marcel Dekker, New York, 1991.
8. *Metals Handbook,* 10th ed., ASM International, Materials Park, Ohio, 1990.
9. B. S. Rawal et al., "Factors Responsible for Thermal Shock Behavior of Chip Capacitors," *Proc. 37th Elect. Comp. Tech. Conf.,* Boston, pp. 145–156, 1987.
10. R. O. Carlson et al., "Thermal Expansion Mismatch in Electronic Packaging," *Proc. Matls. Res. Soc. Symp.,* vol. 40, pp. 177–190, 1985.
11. C. A. Harper (ed.), *Handbook of Thick Film Hybrid Microelectronics,* McGraw-Hill, New York, 1974.
12. J. R. Larry et al., "Thick Film Technology: An Introduction to the Materials," *IEEE Trans. on Comp. Hybrids and Mfg. Tech.,* vol. CHMT-3, no. 2, pp. 211–225, 1980.
13. D. T. Novick and A. R. Kroehs, "Gold Scavenging Characteristics of Bonding Alloys," *Solid State Tech.,* vol. 17, no. 6, pp. 43–47, 1974.
14. J. C. Fu, "A Study of Structure of Gold-Platinum in Thick Film Conductors," *Precious Metal,* E. D. Zysk and J. A. Bonucci (eds.), pp. 469–483, 1985.
15. P. J. Holmes and R. G. Loasby, *Handbook of Thick Film Technology,* Electrochemical Publishers, Scotland, 1976.
16. J. Savage, "Silver Transfer Through Thick Film Insulants," *Joint Conf. on Hybrid Microelec.,* IERE, Canterbury, pp. 335–340, 1973.
17. J. H. Alexander et al., "A Low Temperature Cofiring Tape System Based on a Crystallizing Glass," *Proc. Intl. Symp. Microelec.,* Orlando, pp. 414–417, 1991.
18. J. H. Alexander and A. S. Shaikh, "The Use of Silver and Silver Plus Gold Conductors with the Ferro Low Temperature Tape," *Proc. 43d ECC Conf.,* Orlando, pp. 888–892, 1993.

19. G. Sarkar et al., "High Reliability Silver-Based Multilayer System," addendum to *Proc. IEMT Symp.,* Japan, 1991.
20. B. D. Cullity, *Elements of X-Ray Diffraction,* 2d ed., Addison-Wesley, Reading, Mass., 1978.
21. L. R. Lichtenberg and M. J. Roll, "Palladium-Silver as an Alternative Metal for Wire Bonding Metallization," *Proc. Intl. Symp. Microelec.,* pp. 229–233, 1988.
22. R. W. Vest, "Materials Interactions During Firing of PdAg Conductors," *Proc. ASM Conf. on Thick Films,* Atlanta, 1988.
23. E. T. Turkdogan, *Physical Chemistry of High Temperature Technology,* Chap. 1, Academic Press, New York, pp. 1–26, 1980.
24. Y. Kudoh et al., "Investigations of Resistance to Solder Leaching of 4 Pd 96 Ag Thick Film Conductors with Ag/Sn/Pd Solders," *Proc. Intl. Symp. Microelec.,* Minneapolis, pp. 51–56, 1987.
25. P. S. Tong and C. E. Falleta, "Silver Palladium Conductors with Excellent Leach Resistance and Solderability but Containing Less Than 1% Palladium," *Proc. Intl. Symp. Microelec.,* Minneapolis, pp. 8–10, 1987.
26. W. Borland et al., "High Conductivity Materials Systems for Advanced Hybrids," *Proc. ECC Conf.,* Houston, pp. 704–713, 1989.
27. G. D. O'Clock, Jr., et al., "PbSn Microstructure: Potential Reliability Indicator for Interconnects," *IEEE Trans. CHMT,* vol. 10, pp. 82–88, 1987.
28. P. W. Dehaven, "The Reaction Kinetics of Liquid 60/40 SnPd Solder with Copper and Nickel: A High Temperature X-Ray Diffraction Study," *Proc. Matls. Res. Soc. Symp.,* vol. 40, pp. 123–128, 1985.
29. D. S. Dunn et al., "Dependence of CuSn and Cu_6OSn_4OPb Solder Joint Strength on Diffusion Controlled Growth of Cu_3Sn and Cu_6Sn_5," *Proc. Matls. Res. Soc. Symp.,* vol. 40, pp. 120–138, 1985.
30. C. W. Allen et al., "A Study of Intermetallic Compound Development in Nickel-Tin Interfacial Zones," *Proc. Matls. Res. Soc. Symp.,* vol. 40, pp. 139–144, 1985.
31. B. S. Chiou et al., "Intermetallic Formation on the Fracture of SnPb Solder and PdAg Conductor Interface," *IEEE Trans. CHMT,* vol. 13, no. 2, pp. 267–274, 1990.
32. P. S. Palanisamy and D. H. R. Sarma, "Thermodynamics of Processing Copper Thick Film Systems in a Reactive Atmosphere," *Hybrid Circuits,* no. 13, pp. 13–20, May 1987.
33. D. R. Gaskell, *Introduction to Metallurgical Thermodynamics,* McGraw-Hill, New York, 1973.
34. D. E. Pitkanen et al., "Materials Compatibility and Processing Relationships for Copper Thick Film Hybrids," *Intl. J. Hybrid Microelec.,* vol. 2, no. 2, pp. 45–51, 1979.
35. D. E. Pitkanen et al., "Part II—Materials Compatibility and Processing Relationships for Copper," *Intl. J. Hybrid Microelec.,* vol. 3, no. 1, pp. 1–6, 1980.
36. R. F. Lund, "Circuit Shorting Mechanisms in Copper Multilayer Systems," *Proc. Intl. Symp. Microelec.,* Anaheim, Calif., pp. 463–471, 1985.
37. P. L. Toch et al., "Failure Mechanism and Process Control Requirements in the Production of Copper Multilayer Interconnect Boards," *Proc. 36th Elect. Comp. Tech. Conf.,* pp. 602–608, 1986.
38. D. H. R. Sarma et al., "An Accelerated Lot Acceptance Test for Adhesion Degradation of Soldered Copper Thick Films in Temperature Cycling," *Proc. Intl. Symp. Microelec.,* Minneapolis, pp. 554–561, 1987.
39. F. N. Patterson, "Conductor Compositions Compressing Nickel Borides," U.S. patent no. 3,943,168, March 9, 1976.
40. M. Kageyama and T. Yamaguchi, "Mechanism of Reaction Between AlN and Glasses," *Proc. 5th Intl. Microelec. Conf.,* Tokyo, pp. 161–164, 1988.
41. K. Allison et al., "Thick Film Materials for Application on Aluminum Nitride Substrates," *Proc. 5th Intl. Microelec. Conf.,* pp. 153–160.
42. K. Allison et al., "Thick Film Conductor Compositions for Use with an Aluminum Nitride Substrate," U.S. patent no. 5,089,172, February 18, 1992.
43. A. Shaikh, "Thick Film Pastes for AlN Substrates," *Advancing Microelec.,* vol. 21, no. 1, pp. 18–21, 1994.
44. D. L. Hankey et al., "Scanning Electron Microscopy Techniques for Thick Film Microstructural Characterization," *Advances in Ceramics,* vol. 11, *Processing for Improved Productivity,* K. M. Nais (ed.), ACS, Columbus, Ohio, pp. 117–130, 1984.
45. M. Hrovat and D. Kolar, "Investigation in the Al_2O_3–Bi_2O_3–CuO System," *J. Matl. Sci. Letters,* vol. 3, pp. 659–662, 1984.
46. W. D. Kingery et al., *Introduction to Ceramics,* 2d ed., John Wiley & Sons, New York, 1976.

47. *Metals Handbook,* 9th ed., *Powder Metallurgy,* vol., ASM, Metals Park, Ohio, 1984.
48. T. V. Nordstrom and F. G. Yost, "Sintering Behavior of a Reactively Bonded Thick Film Gold Ink," *J. Elec. Matls.,* vol. 7, no. 1, pp. 109–122, 1978.
49. D. W. Pashley et al., "The Growth and Structure of Gold and Silver Deposits Formed by Evaporation Inside an Electron Microscope," *Philosophy Mag.,* vol. 10, pp. 127–158, 1965.
50. R. M. German and Z. A. Munin, "Identification of the Initial Stage Sintering Mechanism Using Aligned Wires," *J. Matl. Sci.,* vol. 11, pp. 71–77, 1976.
51. S. S. Cole, "The Sintering Mechanism in a Silver-Palladium Film," *Proc. Intl. Symp. Microelec.,* Washington, paper 2A1, 1972.
52. R. W. Vest, "Conduction Mechanics in Thick Film Microcircuits," final technical report, grant nos. DAHC-15-70-G7 and DAHC-15-73-G8, ARPA, December 1975.
53. H. Z. Wee and R. W. Vest, "Densification and Adhesion of Silver Thick Film Conductors," *Intl. J. Hybrid Microelec.,* vol. 10, no. 4, pp. 20–24, 1987.
54. R. O. Cooper and E. M. Monahan, "The Characterization of the Solderability of Thick Film Conductors Using a Meniscograph," *Proc. Intl. Symp. Microelec.,* Dallas, pp. 8–12, 1984.
55. "Solderability Tests for Printed Boards," Joint Industry Standard, EIA/IPC, ANSI/J-STD-003, pp. 11–15, April 1992.
56. H. L. Pinch and D. J. Sjostak, "Solderability, Chemical Composition, and Surface Characterization of Fritless Copper Thick Film Inks," *Proc. Intl. Symp. Microelec.,* Seattle, pp. 398–404, 1988.
57. T. T. Hitch, "Adhesion Measurements on Thick Film Conductors," Adhesion Measurement of Thin Films, Thick Films, and Bulk Coatings, ASTM STP 640, K. L. Mittal (ed.), ASTM, pp. 211–232, 1978.
58. G. P. Anderson et al., "The Influence of Loading Direction upon the Character of Adhesive Bonding," *J. Coll. and Interface Sci.,* pp. 600–609, 1974.
59. K. J. Williams, "Adhesion Test Methods for Solderable Thick Film Conductors," Brazing and Soldering, no. 15 (autumn), pp. 10–18, 1988.
60. R. H. Zeien, "Characterization of Thick Film Fritless Metallization," *Proc. Intl. Symp. Microelec.,* Boston, pp. 7–15, 1974.
61. W. A. Crossland and L. Hailes, "Thick Film Conductor Adhesion Reliability," *Solid State Tech.,* vol. 14, no. 2, pp. 42–47, 1971.
62. E. A. Webb, "Effects of Copper Thick Film Processing on Adhesion and Bondability," *Proc. 6th Euro. Microelec. Symp.,* Bournemouth, U.K., pp. 128–135, June 1987.
63. R. Senkalski et al., "Large-Area Non-Warp Dielectric for Multilayer Structures," *Proc. Intl. Symp. Microelec.,* Atlanta, pp. 132–135, 1986.
64. E. Liang et al., "Thick Film Materials System for Air and Nitrogen Firing Applications," *Proc. IEEE 36th Elec. Comp. Conf.,* pp. 493–500, 1986.
65. J. Barrett and P. Moran, "Laser Drilling of Vias in Unfired Thick Film Dielectrics," *Proc. Intl. Symp. Microelec.,* Dallas, pp. 1–7, 1984.
66. A. Shaikh et al., "Thick Film Interfacial Interactions and Resultant Effects on Fired Film Properties," *Proc. Intl. Symp. Microelec.,* Baltimore, pp. 568–576, 1989.
67. J. M. Herbert, *Ceramic Dielectrics and Capacitors,* Gordon and Beach, 1985.
68. A. Shaikh et al., "A Hermetic Low K Dielectric for Alumina Substrate," *Proc. Intl. Symp. Microelec.,* pp. 189–194, 1988.
69. S. J. Stein et al., "Interconnection and Packaging of Advanced Electronic Circuitry," *Proc. Intl. Symp. Microelec.,* Orlando, pp. 130–134, 1991.
70. Gary P. Shorthouse et al., "A Hermetic Very Low K High Resolution Thick Film Dielectric for High Speed High Density Interconnections," *Proc. Intl. Symp. Microelec.,* pp. 528–532, 1991.
71. A. Shaikh et al., "New Materials for High Speed Multilayer Circuits," *Proc. 8th Euro Hybrid Microelec. Conf.,* Rotterdam, 1991.
72. D. Kellerman, "The Development and Characterization of a Low Dielectric Constant Thick Film Material," *Proc. 37th Elec. Comp. Conf.,* Boston, pp. 316–327, 1987.
73. M. K. Rao et al., "Effects of Infra-Red Firing on the Properties of Low K Thick Film Dielectrics Compositions," *Proc. Intl. Symp. Microelec.,* Atlanta, pp. 119–123, 1986.
74. C. M. Val et al., "Copper-Dielectric Interactions—A Comprehensive Study," *Proc. Intl. Symp. Microelec.,* pp. 37–48, 1980.
75. V. P. Suita and R. J. Bacher, "Firing Process-Related Failure Mechanisms in Thick Film Copper Multilayers," *Proc. 36th Elec. Comp. Conf.,* Seattle, pp. 471–480, 1986.
76. C. R. S. Needs, "The Accelerated Life Testing of Copper Thick Film Multilayer Materials," *Proc. Intl. Symp. Microelec.,* Atlanta, pp. 840–847, 1986.

77. R. Gardner et al., "Materials Science Aspect of a Thick Film Copper/Dielectric System," *Proc. Intl. Symp. Microelec.*, pp. 285–294, 1990.
78. R. R. Sutherland et al., "A Comparison of Reliability of Copper and Palladium Silver Thick Film Crossovers," *Proc. 35th Elec. Comp. Conf.*, pp. 498–504, 1985.
79. S. S. Tamhankar and R. Keusseyan, "Optimization of Atmosphere Doping for Firing Photoformable Copper Thick Film Materials," *Proc. Intl. Symp. Microelec.*, Chicago, pp. 8–83, 1990.
80. E. S. Dettmer et al., "Hybrid Design and Processing Using Aluminum Nitride Substrates," *Proc. Intl. Symp. Microelec.*, pp. 545–553, 1988.
81. S. Chitale et al., "ESL Thick Film Materials for AlN," *J. of Adv. Microelec.*, vol. 22, no. 1, pp. 21–23, 1994.
82. V. Komarov et al., "Classification of Dopants for Barium Titanate," *Izv. Akad. Nauk SSSR, Ser. Fiz.*, vol. 29, pp. 1873–1875, 1975.
83. O. I. Prokopalo et al., "Classification of Modifier for Barium Titanates," *Izv. Akad. Nauk SSSR, Ser. Fiz.* (English trans.), vol. 33, pp. 1075–1079, 1969.
84. Leppavuori et al., "Electrical Properties of Thick Film Capacitor Based on Barium Titanate Glass Formulation," *Thin Solid Films*, vol. 86, pp. 287–295, 1981.
85. A. Ikegami, "Thick Film Capacitor Materials of the Powder-Glass Binary Systems and Their Dielectric Properties," *Electrocom. Sci. and Tech.*, vol. 9, pp. 147–152, 1981.
86. L. C. Hoffman and T. Nakayama, "Screen Printed Capacitor Dielectrics," *Microelec. and Reliability,* Paragon Press, vol. 7, pp. 131–135, 1968.
87. K. Abe et al., "Development of the Thick Film Capacitor and Its Application for Hybrid Circuit Modules," *Proc. 29th Elec. Comp. Conf.*, pp. 277–285, 1979.
88. J. V. Biggers et al., "Thick Film Glass Ceramic Capacitor," *Solid State Tech.*, pp. 63–66, 1970.
89. R. A. Delaney et al., *IBM J. of R&D*, vol. 11, pp. 511–519, September 1967.
90. J. P. Holden, *Radio Electron. Eng.*, pp. 381–387, December 1968.
91. R. L. Stermer, *IEEE Intl. Conv. Rec.*, vol. 12, pt. 9, pp. 47–53, 1964.
92. H. Stetson et al., "Development of a High K Thick Film Capacitor with Copper Electrodes for Nitrogen Firing," *Proc. 36th IEEE Elec. Comp. Conf.*, pp. 488–492, 1986.
93. J. W. Asher and C. R. Pratt, *Proc. Electron. Comp. Conf.*, pp. 239–245, 1968.
94. M. Anliker et al., *Helv. Phys. Acta.*, vol. 27, pp. 99–124, 1954.
95. Shen-Li Fu and Gung Fu Chen, "Low Temperature Firing Thick Film Dielectrics in the System $Pb(Fe_{2/3}W_{1/3})_x(Fe_{1/2}Nb_{1/2})_{0.86-x}Ti_{0.14}O_3-Bi_2O_3-Li_2O$," *Am. Cer. Soc. Bull.*, vol. 66, no. 9, pp. 1397–1400, 1987.
96. S. L. Fu and G. F. Chen, "Low Firing $Pb(FeW)(FeNb)TiO_3$ Based Thick Film Capacitor Materials," *Proc. Intl. Symp. Microelec.*, Anaheim, Calif., pp. 48–51, 1985.
97. T. C. Reiley et al., "A Low Temperature Firing Thick Film Capacitor Material Based on Lead Iron Niobate Tungstate," *Matl. Res. Bull.*, vol. 19, no. 12, pp. 1543–1549, 1984.
98. S. J. Stein et al., "High Reliability Thick Film Capacitor Dielectrics," *Proc. Intl. Symp. Microelec.*, Dallas, pp. 433–440, 1984.
99. J. Steinberg, "Materials and Applications for Thick Film RC Networks," *Proc. Intl. Symp. Microelec.*, Chicago, pp. 276–283, 1990.
100. Z. Strand, *Glass-Ceramic Materials,* Elsevier, New York, 1986.
101. C. R. S. Needs and D. P. Button, "Reliability Testing of Thick Film Multilayer Materials," *Proc. Elec. Comp. Conf.*, pp. 505–511, 1985.
102. Hewlett-Packard, "Basics of Measuring the Dielectric Properties of Materials," HP application note 1217-1, HP lit. no. 5091-3300E, February 1992.
103. IEC, "Recommended Method for the Determination of the Permittivity and Dielectric Dissipation Factor of Electrical Insulating Materials at Power, Audio and Radio Frequencies Including Meter Wavelengths," IEC pub. no. 250.
104. MIL-M-28787 C Test Method, app. B, par. 40.8.3.7, January 29, 1988.
105. M. E. Deunesnil, "Resistor and Resistor Composition," U.S. pat. no. 3052573, September 4, 1962.
106. E. M. Davis et al., "Solid Logic Technology Versatile High Performance Microelectronics," *IBM J. of R&D*, vol. 8, no. 2, pp. 102–114, 1964.
107. E. H. Melan and A. H. Mones, "The Glaze Resistor—Its Structure and Reliability," *Proc. IEEE Comp. Conf.*, p. 76, 1964.
108. E. H. Melan, "Stability of Palladium Oxide Resistive Glaze Films," *Microelec. Reliability,* vol. 6, p. 53, 1967.
109. P. R. Van Loan, "Conductive Ternary Oxides of Ruthenium and Their Use in Thick Film Resistor Glazes," *Amer. Cer. Bull.*, vol. 51, no. 3, pp. 231–234, 1972.

110. R. E. Cote et al., "Factors Affecting Laser Trim Stability of Thick Film Resistors," *Proc. Intl. Symp. Microelec.,* Vancouver, pp. 128–137, 1976.
111. S. Vasudevan and A. Shaikh, "Structure Property Modeling of Low Ohm Thick Film Resistors," *Proc. Intl. Symp. Microelec.,* Dallas, pp. 685–694, 1993.
112. D. Bender and R. Lathroy, "Novel High Performance Resistor Design for Telecommunications A-C Surge Protection," *Proc. ISHM Adv. Matl. Tech. Conf.,* Orlando, 1989.
113. C. Y. Kuo, "Thick Film Circuits," *Engineered Materials Handbook,* vol. 4, *Ceramics and Glasses,* ASM, 1991.
114. B. S. Lee and R. W. Vest, "Peak Temperature Glass Viscosity Effects on the Properties of Thick Film Resistors," *Cer. Trans.,* vol. 11, pp. 405–416, 1990.
115. S. Vasudevan and R. W. Vest, "The Effect of Conductive Particle Size on the Temperature Dependence of Resistance in Thick Film Resistors," *Proc. Am. Cer. Soc. Symp.,* pp. 417–428, Indianapolis, 1989.
116. Osamu Abe et al., "The Effect of Various Factors on the Resistance and TCR of RuO_2 Thick Film Resistors," *Active and Passive Elec. Comp.,* vol. 13, pp. 67–83, 1988.
117. P. F. Garcia et al., "Particle Size Effects in Thick Film Resistors," *J. Appl. Phys.,* vol. 53, no. 7, pp. 5282–5287, 1982.
118. B. Morten et al., "Thick Film Technology and Sensors," *Sensors and Actuators,* vol. 16, p. 109, 1983.
119. L. Golonka et al., "Influence of Composition and Construction Parameter on the Basic Properties of Thick Film Thermistors," *Hybrid Cir.,* no. 28, pp. 9–12, 1992.
120. P. Palanisamy and K. E. Ewing, "Self Sealing Thermally Sensitive Resistor and Method of Making Same," U.S. patent no. 4,452,726, June 5, 1984.
121. Jayanth S. Shah, "Strain Sensitivity of Thick Film Resistors," *IEEE Trans. CHMT,* vol. 3, no. 4, pp. 554–564, 1980.
122. M. Prudenziati and B. Morten, "Piezoresistive Properties of Thick Film Resistors: An Overview," *Hybrid Cir.,* vol. 10, pp. 20–25, 1986.
123. M. Purdenziati et al., "Very High Strain Sensitivity in Thick Film Resistors: Real and False Super Gauge Factors," *Sensors and Actuators,* vol. 19, pp. 401–414, 1989.
124. P. R. Garcia et al., "Electrical Conduction and Strain Sensitivity in RuO_2 Thick Film Resistors," *J. Appl. Phys.,* vol. 54, no. 10, pp. 6002–6008, 1983.
125. R. T. Hopper, "How to Apply Noble Metals to Ceramics," *Cer. Ind.,* June 1963.
126. K. Bab et al., "Formation of Thin Film Resistors By Use of Metallo-Organic Deposition Method and its Application for Thermal Print Head," *Proc. Intl. Symp. Microelec.,* Seattle, pp. 381–386, 1988.
127. S. Taguchi and T. Watanabe, "The Development of Thick Film Resistors for Thermal Print Head," *Proc., Intl. Symp. Microelec.,* Minneapolis, pp. 221–230, 1987.
128. S. M. Chitale and R. W. Vest, "Critical Relationships Between Particle Size, Composition, and Microstructure in Thick Film Resistors," *IEEE Trans. CHMT,* vol. 11, no. 4, pp. 604–609, 1988.
129. A. N. Prabu and R. W. Vest, "Investigation of Microstructure Development in RuO_2 Lead Borosilicate Glass Thick Film," *Matl. Sci. Res.,* vol. 10, pp. 399–401, 1978.
130. A. Kushy, "Chains of Conducting Particles that Determine the Resistivity of Thick Resistive Films," *Thin Solid Films,* vol. 121, p. 263, 1984.
131. Mathey Biship, Inc., *Application Data Index,* EMG 2.2, June 1972.
132. M. V. Coleman, "Evaluation Methods for the Examination of Thick Film Materials," *Radio and Electronic Engineer,* vol. 45, no. 3, pp. 121–130, 1975.
133. G. E. Pike and C. H. Seager, "Electrical Properties and Conduction Mechanisms of Ru-Based Thick Film (Cermet) Resistors," *J. Appl. Phys.,* vol. 48, pp. 5122–5168, 1977.
134. D. P. H. Smith and J. C. Anderson, "Electrical Conduction in Particulate Thick Films," *Philosophy Mag. B,* vol. 43, no. 5, p. 811, 1981.
135. P. J. S. Ewen and J. M. Robertson, "A Percolation Model of Conduction in Segregated Systems of Metallic and Insulating Materials: Application to Thick Film Resistors," *J. Appl. Phys.,* vol. 14, pp. 2253–2268, 1981.
136. R. W. Vest, "A Model for Sheet Resistivity of RuO_2 Thick Film Resistors," *IEEE Trans. CHMT,* vol. 14, no. 2, pp. 396–406, 1991.
137. Toshio Inokuma et al., "The Microstructure of RuO_2 Thick Film Resistors and the Influence of Glass Particle Size on Their Electrical Properties," *IEEE Trans. CHMT,* vol. 7, no. 2, pp. 166–175, 1984.
138. J. A. Loughran and R. A. Sigsbee, "Termination Anomalies in Thick Film Resistors," *Proc. Intl. Symp. Microelec.,* 1969.

139. R. M. Hill, *Proc. IEEE Conf. 402,* London, 1977.
140. D. P. H. Smith and J. C. Anderson, "Electrical Conduction in Thick Film Paste Resistors," *Thin Solid Films,* vol. 71, p. 79, 1980.
141. N. C. Halder, "Electron Tunnelling and Hopping Possibilities in RuO_2 Thick Films," *Electro. Comp. Sci. and Tech.,* vol. 110, pp. 21–34, 1983.
142. M. Prudenziati, "Electrical Transport in Thick Film (Cermet) Resistors," *Electro. Comp. Sci. and Tech.,* vol. 10, pp. 285–293, 1983.
143. R. W. Vest et al., "The Dependence of Charge Transport on Microstructure in Thick Film Resistors," *Proc. 5th Euro. Hybrid Microelec. Conf.,* Italy, pp. 406–412, 1985.
144. S. Vasudevan, "Charge Transport Mechanisms in Thick Film Resistors," Ph.D. dissertation, Purdue Univ., Ind., 1990.
145. David E. Hayes and Donald R. Emmons, "Laser Pulse Measurements for Safe and Proper Trimming System Alignment," *Hybrid Circuit Tech.,* pp. 35–40, November 1987.
146. M. A. Barnes et al., *An Introduction to Rheology,* Elsevier, 1989.
147. F. R. Eirich (ed.), "Rheology—Theory and Applications," vol. 3, Academic Press, 1960.
148. F. R. Eirich (ed.), "Rheology—Theory and Applications," vol. 4, Academic Press, 1960.
149. P. Sherman, *Industrial Rheology,* Academic Press, 1960.
150. K. M. Oesterle and M. B. Palmer, "Viscometry of Coating Material," *Treatise on Coatings,* vol. 2, Marcel Dekker, New York, pp. 123–248.
151. T. R. Tadros, "Control of the Properties of Suspensions," *Colloid and Surfaces,* vol. 18, pp. 137–173, 1986.
152. J. W. Goodwin, "The Rheology of Dispersions," *Colloid Sci.,* vol. 2, The Chemical Society, pp. 246–293, 1975.
153. T. C. Patton, "Paint Flow and Pigment Dispersion," 2d ed., Wiley Interscience, 1979.
154. W. H. Bauer and E. A. Collins, "Thixotropy and Dilatancy," *Rheology—Theory and Applications,* vol. 4, Academic Press, pp. 423–466, 1960.
155. R. J. Hunter, "Foundations of Colloid Science," vol. 1, Clarendon Press, 1987.
156. J. Th. G. Overbeck, *Adv. Colloid Interface Science,* vol. 15, p. 251, 1982.
157. T. Sato, *Coatings Tech.,* vol. 65, no. 825, pp. 113–121, 1983.
158. G. Y. Onoda, Jr., *Ceramic Processing Before Firing,* John Wiley and Sons, p. 235, 1978.
159. J. N. Israelachvili and B. W. Ninham, *J. Colloid Interface Science,* vol. 58, p. 14, 1977.
160. A. S. Michaels and J. C. Bolger, *Ind. Eng. Chem. Fund.,* vol. 1, no. 3, p. 153, 1962.
161. R. Buscall et al., "Scaling Behaviour of the Rheology of Aggregate Networks Formed from Colloidal Particles," *J. Chem. Soc., Faraday Trans. 1,* vol. 84, no. 12, pp. 4249–4260, 1988.
162. R. C. Sunntag and W. B. Russel, *J. Colloid Interface Science,* vol. 17, p. 33, 1987.
163. D. C.-H. Cheng, "Yield Stress: A Time-Dependent Property and How to Measure It," *Rheolog. Acta,* vol. 25, pp. 542–554, 1986.
164. C. S. Khadilkar, unpublished work.
165. R. M. Stanton, *Proc. Intl. Symp. Microelec.,* pp. 419–432, 1976.
166. C. S. Khadilkar and M. D. Sacks, "Effects of Poly (Vinyl Alcohol) on the Properties of Model Silica Suspensions," *Cer. Trans.,* vol. 1, pt. A, *Cer. Powder Sci.,* pp. 397–409, 1988.
167. D. R. Kobs and D. R. Voight, "Parametric Dependencies in Thick Film Screening," *Solid State Tech.,* pp. 34–41, February 1971.
168. C. Missels, *Hybrid Circuit Tech.,* pp. 11–13, May 1985.
169. D. E. Riemer, "The Theoretical Fundamentals of the Screen Printing Process," *Hybrid Cir.,* vol. 18, pp. 8–17, January 1989.
170. J. A. Owczarek and F. L. Howland, "A Study of the Off-Contact Screen Printing Process," pts. I and II, *IEEE Trans. CHMT,* vol. 13, no. 2, pp. 358–375, June 1990.
171. D. E. Riemer, "Ink Hydrodynamics of Screen Printing," *Proc. Intl. Symp. Microelec.,* Anaheim, Calif., pp. 52–58, 1985.
172. D. E. Riemer, "The Shear Flow Experience of Ink During Screen Printing," *Proc. Intl. Symp. Microelec.,* Minneapolis, pp. 335–340, 1987.
173. D. O. Brown, "Screen Printing—An Integrated System," *Proc. Intl. Symp. Microelec.,* Atlanta, pp. 582–590, 1986.
174. M. Chi Rang et al., "A Model for Deposition of Thick Films by the Screen Printing Technique," *Proc. Intl. Symp. Microelec.,* Dallas, pp. 604–609, 1984.
175. R. E. Trease and R. L. Dietz, "Rheology of Pastes in Thick Film Printing," *Solid State Tech.,* vol. 1, pp. 39–43, 1972.
176. A. C. Zettlemoyer and R. R. Myers, "The Rheology of Printing Inks," *Rheology,* vol. 3, Academic Press, pp. 145–188, 1960.
177. J. R. Larry, "Influence of Surface Energies on Line Resolution in Screen Printing," *Solid State Tech.,* pp. 48–58, June 1972.

178. K. Gilleo, "Rheology and Surface Chemistry for Screen Printing," *Screen Printing,* pp. 128–132, February 1989.

179. S. Wu, "Rheology of High Solids Coatings," "Analysis of Combined Sagging and Levelling," *J. Appl. Polymer Sci.,* vol. 22, pp. 27–83, p. 2791, 1978.

180. H. L. Beeferman and D. Bergren, "Practical Applications of Rheology in the Paint Industry," *J. Paint Tech.,* vol. 38, no. 492, January 1966.

181. J. D. Mackenzie and R. Shuttleworth, "Phenomenological Theory of Sintering," *Proc. Phys. Soc.,* London, vol. 62, no. 12B, pp. 833–852, 1949.

182. G. Scherer and T. Garino, "Viscous Sintering on Rigid Substrate," *J. Am. Cer. Soc.,* vol. 68, no. 4, pp. 216–220, 1980.

183. M. D. Sacks and T. Y. Tseng, "Preparation of SiO_2 Glass From Model Powder Compacts," "Sintering," *J. Am. Cer. Soc.,* vol. 67, no. 8, pp. 532–537, 1984.

184. J. P. Bradley, "Copper Thick Film Nitrogen Atmosphere Furnace Design and Firing Process Considerations," *Proc. Intl. Symp. Microelec.,* pp. 435–453, 1985.

185. I. Strobeck et at., "Substrate Bowing of Multilayer Thick Film Circuits," *Hybrid Circuits,* no. 11, pp. 21–23, 1986.

4

Thin Film Materials and Processing

Aicha Elshabini-Riad and Fred D. Barlow

4.1 Introduction

This chapter describes the basic processes in thin film technology, the properties of materials used in practical thin film applications for the formation of resistors, capacitors, inductors, conductors, and semiconductors, as well as design guidelines for thin film realization in microelectronics. In addition, characterization techniques to evaluate thin films and emerging thin film materials will be discussed.

Thin film circuits are composed of films with a thickness varying between 50 and 20,000 Å (0.2 to 80 μin) formed by molecular deposition techniques such as evaporation, sputtering, anodization, or chemical vapor deposition. The selection of specific materials to realize a particular thin film component is determined by the desired electrical and mechanical properties as well as the available deposition techniques and compatibility with other materials on the same substrate. Since the film in this technology is grown by a process involving individual atoms or molecules, the deposition conditions are critical because they determine the properties of the resulting film.

4.2 Thin Film Deposition Techniques

This section provides a basic explanation of essential thin film deposition techniques necessary to enable the designer to understand the thin film realization process. For more extensive reading on this section, the reader is referred to Vossen and Kern.[1]

4.2.1 Vacuum evaporation

Vacuum evaporation is one of the most widely used methods for thin film deposition. The technique consists of evaporation or sublimation of a solid or liquid material by heating it to a sufficiently high temperature under high vacuum conditions. A gas or vapor of the material is produced and condenses onto a cooler substrate to form the required film. The deposited films achieved vary in thickness; the film deposited can be as thin as a monolayer or as thick as a few micrometers. The gaseous state of the material is dependent upon the conditions in the vacuum chamber such as the pressure, the nature of the ambient gas, the temperature of the substrate during deposition, and the volume and mass of the source material. Basically three evaporation methods are used: electron beam evaporation, thermal evaporation, and flash evaporation. All these processes utilize virtually identical vacuum chambers, substrate holders, and deposition monitoring equipment. The techniques differ in the way in which they heat and consequently evaporate the source material.

The evaporation process is commonly used for thin film deposition of metals such as gold, copper, chromium, nickel-chromium alloys, and aluminum. These metals have a relatively high vapor pressure at low temperature. Therefore, their evaporation temperatures do not exceed their melting temperature. For example, the pressure for evaporation of chromium is reached at 1400°C while its melting temperature is 1900°C. The theory of evaporation can be explained in terms of the kinetic theory, the thermodynamic theory, and the solid-state theory.

The source is heated until its vapor pressure reaches a sufficient high value, resulting in the evaporation or sublimation of the material. The evaporation rate cannot exceed a certain value that is a function related to the equilibrium vapor pressure at a specific temperature, and can be estimated from the enthalpy and entropy of the source material. In fact, the evaporation rate observed can be smaller than actually permitted by the equilibrium pressure. The gas produced by the heated source material condenses onto the cool substrate, resulting in the growth of a film. This process is highly directional, resulting in the formation of a uniform film on the surface of an object but not on the side edges of the object. A high deposition rate can very easily be obtained with this process (about 100 to 1000 Å/min). Normally, a thickness monitoring device is included in an evaporation unit that provides a real time indication of the deposition thickness and rate. The vacuum required in the evaporation process is on the order of 10^{-5} to 10^{-9} torr. One torr is equal to the pressure of 1 mmHg.

Thermal evaporation is normally accomplished by heating a source material with an electric filament. A vacuum chamber contains a platform on which the substrates to be coated are placed, as shown in Fig. 4.1. The chamber also houses a pair or several pairs of electrodes linked to the main power supply. Tungsten filaments, boats, or coils can be connected between each pair of electrodes, and the current through them can be separately controlled by external knobs. The current required to produce a given evaporation rate

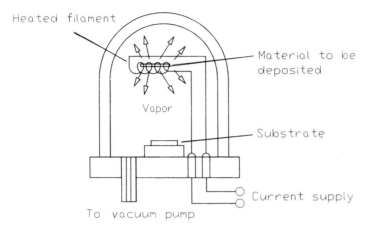

Figure 4.1 A thermal vacuum evaporation system.

depends upon the source material, the quality of the vacuum, and the filament configuration. Thermal evaporation units are often supplied with power supplies capable of delivering several hundred amperes of current. The advantage of this process is its simplicity and the relatively low cost of the equipment. The disadvantage is that materials with high melting points cannot be effectively evaporated, since very large amounts of power are required.

Electron beam (E-beam) evaporation is similar to thermal evaporation except for the heating mechanism. E-beam evaporation utilizes a high energy beam of electrons to heat a small area of source material that is located in a water cooled crucible, as illustrated in Fig. 4.2. The electron beam is focused onto a small area so that very high, localized temperatures can be achieved .

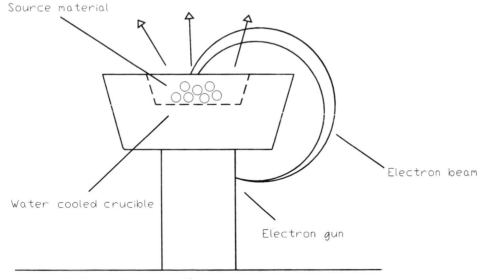

Figure 4.2 An electron beam evaporation source.

This method has the advantages of higher power densities so that a wider range of materials can be evaporated, as well as eliminating any source and crucible reactions common with thermal evaporation. Its primary disadvantage is the relatively high cost of equipment.

One of the problems with evaporation is that complex materials or alloys are very difficult to evaporate since their components or constituents usually have different vapor pressures. The result is that the films do not share the same stoichiometry as the source, since the rate of evaporation is not constant for all the components. Also, some source materials react with the boat or crucible or decompose when heated in the conventional manner. Flash evaporation is a technique which overcomes these problems by depositing a small quantity of source material onto a preheated surface. The material is heated and evaporates instantly, thus preserving the desired composition. Typically, a conventional evaporation boat is heated to the desired temperature and the source material is slowly dropped onto the boat at a constant rate. Various systems have been devised to deliver the source material to the boat. Most of these systems employ a tray which vibrates and delivers the material to the boat via a chute or tube, as illustrated in Fig. 4.3. Applications of this flash evaporation technique include high temperature superconductors[2,3] and optoelectronic materials.[4]

A wide range of evaporated materials such as metals (pure and alloys), semiconductors, dielectrics, chemical elements, and compounds are used in thin film applications. Each of these materials has a particular temperature that must be maintained to produce the required rate of evaporation. In tantalum technology, thin film conducting material and capacitor counterelectrodes are deposited by vacuum evaporation. Appendix A, a table titled

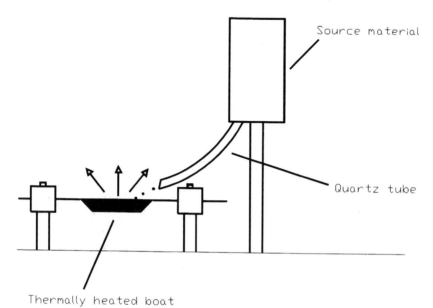

Figure 4.3 A flash evaporation source.

"Deposition Techniques,"[5] provides various conditions of evaporation for common thin film materials. This table is found at the end of this chapter.

4.2.2 Sputter deposition[6,7]

Sputtering is a cold process of physical vapor deposition that involves the removal of material from a solid cathode (a metal or insulator) by bombarding the cathode with positive ions from a rare gas discharge. The sputtering process involves the transfer of momentum from impacting ions to surface molecules. As a result, high melting point materials (such as tantalum, tungsten, and molybdenum) can be deposited onto virtually any substrate material. An evacuated chamber (10^{-6} to 10^{-8} Torr) containing a source material as a cathode, connected to negative potential, and a glass or other inert type of substrate as an anode, connected to ground, is pressurized to a few millitorr with argon, as in Fig. 4.4. A plasma is generated between the two plates by a high voltage and positive argon ions form, bombarding the cathode source and dislodging its atoms. These source atoms form a thin layer of the sputtered material on the substrate surface. The factors influencing the sputtering deposition rate include current, voltage, pressure, source-to-substrate distance, gaseous impurities, and substrate temperature. An optimization of the various sputtering parameters through robust design technique must be performed with each sputtering chamber design in order to produce a film with the desired properties.

Sputtering can be used to produce films with very uniform thickness since the sputtering source can be made broad and flat. Sputtering may be achieved at temperatures between 400 and 500°C for low vapor pressure and high melting point materials (usually 1500 to 3000°C is needed for depositing these materials by the evaporation process). Also, various useful forms of film materials can be produced by altering the sputtering process in appropriate ways. For tantalum, for example, there exist beta tantalum film, body-cen-

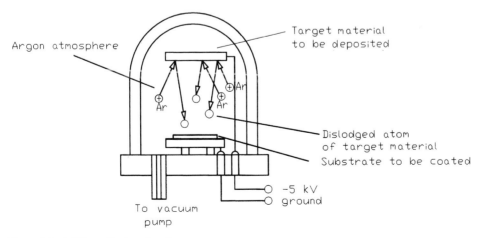

Figure 4.4 A DC sputtering chamber.

tered cubic tantalum, and a low-density form of tantalum film, all of which can be produced by controlling the sputtering voltage or the chamber pressure. In addition, the amount of reactive gas in the solution produces a film with different characteristics. For example, adding large amounts of oxygen during sputtering produces the compound tantalum pentoxide, an insulator for capacitor fabrication. Cosputtering can also be used, allowing more than one material to be sputter deposited simultaneously from several targets.

Reactive sputtering is obtained when a compound of the sputtered material and the sputtered gas may actually be the desired end product. Reactive gases may be deliberately added to the sputtering system to produce this compound in film form. Reactive sputtering can involve oxides, nitrides, sulfides, and other materials. RF sputtering for insulators requires an applied rf field appearing mainly between the two electrodes and a magnetic field which can perform a valuable function by constraining the electrons and reducing their chances of being scattered at these low pressures.

4.2.3 Electroplating and anodization

Electroplating and anodization are both electrochemical processes which are based on the same principle. However, anodization is used to oxidize a material such as a pure metal, while electroplating is used to deposit metals.

As an example of anodization, consider the formation of tantalum pentoxide from pure tantalum. Positive ions (cations) in the electrolytic liquid filling the anodization tank are attracted to the negatively charged cathode and move toward it, and the negative ions (anions) are attracted to the positively charged anode and move toward it. Thus an electrolytic conversion of the material occurs. In a tank filled with an electrolyte, both the tantalum thin film and the tantalum cathode are immersed. The tantalum thin film acts as the anode with positive voltage applied to it. Electron current flows in the external circuit and ions flow in the electrolyte. Conductive tantalum film is converted into tantalum pentoxide at the anode location. Tantalum oxide is an excellent electric insulator (most of the anodic potential drop is across this film); its resistance increases as the thickness increases (which can be controlled to a great precision). Therefore, tantalum pentoxide is formed uniformly over the resistor surface. Anodization is used to protect thin film resistors, to adjust their values precisely, to form the dielectric material for thin film capacitors, or to produce protective coating (hard and glassy uniform film) on thin film elements such as resistors to alleviate deterioration of the resistive material by heat and other factors. Thus the oxide grows on the metal anode surface, while hydrogen results at the cathode site. Examining the chemical reactions, one can determine the vital necessity of water presence; therefore, anodization is usually performed in an aqueous electrolyte. The tantalum oxide films produced by anodic oxidation are amorphous, pore-free, and chemically resistant. Anodic oxide growth is limited by film breakdown, an occurrence of sparking, and a change in color (from a bright uniform interference color of the amorphous oxides to a dull gray color). This phenomenon can also occur at low voltages, below the oxide breakdown, and can be attributed to impurities and/or surface inhomogeneities.

Electroplating works just the opposite of anodization in that a negative voltage is applied to the part (the cathode) and a positive voltage to an anode, both of which are immersed in a suitable plating solution. The part to be plated must be electrically conductive, which is often accomplished by depositing a seed layer by evaporation or sputtering. The electroplated film is then added to the seed layer. Anodes commonly used are either stainless steel or are composed of the material to be deposited. For example, stainless steel or pure silver electrodes can be used for silver plating, while stainless or copper anodes would be appropriate for copper plating. The solutions used to plate a given material are electrolytic solutions composed of the material to be deposited. As a result, when a suitable voltage is applied across the cathode and the anode material is deposited onto the cathode and the anode is slowly consumed.

Conditions used for plating must be carefully controlled in order to produce high quality films. Typically current densities on the order of 10 A/ft^2, based on the surface area of the part, are used while the anode surface area must be the same or larger than that of the part. The value of current density is critical since it controls the deposition rate as well as film quality. Higher current densities result in greater deposition rates but lower quality films. As a result, the current density must be optimized to produce the highest quality films in the shortest period of time. Plating is commonly used for gold, copper, silver, nickel, and a few other materials for thickness up to tens of microns.

4.2.4 Laser ablation

Laser ablation utilizes a high power laser pulse to vaporize a small area of a target material. The cloud of target material generated is then deposited onto a substrate which is located in the vicinity. This type of process is usually performed in a very high vacuum chamber which provides a window for the laser pulse from an external laser source. Ultraviolet excimer lasers delivering 0.1 to 1 J pulses 15 to 45 ns in duration are commonly used with 1 to 100 Hz repetition rates, to produce deposition rates of 0.1 to 100 nm/s.[8] By scanning the laser beam across the target and rotating the substrate, a uniform film can be produced.

The primary advantage of this type of process is its ability to reproduce the composition of the target in the vapor and subsequently in the resulting films. In other words, dissimilar components of a target can be liberated from the target at the same rate. This property makes this type of process ideal for the deposition of complex multicomponent materials. Another advantage of this technique is the ability to deposit virtually any material.

To date, this process has been used to deposit a number of materials including high-temperature superconductors,[9–11] graphite,[12] and electro-optical materials.[13]

4.2.5 Sol-gel coatings

Sol-gel coatings result from spin coating, dipping, or spraying a chemical solution onto a substrate. This solution is a stable mixture of suspended precursor particles known as a sol gel. Once applied to a substrate, a transition or destabilization of the sol occurs, which is marked by a significant increase in the viscosity of the coating, resulting in the formation of a tacky gel. The gel is hardened to a film by drying, typically in air at 100 to 125°C. Although the films are deposited at room temperature, a heat treatment at a high temperature is usually required to create a dense adherent film. Sintering temperatures are typically in excess of 300°C. Final film thickness is determined by the application method, but films are typically 50 nm to 1 mm thick.

The chemical composition of the sol is determined by the material to be deposited, since this process relies on a chemical reaction. One of the more common types of sols is based on a metal alkoxide $M(OR)_x$, with the R representing an alkoxy group. The metal alkoxide is dissolved in a solvent with water and is destabilized by hydrolysis, resulting in the formation of a metal oxide film. Various catalysts may be added to modify the rate of the reaction and hence the properties of the resulting films.

The advantages of this type of process include a very high degree of homogeneity over large areas, as well as high purity. The primary disadvantages are the high cost of the materials as well as the narrow range of materials which can be deposited with this technique.

A number of applications have been demonstrated for this process. Several of these applications and materials are listed in Table 4.1. A great deal of interest has been sparked by the discovery of high temperature superconductors, since this process is ideal for the deposition of these materials. Further discussion is included in Sec. 4.8.1. Several reviews of this technique include a more extensive discussion of this topic.[1, 16–20]

4.3 Substrate Materials

Commonly used substrate materials for thin film deposition include alumina, glass, beryllia, aluminum nitride, and silicon. For silicon use, the substrate surface is covered with a silicon dioxide dielectric layer. Ceramic materials are often used for thin film applications since ceramics possess high thermal conductivity and good chemical stability and are also resistant to thermal and mechanical shocks. Glazed substrate surfaces provide good surface finish

TABLE 4.1 Sol-Gel Material Applications

Material	Application
Indium tin oxide (ITO)[13] Cadmium stannate[14]	Transparent conductors
Y-Ba-Cu-O	High speed interconnections
Barium titanate	Dielectric for capacitors
Titanium	Electrodes

and low porosity and possess a low dielectric constant value. Glazing typically yields a surface roughness of less than 1 μin (about 250 Å).

The properties of the substrates used to grow defect-free films for thin film circuits include good surface smoothness, proper thermal expansion coefficient, good mechanical strength, high thermal conductivity, chemical stability, porosity, low cost, high electrical resistance, and reasonable uniformity of material composition.

Substrate flatness directly influences the minimum achievable linewidth and spacing. In general, films are tolerant of surface roughness less than an order of magnitude of the film thickness. In order to maintain uniform characteristics of thin film elements, it is important to minimize surface roughness. The thermal expansion coefficient of the substrate should be similar to the deposited film in order to minimize mechanical and residual stresses occurring in the films. High mechanical strength and thermal shock resistance are required in order to enable the substrate to withstand the rigors of processing and normal usage. Ceramic materials are superior to fulfill this function effectively. High thermal conductivity is required in order to remove heat, allowing circuits with high component densities. Inertness to chemicals used in circuit processing is a necessary requirement. Ceramic materials possess better chemical stability than glasses, especially at higher temperatures. These materials are not attacked by etchants used in processing thin film circuits.

Low porosity is required to prevent the entrapment of gases (causing film contamination). These substrate materials must be good insulators at room temperature. Finally, uniformity of substrate properties must be maintained. Therefore, control of electrical properties is very important. Similarly, control of substrate purity, density, and surface properties has to be ensured.

Substrates for thin films have high resistivity and high dielectric strength and are chemically resistant. For example, tantalum film processing requires hydrofluoric acid as the etchant. Therefore, glass or glazed substrates must be covered with a thin layer of tantalum oxide (to resist the acid). The volume resistivity of alumina, beryllia, and aluminum nitride is on the order of 10^{15} (or higher), 10^{14}, and 10^{13} Ω cm, respectively. The relative dielectric constants measured at 1 MHz of alumina, beryllia, and aluminum nitride are 9.7, 7, and 8.9, respectively. The thermal conductivity of ceramics is superior compared to that of glass [about 30 to 33 W/(m•K)]. A glazed ceramic will have slightly lower thermal conductivity than raw ceramics owing to the glasslike glaze layer on its surface. The thermal conductivity of aluminum nitride is on the order of 130 to 170 W/(m•K), and that of beryllia is on the order of 180 to 260 W/(m•K). A high thermal conductivity is needed for heat dissipation from active components and resistive components deposited on the surface.

The fluctuations in the substrate surface can be described in terms of roughness and waviness. Roughness is a short-range fluctuation (average deviation of the surface from some arbitrary mean value). The surface roughness can basically result in an uneven film thickness. Resistive films on rough substrates exhibit a high sheet resistance owing to an apparent increase in the length between two points and also because of thin spots in the film. These films are also less stable during thermal aging process.

Additionally, the performance of capacitors is affected by surface roughness, since a premature breakdown of the dielectric film may occur. Raw ceramic materials are not smooth enough for very fine linewidths or for applications requiring extreme stability. Therefore, capacitors and high stability resistors are deposited on polished and/or glazed ceramic substrates. Waviness is a periodic variation characterized by a peak-to-peak amplitude and a certain repetition interval. Flatness, or the lack of waviness, is an important requirement, especially if thin (<50 μm) and well-defined linewidths are to be achieved. The linewidth of a resistor is widened by the waviness of the substrate surface (even when controlled by the surface mask). As a result, the ohmic value of resistors formed on rough surfaces is lower than expected.

4.4 Conductor Materials

Conductor films are typically formed of several metals in order to achieve the desired properties of a thin film metallization. These properties include high conductivity, good adhesion to both the substrate and the other deposited films, resistance to etchants, capability of being etched to linewidths on the order of a few micrometers using conventional photolithographic technology, ease of bonding, and compatibility with other materials and processes. Metals with good conductivity usually exhibit poor adhesion to substrate material. One can enhance the adhesion property of these high conductivity metals by providing intermediate layer(s) of titanium, chromium, or nichrome.

Copper is often evaporated to form the required metallization. Copper possesses a high electrical conductivity and an excellent solderability. Usually copper is covered with an evaporated layer of a noble metal in order to protect the surface from oxidization. An adhesive underlayer such as titanium or chromium is often used to promote the adhesion of the copper to the substrate.

Gold can be either evaporated or sputtered and is used as a top layer to protect lower evaporated (or sputtered) layers. It can also be used to achieve the required metallization and/or to form the capacitor electrodes. An adhesive underlayer of titanium or chromium is often used since gold does not adhere well to most substrate materials. Gold is more difficult to solder to, since it usually alloys with soft solders, causing mechanically poor connections.

Aluminum can be evaporated and is a common metal often used for conductors and the realization of capacitor electrodes. It exhibits an acceptable degree of adhesion to various substrate materials. One major drawback of aluminum is its reaction with many crucibles during the evaporation process.

Palladium alloy can also be evaporated and is used as a protective layer to prevent surface oxidization of other metals. An adhesive underlayer of titanium or chromium is usually used to improve the adhesion of palladium to glass and ceramic substrates. Titanium can be evaporated and is mainly used as an adhesive underlayer material.

Nickel is often alloyed with chromium to form nichrome material. Nickel–chromium, or nichrome, is used as an interface material to promote the adhesion of some metals or as a thin film resistor material. Nichrome possesses a high resistivity on the order of 100 $\mu\Omega$ cm or higher. Nichrome exhibits poor solderability and can be easily oxidized. Chromium is used as an adhesive, since its adhesion to the substrate and other films is very high. Finally, silver metal can be evaporated and is often used to presensitize the dielectric material surface. Table 4.2 provides a list of several metals with their respective most important properties.

4.5 Resistor Materials

The selection of thin film resistor materials depends on basic considerations such as temperature coefficient of resistance (TCR), voltage coefficient of resistance (VCR), electrical noise figure, resistance stability, achievable sheet resistivity R_S, allowable power density, compatibility with other materials, and film deposition method used. The bulk resistance R_B of a thin film resistor is a basic material property. The sheet resistance (R_S, in Ω/sq) refers to the volume or bulk resistivity normalized with respect to the thickness of the film.

The four-point probe measurement technique is used to measure the sheet resistance of a thin film accurately without any etching process. Typical thin film sheet resistances lie in the range 10 to 2000 Ω/sq with an initial tolerance on the order of $\pm 10\%$. Trimming can be performed to bring these resistance values to tighter tolerances (0.01%) if desired. The temperature coefficient of resistance (TCR) provides useful information in characterizing the properties of the deposited thin films (very thin films in particular). TCR (in ppm/°C) is defined as $(1/R)(dR/dT) \times 10^6$, where dR and dT refer to a change in the resistance value R due to a change in the ambient temperature T (room temperature represents the reference point in this case). This parameter is especially important in some applications which require circuit operation in unstable ambient temperature, thus affecting considerably the circuit performance. In general, thin films have very low TCR values as compared to bulk materials. Very low values of TCR, approaching zero, can also be achieved in some cases. Typical TCR values are well within ± 100 ppm/°C. Resistor stability is closely related to substrate inertness, substrate surface smoothness, and substrate thermal conductivity. Therefore, thin film resistor design and performance are greatly affected by the choice of substrate mater-

TABLE 4.2 Properties of Metal Films

Metal	Resistivity, $\mu\Omega$ cm	Melting point, °C	Temp. for 10^{-2} torr vapor pressure, °C
Copper (Cu)	1.7	1084	1257
Gold (Au)	2.4	1063	1397
Aluminum (Al)	2.8	660	1217
Palladium	11	1550	1462
Palladium–gold	21	1450	1300
Titanium (Ti)	55	1667	1737
Nickel (Ni)	29	1452	1527
Chromium (Cr)	170	1900	1397

ial. Thin film resistors are known to exhibit good stability over time under different environmental conditions. The voltage coefficient of resistance (VCR) describes the dependence of the resistance value on the voltage applied across the thin film resistor. Small VCR values are normally observed for the case of thin films (on the order of 10 ppm/V or less). The improved performance of thin film resistors as compared to wirewound resistors has made these elements very reliable.

A typical thin film resistor structure is illustrated in Fig. 4.5. A thin resistive film on the order of a few hundred angstroms is deposited on an insulating substrate by means of vacuum evaporation or sputtering. Metallic end contacts are formed through a similar procedure. Thin film resistor materials can be metals, metal alloys, and metal compounds. Metals in bulk form possess low resistivity values ranging between 0.1 and 200 $\mu\Omega$ cm. When deposited in film form, these metals achieve significantly higher resistivities, depending on the thickness achieved.

Tantalum films (tantalum nitride and tantalum oxynitride resistors), nickel–chromium (nichrome) alloy, chromium films, chromium silicon monoxide films, and chromium–cobalt films are covered in this section. Common properties of interest to the design engineer include sheet resistivity, temperature coefficient of resistance, stability, and stability after trimming.

4.5.1 Tantalum films

Tantalum is known as a "valve" metal, a name given to metals forming tough self-protective oxides through anodization (anodic oxidation) or heat treatment in the presence of oxygen. Tantalum is a commonly used metal for thin film resistors. Tantalum films possess high stability with respect to aging and tem-

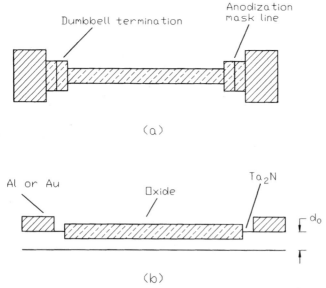

Figure 4.5 Typical thin film resistor structure. (a) Top View. (b) Cross section.

perature. Bulk tantalum possesses a resistivity on the order of 15 $\mu\Omega$ cm, while tantalum thin films possess a higher resistivity at the same temperature.

Thin films of tantalum are usually deposited by sputtering because of refractory nature of tantalum. Special precautions are taken during deposition since the sputtered films have a tendency to be contaminated. Some doping impurities such as nitrogen and/or oxygen are allowed in the tantalum films in order to achieve specific properties. These sputtered films possess a high resistivity. Anodizing them provides also a means of trimming these resistors to a tight tolerance. Many forms of these films exist.

4.5.1.1 Tantalum nitride films Tantalum nitride films have the composition Ta_2N or TaN depending on nitrogen concentration. Nitrogen-doped tantalum films have resistivities on the order of 400 $\mu\Omega$ cm and temperature coefficients of resistance (TCR) on the order of -100 to -200 ppm/°C. The doping can be achieved by introducing a controlled amount of nitrogen into the sputtering chamber. These films also possess higher stability under various environmental conditions than do tantalum films. Resistors composed of Ta_2N are more stable than those formed with TaN. A stabilizing treatment consisting of film anodization, followed by heat treatment and another anodization at higher voltage, promotes the stability of these elements. Tantalum nitride films with thicknesses on the order of 200 to 1000 Å produce sheet resistances on the order of 300 to 50 Ω/sq. Also, these nitrogen-doped films can be anodized in a similar manner as pure tantalum. The sputtering voltage, bias voltage, N_2 partial pressure, and temperature are critical parameters that affect the resistivity of tantalum films.

4.5.1.2 Tantalum–oxynitride films (TaO_2N) Tantalum–oxynitride films can be obtained by adding oxygen as well as nitrogen in the reactive sputtering process of tantalum. The additional nitrogen and oxygen are used to provide higher resistivity films. The TCR can be greatly modified by varying the pressure of both the oxygen and the nitrogen in the sputtering chamber during the sputtering process.

4.5.2 Nickel–chromium films (Ni–Cr)

A common resistive film composed of 80% nickel and 20% chromium alloy provides the desired properties of TCR and sheet resistivity for some applications. In general, the ratio of nickel to chromium can vary from 40:60 to 80:20. Bulk resistivity on the order of 110 $\mu\Omega$ cm and TCR on the order of 100 ppm/°C have been cited. The TCR obtained depends on the nickel–chromium ratio as well as specific deposition conditions. The TCR is found to be more negative for films with high chromium content and high sheet resistivity. The 80:20 composition exhibits an increased TCR and a higher sheet resistivity. Nickel–chromium films are commonly obtained by sputtering but can also be deposited by vacuum evaporation. Sputtering provides improved composition control of nickel–chromium films over a wide range of sheet resistance. Special care should be taken in the evaporation process to provide a film with uniform composition. The difficulty arises from the fact that chromium and nickel possess different vapor pressures, chromium being higher than nickel. The difference in vapor pressure decreases as the temperature increases. Some of the techniques to obtain a uniform film composition are utilizing a different source composition from the desired film deposition, a large source, or flash evaporation. If evaporation is elected as the technique for nickel–chromium film deposition, several requirements should be satisfied in order to produce successful results, including relief of internal stresses in substrates by keeping them within a certain temperature range during deposition, adjustment of residual gas pressure and rate of deposition to control oxidation, and postannealing in air. Partial oxidation of chromium constituent during deposition represents one serious problem. Nickel–chromium resistors possess sheet resistivities in the range 1 to 500 Ω/sq. The incorporation of additives such as copper and aluminum has modified the characteristics of these films. In addition, heat treatment also has an effect on the sheet resistance. Maissel[21] has illustrated the percent resistance change due to heat treatment as a function of sheet resistance for both vacuum-evaporated and sputtered nichrome. The sputtered films show larger resistance increase than evaporated films. Nickel–chromium films are also known to be unstable under high humidity condition; therefore, a coating layer, such as silicon monoxide, is provided to increase the stability of these films. This protective technique improves the resistor stability under wide variations of atmospheric conditions. Polyimide coatings are also used to improve the stability and passivate resistors.

4.5.3 Chromium films (Cr)

Chromium is a simple single-component system. It is not a particularly refractory metal. Therefore, temperature limitations for continuous operation are not relevant. Heat treatment of chromium films in vacuum may cause a decrease in resistance values. Pure chromium films possess higher resistivities than nichrome films with achievable sheet resistances on the order of 200 Ω/sq within a tight tolerance without trimming. Much higher sheet resistances can be achieved with thinner layers but with an increased tolerance. Chromium films are used for thin film resistor realization as well as underlays for thin film conductor metals. Chromium films are deposited by either vacuum evaporation or sputtering. The film properties are extremely sensitive to deposition conditions. The TCR of chromium films is not as good as that of nickel–chromium films. However, the adhesive strength of chromium to glass substrates is superior, making it an attractive metal for resistive films.

4.5.4 Ruthenium films

High resistivity can be achieved with ruthenium film material. The films are deposited by evaporation onto ceramic, glass, or silica substrates. Sheet resistance values on the order of 250 to 300 Ω/sq with a TCR close to zero may be achieved. Higher sheet resistances on the order of 10 000 Ω/sq with a TCR around -400 ppm/°C may also be obtained. A silicon monoxide passivation layer ensures the stability of this type of resistor.

4.5.5 Cermets

Cermets constitute an important class of materials for resistor fabrication (metal-dielectric systems such as the chromium–silicon monoxide system, Cr–SiO). Several techniques or processes are used to deposit films of these materials, such as flash evaporation and rf sputtering. The cermet composition greatly affects the film resistivity. Cermets of 20% SiO (silicon monoxide) and 80% Cr (chromium) possess a resistivity on the order of 550 $\mu\Omega$ cm (TCR -180 ppm/°C) after deposition (200°C), and 400 $\mu\Omega$ cm (TCR 0 ppm/°C) after annealing at 400°C. High resistivities on the order of 10^3 to 10^5 $\mu\Omega$ cm and good stability can be obtained with reasonable TCR values. Varying the percentage of SiO and Cr greatly affects both the resistivity and the TCR values.

4.5.6 Other resistor materials

Table 4.3 summarizes the various important materials used for thin film resistors.

4.6 Capacitor Materials

A thin film capacitor usually consists of three layers, a bottom electrode, a dielectric material, and a top electrode. The insulating material may contain two different types of dielectrics. Different capacitor structures are illustrated in Fig. 4.6. Characteristics of capacitor elements include capacitance per unit area, temperature coefficient of capacitance (TCC), dissipation factor, dielectric constant, and breakdown voltage. Other properties of interest are leakage current and maximum working voltage stress. High-quality thin film capacitors possess a low pinhole density, a small leakage current, low dissipation factor, and low temperature coefficient of capacitance, in addition to reasonably high capacitance per unit area. Because of manufacturing low yields, few companies utilize parallel plate thin film capacitors. Low yields are due to the inherent difficulties in forming pinhole-free dielectric films. Thin film capaci-

TABLE 4.3   Characteristics of Thin Film Resistor Materials

Film material	Deposition technique	Sheet resistance R_S, Ω/sq	TCR, ppm/°C
Cr-Ni (20-80)	Evaporation	10–400	100–200
Cr-Si (24-76)	Flash evaporation	100–4000	± 200
Cr-Ti (35-65)	Flash evaporation	250–600	± 150
Cr-SiO (70-30)	Flash evaporation	Up to 600	-15 to -200
SnO$_2$	Spray pyrolysis/CVD	Up to 10^4	-50 to -200
W, Mo, Ru	Sputtering	10–500	-20 to -100
Ta	Sputtering	Up to 100	± 100
Ta$_2$N, TaN	Reactive sputtering	10–100	-85

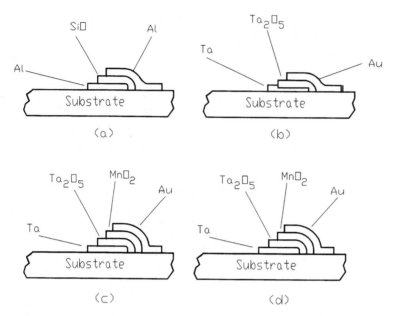

Figure 4.6 Thin film capacitors. (*a*) SiO capacitor. (*b*) Ta$_2$U$_5$ capacitor. (*c*) MnO$_2$–Ta$_2$O$_5$ capacitor. (*d*) SiO–Ta$_2$O$_5$ capacitor.

tors fabricated using interdigitated conductor geometries are often used in the fabrication of high frequency hybrid microcircuits.

Common dielectric materials used for thin film capacitor formation include silicon monoxide (SiO), tantalum oxide (pentoxide Ta$_2$O$_5$), and organic polymers. Capacitor electrodes are usually made from aluminum, gold, copper, or tantalum, depending on the dielectric material used.

4.6.1 Silicon monoxide capacitors

Deposition of amorphous silicon monoxide (SiO) films by vacuum evaporation results in capacitance values between 10 and 1.8 nF/cm^2 for dielectric film thicknesses between 0.5 and 3 μm. The tolerance of the capacitance value can be controlled to within ±5%. The properties of the deposited dielectric film depend on the degree of oxidation during deposition and during annealing (around 400°C). To ensure the proper amount of oxygen to be added, special designed evaporation sources are used in the process as well as an oxygen partial pressure in the evaporation system. The vapor pressure of silicon monoxide dielectric increases considerably as the temperature increases. This property enables the evaporation of the material at much lower temperatures than either silicon or silicon dioxide.

The silicon monoxide dielectric material has been found to give a relative dielectric constant of about 6, a leakage current less than 1 nA, a TCC between + 100 and + 300 ppm/°C, a dissipation factor of approximately 0.01

to 0.1 at 100 kHz, and a maximum allowable stress of 2×10^6 V/cm. These properties are dependent on the deposition conditions and the nature of the electrode material.

Aluminum is commonly used for both the base electrode and the counter-electrode for silicon monoxide capacitors since it forms a smooth layer and has the advantage of being able to vaporize around pinholes by discharge currents. Copper metallization is another alternative used for the base electrode, with copper, aluminum, or gold for the counterelectrode.

4.6.2 Tantalum oxide capacitors

Tantalum pentoxide (Ta_2O_5) dielectric is formed by anodizing tantalum metallization. The resulting dielectric material is amorphous and chemically resistant. The tantalum metallization for the base electrode is achieved through sputtering (depositing a layer of tantalum or tantalum nitride of a thickness of about 0.5 μm). After anodization, a counterelectrode of aluminum or gold is deposited by evaporation or sputtering. The dielectric constant of the resulting oxide is on the order of 25, the capacitance per unit area is about 0.3 μF/cm^2 (depending on the thickness of anodization), the maximum working stress is 1.5×10^6 V/cm, and the dissipation factor is on the order of 0.002 at frequencies in the kilohertz range. The temperature coefficient of capacitance falls in the range of $+160$ to $+250$ ppm/°C. A process consisting of using a constant current anodization followed by constant voltage anodization, in addition to the use of an anodic etch, may be utilized to eliminate pinhole formation in the oxide.

4.6.3 Manganese oxide tantalum oxide capacitors

A high conductivity semiconductor manganese oxide (MnO_2) film is deposited by reactive sputtering or evaporation on top of an anodized tantalum film. In this capacitor structure, tantalum is used as the base electrode, and gold or aluminum is used as the counter or top electrode. The composite dielectric structure enhances the dielectric properties of the capacitor element, eliminating weak spots and thus enabling the use of thinner tantalum oxide layers. The resulting dielectric constant is on the order of 22, with a maximum working stress of 2×10^6 V/cm and a capacitance per unit area of 1 μF/cm^2.

4.6.4 Silicon monoxide–tantalum oxide duplex capacitors

The combination of these two dielectric materials enables one to use thinner dielectric thicknesses without a high probability of pinhole formation. A silicon monoxide film on the order of 2000 Å thick and a Ta_2O_5 film on the order of 500 Å thick may be used. Tantalum or tantalum nitride is used as the base electrode and aluminum is used as a counterelectrode for this capacitor structure. A dielectric constant of 6, a maximum allowable working stress of 2.5×10^6 V/cm, and a capacitance per unit area of 0.02 μF/cm^2 typically result from this type of structure.

4.6.5   Other thin film capacitor materials

Other dielectric materials for thin film capacitor formation include Al_2O_3, Si_3N_4, MgF_2, and $BaTiO_3$. Table 4.4 summarizes the dielectric properties of thin insulating films of various materials for various capacitor applications.

4.7   Inductor Materials

The use of thin film inductors is limited to high frequency circuits owing to the low inductance value achievable on the substrate. Practical physical dimensions of the spiral inductor have limited the inductance value to 10 μH with a quality factor on the order of 50 to 100 at about 15 MHz. Higher inductance values can be achieved by depositing the inductor on a ferrite substrate or using multilayer structures.

4.8   Emerging Thin Film Materials

4.8.1   Superconductor thin film materials

The pioneering discovery by Bednorz and Muller[22] of superconductivity above 30K in the La–Ba–Cu–O system has generated an unprecedented amount of research on superconducting oxides, including the discovery of a superconducting transition above 77K in the Y–Ba–Cu–O, Bi–Sr–Ca–Cu–O, and Ti–Ba–Ca–Cu–O systems.

Although other materials exist, these three systems are the most important for many applications, since they have the highest critical temperatures. These material systems are very complex and possess many phases. For example, $Y_1Ba_2Cu_3O_7$ is a high temperature superconductor (HTS) with a critical temperature of 91 K, while $Y_2Ba_1Cu_1O_x$ is an insulator. The principal phases of these systems and their associated transition temperatures are listed in Table 4.5.

The most common methods used to create superconducting films include sputtering, coevaporation, and laser ablation. The most popular method to produce high temperature superconducting films is rf sputtering. The as-deposited films possess little structure and are generally insulating. Postdeposition annealing treatments in an oxygen atmosphere at temperatures above 800°C are normally necessary to transform the films into superconductors with high critical temperatures. Inadequate heat treatments, such as annealing at low temperatures below 750°C, usually give films a semiconducting behavior before the superconducting transition. Annealing treatment is an important process which determines the final oxygen content and the resulting properties of the film. The critical processing variables in this technique are annealing temperature, oxygen partial pressure, and quenching or cooling rate of the sample.

Substrate reactions to the superconducting films are a major factor of degradation. Diffusion of substrate material into an HTS results in the cre-

TABLE 4.4 **Properties of Thin Film Dielectric Materials****

Film material	Dielectric constant	Dissipation factor	Frequency, kHz	Breakdown voltage, V/cm	TCC, ppm/°C	Thickness, μm	Deposition technique†
Al_2O_3	9	0.008	1	3×10^5	300	1.5	Ev
$BaSrTiO_3$	3–5	0.004–0.02	0.1–1000		300	2	Ev
SiO	6	0.015–0.02	1–1000			1	Ev
SiO_2	3–4	0.001		$\sim 10^6$		1	RS
Si_3N_4	5.5			10^7		0.03–0.3	CVD
Ta_2O_5	25	<0.01	0.1–50	6×10^6	250	1	An
TiO_2	50	0.01	1		300	1	An
W_2O_5	40	0.6				1	An
CaF_2	~3	0.05	0.1			1	Ev
LiF	~5	0.03	0.1			1	Ev
Nb_2O_5	39	0.07	1			1	RS
Polystyrene	2.5	0.001–0.002	0.1–100			1	GDP
Polybutadiene	2.5	0.002–0.01	0.1–100			1	UVP
$BaTiO_3$	200	0.05	1			1	Ev

*In this table, Al or Au films are generally used as electrodes for these measurements.
†Ev, evaporation; RS, reactive sputtering; CVD, chemical vapor deposition; An, anodization; GDP, glow-discharge polymerization; UVP, ultraviolet polymerization.
Source: Reference 45.

TABLE 4.5   High Temperature Superconductors and Transition Temperatures

Material	Critical temp., K
$Y_1Ba_2Cu_3O_7$	91
$Bi_2Sr_2Ca_2Cu_3O_x$	110
$Ti_2Ba_2Ca_2Cu_3O_{10}$	125

ation of insulating phases that reduce the current density of the overall film and may result in an insulating film if the reaction is severe. Certain substrate materials, such as yttrium stabilized zirconia (YSZ), do not react with most of the superconductors, while other materials require a buffer layer to prevent substrate reactions. These buffer layers prevent a physical contact between the superconductive material and the substrate. Commonly used buffer layer materials include YSZ, magnesium oxide (MgO), and cerium oxide (CeO_2).

Most of the applications for these materials are a result of their very low loss characteristics. Particularly for high frequency applications, they have significantly lower loss than conventional metals. Additionally, these materials normally exhibit lower propagation delays than conventional metals and are prime candidates for use as high speed circuit interconnections.

4.8.2   Diamond films

Diamond is a very desirable material for many electronic systems since it possesses the largest thermal conductivity of any material, has one of the lowest coefficients of expansion, is very hard, and is unaffected by most chemical etchants. The high thermal conductivity coupled with the low electrical conductivity of pure undoped diamond makes it very desirable for use in applications where high power or large packing densities are required.

The exact definition of what constitutes a true diamond film has not yet been standardized. Carbon films with a nearly crystalline diamond structure, those films with a polycrystalline structure, and amorphous carbon films can all be produced. These materials are normally referred to as diamond films in the case of single crystalline or large grain size polycrystalline films, and diamond like carbon (DLC) in the case of less well ordered structures. The properties of these films differ greatly and depend on the deposition process. For example, the thermal conductivity of diamond films may be as much as ten times greater than that of DLC films, while the DLC films are significantly easier to produce.[23] DLC films are often contaminated by hydrogen, which is used in the production process. Hydrogen content ranges from ~0% for very high purity diamond films to as much as 25% for DLC materials.[24] DLC films with very large hydrogen content are often referred to as hydrogenated DLC (HDLC).

A large number of techniques have been developed to produce diamond and DLC films. Originally, diamond films were produced under very high temperatures and pressures which produce an environment where diamond is thermodynamically stable rather than graphite. These high pressure–high temperature (HPHT) methods are very costly and subject the substrate material to extreme environmental conditions. In recent years, many low temperature–low pressure techniques have been developed which allow the production of diamond films at much lower temperatures and are more economical than HPHT methods. A large number of methods exist for the low temperature production of diamond and DLC films. The majority of these techniques fall into one of three categories: assisted chemical vapor deposition (CVD), flame combustion, and beam methods.

Some of the more common CVD techniques include hot filament CVD (HFCVD), microwave plasma assisted CVD (MPACVD), rf plasma assisted CVD (RFCVD), electron assisted CVD (EACVD), and DC current discharge CVD. The basic process employed by these techniques is to decompose a gas mixture under high temperature. Typically, a hydrocarbon hydrogen mixture is used. The result is a release of carbon and atomic hydrogen. The carbon is deposited on a cooler substrate and thus produces a diamond or DLC structure. The hydrogen selectively etches away graphite formations without damaging the diamond structure. The manner in which the gases are decomposed is what differentiates these techniques. These processes are typically characterized by a high film purity, a high degree of crystallinity, a good homogeneity, a relatively fast growth rate, and a moderate substrate temperature (800 to 1000°C).[25]

Flame combustion methods utilize a flame to decompose a hydrocarbon material at atmospheric pressure. Carbon is supplied to a substrate by the flame, resulting in the formation of a diamond or DLC film. Acetylene, ethane, propane, and several other materials have been used to produce these diamond or DLC films.[26] Films produced with this technique tend to be lower in quality than CVD films since the uniformity of the growth is difficult to control. The advantage of this technique is its simplicity as well as the fact that very high growth rates can be obtained.

Ion beam techniques have the decisive advantage of very low substrate temperatures. This allows the production of films on materials which cannot withstand the high temperatures required by other methods.[27] These techniques generate the high temperature and pressure required to produce pure diamond on a microscopic scale. Typically, a high energy beam of carbon atoms is focused onto a substrate to generate the conditions required for film growth. This technique requires very complex equipment and suffers from low growth rates but allows the formation of diamond and DLC films on a very wide range of materials.

To date, the majority of electronic applications for diamond are in the area of electronic packaging. A number of researchers are developing metallization systems and die attachment methods which use diamond films or substrates as a heat spreader, particularly in the area of multichip modules, where very high power densities create great difficulties in achieving good thermal management.

4.8.3 Thin film optical materials

Applications of thin films, including transparent films, in microelectronics include thin film reflection filters, photonic devices (mainly light emitting diodes, lasers, and detectors), antireflection coatings for solar cells, photodetectors, and nearly all optical equipment including cameras, microscopes, and telescopes, high reflectance or beamsplitter coatings, mirrors for lasers, and integrated optical waveguides. A deposited thin film of a metal element, possessing a certain refractive index different from the material or substrate it is deposited on, can change the fraction of radiation reflected from the top surface as well as the portion of radiation being absorbed in the material. Factors such as surface roughness, grain boundaries, pinholes or pits, appearance, porosity, presence of impurities and dopants, and presence of microcracks resulting from stresses depending on the heat treatment may affect these fractions of radiation. These metals include aluminum, copper, silver, platinum, and gold. The amounts of reflected and absorbed radiation in the deposited thin film layers can be described in terms of two optical constants, the refractive index and the index of absorption, describing the interaction of electromagnetic radiation with electron carriers in the thin film material. These two optical constants are dependent on each other through dispersion relations governed by electromagnetic theory, and they are highly sensitive functions of the radiation wavelength, as well as sensitive to the thin film deposition conditions and the surface conditions of the material. Adhesion to the surface of the material and long term stability under various environmental conditions are issues of concern associated with these thin metal films. Thus protective coatings of various thin oxide layers are usually used, slightly affecting the radiation absorbance of these thin films. Care is usually administered in order to prevent degradation of reflectance due to oxidation, tarnishing, wear, and scratching.

Dielectric materials used in optical coatings range from compounds, sulfides, glasses, and oxides, finally to fluorides. A thin layer of a dielectric material, mostly applied by a sputtering process, a reactive evaporation, or perhaps a chemical deposition process, is usually applied at the surface. Therefore, the dielectric material is considered transparent to the radiation wavelength with presumably a low absorption coefficient to the light. Finally, optical properties in semiconductor materials can be described in terms of the absorption coefficient, a strong function of radiation wavelength. A long as well as a short wavelength cutoff of the radiation energy exists in a semiconductor material to describe the amount of photon absorption. This photon absorption is dependent on the excitation energy of a semiconductor material, equivalent to the energy bandgap in band to band transition. Photomultiplication within the active region of the semiconductor material can occur under a large reverse bias condition. The main criterion is to guide the incident radiation to be detected to the depletion region with minimum attenuation, where it can be absorbed and photogenerated carriers can move faster under the influence of an electric field. Realization of semitransparent layers with antireflection coating, realization of buffer and matching interfaces through multilayer structures, the use of heterojunction interfaces and back illumination are viable techniques to ensure this condition.

4.9 Thin Film Design Guidelines

4.9.1 Introduction

The design involves hybrid thin film circuits utilizing insulating substrates on which conductor paths and/or passive components have been deposited by means of vacuum evaporation, sputtering, or chemical vapor deposition techniques. Add-on or attached components may include resistors, capacitors, inductors, and various semiconductor devices.

Thin film hybrid circuits consume very little space on the substrate surface. They also possess a high performance as compared to the etched printed circuit boards and the custom monolithic integrated circuits. This is due to the shorter circuit paths, the tighter thermal coupling, the superior control of parasitics, and the tighter component tolerances that can be achieved. In addition, the thin film technology is highly reliable owing to the minimum number of interconnections, and is readily adaptable to any design modifications. Thin film hybrids are very advantageous to use in high frequency applications (exceeding 500 MHz).

4.9.2 Thin film integrated components: resistors and capacitors

In this section, the four fundamental thin film processes are used to make thin film resistors and capacitors using tantalum technology as a demonstration example.

4.9.2.1 Resistors   The resistor material is tantalum nitride. The oxide layer resulting from anodization of the resistor to obtain the final desired value grows at the expense of the underlying film. It is the consumption of the tantalum material that decreases the resistor cross section, increasing the resistance value to the desired amount.

The transition section between the film resistor and its associated conductor pattern is wider than the resistor path. This widened section, referred to as a "dumbbell" termination, is used to minimize the effect on resistor stability of the narrow area of resistor material at each end of the resistor, which must be left unanodized during processing.

One of the most critical factors in a thin film circuit is the resistor film temperature. It is affected by the resistor geometry, the total circuit power dissipation, the influence of other dissipative elements, the substrate used in the application, and the type of external heat sink used. If insufficient information is provided for power dissipation, it is reasonably safe to assume a value of 50 W/in^2 in designing these thin film resistors, as well as a maximum film operating temperature of 150°C.

4.9.2.2 Capacitors A reasonably high capacitance per unit area can be obtained with the tantalum material. Anodically grown tantalum pentoxide, with a dielectric constant of about 25 and a very high dielectric strength, is used as the capacitor dielectric. Very thin films of tantalum pentoxide are produced (less than 1000 Å). The tantalum pentoxide dielectric is sandwiched between two tantalum electrodes. The tantalum adheres well to the substrate. However, these capacitors cannot be used in precision networks because of their high sensitivity to humidity. The problem is solved by using a counterelectrode consisting of a thin layer of nichrome followed by gold deposition. Figure 4.7 illustrates a typical capacitor structure.

4.9.3 Resistor design rules

A thin film resistor consists of a resistive path deposited on an insulating substrate. Deposited metallic end terminations form the required contacts. The resistance value R of a rectangular or bar pattern in ohms is of the form

$$R = \frac{\rho L}{tW}$$

where ρ refers to the specific or bulk resistivity of the material in Ω cm. L, W, and t represent the length, the width, and the thickness of the resistor film in

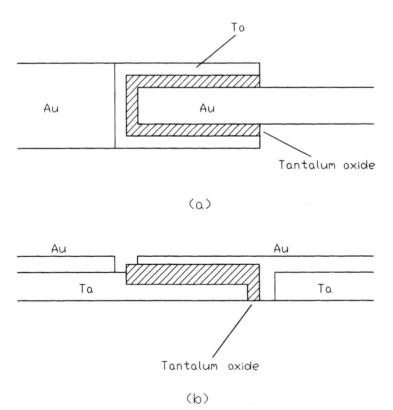

Figure 4.7 Typical thin film capacitor structure. (*a*) Top view. (*b*) Cross section.

cm, respectively. The thickness t that can be achieved varies typically from 50 to 10 000 Å. Uniform thickness can be obtained on smooth substrate surfaces such as glass as compared to ceramic surfaces. The thickness t is the third dimension into the paper. The above equation can be rewritten as

$$R = \frac{R_s L}{W}$$

where R_s refers to the sheet resistivity of the resistor ($R_s = \rho/t$) in Ω/sq and L/W refers to the aspect ratio of the resistor and defines the number of squares needed to realize a certain resistance value (Fig. 4.8).

For a certain substrate, one sheet resistivity is normally used to achieve all resistance values with different numbers of square counts, thus determining different lengths of resistor lines (straight or serpentine). The sheet resistivity is a widely used parameter for comparing films, in particular those films of the same material deposited under similar conditions.

In order to achieve a high resistance value in a small area, it is desirable to have resistor films with high sheet resistance values. On the other hand, one can observe that the higher the sheet resistivity value of a thin film resistor material, the more difficult it is to control the stability. Table 4.6 indicates the specific resistivity of the different tantalum systems.

Resistor patterns can be of rectangular shape or meander shape or a combination of the two patterns. The resistor design must take into consideration the trimming technique that is subsequently used; otherwise an adverse effect can occur on the resistor stability and also the probe pad locations. Figure 4.8 illustrates both the rectangular (or bar) and the meander or zigzag pattern. The meander pattern is characterized by a high square count. Each

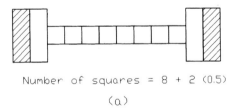

Number of squares = 8 + 2 (0.5)

(a)

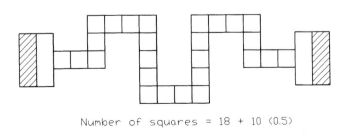

Number of squares = 18 + 10 (0.5)

(b)

Figure 4.8 Resistor designs. (a) Rectangular or bar pattern. (b) Meander pattern.

TABLE 4.6   Specific Resistivity of Various Tantalum Films

Ta film	bcc Ta	$-$Ta	Ta_2N	Ta + O	Ta + O + Ta_2O_3
ρ, $\mu\Omega$ cm	24–50	180–220	240–300	40–300	250–2000

corner of the meander pattern can be considered one-half square owing to current crowding, and the dumbbell terminations each contribute half a square.

With the proper control of deposition processes, thin film resistors can be achieved to a $\pm10\%$ tolerance without trimming. Control of final dimensions is necessary and critical in order to achieve a specific resistor value to tight tolerances. Resistor tolerances as tight as ±0.1 can be attained using thermal stabilization and trim anodization or laser trimming action. Conductor lines possess a minimum linewidth of 50 μm and minimum spacing of 50 μm. The minimum resistor dimensions are on the order of 2 by 2 mils. Conductor-resistor overlap is on the order of 1 mil minimum and resistor underlap is on the order of 2 mils minimum from each side of the conductor. Figure 4.9 and the corresponding Table 4.7 illustrate various thin film design guidelines.

Substrate flatness directly influences the minimum achievable linewidth and spacing. In general, films are tolerant of a surface roughness which is less than the film thickness. When noticeable surface roughness exists, the characteristics of the thin film elements in turn vary. Thin film resistors exhibit an apparent higher sheet resistance value when deposited on rough substrate surfaces. In addition, resistor elements demonstrate more stability on flat surfaces. Rough substrate surfaces result in variations in dielectric films, leading to premature dielectric breakdown.

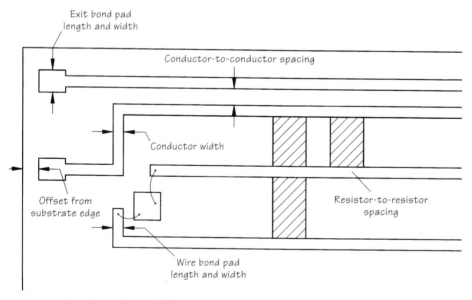

Figure 4.9   Thin film design guidelines.

TABLE 4.7 Thin Film Design Guidelines

Dimension	Minimum value, mils	Typical value, mils
Conductor width	2	10
Conductor spacing (edge to edge)	2	5
Offset from substrate edge	10	12.5
Conductor to resistor spacing	2	5
Exit bond pad width	10	12.5
Exit bond pad length	10	12.5
Wire bond pad (width and length)	5	10
Resistor to resistor spacing	2	5
Resistor width	5	10
Resistor length	5	Can vary

The temperature coefficient of resistance TCR is defined by

$$\text{TCR} = \frac{R(T_2) - R(T_1)}{R(T_1)} \frac{1 \times 10^6}{T_2 - T_1} \quad (\text{ppm/}^\circ\text{C})$$

where $R(T_i)$ is the resistance at temperature T_i. T_1 and T_2 denote two different temperatures. In thin films, the TCR depends on the thermal expansion mismatch between the substrate and the thin film metal. TCR are negative in value, with desired values around -100 ppm/$^\circ$C. Other parameters of interest include the voltage coefficient of resistance and the noise coefficient. Thin films can be thermally stabilized and protected against environmental effects by depositing the proper protective layers to cover the elements and minimize their drift in values.

The design of thin film resistors involves the choice of a suitable film, the determination of the film thickness, the choice of the pattern with the required number of squares, and the selection of the linewidth and spacing that results in an acceptable power density. The choice of film material is based on considerations such as the temperature and voltage coefficients of resistance, the noise coefficient, the stability of resistance with time, the allowable power density, and the method of film deposition.

The power density P_d, of the resistor is a function of the resistor material, the substrate properties and dimensions, any conductive films deposited on the substrate, and external heat sinks. As the power density increases, so does the temperature of the film. This rise in film temperature results in thermal oxidation of the material leading to an increase in the resistance value. To add more complexity to the problem, it is found that the thermal conductivity for most substrate materials is temperature-dependent. In addition, nonuniform deposition of film material results in nonuniform power dissipation in the film. The allowable maximum power density and the power-dissipation requirement of the resistor are used to place a limit on the physical size of the resistor. The required resistor precision puts a lower limit on the linewidth. The variations in the pattern geometry during the deposition process affect the final resistor value.

4.9.4 Resistor passivation

Thin film resistor materials include tantalum films, nickel–chromium (nichrome) films, chromium films, ruthenium films, and cermets. Tantalum nitride films possess higher stability under various environmental conditions. A stabilizing treatment consisting of film anodization, followed by heat treatment and a second anodization at higher voltage, further promotes the stability of these elements. Although nickel–chromium films provide the desired properties of temperature coefficient of resistance, sheet resistivity value, and good adhesion, theses films are known to be unstable under high humidity condition. Therefore, a coating layer (usually silicon monoxide) is provided to maintain acceptable stability of these films.

A technique for passivating thin film hybrids utilizing nickel–chromium resistors includes the use of polyimides.[39] The method consists of spinning the polyimide on the substrate surface and then baking it at a low temperature to form the required film. The low temperature cure is necessary in order to remove the solvents before the high temperature cure. Forming a pattern in this passivation material takes place through the application of positive photoresist, exposure, and development in order to etch the resist as well as the polyimide successfully. The film is then baked at moderate temperature (250 to 300°C) to form the polyimide film. The polyimide can protect thin film resistors satisfactorily under different operating and environmental conditions (under load, high humidity, and high temperature). The polyimide material is found to have no effect on the microwave performance of these thin film hybrid circuits. For ruthenium films, a silicon monoxide protective overcoating helps to assure good resistor stability.

4.9.5 Capacitor design rules

Figure 4.7 illustrates a top view and a cross-sectional view of a thin film parallel plate capacitor structure. The structure is comprised of a dielectric film sandwiched between two conductive layers. The capacitance value C is of the form

$$C = \frac{\varepsilon_0 \varepsilon_r A}{t}$$

where A is the common area of the overlapping electrodes, ε_0 is the permittivity of the free space ($\varepsilon_0 = 8.85 \times 10^{-14}$ F/cm), ε_r is the dielectric constant (relative permittivity), and t is the thickness of the dielectric film. The capacitance density C_D is defined as

$$C_D = \frac{C}{A} = \frac{\varepsilon_r \varepsilon_0}{t}$$

and a maximum value of C_D is desirable. The capacitance density is used to determine the area needed to achieve a certain capacitance value. Fringing effects at the edges of the thin film capacitor are negligible owing to the small thickness of the dielectric compared to the lateral dimensions. An upper limit exists for the capacitance density value achievable for a particular dielectric material. The reason can be attributed to a very thin dielectric film (below 1000 Å) affecting severely the dielectric strength due to pinhole formation and other defects appearing in the thin film. Thus the upper limit for capacitance density is on the order of 0.01, 0.1, and 0.2 $\mu F/cm^2$ for silicon monoxide (SiO), tantalum pentoxide (Ta_2O_5), and titanium oxide (TiO_2) capacitors, respectively.

For most dielectric film materials, the dielectric strength ranges between 1×10^5 and 1×10^7 V/cm. For SiO film, capacitor breakdown has been observed when electric fields ranging between 4×10^5 and 4×10^6 V/cm are applied even for several thousand angstroms dielectric thickness. The presence of defects in the oxide film results in a certain variability of the breakdown field for capacitors deposited on the same substrate. The breakdown field depends also on the substrate material and can be increased significantly by precoating the substrates with a 1-μm layer of SiO. In addition, the metal selected for the capacitor electrodes formation affects the breakdown characteristics of the capacitor structure.

The temperature coefficient of capacitance (TCC) is an important characteristic of the capacitor element. It is of the form

$$\text{TCC} = \frac{1}{C}\frac{\partial C}{\partial T}$$

The dielectric film must be continuous and free of pinholes, thus dictating a lower limit on achievable film thickness. The lower limit for practical evaporated dielectric film is on the order of 3000 Å and for anodic film is on the order of 100 Å. Other capacitor characteristics of interest include stored energy, leakage current, dielectric breakdown, maximum working stress, and dielectric loss. Polarity is a factor of concern in the thin film capacitor design in order to prevent dielectric breakdown under large reverse bias. To realize a nonpolar capacitor structure, two capacitors are formed in series, requiring a large area on the substrate.

Three principal sources of loss exist in thin film capacitors; dielectric loss, electrode resistance, and lead-in resistance. The dielectric loss is independent of the dielectric area and the dielectric thickness. It is also independent of frequency (first degree of approximation). The electrode resistance and lead-in resistance are in series with the dielectric layer. The lead and the electrode resistance can be reduced substantially by contacting the base electrode material on three sides. The dielectric loss is measured in terms of the dissipation factor, defined as tan δ. The angle represents the net phase shift occurring between the voltage applied across the capacitor and the current flowing through the capacitor, with the dielectric material as the medium between the two electrodes. The dissipation factor can be represented by the expression

$$\tan \delta = \omega\, C_s R_s$$

where ω is the angular frequency and C_s represents an ideal capacitor with a zero dissipation factor in series with a resistor R_s. R_s represents the summation of the dielectric loss in addition to electrode resistance and lead-in resistance.

In addition to the parallel plate capacitors, interdigitated capacitors are also commonly used. In this design type, the separation width W is on the order of 25 μm for finger width and spacing. The capacitance expression C is of the form

$$C = (2N_f - 1)K_1(H + 2.1W)$$

where N_f refers to the number of adjacent finger cells. W is the width and spacing of fingers, K_1 is equal to $(\varepsilon_r + 1)(0.05)$ pF/cm, and ε_r is the dielectric constant of the substrate material. Finally, H is the finger cell length. H may be conveniently selected to provide an integer value of N_f. Simple crossovers are achieved by thermocompression wire bonding or performing a ribbon between bonding pads typically 120 by 120 μm.

4.9.6   Inductor design rules

Thin film spiral inductors can be either square or circular in shape. These inductors can be realized by depositing metallization by evaporation or sputtering or even screen printing. The inductance value L of a square spiral is of the form

$$L = 0.0216S^{1/2}N^{5/3} \qquad \text{nH}$$

where S refers to the surface area of the coil in mils^2 and N is the number of turns. The inductance value L of a circular spiral is of the form

$$L = \frac{0.8a^2 N^2}{6a + 10c} \qquad \text{nH}$$

where N denotes the number of turns of the coil, a represents the average radius ($a = d_o/4 + d_i/4$), and c represents the breadth of the coil ($c = d_o/2 - d_i/2$). The terms d_o and d_i refer to the outer and the inner diameters of the coil, respectively. The quality factor Q of a thin film spiral inductor is of the form

$$Q = \omega L_s\left[1 - \left(\frac{f}{f_r}\right)^2\right]\Big/R_s$$

where $\omega = 2\pi f$, f is the frequency of interest, ω is the low-frequency inductance value, R_s is the series resistance of the spiral inductor, and f_r is the self-resonant frequency given by

$$f_r = \frac{1}{2\pi \sqrt{L_s C_s}}$$

where C_s is the distributed capacitance value, and L_s is the distributed inductance value.

4.9.7 Transmission lines

Transmission lines are used to carry information or energy from one location to another in microwave circuits. Passive elements in conventional microwave circuits are commonly distributed and use sections of transmission lines and waveguides. Transmission lines are characterized in terms of the characteristic impedance Z_0, the ratio of the voltage to the current at any point along a perfectly terminated line, the complex propagation constant γ, which is comprised of a real part representing the attenuation constant α, and an imaginary part representing the phase constant β. Bahl and Bhartla[28] illustrate various transmission structures for microwave integrated circuits (MICs). The stripline, the microstrip line, the slot line, and the coplanar line represent planar transmission lines. Expressions for the characteristic impedance and the attenuation constant of these lines are summarized in Ref. 28.

4.10 Fabrication Sequence for Thin Film Resistor-Conductor Circuits

The realization of planar resistors consists of depositing thin films of lossy metal on a dielectric base. Nickel–chromium (nichrome) and tantalum nitride are the most popular and useful film materials to realize thin film resistors. Consider a structure consisting of a tantalum–nitride resistive film and a conductor structure composed of titanium, palladium, and gold, respectively.

4.10.1 Substrate preparation

For glass and ceramic glazed substrates, a tantalum oxide film is placed between the substrate and the resistive film (resistant to hydrofluoric acid as tantalum etchant). This is achieved by sputtering a thin layer of tantalum which is then oxidized in an oven under oxygen atmosphere at $\sim 400°C$.

The substrate is then coated with resistive and conductive films. A tantalum–nitride resistive film is sputtered onto the ceramic substrate (300 to 1500 Å thick). The resistive film is defined through photolithography to obtain various resistor patterns. Successive evaporation of 1000 to 2000 Å of titanium (to provide good adhesion of the palladium–gold conductor film to the tantalum–nitride resistive film), 2000 Å of palladium (to act as a diffusion barrier to separate the titanium and gold), and 13,000 to 18,000 Å of gold follows. Again, definition of the metallization is realized. Sometimes, electroplating is used to deposit the gold when a thickness on the order of 50,000 Å is needed for high conductivity metallization runs. An alternate method commonly used would be to deposit both the resistor and metallization layers successively followed by etching these layers using photolithographic technique.

4.10.2 Photolithographic techniques

The selective removal of a thin film, in order to form a structure or pattern in the film, involves the use of photolithographic techniques. The film to be patterned is deposited over the entire substrate and a photosensitive polymer is applied to the surface by immersion, spraying, or spinning. There are two types of photosensitive polymer or photoresist: positive and negative. For a positive photoresist, the exposed regions can be removed in a developer solution. For a negative photoresist, the unexposed regions are removed by the developer. The photoresist undergoes a chemical change when exposed to light. A pattern is developed in the resist and the exposed or unexposed film on the substrate is then etched away in a suitable solution. A mild heat treatment, called a soft bake, is normally used to improve the etch resistance of the photoresist. Appendix B, a table titled "Metal Etchants for Thin Film Materials," includes a list of etchants for some common film materials. This table is found at the end of this chapter.

The master artwork of the desired pattern can be generated using a computer-aided design (CAD) package that can be developed reasonably in house. Precise photographic techniques are required to generate the positive or negative copy of the desired pattern. The layout of the desired circuit is cut out by a plotter with a diamond scriber of a two-layer Mylar material (the lower clear Mylar layer acts as a support and the upper layer which is optically opaque is a ruby-colored Mylar film cut and peeled where the thin film components are to be located). The master artwork is 20 to 40 times larger than the final desired configuration. High precision photoreduction methods are used to obtain final positive or negative glass or film masks to the desired dimensions.

4.10.3   Pattern generation and etching

1. The photoresist (either positive or negative) is applied and the conductor pattern is defined. The exposed conductor material is removed by selective etching.
2. Next, the photoresist is stripped. A new layer of photoresist is applied and exposed to define the combined resistor-conductor pattern.
3. The resistive film is then etched and the photoresist is stripped.

The selective removal of layers of the composite thin film is achieved by either immersion or spray etching. Etchants used for thin films are usually more dilute than those for bulk materials (to reduce etch rate and to minimize attack of the photoresist and the film under it).

4.10.4   Adjustment of final value or trimming

Resistor trimming can be achieved using several methods. These methods include anodization, heat treatment, electrical method, and mechanical method, chemical method, and laser method.

1. *Anodization:* In trim anodizing or electrochemical technique in a thin film resistor, the thickness of the conductive metal is uniformly reduced by the growing a layer of insulating tantalum oxide at the expense of the parent

film in the tantalum system. Thus the effective thickness of the film is reduced. The anodic film is also used as the dielectric film for thin film capacitors realization. The minimum anodization voltage for tantalum resistors is 30 V. Tantalum oxide grows at the rate of 17 Å/V while consuming 6.3 Å/V of tantalum. The anodization is conducted at a constant current until the proper or exact resistance value is measured. The resistor is anodized at one value of current until the resistance value is close to the final value. Then the current is reduced to a lower value to reduce the anodization rate. The process is repeated until the desired tolerance is achieved.

Successive anodizing and measurement of the resistor value are achieved. The resistance of the film must be measured between anodization steps. Resistive films are made at least 5% lower in value than specified. Low tolerances of 0.01% can be obtained. Anodization is the preferred method for trimming tantalum film resistors since it possesses the advantage of providing a protective layer on top of the resistor surface. The interference color of the film changes as the anodization progresses. Thus, if the oxide layer formed uniformly, it shows an even color. Large resistance values may cause anodization to form a thicker oxide layer near the termination connected to the power supply. A lower current density and slower rate of anodization usually overcome the occurrence of the problem.

2. *Heat Trimming:* Heat trimming involves localized heating to adjust the resistance of the film. The process involves heating in the air at temperatures of several hundred degrees Celsius. This process is a treatment used for the stabilization of resistance values as well as trimming. The method is not a very precise one. Heat sources include pulsed lasers, infrared lamps, and carbon arcs.

3. *Electrical Trimming:* Electrical trimming involves an appropriate current passing through the resistor film generating heat, which trims the resistor to a closer resistance value.

4. *Mechanical Trimming:* Mechanical trimming involves removal of a portion of the resistor material by sandblasting or vaporization or cutting without causing any weak spots where local heating may occur. Usually the mechanical trim involves the reduction of the linewidth of the resistor uniformly along its length.

5. *Chemical Trimming:* Chemical trimming includes painting with various conductive inks. In this case, the resistance value decreases in value by this trimming process.

6. *Laser Trimming:* In this trimming method, the film properties are unchanged, but the geometry varies. The bar, top hat, ladder, and loop resistor designs represent different shapes for laser trimming.

Trimming a thin film capacitor structure is concerned with the ability of controlling the capacitance density of the structure and the overlap area of the electrodes. These two factors constitute the determining factors deciding an achieved capacitance value. As an example, the capacitance density of a tantalum oxide structure can be controlled accurately by controlling the thickness of the anodic oxide through the anodization voltage control. The overlap area of the capacitor electrodes depends upon the control of the width of each of the electrodes and their precise alignment with respect to each other.

Tolerances as tight as ±0.1% can be achieved for thin film capacitors by adjusting the capacitance value in discrete steps. The pattern type consists of a main capacitor in parallel with a set of binary-weighted trimming capacitors which can be removed by the trimming process. Laser trimming of the final value of an interdigitated capacitor may be accomplished by cutting through an exact number of fingers.

4.10.5   Thermal stabilization

The discontinuous transition between the different materials is normally proved to be unstable. By thermal oxidation consisting of heating and aging, oxygen atoms diffuse into the tantalum at the metal–oxide interface, producing oxygen-doped tantalum and tantalum-rich tantalum oxide. Thus a continuous transition results between the different materials. A relatively small increase in the resistance value occurs (the conductivity of tantalum at the interface is reduced because of the oxygen diffusion into the tantalum). A trim-anodization step is necessary in order to adjust the value to a tight tolerance.

4.10.6   Protection of thin film resistors

Tantalum pentoxide, Ta_2O_5, grown by anodization, forms a hard, glassy, uniform film over the resistor element in order to provide protection from damage during handling and from deterioration of the resistor material by heat and/or other factors.

4.11   Characterization of Thin Films

The extensive use of thin films in microelectronics applications has necessitated a careful study of the intrinsic nature of these films. Available characterization techniques range from providing information on the first few atom layers of the surface to probing more deeply in the film.

4.11.1   Film thickness determination

Thin film properties and performance depend greatly on the thickness of the thin film produced during various deposition techniques as well as the deposition conditions. Different optical and mechanical methods are available for film thickness measurement. Optical techniques usually rely on the interface of two light beams with different optical paths, owing to the resultant film thickness, and these techniques are usually nondestructive in nature. These films can be either opaque or transparent in nature.

Ellipsometry is used to measure the thickness of thin films (as thin as 10 Å) as well as to determine the optical constant of both films and substrates. The technique is based on measuring the state of polarization of light reflecting at an incidence (nonnormal) from the surface of the film.

The mechanical method for thickness measurement depends on a stylus undergoing a vertical movement to trace the film surface and to relate that surface to the film-substrate step. This technique is helpful with very smooth substrate surfaces possessing minimum surface roughness. The film thickness to be measured lies in the range 100 Å to 100 μm, with a limited resolution of 10 Å.

4.11.2 Adhesion measurement

The film must possess adhesion to the substrate; that is, it must develop a strong interfacial atomic bond, and this electronic and chemical interaction at the film–substrate interface generally affects the film properties and stability. The adhesion depends on the substrate morphology, the various chemical interactions occurring at the interface, the diffusion rates, the nucleation processes, and certainly the nature of the interface. Film adhesion ranges from a low adhesion strength as in the case of an abrupt interface (with a change from the substrate material to the film occurring within a few angstroms) to a medium adhesion strength as in the case of a compound interface (an interface characterized by a chemical reaction and diffusion occurring between the substrate and the film within a certain distance exceeding a few angstroms), finally to strong adhesion strength as in the case of a diffusion interface (an interface with a gradual change in the composition occurring between the film and the substrate). Adhesion tests include tensile, shear, and scratch methods.[29,30] Tensile testing consists of a direct pull of an adhesive material, which is attached to the film, with a certain measurable applied force to result in an interfacial separation. Shear testing is conducted with some sort of adhesive tape which is pulled away from the film with a certain rate and angle. Scratch tests determine the adhesion by measuring how easily the film can be scratched away.

4.11.3 Structural characterization

The structural characterization includes both lateral and vertical dimension determination, thickness uniformity, etching uniformity, grain size, existence of film voids, lack of adhesion, presence of microcracks, and interfacial regions for multilayer structures.

The scanning electron microscope (SEM) offers high magnification with good resolution combined with a decent depth of field. The SEM uses an electron beam to produce an enlarged picture of the sample in order to examine the surface or to conduct cross-sectional analysis to determine all dimensions accurately. The SEM is based essentially on three principal types of electron microscopes: scanning, transmission, and emission. In both the scanning and transmission electron microscope, an image is produced as a result of an electron beam incident on the sample. On the other hand, in the emission micro-scope, the sample itself constitutes the electron source.[31, 32] Most energy-dispersive x-ray analyses are interfaced to SEMs. In this case, the electron beam can serve to excite x-rays from the probed specimen area.

4.11.4   Chemical characterization

A focused accelerated electron beam, of certain energy incident on a sample surface, can result in incident electrons being absorbed, emitted (an example is Auger electron spectroscopy, or AES), reflected (an example is scanning electron microscopy, or SEM), or transmitted (an example is transmission electron microscopy, or TEM) and can cause light or x-ray emission. Electrons emitted can be categorized as low energy or secondary electrons, or low yield or backscattered electrons. The environmental surrounding for these various techniques is typically a high vacuum level. The detection limit or sensitivity varies a great deal with the beam current and is different from one technique to another. Ohring[33] provides a summary of major chemical characterization techniques.

Auger electron spectroscopy (AES), x-ray photoelectron spectroscopy (XPS), also known as electron spectroscopy for chemical analysis (ESCA), and secondary ion mass spectrometry (SIMS) or ion microscopy are powerful surface analytical techniques (typical 50 Å depth) to enable researchers to study the chemical and compositional properties of thin film materials.[38] Deeper probing is possible with these techniques by sputter etching the film and analyzing the resultant exposed surface. These techniques are capable of detecting the presence of all known elements in nature (except hydrogen and helium with the AES technique). In the AES technique, an electron from the K shell is injected by a primary electron. An outer shell electron L_1 fills the resultant K-shell vacancy. The energy difference $E_{L1} - E_K$ is transferred to a third electron, which is referred to as the Auger electron. It appears at the $L_{2,3}$ level. The Auger electrons emitted from the sample upon which an electron beam was incident are plotted as a function of energy, or in some cases the signal differentiation is plotted in order to improve the detection limit to 0.1%. Data interpretation to identify the various elemental species or elements is possible by available database published in the literature.[34,35] The Auger technique is a three-electron process, referring to the $KL_1L_{2,3}$ transition occurring in the sample.

Ion microscope, or SIMS, is another powerful and highly sensitive analytical technique capable of detecting all elements and molecular species, with resolution depth on the order of 50 Å. A destructive removal of material from the sample by ion bombardment or sputtering neutral atoms to the surface occurs, followed by a thorough analysis of this removed material by a mass spectrometer or analyzer.[36] This technique is mainly used to identify dopants in semiconductor materials, or inclusion in thin films. Dynamic SIMS enables a depth profiling of surface layers.

XPS is used to identify and detect virtually all chemical species at the sample surface (except hydrogen and helium). Photoemission from the sample occurs when x-ray energy exceeds the excitation energy. Although x-rays penetrate deeper in the sample, emitted photoelectrons result from the surface layer of that sample (10 to 50 Å). Depth profiling is realized by ion beam sputtering.

Rutherford backscattering spectrometry (RBS) presents different surface analytical techniques. The RBS technique is based on backscattering of energetic ions such as helium ions incident on a sample. This technique allows researchers to measure the mass of the different elements as a function of depth into the sample. This information allows the determination of the structure or stoichiometry of the film.[37]

4.11.5 Modifications of surfaces and films

The objective of surface and film modification is to improve the surface and film properties. This objective can be achieved in terms of film and coating deposition or modification of existing surfaces through the use of directed energy sources.

The chemical vapor deposition (CVD) process permits the deposition of thin films of diamond on a variety of materials used as cutting tools and tribology, thus benefiting from diamond superior resistance to wear and corrosion as well as its low coefficient of friction.[42,43] In addition, a deposited diamond film on the back of wafers and substrates can greatly improve the thermal management of thin film hybrid circuitry built on these surfaces, owing to the high thermal conductivity of diamond.[40]

Banks[41] illustrates the effects of ion beam techniques on a variety of materials. He provides a summary of surface texturing technologies or surface modification techniques such as ion beam sputter texturing, sputter etching, etching, and simultaneous deposition and etching of surfaces, arc texturing technology, and atomic oxygen texturing in order to improve the surface properties of these materials. Ion surface interactions through an ion implantation process can be used to accelerate nitrogen or carbon ions toward metal surfaces in order to improve their properties.[42] These charged atoms are embedded beneath the surface at a certain depth upon colliding with the surfaces, producing a new alloy in the implanted region with desired properties of enhanced wear resistance and improved corrosion resistance. Steels, stainless steels, carbides, and aluminum are among the metals that can experience an ion implantation process for better performance. Ion implantation has been also successful with semiconductors to produce a film layer resistant to formation of defect-free materials. These implants can be followed by thermal annealing.

4.12 Emerging Technology: Multichip Module Deposited (MCM-D)

4.12.1 Concept

MCM-deposited (MCM-D) has been defined as multichip modules that are constructed using unreinforced low dielectric constant (k<5) materials that are adjacent to the signal plane, with base substrates required for dimensional stability that include ceramic, silicon, copper, glass-reinforced laminate, metal, or metal composites. Conductors are fabricated by sputtering or plating metals such as aluminum, copper, gold, or palladium onto the low dielectric materials, and the metals are subsequently patterned with photolithography techniques. Cost, conductivity, and material compatibility are the prime factors defining a specific metallization. Currently, MCM-D offers the capability of producing the highest circuit density and smallest size than other multichip module techniques. On the other hand, the processing techniques and materials used to construct these modules make it a costly technology. The following sections describe the existing materials and processes for MCM-D technology, as well as recent advances in materials and processes that include the use of photodefineable polyimides, integrating components within the base substrate, as well as the use of advanced materials such as diamond and high temperature superconductors.

4.12.2   Overall process sequence

MCM-D can be fabricated using either the "chips-first" technique or the "chips-last" technique. Chips are mounted in cavities in the base substrate using the chips-first technique, then dielectric and metallization layers are subsequently deposited to form interconnects. With the chips-last technique, the dielectric layers and metallization layers are deposited first on the base substrate, and chips are mounted last to interconnects on the top layer.

Minimum linewidth and spacing is on the order 0.06 mm (0.0025 in) with a nominal value on the order 0.25 mm (0.010 in), and the metallization is kept at a minimum distance on the order 0.25 mm (0.010 in) from the edge of substrate. The dielectric constant of the substrate material is on the order of 8, while the dielectric constant of the separating dielectric layers in the range 3–6. In general, a final thickness of gold metallization ranging between 0.0025 and 0.005 mm (0.0001–0.0002 in) is needed to provide the proper conductivity figure and the wire bonding capability. The dielectric layers are deposited by spray or spin coating. Vias can be produced by wet etch, laser patterning, or through the use of photosensitive polyimides. Vias can be filled with metals using wet chemical plating. Metal layers can be applied by electroplating, sputtering, evaporation, and enhanced ion plating (EIP), and are patterned using reactive ion etch (RIE), or wet etch (for nonphotosensitive materials). Photosensitive dielectrics can be used to eliminate many processing steps, and only involve applying and developing the dielectric material through photolithography and then applying metallization.

4.12.3   Substrates

Base substrates should possess such properties as high thermal conductivity, good electrical insulation, high mechanical strength, high temperature resistance, low weight, and minimal cost. A substrate with a coefficient of thermal expansion (CTE) close to that of silicon is usually desirable for flip chip bonding or attachment of large dice. The most commonly used substrates tend to be ceramics, metals, silicon, copper, glass reinforced laminates, or metal composites.

Ceramic substrates include alumina (Al_2O_3), aluminum nitride (AlN), and beryllia (BeO). Alumina is widely used and can be found in many sizes and thicknesses. Aluminum nitride offers improved heat dissipation over alumina; also, it offers CTE matched materials and possesses a high bending strength and a high strength. Aluminum nitride is not as readily available as alumina and it is also at a higher cost. Similar to aluminum nitride, beryllia also offers an improved heat dissipation over alumina for efficient heat removal from the substrate. Beryllia is also harder to obtain than alumina and is at significantly higher costs, but the most significant disadvantage is that it is highly toxic if handled incorrectly.

Metal substrates are mostly suitable for harsh environments. They possess high strength, are resistant to shock, and also act as an effective ground plane. Metal substrates also possess a low cost and a good thermal conductivity. Copper is as a good metal substrate for MCM-D applications. Copper is usually alloyed with other materials to obtain a good CTE match to silicon, and it can be coated with an insulating layer to overcome reactive behavior with most metal etchants.

Recent trends indicate that silicon is the substrate of choice for applications requiring extremely high circuit density. Higher circuit density is acquired by incorporating components into the substrate itself. Modules have been fabricated with resistors, capacitors, diodes, npn transistors, and pnp transistors. Having the capability to incorporate components into the substrate increases significantly circuit density and frees up vital surface areas for conductor line patterns. Surface areas is crucial because of the limited number of conductor layers that can be processed (~5–10 layers). In addition, silicon has a high thermal conductivity and it is stable, allowing extreme fine lines. The silicon low coefficient of thermal expansion can result in warpage when polyimides are used.

Advanced research in substrate materials has been in the area of chemical vapor deposition (CVD) diamond films.[43,44] Synthetic diamond substrates have the capability of offering improved properties over alumina, aluminum nitride, and beryllia substrates; these properties include a lower dielectric constant, higher thermal conductivity, reduced chemical susceptibility, and a good CTE match with silicon.

The thermal conductivity is the greatest asset of diamond substrates, with values in excess of 2000 W/m•K. Incorporating diamond substrates into MCM-D technology can greatly improve the thermal management performance by enhancing heat removal from chips.

Synthetically produced chemical vapor deposition (CVD) diamond substrates with high figure of thermal conductivity have addressed potential problems in large size MCM applications, namely thermal management problem with heat spreading and dissipating large transient power pulses as well as removing large steady heat loads. Reliable metallization processes for chemical vapor deposition of diamond (CVDD) have been realized. These methods include transition metal bond coats, diffusion barriers, patterning, and solder die attach. Diamond substrates can also be doped to have semiconductor properties that provide the capability to integrate components directly into the substrate. Because diamond is expensive, diamond thin films can be used as an alternative at a reduced cost. Diamond thin films can be applied to a highly thermal conductive substrate, such as AlN, in order to improve heat dissipation, but at a much smaller cost than using diamond substrates.

Diamond materials have disadvantages that are currently under investigation. The coefficient of thermal expansion for diamond is considerably different than gallium arsenide or indium phosphide and structural integrity problems can occur when a module fabricated with these materials is subjected to temperature extremes. Despite this disadvantage, diamond possesses characteristics that can greatly enhance the MCM-D technology performance.

4.12.4   Dielectrics

Dielectrics for use in MCM-D applications should possess a low dielectric constant, low loss tangent, low moisture absorption, good adhesion to the metallization and base substrate, good planarization, high thermal conductivity, and a coefficient of thermal expansion comparable to that of the die and base substrate. Polyimides and benzocyclobutenes (BCBs) possess these vital requirements and are often used for MCM-D applications.

Most of recent research work has involved polyimides; fluorinated polyimides; low stress polyimides with low CTE value in the xy direction, low dielectric constant value, and low water absorption, but unfortunately low adhesion to other materials; photosensitive polyimides to reduce the number of processing steps; and polyphenylquinoxaline (PPQ). Conventional processing techniques, including dry etching, wet etching, and reactive ion etching (RIE), have been simplified by the development of photosensitive polyimides.

This comparison indicates that photosensitive polyimide processing requires fewer steps than other techniques, therefore reducing considerably processing time and cost. To address current problems including corrosion caused by direct use of copper over polyimides when deposited from acid base, five layers of metallization deposition is used. An additional problem involves proper adhesion of copper to polyimides, and an adhesion promotion is necessary in this case.

Benzocyclobutane (BCBs) have basically the same properties as polyimides. BCBs possess low dielectric constant values and low loss factors; they also have a low moisture absorption and adhere well to ceramic, copper, and aluminum, and have superior surface smoothness. High density multilayer interconnect is usually formed over a silicon substrate with BCB as the dielectric material. Metal layers are deposited, followed by photoresist patterning, and etch, and then a coating of BCB polymer, and etching of vias in the polymer by reactive ion etch after photolithographic definition of the vias. Conventional processing techniques, however, have limited BCB advances in MCM-D applications. Limitations include methods with processing vias, which require the use of metal or inorganic masks, and processing methods that use photoresist or wet etch methods, which can be complicated, time consuming, and costly. Recent improvements with processing techniques have been accomplished, reducing considerably the curing time.

The most recent advances have been in the development of photosensitive BCBs. Photosensitive BCBs transcend some of the limitations that exist with previous BCBs by incorporating photolithography and wet etching techniques, and even have improved properties over polyimides, such as a lower conductor resistance, lower propagation delay, and improve planarization.

4.12.5 Metallization

Three types of metallization are used in conjunction of MCM-D technology: copper, gold, and aluminum. Copper is the most widely used because of low cost, high stability, and high conductivity. The disadvantage of copper is that it is not compatible with many polyimides, therefore a buffer layer must be applied between the copper and the polyimide for good adhesion and good interface. This additional layer increases the number of processing steps in order to fabricate the module. Aluminum is also widely used, although not as frequently as copper. Aluminum is compatible with most polyimides, but its conductivity is lower than that of copper. Gold is an alternative to copper and aluminum, possessing a good conductivity and good stability, but at higher cost.

4.13 Conclusion

This chapter describes the basic processes in thin film technology and the properties of materials used in thin film applications for the formation of passive electrical elements, as well as providing design guidelines for thin film realization in microelectronics. It also provides a brief description of various characterization techniques to evaluate thin films, and a description of emerging thin film materials. One important conclusion associated with this technology is that deposition conditions are critical since they determine the properties of the resulting film, which is grown by a process involving individual atoms or molecules.

For multichip module deposited (MCM-D) technology, the high cost of fabricating MCM-Ds has limited its use to high performance and high speed applications such as computer workstations, and military applications. Current applications mostly use photodefineable polyimides for the dielectric material with copper metallization although aluminum metallization is occasionally used as an alternative. A variety of substrates are being used in current applications that include aluminum, alumina, aluminum nitride, and silicon. Silicon substrates have been the choice for high speed applications because of the capability of integrating components into the substrate.

Appendix A, a table titled "Deposition Techniques,"[5] and Appendix B, a table titled "Metal Etchants for Thin Film Materials," follow on pages 4-42 through 4-53.

MATERIAL	SYMBOL	MP°C	S/D	g/cm³	10^{-8}T	10^{-6}T	10^{-4}T	E-BEAM	CRUCIBLE	COIL	BOAT	BASKET	SPUTTER	COMMENTS
Aluminum	Al	660		2.70	677	821	1010	Ex	TiB2BN	W	TiB2BN Al2O3	W	RF DC	Alloys and wets. Stranded W is best.
Aluminum Antimonide	AlSb	1080		4.3	–	–	–	–	–	–	–	–	RF	–
Aluminum Arsenide	AlAs	1600		3.7	–	–	1300	–	–	–	–	–	RF	–
Aluminum Bromide	AlBr3	97		3.01	–	–	50	–	Gr	–	Mo	–	RF	–
Aluminum Carbide	Al4C3	1400		2.36	–	–	800	F	–	–	–	–	RF	n = 2.7
Aluminum 2% Copper	Al2%Cu	640		2.82	–	–	–	–	–	–	–	–	RF, DC	Wire feed and flash. Difficult from dual sources.
Aluminum Fluoride	AlF3	1257	S	3.07	410	490	700	P	Gr	–	Mo,W,Ta	–	RF	Decomposes. Reactive evap. In 10^{-3} N2 with glow discharge.
Aluminum Nitride	AlN		S	3.26	–	–	1750	F	–	–	–	–	RF RF-R	
Aluminum Oxide	Al2O3	2045		3.97	–	–	1550	Ex	–	–	W	W	RF-R	Sapphire excel. In EB, forms smooth hard films. n = 1.66
Aluminum Phosphide	AlP	2000		2.42	–	–	–	–	–	–	–	–	RF	
Aluminum 2% Silicon	Al2%Si	640		2.69	–	–	1010	–	TiB2BN	–	–	–	RF DC	Wire feed and flash. Difficult form dual sources.
Antimony	Sb	630		6.68	279	345	425	P	BN, C, Al2O3	Mo,Ta	Mo*** Ta***	Mo Ta	RF DC	Toxic. Evaporates well.
Antimony Telluride	SB2Te3	619		6.50	–	–	600	–	C	–	–	–	RF	Decomposes over 750°C.
Antimony Trioxide	Sb2O3	656		5.2 or 5.76	–	–	300	G	BN, Al2O3	–	Pt	Pt	RF-R	Toxic. Decomposes on W. n = 1.85
Antimony Triselenide	Sb2Se3	611		–	–	–	–	–	C	–	Ta	–	RF	Stoichiometry variable.
Antimony Trisulphide	Sb2S3	550		4.64	–	–	200	–	–	–	Mo,Ta	Mo, Ta	–	No decomposition. n = 3.01
Arsenic	As	814	S	5.73	107	150	210	P	Al2O3, BeO, VC	–	C	–	–	Toxic. Sublimes rapidly at low temperature.
Arsenic Selenide	As2Se3	360		4.75	–	–	–	–	–	–	–	–	RF	–
Arsenic Trisulphide	As2S3	300		3.43	–	–	400	F	Q	–	Mo	–	RF	n = 2.7
Barium	Ba	710		3.5	545	627	735	F	Metals	W	W,Ta,Mo	W	RF	Wets w/o alloying–reacts with ceramics.
Barium Chloride	BaCl2	962		3.86	–	–	–650	–	–	–	Ta	–	RF	Gentle preheat to outgas.
Barium Fluoride	BaF2	1280		4.83	–	–	–700	–	–	–	–	–	RF	n = 1.47
Barium Oxide	BaO	1923		5.72 or 5.32	–	–	1300	P	Al2O3	–	Pt	Pt	RF,RF-R	Decomposes slightly. n = 1.98
Barium Sulphide	BaS			4.25	–	–	1100	–	–	–	–	–	RF	n = 2.16
Barium Titanate	BaTiO3		D	6.0	–	–	–	–	Q	–	Mo	–	RF	Gives Ba. Co-evap from 2 sources or sputter.
Beryllium	Be	1278		1.85	710	878	1000	Ex	BeO, C, VC	W	W,Ta	W	RF DC	Wets W/Mo/Ta. Metal powder & oxides toxic. Evap. easily.
Beryllium Chloride	BeCl2	440		1.90	–	–	–150	–	–	–	–	–	RF	
Beryllium Fluoride	BeF2	1280	S	1.9	–	–	–200	G	–	–	–	–	–	Toxic
Beryllium Oxide	BeO	2530		3.01	–	–	1900	G	–	–	–	W	RF, RF-R	Toxic. No decomp. from EB guns. n = 1.72
Bismuth	Bi	271		9.80	330	410	520	G	Al2O3, VC	W	W,Mo,Ta	W	DC RF	Toxic vapor. Resistivity high. No shorting of baskets.

*Influenced by composition **All metals alumina coated ***C-plated rod or strip S = Sublimes D = Decomposes
Q = Quartz Incl = Inconel VC = Vitreous Carbon SS = Stainless Steel C = Carbon Gr = Graphite P = Poor
RF = RF sputtering is effective RF-R = Reactive RF sputtering is effective DC = DC sputtering is effective Ex = Excellent F = Fair DC-R = Reactive DC sputtering is effective

MATERIAL	SYMBOL	MP°C	S/D	g/cm³	Vap. Press. @ temp°C			EVAPORATION TECHNIQUES					SPUTTER	COMMENTS
					10^{-8} T	10^{-6} T	10^{-4} T	E-BEAM	CRUCIBLE	COIL	BOAT	BASKET		
Bismuth Fluoride	BiF3	727	S	8.75	-	-	-300	-	-	-	-	Pt	RF	Sublimes
Bismuth Oxide	Bi2O3	820		8.9	-	-	-1400	P	-	-	Pt	Pt	RF, RF-R	Toxic vapor n = 1.97
Bismuth Selenide	Bi2Se3	710		7.66	-	-	-650	G	Gr, Q	-	-	-	RF	Co-evap. from 2 sources or sputter.
Bismuth Telluride	Bi2Te3	585		7.85	-	-	-600	-	Gr, Q	-	W,Mo	-	RF	Co-evap. from 2 sources or sputter.
Bismuth Titinate	Bi2Ti2O7	-	D	-	-	-	-	-	-	-	-	-	RF	Sputter or co-evap from 2 sources in 10^{-2}T O2.
Bismuth Trisulphide	Bi2S3	685		7.39	-	-	-	-	-	-	-	-	RF	n = 1.5
Boron	B	2100	S	2.36	1278	1548	1797	Ex	C, VC	-	C	-	RF	Explodes w/rapid cooling. Forms carbide with container.
Boron Carbide	B4C	2350		2.50	2500	2580	2650	Ex	-	-	-	-	RF	Similar to chromium.
Boron Nitride	BN	2300	S	2.20	-	-	1600	P	-	-	-	-	RF, RF-R	Decomposes. Reactive sputter prefered.
Boron Oxide	B2O2	460		1.82	-	-	1400	G	-	-	Pt,Mo	-	-	n = 1.46
Boron Trisulphide	B2S3	310		1.55	-	-	800	-	Gr	-	-	-	RF	—
Cadmium	Cd	321		8.64	64	120	180	P	Al2O3, Q	-	W,Mo,Ta	W,Mo,Ta	DC RF	Bad for vacuum systems. Low sticking coeff.
Cadmium Arsenide	Cd3As2	721		6.21	-	-	-	-	Q	-	-	-	RF	—
Cadmium Fluoride	CdF2	1070		6.64	-	-	-500	-	-	-	-	-	RF	n = 1.56
Cadmium Oxide	CdO	900		6.95	-	-	-530	-	-	-	-	-	RF-R	Disproportionates. n = 2.49
Cadmium Selenide	CdSe	1264		5.81	-	-	540	G	Al2O3, Q	-	Mo,Ta	-	RF	Evaporates easily. n = 2.4
Cadmium Sulphide	CdS	1750	S	4.82	-	-	500	G	Al2O3, Q	-	W,Ta	W	RF	Sticking coeff. affected by substrate temp. Stoichiometry variable. n = 2.4
Cadmium Telluride	CdTe	1098		6.20	-	-	450	-	-	-	W,Mo,Ta	Ta,Mo	RF	Stoichiometry depends on substrate temperature. n = 2.6
Calcium	Ca	842	S	1.55	272	357	459	P	Al2O3, Q	W	W	W	-	Corrodes in air.
Calcium Fluoride	CaF2	1360		3.18	-	-	-1100	-	Q	W,Mo,Ta	W,Mo,Ta	W,Mo,Ta	RF	Rate control important. Preheat gently to outgas. n = 1.23
Calcium Oxide	CaO	2580		3.35	-	-	-1700	-	ZrO2	-	W	-	RF RF-R	Forms violatile oxides with W & Mo. n = 1.84
Calcium Silicate	CaOSiO2	1540		2.90	-	-	-	G	Q	-	-	-	RF	n = 1.61
Calcium Sulphide	CaS	-		2.18	-	-	1100	-	-	-	Mo	-	RF	Decomposes. n = 2.14
Calcium Titanate	CaTiO3	1975		4.10	1490	1600	1690	P	-	-	-	-	RF	Disproportionates except in sputtering.
Calcium Tungstate	CaWO4	-		6.06	-	-	-	G	-	-	W	-	RF	n = 1.92
Carbon	C	-	S	1.8-2.3	1657	1867	2137	Ex	Al2O3, BeO, VC	-	-	-	RF	EB preferred. Arc evaporation. Poor film adhesion.
Cerium	Ce	795		8.23	970	1150	1380	G	-	W	W	W,Ta	DC,RF	—
Ceric Oxide	CeO2	2600	S	7.3	1890	2000	2310	G	-	-	-	-	RF, RF-R	Very little decomposition.
Cerium Fluoride	CeF3	1418		6.16	-	-	-900	G	-	-	W,Mo,Ta	Mo,Ta	RF	Preheat gently to outgas. n = 1.7
Cerium Oxide	Ce2O3	1692		6.87	-	-	-	F	-	-	W	-	-	Alloys with source. Use 0.015"-0.020" W boat. n = 1.95
Cesium	Cs	28		1.87	-16	+22	+30	-	Q	-	SS	-	-	n = 1.95

* Influenced by composition   ** Cr-plated rod or strip   *** All metals alumina coated   C = Carbon   Gr = Graphite   S = Sublimes   D = Decomposes
Q = Quartz   Incl = Inconel   VC = Vitreous Carbon   SS = Stainless Steel   Ex = Excellent   F = Fair   P = Poor
RF-R = Reactive RF sputtering is effective   RF = RF sputtering is effective   DC = DC sputtering is effective   DC-R = Reactive DC sputtering is effective

MATERIAL	SYMBOL	MP°C	S/D	g/cm³	Vap. Press. @ temp°C 10⁻⁸T	10⁻⁶T	10⁻⁴T	EVAPORATION TECHNIQUES E-BEAM	CRUCIBLE	COIL	BOAT	BASKET	SPUTTER	COMMENTS
Cesium Bromide	CsBr	636	–	4.44	–	–	~400	–	–	–	W	–	RF	n = 1.70
Cesium Chloride	CsCl	646	–	3.97	–	–	600	–	–	–	W	–	RF	n = 1.64
Cesium Fluoride	CsF	684	–	3.59	–	–	500	–	–	–	W	–	RF	–
Cesium Hydroxide	CsOH	272	–	3.67	–	–	550	–	–	–	Pt	–	–	–
Cesium Iodide	CsI	621	–	4.51	–	–	500	–	–	–	W	–	RF	n = 1.79
Chiolote	Na$_3$Al$_3$F$_{14}$			2.9	–	–	~800	–	–	–	Mo,W	–	RF	n = 1.33
Chromium	Cr	1890		7.20	837	977	1157	G	VC	W	**	W	RF, DC	Films very adherent. High rates possible.
Chromium Boride	CrB	2760		6.17	–	–	–	–	–	–	–	–	RF, DC	–
Chromium Bromide	CrBr$_2$	842		4.36	–	–	550	–	–	–	Ind	–	RF	–
Chromium Carbide	Cr$_3$O$_3$	1890		6.68	–	–	~2000	F	–	–	W	–	RF, DC	–
Chromium Chloride	CrCl$_2$	824		2.75	–	–	550	–	–	–	Fe, Ind	–	RF	Sublimes easily.
Chromium Oxide	Cr$_2$O$_3$	2435		5.21	–	–	~2000	G	–	–	W, Mo	W	RF / RF-R	Disproprotonates to lower oxides, reoxidizes @ 600°C In air. n = 2.4
Chromium Silicide	Cr$_3$Si$_2$			5.5	–	–	–	–	–	–	–	–	RF, DC	–
Chromium-Silicon Monoxide	Cr-SiO		S	*	*	*	*	G	–	–	W	W	DC, RF	Flash
Cobalt	Co	1495		8.90	850	990	1200	Ex	Al$_2$O$_3$, BeO	–	W	W	DC, RF	Alloys with refractory metals.
Cobalt Bromide	CoBr$_2$	678	S	4.91	–	–	400	–	–	–	Ind	–	RF	–
Cobalt Chloride	CoCl$_2$	740	S	3.36	–	–	472	–	–	–	Ind	–	RF	–
Cobalt Oxide	CoO	1935		6.4	–	–	–	–	–	–	–	–	DC-R / RF-R	Sputter preferred.
Copper	Cu	1083		8.92	727	857	1017	Ex	Al$_2$O$_3$, BN	W	Mo	W	DC, RF	Poor adhesion. Use inter layer, e.g. Cr. Evap. using any source material.
Copper Chloride	CuCl	422		3.53	–	–	~600	–	–	–	–	–	RF	n = 1.93
Copper Oxide	Cu$_2$O	1235	S	6.0	–	–	~600	G	Al$_2$O$_3$	–	Ta	–	DC-R / RF-R	n = 2.70
Cryolite	Na$_3$AlF$_6$	1000		2.9	1020	1260	1480	Ex	VC	–	W,Mo,Ta	W,Mo,Ta	RF	Large chunks reduce splitting. Little decomp.
Dysprosium	Dy	1409	S	8.54	625	750	900	G	–	–	Ta	–	RF / DC	–
Dysprosium Fluoride	DyF$_3$	1360	S	–	–	–	800	G	–	–	Ta	–	RF	–
Dysprosium Oxide	Dy$_2$O$_3$	2340		8.16	–	–	~1400	G	–	–	Ir	–	RF, RF-R	Loses O$_2$.
Erbium	Er	1497	S	9.06	650	775	930	G	–	–	W,Ta	–	DC, RF	–
Erbium Fluoride	ErF$_3$	1350		–	–	–	750	–	–	–	–	–	RF	–
Erbium Oxide	Er$_2$O$_3$	2400		8.64	–	–	1600	–	–	–	Ir	–	RF, RF-R	Loses O$_2$.
Europium	Eu	822		5.26	360		480	F	Al$_2$O$_3$	–	W,Ta	–	RF / DC	Low tantalum solubility.
Europium Fluoride	EuF$_2$	1380		6.5	–	–	950	G	–	–	Mo	–	RF	–
Europium Oxide	Eu$_2$O$_3$	2056		8.18	–	–	1600	G	ThO$_2$	–	Ir,Ta,W	–	RF, RF-R	Loses O$_2$. Films clear and hard.
Europium Sulphide	EuS			–	–	–	–	G	–	–	–	–	RF	–

* Influenced by composition   ** Cr-plated rod or strip   *** All metals alumina coated   C = Carbon   Gr = Graphite   S = Sublimes   D = Decomposes
Q = Quartz   Incl = Inconel   VC = Vitreous Carbon   SS = Stainless Steel   Ex = Excellent   F = Fair   P = Poor
RF = RF sputtering is effective   RF-R = Reactive RF sputtering is effective   DC = DC sputtering is effective   DC-R = Reactive DC sputtering is effective

MATERIAL	SYMBOL	MP°C	S/D	g/cm³	Vap. Press. @ temp°C			EVAPORATION TECHNIQUES					SPUTTER	COMMENTS
					10⁻⁸ T	10⁻⁶ T	10⁻⁴ T	E-BEAM	CRUCIBLE	COIL	BOAT	BASKET		
Gadolinium	Gd	1312	—	7.89	760	900	1175	Ex	Al₂O₃	—	Ta	—	RF DC	High Ta solubility.
Gadolinium Carbide	GdC₂	—	—	—	—	—	1500	—	C	—	—	—	RF	Decomposes
Gadolinium Oxide	Gd₂O₃	2310	—	7.41	—	—	—	F	C	—	Ir	—	RF, RF-R	Loses O₂.
Gallium	Ga	30	—	5.90	619	742	907	G	Al₂O₃, BeO, Q	—	—	—	—	Alloys with refractory metals. Use EB gun.
Gallium Antimonide	GaSb	710	—	5.6	—	—	—	F	C	—	W,Ta	—	RF	Flash evaporate.
Gallium Arsenide	GaAs	1238	—	5.3	—	—	—	G	C	—	W,Ta	—	RF	Flash evaporate.
Gallium Nitride	GaN	—	S	6.1	—	—	-200	—	Al₂O₃	—	—	—	RF RF-R	Evaporate Ga in 10⁻³ N₂.
Gallium Oxide	Ga₂O₃	1900	—	5.88	—	—	—	—	—	—	—	—	RF	Loses O₂.
Gallium Phosphide	GaP	1540	—	4.1	—	770	920	Ex	Q	—	P, W	W	RF	Does not decompose. Rate control important.
Germanium	Ge	937	—	5.35	812	957	1167	Ex	Q, Al₂O₃	—	W, C, Ta	—	DC, RF	Excellent films from EB guns.
Germanium Monoxide	GeO		—	—	—	—	500	—	Q	—	—	—	RF	—
Germanium Nitride	Ge₃N₂		S	5.2	—	—	~650	—	—	—	—	—	RF-R	Sputtering preferred.
Germanium Oxide	GeO₂	1086	—	6.24	—	—	625	G	Q, Al₂O₃	—	Ta	W Mo	RF-R	Similar to SiO film, predominately GeO.
Germanium Telluride	GeTe	725	—	6.20	—	—	381	—	Q, Al₂O₃	—	W, Mo	W	RF	—
Glass, Schott 8329	—	—	—	2.20	—	—	—	Ex	—	—	—	—	RF	Evaporable alkali glass. Melt in air before evap.
Gold	Au	1062	—	19.32	807	947	1132	Ex	Al₂O₃, BN, VC	W	W***, Mo***	W	DC RF	Films soft, not very adherent.
Hafnium	Hf	2230	—	13.09	2160	2250	3090	G	—	—	—	—	DC, RF	—
Hafnium Boride	HfB₂	3250	—	10.5	—	—	—	—	—	—	—	—	DC, RF	—
Hafnium Carbide	HfC	4160	S	12.2	—	—	—	—	—	—	—	—	DC, RF	—
Hafnium Nitride	HfN	2852	—	13.8	—	—	2600	—	—	—	—	—	RF, RF-R	—
Hafnium Oxide	HfO₂	2812	—	9.68	—	—	2500	F	—	—	W	—	DC, RF, RF-R	Film HfO.
Hafnium Silicide	HfSi₂	1750	—	7.2	—	—	—	—	—	—	—	—	RF	—
Holmium	Ho	1470	S	8.80	650	770	950	G	—	W	W, Ta	W	—	—
Holmium Fluoride	HoF₃	1143	—	—	—	—	~800	—	Q	—	—	—	DC, RF	—
Holmium Oxide	Ho₂O₃	2370	—	8.41	—	—	—	—	—	—	Ir	—	RF, RF-R	Loses O₂.
Inconel	Ni/Cr/Fe	1425	—	8.5	—	—	—	G	—	W	W	W	DC, RF	Use fine wire pre-wrapped on W. Low rate req'd for smooth films.
Indium	In	157	—	7.30	487	597	742	Ex	Gr, Al₂O₃	—	W, Mo	W	DC, RF	Wets W & Cu. Use Mo liner.
Indium Antimonide	InSb	535	—	5.8	—	—	—	—	—	—	W	—	RF	Decomposes; sputter preferred; or co-evap. from 2 sources; flash.
Indium Arsenide	InAs	943	—	5.7	780	870	970	—	—	—	W	—	RF	—

* Influenced by composition   ** C-plated rod or strip   *** All metals alumina coated   C = Carbon   Gr = Graphite   S = Sublimes   D = Decomposes
Q = Quartz   Incl = Inconel   VC = Vitreous Carbon   SS = Stainless Steel   Ex = Excellent   F = Fair   P = Poor
RF = RF sputtering is effective   RF-R = Reactive RF sputtering is effective   DC = DC sputtering is effective   DC-R = Reactive DC sputtering is effective

MATERIAL	SYMBOL	MP°C	S/D	g/cm³	Vap. Press. @ temp°C			EVAPORATION TECHNIQUES					SPUTTER	COMMENTS
					10^{-8} T	10^{-6} T	10^{-4} T	E-BEAM	CRUCIBLE	COIL	BOAT	BASKET		
Indium Oxide	InO	—	S	6.99	—	—	650	—	—	—	—	—	RF	Decomposes
Indium Phosphide	InP	1058		4.8	—	630	730	—	Gr	—	W, Ta	W, Ta	RF	Deposits P rich.
Indium Selenide	In₂Se₃	890		5.7	—	—	—	—	Gr	—	—	—	RF	Sputtering preferred; or evap. from 2 sources; flash.
Indium Sesquisulphide	In₂S₃	1050	S	4.90	—	—	850	—	Gr	—	—	—	RF	Film In₂S.
Indium Sulphide	InS	653		5.87	—	—	650	—	Gr	—	—	—	RF	—
Indium Telluride	In₂Te₃	667		5.8	—	—	—	—	—	—	—	—	RF	Sputtering preferred; or co-evap. from 2 sources; flash.
Iridium	Ir	2459		22.4	1850	2080	2380	F	ThO₂	—	—	—	DC, RF	—
Iron	Fe	1535		7.86	858	998	1180	Ex	Al₂O₃, BeO	W	W	W	DC, RF	Attacks W. Films hard, smooth. Preheat gently to outgas.
Iron Bromide	FeBr₂	689		4.64	—	—	561	—	Fe	—	—	—	RF	—
Iron Chloride	FeCl₂	670	S	2.98	—	—	300	—	Fe	—	—	—	RF	—
Iron Iodide	FeI₂	592		5.31	—	—	400	—	Fe	—	—	—	RF	—
Iron Oxide	FeO	1425		5.7	—	—	—	P	—	—	—	—	RF, RF-R	Decomposes; sputtering preferred.
Iron Oxide	Fe₂O₃	1565		5.24	—	—	—	G	—	—	W	W	—	Disproportionates to Fe₃O₄ at 1530°C. n = 3.0
Iron Sulphide	FeS	1195		4.84	—	—	—	—	Al₂O₃	—	—	—	RF	Decomposes
Kanthal	FeCrAl			7.1	—	—	—	—	—	W	W	—	DC, RF	—
Lanthanum	La	920		6.17	990	1212	1388	Ex	Al₂O₃	—	W, Ta	—	RF	Films will burn in air if scraped.
Lanthanum Boride	LaB₆	2210		2.61	—	—	—	G	—	—	—	—	RF	—
Lanthanum Fluoride	LaF₃	1490		~6	—	—	900	G	—	—	Ta, Mo	Ta	RF	No decomposition. n ≈ 1.6
Lanthanum Oxide	La₂O₃	2250		5.84	—	—	1400	G	—	—	W, Ta	—	RF	Loses O₂. n ≈ 1.73
Lead	Pb	328		11.34	342	427	497	Ex	Al₂O₃, Q	W	W, Mo	W, Ta	DC, RF	Toxic
Lead Chloride	PbCl₂	678		5.85	—	—	~325	—	Al₂O₃	—	Pt	—	RF	Little decomposition.
Lead Fluoride	PbF₂	822		8.24	—	—	~400	—	Al₂O₃	—	W, Pt, Mb	—	RF	n = 1.75
Lead Oxide	PbO	890		9.53	—	—	~550	G	Q, Al₂O₃	—	Pt	—	RF-R	No decomposition. n ≈ 2.6
Lead Stannate	PbSnO₃	1115		8.1	670	780	905	P	Al₂O₃	—	Pt	Pt	RF	Disproportionates
Lead Selenide	PbSe	1065		8.10	—	—	~500	—	Gr, Al₂O₃	—	W,Mo	W	RF	—
Lead Sulphide	PbS	1114	S	7.5	—	—	~500	—	Q, Al₂O₃	—	W	W, Mo	RF	Little decomposition. n = 3.91
Lead Telluride	PbTe	917		8.16	780	910	1050	—	Al₂O₃, Gr	—	Pt	—	RF	Vapors toxic. Deposits Te rich. Sputtering preferred, or co-evap. from 2 sources.
Lead Titanate	PbTiO₃			7.52	—	—	—	—	—	—	Ta	—	RF	—
Lithium	Li	179		0.53	227	307	407	G	Al₂O₃, BeO	—	Ta,SS	—	—	Metal reacts quickly in air.
Lithium Bromide	LiBr	547		3.46	—	—	~500	—	—	—	Ni	—	RF	n = 1.78
Lithium Chloride	LiCl	613		2.07	—	—	400	—	—	—	Ni	—	RF	Preheat gently to outgas. n = 1.66
Lithium Fluoride	LiF	870		2.60	875	1020	1180	G	—	—	Ni,Ta Mo,W	—	RF	Rate control important for optical films. Preheat gently to outgas. n = 1.33

*Influenced by composition   ** Cr-plated rod or strip   *** All metals alumina coated   C = Carbon   Gr = Graphite   S = Sublimes   D = Decomposes
Q = Quartz   Ind = Inconel   VC = Vitreous Carbon   SS = Stainless Steel   Ex = Excellent   F = Fair   P = Poor
RF = RF sputtering is effective   RF-R = Reactive RF sputtering is effective   DC = DC sputtering is effective   DC-R = Reactive DC sputtering is effective

MATERIAL	SYMBOL	MP°C	S/D	g/cm³	Vap. Press. @ temp°C			EVAPORATION TECHNIQUES					SPUTTER	COMMENTS
					10⁻⁸ T	10⁻⁶ T	10⁻⁴ T	E-BEAM	CRUCIBLE	COIL	BOAT	BASKET		
Lithium Iodide	LiI	446		4.06	-	-	400	-	-	-	Mo,W	-	RF	-
Lithium Oxide	Li₂O	1700		2.01	-	-	850	-	-	-	Pt, Ir	-	RF	n = 1.64
Lutetium	Lu	1652	S	9.84	-	-	1300	Ex	Al₂O₃	-	Ta	-	RF, DC	-
Lutetium Oxide	Lu₂O₃			9.41	-	-	1400	-	-	-	Ir	-	RF	Decomposes
Magnesium	Mg	651	S	1.74	185	247	327	G	Al₂O₃, VC	W	W,Mo,Ta	W	DC, RF	Extremely high rates possible.
Magnesium Aluminate	MgAl₂O₄	2135		3.6	-	-	-	G	-	-	-	-	RF	Natural spinel.
Magnesium Bromide	MgBr₂	195		3.72	-	-	250	-	-	-	Ni	-	RF	Decomposes
Magnesium Chloride	MgCl₂	116		2.32	-	-	400	-	-	-	Ni	-	RF	Decomposes. n = 1.6
Magnesium Fluoride	MgF₂	1266		2.9-3.2	-	-	925	Ex	Al₂O₃	-	-	-	RF	Rate control and substrate heat important for optical films. Reacts with W. Ex with Mo. n = 1.38
Magnesium Iodide	MgI₂	700		4.24	-	-	200	-	-	-	Pt	-	RF	-
Magnesium Oxide	MgO	2800		3.58	-	-	1300	G	C, Al₂O₃	-	-	-	RF, RF-R	Evap. In 10⁻³ T O₂ for stoichiometry. W gives volatile oxides. n = 1.7
Manganese	Mn	1244	S	7.20	507	572	647	G	Al₂O₃, BeO	W	W,Ta,Mo	W	DC, RF	-
Manganese Bromide	MnBr₂	695		4.38	-	-	500	-	-	-	Ind	-	RF	-
Manganese Chloride	MnCl₂	650		2.98	-	-	450	-	-	-	Ind	W	RF	-
Manganese Dioxide	MnO₂			5.02	-	-	-	P	-	-	W	W	RF-R	Loses O₂ @ 535°C
Manganese Sulphide	MnS			3.99	-	-	1300	-	-	-	Mo	-	RF	Decomposes. n = 2.7
Mercury	Hg	-39		13.55	-68	-42	-6	-	-	-	-	-	-	-
Mercury Sulphide	HgS		S	8.10	-	-	250	-	-	-	-	-	RF	Decomposes
Molybdenum	Mo	2610		10.22	1592	1822	2117	Ex	Al₂O₃	-	-	-	DC, RF	Films smooth, hard. Careful degas req'd.
Molybdenum Boride	MoB₂	2100		7.12	-	-	-	P	-	-	-	-	RF, DC	-
Molybdenum Carbide	Mo₂C	2687		9.18	-	-	-	F	-	-	-	-	RF, DC	Evaporation of Mo (CO)₆ yields Mo₂C.
Molybdenum Disulphide	MoS₂	1185		4.80	-	-	50	-	-	-	-	-	-	-
Molybdenum Silicide	MoSi₂			6.3	-	-	-	-	-	-	-	-	RF	Decomposes
Molybdenum Trioxide	MoO₃	795		4.70	-	-	900	-	Al₂O₃, BN	-	W	Mo	RF	Slight O₂ loss. n = 1.9
Neodymium	Nd	1024		7.00	731	871	1062	Ex	Al₂O₃	-	Ta	-	DC, RF	Low Ta solubility.
Neodymium Fluoride	NdF₃	1410		6.5	-	-	900	G	Al₂O₃	-	Mo,W	Mo, Ta	RF	Very little decomposition. n = 1.6
Neodymium Oxide	Nd₂O₃	2272		7.24	-	-	~1400	G	ThO₂	-	Ta,W	-	RF, RF-R	Loses O₂, films clear. EB preferred. n = 1.79
Nichrome IV	Ni/Cr	1395		8.50	847	987	1217	Ex	Al₂O₃, BeO,VC	W	•••	W, Ta	DC, RF	Alloys with refractory metals.
Nickel	Ni	1453		8.90	927	1072	1262	Ex	Al₂O₃, BeO, VC	W	W	W	DC, RF	Alloys with refractory metals. Forms smooth adherent films.
Nickel Bromide	NiBr₂	963	S	4.64	-	-	362	-	-	-	Ind	-	RF	-
Nickel Chloride	NiCl₂	1001	S	3.55	-	-	444	-	-	-	Ind	-	RF	-
Nickel Oxide	NiO	1990		7.45	-	-	1470	-	Al₂O₃	-	-	-	RF-R	Dissociates on heating. n = 2.18

* Influenced by composition   ** C-plated rod or strip   *** All metals alumina coated   C = Carbon   Gr = Graphite   S = Sublimes   D = Decomposes
Q = Quartz   Ind = Inconel   VC = Vitreous Carbon   SS = Stainless Steel   Ex = Excellent   F = Fair   P = Poor
RF = RF sputtering is effective   RF-R = Reactive RF sputtering is effective   DC = DC sputtering is effective   DC-R = Reactive DC sputtering is effective

MATERIAL	SYMBOL	MP°C	S/D	g/cm³	Vap. Press. @ temp°C			EVAPORATION TECHNIQUES					SPUTTER	COMMENTS
					10^{-8}T	10^{-6}T	10^{-4}T	E-BEAM	CRUCIBLE	COIL	BOAT	BASKET		
Niobium	Nb	2468	–	8.55	1728	1977	2287	Ex	–	–	W	–	DC, RF	Attacks W source.
Niobium Boride	NbB2	3050	–	6.97	–	–	–	–	–	–	–	–	RF, DC	–
Niobium Carbide	NbC	3800	–	7.82	–	–	–	F	–	–	–	–	RF, DC	–
Niobium Nitride	NbN	2573	–	8.4	–	–	–	–	–	–	–	–	RF, RF-R	Reactive. Evaporate Nb in 10^{-3} T N2.
Niobium Oxide	NbO		–	6.27	–	–	1100	–	–	–	Pt	–	RF	–
Niobium Pentoxide	Nb2O3	1530	–	4.47	–	–	–	–	–	–	W	W	RF, RF-R	n = 1.95
Niobium Telluride	NbTex		–	7.6	–	–	–	–	–	–	–	–	RF	Composition variable.
Niobium-Tin	Nb3Sn		–	–	–	–	–	Ex	–	–	–	–	RF, DC	Co-evaporate from 2 sources.
Niobium Trioxide	Nb2O3	1780	–	7.5	–	–	–	–	–	–	W	W	RF, DC, RF	–
Osmium	Os	1700	–	22.5	2170	2430	2760	F	–	–	–	–	–	Deposits Os in 10^{-3} T O2.
Osmium Oxide	Os2O3		D	–	–	–	–	–	–	–	–	–	–	–
Palladium	Pd	1550	S	12.40	842	992	1192	Ex	Al2O3 BeO	W	W	W	DC, RF	Alloys with refractory metals. Rapid evaporation suggested.
Palladium Oxide	PdO	870	–	8.31	–	–	575	–	Al2O3	–	–	–	RF-R	Decomposes
Parylene	C8H8	300 / 400	–	1.1	–	–	–	–	–	–	–	–	–	Vapor depositable plastic.
Permalloy	Ni/Fe	1395	–	8.7	947	1047	1307	G	Al2O3 VC	–	W	–	DC, RF	Film low in Ni.
Phosphorus	P	41.4	–	1.82	327	361	402	–	Al2O3	–	–	–	–	Material reacts violently in air.
Phosphorus Nitride	P3N5		–	2.51	–	–	–	–	–	–	–	–	RF, RF-R	–
Platinum	Pt	1769	–	21.45	1292	1492	1747	Ex	C ThO2	W,Pt	W	–	DC, RF	Alloys with metals. Films soft, poor adhesion.
Platinum Oxide	PtO2	450	–	10.2	–	–	–	–	–	–	–	–	RF-R	–
Plutonium	Pu	635	–	–	–	–	–	–	–	–	W	–	–	Toxic, radioactive.
Polonium	Po	254	–	9.4	117	170	244	–	Q	–	–	–	–	Radioactive
Potassium	K	64	–	0.86	23	60	125	–	Q	–	Mo	–	–	Metal reacts rapidly in air. Preheat gently to outgas.
Potassium Bromide	KBr	730	–	2.75	–	–	450	–	Q	–	Ta, Mo	–	RF	Preheat gently to outgas. n = 1.49
Potassium Chloride	KCl	776	–	1.98	–	–	510	–	–	–	Ta, Ni	–	RF	Preheat gently to outgas. n = 1.49
Potassium Fluoride	KF	880	–	2.48	–	–	~450	–	Q	–	–	–	RF	Preheat gently to outgas. n = 1.35
Potassium Hydroxide	KOH	360	–	2.04	–	–	~400	–	–	–	Pt	–	RF	Preheat gently to outgas.
Potassium Iodide	KI	723	–	3.13	–	–	~500	–	–	–	Ta	–	RF, DC	Preheat gently to outgas. n = 1.68
Praseodymium	Pr	931	–	6.78	800	950	1150	–	–	–	Ta	–	RF, RF-R	–
Praseodymium Oxide	Pr2O3	2125	–	6.88	–	–	1400	G	ThO2	–	Ir	–	RF, RF-R	Loses O2.
Radium	Ra	700	–	5.0	246	320	416	–	–	–	–	–	–	–
Rhenium	Re	3180	–	20.53	1928	2207	2571	Ex	–	–	–	–	DC, RF	Fine wire will self-evaporate.
Rhenium Oxide	ReO3		–	~7	–	–	–	–	–	–	–	–	RF	Evaporate Re in 10^{-3} T O2.
Rhodium	Rh	1966	–	12.4	1277	1472	1707	G	ThO2 VC	W	W	W	DC, RF	EB gun preferred.
Rubidium	Rb	38.5	–	1.47	-3	37	111	–	Q	–	–	–	DC, RF	–

*Influenced by composition   ** C-plated rod or strip   *** All metals alumina coated   C = Carbon   Gr = Graphite   S = Sublimes   D = Decomposes
Q = Quartz   Incl = Inconel   VC = Vitreous Carbon   SS = Stainless Steel   Ex = Excellent   F = Fair   P = Poor
RF = RF sputtering is effective   RF-R = Reactive RF sputtering is effective   DC = DC sputtering is effective   DC-R = Reactive DC sputtering is effective

MATERIAL	SYMBOL	MP°C	S/D	g/cm³	Vap. Press. @ temp°C 10^{-8} T	10^{-6} T	10^{-4} T	EVAP. E-BEAM	CRUCIBLE	COIL	BOAT	BASKET	SPUTTER	COMMENTS
Rubidium Chloride	RbCl	715	–	2.76	–	–	-550	–	Q	–	–	–	RF	n = 1.49
Rubidium Iodide	RbI	642	–	3.55	–	–	-400	–	Q	–	–	–	RF	–
Ruthenium	Ru	2700	–	12.6	1780	1990	2260	F	–	–	W	–	DC, RF	–
Samarium	Sm	1072	–	7.54	373	460	573	G	Al₂O₃	–	Ta	–	RF, DC	–
Samarium Oxide	Sm₂O₃	2350	–	7.43	–	–	–	G	ThO₂	–	Ir	–	RF, RF-R	Loses O₂. Films smooth, clear.
Scandium	Sc	1539	S	2.99	714	837	1002	Ex	Al₂O₃, BeO	–	W	–	RF	Alloys with Ta.
Scandium Oxide	Sc₂O₃	2300	–	3.86	–	–	–	–	–	–	–	–	RF, RF-R	–
Selenium	Se	217	–	4.2-4.8	89	125	170	G	Al₂O₃, VC	W,Mo	W,Mo	W,Mo	RF, DC	Toxic. Bad for vacuum systems.
Silicon	Si	1410	–	2.42	992	1147	1337	Ex	BeO Ta VC	–	W,Ta	–	DC, RF	Alloys with W; use heavy W boat. SiO produced above 4x10⁻⁶ Torr. EB best.
Silicon Boride	SiB₆	2700	–	–	–	–	–	P	–	–	–	–	RF	Sputtering preferred.
Silicon Carbide	SiC	2700	–	3.22	–	–	1000	–	–	–	–	–	RF	–
Silicon Dioxide	SiO₂	1610 -171	–	2.2-2.7	*	*	1025*	Ex	Al₂O₃	–	–	–	RF	Quartz excellent in EB. n = 1.47
Silicon Monoxide	SiO	1702	S	2.1	–	–	850	Ex	Ta	W	Ta	W	RF, RF-R	For resistance evap. use baffle box and low rate. n = 1.6
Silicon Nitride	Si₃N₄		S	3.44	–	–	800	–	–	W	–	–	RF, RF-R	–
Silicon Selenide	SiSe				–	–	550	–	Q	–	–	–	RF	–
Silicon Sulphide	SiS		S	1.85	–	–	450	–	Q	–	–	–	RF	–
Silicon Telluride	SiTe₂			4.39	–	–	550	–	Q	–	–	–	RF	–
Silver	Ag	961	–	10.49	847	958	1105	Ex	Al₂O₃, Mo	W	Ta,Mo	W	DC, RF	Evaporates well from any source.
Silver Bromide	AgBr	432	–	6.47	–	–	380	–	Q	–	Ta	–	RF	Preheat gently to outgas. n = 2.25
Silver Chloride	AgCl	455	–	5.56	–	–	520	–	Q	–	Mo,Pt	Mo	RF	Preheat gently to outgas. n = 2.07
Silver Iodide	AgI	558	–	5.67	–	–	-900	–	–	–	Ta	–	RF	Preheat gently to outgas. n = 2.21
Sodium	Na	97	–	0.97	74	124	192	–	Q	W	Ta,SS	W	–	Preheat gently to outgas. Metal reacts quickly in air.
Sodium Bromide	NaBr	755	–	3.20	–	–	-400	–	Q	–	–	–	RF	Preheat gently to outgas. n = 1.64
Sodium Chloride	NaCl	801	–	2.16	–	–	530	P	Q	–	Ta,W,Mo	–	RF	Cu oven, little decomposition. Preheat gently to outgas. n = 1.54
Sodium Cyanide	NaCN	563	–	–	–	–	-550	–	–	–	Ag	–	RF	Preheat gently to outgas. n = 1.45
Sodium Fluoride	NaF	988	–	2.79	945	1080	1200	P	BeO	–	Mo,Ta,W	–	RF	Preheat gently to outgas. No decomposition. n = 1.36
Sodium Hydroxide	NaOH	318	–	2.13	–	–	-470	–	–	–	Pt	–	–	Preheat gently to outgas. n = 1.36
Spinel	MgO 5Al₂O₃		–	8.0	–	–	–	G	–	–	–	–	RF	n = 1.72
Strontium	Sr	769	–	2.6	239	309	403	G	VC	W	W,Ta,Mo	W	RF, DC	Wets but does not alloy with refractory metals. May react in air.

* Influenced by composition   ** Cr-plated rod or strip   *** All metals alumina coated   C = Carbon   Gr = Graphite   S = Sublimes   D = Decomposes
Q = Quartz   Incl = Inconel   VC = Vitreous Carbon   SS = Stainless Steel   Ex = Excellent   F = Fair   P = Poor
RF = RF sputtering is effective   RF-R = Reactive RF sputtering is effective   DC = DC sputtering is effective   DC-R = Reactive DC sputtering is effective

MATERIAL	SYMBOL	MP°C	S/D	g/cm³	Vap. Press. @ temp°C			EVAPORATION TECHNIQUES					SPUTTER	COMMENTS
					10^{-8}T	10^{-6}T	10^{-4}T	E-BEAM	CRUCIBLE	COIL	BOAT	BASKET		
Strontium Fluoride	SrF_2	1450	–	4.24	–	–	–1000	–	Al_2O_3	–	–	–	RF	n = 1.44
Strontium Oxide	SrO	2460	S	4.7	–	–	1500	–	Al_2O_3	–	Mo	–	RF	Reacts with Mo and W. n = 1.87
Strontium Sulphide	SrS	–	–	3.70	–	–	–	–	–	–	Mo	–	RF	Decomposes. n = 2.11
Sulphur	S_8	115	–	2.0	13	19	57	–	Q	–	W	W	–	Bad for vacuum system.
Supermalloy	Ni/Fe/Mo	1410	–	8.9	–	–	–	G	–	W	W	–	RF, DC	Sputtering preferred; or co-evap. from 2 sources, Permalloy and Mo.
Tantalum	Ta	2996	–	16.6	1960	2240	2590	Ex	–	–	–	–	DC, RF	Forms good films.
Tantalum Boride	TaB_2	3000	–	12.38	–	–	–	–	–	–	–	–	RF, DC	–
Tantalum Carbide	TaC	3880	–	14.65	–	–	–2500	–	–	–	–	–	RF, DC	–
Tantalum Nitride	TaN	3660	–	16.30	–	–	–	–	–	–	–	–	RF, RF-R, DC	Evap. Ta in 10^{-3} T N_2.
Tantalum Pentoxide	Ta_2O_5	1800	–	8.74	1550	1780	1920	G	VC	W	Ta	W	RF, RF-R	Slight decomposition. Evap. in 10^{-3} T O_2. n = 2.6
Tantalum Sulphide	TaS_2	1300	–	–	800	950	1150	–	–	–	Ta	–	RF	–
Technetium	Tc	2200	–	11.5	1570	1800	2090	–	–	–	–	–	–	–
Teflon	PTFE	330	–	2.9	–	–	–	–	–	–	W	–	RF	Baffled source. Film structure doubtful.
Tellurium	Te	452	–	6.25	157	207	277	P	Al_2O_3, Q	W	W	W,Ta	RF	Toxic. Wets without alloying.
Terbium	Tb	1357	–	8.27	800	950	1150	Ex	Al_2O_3	–	Ta	–	RF	–
Terbium Fluoride	TbF_3	1176	–	–	–	–	–800	–	–	–	–	–	RF	–
Terbium Oxide	Tb_4O_3	–	–	7.87	–	–	1300	–	–	–	Ir	–	RF	Partially decomposes.
Terbium Oxide	Tb_4O_7	–	–	–	–	–	–	–	–	–	Ta	–	RF	Films Tb_2O.
Thallium	Tl	302	–	11.85	280	360	470	P	Al_2O_3, Q	–	W,Ta	W	DC	Very toxic. Wets freely.
Thallium Bromide	TlBr	480	S	7.56	–	–	–250	–	Q	–	–	–	RF	Toxic
Thallium Chloride	TlCl	430	S	7.00	–	–	–150	–	Q	–	–	–	RF	n = 2.25
Thallium Iodide	TlI	440	S	7.09	–	–	–250	–	Q	–	–	–	RF	n = 2.78
Thallium Oxide	Tl_2O_3	717	–	9.65	–	–	350	–	–	–	–	–	RF	Goes to Tl_2O at 850°C
Thorium	Th	1700	S	11.7	1430	1660	1925	Ex	–	W	W,Ta,Mo	W	–	Toxic, radioactive.
Thorium Carbide	ThC_2	2773	–	8.96	–	–	–2300	–	C	–	–	–	RF, DC	Radioactive
Thorium Dioxide	ThO_2	3050	–	10.03	–	–	–2100	G	–	–	W	–	RF, RF-R	Radioactive
Thorium Fluoride	ThF_4	90	–	6.3	–	–	–750	F	VC	–	Ni	W	RF	Radioactive
Thorium OxyFluoride	$ThOF_2$	900	–	9.1	–	–	–	–	–	–	Mo,Ta	–	–	Radioactive. n = 1.52
Thorium Sulphide	ThS_2	–	–	6.80	–	–	–	–	Q	–	–	–	RF	Sputtering preferred; or co-evap. from 2 sources.
Thulium	Tm	1545	S	9.32	461	554	680	–	Al_2O_3	–	Ta	–	DC	–
Thulium Oxide	Tm_2O_3	–	–	8.90	–	–	1500	–	–	–	Ir	–	RF	Decomposes
Tin	Sn	232	–	7.75	682	807	997	Ex	Al_2O_3	W	Mo	W	DC, RF	Wets Mo. Use Ta liner in EB guns.
Tin Oxide	SnO_2	1127	S	6.95	–	–	–350	Ex	Q Al_2O_3	W	W	W	RF, RF-R	Films from W are O_2 deficient, oxidize in air. n = 2.0
Tin Selenide	SnSe	861	–	6.18	–	–	–400	–	Q	–	–	–	RF	–
Tin Sulphide	SnS	882	–	5.08	–	–	–450	–	Q	–	–	–	RF	–

MATERIAL	SYMBOL	MP°C	S/D	g/cm³	Vap. Press. @ temp°C			EVAPORATION TECHNIQUES					SPUTTER	COMMENTS
					10^{-8} T	10^{-6} T	10^{-4} T	E-BEAM	CRUCIBLE	COIL	BOAT	BASKET		
Tin Telluride	SnTe	780		6.44	–	–	~450	–	Q	–	Ta	–	RF	Alloys w/ refractory metals; evolves gas on 1st heating.
Titanium	Ti	1675		4.50	1067	1235	1453	Ex	VC TiC	–	Ta	–	DC, RF	–
Titanium Boride	TiB₂	2980		4.50	–	–	~2300	P	–	–	–	–	RF, DC	–
Titanium Carbide	TiC	3140		4.93	–	–	~2300	–	–	–	–	–	RF, DC	–
Titanium Dioxide	TiO₂	1640		4.29	–	–	~1300	F	–	–	–	W	RF, RF-R	Sub-oxide, must be reoxidized to rutile. Ta reduces TiO₂ to TiO and Ti. n = 2.4
Titanium Monoxide	TiO	1750		4.93	–	–	~1500	G	VC	–	W, Ta	–	RF	Preheat gently to outgas. n = 2.2
Titanium Nitride	TiN	2930		5.43	–	–	–	G	–	–	Mo	–	RF, RF-R, DC	Sputtering preferred. Decomposes with thermal evaporation.
Titanium Sesquioxide	Ti₂O₃	2130		4.6	–	–	–	G	–	–	W	–	RF	Decomposes
Titanium Silicide	TiSi₂		S	–	–	–	–	–	–	–	–	–	RF, DC	–
Tungsten	W	3410		19.3	2117	2407	2757	G	–	–	–	–	RF, DC	Forms violatile oxides. Films hard & adherent.
Tungsten Boride	WB₂	2900		12.75	–	–	–	P	–	–	–	–	RF	–
Tungsten Carbide	W₂C	2860		17.15	1480	1720	2120	Ex	C	–	C	–	RF, DC	–
Tungsten Disulphide	WS₂	1250		7.5	–	–	–	–	–	–	–	–	RF	–
Tungsten Selenide	WSe₂			9.0	–	–	–	–	–	–	–	–	RF	–
Tungsten Silicide	WSi₂	2900		9.4	–	–	–	–	–	–	–	–	RF, DC	–
Tungsten Telluride	WTe₃			9.49	–	–	–	–	Q	–	–	–	RF	–
Tungsten Trioxide	WO₃	1473		7.16	–	–	980	G	–	–	W, Pt	–	RF-R	Preheat gently to outgas. W reduces oxide slightly. n = 1.68
Uranium	U	1132		18.7	1132	1327	1582	G	–	W	Mo, W	W	–	Films oxidize.
Uranium Carbide	UC₂	2260		11.28	–	–	2100	–	C	–	W	–	RF	Decomposes.
Uranium Dioxide	UO₂	2176		10.9	–	–	1800	–	–	–	W	–	RF	Ta causes decomposition.
Uranium Fluoride	UF₄	1000		–	–	–	300	–	–	–	Ni	–	RF	–
Uranium Oxide	U₃O₈		D	8.30	–	–	–	–	–	–	W	W	RF-R	Decomposes at 1300°C to UO₂
Uranium Phosphide	UP₂			8.57	–	–	1200	–	–	–	Ta	–	RF	Decomposes
Uranium Sulphide	U₂S₃	–		–	–	–	1400	–	–	–	W	–	RF	Slight decomposition.
Vanadium	V	1890		5.96	1162	1332	1547	Ex	–	–	Mo	–	DC, RF	Wets Mo. EB evaporated films preferred.
Vanadium Boride	VB₂	2400		5.10	–	–	1800	–	–	–	–	–	RF, DC	–
Vanadium Carbide	VC	2810		5.77	–	–	1800	–	–	–	–	–	RF, DC	–
Vanadium Dioxide	VO₂	1967		–	–	–	575	–	–	–	–	–	RF, RF-R	Sputtering preferred.
Vanadium Nitride	VN	2320		6.13	–	–	–	–	–	–	–	–	RF, RF-F, DC	–
Vanadium Pentoxide	V₂O₅	690		3.36	–	–	~500	–	Q	–	–	–	RF	–
Vanadium Silicide	VSi₂	–		4.42	–	–	–	–	–	–	–	–	RF	–
Ytterbium	Yb	824		6.98	520	590	690	G	–	–	Ta	–	DC, RF	–
Ytterbium Fluoride	YbF₃	1157		–	–	–	~800	–	–	–	–	–	RF	–
Ytterbium Oxide	Yb₂O₃	2346		9.17	–	–	~1500	–	–	–	Ir	–	RF, RF-F	Loses O₂.

* Influenced by composition   ** Cr-plated rod or strip   *** All metals alumina coated   C = Carbon   Gr = Graphite   S = Sublimes   D = Decomposes
Q = Quartz   Incl = Inconel   VC = Vitreous Carbon   SS = Stainless Steel   Ex = Excellent   F = Fair   P = Poor
RF = RF sputtering is effective   RF-R = Reactive RF sputtering is effective   DC = DC sputtering is effective   DC-R = Reactive DC sputtering is effective

MATERIAL	SYMBOL	MP°C	S/D	g/cm³	Vap. Press. @ temp°C			EVAPORATION TECHNIQUES					SPUTTER	COMMENTS
					10^{-8}T	10^{-6}T	10^{-4}T	E-BEAM	CRUCIBLE	COIL	BOAT	BASKET		
Yttrium	Y	1509		4.48	830	973	1157	Ex	Al_2O_3	–	W,Ta	W	RF, DC	High Ta solubility.
Yttrium Aluminum Oxide	$Y_3Al_5O_{12}$	1990		–	–	–	–	G	–	W	–	–	RF	Films not ferroelectric.
Yttrium Fluoride	YF_3	1387		4.0	–	–	–	–	–	–	–	–	RF	–
Yttrium Oxide	Y_2O_3	2680		4.84	–	–	2000	G	C	–	W	–	RF, RF-R	Loses O_2, films smooth and clear. n = 1.79
Zinc	Zn	419		7.14	127	177	250	Ex	Al_2O_3, Q	W	Mo,W,Ta	W	DC, RF	Evaporates well under wide range of conditions.
Zinc Antimonide	Zn_3Sb_2	546		6.3	–	–	–	–	–	–	–	–	RF	–
Zinc Bromide	$ZnBr_2$	394	D	4.22	–	–	–300	–	C	–	W	–	RF	Decomposes
Zinc Fluoride	ZnF_2	872		4.84	790	905	1035	–	Q	–	Pt Ta	–	RF	–
Zinc Nitride	Zn_3N_2			–	–	–	–	–	–	–	Mo	–	RF	Decomposes
Zinc Oxide	ZnO	1975		5.61	–	–	–1800	–	–	–	–	–	RF-R	–
Zinc Selenide	ZnSe	1526		5.42	–	–	660	–	Q	W,Mo	Ta,W,Mo	W	RF	Preheat gently to outgas. Evaporates well. n = 2.6
Zinc Sulphide	ZnS	1830	S	4.09	–	–	–800	G	Q	–	Ta,Mo	–	RF	Preheat gently to outgas. Films partially decompose. Sticking coeff. varies with substrate temp. n = 2.3
Zinc Telluride	ZnTe	1238		6.0	–	–	–600	–	–	–	Mo,Ta	–	RF	Preheat gently to outgas. n = 2.8
Zircon	$ZrSiO_4$	2550		4.56	–	–	–	–	–	–	–	–	RF	–
Zirconium	Zr	1852		6.40	1477	1702	1987	E	–	–	W	–	RF, DC	Alloys with W. Films oxidize readily.
Zirconium Boride	ZrB_2	3040		6.08	–	–	–	G	–	–	–	–	RF, DC	–
Zirconium Carbide	ZrC	3540		6.73	–	–	–2500	–	–	–	–	–	RF, DC	–
Zirconium Nitride	ZrN	2980		7.09	–	–	–	–	–	–	–	–	RF, RF-R, DC	Reactively evaporate in 10^{-3} T N_2.
Zirconium Oxide	ZrO_2	2700		5.6	–	–	–2200	G	–	–	W	–	RF, RF-R	Films O_2 deficient, clear and hard. n = 2.05
Zirconium Silicide	$ZrSi_2$			4.88	–	–	–	–	–	–	–	–	RF, DC	–

*Influenced by composition   ** C-plated rod or strip   *** All metals alumina coated   C = Carbon   Gr = Graphite   S = Sublimes   D = Decomposes
Q = Quartz   Incl = Inconel   VC = Vitreous Carbon   SS = Stainless Steel   Ex = Excellent   F = Fair   P = Poor
RF = RF sputtering is effective   RF-R = Reactive RF sputtering is effective   DC = DC sputtering is effective   DC-R = Reactive DC sputtering is effective

Appendix B   Metal etchants for thin film materials

Elemental metals	Etching solution	Etch rate or etch time
Gold	1.4 g KI, 1 g I_2, 40 ml H_2O	0.5–1 μm/min
Indium	Mineral acids	—
Nickel	$FeCl_3$, 42-49°C Be; 43–54°C	12–25 μm/min
Palladium	1HCl, 10HNO_3, 10CH_3COOH	1000 Å/min
Platinum	8H_2O, 7HCl, 1HNO_3, 85°C	400–500 Å/min
Silver	5–9 HNO_3, 1-5 H_2O; 39–49°C	12–25 μm/min
Titanium	9H_2O, 1HF; 32°C	12 μm/min
Aluminum	1HCl, 4H_2O; 80°C	—
Chromium	164.5 g	—
	$Ce(SO_4)_2 \cdot 2(NH_4)_2SO_4 \cdot$ 2H_2O, 43 ml $HClO_4$, H_2O to make 1 L; 25–50°C	
Copper	$FeCl_3$, 42°C Be; 40°C	50 μm/min

References

1. J. L. Vossen and W. Kern (eds.), *Thin Film Processes II,* pp. 501–522, Academic Press, San Diego, 1991.
2. C. Stolzel, M. Huth, and H. Adrian, "C-Axis Oriented Thin $Bi_2 Sr_2 CaCu_2C_{8+\&}$ Films Prepared by Flash Evaporation," *Physica,* vol. C204, pp. 15–20, 1992.
3. T. Hato, Y. Takai, and H. Hayakawa, "High Tc Superconducting Thin Films Prepared by Flash Evaporation," *IEEE Transactions on Magnetics,* vol. 25, no. 2, pp. 2466–2469, March 1989.
4. C. De Las Heras and C. Sanchez, "Characterization of Iron Pyrite Thin Films Obtained by Flash Evaporation," *Thin Solid Films,* vol. 199, pp. 259–267, 1991.
5. This table is from *Deposition Techniques,* Kurt J. Lesker Co., Clairton, Pa., 1992, pp 25-15 through 25-25.
6. H. E. Horng, J. C. Jao, H. C. Chen, H. C. Yang, H. H. Sung, and F. C. Chen, "Critical Current in Polycrystalline Bi-Ca-Sr-Cu-O Films," *Physical Review B,* vol. 39, no. 13, pp. 9628–9630, May 1989.
7. P. Wagner, H. Adrian, and C. Tome-Rosa, *Physica C,* vol. 198, p. 258, 1992.
8. J. Dieleman, E. Van de Riet, and J. C. S. Kools, *Japanese Journal of Applied Physics,* pt. 1, vol. 31, no. 6B, pp. 1964–1971, 1992.
9. A. J. Basovich et al., *Thin Solid Films,* vol. 228, pp. 193–195, 1993.
10. F. Sanchez et al., *Applied Surface Science,* vol. 69, pp. 221–224, 1993.
11. A. Giardini Guidoni and I. Pettiti, *Applied Surface Science,* vol. 69, pp. 365–369, 1993.
12. C. Germain, C. Girault, J. Aubreton, and A. Catherinot, *Applied Surface Science,* vol. 69, pp. 359–364, 1993.
13. G. A. Petersen and J. R. McNeil, *Thin Solid Films,* vol. 220, pp. 87–91, 1992.
14. N. J. Arfsten, *Journal of Non-Crystalline Solids,* vol. 63, p. 243, 1984.
15. H. Dislich, *Journal of Non-Crystalline Solids,* vol. 57, p. 371, 1983.
16. C. J. Brinker and G. W. Scherer, *Sol-Gel Science,* Academic Press, Orlando, 1990.
17. D. Segal, *Chemical Synthesis of Advanced Ceramic Materials,* Cambridge University Press, Cambridge, 1989.
18. L. C. Klein (ed.), "Sol-Gel Technology for Thin Films, Fibers, Preforms, Electronics and Specialty Shapes," Noyes Publications, Park Ridge, N.J., 1988.
19. L. C. Klein, *Ceramic Engineering Science Proceedings,* vol. 5, p. 379, 1984.
20. H. Dislich, *Journal of Non-Crystalline Solids,* vol. 80, p. 115, 1986.
21. Maissel and Glang (eds.), *Handbook of Thin Film Technology,* McGraw-Hill, New York, 1970.
22. J. G. Bednorz and K. A. Muller, *Zeitschrift für Physik,* B, vol. 64, p. 189, 1986.
23. L. Kempfer, *Materials Engineering,* vol. 107, no. 5, pp. 26–29, May 1990.

24. R. Messier et al., *Journal of Metals,* September 1987, pp. 8–11.
25. W. Zhu, B. R. Stoner, B. E. Williams, and J. T. Glass, *Proceedings of the IEEE,* vol. 79, no. 5, May 1991.
26. Y. Hirose and M. Mitsuizumi, *New Diamond,* vol. 4, pp. 34–35, 1988.
27. A. H. Deutchman and R. J. Partyka, *Advanced Materials and Processes, ASM Int.,* June 1989, pp. 29–33.
28. I. Bahl and P. Bhartla, *Microwave Solid State Circuit Design,* Wiley, New York, 1988.
29. P. A. Stenmann and H. E. Hintermann, *Journal of Vacuum Science and Technology,* vol. A7, p. 2267, 1989.
30. J. Valli, *Journal of Vacuum Science and Technology,* vol. A4, p. 3007, 1986.
31. V. E. Cosslett, "Fifty Years of Instrumental Development of the Electron Microscope," in R. Barer and V. E. Cosslett (eds.), *Advances in Optical and Electron Microscopy,* vol. 10, Academic Press, San Diego, pp. 215–267, 1988.
32. J. I. Goldstein, D. E. Newbury, P. Echlin, D. C. Joy, C. Fiori, and E. Lifshin, *Scanning Electron Microscopy and X-Ray Microanalysis,* Plenum, New York, 1981.
33. G. Thomas and M. J. Goringe, *Transmission Electron Microscopy of Materials,* Wiley, New York, 1979.
34. Milton Ohring, *The Materials Science of Thin Films,* Academic Press, Orlando, Table 6-3, p. 276, 1992.
35. L. E. Davis, N. C. MacDonald, P. W. Palmberg, G. E. Riach, and R. E. Weber, *Handbook of Auger Electron Spectroscopy,* Physical Electronics Industries, Inc., Eden Prairie, Minn., 1976.
36. G. E. McGuire, Auger Electron Spectroscopy Reference Manual, Plenum, New York, 1979.
37. M. T. Bernius and G. H. Morrison, "Mass Analyzed Secondary Ion Microscopy," *Review of Scientific Instruments,* vol. 58, pp. 1789–1804, October 1987.
38. J. R. Bird and J. S. Williams (eds.), *Ion Beams for Materials Analysis,* Academic Press, Sydney, 1989.
39. S. C. Miller, "A Technique for Passivating Thin Film Hybrids," *Proceedings of the 1976 International Symposium of Hybrid Microelectronics (ISHM),* 1976, p. 39.
40. Grant Lu, "CVD Diamond Films Protect against Wear and Heat," *Advanced Materials and Processes, ASM Int.,* vol. 144, no. 6, pp. 42–43, December 1993.
41. Bruce Banks, "Modify Surfaces with Ions and Arcs," *Advanced Materials and Processes, ASM Int.,* vol. 144, no. 6, pp. 22–25, December 1993.
42. Brian Holtkamp, "Ion Implantation Makes Metals Last Longer," *Advanced Materials and Processes, ASM Int.,* vol. 144, no. 6, pp. 45–47, December 1993.
43. H. A. Naseem et al., "Metallization of Diamond Substrates for Multichip Module Applications," *Proceedings of the 1993 International Conference and Exhibition on Multichip Modules,* Apr. 14–16, 1993, pp. 62–67.
44. C. D. Iacovangelo and E. C. Jerabek, "Metallizing CVD Diamond for Electronic Applications," *Proceedings of the International Symposium of Hybrid Microelectronics (ISHM),* 1993, Dallas, pp. 132–138.
45. L. Eckertova, *Physics of Thin Films,* Plenum, New York, 1986.

Pure Copper Metallization Technologies

Thomas C. Evans

5.1 Introduction

At the close of the twentieth century, new methods of ceramic metallization have become extremely important to designers, manufacturers, and users of hybrid microelectronics. They are dealing with technological advances at all levels of the hybrid environment, finding that interactions between materials have become a key element to proper design of new products. Recognizing that the performance of many existing hybrid products is limited by thin metallization layers of low electrical and thermal conductivity, there is a concern that the potential of future products, imagined to have even higher performance, may not be realized. Material engineering developments in the field of ceramic to copper bonding are providing metallized substrates that have superior properties for certain hybrid applications, with the promise of meeting performance issues for future designs. Table 5.1 outlines some of the requirements driven by technological advances which apply to copper metallized substrates.[1,2]

TABLE 5.1   System/Performance Requirements for Metallized Substrates

Power management	High voltage, high current
	High speed/frequency
	High efficiency
	High thermal loads
Information management	High speed/frequency
	Low signal loss
	Low noise

To provide greater material performance in hybrid applications, many variations of copper and ceramic, and bonding mechanisms between the two, have been investigated for use in metallized substrates. Unique limitations of most combinations have been found to restrict the potential applications to just a few specialized cases. The three different approaches presented here, direct bond copper (DBC), plated copper, and active metal bond (AMB), have found more general application and acceptance because of their ability to solve performance problems. Pure copper on ceramic is a successful metallization technology, providing good thermal and electrical conductivity, adhesion strength, and reliable performance for microelectronic products.

Pure metals are generally strong, yet malleable, with good heat transfer and high electrical conductivity. However, most metals expand or contract significantly with temperature change, corrode, and wear relatively easily. The combination of pure metal and ceramic, which have very different properties, to create a hybrid substrate is not an obvious choice. Ceramics are hard, brittle insulators, chemically inert, and thermally very stable, but are generally poor heat conductors. Although it is difficult to imagine two more dissimilar electronic materials, their complementary nature in a metallized substrate makes them extremely useful. Pure copper metallization of ceramic applies the best properties of each material to enhance the hybrid performance.

Copper metal is recognized as the preferred conductor for interconnection, especially for the printed circuit board (PCB) industry, where etching a circuit pattern in copper bonded to an organic substrate is a standard practice. The advantages of using very thin layers of copper on PCBs also apply to the thicker layers bonded to ceramic. From either point of view, copper is second only to silver metal for electrical and thermal conductivity when considered on a volume basis. Yet, the advantages of using silver are accompanied by the risk of an unstable and high cost for a relatively small increase in performance. Pure copper is quite soft, easily etched or machined, is not a significant environmental threat, and is also nonmagnetic. But, even with the advantages, designers will always find performance limitations. It is evident that copper has significant advantages in certain applications when compared to standard hybrid metallization technologies.

The early hybrid applications of copper metallization were driven by individual products requiring extreme levels of performance that eliminated the option of using existing metallization technologies. Logic applications found speed limited by the resistivity of the trace, and power applications found the thermal conductivity inadequate for good thermal management.[3] Solving thermal issues of high power and high current led the innovators to adhere, pattern, and evaluate hybrids with copper on ceramic. Of lesser concerns were the price and unknown reliability.[60] Their incentive to use copper came from its advantageous properties and the realization that the product would not be possible otherwise. For many designs, the operating temperatures of active devices were too high when mounted to the standard metallized hybrid substrates. Mean time between failures was a hot topic of conversation between manufacturers and customers, especially when mounted power hybrid components failed. Substrates using copper metallization could not only conduct the heat better, but spread the heat to a greater surface area. This spreading is especially important with thicker copper on ceramics having low thermal conductivity to effectively reduce the operating temperature of the device and increase the reliability.

Which ceramic substrate and copper metallization technique are going to be used for the new product? Shorter product life cycles, increasing competitive pressures, and rapid technological improvement all lead designers to consider more of the material constraints. The ability of the selected materials to meet future, and many times unknown, requirements, must now be evaluated. Beginning with a step by step determination of fundamental performance limitations due to the metallized substrate material, a specific choice is made to relieve as many constraints as possible. However, that go, no-go approach to substrate material selection is changing. Recent hybrid product development programs for new substrate materials have been evaluated not just for material properties important today, but properties necessary to meet the future challenge. One important aspect of this new evaluation method is the examination and documentation of the upper performance limits of DBC, plated copper, and AMB for better decision making in the design process. Meeting the present and future performance requirements is more likely with a thorough characterization of the metallized substrate.

Typical hybrid assembly processes using pure copper metallized substrates have little difficulty in producing a quality finished product. Processing from device attachment and interconnection through rework is accomplished with the existing capability of the user. Plating of the copper surface may be desired, with many options possible to support a great variety of assembly and reliability needs.

Copper on ceramic hybrid applications have three major markets with the largest users being in Japan. The United States and Europe are smaller markets, but all three are growing rapidly. A rough estimate of the total world market for pure copper–ceramic substrates may not be very accurate, but a breakdown would show Japan using 60%, with Europe and the United States at 20% each.[4]

From a global perspective, sources of copper metal are readily available and relatively secure with open markets for the foreseeable future. Environmentally, copper has the advantage of very low toxicity, with a potential for recycling in higher volumes. This is a significant advantage as a result of the new focus on issues of electronic product disposal, the "deengineering" of populated hybrid substrates. European sources of copper metallized ceramic must already deal with this issue. Therefore, use of pure copper on most ceramic types should be classified as environmentally friendly and available in the future, even if classifications and use of other metallizations become restricted for environmental reasons.

Copper metallization technology is viable for many hybrid processes and products, but competitive technologies still have significant advantages. There may need to be a blend of metallization technologies, taking some of the advantages from each. Thick and thin film materials have already been combined with pure copper metallization to produce a more functional hybrid. This is indicative of the emphasis that should be placed on the tradeoff of design, materials, and processes for performance-driven products. No one material is likely to satisfy circuit demands on all counts. Therefore, a thorough understanding of the circuit performance and customer or market requirements with each combination of materials is absolutely required. The choice of ceramic and copper metallization technology should not begin until all factors affecting the product are fully defined. Selection of the best option will occur if a careful comparison is performed and each selection completely evaluated over all operating conditions. More options exist for the choice of ceramic and metallization than ever before, forcing the designer to do a much more thorough evaluation before a final selection, but producing a more functional product in the end.[5]

5.2   Substrates

Ceramic materials have been a cornerstone of hybrid microelectronics, making it possible to add many functional elements to a product. The traditional ceramics have met the two basic hybrid requirements of performance and cost while successfully competing with organic substrate materials in performance. Ceramics commonly used in the past will continue to have broad application with standard hybrid metallizations, but, for reasons of performance, their fundamental limitations will limit their future role unless other metallization technologies can extend the useful range of substrate properties. Pure copper on ceramic will enhance traditional ceramic performance, especially those ceramics with high thermal resistance.

Copper metallization technologies have also been developed to meet the demands of new hybrid products in conjunction with advanced ceramics. Usually, these advanced ceramics (frequently referred to as technical ceramics) have been formulated to meet a specific need and possess one or more superior material properties. Many commercial and academic technologists have found an opportunity to apply pure copper metallization processes that complement and strengthen the advantages present in one of the technical ceramics as a method of increasing the overall utility of the metallized substrate. Table 5.2 lists the typical substrate types involving pure copper as an illustration of the traditional and advanced ceramics being developed for future hybrid products.[41,44]

Two classifications of ceramic materials have important implications for selection of a pure copper metallization technology. Oxide ceramics such as alumina and beryllia (Al_2O_3 and BeO, respectively) have oxygen within the chemical composition, and the bonding processes between copper metal and ceramic are relatively well understood. Nonoxide ceramics, including aluminum nitride and silicon carbide (AlN and SiC), are newer materials competing with BeO for advanced hybrid applications. Consequently, bonding mechanisms are less understood. In both oxide and nonoxide ceramic metal-

TABLE 5.2   Comparison of Substrate Materials for Pure Copper Metallization

Ceramic	Temperature coefficient of expansion, $10^{-7}/°C$	Thermal conductivity, W/m•°C	Dielectric constant (@ 1 MHz)	Strength, MPa	Dielectric strength, kV/cm
99.5% Al_2O_3 (alumina)	6.5	30	8–9.5	350	87
96% Al_2O_3	6.1	25	9.3	317	83
BeO (beryllia)	6.3	250	6.7	490	90
AlN (aluminum nitride)	4.2	320 (150–170 typical)	8.8	320	155
SiC (silicon carbide)	3.8	270	40.0	450	4000
Diamond	1.5	1200	5.2		>200
Silicon	3.5	120	12.0	200	100
Copper metal	16	400			

lization development, work continues to better define the actual chemical processes and stoichiometry of the interfacial bond layer between ceramic and copper. This is not to infer that complete understanding is required before wider usage of pure copper metallization as a reliable hybrid substrate can occur, but to note that questions still must be answered.

Alumina is the oxide ceramic that has had the benefit of more metallization research than any other and is used extensively within the hybrid industry. Three factors have made alumina a common starting point against which alternative ceramic materials are measured: the widespread experiences of using alumina, the existing infrastructure, and lower costs. These factors combine to give hybrid users a powerful incentive to continue to use it as the ceramic base for metallized substrates. Each of the pure copper metallization technologies—DBC, plated copper, and AMB—has started with alumina as the initial ceramic material for evaluation. For refinement of metallization technologies, there is good control of the ceramic composition and surface condition of most alumina sources, making it an excellent choice as a development vehicle.[6]

For all the advantages of alumina, there are limitations to performance that affect more hybrid products with each new generation of products. The most basic limitation for design evaluations of alumina is poor thermal conductivity relative to other types of ceramic. Alumina has much better thermal performance than printed circuit boards, but there are other traditional and technical ceramic materials available that have several times the thermal conductivity. To continue to enjoy the economic advantages of using alumina, users have examined the possibility of increasing thermal performance by several different methods. Thermal vias and additional cooling apparatus have been tried, but a more effective solution has been pure copper metallization. Copper increases the thermal performance by effectively increasing the thermal load capacity for higher power applications. In terms of cost and reliability, the use of alumina has been extended beyond past limits, making it a good combination with pure copper metallization for many hybrid products.

BeO is the other oxide ceramic used with all copper metallization technologies. It has the advantage of a lower temperature coefficient of expansion (TCE), a lower dielectric constant, and much higher thermal conductivity compared to alumina. Bonding copper to BeO has been successfully done for some time, and many applications will have BeO substrates for the foreseeable future. Like alumina, BeO enjoys a large supporting infrastructure, giving assurance of consistent and reproducible substrate properties. There are disadvantages, however: higher cost, environmental hazards of certain assembly processes, and the fact that the thermal conductivity declines significantly with increasing temperature. No matter what metallization is applied, the properties of the BeO are both an invitation and a limitation.

The many new types of technical ceramics present a difficult challenge to bonding by the three methods of pure copper metallization. Understanding and control of the mechanisms of chemical and mechanical bonding have been elusive as ceramic manufacturing processes and materials develop. Especially difficult has been AlN ceramic and the raw material formulations that were constantly modified for better ceramic material characteristics. In general, the chemical composition is determined by the pure ceramic and by small amounts of oxide added to facilitate sintering. After the ceramic is processed, the surface composition reflects the overall chemical makeup, but the physical surface has different phases, or crystalline structures, such as grain boundaries typical of most solid ceramic materials at room temperature. Problems arise when a chemical and mechanical bond depends on a particular chemical and physical composition of the ceramic surface. Without a thorough understanding of the topography and the need for consistency, ceramics of differing chemical composition and structure make consistent and repeatable bonds almost impossible. Also, physical surface conditions such as roughness and flatness may play a significant role in adhesion because of secondary effects of mechanical bonding and actual contact with a solid sheet of copper. Poor adhesion test results are often traced to some surface condition previously unknown, or uncontrolled, on the technical ceramic. Special concern should be paid to very small changes in chemical or physical properties that can drastically alter adhesion properties of a given bonding process.[9,35,61,62]

Surface conditions play an important part in determining the optimum process for copper metallization for all ceramic types, not just technical ceramics. To get the most stable and reproducible surface, as-fired ceramic substrates are preferred for several reasons. Sources of contamination are much more limited, and the surface is generally contamination-free with proper handling. Some hybrid products require surface treatment of the as-fired substrate, such as polishing to meet the tight mechanical design tolerances of large substrates or extremely high frequency applications. However, processes such as grinding, lapping, and polishing can leave cracks, particles, and residual stresses within the body of the ceramic. Grinding, for example, is known to significantly increase the amount of residual stress within the ceramic, leaving microcracks that may be driven by the stress to propagate later beneath the metallization.[7,17] Lapping and/or polishing to an appropriate depth into the ceramic surface can remove microcracks, but the extra handling may introduce contamination such as small ceramic particles.

Interstitial voids can cause many problems as well. Any of these variations has been found to decrease the adhesion strength of the bond.[7] Preventing an adhesion problem may make it necessary to develop the specific surface conditions that are allowed before pure copper can be consistently bonded to the ceramic. Fortunately, resintering to an original as-fired surface finish works well to eliminate adverse side effects of physical ceramic modification for many of the pure copper metallization processes.[10] Surface finishes of as-fired or resintered ceramics are generally in the range of 15 to 30 μm, and have been acceptable for DBC, plated copper, and AMB technologies.

From the above discussion, it is clear that control of surface conditions is critical to the success or failure of bonding pure copper to ceramic. Scanning electron microscope (SEM) photos of the substrate surface before metallization will usually give some clue to the nature of an adhesion problem caused by surface conditions, especially if a subtle change has occurred. The permissible ranges of each surface condition, such as chemistry, roughness, grain size and shape, microstructure, and boundary phase, are variables to be understood when adhesion problems arise.[8,9] It is much more desirable to establish design parameters and rugged metallization processing using ceramic material as delivered from the vendor, without special and costly preparation. With a careful and deliberate approach to the surface condition of the ceramic substrate, the process of applying pure copper metallization will produce higher yields and greater reliability and will be more economically viable.

5.3   Direct Bond Copper Technology

The direct bond copper on alumina process was developed and patented by General Electric engineers early in the 1970s. Within the next few years, it was incorporated into the design of several power hybrid modules used by the military.[11] DBC is a unique method of joining thick, pure copper layers to oxygen-containing ceramic substrates without a separate processing step or additional material for creation of the interfacial adhesion layer. In addition to the obvious advantage of good thermal transfer between copper and oxide ceramic, high current applications require the high conductivity of pure copper, and can take advantage of integral leads extending from the substrate. Interconnection can be done across several substrates, and vias may be created to meet the necessary levels of circuit density for a multilayer approach. To meet even higher levels of hybrid performance requirements, DBC technology has focused on copper metal bonded to a variety of advanced ceramic materials, including nonoxide types.

Mating of two dissimilar electronic materials without the addition of another element to the interface between them summarizes the DBC process. A tough interface is generated between just the pure copper and the ceramic substrate, one that can reliably pass through assembly processes and temperature cycles during use. To make this interface, DBC technology employs a unique chemistry that takes advantage of the lower-melting-point copper and oxygen eutectic, below that of either pure copper or oxide ceramic. At that melting temperature, the eutectic is the only liquid present, wetting and bonding to both surfaces.

Explanation of the direct bond process between copper and alumina begins with a review of the partial phase diagram in Figure 5.1, detailing physical and chemical states at different percentage compositions and temperatures.[13] Several features of the phase diagram pertain to aspects very important to the DBC process. Pure copper melts at 1083°C but will not bond with any ceramics in the liquid state. Just below the melting point of pure copper, very small amounts of oxygen will dissolve into the copper metal and exist as a solid solution, termed the *alpha state*. Higher concentrations of oxygen and different temperatures give rise to mixtures of copper, cuprous oxide (Cu_2O), and alpha material. Copper and oxygen, at the proper concentrations and temperature, have a eutectic mixture with a melting point beginning at a little over 1060°C. With careful control of the temperature at 1065°C, 0.39% by weight oxygen and copper metal form a liquid that can melt, wet, and bond tightly to the surfaces in contact with it when cooled to room temperature.

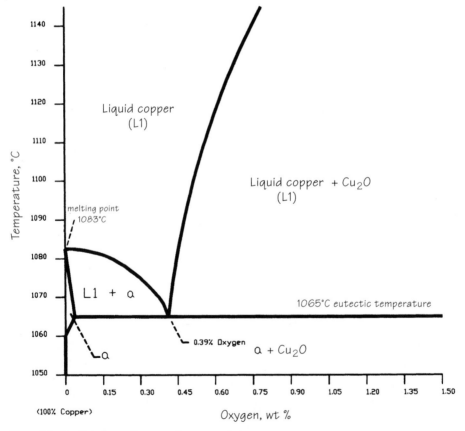

Figure 5.1   Partial phase diagram of copper and oxygen.

Since it does not melt the bulk copper layer, the eutectic is a convenient and useful way to avoid the introduction of another interfacial layer.

Before the copper, aluminum, and oxygen interfacial layer is discussed in detail, the separate steps of DBC processing will be described to give the necessary background. Figure 5.2 shows the steps necessary to form the DBC hybrid metallized substrate. For the second and third steps, several different methods of introducing the oxygen necessary for the bond have been tried. Oxygen has been added through control of the furnace atmosphere, and copper oxide powder has been applied to the ceramic surface. These methods have made it difficult to control the condition of the final bond, making the results unacceptable.[12,13] A more favorable method begins with a thermal or chemical oxidation of the copper sheet to be placed onto the ceramic substrate. For example, a thin, dimensionally controlled layer of cupric oxide (CuO) grown onto the copper sheet can supply oxygen for the generation of eutectic material. This provides a very uniform and controlled amount of oxygen, which addresses the concern that the cupric oxide present at the beginning of the DBC process must be completely consumed, so that it is not present in the interfacial layer after bonding to create a brittle layer.

After alignment of the oxidized copper sheet to the clean ceramic substrate, the assembly is sent through a belt furnace for high temperature processing at

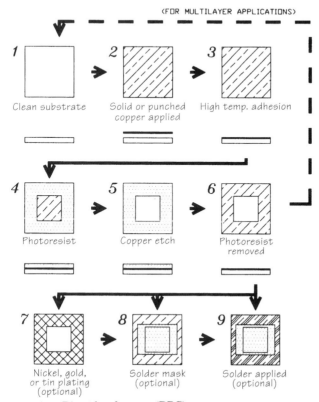

Figure 5.2   Direct bond copper (DBC) process.

1065°C. Fixturing to put pressure on the assembly is not necessary to promote the bonding process, so it is easier to find an appropriate temperature profile based on the mass of the substrates alone. An inert atmosphere inside the furnace provides the last condition necessary for bonding, as the melting point of the eutectic is reached. Liquid eutectic wets both surfaces, chemically and mechanically adhering to them. Slow cooling is necessary to accommodate the interfacial stress of mismatched TCE between copper and ceramic. The cooling time depends on substrate size and the thickness of copper and ceramic.

There is a lack of agreement regarding the actual chemical reactions and crystalline structures present in the interfacial layer both during and after formation. As a consequence, the explanations of good or bad properties of the interfacial/bond layer are not universally accepted. In spite of the more technical disagreements, several aspects of the chemical reactions have become accepted as present in the solid interfacial layer. Referring to the phase diagram in Fig. 5.1, oxygen at the alumina–copper interface at a temperature of 1065°C will cause a liquid copper–oxygen eutectic to form. This eutectic is a mixture of Cu_2O (cuprous oxide) formed from the CuO (cupric oxide) on the copper surface, alpha material, and liquid copper. As the bonding process begins, the Cu_2O first wets and then reacts with the Al_2O_3, forming very small concentrations of copper aluminates ($CuAl_2O_4$ and, more important, $CuAlO_2$). Other grain boundary constituents from sintering agents are dissolved into the eutectic liquid at this same time. Upon cooling, Cu_2O becomes segregated as crystals within the interfacial layer, bonding to the pure copper, alumina, and copper aluminates. This is the interfacial layer bonding pure copper metal to alumina; and it is used as a model to describe the existing process.[13,14]

Cooling the liquid interfacial layer to the solid state and continuing down to room temperature produces interfacial stress between the copper and ceramic. Excessive stress after bonding, or preexisting defects which weaken the ceramic, will generally cause ceramic defects such as bowing, cracking, or delamination. Most of the interfacial stress is caused by copper metal contracting at a rate almost twice that of ceramic, placing a tensile stress on the copper and compressive stress on the ceramic. As a broad statement of the stress accommodation between the two materials, there are three factors:

1. Plasticity of malleable copper in an annealed condition
2. Slippage between copper grains
3. Strength of the adhesive bond

Slow cooling of the copper sheet will anneal it into what has been termed a *dead soft* condition. With minimal hardness leading to plastic flow of the metal, this annealing affects the grain size, reducing stress by allowing individual grains to slip past each other as the copper shrinks. Stress is lessened by these factors, but not completely eliminated. Adhesion strength must make up the difference in stress accommodation between copper and ceramic, with mechanical and chemical surface conditions of the alumina having the greatest effect on the quality of the resulting bond. The surface chemistry affects formation and composition of the eutectic mixture, both as a liquid and as the solid interfacial layer at room temperature. Minute changes to chemical composition, usually the concentration of magnesium, calcium, or other sintering agents, can dramatically change the adhesion strength.[61,62] Physical surface imperfections have a detrimental effect on mechanical properties of the bond, especially particles and microcracking as discussed in Sec. 5.2, "Substrates." Thorough understanding of the stress and composition of the interfacial layer is the most important part of DBC technology, providing the adhesion mechanism and accommodating the stresses for reliable performance.

The interfacial layer formed by DBC has one other characteristic worthy of note. Blisters or voids between copper and ceramic may appear after the direct bond process, especially with a thinner copper sheet of 0.005 in (127 μm) or less. Larger metallized areas completely covering the ceramic are also prone to the phenomenon, having an accumulation of camber tolerance for the ceramic to keep the copper sheet from making complete contact. One method of relieving this problem has been a special preparation of the copper sheet, involving grooves in the copper of various patterns and depths. These have been found effective in moving the gas pockets away from the interface to the edges, but reduce the copper surface area directly contacting the ceramic surface and reduce the adhesion strength. In addition, groove edges along the bottom of the copper layer can collect other materials such as flux from soldering operations, which may cause reliability problems if they are released later as contamination. Refinements to DBC technology indicate that the grooves are not required for well-controlled processes with high temperatures and attention to surface cleanliness. Very small changes in the low levels of oxygen in a furnace atmosphere will cause slight but critical variations of the liquid eutectic melting point. Complete absence of oxygen in the furnace may allow the oxidized copper surface to be reduced, leaving not enough CuO for the eutectic reaction. Each of these problems may lead to areas of slightly higher melting points because of too much or too little oxygen, and the surfaces are not wet when the normal eutectic point is reached. Water vapor is another contaminant to control for the same reason, making it important to use very low dew point gases.

Voids most likely come from contamination of either surface to be bonded. Among the many sources of contamination, human contact has been known to cause voids for some time. Proper cleaning and handling of surfaces before bonding is therefore essential. Voids may also be caused by a lack of flatness between copper and ceramic areas, so that the eutectic mixture cannot flow between them and wet them. Although the copper sheet does become soft at 1065°C and conform somewhat to the ceramic surface, it may not be able to adapt to an uneven ceramic surface. This may be a contributing factor to the problems with very thin copper sheet, noted earlier. Copper layer thickness below 0.005 in (127 μm) are the primary problem; voids may form over uneven ceramic with less weight pressing the copper down for contact.

The next step of the fabrication process is chemical etching of the copper sheet. (Although prepunched copper can be used, prepunching is done only with designs that can tolerate poor dimensional stability.) General design rules call for copper metallization less than half the thickness of the ceramic, with solid copper sheets preferred for the DBC process. Chemical etching after bonding creates the desired pattern, with careful attention to the dimensional effects of etching thick copper.

Before the etching process can begin, however, photolithography applies the circuit pattern to the copper surface. Etching thick copper requires artwork that has been adjusted by an *etch factor*. The factor is necessary because the etch process removes copper from beneath the photoresist layer, while it removes copper across the substrate surface and separates the individual patterns. Copper thickness is the determining factor in what etch factor will be used. Figure 5.3 illustrates the etch process and details some of the photolithographic compensations that are necessary. Copper is initially removed by the etching solution from the top surface of the copper, through areas exposed by openings in the photoresist lamination. As the etch process continues, however, copper is removed in all directions, not just through the copper to the substrate surface. When the etch process has separated the copper areas across the substrate surface to the required dimensions, the etch process is stopped. A curved sidewall in the copper thickness dimension leaves different top and bottom dimensions to the etched copper pattern. Etch factor B accounts for this difference, and the difference between the bottom width and the edge of the artwork image/photoresist lamination is termed etch factor A. Depending on the rate of etching, copper thickness, and photolithography process, artwork must be adjusted to leave the required copper pattern on the ceramic surface. Etch factor A of Fig. 5.3 is generally 0.004 in (102 μm) for 0.008-in-thick (203-μm) copper, meaning that the artwork image created by the designer would be wider than the copper trace width after etching. Minimum feature sizes of lines and spaces are typically

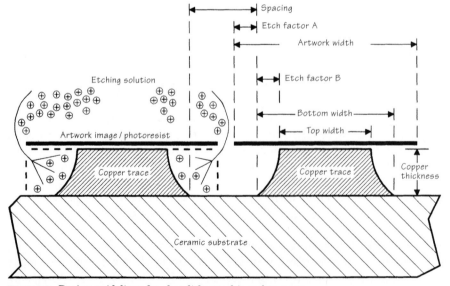

Figure 5.3 Design guidelines for photolithographic etch process.

twice the thickness of the copper, but with accurate design allowances for spacing and etch factors A and B, patterned features the same size as the copper thickness are possible. Consideration should also be given to allowing openings in the photoresist for etching solution to reach the copper surface. This dimension is usually one half the thickness of the copper, causing a further spreading of the pattern, and increasing the minimum spacing of copper feature sizes. As a final note on circuit density, tolerances of the copper trace width, both top and bottom, and spacing between them is approximately ±0.003 in (±76 μm), as a result of different etching rates along the edge of the copper.

Table 5.3 includes a summary of DBC design rules, and provides a comparison to the other copper metallization technologies. Standard ceramic substrate thicknesses can be used, but less than 0.025 in (635 μm) should be carefully considered for limiting the overall size and stresses that may cause ceramic fracture. Substrate size is another design consideration, up to 4.000 in sq. (10 cm) being the upper limit in most cases. If integral leads are not required, multi-up substrates should be considered. Prescoring a large ceramic substrate matching the final size required is a cost-saving step, but design guidelines for multi-up substrates are specific to the scoring technique, with ceramic type and thickness important to final yields.

Choosing the thickness of the copper is a major design decision. The choice must balance electrical and thermal performance with size and weight constraints. Designers should bear in mind that, as the copper thickness increases, the ceramic thickness must increase as well to withstand the TCE mismatch. For a 0.005- to 0.008-in (127- to 203-μm) minimum, with the possibility of going to 0.500 in (12.7 mm) as a maximum, 0.005-in-thick (127-μm) copper is the accepted lower limit, before the onset of increased numbers of blisters or voids in the interfacial bond layer. The preferred minimum is 0.008 in; this is the maximum copper thickness possible on both sides of ceramic less than 0.025 in (635 μm) thick. Substrates are available with 0.008-in copper layers on each side of a 0.010-in(254-μm)-thick alumina ceramic, but handling should be well-controlled.

The advantages of thinner copper are several. Etching is possible down to smaller geometries, which become more important in striving for the elusive goal of higher input/output (I/O) on the hybrid product. Sidewalls on either side of the thinner copper trace are more perpendicular, or rectilinear, providing better electrical performance at higher frequencies. TCE is closer to that of the ceramic, which improves reliability of components mounted to the surface and reduces the compressive stress on ceramic. Finally, the larger-size hybrids will have savings in size and weight, depending on the coverage of copper over the total surface area. Thicker copper is usually required for current-carrying capacity, and circuit density must be sacrificed for better electrical and thermal performance. A last design consideration for the copper layer is the tolerance of the thickness dimension, with ±0.0025 in (63 μm) being typical without lapping the surface to a closer tolerance.

Double-sided copper is generally required on all ceramic substrates with DBC metallization. Exceptions may be applications with small surface areas,

TABLE 5.3 Comparison of Design Rules for Pure Copper Metallization

Technology	Line width, in		Etch factors,* in	Registration,† in	Copper thickness,‡ in			Via diameters, in		Integral leads from edge of ceramic	Camber,§ inch per linear inch
	Min.	Typical			Min.	Typical	Max.	Min.	Typical		
Direct bond copper	0.015	0.020	0.004 to 0.008 with 0.020 pullback required from ceramic edge	±0.008	0.005	0.012	0.500	0.016	0.064	0.100 in sq. required on ceramic for integral leads	±0.004
Plated copper	0.002	0.005	0.0005	±0.005	0.001	0.002	0.005	0.005	0.025	Leads not possible	±0.003
Active metal bond copper	0.015	0.020	0.004 to 0.008 with 0.020 pullback required from ceramic edge	±0.010	0.008	0.010	0.012	Vias not possible		0.100 in sq. required on ceramic for integral leads	±0.004

*The width of the artwork pattern determines final conductor width. Etching will reduce the original artwork width by several mils, depending upon copper thickness. Therefore the artwork must have a wider dimension to compensate. This is called the etch factor, as shown in Fig. 5.3.
†The artwork is registered to a feature on the ceramic substrate, usually a corner of the ceramic surface.
‡Copper layers can be lapped to a lesser thickness after metallization.
§Assume that both ceramic sides have copper metallization. Then camber depends on differences in percentage of ceramic surface area coverage by copper.

or very thick ceramic, but the bowing of the metallized substrate is very difficult to deal with through assembly, test, and in the final product. Although
this may be a penalty of additional size and weight, there are benefits to
higher thermal conductivity with alumina and the need for ground planes in
high power and/or high frequency designs.

Integral leads are designed to extend from the substrate edge, bonding a
copper sheet larger than the ceramic substrate. After etching, there must be
some minimum coverage of the lead to the surface of the ceramic to ensure
good adhesion. A bonded area of 0.100 in^2 (2.54 mm^2) will provide the necessary strength for ease of handling and later assembly processing. These leads
have many electrical advantages for power applications, but the assembly
design considerations especially benefit from the ability to form leads if
desired. Ease of assembly, better performance, and more reliable interconnection usually result. Forming also makes possible three-dimensional hybrid
structures, with the unique ability of having copper leads join more than one
ceramic substrate. Space constraints may be apparent in two dimensions, but
with design space in three dimensions, copper on ceramic technology becomes
a solution to more design problems.

Multilayer DBC substrates have alternating layers of copper and ceramic,
up to four layers of each. Metallized vias in the ceramic for electrical interconnection between layers are down to a diameter of 0.016 in (406 μm), with
larger sizes possible if current density demands it. By placing copper pellets
of a slightly smaller diameter into the via hole, contact is made between
metallized layers. Vias require a pad area around the hole of 0.040 in (1 mm)
or more to ensure reliable electrical contact, and hermeticity if required.

A new technology such as DBC rarely stands alone as a solution. Thick and
thin film technologies have both been combined with the DBC approach to
enhance the capability of the substrate and make use of the existing facilities
of the user. The addition of resistors and localized high density areas are two
more tools to increase the capability of the substrate, making possible combinations of DBC with other metallization technologies. In particular, copper
thick film conductors and resistors have been successfully combined with
DBC, the thick film and pure copper overlapping by 0.080 to 0.100 in (2 to 2.5
mm) on each end of the resistor network. Designs requiring higher interconnection density can also benefit from multilayer copper thick film, combining
the thermal and electrical advantages of pure copper conductors with small
components which require the large number of I/Os usually associated with
the control circuitry of power hybrids.

Aluminum nitride ceramic is a relatively new substrate material for use
with DBC, and is still struggling with its own problems of early introduction
as a technical ceramic. Some of these problems have contributed to the
mixed results with DBC technology, such as different grain structures,
surface degradation from cleaning, and variable amounts of different sintering agents.[8] Finding an optimum DBC process requires a consistent and stable ceramic product, so that pertinent development issues can be addressed.
The most obvious issue is that direct bond processing does not occur without

the surface presence of solid oxide ceramics. To make a suitable chemical and mechanical bond with DBC, additional preparation of the AlN surface is required in the form of oxidation of the AlN crystals at the surface of the substrate, forming a thin layer of alumina ceramic. This is then the ceramic oxide to be direct bonded with pure copper by the methods already discussed.[15]

Pretreatment of the AlN surface and then forming the oxide layer has been the greatest challenge, requiring many different combinations of process steps. Some pretreatment methods are both mechanical and chemical in nature, polishing the surface and lightly etching it in aqueous solution to obtain the proper characteristics according to the manufacturer's process. Oxidizing the AlN surface to a thin layer of alumina has generally been done by thermal diffusion, heating the substrate to temperatures of around 1250°C for 5 to 20 min in the presence of oxygen gas. Good control of the alumina thickness at anywhere from 1 to 6 μm determines the time and temperature required. Cracks on the new alumina surface seem to degrade the adhesion strength after DBC processing, making the understanding of the cracking mechanism and its elimination from the surface very important. Work has been done to anneal the resulting oxide layer in the presence of argon at high temperature, to create a smooth and uniform surface for better adhesion to copper metal. While modifications to the surface of the alumina have been marginally beneficial, control of the alumina thickness and overcoming the thermal mismatch with AlN has been deemed critical to the adhesion strength.[8,12,16,18]

Successful vendors of DBC with AlN are overcoming several problems and setting design guidelines for power hybrid circuits. There are still concerns regarding DBC of AlN, however. AlN microcracking at the alumina interface is a problem that arises whenever materials or fabrication process parameters change, such as properties or thickness of the alumina interfacial layer.[17] Propagation of the crack seems to be assisted by the TCE mismatch of the alumina, AlN, and the copper metallization. Microcracks at the edges of copper to alumina bonds have also been observed, from the same stresses. To address this problem, grooves have been put into the copper sheet, but this leads to a reduction of surface area in direct contact, and increases the thermal impedance already present as a result of the alumina interface between the copper and AlN. Good process controls of AlN chemistry, structure, and preparation, alumina thickness and porosity, and DBC processing have yielded good AlN copper metallized substrates. Design guidelines generally have the AlN ceramic thickness at least 2 times the pure copper metallization thickness. Multilayer structures of up to four layers are possible, using large vias for interconnection that can be formed with a hole diameter of approximately 0.080 in (2 mm) and placing a slightly smaller surface-oxidized copper plug inside for bonding to the surrounding materials. An excellent adaptation of AlN is the combination with alumina in a single product, using the additional performance of AlN only where needed and benefiting from the lower cost of alumina.

5.4 Plated-Copper Technology

A well-known application of pure copper in electronics is the printed circuit board industry, where copper plating and etch technologies are commonly used. Firmly established for many years as the preferred material and process for interconnection on PCBs with through-hole components, copper conductors have met the challenge of increased circuit performance and fine pitch technologies. In attempting to duplicate this success on ceramic substrates, plated-copper metallization of ceramic started as a spin-off of PCB processing. Because of the ease of plating compared to other pure copper metallization methods and the abundance of existing plating capability, copper-plated substrates are an evolutionary step in the search for higher levels of hybrid performance.

Plated copper is bonded to ceramic surfaces primarily by mechanical bonding mechanisms, typically using a semiadditive process in aqueous solutions. There have been many variations of the plating chemistries and combinations with other metallization technologies in the search for an economical method of reliable adhesion and better understanding of the ceramic surface conditions. The difficulty for every approach lies in achieving a reliable bond between two dissimilar materials with a plating technology of low thermodynamic energy.[19—22, 60]

As already mentioned, ceramics are generally inert and are unlikely to chemically bond with other materials at room temperature. Methods related to the science of plating metals on PCBs have been used to similarly prepare the ceramic surface and change the stable nature of the material to facilitate bonding. Most plated-copper technologies use specific pretreatments of the ceramic substrate before electroless deposition of pure copper. This is the most important process step, leading to adhesion at the copper-to-ceramic interface independent of the oxide or nonoxide nature of the ceramic. The properties of an adhesive layer deposited over the ceramic surface are critical to the reliability of the finished copper metallization.

Three categories of pretreatment will be discussed:

1. Modification of the ceramic surface for a rougher finish
2. Modification of the ceramic surface with a catalyst
3. Vacuum deposition of an adhesion layer

The first method relies exclusively on mechanical bonding of copper metal to the ceramic surface. Mechanical bonding is facilitated by a rougher surface, with pores for copper to lock into the ceramic crystal structures. A rough surface on the ceramic may be thought of as an adhesion layer, allowing the copper to surround and mechanically interlock with the crystalline structures. The surface is modified by a strong acid or base in an aqueous solution for most ceramic materials. The strength of the etch solution and time of exposure are adjusted to expose crystalline structures and enlarge grain boundaries without reducing their adhesion to the bulk ceramic material. This is necessary to open up the tightly bound and closely packed structures present at the surface. Surface roughness of the ceramic surface is measured and adjusted to provide the best adhesion with copper. Wax adheres more strongly to a textured cloth than to smooth glass; the same principle applies to finding the optimum surface finish for copper plating of a ceramic surface. Adhesion strength has been adequate for some applications, but the search continues for methods to increase the strength of the bond. One method is particularly successful: sending the metallized substrate through a thick film firing profile of 800 to 950°C. Firing of the plated ceramic during or after the overall plating process increases adhesion strength, potentially creating a chemical bond, but requires careful attention to internal stresses that may be introduced.[23,35]

For the second copper plating approach, pretreatment methods have included activation of the ceramic surface, usually by a tin–palladium catalyst that assists with the deposition chemistry of electroless-plated copper. This process begins with a mild etch of the ceramic, leaving a clean, particle-free surface for application of seed layers. Typically a screen printing process applies the catalyst-containing seed layer for activation and adhesion of the ceramic material and copper in solution. The nature of a seed layer is usually proprietary, providing catalytic activity to remove cupric ions from solution and deposit them as pure copper metal. A proposed mechanism uses high pH aqueous solutions containing formaldehyde as the reducing agent, and cupric ions coming into contact with the activated ceramic substrate:[23]

$$2Cu^{+2} + HCHO \text{ (formaldehyde)} + 3OH^{-} \xrightarrow{Pd} 2Cu^{+} + HCOO^{-}$$

$$+ H_2O \rightarrow Cu \text{ (deposited)} + Cu^{+2}$$

The catalyst is deactivated during the electroless process, and process parameters must be optimized for an acceptable rate of electroless copper deposition. The concentrations are important to control to avoid too low a copper density, voids, blisters, or unusual copper grain structures in the copper flash plating. Adhesion is primarily a mechanical bond at this point. However, there may be catalytic activity which allows some form of chemical reaction, but no definitive answer has been found.[24]

The other method of ceramic surface preparation uses thin film vacuum processes to deposit a metal adhesion layer. The same approach can also deposit the pure copper layer, but is limited in the thickness that can be applied. This is a costly and technologically sophisticated process, and applications are limited to expensive products requiring very low throughput. Economically, it is generally more practical to plate the copper to the desired thickness electrolytically, over the adhesion layer. Many types of metal adhesion layers are deposited, depending on the ceramic type, existing capability, and expected use. Other drawbacks to this approach apply to high frequency applications, where the excessive electrical and thermal resistance of the interfacial layer may cause unacceptable losses. The adhesion between ceramic and thin film layer is typical of the thin film characteristics described in Chap. 4 of the handbook.

With a thin layer of electroless copper applied to the ceramic surface, standard photolithographic techniques of the PCB industry can begin. Photoresist is applied, and the choice of type may be based solely on price, but more likely will be based on past experience and existing capability. Dry film or liquid resist may be of the positive or negative exposure type; each approach has advantages for particular applications. Substrates requiring a very fine line geometry, down to 0.002 in (50 μm), may need to use liquid photoresist and forgo a thick copper layer. Dry photoresist allows a thicker copper plated layer of 0.005 in (125 μm) and vertical side walls, which are important at higher frequencies, but cannot produce the fine geometries that a liquid photoresist can.[25]

After exposure, the photoresist is patterned and is ready for the electrolytic copper plating process. The copper flash plating can carry current across the surface of the entire substrate, for deposition of the bulk copper layer. Electrolytic baths immerse the prepared ceramic substrate in copper solution, providing electrical contact for deposition of pure copper onto the substrate. Bath chemistry should be constantly monitored, especially during plating of technical ceramics that may contaminate the bath. The plating current density has an important effect on the copper structure of the plated layer, yielding large grain size and low density if deposition occurs too rapidly. The larger grains have poor adhesion between grains and additional problems of internal stress causing voids, cracks, and decreased adhesion to the ceramic surface. After plating to a desired thickness, no more than the thickness of the photoresist, the metallized substrate is ready for the next process step: removal of the photoresist, which leaves the circuit pattern as well as the original electroless copper and/or other adhesion layers. A quick etch step removes the adhesion layers, and electrically isolates the pattern. As shown in Fig. 5.4, other processes may be required, such as nickel and gold plating or the application of a solder mask. The pure copper metallized substrate is now ready for testing and assembly.

Plated copper has advantages derived from two basic characteristics, the low temperature processing and the thinner copper metal layer (less than 0.005 in), both contributing to reduction of stress within the interfacial layers of metallized substrate. Room temperature joining of the copper and ceramic results in less TCE stress than high temperature processing. Thin copper layers formed at low temperature with less stress at the interface and less mechanical strength will not cause as much warp or bow in the ceramic substrate. This is an important consideration for thin ceramic substrates where mechanical strength is poor, and for designs allowing only one side of the substrate to be covered with copper metallization. Depending on copper thickness, ceramic thickness, and area coverage, two-sided copper may not be required for metallized substrate integrity.

The high TCE of pure copper below 0.005 in (125 μm) thickness is significantly reduced in the x and y dimensions, strongly influenced by the lower TCE of the much stronger ceramic. Thin copper on ceramic will be advantageous to the TCE mismatch of attached components, reducing the stress on

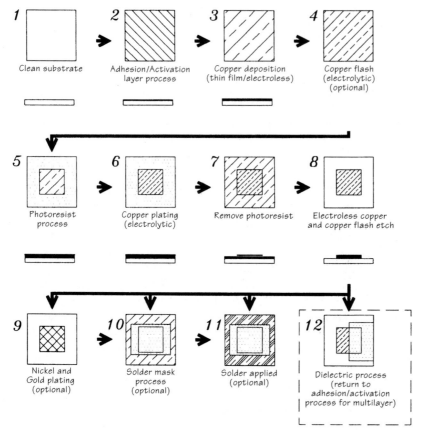

Figure 5.4 Plated-copper process.

components with changes in temperature. This is important for application of bare silicon chips to the metallized surface, making more reliable interconnections with either wirebond or flip chip processing.

Thinner copper metal is easier to etch, and smaller traces are the result. Trace width and spacing of 0.005 in (125 μm) is typical, with 0.002 in (50 μm) possible. Unlike the DBC and AMB processes, the plating into the photoresist openings makes dimensional control a function of the photoresist, not etching through copper. Design guidelines for the processes are compared in Table 5.3.

Printed circuit board design rules can be used for layout, but cannot be applied to multilayer products. Multilayer structures are possible, with the application of dielectric that has adhesion to the pure copper and ceramic. Several types of dielectric have been used, in both high and low temperature applications. Dielectric layers usually found in thick film processing have the potential of increasing density and take advantage of high temperature firing to increase adhesion. This is a new and potentially useful approach, and work continues to improve its capability.[19,26]

Via metallization coverage in both ceramic and dielectric materials is an advantage of plated-copper metallization. The via walls in ceramic are usually a problem, having loose, ragged surfaces or laser-melted debris in them. Thick film has difficulty filling the hole even with vacuum pulling from below, and thin film technology has shadow effects which demand larger via hole sizes. Other pure copper metallization technologies must put copper into the via as a separate process step, with large sizes required. With copper plating, the through hole via in ceramic can be thoroughly metallized down to relatively small diameters regardless of the method used to create it.[21] The smaller via diameter is not just a saving of real estate, but an important consideration for reduction of the TCE mismatch to the ceramic. With less copper material, the via will not expand with as much force, potentially breaking the ceramic surrounding it.

Vias are not the only method of connecting both sides of a metallized ceramic substrate. Copper metal can be plated along an edge of the ceramic, creating electrical contact to each side.[20] This is less expensive than forming the via and metallizing the interior, but the ceramic surface conditions along an edge must be considered. Laser processing may cause glassy surfaces, and the adhesion along the edge is much more important than the adhesion inside the via since the edge is exposed to potential handling damage. Grinding the edge after laser cutting is one method of overcoming the adhesion problem. The two best solutions for copper plating ceramic edges are cutting substrates with a diamond-embedded blade or using pressed ceramic substrates, both of which leave an adherable surface.

With finer pitch a possibility, assembly techniques of wirebond and small outline surface mount technology (SMT) components become more important. These techniques avoid many of the issues related to bare copper and its oxidation with selective plating for protection of the copper surface in sensitive areas. Whether it provides selective or complete coverage, plating is typically gold over nickel for the greatest variety of later assembly processing. For mass solder reflow and greater throughput, a solder mask can be applied to the substrate if required, and solder can be reflowed to the copper surface as a substitute for nickel/gold plating.

Alumina and BeO have been the ceramic materials of choice for evaluations of plated-copper technology. They are widely used, and have well-understood properties resulting from many years of use by the hybrid industry. Other ceramic materials have been investigated, but most involve a thin film metallization approach to deposit a compatible adhesion layer which is then plated with copper metal.[28,29] Even without the broad base of experimental results, special mention should be made of aluminum nitride ceramic. Using the activation pretreatment technique of tin and palladium catalysts, electroless copper has been plated on AlN. Results were reported as acceptable, with the unusual effect of hydrogen gas appearing at the ceramic surface and having a detrimental effect on pure copper density.[22,27]

5.5 Active Metal Brazing Copper Technology

Metal brazing compounds are usually an alloy of two or more metals, with combinations selected for a set of desired properties. Melting points above

500°C differentiate braze materials from lower melting point solders, along with much greater strength and adhesion when bonded. For processing, braze compounds are heated above the melting point for wetting and bonding to many different types of glass, ceramic, and metal. They must promote intimate contact between surfaces, joining them into a chemically and mechanically bonded structure having the ability to accommodate thermal expansion stresses from cooling after fabrication or temperature changes during operation. The performance of any active metal braze compound for creation of hybrid substrate is judged by its ability to react and reach chemical equilibrium at the interface of copper and ceramic. Ceramics have presented the greater challenge to good braze joints because of the many different types, TCE mismatch with metal, and inert nature of the material preventing wetting and adhesion. Active metal braze compounds have been developed to meet this challenge, using titanium to start a reaction with ceramic during the braze process, resulting in wetting and diffusion bonding of the braze material to both copper and ceramic. With this capability it is possible to copper-metallize a ceramic surface, create hermetic enclosures, and fasten both metal and ceramic structures to the hybrid substrate surface for better utility.[30,31,36]

Brazing of ceramic has long been the province of refractory metallization processes using thick film screening or spray application to put refractory metal onto the ceramic surface. A slurry of tungsten, or molybdenum, and manganese is dried and sintered to the ceramic in a reducing atmosphere at around 1200 to 1500°C. This metallized ceramic surface has good wetting and adhesion characteristics, but requires plating or other braze materials to create a reliable bond.[7,31,32] In addition, the metals used are expensive and poor conductors of heat and electricity.

Active metal brazing technology has several advantages over refractory metallization, for reasons of simplicity and economy. The braze compound containing an active metal species is processed in a vacuum or inert atmosphere, without separate steps or addition of other braze compounds. AMB has the potential for use across a wide variety of newly developed ceramics, and as the types of ceramic material increase, a standard technique to metallize them becomes very attractive. Table 5.4 lists many of the active metal braze compounds used today.[36] Other metallization techniques may require extraordinary measures, depending on the type of newly developed ceramic, and produce an unknown interface of ceramic to metal. The flexibility and simplicity of AMB are important options for the hybrid designer requiring copper-metallized ceramic substrates.[7,33]

The active metal of the braze compound is usually titanium, although other transition elements from column IV-B of the periodic table such as zirconium or hafnium can be used. Titanium is characterized extensively in the research literature and is used in the majority of brazing compounds for AMB applications. At temperatures of 700 to 1050°C in a vacuum of 10^{-5} torr, or low pressure and dry ($-70°C$ dew point) inert gas environment, active metals exhibit strong chemical activity within the alloy. This promotes the brazing process of ceramic materials without compression. An alloy containing 1 to 2 wt %

TABLE 5.4   Active Metal Brazing Alloys

Alloy	Composition, wt %	Brazing temperatures, °C	TCE, ppm/K
Au–Ni–Ti	96.4–3–0.6	1025–1030	16.1
Cu–Al–Si–Ti	92.75–2–3–2.25	1025–1050	19.5
Pb–In–Ti	92–4–4	850–950	
Ti–Cu–Ni	70–15–15	960–1000	
Ti–Cu–Ag	68.8–26.7–4.5	810–900	
Sn–Ag–Ti	86–10–4	850–950	
Ag–Ti	96–4	1000–1050	
Ag–Cu–Ti	70.5–26.5–3	850–950	
Ag–Cu–Ti	64–34.5–1.5	850–950	
Ag–Cu–Ti	63–35.2–1.75	830–850	18.5
Ag–Cu–In–Ti	72.5–19.5–5–3	850–950	
Ag–Cu–In–Ti	61.5–24–14.5–(Ti)*	850 –950	
Ag–Cu–In–Ti	59–27.2–12.5–1.25	700 –750	18.2
Ag–Cu–Al–Ti	92.75–5–1–1.25	900–950	20.7
Ag–Cu–Sn–Ti	63–34.2–1–1.75	810–860	8.7
Ag–Cu–Ni–Ti	71.5–28–0.5–(Ti)*	850–950	
Ag–Cu–Ni–Ti	56–42–2–(Ti)*	900–1000	

*Titanium concentrates below 0.1%.

titanium becomes liquid and diffuses into the ceramic, chemically reacting with other metallic elements. The chemical reaction is of the redox type, driven by high temperature and differences in electromotive force between metals. For AMB of copper to alumina, the high oxidation potential of titanium causes reduction of aluminum. The titanium then bonds with both the oxygen and aluminum to allow the ceramic surface to wet to the bulk braze material. The net redox reaction is $Al_2O_3 + 5Ti = 3TiO + 2AlTi$.

This process is termed *gettering of oxygen* by anion absorbing active metal species of cathodic charge. The amount of surface oxidation or the extent of the redox reaction is important to control; the activity level must be balanced to adequately wet the ceramic without producing a brittle interface from the presence of excess titanium. Higher concentrations lower the malleability of the interfacial layer, making it more brittle. This increases the stress from cooling to room temperature and the possibility of ceramic fracture. The ability of the interface to overcome the TCE stress between copper and ceramic is determined by the ductility of the interfacial layer, flow, and crystalline structure. Process variables, such as time and temperature, also determine the microstructure, which is important to the stress profile and eventual adhesion strength of the bond. Some of the other compounds that appear in the interfacial area and have crystalline structures that are affected by the time and temperature include titanium oxides (TiO_2 and Ti_2O_3), aluminum titanate (Al_2TiO_5), $CuAlO_2$, and Cu_2Al_4O, all reacting at the interfacial area for chemical and mechanical bonds between copper and ceramic.[35,36]

Work has been done to coat ceramic surfaces with pure titanium before brazing, for example, by fusing titanium hydride slurries or depositing thin films, but the ability to apply filler braze material with the active metal component already in it has made these approaches less practical. In combination with silver, copper, and other filler metals such as tin or indium, the braze material can be uniformly applied to the ceramic surface as a paste, or in a separate metal preform (sometimes called a *foil*) between the surfaces to be joined. Four methods are typically used:

1. Powders of titanium hydride and filler braze metals mixed with an organic vehicle are applied to surfaces and dried.
2. Individual preforms of titanium and braze metals are stacked between surfaces.
3. Preforms are laminated together, with the titanium between layers for better protection from contamination.
4. A true alloy of braze metal is formed into a powder or foil.[34]

In the first three applications listed above, care must be taken to keep oxygen, nitrogen, and water vapor from reacting with the titanium. Keeping the adverse titanium reactions from occurring depends on the process environment and the physical state of the braze material. With the powder of titanium hydride, large amounts of surface area are exposed during the braze process; surface outgassing may therefore deactivate the titanium metal and not enough will remain to wet with the ceramic surface. A more expensive but better quality approach is to form a true alloy of titanium with other metals, and then select one of the application methods. A true alloy will protect the titanium within the alloy until it reaches the liquid state. This has the benefit of reducing the amount of titanium required, and the resulting interfacial layer is more ductile and can form strong reliable bonds.

There are several important variables in active metal brazing for the user to be aware of, and each will be discussed as the process is described. The AMB process begins as outlined in Fig. 5.5, with application of the braze compound to a clean ceramic surface in one of the previously discussed forms. The condition of the ceramic surface must be understood and controlled, especially surface modification by grinding, which has already been discussed.[7,37] Active metal content of the filler metal must be at the appropriate concentration, but as important is equal distribution throughout the braze material. Otherwise, localized areas of the interface will be very different in wetting and adhesion strength. Next, a pure copper sheet is placed on the surface of the ceramic with active metal braze, and the assembly is put into a high temperature vacuum chamber or inert gas furnace. Processing the copper through vacuum or inert atmosphere prevents surface oxidation, and facilitates later plating or assembly to the clean, pure copper surface.

While the components are heated, outgassing takes place and reduces the amount of contamination on the surfaces, but can passivate the active metal, especially in the pure forms. Clean surfaces and use of titanium alloys reduce the negative effects. The heating rate is a critical factor, especially when the filler braze metal containing silver readily reacts with the copper foil. This

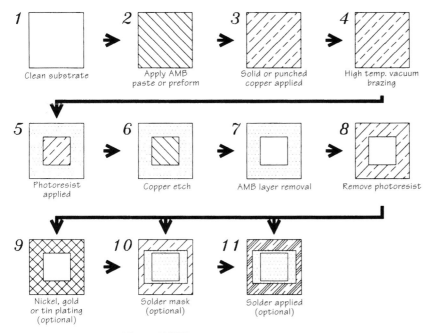

Figure 5.5 Active metal braze (AMB) process.

leads to an increase in the melting temperature of the braze alloy at the point of contact and may leave interfacial voids. Although a more rapid heating rate is desirable for this reason, a large thermal mass such as a ceramic substrate over 1 in^2 (6.5 cm^2) will not reach the braze temperature at the same time as the copper foil, causing the braze compound to preferentially wet to the copper and leaving the ceramic unbonded.[34] Preforms are more prone to premature brazing on the copper foil than are powders applied to the ceramic surface. Therefore, spending a short time just below the melting point of the braze compound balances the two factors, before a rise of 5°C/min to the peak temperature, which is held for approximately 10 min. At the melting point of the active metal braze compound, titanium initiates the braze process by activating the ceramic surface. A well-controlled braze profile with hold time at an optimum temperature is required for thorough completion of the braze reaction without overreaction, and degradation of the interfacial microstructure. Cooling to room temperature, the interfacial layer has enough ductility to reduce the residual stress of mismatched TCE, yet remains strong enough to adhere to the copper and ceramic substrate. Figure 5.6 shows the temperature vs. time relationships for heating of the copper, braze, and ceramic materials.[7,38]

Variables such as thickness of the finished joint influence the amount of ductility and significantly affect the thermal performance. Braze interfacial layers after bonding range from 0.4 to 0.6 mil (10 to 15 μm), but are not thoroughly characterized for mechanical or thermal performance. Along with atmospheric conditions during brazing and the roughness of the surfaces being joined, the variables of the AMB process are infrequently studied and depend on the skill of the user for good process development.[7,32,38]

A phenomenon unique to AMB is blushing of the ceramic, observed after brazing. *Blushing* is the excess flow of braze compound in the liquid state during the braze process. Too much blushing onto areas not requiring braze compound can deplete the amount of active metal available for the bond and contaminate other areas of the circuit with conductive material. A method of painting or placing a removable layer of material to protect surfaces is used to prevent contamination. Changes to the contents of braze metals can help reduce blushing, but care must be taken that enough braze flows. Insufficient flow will minimize wetting, reducing the bonded area and adhesion strength by creating voids. Control of the braze material content and temperature profile are critical to preventing a blushing problem and ensuring that a hermetic seal between copper and ceramic is made. However, such close control of the wetting behavior prevents via formation, since braze flow in the liquid state is so limited. Multilayer structures are therefore not in electrical contact, since copper plugs in via holes have not been wet and bonded.

Grooves in the copper foil have been used in the past to improve the properties of the finished substrate, letting gases escape and eliminating the potential for blisters to form. They have not worked very well in AMB processing

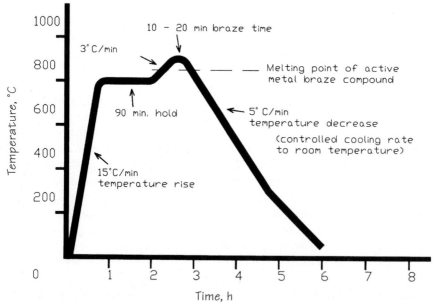

Figure 5.6 Active metal brazing profile.

because of the lack of flow of the braze material, producing a thicker interface of less overall strength between copper and ceramic.

To create the metallized circuit pattern, there are two choices. Prepunching the pure copper sheet prior to brazing is usually done for larger geometries, and may require bracing or tie bars to keep the pattern from changing shape. The other option is to etch the pure copper sheet after brazing to the ceramic, as discussed in "Direct Bond Copper Technology," with one important exception. Photolithographic processing and copper etching leave no metal over the openings of ceramic surface, but does leave some of the braze material and activated ceramic, which is conductive. If the photoresist is left on the unetched copper surface, bead blasting the surface of the ceramic will remove the thin activation layer, isolating the copper circuitry without damaging the conductor.

The AMB process is also finding much wider application for other ceramic materials such as nitrides, carbides, and other nonoxide ceramics. The bonding mechanism of braze materials containing titanium with AlN is typically TiN with trace amounts of free aluminum and quaternary intermetallics such as Ti–Al–Cu–N dispersed within the braze joint.[39] A significant challenge is ensuring the amount and homogeneity of the active metal in the interfacial layer, since a higher titanium content of approximately 7% is usually required for adequate wetting to begin the braze process. Excess amounts of active metal can cause embrittlement and promote further chemical activity after the joining process is complete. Limiting temperature exposure of the metallized substrate to 400°C or less reduces this effect, however.[31]

Titanium is known to produce interfacial reaction products of titanium carbides and borides as well as nitrides. Government funding of work to evaluate Si_3N_4 and SiC is continuing, and has led to some promising results. The applicability to new and diverse ceramic materials is a significant benefit to the development of high power and high frequency hybrid products,[31,37,40,73] but the selection of active metal, the content, and braze conditions will require more development. Driving this work will be the increased hybrid performance requirements, with users looking for the benefits of copper bonded to many different ceramic types.

5.6 Comparisons

Hybrid products have depended on metallized substrates primarily to interconnect active elements of the circuit design. Doing this in a cost effective and reliable way over the past 30 years or more brought about the term *standard* metallization technologies, which includes both thick and thin film applications. Thermal dissipation is probably the requirement noticed most, but high speed requirements on metallized substrates are forcing designers to make comparisons beyond just the typical resistivity of the conductor to DC signals. As the new designs force performance requirements to grow and change, the advantages of copper on ceramic have become more evident. Especially important to the understanding of the advantages and limitations of each hybrid substrate is the trend toward combinations of metallization technologies, bringing flexibility and a better compromise of performance to the final product. Table 5.5 lists the properties of pure metals and their standard hybrid metallization equivalents.[42,43]

Thick film technology is the foundation of the hybrid market as we know it today. Typical thickness is 0.0005 in (0.0127 mm), and gold, silver, silver alloy, and copper thick films are predominant. An easily implemented and understood technique, thick film technology has several advantages[1,3,6,45,56]:

- Design cycle of new and existing products can be extremely short
- Reliability is very good for most hybrid applications, and well-understood
- Capital investment is low
- Volume capability is high

Thick film metallization is a mixture of binders and glass with the metal, bonded to the ceramic substrate. The conductor will not have the electrical and thermal properties of the pure metal, but will perform at a lesser level depending on the metal content and the thickness of the trace. Electrical losses occur because of the higher resistance and excess heat dissipation. As the speed of electrical operation increases, additional thick film characteristics become a problem. Poor edge and top surface definition due to the screen print process can cause the electrical signal to be reduced in strength, delayed, and distorted.

It is interesting to compare soldering to thick films to soldering to pure copper metallization. Solderability testing is usually done as an assembly process evaluation to determine the acceptability of a thick film metallized substrate. There is the tradeoff of good wetting for ease of solder processing and the concern that, with rework, the metal will be removed by leaching. Pure copper metallization cannot be leached from the ceramic, and has good wettability to the solder.

Thin film technology is essentially a vacuum process with much greater equipment and tooling cost and coincidental difficulty of process implementation. Yet it has met the needs of many hybrid markets requiring the unique capabilities of a standard metallization technology. Thin film has a thickness of less than 0.0005 in (0.013 mm or 13 μm), deposited with ion implantation,

TABLE 5.5 Comparison of Metals and Conductors

Metal	Pure metal thermal conductivity, cal/(s•cm•°C)	Pure metal CTE, 10^{-6}/°C	Pure metal electrical resistivity, $\mu\Omega$•cm	Thick film electrical resistance, mΩ/sq	Thin film electrical resistance, mΩ/sq
Aluminum (Al)	0.566	23	2.7		10
Copper (Cu)	0.951	16	1.7	2.5	5
Gold (Au)	0.753	14	2.2	5	10
Platinum (Pt)	0.150	9	10.5		40
Silver (Ag)	1.02	19	1.6	2.5	

sputtering, evaporation/condensation, or chemical vapor deposition (CVD) techniques. Many different types of noble and transition metals can be deposited on the substrate. Advantages include[1,2,41,58]:

- Excellent line definition with edge and surface uniformity
- Fine pitch capability
- Good resistor performance
- High frequency response

Thin films are just that: very thin layers of material (as thin as 1000 Å or 0.1 μm) with metal traces having relatively low thermal conductivity. Pure metal layers can be applied to the substrate, but the electrical resistance is high because of a lack of metal volume in the conductor. High costs of operation, sensitive process parameters, and environmental concerns are other drawbacks in comparison to pure copper metallization. The adhesion mechanisms of thin film are important to note as well, since they constitute a mixture of van der Waals and short-order chemical bonding. The ability to assemble and bond materials of mismatched TCE is not as good as that of other metallization technologies.

Thin films find most of their applications where cost and throughput are not a major concern. They are limited to batch processing, which, with more expensive metals, contributes to a more costly substrate. They are not generally made in multilayers, which are technically possible but not practical. A final limitation noted here is that standard thin films are not readily solderable; again the capability exists but is not readily accessible by most metallized substrate users.

Design analysis of which metallization technique to use for a particular hybrid product will focus more on the performance issues, in concert with the layout parameters. All of the pure copper metallization techniques provide better electrical and thermal performance, but with a thicker copper layer in AMB and DBC, there is less capability for fine pitches. Designers must weigh the advantages and disadvantages of each metallization approach, leading in some cases to mixtures of metallization technologies for the best properties of each. A comparison of the standard metallization technologies with pure copper metallization is presented in Table 5.6.

Before justifying the use of copper metallization, it is important to understand the performance of standard metallization technologies if intelligent comparisons are to be made regarding a given application. This perspective leads to a better understanding of the differences between ideal design performance, without any effect of the metallized substrate, versus actual performance. Substrate metallization will always degrade the ideal, but good design analysis minimizes the effect within a specified cost and reliability guideline. Modeling and established design guidelines are useful tools to gain the understanding and perspective necessary. As pure copper metallization matures and desktop computational power increases, these tools will be more generally available.[49,50,51,68]

The comparison of greatest interest here is between each of the pure copper metallization technologies. General statements of advantages and limitations for any newer technology will not be constant, and can be rapidly changed as improvements occur. Table 5.3 lists the design guidelines typical of the industry today, but without judgment of future capabilities due in the near future.

5.7 Assembly

Hybrid assembly techniques on pure copper metallization are similar to existing methods for standard metallization technologies, but there are special considerations to be noted. Advantages of thermal performance to the finished product may be a hindrance to assembly with reflowed solders, and attention to thermal shock through assembly is important to keep TCE stresses under control. With the diversity of components and assembly processes now available, it is important to understand the impact of pure copper metallization technology on assembly techniques and reliability.

As an incoming material to the assembly process, pure copper substrates should be controlled by written documentation that has been agreed on by the user and supplier. This is true of in-house as well as outside substrate fabrication. In verifying the controlled substrate properties before assembly, later processing and reliability should not be affected by performance of the copper surface or substrate materials beneath the metallization. Throughout the hybrid industry, inspections to prevent deviant material from getting into the assembly process vary greatly in philosophy, form, and substance. For that reason, there are three methods to be discussed for controlling the quality and properties of incoming assembly materials, each applicable to the pure copper substrate. Visual criteria, drawings with dimensional controls, and physical test parameters are the primary methods of inspection, usually having procedures associated with each. Various combinations can be used, but the most easily implemented is an inspection procedure for visual criteria to correlate the outgoing inspection of the substrate producer with the incoming examination of the user. Table 5.7 lists some of the copper metallization visual criteria definitions that may be used for inspection or testing of all types. A procedure that notes the size and number of each feature as defined is a good approach for any user to take.[46]

Drawings are another way of controlling the copper metallized substrate; they denote the size and tolerance of features important to the function of the device. Notes on the drawing are sometimes used to control both the visual inspection criteria and the physical test methods, providing the advantage of just one document. In addition, notes designating a particular feature of the drawing are helpful in making the interpretation straightforward. They also can control the basic construction materials of the metallized substrate, leading to better interpretation of all the assembly constraints. The following notes are a partial list, and have general application to pure copper substrates.[47]

TABLE 5.6 Comparison of Hybrid Metallization Technologies

Technology	Ceramic selection	Metallization description	Adhesion mechanism	Geometry (typical)	Electrical and thermal	Hybrid assembly	Reliability	Cost
Thick film	Usually oxide, most types possible	Metal + glass	Chemical + mechanical	0.010 in (250 μm) lines and more than 0.0005 in (12 μm) thickness	Poor	Good, solder rework can be a problem	Good, well-understood	Variable; can use expensive precious metals
Thin film	All types	Pure metal, require adhesion layers	Mechanical + chemical	0.002 in (50 μm) lines and less than 0.0005 in (12 μm) thickness	Adequate	Adequate, not easily soldered	Good, well-understood	High for equipment, process, and materials
Direct bond copper	More selective, oxide based	Pure copper	Chemical + mechanical	0.020 in (400 μm) lines and 0.008 to 0.020 in (200 to 500 μm) thickness	Good	Good	Good, well-understood	Reasonable
Plated copper	All types	Pure copper, usually thin adhesion layer	Mechanical (chemical?)	0.004 in (100 μm) lines and 0.0005 to 0.005 in (12 to 125 μm) thickness	Good	Good	Less understood	Reasonable
Active metal braze copper	All types	Pure copper, with braze adhesion layer	Chemical + mechanical	0.020 in (400 μm) lines and 0.008 to 0.020 in+(200 to 500 μm+) thickness	Good	Good	Adequate	Reasonable

5-31

TABLE 5.7 Assembly Criteria

Criterion	Description
Bleedout	Occurs when the copper metal and ceramic substrate interfacial material leaves a residue just at the edge of the copper metal
Blister	A raised area of metal surface showing adhesion loss below
Bridging	The movement or other mode of relocation of any material which connects two areas not normally connected
Burr	Attached material from mechanical processing, extending from usual surface
Corrosion	Discolored area of copper or ceramic, causing surface texture change
Crack	Irregular fissure appearing as a jagged dark line in surface, separating material on either side
Delamination	Separation with ceramic at edge of copper layer
Dielectric	An insulator or material which will not conduct electricity, used primarily for insulation of more than one conductor layer on a hybrid
Foreign material	Any material from outside the microcircuit, or any nonforeign material that is displaced from its original or intended location. It shall be considered *attached* when it cannot be removed by a nominal gas blow (no more than 10 psig).
Metallization	A pattern of single or multilayer copper conductors; a metallization path or conductor is usually for the purpose of conducting electricity within the microcircuit
Operating metallization	Copper or any other material used for interconnection, except metallized scribe lines, test patterns, unconnected functional circuit elements, unused bonding pads, and identification markings
Original dimension	The dimension or distance that is intended by design (e.g., original metal width, original pattern area)
Scratch	A tearing defect on any surface of the substrate, either copper or ceramic; has much greater length than width (dents, mechanical impact, scrapes, or other types of mechanical damage are usually considered scratches)
Stain	Discolored area of copper or ceramic, not causing surface texture change
Void	The absence of a material normally found at a location; a gap or opening which is an unintentional defect in the material

1. Copper metal; for example CDA 102 OFHC, or ETP
2. Changes to dimensional tolerance of entire drawing, providing the new tolerance for a specific portion of the copper pattern; for example 0.020 ± 0.001 in (508 ± 25 μm) in conductor width for a critical area
3. Copper thickness = 0.008 ± 0.001 in (203 ± 25 μm)
4. Critical areas that must be free of any defects for assembly processing, such as scratches in wirebond area and bleed-out of copper/interfacial material
5. Plating specifications for tin, nickel, and gold, if applicable
6. Physical test methods, such as minimum peel forces and temperature testing
7. Must conform to engineering specification and purchase order
8. This part is not repairable or reworkable, unless written procedure is used

A more complicated and expensive method of controlling the incoming copper substrate is through testing. Physical test methods assure reliability of copper substrates entering assembly processes, determining adhesion strength between copper and ceramic, solderability, and surface finish, among other physical properties. Adhesion strength testing is discussed in Sec. 5.9, "Reliability." It assures the integrity of a hybrid product not just through the assembly process, but over the long term. Solderability data are useful in determining the wetting of the metallized substrate to solder and fluxes. Standard hybrid metallization techniques for evaluation should be used. The expectation is for better performance from pure copper metallized substrates.

The assembly processes necessary to interconnect the various components determine the structure of the copper substrate. There are two particular assembly techniques—soldering and wirebonding—that must be compatible with whatever copper metallization technology used. A partial list of assembly processes is given below:

- Tab
- Wirebond
  - Gold
  - Aluminum, fine wire
  - Aluminum, large diameter
- Flip chip
- Soft solder
  - Reflow technique
  - Hand solder
- Hard solder
- Epoxy/polyimide
- Thick/thin film resistor
  - Trimming
- Welding
- Brazing

For most applications, plating with nickel and gold, tin, or solder is recommended as a minimum precaution, unless the lowest cost is extremely important and assembly processing provides a well-controlled reducing atmosphere. Passivation of the copper surface is readily apparent as discoloration, and prevents reliable attachment of components. Direct bond copper in particular has a thin oxide coating from the process itself, which must be treated before any assembly can begin. For any copper surface, however, oxidation will occur rapidly even at room temperature, creating difficult assembly problems for many processes. Substrates may be solder-plated or may use organic surface treatments to minimize copper oxidation, but, while some later assembly processes may benefit, others become difficult or impossible.

Plating is advantageous to all assembly processing, and pure copper has good density at the surface for either electroless or electrolytic nickel, making thorough coverage of the copper surface possible. Nickel/gold plating is recommended for three reasons:

1. It creates a barrier to copper diffusion into bare semiconductor materials and thus prevents changes in properties of semiconductor components from this source.
2. It prevents exposure of pure copper metal to the environment.
3. It provides the clean and consistent gold surface that wirebond processes need for good adhesion of wire to metallization.

To test the integrity of the plating over copper metallization, a bake test is recommended. Expose the test substrate to 350°C for 15 min and observe it at room temperature for blisters or corrosion. Discoloration may occur, but each lot of copper substrates should have a characteristic appearance after heating. Significant changes to the appearance may be cause for further investigation. An effective test of plating adhesion is pressing scotch tape to the plated surface after heating, and peeling directly upward. Lifting or peeling of the plated surface is not acceptable.

Solderability is excellent compared to other metallization technologies, with pure copper or its plated surfaces easily wet by lead/tin solders. Leaching of the copper by most solders is not a concern, however passivation can occur and solders will no longer adhere. This is especially true of the solder flux formulations now available. Problems with soldering or plating to the copper surface generally are not due to the properties of the copper metal. Rather, they are due to the low flux activity and, occasionally, to excess surface contamination from a small amount that is almost always present.

Rework can be performed without excessive concern for the reliability of the bond to copper metallization or the solderability of the copper surface. Bear in mind, however, that each rework process is a temperature cycle, and careful attention should be paid to the differences in temperature across the surface of the hybrid device and the possibility of thermal shock. Slow preheating and cooling down are the ways to minimize the adverse stress effects of different TCEs in materials such as pure copper and ceramic. Heating the entire hybrid device during rework minimizes the stress, but unfortunately reflows solder on all of the components present. In fact, a problem with copper metallization is that not enough heat can be generated and localized to perform the reflow process at all. Insulation of the substrate is usually necessary.

Epoxy and eutectic die attachment materials are usually best used on a plated copper surface, unless the specific materials are well-understood. Nickel and/or gold plating reduces the chances of a reliability problem. Wirebond processing is good on a copper surface for aluminum wire of all diameters and aluminum ribbon. Gold wire requiring a thermosonic process can be bonded as well, but heating a large substrate can be difficult. For either wire type, but especially gold, nickel or nickel/gold plating is advisable in areas of wirebond connection, to reduce the risk of an oxidized copper surface interfering with formation of the bond. Wirebond process parameters are usually different from those used on thinner metallizations, copper being a fairly soft material and thick. Matching wire characteristics, such as tensile strength and hardness, bond tool shape, and process parameters of time and power to pure copper metallization are possible, but careful evaluation is required.[60] An advantage of pure copper metallization comes from the flatness of the top surface, compared to thick films. Wirebonding to a flat surface applies equal amounts of energy to the entire wirebond foot area. The resulting bond is better attached to the metallization.

Flip chip attachment is the interconnection of a discrete silicon semiconductor chip or die to the metallized ceramic surface. As opposed to typical die attachment, with the semiconductor die bonded on its back and wirebond pads facing away from the substrate, the die is flipped over and attached to wirebond pads directly on the substrate with solder. This method of attachment is sensitive to mismatched TCE between substrate and the semiconductor die, and as the temperature rises and falls the stress on the interconnections may cause failure. Pure copper metallization should be evaluated to determine if a minimum thickness will limit the copper expansion and contraction in the x and y directions more closely to that of the ceramic and provide a reliable flip chip attachment method.

Power hybrid assembly is an important use of pure copper metallized substrates, and requires attachment to a broad variety of components. Discrete semiconductors, transformers, and large and variable capacitors, along with more typical surface mount devices, require excellent contact with copper and adequate capacity at the attachment points to withstand significant thermal and electrical stresses. A lack of standardization, the importance of the final appearance, and the unique geometries necessary lead to hand soldering as a common assembly technique. Pure copper metallization has the advantage of withstanding much longer solder reflow times, but demands much higher thermal input to attain reflow temperature. Integral leads save time and cost during assembly, with no stamped parts or wires, less assembly time, less scrap, and elimination of some of the more exotic interconnections necessary for high power applications.

Encapsulation is the last assembly process to be considered here. As protection against environmental contamination, covering the surface of the completed hybrid assembly improves reliability but adds to cost. The encapsulation is sometimes screen-printed over metallization, before component attachment, but more often fills the hybrid cavity after assembly and test. The variety of encapsulants and the properties of each are many, but adequate shelf life, pot life, no mixing, and low toxicity are desirable characteristics to find in any product.

There may be limitations in later processing of components onto the copper metallized substrate, especially if a reducing atmosphere is used. Hydrogen embrittlement has been observed when copper materials containing cuprous oxide are exposed to a reducing atmosphere.[43] With the loss of any oxide-containing material from the adhesion layer between copper and ceramic, the strength of the bond is compromised. Another consideration involves processing at temperatures over 600°C; exposure to such high temperatures should be carefully reviewed to understand its effect on reliability for a particular pure copper metallization technology. The length of time and the environment of the high temperature exposure are important factors in the reliability of the device after processing. They may be a particular concern for later assembly of AlN substrates, which are particularly susceptible to the TCE stresses and reducing atmospheres of some copper metallization technologies. At this point in time, the user should process samples of the copper metallized AlN substrate through the expected assembly process conditions and thoroughly evaluate the effects.

5.8   Thermal Performance

The high thermal conductivity of pure copper metallization is a fundamental and obvious advantage for hybrid designs, no matter what ceramic is used. A metallized substrate with copper metal in close proximity to small sources of steady or transient heat generation conducts more heat over a greater surface area, thereby lowering the operating temperature and improving the reliability. Thus, with copper's benefits of high thermal capacitance and low thermal resistance, many different types of waste heat can be dissipated without the addition of special cooling methods.

A primary goal of thermal analysis of any metallized hybrid substrate is restriction of the operating temperature below some critical value for active devices, above which long-term operation is not guaranteed. In thermal analysis of copper metallized substrates, it is important that the total thermal path be considered for a particular design. Consideration of all the pathways, not just those through copper and through ceramic, will ensure a design meeting thermal performance goals. The larger perspective may be crucial to ensuring that the excellent thermal conductivity in a material such as copper is not wasted by inadequate dissipation imposed by other hybrid materials or system level constraints.

In the event that the thermal pathway is found adequate, the designer should look at materials that are closer to the heat source and appreciate the greater effect these conducting materials have on the thermal transfer of energy. Immediately below the active device that is creating thermal energy, an important junction in the thermal path exists at the plane of contact between copper and ceramic. Each of the copper metallization technologies has a unique approach to creation of an adhesion mechanism. Variations in materials and processes exist even within one technology, such as different braze compounds for active metal bonding, or seed layer vs. thin film metal adhesion layer for plated-copper processes. The different methods will form quite different interfacial layers important to adhesion, and affect other metallized substrate properties such as thermal conductivity. Most obvious is the effect of layer thickness on thermal resistance, but the actual conductivity value of the material, variations in thickness, voids, and other physical imperfections can significantly degrade thermal performance from a much higher theoretical value. Understanding interfacial properties and variations are fundamental to a good definition of the total thermal path and proper thermal design of a pure copper metallization hybrid.

Other metallization technologies such as thick and thin film are not usually designed in conjunction with the thermal issues, but simply introduce basic thermal limits for the designer to accommodate by other means. By approaching the thermal management issue with an expectation of modifying the copper metallization for better thermal performance, hybrid designs can be made more flexible and cost effective.

Choosing a copper layer thickness to control the temperatures of components mounted to the surface should be considered for lower thermal conductivity ceramics. Consideration should also be given, however, to the increase in TCE and adverse stress on mounted components when thicker copper is used. In looking at the question of optimum thickness, development work has shown that the thermal conductivity of the ceramic determines which dimension to use. Lower thermal conductivity ceramics benefit from thicker copper, where higher value ceramics do not. There is a tradeoff of increased thermal resistance due to the greater distance traveled by the heat flux through a thicker copper layer vs. a decrease in thermal resistance due to the additional spreading within the copper layer to a larger surface area of ceramic. Ceramics with low thermal conductivity benefit more from a larger surface area, which outweighs the penalty of greater distance through the thicker copper. In more thermally conductive ceramics, thicker copper produces a thermal penalty of additional material for heat to pass through and raises the overall thermal resistance, while the ceramic has the ability to absorb more of the heat before it expands to a greater surface area. For these substrates, thicker copper can actually degrade thermal performance. For instance, work has been done to demonstrate that 0.005-in (125-μm) copper is the maximum required for BeO and AlN; thicker copper layers do not appreciably reduce the thermal resistance of the circuit.[51] On alumina, however, the thermal performance improves with copper thicknesses up to 0.012 in (300 μm).[49,70]

One other method of increasing copper thickness is already a standard configuration: bonding copper to each side of the substrate, which balances the stress of mismatched TCE. Copper metallizing both sides of ceramic with poor thermal conductivity increases the surface area of thermal transfer through each layer, again overcoming the drawback of a longer thermal path through the copper. Alumina ceramic 0.025 in (625 μm) thick has been evaluated to show a 40% reduction of thermal resistance with 0.010-in-thick (250-μm) copper applied to both sides instead of just one side. The same evaluation of BeO showed only a 7% thermal improvement with double-sided copper metallization.[48]

Increasing the copper thickness and applying copper to both sides of the ceramic are not the only methods of increasing the thermal performance. For any given thickness, increasing the surface area of copper surrounding an active device will also lower the thermal resistance. This is especially true if there are limits to the thickness of copper and low thermal conductivity ceramic is the material of choice. Again, increased surface area to transfer heat from copper to ceramic is the reason for improvement. There is no getting something for nothing, however; depending on the number of devices to be attached, the additional surface area may limit the circuit density for a specified substrate size. Nevertheless, by increasing the area and thickness of copper beneath the active device as necessary, enough thermal performance may be obtained to prevent more costly solutions such as external heat sinks.

Thermal evaluations for some hybrid products may have another factor to consider in pure copper substrates. Transient thermal loads can be accommodated by a physical property of copper that is related to thermal conductivity, but important in its own right. The large value of heat capacity, sometimes called *thermal capacitance,* can absorb sudden thermic events, giving a designer the dynamic thermal response required to keep active devices below critical temperatures. Therefore, peak thermal loads over short periods of time can be more easily accommodated without special design modifications or expensive packaging of the active device.[50] The high thermal capacity and conductivity of copper also benefit temperature-sensitive hybrid circuits and their mounted components, enabling them to quickly reach an equilibrium temperature when power is applied. Drift of electrical parameters due to slow temperature increases on metallized substrates with poor thermal conductivity, a vexing problem for test engineering, can thus be eliminated.[68]

A better understanding of the thermal issues for a specific hybrid design usually begins with rough calculation of the heat output by any number of sources on the substrate, the spreading angle through different values of thermally resistant material, and estimation of junction resistance. Calculations are a good approximation of basic thermal performance, but they are not realistic in many instances. Most analyses take place with single devices, operated in isolation and having ideal interfaces in the total thermal path. Real world factors such as variations in the electrical performance of the devices with temperature, the spacing between them, and the total number of closely spaced devices both on and off the substrate can all interact, challenging the designer during thermal evaluations. Although these issues will not be discussed in detail here, it should be noted that components with smaller surface areas generating high concentrations of heat flux, over 1000 W/in^2 (200 W/cm^2), can have other thermal effects very difficult to model or calculate. Other difficulties arise from various spreading angles and material thermal conductivity values that give disparate answers, especially for newer materials such as copper on advanced ceramics. These problems indicate there is not a broad practical base of knowledge for thermal analysis of thick, pure copper hybrids. Copper on ceramic metallization has changed the way thermal calculations need to be done. The past assumption by hybrid designers that the total thermal resistance is a function of area based on angles of thermal transfer though materials below the heat source is still valid, but thermal spreading of copper effectively increases the angle and transfers significant amounts of heat along the copper layer itself, as already discussed. The effects on other hybrid materials and at the system level may not be included in a theoretical evaluation. Thermal performance based on past experience and calculation is a good idea, but the empirical approach of experimentation, even when limited to basic structures, may prevent lost time and effort if a product is destined to be unsuccessful for thermal reasons. Experimentation is a key element of any evaluation, and thermal testing of pure copper metallized prototypes should be a part of more complex designs.[51-56]

Plated-copper technology has one thermal advantage worth noting. Discussed in Sec. 5.4, "Plated-Copper Technology," vias in plated copper are an advantage for multilayer structures, having small diameters and high density. Plated through-hole vias of pure copper can also make an improvement in thermal performance, especially that of low thermal conductivity ceramic substrates such as alumina. Thermal vias are filled with pure copper, from 0.003 to 0.050 in (0.08 to 1.3 mm) in diameter, for ceramic thickness from 0.015 to 0.100 in (0.38 to 2.5 mm). This may be a lower-cost alternative than using advanced technical ceramics.

5.9 Electrical Performance

As semiconductor technology on silicon advances, electrical performance requirements from higher level systems down to the individual components become more difficult to achieve. Selecting pure copper metallization for appropriate hybrid products will permit meeting existing and future electrical requirements, often providing more performance capability than will be utilized. This is a significant advantage with the increasing rate of technological change shortening product life cycles, and each new generation demanding more electrical performance of the conductor. Unused capacity in today's product may be required tomorrow, giving the design analyst reason to consider and justify implementation of a newer metallization technology.

Power hybrid applications have been the driving force behind decisions to use copper on ceramic; they are the motivation for development of many advances and guidelines for use. At such high levels of power management, the optimum design of power hybrids for different types of power supplies, switches, regulators, and converters is usually done with derating of theoretical limits, a necessary step due to the physical imperfections in geometry and material properties. For the purposes of comparison, copper conductors should be evaluated for true performance in conjunction with other system components, with their advantages and disadvantages compared to those of other conductor metallizations. From the theoretical to actual design characterizations that have been performed, copper substrates should be evaluated in several categories that might include:

- High voltage
- High current

- Efficiency (small losses)
- Faster propagation times (higher speed/clock rates)
- Skin effects
- Electromigration

In addition to these high power and speed categories, a recent area of interest for electrical performance of pure copper is application to very low power products, those that have rapidly evolved to less than a standard 5-V operation. Extremely low voltage and current losses, high speed, and less noise interference have become important to consumer products operating at the 3.3-V level, with the goal of extending battery life as much as possible. As applications at lower voltages are developed, the electrical performance advantages of copper become even more important.

The two characteristic requirements of power hybrid applications, high voltage and high current, are met by using thick, pure copper metallization. Definitions are helpful to begin understanding the reasons for this use. High voltage is generally at or above 50 V, with an upper limit of approximately 1200 V. Current levels above 20 A would be classified as high current, but designs requiring constant 400 A or more have been produced. In reality, the ability of copper conductors to support high voltage and high current is closely linked to the physical properties of low electrical resistance and high thermal conductivity. With low resistance, current losses are reduced and the voltage level is higher for better power distribution at all points along the copper conductors. Good thermal conductivity keeps the temperature lower, and components with high TCE are therefore more reliable and have less resistance to the flow of electricity. DBC and AMB have a significant advantage for power hybrids in the capability of producing wide, flat integral leads, greatly reducing the lead resistance, inductance, and voltage drop normally present with soldered interconnections.

With each new product or technological improvement, new parameters need to be defined for evaluation of the various types of metallization on ceramic. A parameter that has become very important to the success of new hybrid products is high speed performance. As operating frequencies have increased from kilohertz to megahertz to gigahertz, the impact of conductors on electrical and thermal performance has become critical. Individual technological improvements such as higher frequency operations rarely happen alone, however, and other factors must be taken into account. The simultaneous occurrence of high voltage, high current, and high frequency signals causes unique interactions with substrate materials, usually coming together in the material evaluation as thermal problems.[60]

Basically, there are two causes of degradation to the signal path at high frequencies: reflections and the adverse effects of line resistance.[58] One advantage of pure copper conductors in high speed applications is low resistance (less than 1.0 mΩ/sq) and its relationship to a phenomenon called the *skin effect*. The current of a high speed signal traveling down the length of a metal trace exists primarily on the surface, and the depth of penetration into the trace depends on frequency, electrical conductivity of the metal, and the thickness of the conductor. The depth of penetration of current into the conductor is called the *skin depth*. Hybrid substrate metallizations are usually 1 mil (0.025 mm) or less in thickness, and range from 3 to 30 mΩ/sq in trace resistance. With thinner metallizations and higher electrical resistance, transmission-line impedance models take these and other factors into calculations to determine the efficiency loss, since the skin depth may be much greater than the metallization thickness. Pure copper metallizations of 0.005 in (125 μm) or more allow the high frequency current to pass with greater efficiency, since the skin depth is less than the thickness. Different frequency ranges change the fundamental skin depth relationships and relative advantages, but the low electrical resistance properties of pure, thick copper still apply.

All hybrid materials exposed to high voltage and high current are stressed, but not necessarily in the mechanical sense. Just as high temperature or high pressure can drive chemical reactions thermodynamically prohibited at room temperature and pressure, high voltage and current can change the material properties of the conductor surface through chemical reactions normally prohibited. Even if an unwanted reaction occurs slowly at room temperature, the electrical stress can accelerate it significantly. Copper is usually plated with nickel to inhibit this process, but can still be affected if the plating is porous or has openings. Corrosion of a copper surface is a typical example of stress-induced change to a conductor, especially with exposure to moisture or other contaminants at the same time. If the corrosion were as conductive and smooth as the pure copper, and did not migrate to other parts of the substrate surface, there would be no problem. However, as the preceding discussion noted, the skin effect is an important consideration at high frequency, and conductors having a rough and nonconductive surface will interfere with transmission of the electrical signal. Much higher losses and lower efficiency can hurt the reliability of the device.[59]

A very important factor in the use of pure copper metallization is selecting an appropriate copper metallization technology to bond with many types of lower dielectric constant ceramic substrates, which will allow better high frequency performance. Beryllia has the lowest dielectric constant of the standard ceramics, but other materials appear ready to meet high frequency needs. Refer to the substrate chapter for the dielectric constants of newer materials. Copper metallization technology holds the promise of metallizing almost all advanced ceramics, but especially those with lower dielectric constants. The flexibility of combining a substrate with a low dielectric constant with thick, pure copper metallization is one of the reasons for the rapid pace of development and application of this technology.

5.10   Reliability

As discussed in this chapter, the joining of ceramic and pure copper metal to form a metallized substrate can be done in several different ways. Regardless of the method of fabrication, substrates are expected to have the following ideal properties:

- Adhesion strength unaffected by environmental conditions
- Thermal conductivity
- Electrically conductive pattern
- Inertness
- Compatibility with standard assembly processes
- Economy

Fundamental to acceptance of pure copper metallized substrates is an accommodation of thermal expansion differences between the materials to obtain a sound, reliable bond when subjected to thermal and/or mechanical stresses. In simplest terms, the copper and ceramic must have adequate adhesion before any thermal or electrical properties are useful. Each property listed above is predicated on the adhesion characteristics of the interfacial layer. Several concepts are helpful in understanding the adhesion mechanisms. Chemical reactions are the strongest binding mechanisms because of short order effects, causing permanent changes to the electronic structure of elemental and molecular species present. Secondary adhesion effects are termed *mechanical bonding,* such as van der Waals (dipole) attraction and physical interlocking in a matrix.[36] Together, the adhesion mechanisms for each pure copper metallization approach must have sufficient strength over a wide range of environments to overcome the interfacial stresses and reliably bond the copper to the ceramic.

Stress within the metallized substrate interface comes from two sources: stress present in the ceramic before copper metallizing and stress created from the process of metallization itself. Residual stress in the ceramic is usually due to grinding or laser processing. As previously discussed in Sec. 5.2, "Substrates," resintering the ceramic is the preferred method of relieving this induced stress. Laser scribing or via formation has been shown to be a cause of residual stress which can appear later in the hybrid processing as warpage, cracks, or shifts in resistor value. Ceramic substrate modification and preparation thus affects the copper-metallized substrate, adding to the stress that comes from bonding two materials of different TCE.

Traditional and advanced ceramics for hybrid applications have a TCE range of 4 to 10 ppm, while copper has a TCE of 17 ppm, causing a high rate of volumetric expansion with temperature for copper that is constantly opposed by the lower rate for ceramic. For high temperature DBC and AMB formation, initial stress of the ceramic is compressive after cooling, and fortunately the strength of ceramic is good in compression. So long as the compressive strength for a given thickness of ceramic is not approached, the substrate will perform reliably. Smaller substrate sizes are less affected by this mismatch, but as the size increases to 2 in^2 (25 cm^2) and above, interfacial stress grows very large between copper and ceramic at the perimeter of the substrate during any temperature cycle.

Single-sided copper on ceramic requires thicker ceramic and smaller substrate sizes. Even with those precautions, there will be some bowing of the substrate to cause assembly, thermal transfer, and reliability problems. Balancing the mismatched TCE with copper on each side of the substrate may increase the total compressive stress, but it is then symmetrical on each side of the ceramic, making the substrate planar and less prone to cracking or delamination. Designers still should account for differences in ceramic surface area coverage on each side that may cause asymmetric compressive stress, but should be especially aware that ceramic strength, which is normally tolerant of pure copper metallization process stress, may become unreliable if weakened by residual stress.

Each of the pure copper to ceramic metallization technologies relies on one fundamental way to meet the challenge of interfacial stresses: they use demonstration of adhesion strength as a proof of reliability. Good values of strength are important, but without examining the effect of environmental factors over time and using a standard adhesion test procedure, there is some question about the conclusions that are often drawn from the data. At first glance, tests to measure adhesion strength are an easily understood concept, and should be the most important tool in making design and reliability decisions. In reality, however, adhesion tests to demonstrate the various properties of the bond between copper and ceramic are not standardized to any great degree, and thus good correlation between sources of data on adhesion strength is lacking. The test data have been formulated by different manufacturers and users to give confidence in the adhesion performance of the final product, but for many manufacturers the data have been derived from several approaches to pure copper metallization of ceramics. Proper attention should be paid to a consistent and realistic test procedure to measure the adhesion strength.

Testing is critical to achieve the highest adhesion strength possible for specific materials and designs, guiding the judicious selection of materials and fabrication parameters such as surface conditions, process environments, time, and temperature. A standardized procedure is also required for evaluation of adhesion degradation, measuring the relative amounts and rates during environmental testing. If adhesion is poor, changes in process parameters are required, and with further experimentation make it possible to identify just what variable was at fault. Adhesion strength testing in essence pulls the copper metal from the surface of the ceramic as the fundamental test for the properties of bonding between copper and ceramic, but actually provides information on all of the materials and processes that create the metallized substrate. Tests with poor standardization cause problems not only with comparisons between sources, but also with each set of test data from one source. Therefore, well-performed adhesion testing is an important component of any decision regarding a pure copper metallized substrate.

As part of the overall evaluation of metallized substrates, adhesion testing is a method of comparison between:

- Different types of metallization technologies
- Different types of pure copper metallization technologies

- Different types of ceramic substrates after pure copper metallization
- Process or material changes for improvement of the adhesion strength
- Before and after environmental test conditions

Adhesion strength values of ceramic to copper metallization allow users to make these comparisons and others, evaluating the adhesion performance in a valid way.

Actual testing of metallization involves pulling an area of copper from the surface or peeling a given width along the ceramic surface. Pull testing employs a standard force per unit area, but requires careful preparation of samples before testing. For example, mounting on epoxy or soldering on pins may use different amounts of material at each data point, contributing to variability in the result. The peel test has become a preferred method, being a fast, economical procedure allowing more of the surface to be checked for consistent adhesion strength. For the peel test, a substrate must have traces of a standard and reproducible width, usually from 0.125 to 0.350 in (3.2 to 8.9 mm) to assure that edge effects will not interfere with strength values. Fillets of adhesion/interfacial material on the side of the copper trace at the junction with ceramic are not uniform, and can give false indications of adhesion strength with conductors less than 0.125 in wide. One end of a copper conductor is pulled orthogonal to the surface of the substrate, at a standard pull rate of 0.13 mm/s (0.005 in/s). The constant force to separate the copper from the substrate is then measured and recorded. Figure 5.7 illustrates the arrangement of peel test apparatus to obtain a standardized procedure. By pulling with a free moving point at a longer distance from the ceramic surface, slight variations of the pull angle of 90° and rotational moments are minimized as the copper peels away. A peel force at less than 90° along the trace and pulling behind the point of separation can distort the measurements, since shear force contributions to the measured reading of bond tensile strength may give the appearance of a different bond strength than is actually the case. Peeling to either side of the trace initiates separation along an edge, and does not accurately produce a value of adhesion strength across the true perpendicular width.[62] With a standard method and apparatus to avoid the problems just mentioned, units for the adhesion values should be reported as force per metallized trace width, with the rate of pulling copper from substrate included. Typical values are recorded as 30–40 lb/in or 53–70 N/cm.

Special consideration must be given to the thickness of the copper, which adds to the adhesion strength value. With pure copper traces having thickness above 0.005 in (0.13 mm) the tendency of the trace to resist a 90° bend may erroneously contribute to the strength of the bond. By setting up the peel test apparatus to measure the resistance to bending of a straight piece of copper of identical material, dimensions, and processing as the copper trace of interest, the force to make a 90° copper angle can be determined and subtracted from the nominal values of copper peel strength. This will give the true adhesion force without contributions from the force to bend the thick copper.[61,62]

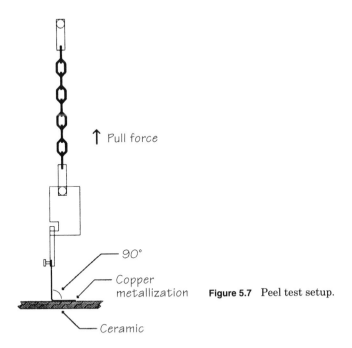

↑ Pull force

— 90°

— Copper
   metallization   **Figure 5.7**   Peel test setup.

— Ceramic

The standard method of adhesion testing just discussed is crucial to the evaluation of stress on copper metallized substrates subjected to simulated environmental conditions. Long term reliability evaluations generally determine the ability of materials to resist fatigue failure from propagation of very small defects. Figure 5.8 shows the cross section of a finished hybrid, and the materials stressed by environmental testing. The predominant simulations of long term reliability are temperature cycling and temperature aging, standard environmental tests used to evaluate the copper to ceramic adhesion and interfacial layer. For temperature cycling, the Mil-Std 883 Method 1010 is a good source of information, providing a straightforward method of performing the test. In making comparisons with other test methods, note the length of time at temperature and duration of the temperature changes, since they lack standardization in some cases. Automotive testing typically uses −40 to +150°C as a temperature range, and as with other test condition limits, may be less stringent than Mil-Std 883.[64,67] Adhesion tests are suggested after 25, 50, and 100 cycles. Temperature aging is usually in air at 150°C, with periodic adhesion testing done at 50, 100, 250, 500, and 1000 h. Plotting the average adhesion strength vs. the total number of cycles or number of aging hours allows the user to intelligently evaluate the long term reliability of the bond.[66,68]

Adhesion testing before, during, and after the environmental tests indicate the condition of the interfacial layer and whether any change has occurred. As a rule, trends are more important than absolute values. A lower value of initial adhesion strength that drops slightly during the environmental test program may be more desirable than a high initial value dropping more than

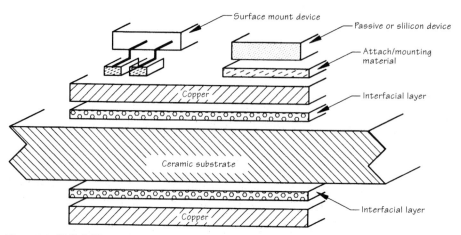

Figure 5.8 Reliability concerns.

50%. One important experimental parameter for accurate and reproducible testing is the implementation of a stabilization time after environmental testing and before peel testing. Waiting 24 h allows the materials to relax and reach a common condition for more uniform data from each test strip.[60,63]

Environmental testing is done to evaluate more than just the reliability of copper and ceramic adhesion. Finished hybrid products are tested to determine the strength of adhesion between copper metallization and components mounted to the surface. As the copper expands and contracts with temperature, the device mounted to the surface must have the strength to withstand the TCE stress acting on it and the mounting material. Environmental testing a particular hybrid configuration determines if the stresses of TCE mismatch will degrade either the component itself or the mounting material holding it to the copper surface. The stresses of TCE mismatch are more of a concern if the hybrid product undergoes large changes in temperature due to its own thermal output or due to the environment. The magnitude of the stress depends on two factors: total surface area covered by the active device and compliance of the mounting material. Consideration must therefore be given to the thickness of the copper beneath the device, the size of the component, the ceramic TCE, and the actual temperature changes during operation.

Copper expands and contracts at a rate of 17 ppm/°C, compared to a lower value of 4.5 ppm for silicon and between 6 and 12 ppm for other ceramic and packaged devices. The value for copper is slightly overstated because it is not clear just how much the ceramic reduces the x and y directions of copper expansion and contraction during temperature changes. For each variation of ceramic thickness, TCE, and flexural strength, there would be variations of the copper thickness and specific coverages of surface area to be considered. Figure 5.9 shows the TCE for different copper thicknesses on 0.040-in-thick (1.0-mm) alumina ceramic as a reference.[70]

The thinner the copper, the smaller the component, the more compliant the mounting material, the lower the ceramic TCE, and the smaller the tempera-

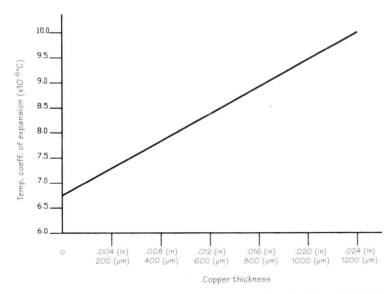

Figure 5.9 TCE change with copper thickness for a 0.04-in-thick 96% alumina substrate.

ture range, the less concern there should be for the die breaking. For a soldered flip chip attachment, the copper metallization technology generally cannot support the fine lines required for interconnection. Plated copper may approach this capability, and if so, the thin copper is much less a concern than the TCE of the ceramic.

For solder applications of pure copper metallization, there is a reliability advantage over thick and thin film. With plated or unplated copper, repeated exposure to solder reflow does not affect the long term reliability by solder leaching. With thick film materials, solder leaching has been one of the limitations, especially if rework is necessary. Of more importance to the reliability question for all copper metallizations, however, is the degraded adhesion to the ceramic substrate which occurs after the formation of copper–tin intermetallics at higher temperature and over time. Copper loses long term adhesion due to an increase of molar volume from the intermetallics and the subsequent changes in TCE between several different materials within the typical solder volume of an interconnection. Thinner copper, typical of plated-copper metallization, is more prone to the reliability problem.[19,30,60,64]

Corrosion of metallization is always a concern in any microelectronic hybrid application. Exposure to stresses from high power applications in a difficult environment, such as an automobile, have led to many of the reliability tests discussed herein. If copper undergoes chemical reactions on the surface caused by these environments, corrosion by-products include copper salts and oxides, which are not corrosive or catalytic but can be corroded themselves. Conditions leading to copper corrosion can also lead to electromigration along the ceramic surface. Silver metal is notorious for moving from the edges of a metallized area into isolated areas of the hybrid surface and introducing leakage or complete failure of the device. Any metal will move under the right conditions of moisture, voltage gradients, and ceramic surface finish. For preventing corrosion of just the copper, an inert plating such as nickel is recommended; lesser used methods compatible with copper metallization include solder mask, organic films such as polyimide, or low temperature dielectric.[60,69] Each will protect the copper surface, but a better solution may be encapsulation of the entire copper metallized substrate after assembly. Silicones are probably the most widely used encapsulant, with good resistance to harsh environments and high power stresses. The coating not only protects the metallization and components, but also can contribute to the thermal performance by providing better dissipation of heat in comparison to air. Power hybrid applications in particular use encapsulants to provide high voltage insulation, moisture protection, chemical resistance, lower dielectric constant, and thermal dissipation. From glob tops and clear silicones to hard-set potting compounds, encapsulants to protect surfaces and components from the adverse impact of contamination are advisable.

5.11 Applications

A collection of active and passive devices placed on the surface of a copper metallized substrate can be classified under any number of hybrid product types. Applications using copper metallization, generally defined as products that process and manage power, have several characteristics in common. High power requires the excellent thermal and electrical properties of copper metallization, but must accommodate a lesser capability for smaller lines and spaces.

Power supplies and power converters have demanded much of the recent development in copper metallization technology. There has been a focus on the inherent advantages of ceramic hybrid technology for the stressful environment inside a power supply, and on those of pure copper in particular. The supply is the heart of many electronic systems, and certain classifications of switching power supplies have even more components, such as bridges, rectifiers, and transistors, that need reliable materials that withstand large electrical and thermal stresses. Especially important are the improvements possible with flat plate inductors and high voltage standoffs. Thermal management is one of the biggest problems, and the thermal conductivity of copper metallization is one the best ways to meet this challenge.

Automotive engineers have used hybrid technology for many years, and now find copper metallization useful for several different applications. The tremendous increase of intelligent functions for the average car are selectively placed into a hybrid format, taking advantage of its reliability in a harsh environment and its capacity for managing more power. Copper metallization, with its ability to reliably handle high current and high power demands, has made the life of an automotive circuit designer a little easier, while relieving environmental concerns.

Table 5.8 lists many of the markets driving usage of pure copper metallization technology today, and the photographs in Figs. 5.10 to 5.15 further illustrate applications of the various copper metallization technologies.

TABLE 5.8   Applications

Linear power power supplies	Switching power supplies	DC/DC converters	AC/DC converters
Voltage regulators	Auto ignition circuits	Traction controllers	
Motor controllers	Frequency converters	Solenoid drivers	Speed controllers
Actuators	Switch drivers	Power op amps	Rectifying bridges
Inverters	Power resistors		
Thyristor modules	IGBT modules	Bipolar transistor modules	
Peltier elements	High temperature sensors	Power coils	
RF power modules	High frequency transmitter modules	MMIC	

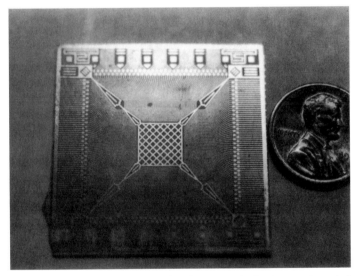

Figure 5.10 Plated copper on alumina ceramic for tape automated bonding (TAB), with 0.002-in (0.05 mm) lines and spaces next to die attach pattern. (*Courtesy of Thomas C. Evans.*)

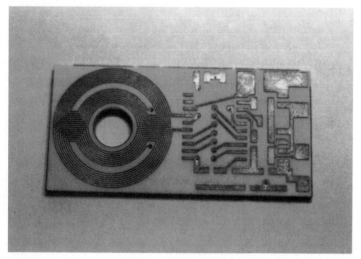

Figure 5.11 Radio-frequency application of plated copper technology, with circuitry on both sides of substrate. (*Courtesy of Thomas C. Evans.*)

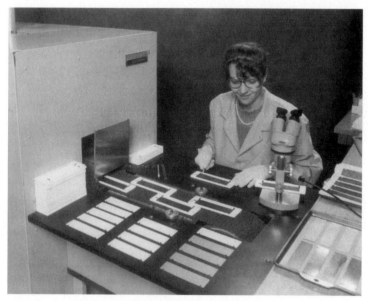

Figure 5.12 After alignment of the preoxidized copper sheet and the ceramic, substrates are loaded onto the belt of a nitrogen atmosphere furnace for bonding. (*Courtesy of Brush Wellman.*)

Figure 5.13 An example of a power control/frequency converter module for "smart power" driver ICs with DBC on alumina, capable of handling 1500 W at 90°C ambient. Copper traces 0.040 in (1 mm) wide and 0.012 in (0.3 mm) thick can handle up to 100 Å of continuous current. (*Courtesy of Thermalloy/Curamik.*)

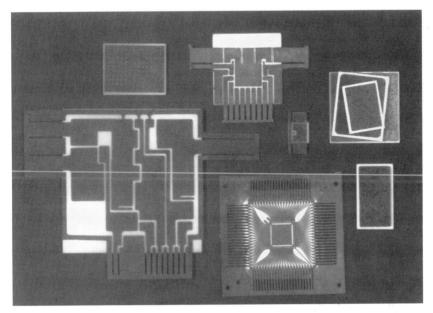

Figure 5.14 Direct bond copper substrates, illustrating various configurations of integral leads. (*Courtesy of Brush Wellman.*)

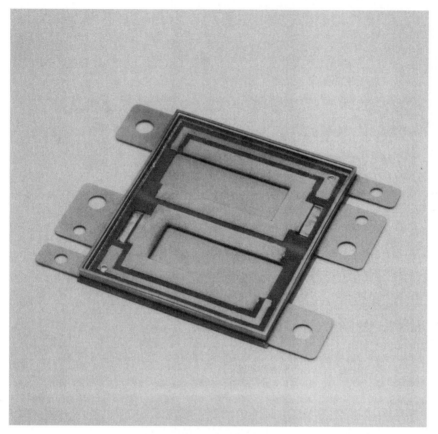

Figure 5.15 This low profile hermetic package, approximately 3.5 in (8.9 cm) square, is designed to handle 400 A through two switching transistors. The flat copper-clad molybdenum power leads are sandwiched between two AlN plates, the upper plate having cutout die cavities. AlN-to-metal joints were made by active metal brazing. (*Courtesy of Ceradyne, Inc.*)

References

1. "Guidelines for Multichip Module Technology Utilization," IPC-MC-790, August 1992, Inst. for Interconnecting and Packaging Electronic Circuits, Lincolnwood, Ill.
2. D. Seraphim, R. Lasky, C.-Y. Li, *Principles of Electronic Packaging,* McGraw-Hill, New York, 1989.
3. C. Harper (ed.), *Electronic Packaging and Interconnection Handbook,* McGraw-Hill, New York, 1989.
4. B. Moody, "Size and Scope of the U.S. Power Hybrid Market," HMRC, 1990 and 1991.
5. T. W. Dekleva and D. B. James, "High Density Electroplated Copper Circuitry Incorporating Air Firing Thick Film Resistors," *Proc. 1988 ISHM Symposium,* pp. 554–566.
6. J. Licari and L. Enlow, *Hybrid Microcircuit Technology Handbook,* Noyes Publications, Park Ridge, N.J., 1988, pp. 26–31.
7. H. Mizuhara, E. Huebel, and T. Oyama, "High-Reliability Joining of Ceramic to Metal," *Ceramic Bulletin,* vol. 68, no. 9, 1989, pp. 1591–1599.
8. J. Jarrige, J. Mexmain, J. Michelet, J. Guinet, B. Guillaume, and J. Hubert, "AlN Substrates Requirement for Copper Metallization," *IMC 1990 Proc.,* Tokyo, pp. 273–278.
9. J. Holowczak, V. Greenhut, and D. Shanefield, "Effect of Alumina Composition on Interfacial Chemistry and Strength of Direct Bonded Copper–Alumina," *Ceram. Eng. Sci. Proc.,* vol. 10, 1989, pp. 1283–1294.

10. N. Schulz, A. Elshabini-Riad, and R. Hendricks, "More on the Role of Residual Stresses in Ceramic Substrate Materials and Metallization," *Proc. 1990 ISHM Symposium,* pp. 321–328.

11. J. Dickson, "Direct Bond Copper Technology: Materials, Methods, Applications," *ISHM 1982 Proc.,* pp. 103–109.

12. W. Chiang, V. Greenhut, D. Shanefield, and L. Johnson, "Substrate Pretreatments and Strength Limiting Factors in Cu–AlN Direct Bonds," *J. Mat. Res. Soc.,* in press.

13. W. Chiang, "Interfacial Structure, Process Control and Mechanical Properties of Cu to Al_2O_3 and AlN Direct Bonds," dissertation, Rutgers U., New Brunswick, N.J., 1991.

14. M. Baldwin, "Copper Metallization of Alumina," dissertation, Colorado School of Mines, Golden, Colo., 1993.

15. F. Miyashiro, N. Iwase, A. Tsuge, F. Ueno, M. Nakahashi, and T. Takahashi, "High Thermal Conductivity Aluminum Nitride Ceramic Substrates and Packages," *IEEE Trans. CHMT,* vol. 12, no. 2, June 1990, pp. 313–319.

16. N. Iwase, K. Anzai, and K. Shinozaki, "Aluminum Nitride Substrates Having High Thermal Conductivity," *Solid State Technology,* October 1986, pp. 135–138.

17. K. Paik, "Evaluation of Various AlN Substrate Materials for High Density Interconnect Applications," *ISHM 1992 Proc.,* pp. 555–560.

18. N. Iwase, K. Anzai, K. Shinozaki, O. Hirao, T. Thanh, and Y. Sugiura, "Thick Film and Direct Bond Copper Forming Technologies for Aluminum Nitride Substrate," *IEEE Trans. CHMT,* vol. 8, no. 2, June 1985, pp. 253–258.

19. W. Kinzy Jones, "Evaluation of Copper Plated Ceramic Substrates," *ISHM 1988 Proc.,* pp. 164–169.

20. J. Muir and J. Williams, "Copper Metallization of Conventional and Alternative Substrates," *ISHM 1988 Proc.,* pp. 196–202.

21. C. L. Lassen and J. R. Williams, "Plated Copper on Ceramic," *EP&P,* May 1989, pp. 52–54.

22. G. J. Shawhan and G. R. Sutcliffe, "Plated Copper Metallization for Power Hybrid Manufacturing," *Hybrid Circuit Technology,* April 1990, pp. 37–42.

23. W. Jorgensen and K. Adam, "Copper Plating Process for Metallizing Alumina," *ISHM 1984 Proc.,* pp. 347–352.

24. M. Capp and K. Zsamboky, "Four Pure Cu Metallization Products, An Adhesion Value and Mechanism Comparison Study," *Proc. 1990 ISHM Symp.,* pp. 229–238.

25. R. Delaney, "Copper Plated Ceramic Hybrids," *ISHM 1984 Proc.,* Dallas, pp. 394–398.

26. C. Park, "Kyocera Thin Film Technology," technical presentation and personal correspondence, 1993.

27. B. Chiou and J. Chang, "Electroless Cu Plated AlN Substrate," *Proc. 1993 IEEE Electronic Components and Technology Conf.,* Orlando, pp. 1085–1089, June 1993.

28. P. Donahue, J. Page, E. Thiele, Y. Hu, M. Saltzberg, and S. Gallo, "System 901: Low *K,* Copper MCM-C Packaging," *Proc. 1993 IEEE Electronic and Technology Conf.,* Orlando, pp. 893–895, June 1993.

29. "Low Temperature Fireable Ceramic Green Sheet System with Copper Conductors," technical data sheet, Shoei Chemical, 1993.

30. J. Intrater, "The Challenge of Bonding Metals to Ceramics," *Machine Design,* November 23, 1989.

31. J. Intrater, "How to Select the Right Metallization/Joining Method," *Ceramic Industry,* February 1991.

32. R. D. Watkins, "Types of Ceramic Joining and Their Uses," *Ceramics and Glasses: Engineered Materials Handbook,* vol. 4, ASM International, Materials Park, Ohio, pp. 478–481, 1991.

33. M. Schwartz, "Ceramic Joining," ASM International, Materials Park, Ohio, 1990.

34. H. Mizuhara and E. Huebel, "Joining Ceramic to Metal with Ductile Active Filler Metal," *Welding Journal,* October 1986.

35. J. Pask and A. Tomsia, "Wetting, Surface Energies, Adhesion, and Interface Reaction Thermodynamics," *Ceramics and Glasses: Engineered Materials Handbook,* vol. 4, ASM International, Materials Park, Ohio, pp. 482–492, 1991.

36. H. Mizuhara and T. Oyama, "Ceramic/Metal Seals," *Ceramics and Glasses: Engineered Materials Handbook,* vol. 4, ASM International, Materials Park, Ohio, pp. 502–510, 1991.

37. N. Anderson and S. Weinshanker, "Brazing of Aluminum Nitride," *Inter. J. Hybrid Microelectronics,* vol. 14, no. 4, pp. 121–128, December 1991.

38. "Brazing with Dow Developmental Aluminum Nitride," Ceramics and Advanced Materials technical datasheet, Dow Chemical, Midland, Mich., 1993.

39. E. Luh, J. Enloe, A. Kovacs, and R. Lucernoni, "Metallization of Aluminum Nitride Packages," *IEPS Journal,* vol. 13, no. 2, pp. 3–5.

40. E. Lillie, P. Ho, R. Jaccodine, and K. Jackson, "Modification and Reactions of Aluminum Nitride Surfaces," *Electronic Packaging Materials Science V,* Materials Research Society, vol. 203, Pittsburgh, pp. 241–252, 1991.

41. R. Cadenhead, "Materials and Electronic Phenomena, Part IV," *Inside ISHM,* June 1985, pp. 9–18.

42. J. King (ed.), *Materials Handbook for Hybrid Microelectronics,* Artech House, Boston, 1989.

43. *ASM Handbook,* vol. 2, *Properties and Selection: Nonferrous Alloys and Special-Purpose Materials,* 2d printing, January 1992.

44. R. C. Weast (ed.), *CRC Handbook of Chemistry and Physics,* 62d ed., CRC Press, Boca Raton, Fla., 1982.

45. C. Harper, *Handbook of Thick Film Hybrid Microelectronics,* McGraw-Hill, New York, 1974.

46. "Inspection of Direct Bonded Substrates (DBCu)," Inspection specification, Stellar Industries Corp.

47. R. Buchanan, *Ceramic Materials for Electronics,* 2d ed., Marcel Dekker, New York, pp. 455–509, 1989.

48. R. Vissar and J. Snook, "Direct Bond Copper (DBCu) Technologies," Stellar Marketing Literature.

49. D. Hopkins, K. Dalal, and S. Bhavnani, "Effect of Metallization Thickness on Thermal Conductance of a First-Level Power Hybrid Structure," *Inter. J. Microcircuits & Electronic Packaging,* vol. 16, no. 3, third quarter, pp. 189–197.

50. N. B. Nguyen, G. DiGennaro, F. Jones, and C. Kerfoot, "Hybrid Packaging Techniques for Very High Power Converters," *PCIM,* pp. 28–34, October 1990.

51. M. M. Hussein, D. J. Nelson, and A. Elshabini-Riad, "Thermal Management of Hybrid Circuits: Effect of Metalization Layer, Substrate Material and Thermal Environment," *ISHM 1990 Proc.,* pp. 389–394.

52. D. Hopkins, S. H. Bhavnani, and L. Tsai, "Numerical Modeling and Experimental Comparison of Copper Bonded AlN, Al_2O_3, and BeO Power Hybrid Structures," *IEPS 1992 Proc.,* pp. 200–212.

53. N. Nguyen, "Using Advanced Substrate Materials with Hybrid Packaging Techniques for Ultrahigh-Power ICs," *Solid State Tech.,* pp. 59–62, February 1993.

54. D. Hopkins, S. Bhavnani, and K. Dalal, "Thermal Performance Comparison and Metallurgy of Direct Bonded AlN, Al_2O_3, and BeO Assemblies," *ISHM 1992 Proc.,* pp. 577–583.

55. N. Nguyen, "Adoption of Advanced Materials in Hybrid Packaging Technique for Ultra High Power Semiconductor Devices," *ISHM 1992 Proc.,* pp. 567–570.

56. J. Sergent, T. Evans, G. Newell, and J. Fudala, "New Challenges in Packaging Power Hybrids," *EP & P,* pp. 46–48, September 1990.

57. King and C. W. Harrison, "Current Distribution and Impedance Per Unit Length of a Thin Strip," *IEEE Trans. Antennas and Propagation,* vol. 14, p. 252, March 1966.

58. R. Tummala and E. Rymaszewski, *Microelectronics Packaging Handbook,* Van Nostrand Reinhold, 1988.

59. J. Steinberg, "Metallizing," *Ceramics and Glasses: Engineered Materials Handbook,* vol. 4, ASM International, Materials Park, Ohio, pp. 542–545.

60. R. W. Johnson, R. Weeks, D. C. Hopkins, J. Muir, and J. R. Williams, "Plated Copper on Ceramic Substrates for Power Hybrid Circuit," *IEEE Trans. CHMT,* vol. 12, no. 4, December 1989, pp. 530–536.

61. W. Chiang, V. Greenhut, D. Shanefield, and L. Johnson, "Processing and Testing Variability in Direct Bonded Copper on Electronic Grade Alumina Substrates," *IEEE Trans. Micro.,* in press.

62. J. Holowczak, V. Greenhut, and D. Shanefield, "Peel Adhesion Bond Strength of Direct Bonded Copper–Alumina as Affected by Alumina Sintering Aids," *Metals–Ceramic Joining,* Minerals, Metals, and Materials Society, 1991.

63. E. Webb, "Effects of Copper Thick Film Processing on Adhesion and Bondability," *ISHM Europe 1987 Proc.,* pp. 128–135.

64. S. Palanisamy, D. Sarma, and D. Weaver, "An Accelerated Lot-Acceptance Test for Adhesion Degradation of Soldered Copper Thick Films in Temperature Cycling," *ISHM 1987 Proc.,* pp. 554–561.

65. S. Tamhankar, R. Wolf, E. Chang, and M. Kirschner, "Defects in Copper Thick Film Conductors Produced Under Different Conditions," *International J. Microcircuits and Electronic Packaging,* vol. 16, no. 1, pp. 71–78, 1st quarter 1993.

66. L. Dolhert, J. Lau, J. Enloe, and E. Luh, "Performance and Reliability of Metallized Aluminum Nitride for Multichip Module Applications," *International J. Hybrid Microelectronics,* vol. 14, no. 4, pp. 113–120, December 1991.

67. K. H. Dalal, "Substrate and Metallization Selection for High Power Hybrid Circuits Based on Thermal Resistance and Temperature Cycling Reliability," *Proc. 6th International Applied Power Electronics Conf.,* Dallas, pp. 347–354, March 1991.
68. A. Kraus and A. Bar-Cohen, *Thermal Analysis and Control of Electronic Equipment,* McGraw-Hill, New York, 1983.
69. C. Fisher, "Copper Technology: A Decade of Development," *Hybrid Circuit Tech.,* pp. 37–40, September 1886.
70. W. Martin, B. Waibel, and W. Lasser, "Thermal Resistance and Temperature Cycling Endurance of DBC Substrates," *Hybrid Circuits,* pp. 29–33, May 1990.
71. P. Maier and J. Jacobsen, "Power Devices for Frequency Converters," PCIM Europe, pp. 292–296, November/December 1993.
72. "Active Bonded Copper on AlN," technical data sheet, Sherritt Inc., Alberta, Canada, 1993.
73. *Aluminum Nitride Bibliography,* Sherrit Inc., Alberta, Canada, 1993.
74. J. Blum, "Aluminum Nitride Substrates for Hybrid Microelectronic Applications," *Hybrid Circuit Tech.,* pp. 7–14, August 1990.

6

Assembly of Hybrid Microcircuits

Roger L. Cadenhead

6.1 Introduction

As integrated circuits increase in complexity the requirements for high density packaging will ultimately lead to novel forms of hybrid assembly. Typically, two types of assembly are used in microelectronics. The first is known as chip-and-wire (since there are now many bare-chip assembly techniques not requiring wires, this method of assembly will be called *bare-chip* assembly). The second method of assembly was originally known as solder assembly but has come to be known as the surface mount technology (SMT). Combining the manufacture of all varieties of multichip modules (MCMs) and SMT shows the use of both bare-chip assembly and SMT.

Figure 6.1 highlights the differences in fabricating microcircuitry using hybrid assembly techniques and polymeric PWB through-hole assemblies. The distinctions between these two types of microelectronic assembly were once very obvious. However, today they are so strongly integrated that they must be taught together. The term *hybrid* was originally applied to an assembly technique to describe a microelectronics circuit that is a hybrid of different assembly methodologies. This is even more true today than it was twenty years ago when the first edition of this handbook was published. For example, the traditional solder assembly (the SMT) and the bare-chip assembly techniques are normally found in the same system.

Different sections of this chapter fully elucidate both assembly methodologies. Figure 6.2 gives a simplified overview of the chip-and-wire (this figure

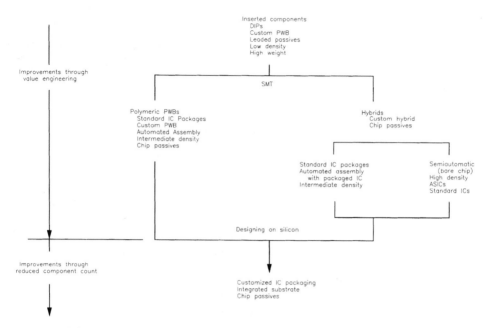

Figure 6.1 Fabrication differences between SMT using hybrids and polymeric PWBs.

really is chip-and-wire) assembly technique; other bare-chip techniques are fully explained in this chapter. In addition, two sections of this chapter on microelectronic assembly concern themselves with materials properties, i.e., intermetallic formation during bare-chip assembly and solder, the *heart* of the SMT. For the first time, factory automation and a proposed decision support system to support hybrid assemblies are both included. Last, statistical process control takes its place in the assembly of hybrid microcircuits in this handbook.

The SMT now addresses flip-chip and chip-on-board issues that were previously addressed only by hybrid microelectronics. Multichip modules now come in three varieties: laminated (MCM-L), ceramic (MCM-C), and deposited (MCM-D). The major issues in MCM and SMT assembly include the use of leaded versus leadless components, single-chip versus multiple chip packages, and the use of solder as an interconnection for both assembly techniques. Packaging necessities, sealing of covers, conformal coatings, and attachment of hybrid assemblies to the package are addressed in Chap. 7.

Bare-chip hybrid assembly will have more materials issues to address than do typical solder assemblies. Several assembly techniques are addressed such as TAB and flip-chip bonding plus simple wire bonding. The majority of bare-chip assemblies are used for high reliability applications. Since reliability is a paramount issue, these assemblies are typically used with higher-cost materials.

As technology comparisons are made in the following sections, it will become readily obvious that the world of *assembly of hybrid circuits* has become more

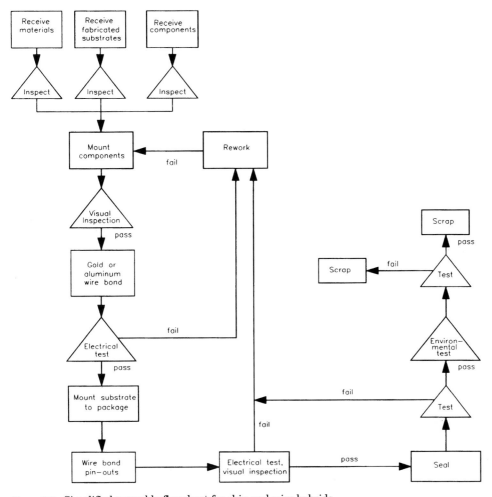

Figure 6.2 Simplified assembly flowchart for chip-and-wire hybrids.

complicated than it was at the time of the first edition of this handbook (1974). Where possible, the same style of address will be used. However, over 80% of this chapter consists of new material, either updated or addressed for the first time. Almost all the listings in the bibliography are dated after the copyright of the first handbook. As new technologies have evolved, old ones have been supplanted. For example, no inclusions have been made for beam lead bonding. The first edition addresses those issues, and this one will neither add to nor replace what was said at that time.

6.1.1 Technology comparisons

Hybrid technology has its position in the middle ground of microcircuit packaging. Neglecting the world of point-to-point wiring, the through-hole PWBs represent the highest volume, largest size, heaviest assembly technique. As hybrid solder assembly techniques began to be applied to the PWB world (surface mount technology), size, weight, and volumes of equivalent circuits plummeted drastically, often by 50% or more. However, with SMT being used as a one-for-one replacement for the through-hole PWB, reliability did not increase, as has been the demonstrated phenomenon of miniaturization in the hybrid world.

The reason is simple. With SMT, the number of interconnections decreased from that of PWB polymer technologies with more of the circuitry being fabricated in the solid state. Although this enhanced reliability, a host of problems not experienced with through-hole PWB technology surfaced. These problems involved optimum solder flow, flux residue removal, coefficient of thermal expansion mismatch, and solder joints cracking under environmental stresses. The solution to these problems had to be obtained during the assembly of the product, resulting in extensive rework and repair, often culminating in product redesign.

However, as more was learned about the surface mount technology, the reliability of this kind of hybrid assembly increased. The expected increase in reliability associated with miniaturization and size reduction should be evident in line with historically achieved levels of SMT densification.

Ascending the density scale of microelectronic circuitry manufacture is traditional thick film hybrid technology, followed by thin film microelectronics and finally the multichip module (wafer-scale assemblies). There is no distinct dividing line from SMT on either polymer or ceramic wiring boards through the multichip modules. To assemble contemporary microcircuits requires the intermingling of all these technologies and manufacturing techniques to the point that no clear division exists today. After the multichip module comes wafer-scale integration, and then the fully integrated circuit (monolithic IC). The smallest method of circuitry manufacture in scope, scale, and even possibility (at least to the current extent of human knowledge) is *molecular electronics.*

In manufacturing minicircuits and microcircuits, the basic unit upon which assembly occurs is the substrate. This is true in all technologies except point-to-point wiring and molecular electronics. In conventional through-hole PWB technology and in commercial SMT assemblies, devices are individually packaged and therefore easily tested before assembly. However, the package consumes most of the volume of SMT, conventional polymeric PWBs, and ceramic wiring boards (CWBs). This is also the case in thick film solder assemblies. In contrast, many hybrid types (thick film, thin film, and multichip modules) possess bare chips and use assembly techniques conducive to the handling of these bare chips. Nevertheless, protection of both assembly types is an issue.

The extreme in today's microcircuit technologies is the monolithic integrated circuit (IC) in which all of the circuit's components are contained within a single structure. The IC is the device contained in the single-chip package or mounted on the surface of substrates used in hybrid technologies. Therefore, the bare-chip hybrid technologies represent a packaging approach that is intermediate between substrate–assembly technologies and true wafer-scale integration. All hybrid technologies were developed to reduce volume and offer flexibility to interconnect a number of different electrical functions on a single supporting member. This chapter concerns itself with the elucidation of the assembly techniques to make the hybrid assembly ready for test prior to packaging.

6.2 The Decision Support System

In order to decide which assembly technique to use, the original decision as to which fabrication technology to choose is addressed. With so many options from which to choose, a decision support system to evaluate the proper micro-electronics technology to choose is paramount. Decision support systems (DSS) are the result of adding structuredness to the field of problem solving starting with unstructured data. These systems culminate from database manipulation during *information management.* In an effort to add structure for problem solving as support to making decisions, a model can be generated as precursive to building a DSS. This model may revolve around the use of a graphic or pictorial such as a data flow or process flow diagram, thereby reducing the data to a binary (yes/no) format or into sets of mathematical equations (linear programming). In any event, a generator has to be developed.

The need for the DSS is easily recognized. The definition of what should or should not be involved (concept) is the difficult part. The DSS must first be recognized as a support system, not the ultimate in the decision-making process. In simpler cases, the DSS can be structured; usually it is semistructured. The DSS focuses on the *effectiveness* rather than the *efficiency* of the decision and is support to all phases of the decision. The historical portion of the system is the data programmed into the system. In properly constructing the system, it must be made self-teaching through cybernetic loops; i.e., the DSS learns from its experience. This will make the DSS knowledge-pervasive, ensuring its effectiveness.

6.2.1 Building the DSS

The DSS represents a special application of the *information management system.* The DSS collects, maintains, and produces an output based on input received through its programming to handle that information. Designing a classical information system and DSS means automating an existing well-designed manual system. When designing the DSS, however, great care must be given to automation of the information available. If the DSS is made too rigid, it will give answers which are too regimented, too closely guarded. The reason is that the system will do as it is programmed; i.e., the DSS will inject the prejudice of the programmers into the design effort. In so doing, the system cannot be dynamic but rather is some aberration of its initial static self.

The DSS must first be prototyped. Evaluating the prototype system against the considered opinions of experts in the microcircuit field will serve cybernetically to determine if the system is producing the desired result. A periodic audit of the results of this DSS will also be mandatory. This DSS will use the model suggested here and be prompted to design the circuit using the various types of microcircuit assembly techniques familiar to experts in the microelectronics industry.

6.2.2 The model design process

In the area of microelectronics, changes are occurring very rapidly, even daily. Choices are at best tentative and must continually be revised. The DSS gives the applications and design engineer (decision maker) access to the history of previous decisions. This aids the human effort, since we are all subject to incomplete or inaccurate statistical thinking based on our particular experience. The DSS eliminates this inequity.

The decision model, if continually updated, becomes a plethora of statistical information. Considering this database, the DSS is then a communications tool (from the past to the present), furthering the decision maker's conceptual skills by completing tedious deductive tasks. By propagating the necessary knowledge, the engineer's time is freed for abstract or innovative thought rather than the rote memorization of historical facts and from becoming embroiled in interminable calculations concerning known data (prevents reinvention of the wheel).

The technology involved in such a system may be viewed in three parts; i.e., the DSS becomes a tool for *work group collaboration,* a *knowledge representation system* for conceptual modeling, and a *truth maintenance system* of reasoning for changes about the model. To begin the process of constructing any model, experts must be consulted. These experts may have open-ended discussions but of necessity must be able to "boil" these discussions into relevant choices that can be made, even if they are based on arbitrary starting points. From this beginning, this project establishes a model that, through computer learning (self-teaching), will result in a DSS.

The use of such *process knowledge* will result in the desired DSS through trial-and-error experimentation. The *old requirements* will thereby evolve into the *new requirements.* By utilizing experts for input to the DSS, model design starts. After a period of model usage, the DSS (based on the model) becomes *the expert.* Flexibility in the system, however, allows the introduction of new expert knowledge when desired. No matter how the DSS comes into being, any system that is knowledge-pervasive and application-specific is expensive to build and expensive to maintain. However, properly designed and used, the DSS will pay for itself time and again. It will then be more expensive to operate without the DSS than with it.

Stability of the model's design will be dependent on the environment in which the knowledge is used. Since this model (and DSS) is proposed for the turbulent microelectronics environment, the design of the system must always be in question. The more flexible the model's design, the more useful will be the DSS. At this point the system can become independent and a priori knowledge more useful.

6.2.3 The microelectronics packaging technology

Choices of packaging technologies for microcircuits have been few, but in the past 15 years, the surface mount technology (SMT) and the development of multichip modules (MCMs) have reached the point that packaging technologies are catching up to the capabilities of the IC technology since these packaging technologies are becoming able to handle the tremendous number of I/Os of the more complicated ICs (ULSI, ASICs). As early as 1992, the integrated circuit industry was capable of 0.3-μm lines and spaces, although that feat will not be industry-common until the end of this decade.

Often designers of microcircuits to be assembled do not care which technology is chosen to package their microelectronic circuitry. Simply stated, designers need *the one that is possible to use (feasible) in the application* and, if there are several, will invariably choose the one with which they are most familiar unless another can be shown to be more cost effective. Therefore, the purpose of this paradigm (model) is to forecast which combination of fabrication technologies is most cost effective given the choice of several.

6.2.4 Decision support requirements

Decision support systems are the bases for decision making and problem solving. Those systems typically take data and act on those *data* to produce *information* from them. If that information is used in the decision making process, it is called *intelligence*. This leads to the objective of the project: *integration of the technology, the model, and the DSS.* The design of any model is made difficult by the components of the model and the degree of expertise of those charged with the responsibility for constructing that model and subsequently the DSS.

Applications will require a maximum likelihood estimation model for the circuit type being designed. Such a likelihood will depend on a number of factors, e.g., cost, time to market, resources, equipment, and process knowledge. These aspects of integration must be understood by the microelectronics engineer as well as the designer of the DSS. This model is presented as a wide-range decision model. Integrating microelectronic knowledge and decision making knowledge are required for the circuit designer to make the assembly technology selection decision. It is at this juncture, conceptualization, that the greatest economy and highest quality can be achieved.

6.2.4.1 Objective.
Today most circuit designers using microelectronics technology do not know when to use one assembly technology or another cost effectively. To complicate matters, that decision is made by the designer, not the packaging expert, who knows the difference between technologies. After the microcircuit is designed using one set of design rules, it is almost always too expensive and time-consuming to change directions and do anything else. Therefore, the design of a paradigm that may be used by designers and management to determine that the most efficacious method is used is specified here.

6.2.4.2 Criteria. Technology is developing at a rapid pace. It behooves the manufacturer to use the latest technology available but, at the same time, to use the technology that is tested, tried, and proved. The selection of the technology to use for packaging microcircuits is becoming increasingly important, since packaging density will increase from the 150-lead packages common today to devices with as many as 300 leads by 1997. The first step in constructing this model was to determine the density of the substrate to be used in packaging the microcircuit. The best measure of density is the number of inches of conductor that can be fabricated on each square inch of board area. From Table 6.1 and Fig. 6.3, the selection of densities for each packaging technology can be used as decision criteria for the model.

Any technology has to be cost effective to be competitive. However, under unusual circumstances, a given technology may be the only way in which the circuit can be packaged owing to either size and/or weight restrictions as in military applications or owing to the customer's demands. In these instances a decision model is not necessary. The application of customer-dictated technology must be used or the microcircuit cannot be made.

6.2.4.3 Paradigm. A paradigm is an abstraction including key variables and relationships pertinent to the problem. That paradigm may be either mathematical, graphical, or pictorial and has, as its purposes, the provision of an understanding of the problem and its ready solution. To be effective, a DSS must be represented by its model. The system itself is a complex interrelation of correlated components to accomplish a task or set of tasks. The model portrays an interaction of inputs with its environment to produce optimum output. The model and its resultant DSS must be well defined and have the following characteristics.

In designing and producing microcircuits, *several determinants* have to be considered. The design of the paradigm must take each desired feature of the DSS into account. The first is the worth of the product both *technically and economically* to the customer. This is the reason for calling this the *quality*

TABLE 6.1 Available Density

	Range	
Technology	cm/cm^2	in/in^2
Single-sided PWB	12–16	30–40
Polymer thick film	16–22	40–55
Double-sided PWB	16–22	40–55
Multilayer thick film	24–32	60–80
Multilayer PWB	24–180	60–450
Cofired thick film multilayer board (double-sided)	80–180	200–450
Thin film	<240	<600
Multichip modules	240–600	800–1500
Wafer-scale integration	600–1600	1500–4000
Integrated circuits	>1600	>4000

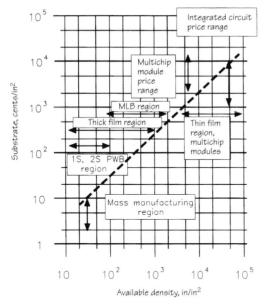

Figure 6.3 Typical price-density estimates (1993 dollars).

model of the product's design. The second consideration is the *economy of producing* the circuit and consequently the *selling price* of the microcircuit. The third important consideration is the *ease* with which the circuit can be manufactured using the chosen assembly technology.

6.2.5 A DSS for assembly technology selection

For the construction of this DSS, a pictorial, heuristically derived model has been developed. The algorithm depicted in the model can be used with only simple inputs that will be available to the designer after breadboarding of the circuit to choose the most cost effective assembly technology. The inputs to make that decision are already available with the CAD input package or are required by the model itself. Further, the model is constructed so that, with original design inputs, a computer-aided software engineering (CASE) tool can be generated. From this information, management and product designers will have an operations tool that can be tailored to the individual enterprise with the simple inclusion of that enterprise's microcircuit layout rules.

As shown in Table 6.1 and Fig. 6.3, there was previously no assembly technology coverage in the "gap" between thin film microelectronic capability and fabrication of the integrated circuit (IC). Although the capability existed to reach 1-μm feature sizes with the IC technology, there was no impetus to do so since the IC would have required too many I/Os (pads) to be effectively packaged. With the advent of the multichip module technology, that shortcoming no longer exists.

Several technology definitions are given throughout this chapter and handbook that must be comprehended in order to understand the model. Also several assumptions had to be made as a starting point to the model and the design of the DSS. Those assumptions are given as prelude to the presentation of the model.

Assumptions

- An optimum combination of technologies can be found.
- There are no technical constraints other than those denoted by this project that make one technology preferable to another.
- The cost per I/O (pad) is optimal for a given technology.
- All substrates are multilayered except thin film ceramic substrates, which are state-of-the-art single-layered.
- The densities quoted in the model fit within the boundaries of densities cited in Table 6.1.
- The densities of the MCM-D will match industry projections.

6.2.6   Summary

The emphasis of decision support systems is and should be on the process rather than the system. The DSS is a support function as the name implies; it is not the ultimate. There are many criticisms, most founded. The DSS must be used to help humans make the final decision. The system will be useful only to the degree it is programmed to be useful. Usually there is not a single model that will do the job as this project might suggest. Extensive system usage will expose intangibles that will add to the process of decision making and expand the need to that of several models.

The importance of this DSS is relieving the decision maker from the mundane. This system will make it easier for the engineer, manager, or executive to concentrate on the intangibles involved in any decision that must be made, particularly in the turbulent, highly charged environment of microelectronics.

With current state-of-the-art multichip module technology, the required assembly technology now exists to package higher density ICs. However, the high cost of practicing the hybrid assembly technology prohibits its *universal* (the correct application of every available hybrid assembly technology) use. With this model, management can readily know when each technology is more cost effective, thereby promoting the use of all hybrid assembly technologies. The substrate selection model is shown in Fig. 6.4.

6.3   The Solder Joint

Two types of microelectronic assembly are studied by this chapter: solder assembly (SMT) and bare-chip assembly. The first of these depends for its integrity wholly upon the solder joint, since it is the weak link in the assembly. The solder joint is the mechanical strength, the electrical connection, and the physical weakness of the assembly. Since the solder joint is the "heart" of the solder assembly, it is examined here. Also to be considered in this chapter are hybrid assemblies on organic wiring boards as well as inorganic wiring boards; both will be called substrates.

The solder joint is examined mechanically, electrically, physically, and metallurgically. The forces exerted on the solder joint depend on the relative coefficients of thermal expansion (CTEs) of the materials involved. Under even the best of circumstances the CTEs of the different materials used will be different. As the microcircuit endures the different thermal stresses during assembly, environmental testing, and/or service life, the solder joint is fatigued. Figure 6.5 illustrates the origin of solder joint fatigue. As can be seen, the steady state affords no stress to the solder joint. The different condi-

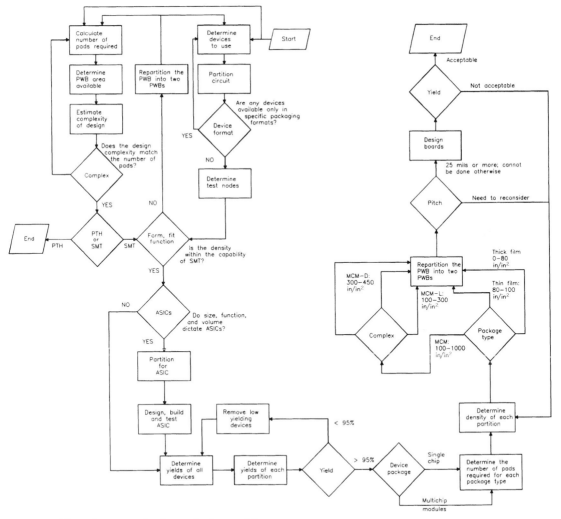

Figure 6.4 The substrate technology selection model.

tions that can exist that stress the solder joints are: the component is heated at a quicker rate than the substrate, the component is heated and the substrate is not, and finally, the substrate is heated while the component is not.

These different heating rates for the differing materials, plus the different CTEs of the materials, lead to the various deformation modes of the solder joint. In-plane displacement results in a shear on the solder joint. In the case of vibration or mechanical flexing, an out-of-plane rotation occurs, causing tension to be placed on the solder joint. Simple tension occurs any time an environmental stress resulting in several g forces is placed on the hybrid assembly. At any given time during the service life one or more of the stresses may occur to the solder joint. Figure 6.6 illustrates these three different deformation modes.

Cyclic fatigue is another of the environmental vagaries faced by the solder joint during service. The causes of such fatigue are many. The first involves temperature variations that occur during the assembly of the microcircuit. The second type of cyclic fatigue faced by hybrid assemblies involves temperature variations during service life. The third type of cyclic fatigue involves the environmental tests the hybrid assembly must face before shipment. The

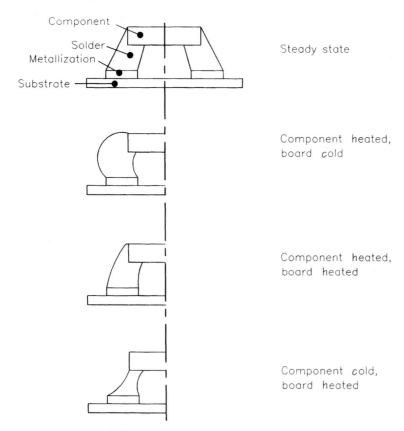

Figure 6.5 Origin of solder joint fatigue.

fourth stress is power cycling, the fifth is mechanical flexing, and the sixth is simple vibration that can readily be experienced during shipment of the microcircuit. As can be seen, solder has to be the most compliant member of the SMT assembly; i.e., solder bears the brunt of cyclic damage. Lastly, during SMT construction, the problem of cyclic fatigue is exacerbated with leadless devices, e.g., chip capacitors, chip resistors, leadless chip carriers, etc.

The failure mechanisms faced by the solder joint under stress involve the metallurgical failure of the solder due to crack propagation until the coherent cross-sectional area of the joint can no longer support the load. Low temperature cracks will start at a free surface and propagate transgranularly. High temperature failures will start anywhere and even travel intergranularly. Often, this knowledge is most applicable during failure analysis, since both types of failure cracks cause microcircuit failures. A good rule of thumb is: Failure is accelerated in accordance with the ratio of ambient temperature (K) to the melting point of the metal (K). Since solders are alloys with relatively low melting points, they are materials that will suffer early failures as indicated by the rule of thumb. Several methods are used to combat fatigue mechanisms. Prevention of oxidation can increase fatigue life by an order of magnitude. Conversely, the presence of a corrosive flux decreases solder joint life by a similar amount.

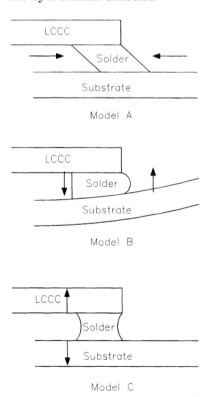

Figure 6.6 Deformation modes of the solder joint. Model A: in-place displacement (simple shear); model B: out-of-plane rotation, as in vibration or mechanical flexing; model C: out-of-plane displacement (simple tension).

There is a well-founded concern by assemblers of SMT hybrids that cyclic testing procedures are not directly related to the service life of the solder joint. If cyclic stress levels are too high, failures in the solder joint are definitely stress-induced. If the frequency of cycling is too high, it is shown that this test condition is unrepresentative of functional service behavior since dwell times are too short for stress relaxation of the solder. Analogously, if the temperature extremes are too high or too low during testing, the material behavior cannot be related to the service conditions of the solder. The following test methods are inappropriate to use when testing the solder joint (Table 6.2).

- *Oven cycling* of assemblies with matched CTEs may not cause sufficient thermal expansion mismatch to produce failures during reasonable test duration.

- *Mechanical cycling* may be too fast to allow stress relaxation of the solder. If stressful relaxation does not occur, the solder joint experiences only a portion of the stresses the test was meant to induce.

- *Thermal shock cycling* may be misleading.

Notice that in comparing the fatigue characteristics of solder alloys used in the surface mount technology, 63:37 and 60:40 are the worst performers of all the alloys listed in Table 6.3. Why then do we use these alloys so often? The reason is that so much is known about them. They are low melting and very pliable. However, their physical characteristics are greatly undesirable.

Figure 6.7 illustrates solder joint fatigue in leadless chip carriers (LCCs). The solder joint in this application is most susceptible to fatigue yet, owing to hermeticity requirements, the LCC is most used in high reliability applications. Figure 6.7 assumes, for mathematical purposes, that the component and substrate are rigid, the solder joint dimensions are small compared with the component size, the solder volume remains constant, and the joint is in shear only.

In using LCCs, there are several methods for producing solder joint standoffs. The LCC may possess a high melting point solder while the substrate is pretinned with a low melting point solder and vice versa. In addition, different melting point solders may be used (one solder on the device, a different one on the substrate) as a method for solder joint standoff. Ceramic particles

TABLE 6.2   Assessment of Accelerated Cyclic Fatigue Testing Methods

Test	Information	Test complexity	Test duration
Functional power cycling	Accurate	High	Very long
Oven temperature cycling	Good, with caution	Moderate	Long–moderate
Mechanical cyclic flexing	Comparative, with caution	Low	Moderate–short
Thermal shock cycling	Misleading	Moderate–low	Short

TABLE 6.3   Comparative Fatigue Characteristics of Solder Alloys Used in SMT

Performance	Composition
Worst	63Sn:37Pb
	60Sn:40Pb
Poor	62Sn:36Pb:2Ag
	65Sn:35In
	42Sn:58Bi
	50Sn:50In
	50Sn:50Pb
Fair	99In:1Cu
	90Sn:10Pb
	99.25Sn:0.75Cu
Good	99Sn:1Sb
	50Pb:50In
	100Sn
Excellent	96Sn:4Ag
	95Sn:5Sb

may be added to the solder, or posts may be built onto the substrate using solder mask or controlled dielectric to place standoffs on the substrate. Increasing the standoff distance between the LCC and the substrate decreases the shear strain on the solder joint and increases its fatigue life. Finally, increasing the solder fillet size beyond the component edges increases the joint area and also decreases shear stress.

The solder joint deforms by both shear and tensile strain; the component and substrate deform to place the solder joint in both types of duress. Inorganic substrates, usually alumina, will easily match the CTEs of components. Organic boards are not so fortunate. They often require the following two methods of combating the degree of deformation, and thereby stress, placed on the solder joint. The first is by the use of a compliant top-layer material, i.e., one in which the top layer is less rigid than the body of the organic substrate, sometimes by a factor of 10:1. The second method of mitigating deformation is by the use of matched CTE cores; i.e., the core material

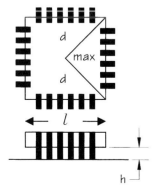

Figure 6.7 Solder joint fatigue in LCCs.

underneath the organic substrate will have a CTE similar to that of the LCC.

An important factor to consider, regardless of the type of stress, is the time it takes for the solder joint to relax or assume its natural state for that given temperature. It may take upward of 1 month for a solder joint to be truly changed by the application of external stress. Figure 6.8 illustrates this principle. The solder used for illustration in this example is 60:40. Note that measurement of the rate of stress relaxation of the shown 60:40 solder is given as a function of temperature. Note also that this curve is universal for all temperatures since the abscissa of the graph in Fig. 6.8 uses the temperature T in Kelvin.

In summary, this chapter explains why the solder joint is the weak link in the SMT. Luckily, it is the easiest material for which to substitute should it become necessary. We have shown that applied stresses must be carefully considered as well as the solder joint microstructure in determining potential failure modes of the solder joint. The cyclic frequency during testing is of great importance, as are the cyclic fatigue extremes used. Various methods of combating CTE differences have also been given before the general discussion of the surface mount technology, hybrid assembly technique.

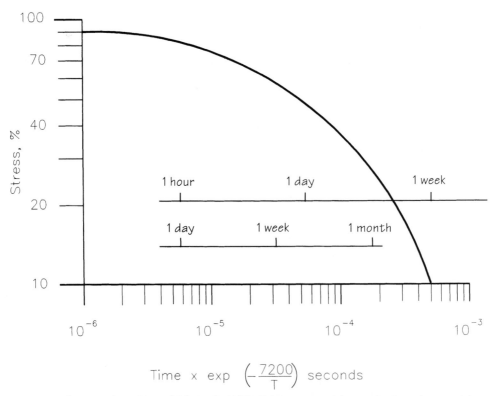

Figure 6.8 Stress relaxation, which in Sn60/Pb40 is expressed by an Arrhennius equation: $E_A = 62$ kJ/mole.

6.4   Hybrid Assembly Using the Surface Mount Technology

Electronic circuitry, as currently constructed, is composed of passive functional blocks such as resistance, inductance, and capacitance plus active components. Tying these functions together into a subsystem or system to execute a predetermined purpose is the job of electronic assembly. Electronic assembly has traversed several generations of interconnect technology to reach today's state of the art. This fourth-generation, state-of-the-art assembly technique is surface mount technology (SMT). SMT is the widespread application of the hybrid assembly technology formerly known as *solder assembly*.

6.4.1   Background

All this activity resulted from the joint efforts of Harry Diamond of the NBS and Harry Rubinstein *et al.* at Centralab in Milwaukee working on the 81-mm mortar fuse. To develop this fuse, Diamond and Rubinstein executed innovations in electronic microcircuit fabrication and assembly due to the very real fear the Allies and the Axis shared toward the end of World War II: The winner conceivably could be determined not on the battlefield, but in the warehouse.

Both sides were rapidly running out of resources (labor and raw materials), specifically resources to make electronics. The fear was simple: the side that ran out of these resources first, lost. To have less labor and fewer materials in a circuit, the circuit's elements requiring those materials had to be made integral to the chassis (substrate) and be produced by bulk methods (reduced labor content). Necessity being the mother of invention, the printed circuit on steatite, forcing some component leads and terminations to be directly attached to surface connections (the SMT), came into being in May 1944.

Initially connection was made by simply stringing conductive material in wire form from one function to another. Because of the unreliable nature of this type of connection, solder was used to secure the connection at either end of the wire. Then technologists like Frank Sprague and Thomas Edison determined that at least the conductive interconnections could be defined on a supportive surface (substrate) by either additive methods (decalcomania—screen printing) or subtractive methods (metallize the surface of the substrate followed by removal of unwanted material). When the wires and other passive components of the circuit were replaced by conductive, resistive, and insulating materials on a ceramic substrate, it became economically sensible to make the soldered connections on the surface of the ceramic. Thus the SMT was born.

After defining the circuitry on the substrate, one layer could be stacked on another (multilayer) with intermediate insulation, thereby compacting the volume an electronic function would occupy (also reducing the weight of the electronics). Now that the interconnective portions of the circuitry were in place, the components had to be added. Initially, on organic substrates, holes were drilled and the leads of the components were inserted and soldered.

Room had to be left on each layer of the multilayer for the lead to penetrate from top to bottom of the substrate.

The thick film technology, culminating from previously referenced efforts on the variable-time mortar fuse in World War II, mounted its components on the surface of a steatite substrate when assembling electronic circuitry instead of using holes for leads (too expensive to drill holes in ceramics). These techniques were later adapted to hybrid assembly in general. This seemingly simple progression in electronic circuitry assembly (the SMT) began a revolution in the hybrid assembly industry.

As the 1980s approached, the common usage of high density integrated circuitry became more common, with feature sizes reaching 0.3 μm (lines and spaces) now possible. This makes high density packaging and interconnection schemes such as the SMT and methods of manufacturing multichip modules absolutely mandatory. Otherwise, the routing requirements around today's high input/output devices would go unanswered by the packaging and interconnect technologies.

6.4.2 Assembly placement equipment

The majority of equipment marketed for surface mount placement is of the chip-shooter or pick-and-place variety (*chip-shooter* is a term denoting the rapid placement of small microdevices on the surface of hybrid assemblies). Some placement machines use an x,y,z traversing head to pick one component at a time and move to place it in a programmed position on the substrate. In like manner, other machines incorporate an x,y,z programmable table to position the substrate under the placement head, which then performs the z-axis motion of component placement.

Usually most hybrid assembly placement machines function with some combination of both types of movement. Equipment incorporating an x,y table is generally faster than equipment with a multiaxis placement head. Very high rates of component placement can be achieved with mass transfer equipment which place some fifty components during one 3-s cycle. However, this type of equipment uses rigid, fixed tooling to achieve speed. The advantages to component placement machinery are programmability, automation, and consistency of product, enabling the manufacturer to build a variety of product designs on one piece of equipment by changing programs and setups with different components.

The important point to stress is that the manufacturing process functions best when the equipment fits the needs of the assembly task. Some of the important assembly features that affect the hybrid assembly plant are placement speed and accuracy, adhesive dispensing, automatic substrate and component handling, off-line versus real-time programming, component verification, setup time, number of different components allowed, size of substrate used, ease of automation, shape of components to be handled, etc. Expect trade-offs in the decision to determine the most appropriate assembly placement machine for the job.

The selection of the appropriate placement equipment to do your hybrid assembly job requires knowledge of systems' controllers, necessary applications to be addressed, CIM exigencies, mechanical positioning systems, and optical design expertise. The increasing use of the SMT as an assembly technology has fortunately resulted in a proliferation of many component styles and sizes. Many new generations of placement systems are necessary to handle the diverse nature of these components, to stock them for hybrid assembly usage, and to visually place them very accurately to accommodate the fine pitches many of these devices now have. One such system is the Sony SS-7$_{II}$ (Fig. 6.9) made to handle odd-shaped, special-purpose components. The SMT often requires the use of a special, separate machine that is very flexible in nature to handle such components as connectors and through-hole devices on

Figure 6.9 The Sony SS-7$_{II}$ odd-shaped component insertion machine.

the same substrate. These machines are made to economically handle both axial-leaded and radial-leaded components.

The relatively high price of many of today's odd-shaped components demands the use of such insertion equipment as the SS-7$_{\text{II}}$. The specially designed, gentle placement method of this machine virtually eliminates component damage while reducing unnecessary component waste due to bent leads. The SS-7$_{\text{II}}$ is only one example of the commercially available machinery that can handle many types of component packaging and several different component types, thereby reducing changeover time and increasing continuous operation.

To address the issue of fine-pitch placement, several companies are offering systems that contain vision as well as other options to increase hybrid assembly yields. Some of these options include fine-pitch handlers with lead spacings as low as 0.016 in and package dimensions up to 1.5 in on a side. All these machines are equipped to connect with other pieces of hybrid assembly equipment such as the screen printer and the solder reflow system to approach full-up automation and eventually CIM (note Sec. 6.12). These placement machines are equipped with both mechanical and vision placement systems that will accurately position the component where it needs to be with relation to the substrate. Such a system is the Zevatech Placemat 560 shown in Fig. 6.10.

There are many fine-pitch, flexible mounting systems to do high precision surface mount placement. Most, if not all, are designed to handle quad flat packs (QFPs) as well as a wide range of otherwise difficult to handle devices. These machines not only must be capable of accurately handling fine-pitch QFPs, they have to be flexible enough to handle a wide range of odd-shaped components and standard chips as well. Figure 6.11 illustrates just such a system, the Sony SS-V$_3$.

Other inclusions of these automated component placement systems are the ability to download directly from a CAD system; i.e., the placement machine can take its direction from the design data to determine where the components must be placed. Usually this is an option, but most companies that make placement equipment offer the software to perform this task automatically, thereby creating an assembly program that will fully automate the machine. In addition, many of these machines are capable of dispensing adhesive to attach components before they are placed. Finally, the vision system of most of these machines will detect bad components and bad substrates, neither placing the bad component nor populating the bad substrate.

One more advantage of using this type of machinery is that they offer a greater latitude when design for manufacturability (DFM) is the question (to be competitive, DFM must always be considered). The placement system may be configured so that it is retrofittable to provide for its address by new manufacturing assembly systems. Through the use of the system's computer, work in process (WIP) may be reduced to the point that all the assembly operations systems of MRP, MRPII, JIT, and CIM are readily achievable. The ingredients that make such placement systems successful are that they be

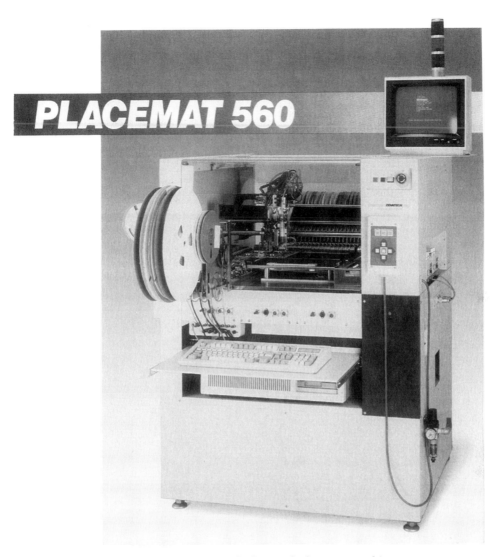

Figure 6.10   The Zevatech Placement 560 flexible, fine-pitch placement machine.

flexible and that they network with the other systems used in high speed hybrid assembly. Figure 6.12 illustrates how such a system may be configured so that it encompasses capacity requirements, facility layout, and assembled parts buffers based on a rigid downtime analysis performed using menu-driven simulation software.

6.4.3 Reflow soldering assembly equipment and processes

In addition to the type of placement equipment selected, the type of reflow process chosen will determine the manufacturing process sequence from

Figure 6.11   The Sony SS-V$_3$ fine-pitch, flexible mounting system.

receiving inspection through cleaning. Because of this, particular attention must be given to the choice of the reflow process and its potential consequences. A number of reflow processes available to reflow hybrid assembly devices are illustrated in Table 6.4.

In vapor phase reflow soldering the major advantages are that the heating of the part is uniform, the heat application is geometrically independent of the sizes and shapes of the substrate and its components, and, finally, vapor phase reflow soldering has a built-in maximum temperature limit. As the part enters the vapor, condensate forms on all exposed surfaces (top, bottom, sides, etc.). As noted, vapor phase reflow soldering uniformly heats the part sufficiently above the solder reflow point to effect formation of the solder joints. This minimizes thermal shock and allows the soldering process to occur in an oxygen-free environment. Conversely, because liquid is condens-

Figure 6.12 SMT production line setup after model simulation. (*After Kiran, Surface Mount Technology.*)

ing on the surface of the workpiece, the surface-mounted components are much more prone to misalignment requiring major rework than the other processes that will be mentioned.

During vapor phase reflow, the great majority of the heat transfer to perform the reflow operation is the result of a saturated vapor yielding its latent heat of condensation to the relatively cool substrate and its components as it is immersed into the heat transfer vapor zones. The amount of energy necessary to bring an assembly to the vapors' temperatures is calculated by

$$\text{Energy required} = \text{SpH avg.} \times (\text{primary vapor temperature} - \text{ambient part temperature})$$

TABLE 6.4 Potential Reflow Solder Processes

Reflow process	Ease of automation	Process complexity	Process control	Process cost
Hot gas	1	1	2	2
Hot oil	3	2	1	2
Hot plate	1	1	2	1
Infrared	1	2	2	1
Laser	1	3	2	3
Vapor phase	2	3	1	2
Wave solder	1	2	1	1

1 = excellent, 2 = good, 3 = poor.

Once vapor condenses on the cool assembly, the still-hot liquid will conduct heat to the cool surfaces. This, combined with heat transfer occurring with the low levels of infrared (ir) radiation from the warm interior of the vapor phase machine, accounts for the total energy invested in solder reflow by vapor phase reflow soldering.

Successful assembly using the vapor phase soldering operation requires a prebake prior to soldering in both batch and in-line systems for two very different reasons. In the *batch vapor phase system,* the parts must be preheated above the secondary fluid's boiling point to prevent condensation of secondary vapor on and under the part. If secondary fluid does form on the part, the rapid volatilization of this liquid as it enters the primary vapor will affect production yields that will be detected at postsolder inspection. With the *in-line vapor phase system,* the assembly enters the hot vapor at a slight angle. This requires a prebake to secure the parts in the dried solder paste to improve process yields during reflow. Both processes require a preheat stage to reduce the thermal shock of the temperature gradients experienced during vapor phase solder reflow. Figure 6.13 schematically represents the two types of vapor phase reflow. Figure 6.14 shows a commercially available vapor phase reflow machine in which the variety of different substrate sizes and shapes does not hinder the effectivity of this type reflow.

Dual-vapor (batch) systems offer advantages over in-line vapor phase reflow systems in that the workpiece enters the heated vapor zone flat (low part movement potential); they are more versatile, allowing a wider product range; and they require simpler tooling. The major disadvantages of the dual-vapor batch systems are that they require maintenance of two vapor layers and an associated acid-stripping mechanism. The in-line system is better suited for a production environment since the in-line process is more easily automatable (leading to automation and ultimately, CIM) and is easier for

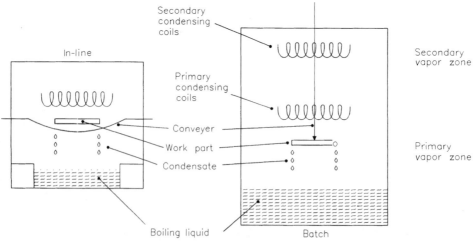

Figure 6.13 In-line versus batch vapor phase process operation.

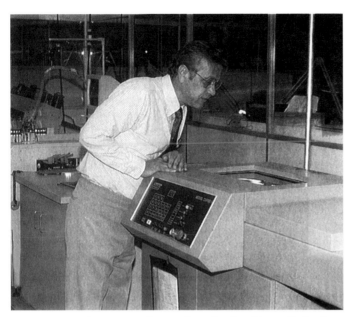

Figure 6.14 Working vapor phase reflow. (*Surface Mount Technology, September 1992.*)

production personnel to operate. The in-line system may also be used for purposes other than solder reflow since the product flows through a continuous assembly environment (Fig. 6.15).

By comparison to the vapor phase reflow soldering process, the infrared solder reflow process transfers heat by radiation. Note the simplified schematic representation of ir systems in Fig. 6.16. The fundamentals of infrared radiation and its interaction with components and substrates must be considered to design an effective infrared reflow system. This interaction is stronger with ir reflow than with vapor phase reflow. A typical ir process requires less power than an equivalent throughput vapor phase soldering system or wave soldering system. The ir process does not require any expensive heat-transfer fluids or an associated acid stripper.

In the ir process, a marked solderability improvement is noted on marginally solderable conductors, i.e., PtAu, PdAg, etc. The principles underlying this observation deal with the activation energies of the activators used in fluxes. Fluxes absorb strongly in the 3- to 4-μm range. These energy absorptions are related to the activator's ability to ionize and thereby become active. In infrared soldering much more energy at this specific wavelength is available at the workpiece, which gives rise to improved solderability.

Infrared solder reflow does not require the prebaking of the components which is typical of the vapor phase process, but prebake is always a positive. Prebake is much easier with ir than with vapor phase since the ir oven itself may perform the prebake in the initial zones on the reflow furnace. This makes ir solder reflow easier to automate. All commercially available ir ovens

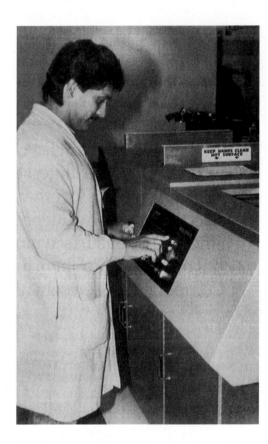

Figure 6.15 Vapor phase used to "snap" cure adhesive. (*After Plapp, Surface Mount Technology.*)

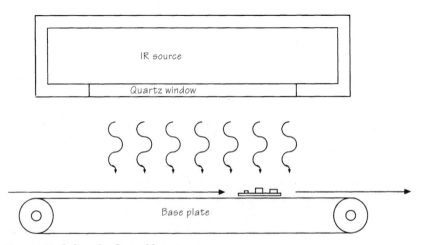

Figure 6.16 Infrared reflow solder process.

utilize an in-line conveyor. This in-line feature makes the ir process attractive for more than one reason. Figure 6.17 shows an ir reflow machine that can be used for more than one purpose.

The radiant energy absorbed by the surface mounted assembly is a function of the absorptivities of the materials on the surface mounted assemblies as well as of the substrate at the specific ir wavelengths used. For hybrid assembly applications an ir panel with low visible output is desired. Major differences in large component absorption will result in *hot spots* during reflow of the assembly. These hot spots can be comprehended but require lowering conveyor speed and platen temperature to allow time for convection to equalize these imbalances. Designing assemblies with large differences in thermal mass causes similar problems in ir reflow. These are the two disadvantages of using ir reflow as compared to vapor phase solder reflow.

Developing an ir system may and probably will determine that some portion of infrared radiation (usually about 60%) is combined with convection heating (about 40%) to design an efficient and effective reflow system. Figure 6.18 illustrates a commercially available ir unit that uses a combination of ir and convection. The type of ir energy used, the relative absorptivities of the assembly and its components, the platen temperature, the conveyor speed, and the platen height are the major variables which must be considered in ir process and equipment design. Adjustable ir panel height is desired to keep panels close to units while still accommodating a wide variety of products and process applications.

Adhesive and wave solder offer the best combination of leaded and leadless assembly technology (mixed technology). In this process, surface mounted devices are attached to the underside of a substrate using a convectionally cured epoxy or a uv-cured adhesive. Once the adhesive has cured, leaded components can be stuffed from the top side of the board. Additional preheat is then required to dry the flux and elevate the components to soldering temperatures. The whole assembly is then simultaneously soldered either using a conventional wave solder machine or with a wave solder machine specifically built for surface mounted devices.

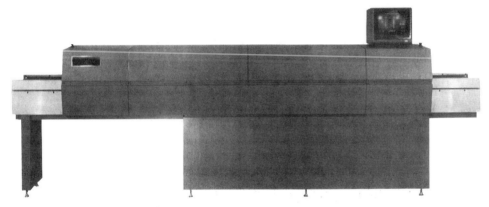

Figure 6.17 The ETS 4813C uv and ir adhesive cure and solder reflow combination.

Figure 6.18   The ETS model 4814 ir reflow apparatus.

The wave solder process does have limitations. Ceramic chip carriers and the plastic J-leaded chip carrier cannot be used on adhesively attached and wave soldered assemblies. Also, the thermal shock to the components during wave soldering is the most severe of the three major reflow processes.

Wave soldering surface mounted devices requires special design considerations. The leadless chip components must be designed such that the length of the component enters the solder wave parallel to the wave. This is because there is a tendency to skip the trailing solder connection if the part enters perpendicular to the wave. This problem has been addressed on wave solder systems designed specifically for surface mount by using a dual-wave design. SOICs should enter the wave perpendicularly since the solder connections are on the sides, not the ends, of the packages. In dual-wave systems, the first wave is a turbulent wave which more evenly distributes the solder around the part. The second wave is a laminar-flow wave which is intended to remove solder defects, i.e., solder bridges caused by the turbulent wave. There must be a physical separation between the waves to allow for flux out-gassing.

The two minor types of reflow soldering to be discussed here are hot plate reflow soldering and forced, heated air soldering (Fig. 6.19). Hot plate soldering is an effective means of soldering one-sided assemblies. This method can be readily adapted to accommodate high volume production. Hot plate soldering works well when using ceramic substrates populated with surface mounted components. The hot air process has an advantage over the hot plate method in that the temperature of the hot air is geometrically independent of the shape and size of the components and can be used in the double-sided population of components. The disadvantage of this system is that it takes longer to affect solder reflow than does the hot plate method.

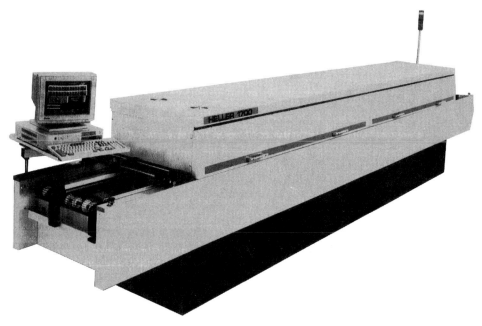

Figure 6.19 The Heller forced-air solder reflow oven.

6.5 Cleaning

Cleaning is treated as a separate section since both types of hybrid assembly may require this process. Primarily, however, only the use of fluxes necessitates cleaning. Cleaning may be done for other reasons in bare-chip assemblies, and those reasons are addressed as well.

6.5.1 To clean or not to clean

The first two questions that will always arise in any given hybrid assembly situation are (1) should I clean at all? and (2) when should I clean? Obviously this question will polarize any group of people deciding the process to manufacture any microcircuit. Most times that argument will be divided along departmental lines; i.e., the design and quality engineers will say to clean while the manufacturing types responsible for P&L will say not to clean. Why does each choose the stand it does? These engineers will feel that it is better to be safe than sorry. Those who watch the bottom line want to make sure that every move they make is value-added. Obviously, the company cannot be competitive if too many non-value-added steps are placed in the process.

Since the Montreal Protocol of 1987, engineers are being told to get rid of CFCs in the workplace. Now both factions are concerned with the issue of cleaning. The pressing issue is simply: When should I clean? The answers seem commonplace, ordinary, but necessary to iterate.

- Clean only when necessary. Do not clean simply for the sake of it, or just to be safe rather than sorry. Perform some tests to be sure cleaning is necessary. Perhaps no steps will have to be taken in this instance. If cleaning seems to be necessary, investigate steps in the process that will alleviate the need for the substance needing to be removed.

- One remarkable thing to note is that some contaminants may be lodged in a position where they will forever remain and will never damage the circuit under its service-life conditions. Poor cleaning technique may dislodge but not dissolve and carry them away. In this circumstance, the contaminant will merely be redeposited in a position where it will cause problems to the microcircuit during its service life.

- Usually it is the flux that needs to be cleaned. With corrosive fluxes, cleaning is a necessity, but with some fluxes such as R, RA, and RMA (Table 6.5), this may not be necessary. Their active ingredients will be innocuous to the circuit's components. Know your process and the materials in it.

- Cleaning also depends on the application and its contractual requirements. The military will almost certainly require cleaning. Determine this fact up front and charge accordingly, preferably with a sliding scale charge depending on the extra costs and restrictions you are going to incur by having to clean. This sliding charge alone may cause the customer to reconsider the contractual exigencies.

- If you do have to clean, do it as soon as possible after the operation that generated the contaminant. Waiting will only make the situation worse, particularly if there is a process operation involving heat after the contaminant-generating operation.

Once you have determined that cleaning will be necessary according to its impact on assembly of the microcircuit (Table 6.6), then learn what types of contamination have to be removed. The design of the equipment as well as the solvents used in the cleaning operation depend on this knowledge. Particulate matter, for example, can be removed with a pressured stream; the type of solvent does not matter unless it causes a secondary problem. Determine if the contaminant is polar (ionic) or if it is nonpolar (organic).

Polar contaminants will always have to be removed. They will degrade performance and promote corrosion, even destroying some of the microcircuit's

TABLE 6.5   Comparison of Flux Types

Flux type	Characteristics
Inorganic	Very active acid or salt, e.g., HCl, HF, $SnCl_2$, KF, $ZnCl_2$; not used in SMT; water soluble
Organic acid (OA)	Stronger than rosin, milder than inorganic acid; not normally used in solder paste (should be); water soluble
Rosin (R, RMA, RA)	Used in solder pastes; contains low halide content for activity; may not be cleaned
Synthetic (SRA, SA)	Mildly corrosive; not usually found in solder pastes

TABLE 6.6   Impact of Contaminants

Contaminant	Impact
Polar or ionic (chloride activators)	Dielectric breakdown Current leakage Component-circuit corrosion
Nonpolar or nonionic (rosin)	Cosmetic May attract ionic contaminants through occlusion Poor electrical contact during bed-of-nails testing Poor electrical contact with surface mount or edge connectors

elements. Nonpolar contaminants may not even have to be cleaned at all unless they caused an aesthetically undesirable product. In that case, let them stay where they are; do not clean at all.

If the residue is rosin (pine extract—naturally occurring), determine what subsequent processing steps will be necessary. Rosin begins to polymerize above 140°C and is completely polymerized at 273°C. If this range of temperature is to be experienced by any part of the circuit in subsequent assembly operations, cleaning is required. Polymerized rosin is difficult to impossible to clean.

In some cases the contaminant may have both a covalent and an ionic nature. Such is the case with all organic acids. If the flux is reacted or if any substance with a recombination nature is on the substrate, then cleaning with an azeotropic polar-nonpolar mixture at elevated temperature may be necessary. In all events, it is necessary to know what the contaminant is and what kind of solvent is necessary to clean the contaminant.

The most prevalent problem in hybrid assembly is the infamous "white residue." It may be any number of things, and untold dollars have been expended determining exactly what it is in any given process. Several answers have been given by different manufacturers, and the makeup of "white residue" may vary depending on the process. Although polymerized rosin is typically black, rosin that has only started to polymerize is white. It may be the activators in the flux that are left on the circuit after evaporation of its solvent. Lead chloride and lead bromide by-products of the solder-flux reaction have also been identified as "white residue." The obvious solution is to clean immediately after the soldering operation is over (the next process step). If that does not work, then seek out a solvent that will clean just after the solder operation terminates. In any event, always clean as soon as possible.

6.5.2 Choosing a solvent

Table 6.7 relates the different types of cleaning solvents with the type of contaminant they will clean or not clean. Further, the table elucidates the positive feature of this class of solvent and enumerates the problems that will be encountered with the solvent. No solvent is immune to this scrutiny. Water may dissolve many substances that will make the solution dangerous, so beware of the problems associated with the use of any solvent. With the use

TABLE 6.7 Cleaning Process vs. Type of Contamination

Type of cleaning	This will clean	This will not clean	Positive features	Problems
Nonaqueous vapor degreasing with CFCs and chlorocarbons	Loose particles and nonpolar organic compounds	Polar organic compounds, embedded particles, inorganic compounds, fingerprints	Nonflammable, cleans in cracks, fresh solvent available, cleaning in vapors and ultrasonics not always necessary	Toxicity, disposal, EPA regulations, decomposition, reactivity, disposal, impurities in rinse water
Aqueous cleaning with water rinsing	Loose particles, both polar and nonpolar organics, inorganic salts, fingerprints	Embedded particles, insoluble inorganic compounds	Best solvent, nontoxic, nonflammable	Dirty solvent, ultrasonics necessary, rust possible
Plasma cleaning	Organics, very thin films	Nonorganics	Final cleaning, surface modification	Thin layers only

of this table it is possible to determine a class of solvent to be used. The next step is to choose the proper solvent that can be used safely and economically.

Remember: *Like dissolves like.* Choose a polar solvent to dissolve and remove ionic contaminants. Water is the best choice of polar solvent. Following that, members of the alcohol family are preferred polar solvents. Nonaqueous solvents dissolve nonpolar contaminants. A plethora of nonpolar solvents are available. The exact choice of solvent will depend of the entire nature of the organic contaminant.

When choosing the solvent several requirements must be met before settling on the final choice of solvent. The chosen material must not alter markings on substrates, components, or plastic packages. The solvent chosen cannot dissolve or disfigure the substrate, components, or plastic packages of the microcircuit. The chosen solvent must not release smog precursors or harmful chlorofluorocarbons into the atmosphere. The solvent will face disposal after doing its job. Therefore, the solvent must not release any kind of harmful contaminant into the water supply. The solvent cannot be a health hazard before, during, or after its use. Finally, use of the solvent (latent heat capacity) must be economical in that the solvent must be energy efficient when used.

6.5.3 Chlorinated solvents

The most commonly used chlorinated solvent is 1,1,1-trichloroethane (TCE). This solvent by itself is effective but decomposes easily in the presence of metals and with heat to hydrochloric acid and other substances which readily attack vapor degreasers. The formerly used trichloroethylene (rather than 1,1,1-trichloroethane) is a photochemical smog precursor. Trichloroethylene and methylene chloride (another popular solvent) are both toxic, suspected carcinogens.

The chlorinated (only) hydrocarbons are not as dangerous to the ozone layer as are the CFCs but are less able to penetrate gaps. Additionally, these solvents are prone to attack both plastics and elastomers. Both these substances are found in microcircuit assembly. A further drawback of using these chlorinated solvents is that they decompose when subjected to prolonged heating. Their decomposition products are both toxic and corrosive; the toxic by-products are detrimental to humans, and the corrosive by-products are detrimental to equipment and products. Finally, ozone depletion is linked to the chlorine content of these solvents. Therefore, they are not acceptable cleaning solvents.

6.5.4 Chlorofluorocarbons (CFCs)

The CFCs have many advantages that have made them popular as solvents to clean microcircuit assemblies. They have a low latent heat of vaporization, which minimizes the energy it takes to operate degreasers with these solvents. CFCs have a low boiling point, which minimizes the thermal damage that can be done to assemblies and their components. Further, CFCs do not dissolve elastomers or plastics. However, there are disadvantages to the use of CFCs.

If the assembly contains any Al, Zn, or Mg, CFCs may cause some damage. However, CFCs are not considered corrosive. The decomposition products of CFCs are both toxic and corrosive, but they are not readily prone to decompose. The depletion of the ozone layer has been linked definitively to the chemical content of the solvent. These solvents figure most strongly in ozone depletion since their use is so widespread.

CFCs are used in refrigeration and air conditioning as well as in microcircuit assembly. In fact, cleaning accounts for only 20% of the problem caused by CFCs. The solvent primarily used in microcircuit cleaning is CFC113. This solvent is not flammable, is extraordinarily stable, and is safe for use by humans. Unfortunately, it is this solvent's extraordinary stability that causes the problem with our ozone layer. Once the solvent has penetrated the atmosphere it can last for up to 100 years before decomposition into innocuous substances.

CFC113 has a very low surface temperature which allows it to readily penetrate under surface-mounted devices. CFC113 also has a high density which allows contaminants to float, making it very easy to get rid of them. In order to stabilize CFC113, it is normally mixed with other substances. As Table 6.8 shows, several different mixtures comprise CFC113. Depending on the degree of polarization needed, members of the alcohol family may be added to trichlorotrifluoroethane. While there are many CFCs, trichlorotrifluoroethane is the one most commonly used as a degreasing solvent.

Table 6.9 shows the chemical composition of halogenated hydrocarbon solvents. Many times this composition is dictated by azeotrope formation. An azeotrope is a solvent blend in which the composition of the vapor is exactly the same as that of the liquid which is constant boiling. One advantage is that both polar and nonpolar solvents are found in the mixture. Another reason for using mixtures as solvents is to prevent interaction between the solvent and reactive metals. The minor constituents found in the hydrocarbon solvents are termed *inhibitors*. Table 6.10 shows the physical properties of many solvents.

TABLE 6.8   CFC113 Cleaning Mixtures

Chemical name	Chemical formula	Density, g/cm^3	Boiling point, °C	TLV, ppm
100% Trichlorotrifluoroethane*	$Cl_2FC\text{-}CF_2Cl$	1.57	47.6	1000
Acetone (11%)	Mixture	1.41	43.6	900
Ethyl alcohol (4%)	Mixture	1.51	44.6	750
Ethyl alcohol (35%)	Mixture	1.17	48.3	750
Isopropyl alcohol (35%)	Mixture	1.15	48.9	500
Methyl alcohol (5%)	Mixture	1.48	39.7	475
Methylene chloride (50%)	Mixture	1.42	36.2	140

*While there are many chlorofluorocarbons, this is the one most commonly used as a degreasing solvent.
Source: *Texas Instruments Tech Journal,* September–October 1990.

TABLE 6.9 Chemical Composition of Halogenated Hydrocarbon Solvents

Solvent	Manufacturer	TTE	TCE	Methylene chloride	Methanol	Ethanol	Isopropanol	Nitro-methane	Isohexene	Acetone
TMS	DuPont	94.05						0.25		
TES	DuPont	95.20			5.7	3.80		1.00		
TMC	DuPont	50.00		50						
DMSA	Allied Signal	94.00			4.0	0.50	0.50	1.00	1.9	
DXF	Allied Signal	91.10			5.9			0.20		0.9
Prelete	Dow Chemical		94.08			6.08				

TABLE 6.10 Physical Properties of Halogenated Hydrocarbon Solvents

Property	TTE	TCE	TMS	TES	TMC	DMSA	DFX	Prelete	Methylene chloride	Trichloro-ethylene	Methanol
Boiling point, °C	47.6	74.1	39.7	44.4	36.2	40.3	39.9	73.3	39.8	86.9	64–66
Density, g/ml	1.57	1.32	1.48	1.50	1.42	1.46	1.43	1.26	1.32	1.46	0.79
Surface tension, dyn/cm at 20°C	17.3	25.9	17.4	17.2	21.4	18.4	17.8	25.2	28.0	31.6	22.6
Latent heat of vaporization, Btu/lb	63.1	102	90.70	76.70	104	89.50	92.60	103.7	142	103	473
Specific heat of liquid, Btu/lb/°F	0.21	0.25	0.24	0.23	0.26	0.24	0.26	0.30	0.28	0.23	0.60
Toxicity TLV, ppm	1000	350	475	770	140	510	480	350	140	50	475
Kauri-butanol value	31	124	45	37	86	49	46	124	115	130	—

6.5.5   Process–materials interaction

Many surface mounted devices may have as little as 0.002 in clearance from the bottom of the device to the substrate. Cleaning under such conditions during assembly is at best difficult. High pressure on the order of 200 psi has been effective, but not always. Additionally, the use of such high pressures has resulted in damage to other of the circuit's components. If cleaning under devices with no more clearance than described is necessary, the use of surfactants is recommended rather than high pressure. Surfactants reduce the surface tension of the cleaning solvent which directly relates to the solvent's ability to clean under SMDs. In fact, the lower the surface tension of the solvent, the better its cleaning action.

Boiling the solvent makes it more aggressive than using that same solvent cold. The reasons are obvious. Cleaning in the vapors allows cleaning with the pure solvent as opposed to cleaning with a solvent that is contaminated with foreign material. Boiling solvent is more aggressive in its attack on contaminants. The agitation experienced with the boiling solvent aids the cleaning process. Ultrasonic agitation also reduces cleaning time by 5X. The use of ultrasonics, however, may be harmful to other of the circuit's elements.

6.5.6   Measurement of cleanliness

The original question concerned whether or not to clean. After making the decision to clean, it is important that the measurement of cleanliness relate to the service requirements of the assembled microcircuit. There are three major methods of cleanliness measurement. The first is a visual inspection which comprehends gross residues on the assembly. The test itself is very subjective, relying heavily on the skill and knowledge of the inspector. Visual inspections are not, by themselves, sufficient. A second, more quantitative measure, is solvent-extract resistivity. This test is performed by measuring the surface resistance of a board and then soaking that same board in a solvent mixture of 75% isopropyl alcohol and 25% water. The resistivity is then measured using an Omegameter. The last measure is insulation resistance. This test is performed by applying voltage to interdigitated fingers, measuring the leakage current, and then calculating the insulation resistance across the contaminated surface.

6.5.7   The Montreal Protocol

Five to fifteen parts per million by volume ozone filters ultraviolet radiation from the sunlight. Several noxious radicals destroy the ozone, thereby allowing dangerous ultraviolet radiation to impinge upon human beings. Many of these radicals are either derived from or precursed by CFCs and other chlori-

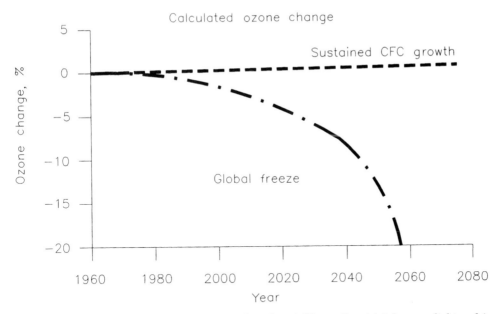

Figure 6.20   The controversy. Ozone (5–15 ppm by volume) filters ultraviolet from sunlight and is destroyed by ClO and NO$_x$ radicals. In the Greenhouse effect CFCs (responbile for 20% of this effect) tend to prevent radiation of heat from the Earth to space. The result is a gradual warming trend.

nated hydrocarbons. In addition to ozone depletion, CFCs tend to shroud the earth, preventing radiation of heat from the earth's surface into space. The result is a gradual warming trend. This problem is known as the *Greenhouse effect*. Figure 6.20 illustrates the problem with our ozone layer that will occur with sustained CFC usage at the current growth rate. If a global freeze is declared on the distribution of this solvent, the effect may also be seen in Fig. 6.20. Since neither instance was desirable to business, the Montreal Protocol was placed into effect in September 1987 to encourage a global freeze.

The Montreal Protocol basically states that CFCs 11, 12, 113, 114, and 115 are regulated, as are the halons. The consumption of CFCs will be capped at the 1986 worldwide levels, reduced by 20% as of mid-1993, and reduced further an additional 30% by mid-1998. Each CFC is not controlled individually but rather as a weighted composite or "chlorine basket." There are provisions in the protocol prohibiting trade in the CFCs and related products in countries who are not parties to the agreement. This is done to provide incentives for all countries to join the protocol. Special provisions are made to allow developing countries to increase their consumption of CFCs while continuing to raise their standard of living if an adjustment is made in the consumption of already developed countries. The effect of the Montreal Protocol can be seen in Fig. 6.21.

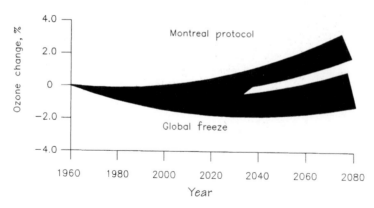

Figure 6.21 Effect of the Montreal Protocol.

6.5.8 Other cleaning possibilities

By simply substituting a hydrogen in the CFC molecule, a safer solvent is derived. However, switching to these solvents will only delay the inevitable. Eventually, all chlorinated cleaning solvents will be banned. Since these solvents are safer, some attention will be paid to them here. The shorthand notation for these solvents is HCFCs.

HCFCs contain hydrogen, carbon, fluorine, and chlorine. They have much shorter lifetimes in the atmosphere than CFCs. For example, HCFC141b will live in the atmosphere for 10 years before decomposition as opposed to CFC113, which exists in the atmosphere for 100 years. The HCFCs have a very low ozone depletion potential, eight times less than CFC113. Further, HCFC141b shows a very low Greenhouse effect, fifteen times less than that of CFC113. Therefore, switching to HCFC as an assembly solvent may buy several years before its usage is banned.

Ultimately, the use of aqueous solvents will become a necessity within the coming few years. However, it is very advantageous to not clean at all unless it is absolutely necessary. It behooves the process engineer to use materials in hybrid assemblies that will not be removed, nor will they do damage to assemblies. One last alternative that might be suggested to substitute for chlorinated solvents is to use a cover gas during the soldering operation. If oxidation is prevented up front, there will be no need to remove it at a later point in the process.

6.5.9 Aqueous cleaning

Aqueous cleaning is becoming extremely popular with the through-hole technology. The high viscosity and surface tension of aqueous solvents make them unattractive for cleaning the tight spaces in most microelectronic assemblies. However, much work is being done to rectify this situation. Therefore, several characteristics and requirements of aqueous cleaning are offered.

Always use deionized water in this type of cleaning. Other additions to water when used as a solvent are neutralizers, saponifiers, surfactants, dispersants, and antifoaming agents. Neutralizers make salts from acidic fluxes, converting them to soluble, corrosive materials. Saponifiers convert rosin fluxes to water-soluble soaps. Saponifiers work in the highly alkaline pH range of 10.5 to 11.8. Surfactants reduce the surface tension of water, which allows this water to penetrate the gaps under surface mounted devices. Dispersants are highly charged materials that attract particulates. Antifoaming agents are added to high pressure sprays.

6.5.10   Cleaning equipment

In order for the equipment (degreasers) used to clean to be effective, the microcircuit assembly should be vertical or slanted during this process operation to permit the solvent to run off the assembly. The substrate should be cool when entering the equipment to allow solvent vapors to condense thereon, dissolving contaminants. After this dissolution of contaminants, the solvent must drain back into the boiling sump.

The vapor zone above the boiling solvent will consist of distilled components of the solvent mixture in some fixed ratio. If the mixture is an azeotrope, that ratio will be the same as the ratio of the boiling liquid. The solvent's components and the ratio of those components must be known to determine the use of the solvent vapors in cleaning and the action of those vapors on the equipment itself.

The best design of any cleaning process and the equipment to execute that process will include several cycles of vapor, spray, and immersion (boiling and rinsing). This may be accomplished by using either in-line or serpentine cleaning equipment.

Two general designs for equipment meet these criteria. The first, known as batch processing, is illustrated in Fig. 6.22. The second type of cleaning equipment is known as in-line and is illustrated in Fig. 6.23. Note in the illustration that the vapor is allowed to condense on the cool assembly, thereby dissolving gross contamination, which is washed away. In using the angled spray, solvent is forced under pressure to impinge on the assembly's surface to facilitate cleaning under surface mounted devices. The assemblies are immersed in the boiling liquid to facilitate further cleaning. Finally, egress cleaning and exit cleaning are done in the presence of vapor.

Inclined and high pressure sprays are often used to facilitate cleaning. An example of the different types of spray pattern that can result is shown in Fig. 6.24. High pressures of 200 psi have been used, but at some detriment to the assembly. The spray pressure usually used is between 20 and 30 psi. During egress and exit the assembly is nominally inclined between 10 and 20° with relation to the vector of the spray pressure.

2 Sump Vapor Degreaser

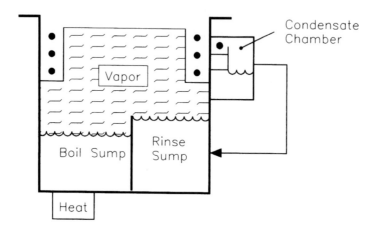

Process Flow:

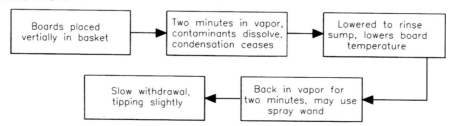

Figure 6.22 Batch vapor degreasing.

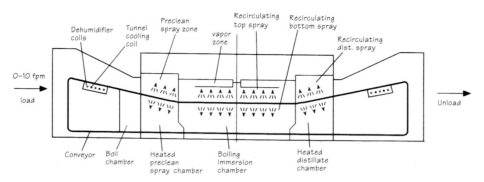

Figure 6.23 In-line vapor degreasing. Vapor condenses on the cool board, dissolves gross contamination that is washed away (egress cleaning). Solvent is forced under pressure at the board surface to facilitate cleaning under SMDs. Liquid is sprayed on component to hold them on conveyor. Cleaning occurs in solvent.

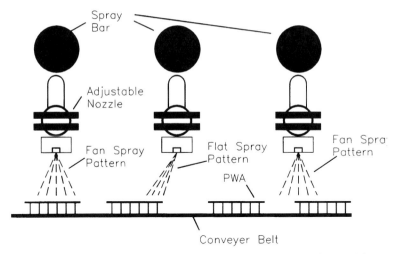

Figure 6.24 Inclined and high pressure spray. High pressures of 200 psi have been used, but at some detriment to the board. Usually, sprays are 20–30 psi. Notice that the board is often inclined 10–20° with relation to the vector of the spray pressure.

6.6 Repair and Rework

Rework and repair are really quite different. Rework is the procedure by which both active and passive devices are corrected on the hybrid assembly, returning that assembly to the requirements of the released production print. Repair returns the errant microcircuit to functionality but does so by changing the physical appearance of the device so that it no longer conforms to the print. Both can be acceptable to the customer but must be contractually allowed. Both the reworked and the repaired hybrid assembly must function electrically.

Neither rework nor repair improves a hybrid assembly; they merely return it to functionality. The best method of *rework and repair* that can be exercised is to "do it right the first time." This can be done by exercising stringent statistical process control in hybrid assembly. The measures described here apply when the assembly is too expensive to discard and even the practice of SPC could not be exercised (incoming bad die) or was not properly exercised.

The major question to be asked by the hybrid assembly operation is: What effect will rework or repair have on the reliability of the device? The immediate benefits to the manufacturer are, of course, increased profits. The long-term situation, if the reworked device fails, is decreased market share and consequent decreased revenue. Repair should be done in such a manner as not to affect the reliability of the microcircuit.

6.6.1 Rework of epoxy bonded devices

This is a method of repair that can be used for both active and passive devices that are attached with either conductive or nonconductive epoxy. This technique, as are most rework and repair techniques, is done manually, preferably by skilled personnel. The following process relates to epoxy-bonded die; simplifying the procedure will make it applicable to passive components as well.

- Remove the wedge bonds from the substrate's conductor.
- Heat the substrate locally where the defective die resides.
- Remove the die by means of tweezers or knife.
- Clean the bonding area.
- Replace the die using repair adhesive.
- Wire bond the die to the substrate contact areas.
- Test and screen the assembly.

6.6.2 Rework and repair of fine-pitched devices (FPD)

Rework and repair require the focusing of energy to one location on the hybrid assembly without deleterious effect to either the substrate or the surrounding components. To accomplish this, specially designed tooling is usually required. Figure 6.25 shows some of these tools that can be used in the rework of FPDs. These tools are designed to deliver localized heat and at the same time not expose adjacent devices to greater than 150°C for more than short durations.

The criteria for selecting a heating method are the same as those discussed in Secs. 6.4 and 6.7. The method to choose during rework and repair, however, has to take into account lead form and fragility, coplanarity, lead pitch and lead count, and device separation as well as the other requirements already mentioned. The methods of heat application that are used in rework include hot gas, focused ir, and resistance-contact soldering with a bar (thermode). These methods are discussed in Sec. 6.7.

Removal and replacement of a TAB device mounted with die-attach adhesive requires not only the replacement of lead attachment but also the removal and replacement of adhesive. TAB devices are typically on 0.012 in lead pitch. As lead pitch decreases, lead fragility increases since leads are narrower and often thinner. How the new leads are attached involves coplanarity difficulties. The same problems and principles of reattachment exist as elucidated in Sec. 6.7. In addition, with TAB devices, the economic decision to rework must be made individually with each case.

6.6.3 Rework of flip chip devices

The use of ir has been described for the removal of flip chips and their replacement. The ease of performing this operation results from the fact that

Flip Chip

TAB

**Single Site Placement-Prototype
SMD Fine Pitch Production**

Figure 6.25 Special tooling for repair and rework.

ir can be readily focused for localized heating. After the area in question is radiated, the chip must be removed just before or immediately after lamp extinction. After removal of the device from the substrate, the area where the new chip is to be placed must be redressed. This may be done by a skilled operator using an iron or other device for solder reflow and wicking the solder from the pad or by solder sucking. Both techniques have long been practiced in the hybrid assembly industry. The reason for removal of the old solder is to prevent solder buildup in the area under the new flip chip that could cause a short and result in further rework.

The replacement chip is held on the site that has been redressed. Moderate pressure is applied to the back of the chip and localized heat once again applied. After reflow has begun, remove the pressure and allow the normal surface tension of the solder to realign the chip to the substrate.

6.6.4 Repair of eutectically bonded devices

Normally the rework of eutectically attached devices is not feasible, although some authors have described the process. Typically this is a repair operation. Even to perform this procedure, the pad has to be there. Usually this will be the case only if the device is known to require eutectic bonding and is suspect (not pretestable). The assembly print will have the device located on one pad. If the repair device is located on an adjacent, redundant pad, this deviates from the print and is classified as a repair.

The steps to follow in the use of redundant pad repair are:

- Do not alter the device that has failed; you cannot improve it or the circuit by anything you do.

- Sever the conductor that connects the failed device from the functionality of the hybrid assembly using a laser or other means; gold is absorptive of Nd:YAG radiation, so this feat is more easily accomplished by overprinting the conductor that may need to be severed by resistor material during the resistor printing sequence.

- Place a die on the redundant pad.

- Wire bond the new die.

- Test and screen.

The one method of replacing eutectically bonded devices that can be classified as rework is the replacement of molytabbed devices, which involves the removal of the defective, molytabbed device and replacement using the rework method needed for replacement of the molytab alone. The two reasons for using molytabbed devices are allowing the possibility of rework and not subjecting the entire hybrid assembly to the extreme temperatures required for eutectic attachment.

6.6.5 Delidding

Delidding is the process by which the lid is removed from the finished microcircuit. This process may not usually be performed without consent of the customer, and must be contractually allowed. This rework method is the province of package sealing, another chapter in this handbook. However, there are a variety of delidding techniques. Milling, engraving, grinding, an "X-acto" knife, and laser machining have been used. The most popular method has been to float the package inverted in a solder pot held around 350°C. When the lid starts to reflow (usually AuSn solder), the package is plucked from the solder pot. Each technique has major risk. The primary problem is the condition of the seal ring after the old lid has been removed. The secondary risk is to the components internal to the microcircuit (see the reliability section in this handbook). The microcircuit should be tested for damage assessment after delidding to determine if any of the methods mentioned here will be workable and if the rework or repair will then be acceptable.

6.6.6 Rework and repair of surface mounted devices

Whereas the repair and rework of bare-chip assemblies is still largely an art form, the technology for repairing surface mounted devices is becoming a science. The reasons are that repair and rework are more readily accomplished, the spacings are more easily addressed, and the strategies have been more fully developed from the repair and rework of through-hole assemblies. Nevertheless, special considerations have to be accorded this hybrid assembly technique.

The first consideration is the solder joint and the effects of repair and rework on the reliability of the overall assembly. Since surface mounted assemblies do have more real estate with which to work, sheathed wires may often be used in the repair of a surface mounted assembly, which is not possible in bare-chip assembly.

Special considerations are accorded the SMT during the design for rework and repair (DFR) such as:

- In DFR it is imperative that the proper clearance be allowed around all packaged devices to permit localized heating for removal and replacement of SMDs.

- In the selection of adhesives, choose an adhesive that will allow the replacement of SMDs, returning the hybrid assembly to its original state.

- Design circuit ground planes in the hybrid assembly to minimize thermal heat-sinking effects.

- The organic substrate is much more susceptible to reheating than is the inorganic substrate; beware the temperatures used in the rework and repair process; design according to the potential for necessary rework and repair.

- Use a maskant on all conductor traces to which a bond is not directly made. Open areas may be left in the maskant to accommodate repair and rework.

After DFR has been addressed, the repair processes closely parallel those already mentioned. The four steps in the repair and rework process are the heating of the localized area where rework will occur, removal of the old device, replacement of the new one, and cleaning of the assembly. Much of this technique may be accomplished semiautomatically with stations such as the one pictured in Fig. 6.26.

The techniques for desoldering and resoldering are the same as discussed in other sections of this chapter. The tooling is normally application-specific and available for use on rework stations that are commercially available.

Hot Gas

- Adjustable Cartridge Heaters
- Simple, Easy-to-Use
  1) Select Cycle 2) Select Operation
  3) Start
- Vertical Viewing
- Computer Controlled
  —9-cycle Storage
  —Automatic Process Control
  —Automatic Sequence Control
  —Learn Mode
- Air Bearing X-Y-θ Table
- Board Support
- Bottom Heater
- Video Monitor

Options:
- Fine Pitch Package
  —Split Prism
  —Fine Adjust Table
  —Force Control
- Computer Interface
- Anomaly Sensing§
- Software
- Keyboard
- Hard Drive
- Floppy Drive

Figure 6.26 Sierra VPG production rework system.

6.7 Component Attachment

The problems of attaching components require that the engineer have familiarity with the different types of component attachment as well as the materials used in that attachment. Most components can be attached with adhesives, either conductive or nonconductive. Whereas the surface mount technology is strongly allied to the solder bond, bare-chip hybrid assemblies may use solder attachment, eutectic die bonding, and/or adhesive attachment.

6.7.1 Organics in microelectronics

The use of organics in the field of hybrid assembly is rapidly expanding since these classes of materials are being developed with characteristics paralleling those of inorganics without the negatives once ascribed to organics. Some of the major recent developments in the organics used in hybrid assemblies are the emergence of tailored surface mount adhesives, the purification of die attach adhesives and other materials coming in close contact with IC dies, and the use of high molecular weight organics in sealing and conformal coatings. In summary, the microelectronics industry has gained faith in the reliability of organics.

In addition to mechanically fastening, adhesives in microelectronic applications are excelling in heat conduction as well as electrical conductivity or isolation. Organics provide shock mounting, seals, substrate protection, and conformal coating for hybrid assemblies. The longevity of such organics required for various applications covers the range from a few seconds to many years to the life of the device.

The organic choice (specifically the adhesive choice) is governed by many considerations including strength, thermal conductivity, application method, cure temperature, and environmental exposure. Ceramics and glasses of many compositions form high temperature hermetic seals with some adhesives. These organics are available as one-part and two-part liquids, powders, dry films, solvent solutions, thermoplastic films, and preformed pellets.

Among the available choices of materials are the *thermosetting resins* and *thermoplastic polymers*. Thermosetting resins, when heated, change irreversibly from fusible and soluble materials into ones which are infusible and insoluble through the formation of a covalently cross-linked, thermally stable network. Thermoplastic polymers soften and flow when heat and pressure are applied; i.e., their polymerization is reversible. Although this is not a desirable attribute of a material for this application, the temperature required for reversal of most thermoplastics' polymerization is high enough to allow their limited use in hybrid assembly.

6.7.2 Component and die bonding adhesives

Adhesives have become popular for such things as chip bonding, passive component bonding, lid bonding, and channel-device attachment. It should always be kept in mind that some military specifications prohibit the use of adhesives because of their organic content.

Adhesive bonding can generally be substituted for solder bonding provided that the adhesive selected will withstand the circuit environments. Naturally if the adhesive will be required to perform a circuit-interconnection function in addition to the bonding function, it must be conductive. The main advantage of adhesive bonding over solder bonding is that it is a room-temperature process. Even though most adhesive formulations must be cured at 100 to 125°C, this is still considerably lower than the 385 to 420°C required for SiAu eutectic bonding or even the 200 to 250°C required for SnPb soldering. Add to this the fact that adhesive operating temperatures range from −55 to 300°C, making adhesive bonding an attractive substitute for the formation of metallurgical bonds.

The choices in conductive adhesives for hybrid-circuit application are essentially two: silver adhesives and gold adhesives. Since this chapter is directly concerned only with assembly processes, it is necessary to exclude the large numbers of adhesives not developed for microelectronic applications and concern ourselves with adhesives used for interconnection and with the properties of adhesives that are harmful to hybrid substrates and components.

There are numerous adhesives for hybrid assembly applications. It would be impossible to cover them all in detail. Fundamentally, there are two-part systems, where the adhesive is mixed with a catalyst, and one-part systems, in which the mixing has already been done by the vendor, and which can be applied directly from the container. Usually, the two-part systems have a longer shelf life prior to mixing and can be mixed to the desired consistency by adding a thinner at the same time that the catalyst is being introduced.

Most adhesives require curing at elevated temperatures (although these temperatures are usually under 125°C), while some will cure at room temperature. Note that room temperature curing formulations have a significantly reduced shelf life. In almost all cases, the shelf life of organics can be extended by storing them in a refrigerator at approximately −40°C. Care should be exercised to prevent freezing.

There are numerous adhesive types, some of which have been formulated for specific purposes rather than general bonding. For example, an adhesive that would be best for active-device bonding might be rather soft, while an adhesive best for passive-device bonding might be thixotropic. The following guidelines can be used in adhesive selection for hybrid assembly:

- Epoxies are the most widely used adhesives because of their versatility, excellent adhesion, compatibility, ease of application, good electrical properties, and resistance to weathering.

- Silicones are used where flexibility, wide temperature range, high frequency, high humidity, and/or atmospheric contamination are encountered.

- Acrylics are used because of their excellent electrical properties, stability, good aging characteristics, optical clarity, and rapid curing.

- Urethane adhesives have flexibility, toughness, and strength from cryogenic temperatures to 125°C.

6.7.3 Adhesive application techniques

The universal tool for applying adhesive (for the purpose of hybrid assembly) is still the needle. The disposable hypodermic syringe is the most popular type of needle to use in dispensing adhesives. Adhesive application is another technique where automation is being instituted in hybrid assembly. In the past, manual dispensing has been used. However, the vagaries of using this method proved to be too wide for accurately reproducing hybrid assemblies. The application of a metered amount of adhesive can be done using a simple desktop apparatus and a foot pedal. Typically dispensing is done on a more automated scale, as shown by the Asymtek Automove 403 (Fig. 6.27).

The features of such a dispenser can easily be tailored to the application. The dispensing heads can be varied for repeatable metering of viscous materials, a trait mandatory to the application of organic adhesives. Like other types of assembly equipment available today, dispensers are typically operated by menu-driven software. The material is encased for longer shelf life. The application is precisely controlled for consistency of product, which ensures a more reliable product once the process has been fully characterized.

Another method of adhesive application is screen printing in the same manner that solder paste is printed. The two problems are (1) being able to screen-print the correct amount of adhesive for a small semiconductor die and (2) having sufficient time to attach the piece parts after printing and before the adhesive starts to tack and cure.

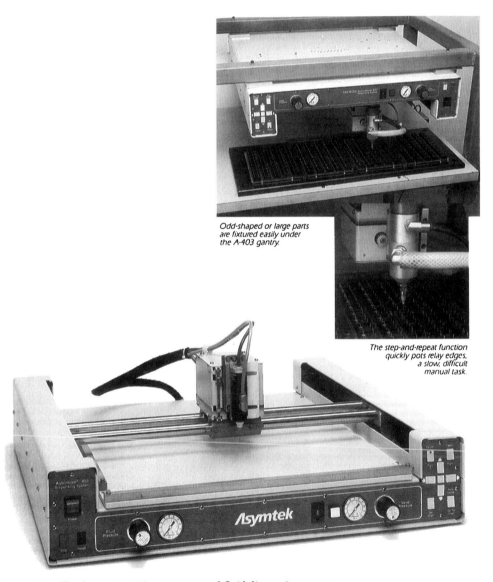

Odd-shaped or large parts are fixtured easily under the A-403 gantry.

The step-and-repeat function quickly pots relay edges, a slow, difficult manual task.

Figure 6.27 The Automove 403, an automated fluid dispensing system.

The room temperature assembly of semiconductor die with adhesive is becoming increasingly popular (except where prohibited by military specification) because of the ease of die attach. The following is the process procedure for bonding components with adhesive (these are the steps executed by the automatic dispenser):

- Clean the substrate to assure that all oils and foreign matter are removed.
- Mix the adhesive in accordance with the manufacturer's specification.
- Apply the required amount of adhesive to the substrate at the place where the die is to be attached (about 0.005 in thick and one-half the die area).
- Place the semiconductor die in position over the adhesive.
- Check for die orientation.
- Seat the die to the substrate surface with slight pressure.
- Reposition the die as necessary to comply with the assembly drawing.
- Place the other die or adhesive-assembled devices on the substrate.
- Cure the adhesive as directed in the manufacturer's specification.

At the end of the cure cycle, the assembled substrate should be ready for wire bonding. While both thermocompression and ultrasonic wire bonding are possible to adhesive-bonded die, the high temperatures associated with thermocompression wire bonding may damage the bond strength of certain adhesives.

Passive components are commonly bonded with adhesive, again because it is a simple process and can be accomplished at room temperature. Since most passive components are heavier than semiconductor dice, enough adhesive must be used for the bonding to prevent loss of adhesion during environmental stresses, i.e., shock, vibration, and centrifugal force. At least three sides of the capacitor ends must be attached. The adhesive must cover the bottom surface of the passive component, mating with the substrate or the substrate's metallization.

6.7.4 Eutectic and soft solder bonding

The most common method of bonding a silicon chip device (transistor, diode, integrated circuit, capacitor, or resistor) to a hybrid thick film conductor is by means of a metallurgical bond, either a eutectic or a solder bond. In bare-chip assembly, the hybrid conductor used is usually gold. The silicon chip forms the second material in the eutectic formation. In a AuSi eutectic bond, no additional material is required to form the metallurgical bond. The AuSi eutectic forms at approximately 370°C. The die bond can be made using a gold-backed silicon die, a bare silicon die, or either type of die with a AuSi solder preform.

The advantage of using a gold-backed die is that the gold backing, if deposited soon enough after wafer fabrication, will prevent the silicon from oxidizing. If the silicon has had a chance to oxidize, the AuSi eutectic bond is difficult to form. Subsonic mechanical motion is required during the chip-bonding process to penetrate any silicon oxide that may have formed on the back of the chip. Once the oxide is penetrated, the eutectic formation proceeds. A proper eutectic formation should occur in greater than 90 percent of the chip's base area to make a reliable mechanical and electrical connection.

If the entire substrate is held at eutectic temperature while more than one or two devices are being attached, the first die will continue a eutectic flow while the succeeding dice are being attached. This continued eutectic flow will destroy the electrical characteristics of the semiconductors. Even if a more sophisticated die-attach machine is used that has a pickup collet and an associated vacuum system, the problem of keeping the heated stage at eutectic temperature still exists. The answer, then, is some form of selective heating. The many forms that selective heating can take include resistance heating, hot gas jets, hot collet, and infrared. With each system the substrate must still be heated, but it can be kept well below the eutectic temperature, leaving the additional temperature increment to the selective heater. Typically, a cooling-gas nozzle is also incorporated to chill the eutectic when the formation is complete.

When the AuSi eutectic temperature requirements cannot be tolerated, a lower-melting solder preform may be used. There is a plethora of soft-solder preform materials from which to choose. The primary method of applying the appropriate temperature is resistance heating under a cover gas. This is accomplished by forcing high current at low voltage through the bonding pad as illustrated by the Crystal Mark DP-4 (Fig. 6.28).

The hot gas method can give excellent results and is very practical for selective heating in die bonding (Fig. 6.29). It is necessary to have a well-constructed mechanical arrangement so that the gas jet is as close to the die as possible and to have a method for controlling the temperature of the gas jet and triggering it on and off. The simplest way to accomplish this is to heat the gas tubes via I^2R heating from a low-voltage, high-current power supply and trigger the gas jet by means of a solenoid valve. Either nitrogen or helium gas can be used, but helium gives superior results owing to its better thermal transfer characteristics.

The hot collet system of selective heating to the eutectic temperature of AuSi applies incremental heat localized directly at the die to be bonded (Fig. 6.30). The collet that picks up the die to be bonded is heated to approximately 390°C. As the die in the heated collet comes in contact with the preheated bonding pad, the die–pad interface temperature rises to the eutectic temperature necessary to cause the AuSi eutectic flow. Even with the hot collet, a cover gas is used to protect the bonding surfaces from oxidation during the bonding process.

Many methods of die attach can be used in assembling hybrid microcircuits. The last example to be elucidated in this chapter details a novel silver-glass dispensing system built by Assembly Technologies, the Argus-SG (Fig. 6.31). This system offers fully automatic operation and changeover,

Figure 6.28 The Crystal Mark die placement–bond system.

vision-controlled die placement, and bond-line thickness accuracies of ± 0.001 in. The advantage of this type of system is its ability to handle very large die (up to 1 in) while still controlling bond-line thicknesses via laser depth determination.

There is a wide choice of die bonding techniques. One technique may work well for one application whereas another may not work at all. Most die bonders are capable of using a variety of bonding techniques and may be adaptable to the hybrid assembler's need through simple configuration changes. There are currently more than 25 suppliers of die bonders worldwide. Each configuration should be studied independently, and the overall process flow coupled with rework exigencies must be taken into consideration before the choice is made.

6.8 Intermetallic Formation

The best way to avoid intermetallic formation problems is to avoid the intermetallic itself; i.e., use only gold-to-gold and aluminum-to-aluminum interconnections. Unfortunately, this cannot always be done. The assembler of hybrid microcircuits must therefore become knowledgeable of Au–Al intermetallics, how they form, why the intermetallics react the way they do, and most importantly, the damage these intermetallics can do.

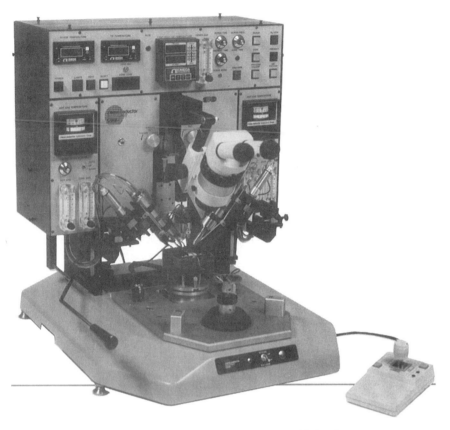

Figure 6.29 The SEC Model 4020 versatile precision eutectic die bonder.

6.8.1 Background

The formation of all gold–aluminum intermetallics has been well document-ed. One of the five intermetallics that is formed, $AuAl_2$, has been tagged "pur-ple plague." It is the one that gets the blame for all Au–Al intermetallic prob-lems simply because it is purple and is highly visible. The name is ominous enough that all gold–aluminum intermetallics are feared, regardless of their nature. These Au–Al intermetallics, including $AuAl_2$, are inevitably formed upon intimate contact of gold and aluminum and are necessary for adequate bond formation. Figure 6.32 graphically depicts the intermetallic growth sequence of the compounds formed during gold wire bonding to Al metalliza-tion on the chip. Note that the intermetallic will interdiffuse into the ball bond to become the mechanism of adhesion between gold and aluminum.

Before the era of microelectronics and Au-to-Al wire bonding, $AuAl_2$ was characterized by Coffinberry and Hultgren in 1938. Since that time, however, many companies have run afoul of the purple plague by using gold-rich inter-connects to Al metallization in addition to "unhealthy" combinations of tem-perature and time, i.e., gold bonds on the aluminum metallization of the chip

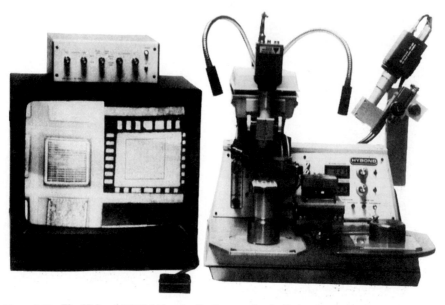

Figure 6.30 The Hybond UDB-140 eutectic–thermoplastic die bonder.

exacerbated by temperature and time. To avoid the problems that result therefrom, the gold–aluminum interface should be made to the substrate by using aluminum bonds on gold metallization. The reasoning will be made clear below.

Since the semiconductor industry has for some time been using aluminum ultrasonic wire bonding to semiconductor devices, there are many millions of hours of reliability data to indicate that the monometallic interface of Al bonding wire contacted to the Al metallization of the chip is not a problem. Au wire bonded to Au metallization has likewise not been a problem. Monometallic joints will not then concern this discussion. The requirements here are investigation of the precautions to be taken when gold and aluminum are coupled in a microelectronic assembly.

6.8.2 Philofsky generalizations

Philofsky reported the manner of *Kirkendall void formation*. Since the aluminum diffuses into the gold-rich side (the ball bond) faster than the gold can diffuse toward the aluminum, microscopic voids or vacancies temporarily exist. The Au and Al react and the layers of Al–Au intermetallics formed diffuse 1.5 × faster into the gold than into the aluminum. The vacancies existing because of the differential diffusion rates of gold and aluminum become the sites for void growth. Then, when the voids are made larger with the interdiffusions of Au–Al intermetallics, they become visible and are known as *"Kirkendall voids."* As these voids grow in size they create electrical and mechanical inequities at the Au–Al interface.

Figure 6.31   Assembly Technologies Argus SG silver glass automated die attach system.

Kirkendall voiding problems are exacerbated by temperature cycling. During temperature cycling, microcracks are formed in the brittle gold–aluminum intermetallics. Continued temperature exposure promotes the growth of these voids depending on the processing temperatures to which the assembly was exposed that incited the voids initially. The use of higher assembly temperatures relates to very low bond failure temperatures under service conditions.

Philofsky has made several generalizations concerning interconnections between gold and aluminum that result from observations of actual phenomena. They are:

- Intermetallic growth continues with time and temperature and will even proceed under ambient conditions.

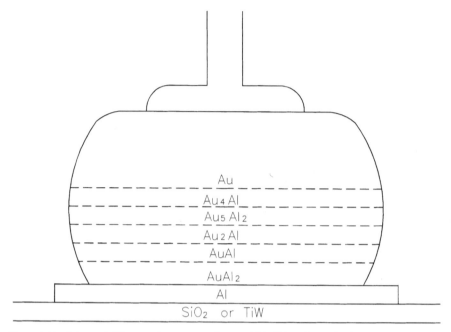

Figure 6.32 Intermetallic growth sequence for a gold ball bond attached to aluminum metallization.

- Using the curve in Fig. 6.33, it can be shown that catastrophic bond failure will occur in 20,000 h (2 years, 16 weeks) at 200°C.

- Thinner aluminum metallization increases the lifetime of the bond if an electrical open does not result due to high assembly temperatures after the Al–Au interface is formed.

- The Au–Al interface failure mechanism is accelerated by temperature cycling.

- Kirkendall voiding results because of the thermal mismatch between the Au_5Al_2 and Au_4Al phases.

- Kirkendall voiding begins at the interface between Au_5Al_2 and Au_4Al.

It is worthwhile to note that failures occur with other than $AuAl_2$ (purple plague) intermetallic.

6.8.3 Time–temperature effects

Aging a bond at 150°C, if no temperature stresses were induced in assembly after the ultrasonic bond was made, will produce only the tan color of four of the Au–Al intermetallics ($AuAl_2$ is purple in color). Continued aging at this temperature will eventually destroy the bond, but normally not within the service life of most device types and usually not before device failures cause failure of the circuit. Any postbond formation processing temperature in

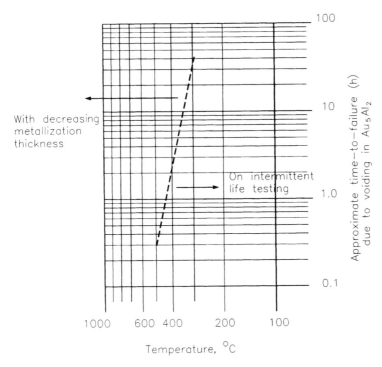

With decreasing metallization thickness

On intermittent life testing

Figure 6.33 Au–Al bond degradation from voiding stress in Au_5Al_2 phases.

excess of 150°C during the assembly of the microcircuit will begin the voiding process, which will cause failure in the bond before the devices fail; i.e., the Au–Al bond interfaces become the weak links if the Au–Al interface is exposed to elevated temperatures during postbonding process operations.

Bushmire reports that Sandia Laboratories performed testing with two types of thin film gold metallization and two types of aluminum wire. The two types of Al wire used were 1% Si and 1% Mg. In both metallization cases, CrAu and TiPdAu, the 1% Si wire proved to make superior bonds. *The conclusion of the work is that electrical resistance can increase to infinity without any degradation in bond strength.* The implication is clear: The problems of resistance do cause circuit failures and those failures will not be detected with a bond pull test.

Therefore, the problem of the differential diffusion rates of either Au-through-Al or Al-through-Au creates major problems for the manufacturer of microcircuit assemblies. Figure 6.34 shows the time of penetration of Au through Al metallization (typical to semiconductor devices). The abscissa of the graph in Fig. 6.34 is temperature while the ordinate is time. The predominant phases that are formed, Au_4Al and Au_5Al_2, are gold-rich. As earlier noted, the high stress that exists at the interface between these two compounds is the site for void formation. To intensify the problem, the large volume increase that serves to propagate these voids is provided by $AuAl_2$ (pur-

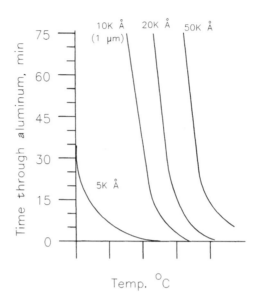

Figure 6.34 Intermetallic behavior in Au-Al system.

ple plague). *When purple plague joins with Kirkendall void formation, the stresses are large enough for bond rupture to occur.*

Several researchers have shown that, owing to the aforementioned phenomena, some disconnects do occur at 150°C after prolonged aging. Increased numbers of disconnects were observed at 200°C, and electrical discontinuities at the Au–Al interface were prevalent at 300°C. These failures occurred regardless of postbond temperature processing stresses given enough time. Interestingly, however, each metallization system produced its own characteristic failure curve, indicating that the problem can be hastened or slowed but cannot be avoided, only delayed. The entire system has to be evaluated before definitive answers can be given concerning the placement of gold wire bonds on aluminum metallization or aluminum wire bonds on the substrate's metallization.

The time–temperature species formation of Au–Al intermetallics is given in Fig. 6.35. Note that the higher the temperature or the longer the time at any given temperature, the more intermetallic compounds are formed. The more compounds that are formed, the greater the number of competing rate constants of diffusion. The more compounds that are moving around at different rates, the more voids and failures will be observed. This quickly becomes a *no-win situation.*

The use of any temperature above 300°C at any time will rapidly promote interdiffusion and reaction at the gold–aluminum interface. The result is the formation of up to five different compounds, $AuAl_2$, $AuAl$, Au_2Al, Au_5Al_2, and Au_4Al (Fig. 6.36 is a phase diagram showing that all five of these compounds do exist as entities). The kinetics of the interdiffusion process are not so important to this discussion as is the fact that these intermetallics are formed and are constantly interdiffusing and reacting with each other. The

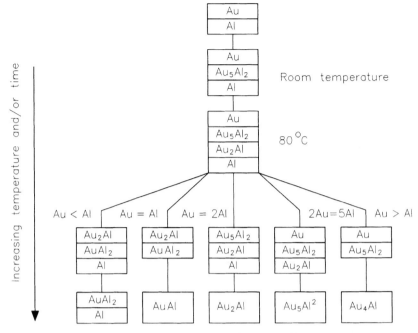

Figure 6.35 Time–temperature relation of Au–Al intermetallics.

relative proportions of Au and Al available for reaction effectively control the amount and type of intermetallics that are formed.

When Au is in excess, Au_4Al is the major phase and if Al dominates, $AuAl_2$ is the predominant phase. AuAl, as might be expected, dominates when Au and Al are present in equal amounts. There are as many permutations and combinations of formation and dissolution of the different compounds as there are elements and compounds in this scenario. All Au and Al compounds will occur with time and temperature. As earlier stated, their presence is not always detrimental to the bond or its electrical characteristics. The loss of bond integrity is directly attributable to Kirkendall void formation, and these voids will form readily. Therefore, the "safe" route for circuits that have assembly temperatures over 200°C is to have Al-to-Al bonds on the chip and Al-to-Au bonds, if they cannot be avoided, on the substrate. If the higher processing temperatures and higher environmental stressing temperatures can be avoided, Au bonds on Al metallization are relatively safe to use.

The tendency toward electrical disconnect (voiding) and "healing" which is occurring simultaneously varies at different rates for different compounds, causing the problem to be indistinctly observable. The only "rules of thumb" to be used here are to make certain that, if gold interconnects are made to Al metallization, the assembly should never exceed 150°C. Conversely, if Al interconnects are to be made to Au metallization, thoroughly investigate the reliability of assembled microcircuits built with the necessary processing temperature hierarchy. If failures occur, investigate the chosen materials, pro-

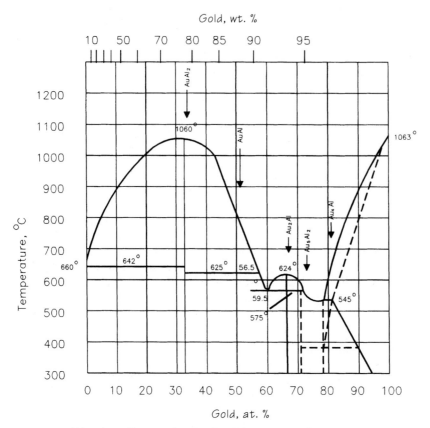

Figure 6.36 AlAu phase diagram showing five AlAu intermetallic compounds.

cessing conditions, or environmental testing requirements to change one or more to meet the circuit's service life requirements.

6.8.4 Au–Al interface degradation

One interesting phenomenon that has been observed is the increase in bond pull strength that occurs in bonds that are formed between gold and aluminum. The reason is that the AuAl compounds that are formed are from 3 to 10 times stronger than either gold or aluminum. Even though such compounds are brittle, they will not lower the bond pull strength and in most cases will increase the bond strength by approximately 10% over time and temperature. The bond *shear* strength (as opposed to bond pull strength) reflects this degradation but does not explain the electrical disconnect propensity of assemblies made with Al–Au interfaces.

Figure 6.37 shows the intermetallic behavior of the AuAl system. Note in Fig. 6.37*a* that zones are beginning to form that are aggravated by the temperature of 300°C. The bottom zone, zone 1, contains AlAu$_2$ (purple plague), which may be seen easily by microscopic examination of the bond area as an

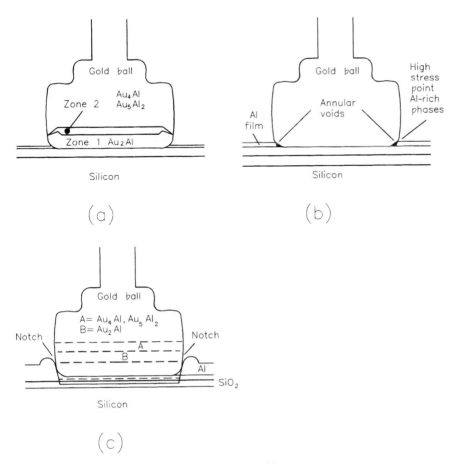

Figure 6.37 Intermetallic formation stages in the AuAl system.

annular ring. Voiding will also start to occur at the Au_5Al_2–Au_4Al interface. Figure 6.37*b* illustrates that bond failures will be accelerated by cyclic stresses in either assembly, environmental testing, or the service environment. Brittle failure mechanics explains the annular voiding that will occur in the bond area. *It is these annular voids that will become responsible for electrical disconnect at the bond interface.*

Figure 6.37*c* shows continued degradation that is induced by volumetric differences in the different phases. All strained interfaces will be susceptible to fracture due to the aforementioned volumetric changes occurring in contiguous phases. Further stresses arise from differences in coefficients of thermal expansion (CTEs). The *notch* formed by cyclic stresses is a propagated annular void. CTE differences, volumetric changes, and phase transformations that are occurring rapidly in these structures serve to aggravate the observation of electrical disconnects in microelectronic assemblies.

6.8.5 Summary

The following points may be made about Au–Al intermetallics that are pertinent to hybrid assemblies:

- Five intermetallics exist in the Au–Al system and are always present in equilibrium at any given temperature—$AuAl_2$, $AuAl$, Au_2Al, Au_5Al_2, and Au_4Al.

- Intermetallics grow continuously; the quantity of Au and Al available determines the type, amount, and sequence of intermetallics formed.

- Intimate contact, even at ambient, is all that is required for initial phase formation to take place. The diffusion and reaction is continuous from that point.

- During the growth of any phase, its thickness increases as the square root of time.

- The intermetallics' structures vary from orthorhombic, monoclinic, fcc, bcc, and distortions thereof; the important thing to remember is that each structure has a different volume.

- With different structures and different volumes coexisting, intrinsic stress always exists at the Au–Al interface.

- Although all Au–Al intermetallics are stronger than either Au or Al in tension, the intermetallic will be more prone to failure due to *brittle mechanics*.

- Based on reaction rate data, 8000 Å Al will be completely transformed to intermetallic in 30 min at 175°C.

- Cracks in the intermetallic layers occur owing to stresses from structural differential expansions.

- With the formation of any particular phase, the diffusion rates of Al or Au through it may be slower or faster than the diffusion rate in the adjacent phase.

6.9 Wire Bonding

The common ways of wire bonding hybrid assemblies are thermocompression (TC) ball bonding, thermosonic (TS) compressive wire bonding, and ultrasonic (US) compressive wedge bonding. TC ball bonding is rarely used at present but will be explained here since its principles are still valid and the method still has much merit. All these wire bonding techniques depend on intimate contact between the bonding wire (either gold or aluminum) and the metallization of either the substrate or component to obtain an atomic interface at the connection.

The most likely long-term wire-bonding reliability problem is *deleterious intermetallic formation*, already discussed in this chapter. The most likely preventative to the *formation* of good bonds is contamination. Contamination may be in the form of oxides on the bonding interfaces or foreign chemical substances attached either on the metal surfaces or on the oxides already present on the metal surfaces. Anything that can be done to prevent the formation of contamination or execute the removal of oxides (and contamination) from the bonding surface will improve the quality of the bond, e.g., the wearing of finger cots, burnishing, plasma etching, ultraviolet cleaning, etc.

The most common form of chemical contamination is finger oils present because of improper handling during processing. While the deformation of the metals that occurs during the bonding process will break the oxide layer that has formed, it will not eliminate the effects of finger oil contamination. Finally, all forms of chemical contamination will require exact knowledge of the type of contamination involved before it can be removed.

Another deterrent to the formation of a good intermetallic bond is the surface irregularities of the metals to be bonded. These surface irregularities must not prevent intermetallic contact over a significant portion of the bond area. Intimate contact is, as earlier stated, mandatory to the making of reliable wire bonds via the formation of intermetallics that are the bonding mechanism of this type of hybrid assembly (chip-and-wire).

6.9.1   Wire bonding techniques

Ball bonding requires that the small wire be fed through a quartz, tungsten carbide, or titanium thick-walled capillary tube. The capillary tube, with one end tapered to a few mils in diameter on the outside, is mounted in a suitable mechanical fixture so that it can be moved both vertically and horizontally. The horizontal positioning must be accomplished by means of precision manipulators with required accuracies of 20 μin while being observed through a microscope.

Before bonding, a small ball is formed on the end of the gold wire by a flame-off mechanism. The metallization to which the wire will be bonded is positioned on the work stage below the bonding capillary. The capillary is brought down over the bonding pad and lowered until the ball is brought into contact with the conductor, where a predetermined amount of force is applied. This deforms the ball and establishes intimate contact between the gold ball and the bonding pad. Wire lead attachment to the other terminal can be accomplished by bonding the wire with the edge of the capillary providing some bonding pressure and ultrasonics supplying the energy. The capillary is then raised, and the flame-off mechanism is used to cut the wire off while forming another ball for the next bonding operation. Figure 6.38 illustrates the ball bonding sequence.

A variation of the ball bonding technique is wedge bonding, accomplished by either thermocompression or ultrasonic techniques. There is a difference in the type of bonding tool used in these two techniques since the wedge bond relies on combinations of time, temperature, pressure, and ultrasonic energy rather than just time, temperature, and pressure. Strictly ultrasonic wedge bonding (no heat) must have bonding tools designed to grip the wire during the vibration sequence. Both the first and second bonds are made using the

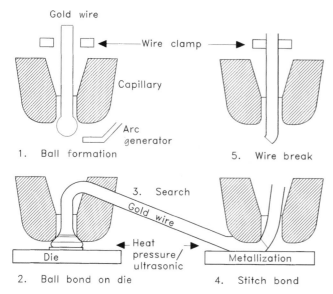

Figure 6.38 Typical ball bonding sequence.

wedge. This type of bond is preferred when bonding to very small, closely placed pads with centers on the order of 0.002 in spacing. Microwave devices have very small bonding pads; wedge bonding is used in this application.

Ball bonding is done with gold wire. Wedge bonding is usually performed with aluminum wire; however, gold may also be bonded with wedge bonding techniques. In ball bonding, the ball and capillary are positioned over the bonding area and then lowered onto the bonding pad. An analogous situation exists in wedge bonding except there is no ball or capillary, only the wedge and the wire. About 100 g pressure is applied to the ball (much less to the wire) for compression-only bonding to take place.

The ball bond is made on the semiconductor die. The capillary is then raised and moved to the metallization on the substrate (usually thick or thin film gold). The wire has no ball in this case and the next bond is made by the edge of the capillary (called a stitch bond). All wedge bonds are stitch bonds. The capillary (or wedge) is then raised in conjunction with some form of wire clipping simultaneously performed. In ball bonding, the wire is then subjected to the flame-off mechanism to form another ball. In wedge bonding, the wire is immediately usable for the next bond.

By varying bonding parameters using properly designed DOEs, improvements in the pull strength of both the ball and wedge bonds can be accomplished. By individual parameter variation a bonding schedule can be established. This bonding schedule will take into account all the independent bonder-controlled variables plus the wire diameter, whether the bond is being made to an aluminum (semiconductor) or gold (substrate) metallization or vice versa.

Wire bonding equipment contains a microscope, a heated stage (not necessary in ultrasonic bonding), and a heated wedge or capillary that will apply pressure (and/or ultrasonic energy) to the wire at the interface of the bonding surface. In addition, a wire feed mechanism is required, as are methods for manipulation reduction as well as equipment and process control.

Typical difficulties that may be experienced in wire bonding are listed below.

- Bonds can be accomplished utilizing any of these bonding techniques which will exceed the wire breaking point in strength; i.e., instead of the bond's breaking, the wire will break during a pull test. The gold wire sizes used for hybrid assembly range from 0.0007 to 0.003 in in diameter. If wire breaking becomes a problem, use a larger diameter wire.

- Semiconductor targets are too small for the wire diameter; bonds cannot be made without shorting to adjacent metallization. Wedge bonding can solve this dilemma.

- Improper stage or capillary temperatures cause active and/or passive devices to fail during assembly. This is the major problem suffered by TC bonding. The stage can be held between 200 and 320°C for very short periods of time. However, the interface temperature between the bonding surfaces will be critical to a good bond unless ultrasonic energy is used; therefore, if many wire bonds are to be made, 320°C temperature for a long bonding period will degrade the semiconductors. In this case, the capillary is heated to 350°C and the substrate lowered to about 200°C.

- The concern with the bonding wire, regardless of its diameter, is the wire's ductility. Bending or cold working of the wire is sufficient to cause poor bonding.

- Contamination of the wire or bonding surfaces by humidity or dust will cause difficulties in bonding.

- The bonding wire may stick to the capillary or to the wedge. The tool used for bonding should have a polished finish (at least 10 to 20 μin) to prevent the wire from sticking to the bonding tool. Even with a glass capillary the wire can stick so that the bond is broken when the tool is withdrawn.

6.9.2 Thermocompression wire bonding

Although this type of bonding is now rarely used, it is the simplest and serves as an introduction to the more commonly used methods of today. Most metals form a metallurgical bond to like or different metals at a temperature considerably below the lower melting point of either of the two metals when pressure is applied normal to the point of contact. TC wire bonding is based on this phenomenon and depends upon time, temperature, and pressure.

The thin layer of oxide present on aluminum wire in air precludes its use for thermocompression bonding. Thus gold wire is the standard wire used in the TC bonding process. The gold wire is fed through a capillary, usually made of tungsten carbide, and a small hydrogen flame or electric discharge is used to cut the wire and form a ball on the end of that wire.

The advantage to TC bonding is that bonding can be made in any direction after the ball has been bonded. Thus bonding can be accomplished over the entire substrate surface without rotating the substrate. The major disadvantage of TC bonding for complex hybrids is the high temperature required for fairly long periods. This temperature extreme results in degradation of the parameters of both active and passive components.

6.9.3   Ultrasonic compressive wire bonding

One of the advantages of ultrasonic compressive aluminum wire bonding is the avoidance of purple plague (Al bonding wire connected to the Al metallization of the chip). However, it is possible to bond gold wire ultrasonically. In general, 0.002 in diameter gold wire or larger can be ultrasonically bonded with relative ease, while the finer 0.001 in diameter gold wire presents many difficulties. The ultrasonic compressive bonding wedge tool configuration is considerably different from that of the thermocompression bonding capillary. The tool design restricts the direction of wire from the first bond to the second bond since the wire must be drawn directly toward the machine operator.

With aluminum wire, the ultrasonic energy and the acoustical high frequency movement of the wire against the conductor pad breaks the refractory oxides surrounding the aluminum wire. Ultrasonic compressive wire bonding usually involves the use of aluminum wire, although as already noted, gold can also be used. The wire is fed through and under a bonding wedge as is done by the WestBond 7400B in Fig. 6.39.

During bonding the wedge presses the wire against the metal termination pad and ultrasonic energy (usually at about 20 to 60 kHz) is applied to the wedge. The wire is rubbed against the contact, causing local heating and a metallurgical weld. The thin oxide coating on aluminum wire is penetrated and the oxide tends to help the frictional heating process, giving a very reliable bond. Wire diameters used are typically 0.0007 to 0.002 in but range as high as 0.20 in in diameter.

There is the need for using large wire diameter bonding for high power hybrid assemblies. Even the applications that require the reliable handling of high current flows at low voltages use large diameter wires. This application may be answered by making several small diameter bonds to the same pad, but this measure is usually stopgap. The bonder shown in Fig. 6.40 not only bonds large diameter wires but also has a rotary head which automatically bonds in any direction and overcomes an inherent disadvantage of many wedge bonders.

After the first bond has been made, the substrate is moved relative to the wedge, pulling the wire through the hole in the wedge. The substrate movement can be in one direction only with most wedge bonders, i.e., the direction of the wire feed. The substrate is positioned until the wire is over the second termination pad and the process is repeated. On completion of the second bond the wire is clamped and pulled away from the bond, leaving a short tail.

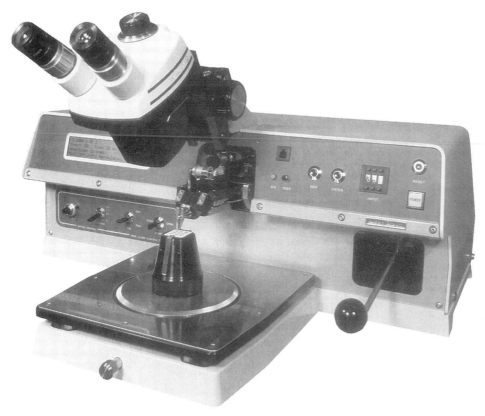

Figure 6.39   The WestBond 7400B deep access, deep reach wedge–wedge wire bonder.

The main difficulty in applying ball and wedge bonding to aluminum had been in forming the ball. Both the hydrogen flame and the electrical discharge method oxidize the aluminum too badly to form an adequate ball for this type of bonding. Capacitive discharge has effectively made the Al ball-and-stitch method work. Therefore, the advantages of this type of bonding can be applied to Al wire bonding, although the disadvantages of such a bonding method have not to date been outweighed by the advantages.

Ultrasonic bonding has several advantages: no heated substrate stage or heated capillaries are required, aluminum is more economical to use than gold, and the bonding area is smaller. A major disadvantage is that the substrate has to be rotated after each bond to set up the direction for the next bond (a disadvantage that is being overcome). These disadvantages are being outweighed by the advantages in such manual bonders as the one shown in Fig. 6.41, the K&S 4123, an aluminum wire or gold wire wedge bonder that is versatile enough to bond simple ICs and discrete devices or complex hybrids with height variations up to 0.2 in. This particular bonder has control over nine variables that make it useful for most applications. However, the bonding platen must be rotated to rotate the substrate, a previously noted disadvantage of this type of bonder.

Figure 6.40 Orthodyne Model 360 rotary head automatic large-wire bonder.

6.9.4 Thermosonic compressive wire bonding

Thermosonic (TS) compressive wire bonding employs the variables of TC bonding, i.e., force, temperature, and time, in addition to the vibratory motion provided by cavitation energy. The ultrasonic energy promotes the bonding process by mechanical displacement of surface contaminants that may exist and by plastic deformation of the surface imperfections. Therefore, this approach offers the advantage of increased surface area without heating the entire substrate to temperatures that will be deleterious to the other on-board components. Ultrasonic vibratory energy causes a temperature rise at the wire–conductor interface that can approach 30 to 50% of the melting point of the metal. Typically, this method of bonding is employed using gold wire.

As wire bonding becomes more sophisticated, machines which once were very simple devices have become sophisticated, as can be seen by the bonders illustrated in this section. In these types of machines, the ultrasonic head is identical to the head seen in all ultrasonic bonders except that a straight wire capillary is used, as on a thermocompression bonder. Also included is the flame-off device necessary to form the ball on the gold wire. Whereas in straight ultrasonic gold wire bonding it is difficult to bond gold wire of less than 0.002 in diameter, on an ultrasonic ball bonder gold wire of 0.001 in diameter (or less) is usually used. With this type of bonder we have the ultimate combination, i.e., a heated stage, a capillary-type tool, and an ultrasonic transducer. The only thing missing is the heated capillary, which is impractical with an ultrasonic transducer.

Figure 6.42, the Hybond model 572, shows a manual thermosonic compressive wire bonder with deep access capability. This bonder offers many options that may be used to change the thermosonic compressive features into a thermocompression wire bonder if the application requires it.

Figure 6.41   The K&S 4123 wedge bonder.

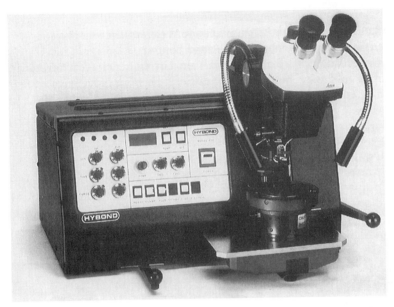

Figure 6.42 The Hybond Model 572 deep access thermosonic wedge bonder, 0.5 to 3.0 mil diameter wire capability.

6.9.5 Automated wire bonding

Wire bonding is one of the main target areas of hybrid assembly for automation. The wire bonding operation is time-consuming, since it involves so many repetitive process steps in the manufacture of one hybrid assembly. Operator training, downtime, and repair/rework due to the lack of consistency in operation are further reasons why the wire bonding operation is prime for automation. The investment cost for using automatic wire bonders is more than offset by the high production rates of these machines, the consistency of product that is obtained, and the reliability of the wire bond once it is made.

Automated wire bonding eliminates individual operator adjustments by automatically bonding all the wires according to a preprogrammed sequence. A typical automated wire bonding system consists of a computer-controlled bonding head and high speed X-Y table, a closed-circuit TV targeting system, a computer programming capability, and a feed mechanism for loading, indexing, and unloading the hybrid substrates. The chips may be aligned manually by the operator, and the bonding is done under computer control after alignment by a computer controlled, high speed pattern recognition system. The FB-118A (Fig. 6.43) from Kaijo is designed for small ball and fine pitch wire bonding in high-pin-count devices such as ASICs by using digital control bonding head characteristics.

Most automated wire bonders can string 10,000 to 15,000 wires per hour compared with about 500 to 1000 wires per hour for semiautomated bonders. Significant savings can be obtained with such a system. Elimination of operator control also results in fewer bonding errors and more consistency in bond-

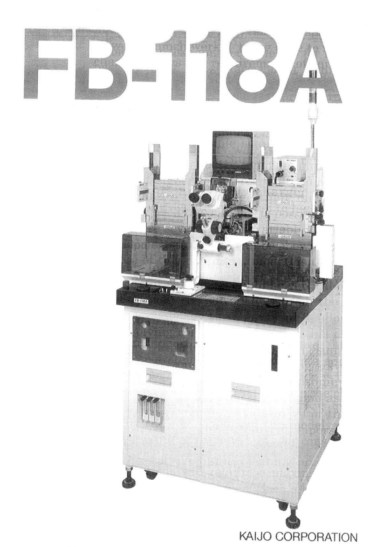

KAIJO CORPORATION

Figure 6.43   The Kaijo FB-118A fully automatic wire bonder.

ing quality. VLSI and ASICs have created the need for an automated assembly process because of high die cost and the requirement for a high reliability wire bonder process that is operator independent.

Automated bonders such as the Hughes Model 2470-III (Fig. 6.44) and the Hughes Model 2460-IV (Fig. 6.45) provide many features not previously available. Their operating systems contain software routines for single-minute exchange of dies (SMED—a JIT concept). A gray-scale vision system and programmable lighting assure repeatable capture of low-contrast substrate features. Most automatic bonders feature multiple operator programmed features that did not exist 20 years ago. The onboard computer provides high speed image processing, information storage, and data retrieval to wire bond

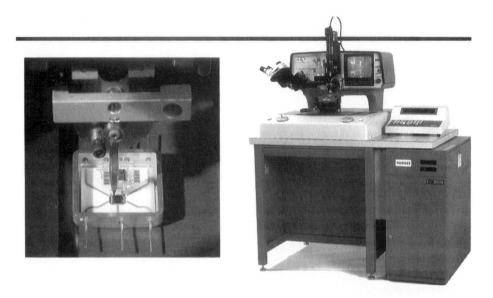

Figure 6.44 At right: the Hughes Model 2470-III automatic wedge bonder. Insert at left: wedge bonding 1-mil gold wire to a microwave device.

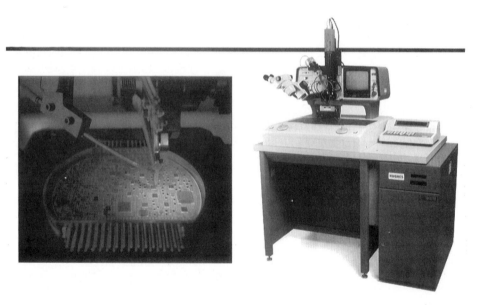

Figure 6.45 The Hughes Model 2460-IV flexible automation for fine-pitch wire bonding. Fully automated program-controlled wire bonding of hybrids and multichip modules.

the fine-pitch, high-pin-count devices standard in the multichip module format. These features may be integrated with the factory to fully realize factory automation.

6.10   TAB and Flip-Chip Bonding

TAB is a semiconductor assembly technology originally developed by General Electric as a possible replacement for wire bonding. TAB is the automated assembly of devices to a substrate using a flexible lead frame known as the tape. Other names for this tape are the *carrier* or the *beam* from which the lead frame is mounted to the device in an automated fashion. Besides the obvious advantage of automation, TAB devices may be pretested by contacting the tape's test pads prior to committing the TAB unit to further processing.

Flip-chip bonding was developed by IBM and perfected by both IBM and Motorola. Flip-chip bonding is another bare-chip bonding technique. The term *flip chip* is used to describe a particular classification of device that has all its contacts on the face of the die. These contacts are in the form of raised bumps, primarily having the bumps' material made of high temperature SnPb metallurgy sometimes alloyed with small amounts of silver.

6.10.1   Tape automated bonding (TAB)

This assembly process is entirely automatic. Using a bonder such as the one shown in Fig. 6.46, IMI's ILB-1207, the inner leads are automatically attached to the semiconductor device. The inner lead bonding machine is very similar to the older reel-to-reel projection machines used in moving pictures. The tapes used in tape automated bonding are typically 35, 48, and 70 mm. The 35 and 48 mm tapes are primarily for small chip sizes with low lead counts. The 70 mm tape is used for VLSI and ASIC devices to accommodate the larger chip size as well as to allow fanout of the test pads to a usable layout.

After the inner leads are bonded, the film is cut and the outer leads are cropped from the film. The chip is then adhesively attached in position and the outer leads are either soldered or bonded to the microcircuit's connecting pads. The primary method of bonding these devices is using an automated outer lead bonder such as the one shown in Fig. 6.47.

The primary reason for the renewed interest in TAB during the past few years has been the move toward chip-on-board technology. There are three basic options for the mounting of TAB devices. The first option is one in which the chip is placed face up and the devices are mounted with the leads bent to the substrate. The second mounting option involves leaving the leads straight and bonding the chip in a face down configuration. The third and final option is called *pocket TAB*. The reason for this name is that the chip is placed in a pocket hewn out of the substrate to accommodate the device's body while the leads remain straight and are connected to the substrate.

Figure 6.46 IMI's ILB-1207 inner lead bonder.

From an assembly point of view, TAB may be done by either single-point or gang methods. The first method is rather obvious, but somewhat slow. The only fast single-point method to date has been using a laser such as that shown in Fig. 6.48, the ESI Model 7100 laser TAB system. The system shown can very accurately and reliably perform outer lead bonding using a pseudo-pulsed laser that is of a Nd:YAG variety. Using laser TAB, the pitch of the device is limited only by the laser beam diameter instead of by the tool size in single-point bonding.

The traditional method of TAB has been gang bonding. This is a technique in which a hot blade (thermode) is used with heat and force to bond all leads to a die or a substrate simultaneously. The problem with using such a technique is coplanarity. By noting the device shown in Fig. 6.49, it is easy to see how difficult it is to have so many leads on one side of any device or on any given substrate coplanar. Therefore, it behooves assembly technologists to find ways of automating single-point tab.

Under any circumstances, TAB has been advantageous for a variety of reasons, including:

- Ability to pretest and burn-in devices prior to final assembly

- Superior high frequency characteristics of TAB leads compared with those of wire-bonded leads (the increased cross-sectional area of TAB leads produces lower electrical and thermal resistances and low, constant inductance)

Figure 6.47 IMI's OLB-1300 outer lead bonder.

- Ability to achieve very dense bonding patterns, such as 0.002 in pads on 0.004 in centers
- Amenability of inner-lead-bonded components to surface mounting
- Essentially flat bond-height profiles to accommodate limited packaging space requirements
- A degree of protection for the normally exposed bond pad afforded by bumped dice

The following are disadvantages of the TAB process as noted by the IPC publication, SMC-TR-001:

- TAB requires specially designed equipment to match each application, at both the chip and substrate interfaces.
- Wafers are expensive in small quantities.
- Lack of commercially available "bumped" tape, wafers, or dice.

- TAB bonding and bumping equipment is more sophisticated than hybrid assembly equipment.
- Chip-on-tape components (TAB carriers) are expensive in small quantities.
- TAB ILB and OLB locations have not yet been fully standardized.
- Dynamic burn-in of TAB packages is more costly in both hardware tooling and burn-in.

Figure 6.48   ESI's Model 7100 laser TAB system.

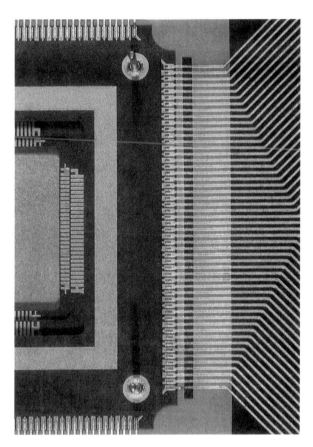

Figure 6.49 A package-to-board second level TAB lead interconnect. The 70 mm 208 I/O lead pattern is Au-to-Au single-point thermocompression bonded to the perimeter of the ceramic chip carrier.

6.10.2 Comparison of interconnect technologies

Fine-pitch technology (FPT) differs in several respects from both through-hole technology and the surface mount technology (SMT) when used for hybrid assembly. Table 6.11 lists the main aspects of package interconnecting and rates the applicability of three different technologies for each aspect. Through-hole technology requires component lead mounting holes in printed wiring board lands; they are not required for SMT or FPT. In the SMT, vias are located away from the land pattern in order to minimize solder flow down the hole and away from the land. This approach wastes real estate. A better alternative is FPT, which uses lands with solder-filled vias.

Through-hole technology commonly uses automated component insertion while FPT and SMT use automated component placement. Manual assembly is possible for each of these approaches but not practical for volume assembly. SMT assemblies commonly use solder paste to add enough solder to assure a quality solder joint. Unfortunately, solder paste complicates the interconnection assembly process and hides yield problems caused by board solderability. Solder bumped lands avoid some of these problems. Solder bumped lands are made by adding fused solder to each land area prior to assembly. Flux holds the component and cleans the metal oxides during reflow.

TABLE 6.11   Process Comparison

	Through-hole technology	Surface mount technology	Fine-pitch technology
Lands with lead mounting holes	5	—	—
Lands with external vias	—	5	3
Lands with internal vias	—	2	4
Component insertion	5	—	—
Component placement	—	5	3
Component lead excise, form, and placement	—	—	3
Use of solder paste	—	5	3
Use of solder mask	3	4	4
Land solder bumping	0	3	4
Use of conductive adhesive	0	1	2
Wave soldering	5	2	0
Mass reflow soldering	—	5	4
Localized reflow soldering	1	3	4
Manual assembly (in volume)	2	1	1
Automated assembly (in volume)	5	5	5
Rework and repair	5	4	4
Process control	3	4	5

5 = essential or common usage
4 = possible and recommended
3 = desirable (but not mandatory)
2 = not recommended or stretching the technology
1 = possible (but impractical)
0 = avoid at all costs

Conductive adhesives can replace the solder joint; however, this approach is not recommended in volume production because of high costs and process difficulties. Wave soldering is commonly used for through-hole technology but is not always suitable for SMT components (see Sec. 6.4) and should not be used for FPT packages.

6.10.3   Flip-chip bonding

The flip-chip concept uses small solder bumps attached to the chip termination lands to be mated with the appropriate lands on the substrate. The chip is placed face down, or flipped. The solder joints are made by reflow at an adequate temperature. Previously, the handling and placement of these miniature die with even smaller balls have been extremely difficult and costly. However, special machinery such as the SEC Model 4150 flip-chip die bonder, shown in Fig. 6.50, allows the versatility for placing the die with the required accuracy.

Modern flip chips are produced with high temperature solder bumps on the appropriate land areas while the chips are still in wafer form. The chips are then excised, placed face down on the substrate, and reflowed by one of the aforementioned reflow techniques. An alternative chip-on-board (COB) flip-chip method is to place the solder bumps on the substrate rather than on the chip. However, the mass techniques of producing flip chips in the wafer form make this method of flip-chip attachment unattractive. The major advantage

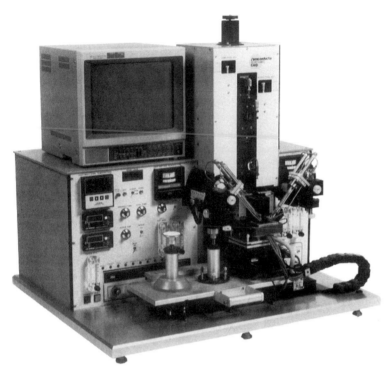

Figure 6.50 SEC 4150 flip-chip die bonder aligns and attaches flip-chip die onto substrates.

offered by using flip chips is the tremendously reduced amount of real estate required with this COB technique. The major disadvantage is that the chips may not be pretested or burned in without being attached to the substrate.

Precision bump placement and singulation of the devices from the wafer are employed to create die uniformity in anticipation of automatic chip handling and placement. Although the placement of devices is commonly done automatically in the surface mount technology, it is still rudimentary in the handling of bare chips. However, automatic handling of chips is one of the advantages of flip-chip technology.

Following placement on the patterned substrate, two requirements must be met. The specific bonding pad area must confine the device by means of a solder dam around the pads. The second requirement is that the solder pad must be pretinned to accept the flip-chip device. After proper placement, the chip and its substrate are placed in a reflow machine that usually uses a nonactivated flux, nitrogen cover gas, and high temperatures. The surface tension of the combining SnPb areas of both the substrate and the device will force the flip chip into precise alignment with the substrate.

To date, only two companies have done much work with the flip chip. However, the tremendous amount of interest generated by chip-on-board manufacturing techniques is increasing the demand for flip-chip manufacturing techniques and equipment to handle these devices.

The reliability data provided for flip-chip devices are still limited even though flip chips have been the subject of many studies over the past 20 years. The only difference that should result from using a flip-chip device and a standard chip-and-wire device is the reliability of the bump under thermal fatigue. IBM has performed much work in this area and shows this soldered bump to be as reliable as any soldered joint used in surface mounting small devices (note Sec. 6.3).

The prime advantage of using flip chips is the small amount of real estate required during substrate preparation. The second advantage offered by using flip chips is that fewer assembly steps are required than in using any of the other bare-chip assembly techniques. These techniques are even superior to most bare-chip handling techniques since automation is coming of age. When considering the use of flip-chip assemblies in commercial environments, they are more reliable than normal chip-and-wire techniques.

In terms of repairing the flip-chip device, it is easier than any of the other bare-chip techniques. Less assembly skill is required in the manufacture of the flip chip, and once the process is developed, assembly yields are higher. Further, the development costs are no more than with other bare-chip techniques. The disadvantages are that the cost of the capital equipment required to assemble these devices is high. They are not as available as had been hoped 20 years ago. They are much more difficult to inspect than are face up die. Lastly, the device cost of flip chips to build most assemblies is still inordinately high.

6.11   Statistical Process Control

While inspection and test equipment are the province of the failure analysis section in Chap. 10, statistical process control (SPC) belongs in the area of assembly of hybrid microcircuits. SPC resulted from initial attempts at statistical quality control (product-centered) into the process-centered SPC we know today. With the advent of automation, SPC is becoming real-time process control (RTPC). The relation between the two is that something has already gone wrong, out-of-control, before action is taken in SPC. Granted the something that is after-the-fact in SPC is small indeed compared to yesterday's method of trying to inspect quality into a product; SPC is still after-the-fact control.

RTPC is not and truly cannot afford to be after-the-fact control. RTPC is connected with automation and automation will produce poor quality product every bit as fast as it will good product. The three major advantages to automation are *reliability, consistency of product,* and *reduced labor content* (see Sec. 6.12). Reliability is possible only if the process is developed fully before automation. Reduced labor content also depends on the process being fully developed. *The product can be consistently made poorly as well as made correctly.* Therefore, RTPC is very necessary for automation. If a process is *beginning* to approach control limits, the machinery needs to "flag" upstream operations to make corrections or to shut down before the disaster becomes paramount. This approach to signaling an imminent error before one happens is RTPC. The expense of programming RTPC is usually avoided in all but fully automated operations. SPC, the initial measure that must be taken, is gaining widespread acceptance in today's hybrid assembly operations.

6.11.1 Principal kinds of control charts

Control charts are designed to focus on the target of any operation (examples of control chart types are in Fig. 6.51). The control chart is constructed by having the target value bounded by its statistical limits to indicate the capability of a given process. Do not mistake the control limits for the specification limits of any process. For example, if the specification limits are lower than the control limits, parts may be inspected that are within specifications but are being built in an out-of-control situation. If the specification limits are larger than the control limits, then even if the part is out of specification, it is being built by an in-control process. If parts are being built properly to a control chart limit, and those limits are within specification limits, then the process is fully capable and in control.

The most commonly used control charts are *X-bar* and *R charts*. These are so called because they are average and range charts. The *X*-bar is calculated by taking 5 parts from the current population and averaging the response of

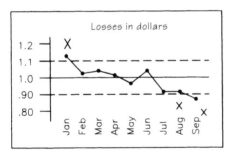

Chart for losses on a scrap assembly line

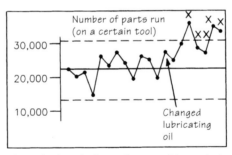

Chart for length of run after a tool is repaired

Chart for inspection ratios

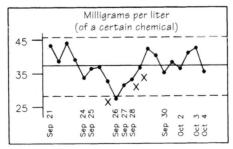

Chart for electroplating bath

Figure 6.51 Typical process control charts.

interest. Since averages are always normally distributed, the X-bar chart is normally distributed and the 3-sigma limit is applicable to the X-bar chart. X-bar charts are usually associated with R charts, the R simply standing for the range or difference between highest and lowest points in the sample.

The next type of control chart is a *p chart*. *p charts* are also called percentage charts or attribute charts. The *p* means proportion. It usually represents a proportion of bad assemblies or assembled parts compared to the total, i.e., the percent defective. Since many microelectronic processes are in excellent control, these are usually parts per million (ppm) charts.

The X-bar, R, and *p* charts are the most common charts used. Other types include *np charts, c charts,* and *u charts.* The *np* charts simply involve the number of defectives found in each sample instead of calculating the percentage. Except for plotting the number defective instead of the percentage or fraction, this chart is in every way the equivalent of a *p* chart. The *c* chart is a special type of attribute control chart which uses the number of defects instead of the number of defectives. The distinction between defects and defective is that a defect is an individual failure to meet a single requirement while a defective is a unit of product which contains one or more defects. Finally, the *u* chart is a variation of the *c* chart. The point plotted in a *u* chart is the average number of defects per unit in a sample of several units.

6.11.2 Why SPC?

Defects cost everyone. Defect costs are very nearly like the iceberg (Fig. 6.52). Only those costs readily visible are usually seen by management. They are scrap, rework, repair, and warranty costs. Below the surface other costs can be associated with defects. For example, the nonassembly workload usually known as overhead is associated with these defect costs. There are other

Figure 6.52 Defects cost everyone.

aspects of this negative cost of quality, e.g., inventory, capital used to generate bad parts, customer dissatisfaction, and loss of market share. Quite often these costs cannot be seen and are not even known. By maintaining control charts, it may be determined where a problem exists (if it does) and to what extent it is affecting the hybrid assembly operation, and through effectively designed experimentation, the corrective action to be taken.

SPC is the first step in striving for excellence on the hybrid assembly floor. It is very important to note that it is the *focus on the process* that accomplishes statistical process control. In order to observe the process properly, all that the process entails must be known. As Fig. 6.53 shows, there are several facets to that process. By maintaining adequate control of the process and everything that attends the process, the assembly will work the first time, every time, as it was designed.

In order to strive for excellence and know what processes to control and what controls to make it is imperative to realize that every work process has a customer and a supplier. *Poor quality is usually planned that way.* The way to avoid this is to understand the concepts of PDCA, i.e., plan, do, check, act. These are the basic building blocks of design of experiments (DOE) and of SPC.

SPC is a philosophy. The old philosophy (Fig. 6.54) was quality based on conformance to the specification. Quite simply, if the part inspected is a conforming one, keep it. If it is not, discard it. *Do not fix the process; do not change anything.* As we have learned, this philosophy is costly and does not even work. The new philosophy (Fig. 6.55) is quality losses based on deviation from the target and is the cornerstone of SPC.

Using this new philosophy and properly strategized DOE, the effort is placed in the front end. At that point, numerous design and process business practices can be placed into effect that will result in a lower cost product to the manufacturer, a more satisfied customer, and an increased market share.

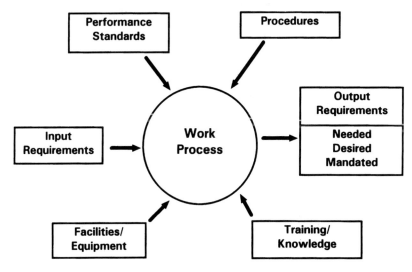

Figure 6.53 The work process, the focus of SPC.

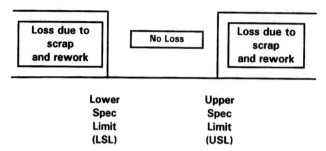

Figure 6.54 The old philosophy.

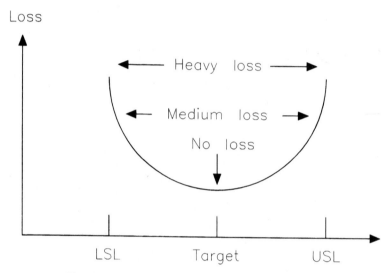

Figure 6.55 The new philosophy. Quality losses are based on deviation from target, not conformance to specifications.

All this must be started at the point of vendor selection and continued throughout the hybrid assembly shop on to total customer satisfaction. Figure 6.56 simplifies this sentiment. However, one might gather from the graph that the cost is the same whether it is paid in the front end or the back end. The cost to the front end is nonrecurring. The cost to the back end is recurring. It will not take long before management recognizes back-end costs and then spends the money to do what should have been done in the first place (usually more than double the cost of *doing it right the first time*).

In order to decide where SPC must be exercised, the problem solving process must be used (Fig. 6.57). The process is a simple one, but one that is normally overlooked; the simplest things usually are. Using this process alerts the assembler of hybrid circuits to the problems and then aids in the decision of whether a pinch point in the process exists or does not exist which may require SPC on a daily basis. To verify this, suitably designed experiments have to be executed. Using the aforementioned problem solving

% Effort

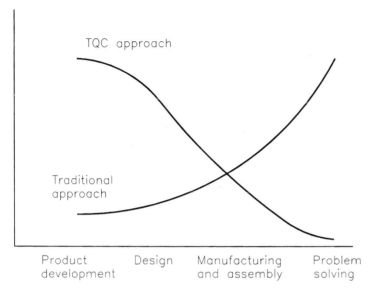

Figure 6.56 Management strategies.

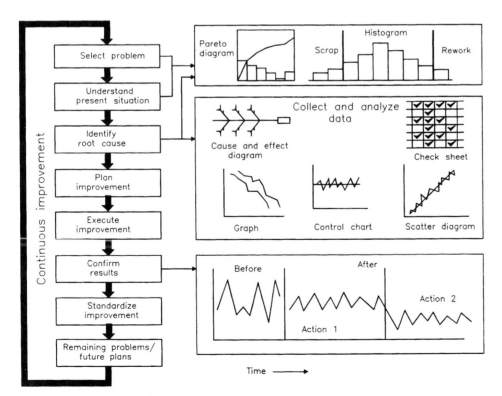

Figure 6.57 The problem solving process.

process results in the effects on SPC shown in Fig. 6.58.

Although C_p and C_{pk} are generally accepted methods of measuring the process, most statisticians will advise the hybrid assembler to be wary of them. Nonetheless, the control charts in Figs. 6.59 and 6.60 show the advantages that can be gained by using some form of process index to control the hybrid assembly floor.

The final point to be made is the use of designed experiments. Yesterday's method of experimentation was to vary one factor at a time while holding all others constant. This would be grand except that the real world does not work that way. Usually everything is varying at the same time. Also, by using the old one-at-a-time method of experimentation, interactions were disregarded and no math models existed to predict the responses for variable combinations. Time was consumed, but little else was accomplished. DOE is the only way.

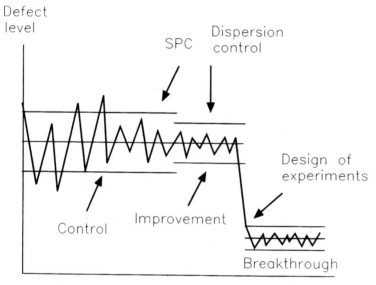

Figure 6.58 Statistical methods, SPC, process improvement, DOE.

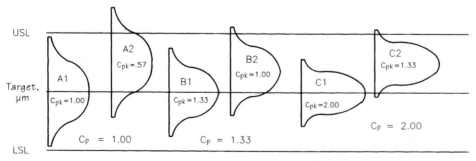

Figure 6.59 Statistical methods, C_p and C_{pk}. C_p = process capability = $(USL - LSL)/6\sigma$. C_{pk} = actual performance of a process = $\hat{C}_p [1 - (2|\bar{x} - \mu|)/(USL - LSL)]$ = smaller of $(USL - \bar{x})/3\sigma$ or $(\bar{x} - LSL)3\sigma$.

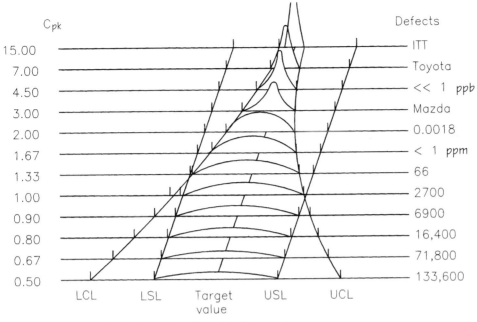

Figure 6.60   Defect reduction is achieved by minimizing variance.

There are several types of experimental design, all with the same objective. They obtain the maximum amount of information using the minimum amount of resources. DOE determines which factors shift the average response, which factors affect the variability, and which factors have no effect at all by following the steps in the process flow diagram (Fig. 6.61). DOE determines factor settings that optimize the response and minimize the cost. From the results of DOE, SPC can be instituted from an empirical process model that will control the response of interest by controlling the input factors, not by simply inspecting out the nonconforming parts.

6.11.3   The arachnoid

The use of SPC and other quality methods must go hand in hand with the business objectives of the organization. The method of accomplishing this is the "arachnoid," a tool that is synonymous with SPC. The arachnoid is a simple way to visualize the performance of an area or, for that matter, an enterprise. The word means "of or pertaining to a spider." The arachnoid looks like a spider web.

An arachnoid (Fig. 6.62) is any regular geometric figure, e.g., square, pentagon, hexagon, septagon, etc. It is used to illustrate performance against industry standards (benchmark) or preset objectives (objectives are *how much* and *when* metrics; goals are open-ended desires). The arachnoid forces the determination and presentation of objectives. It is, in fact, the visual statement (the picture worth a thousand words) of those objectives. The

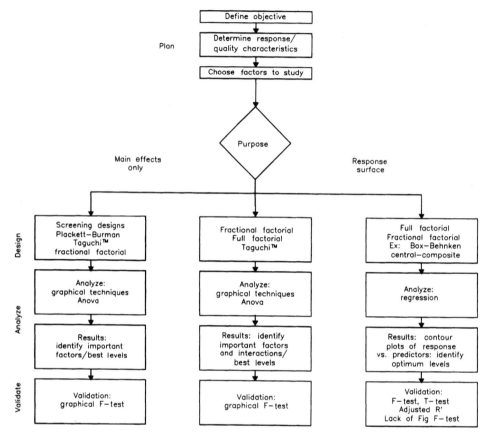

Figure 6.61 DOE methodology.

arachnoid allows the quick and easy projection of performance against those objectives at a glance. Dependent on the creative use of shading or colors on the arachnoid, you may show the major objectives of the project or hybrid assembly area, performance against those objectives, and even how well the industry or rest of the plant is doing on the same scales. The arachnoid shows management where the resources are being applied and if they need to be moved to get the job done (or if more need to be added to meet the schedule). The important point is to meet but not exceed the objectives of the arachnoid.

Note in Fig. 6.62 that no individual spine of the arachnoid necessarily has the same scale or even the same units as any other spine. They may or they may not. The objective is always the spatial location at which the regular geometric figure is executed. The enclosed area may be one color or shade. A plot on the axes of where you are with the project forms another, usually irregular, geometric figure. This enclosed area may be another color or shade. The dollars or worker hours expended on each effort may be reported on the arachnoid if that breakout is known. Therefore, if you are over in one or more spots, it is time to shift resources in order to bring the lagging portion of the project up to acceptable standards.

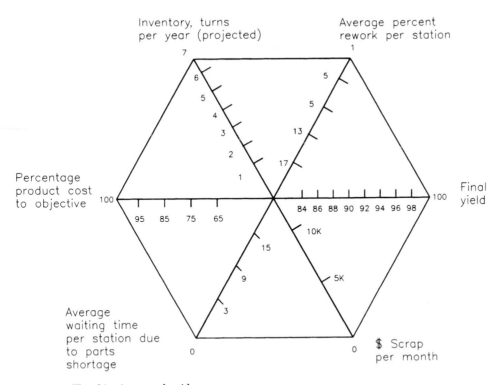

Figure 6.62 The objectives arachnoid.

The arachnoid should be displayed with SPC control charts. It plots the progress of a hybrid assembly area the same as does the control chart. This device may also be used for a group of process efforts that have to work in concert to show the results of SPC. In summary, the arachnoid adds to the use of SPC as a hybrid assembly tool. The advantages far outweigh the disadvantages. Both are "visuals" that aid the engineer and management to know what has been done, what is being done, and what needs to be done.

The disadvantages of SPC are:

- Problem identification takes time.
- Control charts signal problems but do not imply or express cause.

The advantages of SPC are:

- Provides a systematic and efficient method of gathering data and transforming it into information
- Allows the hybrid assembler to make decisions on facts rather than engineering judgment or memory
- Identifies nonrandom impacts to a process
- Warns of process degradations prior to assembling nonconforming hybrids
- Forms a basis for improving a process within the specifications

- Gets line workers involved
- Builds an expert database
- Shows evidence of process improvements

In summary, SPC is the systematic analysis and maintenance of hybrid assembly process performance data collected for the purpose of predicting process trends. SPC can be used in any process environment to prioritize potential opportunities for process improvement, to determine process stability, and to measure the impact of corrective actions.

6.12   Factory Automation

Worldwide competition has increased markedly over the past 30 years. The global nature of competition and the desire of most countries to compete internationally have dictated that microelectronic assembly business concerns, particularly multinational ones, meet this competition. Approaches these organizations usually take consider material requirements planning (MRP) and just-in-time (JIT) manufacturing, more than ever requiring these companies to enforce total quality methodology (TQM). MRP is evolving to manufacturing resource planning (MRP II). The vehicle for using MRP II is computer integrated manufacturing (CIM). Tying these systems together requires an information system making that system as much a part of assembly as bonders and solder reflow equipment.

To be competitive in the assembly of microelectronic circuitry, automation is a contemporary necessity. In order to automate, the assembly process must be fully developed. If it is not, the automated work cell will simply produce rejects at a rate higher than can be done manually. Once the process is developed and automation is exercised, the work cell must be fully integrated with all forms of information management. The difference between the integration of semiautomatic, worker-interface equipment, and fully automated equipment is the materials handling system. The automation of both SMT and bare-chip assembly processes uses the computer to substitute for the redundant tasks that an operator routinely performs.

The automated assembly work cell may integrate components, substrate, and interconnections for both military and commercial applications. To do this, the work cell must be designed with computer-integrated RTPC for controlling the process. Many integrators are available today to provide custom equipment, factory management systems, and integrated materials handling systems to manufacture, control, and provide product in the assembly of microelectronic circuitry. Automation is the first step in interfacing the computer systems already existing in the plant with the assembly area.

6.12.1   Technology

All devices connected to this information system should communicate with a coordinated protocol. Multivendor software must be united via coprocessor before installation into the system. Interface should be done at the software level so that a uniformity of masks will be presented to all users. All backup, recovery, and security modules have to be automatic and transparent to the user. Communications, under these circumstances, will then occur from all devices with equal facility.

Modularity of hardware, firmware, and software will be necessary to execute the design of this system. However, no single topology LAN should be instituted. At the corporate level, the bus may be used. At the department level, the type of LAN employed should adapt to the function that department performs; e.g., R&D might use a web network while many manufacturing areas may use a star LAN. This will allow all users to configure access and menu- driven commands that will best serve the function of that department. The additional processing times required to relate the various functions will be the penalty this system pays until one system becomes the industry standard.

6.12.2 Database

All cell controllers must be responsible for the acquisition and logging of data and updating the database. Because of the nature of the programmable logic controllers, they will add to the database on a real-time basis. They will then be programmed to add data and information based on RTPC as illustrated by the model, both by frequency of occurrence and by event. These data will become part of the product pricing model for both products and services offered by any microelectronics manufacturing concern. Some manual data entry will be necessary to complete the DSS for this system.

The objective of CIM is to develop a cohesive digital database that integrates manufacturing, design, and business functions. Information is sent on demand and as needed to as large a number of intra- and interdisciplinary groups as possible. The database will then be dynamic, but accessible to those with the proper entry credentials, and will become part of the DSS. Penultimately, the database must be programmed with all known product data by engineering. Ultimately, the database will become an accumulation of hybrid assembly experience.

6.12.3 The manufacturing enterprise

Using the information relationship modeling techniques provides a methodology to design a system in which CIM can be effectively applied. The graphic model of this system (Fig. 6.63) shows the relation of all functions involved in the enterprise. The information system links the various activities that are required to service and manufacture microelectronics products. This model is configured to show the interrelation of products, processes, and support services from concept to delivery as well as after-sale support for all microelectronic products. The model for this system consists of manufacturing and support cells, each one designed with feedback and commands that operate to maximize productivity and product differentiation.

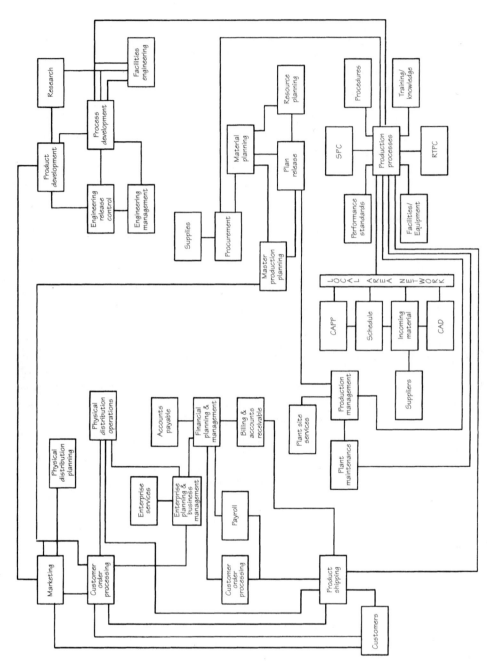

Figure 6.63   The assembly center computer system architecture.

The hierarchy of operation is the same as that of input. The information system is designed to maintain planning and control at the lowest level, diminishing the amount of soft and firmware necessary to institute the system. Workstations must arrange all operations in sequence, removing this responsibility from operations and information management personnel. Controllers and workstation nodes are illustrated by the individual points denoted in Fig. 6.63. The corporate levels of planning and control are not shown. Interactions of the various functions are. Each function is assigned to the level that satisfies the requirements of that function.

Real-time requirements are more quickly needed at operational levels of the organization. They are, however, more prone to the use of knowledge-pervasive and preprogrammed decisions. Therefore, this system will be programmed (expert system) to make these decisions. Top-level decisions require more in-depth information and will be presented by this model as information to that level. By using this model, it is possible to incorporate fault detection, improper input and output, scheduling, and the judicious utilization of any microelectronics enterprise's assets.

Automation of the assembly areas plus detailed information management will easily lead to a combination of material requirements planning plus just-in-time assembly. The result is computer integrated manufacturing (CIM). In order to implement the CIM strategy, the enterprise must be understood completely. All information needs have to be addressed. A model of the enterprise must be defined as a tentative description of the holistic system. The model may be broken into its basic functions, which are shown in Fig. 6.63. Each block represents a support function such as administration, applications development, and decision support systems. The object of CIM is to perform these tasks in such a fashion that consistency of assembly results in reliability of product. In order to implement CIM, the following steps are necessary:

- Develop an enterprise model including your assembly area that establishes all business requirements.

- Establish a common CIM vocabulary.

- Identify all information flows required to achieve assembly automation.

- Outline the basic system requirements for creating, communicating, manipulating, using, and storing CIM information.

- Identify necessary resources and organizations that must contribute in order to achieve CIM objectives.

- Define the CIM architecture that will allow maximum return on necessary information.

- Define process steps to be taken toward CIM implementation in automating the microelectronics operation.

The objectives of installing automated assembly and integrated materials handling systems are the reduction of labor costs (cost-competitive products), consistency of production, and most importantly, increased reliability in the assembled product. To achieve these objectives, the automated assembly center must gather product and process related data during manufacturing. The automated assembly center will then have the capability to schedule and track the product during its manufacture. Since the assembly center will be accurately scheduled and monitored, on-time delivery of product is optimized. The individual pieces of assembly equipment must be tightly monitored and controlled to produce a quality product, to produce any product at all. If quality issues do occur, information will exist in the computer's database that will allow timely problem identification and on-line correction through RTPC due to automated data feedback of product in processing parameters to all areas needing the information.

The traditional solution to such assembly automation is the integration of computer-aided design, computer-aided manufacturing, computer-aided test, and ultimately, computer-aided engineering. The increasing competitive pressures are forcing most microelectronic assembly environments to evaluate day-to-day operations with a view toward CIM. In doing so, enterprises are discovering the need for open, business-driven information systems. It is for this reason that this section is included in this handbook. Systems such as this will enable all departments to work closely together in order to allow engineers to take automated plant-floor, hybrid-assembly equipment into account in the design of microelectronic products. CIM extends communication lines to suppliers and customers as well as all internal departments.

To meet this need, the CIM environment requires a dynamic network of distributed functions. The fact that these functions reside on system platforms makes them a portion of the hybrid assembly process. Many of these platforms are tailored to the specific environment of the enterprise. Many are general purpose. The result is an environment that encompasses the total information requirements of the enterprise from developing of the business plans to the shipment of products.

It is no longer a simple world where the assembly engineer has cognizance of the materials and processes that are necessary to assemble the microelectronic circuit. In order to realize these benefits, the engineer must understand operations systems, flexible manufacturing systems, and the overall system into which the assembled microcircuit will go. In short, CIM supports management efforts to meet the challenges of competing effectively in today's world markets. Several manufacturers have already realized this necessity. Figure 6.64 illustrates an automated line that is offered to produce surface mounted devices. Note that this line manufactured by Panasonic includes die mounters, packaging systems, wire bonders, cure ovens, and coating dispensers. Each machine is made to perform assembly operations in situ. Together, the systems must operate with an integrated material handling system, plus buffers that are provided in the event of failure or to accumulate product where the production rate of one piece of equipment exceeds that of the one in front of it. All systems include RTPC, which must use preprogrammed statistical techniques for determining if any portion of the system is going out of control.

Figure 6.64   The Panasonic COB/MCM fully automated system.

All manufacturers of assembly equipment are gearing their offerings to the concept of automation. Another example is the assembly system shown in Fig. 6.65. Such flexible assembly systems manufacture TAB, C4 flip chip, chip-on-board, fine-pitch devices, QFPs, and regular SMT components. This stand-alone system is modular and may be configured for one or more of these technologies, or is easily retrofitted to any of these assembly technologies. In addition, this system may be included in a more complete computer integrated environment. Most of these systems are made to operate stand-alone or with manual intervention. Figure 6.66 illustrates IMI's contract packaging facility for TAB device development and production. This operation includes wafer bumping, lead pattern design and fabrication, inner-lead bonding, encapsulation, outer-lead bonding, packaging, and final assembly in a controlled environment.

The automated assembly center should be a work cell that is only a portion of the enterprise. The automated assembly center will then include all necessary loading and unloading mechanisms along with transfer systems to facilitate the flow of assemblies through all manufacturing process. Bar code readers are to be used throughout the assembly center to track and identify the progress of the hybrid assemblies. A system computer can then be utilized to direct and track all activities and analyze the production information being supplied by each individual piece of equipment.

6.12.4   Operational concepts

Before institution of automated assembly, it is imperative that a total quality environment be created. Control charts must be integrated with each functional assembly block for detecting unexpected variations in the quality

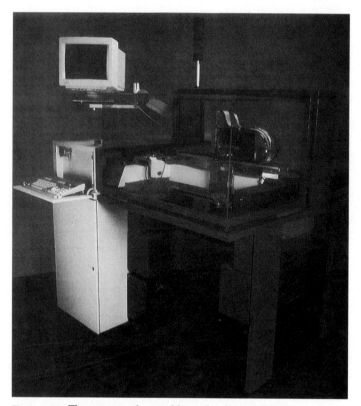

Figure 6.65 The automated assembly work cell by Zevatech.

obtained from each step in the production line. Any variation that exceeds a control limit statistically will cause each piece of assembly equipment to alter its mode of operation or alert supervisory management.

In addition, charts may readily be obtained from this otherwise paperless environment in order to analyze trends, measure continuous improvements, or serve as easily readable documentation of quality checks. The automated assembly center will thereby contain visual, electrical, optic, and mechanical sensors to provide feedback to critical process steps that indicate drifting parameters before they reach an *out-of-control limit*. It is important in the design of all functional hybrid assembly operations that control points are identified, the type of control used at these operations is known, and the type of output that is desired for each process is specified.

Just as important as the assembly equipment itself is the computer system. The computer system will use computer-aided process planning (CAPP) to interface through a local area network in determining exactly what has to happen when it has to happen. Figure 6.67 is offered as an example of a process flow diagram used in the manufacturing of hybrid assembly products and to illustrate the various points at which CAPP will be necessary. Note that this process is an imaginary one for the manufacture of passive components that

Figure 6.66   The International Micro Industries contract assembly facility.

may be used in the surface mount technology. The same type of process diagram can and must be developed for any kind of hybrid assembly process.

From this diagram, hardware may be developed using an assembly center simulation program from which equipment requirements can be developed. The facility requirements follow the development and specification of these equipments and, from the facility requirements, the materials transfer requirements will evolve. The software required to run the system may then be developed. The initial step must be an assembly center operations description. This description must define the products to be assembled and the factory operational rules to be used during production. The manufacturing process employed, inventory guidelines, product identification, inspection requirements, and test points must be defined.

Next, the user requirements should be listed to determine the type of factory control to be exercised. The user requirements can then be established by functional group, and a matrix showing the relationship between the requirements and the groups must be established. Generic functions such as product planning, operator interface, inventory buffers and queues, material identification and handling, and security are included in the assembly center's operational software.

A structured material and information flow can then be generated. The factory layout with all assembly workstations should be mapped to the factory floor. The structured material flow attends product flow relative to the rout-

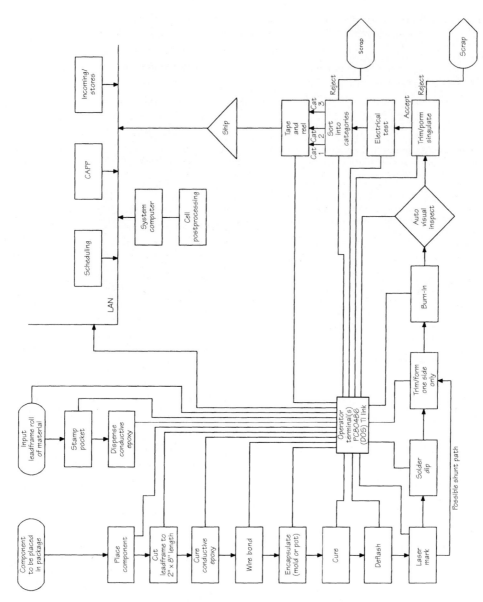

Figure 6.67 Example process flow diagram.

ings (a general form rather than each specific part routing) from workstation to workstation. At each major step the assembly process is identified and the time required is logged by the computer. The machine data for input and desired information output are defined in masks presented by the assembly center's computer-terminal nodes.

The various functions of this assembly center are its hardware configuration, its software, the assembly center interface, the cell controllers' interfaces, the database, the application software, and the performance criteria. Once each of these functions is integrated in the assembly of microcircuits, the manufacturing enterprise may be said to be integrated. Thereby, all data and information will move freely between the functions in the system throughout the assembly of the product. The challenges involved are difficult. The incentives are obvious. To compete effectively on a worldwide scale, the microelectronic enterprise's hybrid assembly environment must use these strategies.

6.12.5 Factory automation remarks

Complete information systems for automated hybrid assembly manufacturing, and any other kind of service or product provision for that matter, are in their infancy. No such system actually exists today and, therefore, direct comparison is not possible. However, the design of this system will work. Standard networks are currently being developed, but at the present, the efforts are somewhat disjointed. In order to compete globally, business concerns are being required to use CIM and MRP II. Information management in some standardized form will be necessary to meet this challenge. This system must and will be integrated into all microelectronic manufacturing plants, if for no other reason than the fact that products containing microelectronics are becoming less and less costly with time and automation will be required to be competitive.

Bibliography

Abbagnaro, L., "Manual Repair: New Techniques and Directions," *Surface Mount Technology,* November 1993.

Appleton, D., "The State of CIM," *Datamation,* Dec. 15, 1984.

AT&T Technologies, *Statistical Control Quality Handbook,* Delmar Printing, Charlotte, N.C., 1984.

Baker, E., "Stress Relaxation in Tin-Lead Solders," *Materials Science and Engineering,* vol. 38, 1979.

Balde, J., "Overview of Multichip Technology," *Electronic Materials Handbook,* vol. 1, ASM Press, Metals Park, Ohio, 1989.

Bender, M., Patterson, F., Kemp, E., and Gantzhorn, J., Jr., "Low Temperature Cofired Ceramic Tape System," *Hybrid Circuit Technology,* vol. 6, no. 2, 1989.

Billmeyer, F., *Textbook of Polymer Science,* 2d ed., Wiley-Interscience, New York, 1971.

Blech, I., and Sello, H., "Some Aspects of Au-Al," *Journal of the Electrochemical Society,* vol. 113, 1966.

Box, G., Hunter, W., and Hunter, J., *Statistics for Experimenters,* Wiley-Interscience, New York, 1978.

Brous, J., and Schneider, A., "Cleaning Surface-Mounted Assemblies with Azeotropic Solvent Mixtures," *Electronics,* April 1984.

Browning, G., Colteryahn, L., and Cummings, D., "Physics of Failure in Electronics," *Reliability Sev.,* vol. 4, RADC, Rome, N.Y., 1966.

Brunetti, C., and Curtis, R., "Introduction," *Printed Circuit Techniques,* National Bureau of Standards, Circ. 468, Nov. 15, 1947.

Bunshah, R. et al., *Deposition Technologies for Films and Coatings,* Noyes Publications, 1982.

Burch, J., and Grudnitski, G., *Information Systems, Theory and Practice,* 5th ed., Wiley, New York, 1989.

Bushmire, D., "Resistance Increases in Gold Aluminum Interconnection with Time and Temperature," *ISHM Proceedings,* 1976.

Cadenhead, R. 1991 chairman of the Hybrid Marketing Research Council, personal observation during working group session.

Cadenhead, R., "Materials and Electronic Phenomena," *Inside ISHM,* November/December 1984.

Cadenhead, R., "The Process Development Philosophy of Characterization," *Inside ISHM,* May–June 1988.

Cadenhead, R., and Bradley, J., "Overview of Hybrid Technology," *Electronic Materials Handbook,* vol. 1, ASM Press, Metals Park, Ohio, 1989.

Cadenhead, R., and DeCoursey, D., "The History of Microelectronics, Part 1," *International Journal for Hybrid Microelectronics,* vol. 8, no. 3, 1985.

Caswell, G. (ed.), *Surface Mount Technology,* ISHM Technical Monograph Series 6984-002, 1984.

Chase, R., and Prentis, E., "Operations Management: A Field Rediscovered," *Journal of Management,* vol. 13, no. 2, 1987.

"Choosing the Right Solvent," *Texas Instruments Technical Journal,* September–October 1990.

The CIM Enterprise, *Advertising Booklet of the IBM Corporation,* White Plains, N.Y., 1989.

Cocca, T., and Pierce, J., "Analysis of Aluminum Wire Bond Degradation to Gold Thick-Films," *ISHM Proceedings,* 1978.

Coffinberry, A., and Hultgren, R., *AIME Proceedings,* 1938.

Crowley, R., "Laser TAB Systems: A New Technology Implementation," *Microelectronics Manufacturing Technology,* March 1991.

Daigle, J., Seidmann, A., and Pimentel, J., "Communications for Manufacturing: An Overview," *IEEE Network,* vol. 2, no. 3, May 1988.

Dhar, V., and Jarke, M., "On Modeling Processes," *Decision Support Systems,* vol. 9, 1993.

Diamond, W., *Practical Experiment Design for Engineers and Scientists,* Lifetime Learning Publications, Belmont, Calif., 1981.

DiNitto, J., and D. Zimmerman (eds.), *Hybrid Microcircuit Design Guide,* ISHM-1402/IPC-H-855, 1982.

Dixon, T., "TAB Technology Tackles High Density Interconnections," *Electronic Packaging and Production,* December 1984.

Dolk, D., "An Introduction to Model Integration and Integrated Modeling Environments," *Decision Support Systems,* vol. 10, 1993.

Ellis, B., "Ionic Contamination Control of Circuits with Surface Mounted Components," *Brazing & Soldering,* no. 9, 1985.

Ellis, B., "Removing Flux Using Batch Water Machines," *Electronic Production,* vol. 16, no. 4, 1987.

Engelmaier, W., "Effects of Power Cycling on Leadless Chip Carrier Mounting Reliability and Technology," *Electronic Packaging & Production,* vol. 23, no. 4, 1983.

Epstein, D., "Rework of Hybrid Microelectronics," *ISHM Proceedings,* 1980.

"Forgiving Reflow Operations Ease Shift to SMT," *Surface Mount Technology,* September 1992.

Gedney, R., and Balde, J., "Future Trends," *Electronic Materials Handbook,* vol. 1, Publication of ASM International, 1989.

Gerry, D., Taylor, B., Sutliff, B., and Miller, W., "Automatic Wire Bonding—Performance Characteristics of a Reduced Thickness Thick Film Gold Conductor," *ISHM Proceedings,* 1978.

Gerwin, D., "Do's and Don'ts of Computerized Manufacturing," *Harvard Business Review,* March–April 1982.

Ginsberg, G., "Chip-on-Board Profits from TAB and Flip-Chip Technology," *Electronic Packaging and Production,* vol. 25, no. 9, 1985.

Gunn, T., "The CIM Connection," *Datamation,* Feb. 1, 1986.

Gunter, B., "The Use and Abuse of C_{pk}, Part 1," *Quality Progress,* January 1989.

Gunter, B., "The Use and Abuse of C_{pk}, Part 2," *Quality Progress,* March 1989.

Gunter, B., "The Use and Abuse of C_{pk}, Part 3," *Quality Progress,* May 1989.

Gunter, B., "The Use and Abuse of C_{pk}, Part 4," *Quality Progress,* July 1989.

Hagge, J., "Predicting Fatigue Life of Leadless Chip Carriers Using Manson-Coffin Equations," *Proceedings IEPS,* 1982.

Hahn, G., "Some Things Engineers Should Know about Experimental Design," *Journal of Quality Technology,* vol. 9, 1977.

Harman, G., and Albers, J., "The Ultrasonic Welding Mechanisms as Applied to Aluminum- and Gold-Wire Bonding in Microelectronics," *IEEE Transactions on Parts, Hybrids, and Packaging,* December 1977.

Harman, G., "Ultrasonic Wire Bonding," *ISHM Proceedings,* 1973.

Harper, C. (ed.), *Handbook of Thick Film Microelectronics,* McGraw-Hill, New York, 1974.

Heller, P., "Introduction," *Surface Mount Technology, A Compendium of Technical Articles Presented at the First, Second, and Third Annual Conferences,* IEPS Publication, 1984.

Hinch, S., *Handbook of Surface Mount Technology,* Longman Scientific & Technical, New York, 1988.

Horowitz, S., Gerry, D., and Cote, R., "Connecting to Gold Thick-Film Conductors," *Electronic Packaging and Production,* November 1977.

Horsting, C., "Purple Plague and Gold Purity," *10th Annual Proceedings, Reliability Physics Symposium,* 1972.

Huang, C., Willwerth, J., Cang, L., and Stein, S., "Large Diameter Aluminum Wire Bonding to Thick Film Conductors," *ISHM Proceedings,* 1982.

Hueners, B., "Absolute Ultrasonic Amplitude Measurement, Calibration, and Troubleshooting of a Wire Bonder Using a Laser Interferometer," *International Journal for Hybrid Microelectronics,* vol. 6, no. 1, 1983.

Hybrid Microelectronics for the 80s, 85260-2476, Integrated Circuit Engineering Corporation.

Inglesby, T. (ed.), "Slaying Dragons with SPC," *Manufacturing Systems,* August 1988.

"An Introduction to Tape Automated Bonding and Fine Pitch Technology," *IPC Publication SMC-TR-001,* January 1989.

Johnson, D., and Chavez, E., "Characterization of the Thermosonic Wire Bonding Technique," *ISHM Proceedings,* 1976.

Johnson, K., Scott, M., and Dawes, C., "Development of Aluminum Ball/Wedge Wire Bonding," *ISHM Proceedings,* 1980.

Jones, R., *Hybrid Circuit Design and Manufacture,* Marcel Dekker, New York, 1982.

Joseph, C., "Using LANs to Automate the Factory Floor," *CIM Review,* summer 1990.

Kay, S., "Making It in Manufacturing," *Computerworld,* Apr. 25, 1988.

Kenyon, W., "Part 1—Water Cleaning Assemblies: Wave of the Future or Washout?" *Insulation/Circuits,* February 1978.

Khadpe, S., "Yield and Throughput Considerations for Automated Wire Bonding of Multi-Chip Hybrids," *ISHM Proceedings,* 1976.

Kilby, J., "Invention of the Integrated Circuit," *IEEE Transactions on Electron Devices,* vol. ED-23, no. 7, July 1976.

Kiran, A., Roberts, D., and Strudler, B., "A `Simulated' SMT Production Line," *Surface Mount Technology,* August 1993.

Kirby, P., and Pagan, J., "The Origins of Surface Mount," *Sixth European Microelectronics Proceedings,* ISHM Europe '87 Conference, June 1987.

Kohl, P., "Multichip Module Size and Yield Considerations," *Hybrid Circuit Technology,* vol. 8, no. 6, 1991.

Koopman, N., Reiley, T., and Totta, P., "High-Density Chip Interconnections," *ISHM Proceedings,* 1988.

Lambert, W., "Ultrasonics Aid Cleaning," *Electronic Production,* vol. 14, no. 10, 1985.

Lassen, C., "Wanted: A New Interconnection Technology," *Electronics,* vol. 52, no. 9, Sept. 27, 1979.

Lau, J., and Rice, D., "Solder Joint Fatigue in Surface Mount Technology: State of the Art," *Solid State Technology,* vol. 28, no. 10, 1985.

Lea, C., *A Scientific Guide to Surface Mount Technology,* Electrochemical Publications, Ayr, Scotland, 1988.

Lee, C., and Chang, J., "Reworkable Die Attachment Adhesives for Multichip Modules," *ISHM Proceedings,* 1991.

Lemond, D., "Key Process Design Factors for Efficient Fluorosolvent Spray Cleaning of Surface Mounted Assemblies," *Proceedings IPC,* Technical Paper IPC-TP-604, 1986.

Lermond, D., "Key Process Design Factors for Efficient Fluorosolvent Spray Cleaning of SMAs," *Printed Circuit Assembly,* May 1987.

Levine, L., and Sheaffer, M., "Wire Bonding Strategies to Meet the Thin Packaging Requirements," *ISHM Proceedings,* 1992.

Liljestrand, L., "Bond Strengths of Inner and Outer Leads on TAB Devices," *Hybrid Circuits,* no. 10, 1986.

Loeb, W., "Hybrid Circuit Market Forecast for 1991," presentation to the HMRC, June 1991.

Lynch, J., "Surface Mount Reliability," *Hybrid Circuits,* no. 11, 1986.

Majni et al., *Journal of Applied Physics,* vol. 52, no. 6, June 1981.

Malloy, G., Koudounaris, A., and Farrel, J., "Evaluation of Hybrid Microcircuit Rework Procedures," *ISHM Proceedings,* 1982.

Manson, S., *Thermal Stress and Low Cycle Fatigue,* McGraw-Hill, New York, 1966.

Marcoux, P. (ed.), "Surface Mount Status of the Technology Industry Activities and Action Plan," *Bulletin of the EIA and IPC,* January 1989.

Marquis, E., and Wallace, A., "Surface Preparation of Thick-Film Gold Automatic Thermosonic Wire Bonding," *ISHM Proceedings,* 1982.

Messner, G., "Price/Density Tradeoffs of Multi-Chip Modules," *ISHM Proceedings,* 1988.

Musselman, R., and Yarbrough, T., "The Fluid Dynamics of Cleaning under Surface Mounted PWA's and Hybrids," *Proceedings of NEPCON/West '86,* February 1986.

Oswald, R., deMiranda, W., and White, C., "Changes in Strength and Resistance of Aluminum to Gold Ultrasonic Bonds After Thermal, Electrical, and Environmental Stresses," *ISHM Proceedings,* 1975.

Page-Walton, J., "PCB Cleaning Systems," *Electronic Production,* vol. 15, no. 6, 1986.

Palmer, D., and Ganyard, F., "Aluminum Wire to Thick Film Connections for High Temperature Operation," *ECC Proceedings,* 1978.

Panousis, N., and Fischer, M., "Ball Bond Shear Testing to Assure High Yields and Reliability on Large Chips," *International Journal of Hybrid Microelectronics,* vol. 6, no. 1, October 1983.

Peck, D., "Heating Technologies in Rework and Repair of TAB and Fine Pitch Devices," *ISHM Proceedings,* 1992.

Philofsky, E., "Intermetallic Formation in Gold Aluminum Systems," *Solid State Electronics,* vol. 13, Pergamon Press, 1970.

Plackett, R., and Burman, J., "The Design of Optimum Multifactorial Experiments," *Biometrika,* vol. 33, 1946.

Plapp, N., "A Vapor Phase Solution for Curing Conductive Epoxy Adhesives," *Surface Mount Technology,* August 1993.

Plunkett, P., and dal Porto, J., "Low Temperature Void Formation in Gold-Aluminum Contacts," *Proceedings of the 32nd IEEE Electronic Components Conference,* May 10–11, 1982.

Puttlitz, K., "Replacing Flip Chips on MLC Multichip Modules Using Focused Infrared (IR): The Basics" *ISHM Proceedings,* 1992.

Ravi, K., and White, R., "Reliability Improvement of 1-Mil Aluminum Wire Bonds for Semiconductors," *Final Report, NASA Contract No.* NAS8-26636, December 1971.

Richardson, D., "A Call for Action: Integrating CIM and MRP II," *Production and Inventory Management Journal,* vol. 29, no. 2, 2Q, 1988.

Ritzman, L., King, B., and Krajewski, L., "Manufacturing Performance—Pulling the Right Levers," *Harvard Business Review,* vol. 62, no. 2, March–April 1984.

Roberts, S., "Systematic Evaluation of Wire Bond Connection Features," *ISHM Proceedings,* 1992.

Rousch, M., "Microwave Hybrids—Frequency Constraints in Materials and Process Techniques," *ISHM Proceedings,* 1981.

Sage, M., "Multichip Modules," *ISHM Proceedings,* 1988.

Saldarini, R., *Analysis and Design of Business Information Systems,* Macmillan, New York, 1989.

Scharr, T., "TAB Bonding a 200-Lead Die," *International Journal for Hybrid Microelectronics,* vol. 6, no. 1, 1983.

Schmidt, S., and Launsby, R., *Understanding Industrial Designed Experiments,* CQG Ltd. Printing, Longmont, Colo., 1989.

Schuessler, P., "Adhesive Die Attach Materials: Their Pros and Cons," *ISHM Proceedings,* 1983.

Sharif, N., and Adulbhan, P., *Systems Models for Decision Making,* Asian Institute of Technology, Bangkok, 1978.

Shimada, W., Kondo, T., Sakane, H., Banjo, T., and Nakagawa, K., "Thermo-Compression Bonding and the Au-Al System in the Semiconductor IC Assembly Process," *Proceedings of the International Conference on Soldering, Brazing, and Welding in Electronics,* Munich, Nov. 25–26, 1976.

Siegal, B., "A More Effective Approach to Evaluating Die Attachment Quality," *ISHM Proceedings,* 1983.

Silverberg, G., "Single-Point Tape Automated Bonding—A Versatile, Efficient Interconnect Technique," *ISHM Proceedings,* 1987.

Skeist, I. (ed.), *Handbook of Adhesives,* 3d ed., Van Nostrand Reinhold, New York, 1990.

Snee, R., Hare, L., and Trout, J., *Experiments in Industry: Design, Analysis, and Interpretation of Results,* ASQC Quality Press, Milwaukee, 1985.

Somerville, D., "Introduction to Surface Mount Technology Part One," *Inside ISHM,* January–February 1986.

Spigarelli, D., "Thermal Separation in Surface Mount Attachment and Rework," *Proceedings, NEPCON West,* February 1989.

Spigarelli, D., and King, D. (cochairmen), "Chip Carriers and Other Surface Mounted Components," *International Journal for Hybrid Microelectronics, Proceedings of the 1982 ISHM Conference,* vol. 5, no. 2, November 1982.

Srubar and Cohen, "The Montreal Protocol," *ISHM Journal,* fall 1988.

Stohr, E., and Konsynski, B., *Information Systems and Decision Processes,* IEEE Computer Society Press, Los Alamitos, Calif., 1992.

Teicholz, E., "Computer Integrated Manufacturing," *Datamation,* March 1984.

Totta, P., "Flip-Chip Solder Terminals," *Proceedings 21st IEEE Electronics Components Conference,* Washington, D.C., 1985.

Traeger, R., "Organics Used i n Microelectronics: A Review of Outgassing Materials and Effects," *Proceedings, Electronics Components Conference,* 1977.

Unger, R., Bycer, M., and Wing, W., "Self Test Wire Bond Technology—Hybrid Automated Lead Tester (HALT)," *ISHM Proceedings,* 1981.

vanHee, K., Somers, L., and Voorhoeve, M., "A Modeling Environment for Decision Support Systems," *Decision Support Systems,* vol. 7, no. 3, 1992.

Vonderembse, M., and White, G., *Operations Management, Concepts, Methods, and Strategies,* 2d ed., West Publishing, St. Paul, 1991.

Vonderembse, M., and White, G., *Operations Management,* 2d ed., West Publishing Company, St. Paul, Minn., 1991.

Wild, R., "Fatigue Properties of Solder Joints," *Welding Journal Research Supplement,* vol. 51, 1972.

Winchell II, V., and Berg, H., "Enhancing Ultrasonic Bond Development," *IEEE Transactions on Components, Hybrids, and Manufacturing Technology,* vol. CHMT-1, no. 3, September 1978.

Wirsing, C., "The Elimination of the Kirkendall Voids by Gold Wire Ultrasonic Bonding," *Proceedings of Southeastcon,* 1974.

Wright, E., and Wolverton, W., "The Effect of the Solder Reflow Method and Joint Design on the Thermal Fatigue Life of Leadless Chip Carrier Solder Joints," *Proceedings IEEE, 34th ECC,* 1984.

Young, R., and Mayer, R., "The Information Dilemma: To Conceptualize Manufacturing as Information Process," *IE,* September 1984.

Electronic Packaging of Hybrid Circuits

A. Elshabini-Riad, Imran A. Bhutta, and Jerry Sergent

7.1 Introduction

Hybrid microelectronic circuits can be realized in thick film, thin film, monolithic, or a combination of these technologies. These circuits perform properly only under the prescribed operating and environmental conditions. Proper operating conditions can be maintained by encasing these hybrid circuits in a protective enclosure. The enclosures are called *electronic packages,* and they play a crucial role in the hybrid circuit's performance.

Looking at the advances in the monolithic circuit technology in the last 50 years, one becomes aware that the amount of resources invested in the development of electronic packages does not begin to compare with the resources spent for the development of electronic devices. Only recently has the electronic community realized that the evolution of microelectronic devices and their packaging are highly interdependent and decided to invest some efforts and resources into research and development of electronic packages. One of the main driving forces behind these recent efforts has been the fact that the electronic devices are reaching their physical limits, and a significant performance improvement can be achieved by optimizing the electronic package design. The smaller and faster devices of today require electronic packages that can match their performance in electrical, mechanical, and thermal aspects.

7.2   Requirements of Electronic Packaging

An electronic package must provide the encased circuit with the proper operating conditions. These conditions include maintaining a proper operating temperature, protection from humidity and contaminants present in the environment, providing a robust mechanical support, and good thermal management with minimum limitation on the circuit bandwidth. These conditions can be summarized in three basic categories:

1. Electrical interconnects and their performance
2. Mechanical support and protection from physical environment
3. Thermal management

Electrical interconnects and resonant modes[1] in high frequency packages can affect the overall performance. In this chapter, only the electrical interconnects and their performance will be discussed in detail. Mechanical and thermal considerations of electronic packages have been considered in other works, including Ref. 2.

7.2.1   Electrical interconnects and performance

An electronic package has two types of interconnects. One type is used primarily to provide connection between the different encased circuits or devices. The other type is used to provide the input/output (I/O) signal to the encased devices. The electrical design considerations for both types of interconnects are very similar, and in the following discussion both types of interconnects will be treated together.

7.2.1.1   Critical issues related to electrical interconnects.   Integrated circuits have evolved from less than one transistor per square millimeter per IC in the 1960s to more than 20,000 transistors per square millimeter per IC.[3] Such large numbers of transistors per unit area are associated with very short on-chip interconnects, and consequently very small signal propagation delay. Compared to on-chip interconnects, the package interconnects are relatively long, and account for most of the overall propagation delay.

In order to operate the packaged device at high frequencies, it becomes important that the propagation delay be reduced to as small a value as possible. This requirement demands that the electrical length of the interconnects be made small and the propagation velocity be increased. The signal delay in an interconnect, t_d, is given by

$$t_d = \frac{l}{v_p} \tag{7.1}$$

where l = length of the interconnect and
v_p = propagation velocity of the electrical signal

For high frequency operations, the electrical interconnects can no longer be considered as simple conductors, instead they must be treated as transmission lines. In a transmission line, in addition to the electrical resistance, other quantities of interest are the inductance and capacitance. Both the inductance and the capacitance depend on the properties of the conducting material as well as the surrounding dielectric. The propagation velocity v_p is determined from the inductance and capacitance per unit length, L and C, respectively:

$$v_p = \frac{1}{\sqrt{LC}} \tag{7.2}$$

At high frequencies, the electrical signal in the interconnects has the characteristics of an electromagnetic wave. Any discontinuity in the path of the signal causes a partial reflection of this electromagnetic wave. The amount of signal reflected from the discontinuity is determined by the reflection coefficient at the discontinuity. Discontinuities in a transmission line's path are represented by a variation in the characteristic impedance of the transmission line. The reflection coefficient at the discontinuity is thus given by

$$\Gamma = \frac{(z_{02} - z_{01})}{(z_{02} + z_{01})} \tag{7.3}$$

where z_{01} = characteristic impedance of the transmission line
z_{02} = characteristic impedance of the discontinuity
Γ = reflection coefficient

Thus, electrical interconnects must be designed so as to reduce the variations in the characteristic impedance along the length of the interconnect. The variation in the characteristic impedance can be attributed to several factors.

1. Variation in the thickness of the dielectric
2. Bends or cutouts in the interconnects
3. Presence of vias or wire bonds
4. Proximity to other signal or ground lines planes

Another important design consideration for electrical interconnects is the presence of electrical cross-talk. Cross-talk is a measure of signal leakage or interference from one transmission line to another. The two adjacent interconnects are coupled together through their mutual inductance and capacitance. Since the electrical interconnects are very dense in many electronic packages, signal interference is a genuine concern.

Cross-talk between interconnects can be effectively reduced either by placing a ground or power plane/line between the two interconnects or by increasing the dielectric separation. Increasing dielectric separation is usually feasible when multilayer technology is used. In such cases, cross-talk can be further reduced by inserting a ground or power plane between the two interconnect layers.

7.2.1.2 Electrical simulation, modeling, and design.

Electrical performance of packages and interconnects has become increasingly more important as the semiconductor devices are operated at higher frequencies and the packaging architecture becomes more and more condensed. As the present electronic packages continue to be reduced in size, the complexity of the interconnects increases. To investigate these complex structures, the designers have to perform three-dimensional electromagnetic simulations to fully characterize all discontinuities and proximity effects.

Electrical simulation. With improving computing resources, the interest in solving electromagnetic problems to obtain greater understanding of the electrical designs has also increased. Several basic approaches exist to solve the integrodifferential equations involved in the electromagnetic analysis. Depending on the solution method, these approaches can be broadly classified as the finite element method, boundary element method, and finite difference method.

In the finite element method, the solution of the field problem is obtained by approximating the field over the domain of interest, using a set of compact basis functions. In the boundary element method, the solution surfaces are divided into small regions with precalculated compact basis functions. In the finite difference method, the equation derivatives are approximated by difference equations.

Apart from the above classification based on the solution methods, the electromagnetic analysis can also be classified on the basis of the electrical analysis approach used to investigate the electrical behavior of electronic packages. The electrical analysis approaches have included electric field integral equation formulation (EFIEF) and its solution based on the method of moment (MoM) approach, geometrical theory of diffraction (GTD), and the finite difference time domain (FDTD) approach. In the past, the above approaches have been used in analyzing electromagnetic problems like radiation, scattering, radar cross section (RCS), induced currents, and potential variations. However, the electrical analysis of microwave packages poses a new set of constraints and limitations on the use of the above approaches; for example, some approaches that provide accurate results for RCS calculation of aircraft [resolution of 30 to 40 ft (9 to 12 m)] in the microwave range of frequencies may not be suitable for geometries of the order of 0.001 in (0.025 mm) that are usually found in microwave packages.

The EFIEF-MoM approach has been used extensively in providing rigorous boundary value analysis of structures. The method requires the solution of the electric field integral in addition to the solution of the weighting function integral. This approach increases the computation time required to perform the necessary analysis and, for today's complex microwave packaging structures, poses a significant challenge for electromagnetic analysis.

The GTD approach can provide the electric field at the point of interest in terms of a spatial attenuation factor multiplied by the electric field at the source. This spatial attenuation factor depends on the medium in which the wave is traveling as well as the shape of the source. In complex dielectric structures where many different mediums are used and where the shape of the source may be different, this poses a problem, since the spatial attenuation factor has to be modified in going from one medium to another.

Electrical modeling. Wideband models of electronic structures can be used to predict the performance of the package over the frequency range of interest. These models in conjunction with models of semiconductor devices may be used to predict the performance of the packaged device. Physically based models also provide the designer with the opportunity to isolate the different discontinuities and predict the circuit's performance without them.

One of the time domain network synthesis (TDNS) approaches used involves iterative network synthesis of the transient response of the electronic structure. TDNS uses the acquired time domain reflectometry (TDR) response of the structure as well as the actual excitation pulse used in the measurement. The acquired excitation pulse is used to excite the wideband model and the simulated TDR response of the model is compared with the acquired TDR response. The network model may include distributed or lumped elements. The elements may include capacitors, inductors, resistors, lossless transmission lines, junction diodes, nonlinear capacitors, and a variety of voltage and current sources. Iterative adjustment of the values of the different elements is needed to obtain a close match between the simulated and the measured response. Once the model has been developed using time domain measurement results, it can be verified by performing frequency domain measurements and comparing the measurement results with the simulated results. Figure 7.1 shows a wideband electrical equivalent model of a microwave monolithic integrated circuit (MIMIC) package. Figure 7.2a shows the comparison between the time domain response of the model and the measurement. Figure 7.2b shows the comparison between the frequency domain response (S_{11}) of the model and the measurement.

Time domain measurements have the advantage of providing the designer with an insight into the electronic structure under consideration. The designer is able to correlate the measured response to the spatial characteristics of the structure. This advantage allows the development of physically based models. The same technique can be used to develop a comprehensive model of the full single chip or multichip modules or it can be used to develop a model for any multilayer structure or one of the interconnects. Values for the different parasitics associated with the interconnects can thus be precisely evaluated.

Electrical interconnect design. For high frequency circuits, the interconnects must be treated as transmission lines. Interconnects can be fabricated as striplines, microstrips, or coplanar lines.[4,5]

Stripline. A stripline can be fabricated by placing the interconnect in a dielectric medium and providing grounded planes at the top and bottom of the dielectric. Striplines have the advantage of tighter coupling between the

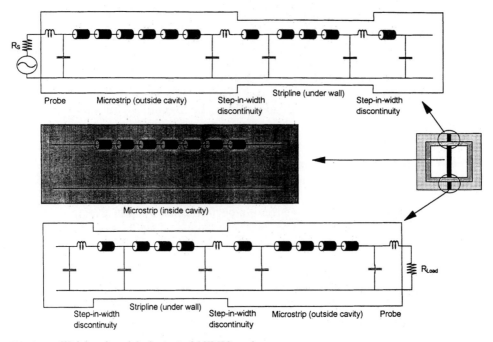

Figure 7.1 Wideband model of a typical MIMIC package.

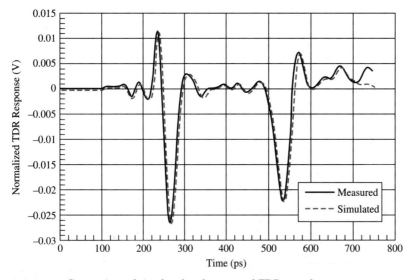

Figure 7.2a Comparison of simulated and measured TDR waveforms.

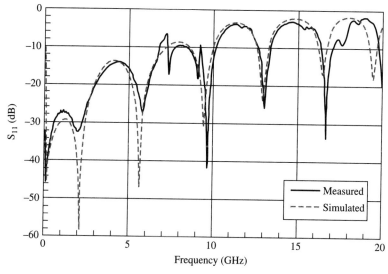

Figure 7.2b Comparison of simulated and measured S_{11}.

interconnects and the ground planes, and, therefore, experience less cross-talk than microstrips. However, access to striplines is difficult and striplines have to be gradually transformed to microstrips or coplanar structures at the contact points.

Microstrip. A microstrip line can be fabricated by placing the interconnects on top of a dielectric and providing a ground plane at the bottom of the dielectric. Microstrips have the advantage of ease of access. However, the bottom ground plane does not provide as effective isolation against cross-talk as the dual ground planes in striplines.

Coplanar line. A coplanar line can be fabricated by placing the ground lines or planes on either side of the signal strip. Coplanar lines provide the designer the advantage of accessing the ground plane on the same surface as the signal strip. The advantage of such a structure becomes clear when coplanar connectors provide I/O to the package. Use of coplanar lines also simplifies the circuit by reducing via structures, which connect the bottom and/or top ground planes to the ground pads.

7.3 Types of Electronic Packages

Electronic packages have been classified on the basis of their design, material, performance, and hermeticity. In terms of design, electronic packages can be classified as the plug-in type having leads protruding from the bottom with a lead spacing of 0.100 in (2.5 mm) or as the surface mount type with either gull-wing or J leads. Depending on their design, electronic packages may be further classified as dual in-line packages, small outline integrated circuit (IC) packages, leadless and leaded chip carriers, pin grid arrays, flat packages, and metal packages.

7.3.1   Dual in-line package (DIP)

A DIP is a through-hole-mounted package with the leads on a pitch of 0.1 in situated on two opposite sides of the package. DIPs can be fabricated by plastic molding, laminated ceramic, or pressed ceramic techniques (CERDIPs). DIPs are sometimes further classified as skinny DIPs, shrink DIPs, and standard DIPs, according to the lead width and lead pitch.

In addition to DIPs, single in-line packages (SIPs) are sometimes used for resistor and capacitor components or networks. For all plug-in packages, leads provide a convenient method for ensuring clearance in assembly and providing a degree of compliance in the mechanical stress established by assembly or expansion coefficient mismatch between the package and the mounting substrate. DIP packages are used primarily for packaging devices with low complexity. However, with the advent of very large scale integration (VLSI) devices and high input/output count, the 0.1-in lead centers resulted in very large packages. A standard 40-pin device requires over 2 in (6 cm) of length and higher pin counts became increasingly difficult to package in DIP form. A DIP is shown in Fig. 7.3.

7.3.2   Small outline package

The limitations of the original DIP package were overcome with the introduction of the small outline (SO) package. Originally conceived by Phillips N.V., these designs have been standardized by the Joint Electron Device

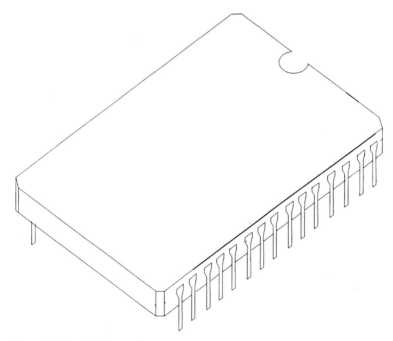

Figure 7.3   Dual in-line package.

Engineering Council (JEDEC) and the Electronic Industries Association (EIA) of Japan. Several differences between the SO and DIP packages are clear and can be identified:

1. The leads on the SO are on 0.05-in centers vs. 0.1 in for DIPs.
2. The SO has low height, occupies less than 50% of the area of a DIP, and has a mass about one-tenth that of a DIP.
3. Like the DIP, it can be soldered to a motherboard and be pretested prior to assembly.

The SO family includes packages for passive devices, packages that contain ICs (SOIC packages), and packages that contain transistors (SOTs). Both plastic and ceramic SO packages are available. Figure 7.4 shows an SO package.

7.3.3 Leadless chip carrier (LLCC) and leaded chip carrier (LCC)

Leaded chip carriers are surface mount packages with gull-wing or J leads. J leads can be spaced more closely than gull-wing leads,[6] since they do not extend beyond the package. J-lead LCCs can also be inserted into specially designed sockets for easier removal and replacement of a defective device.

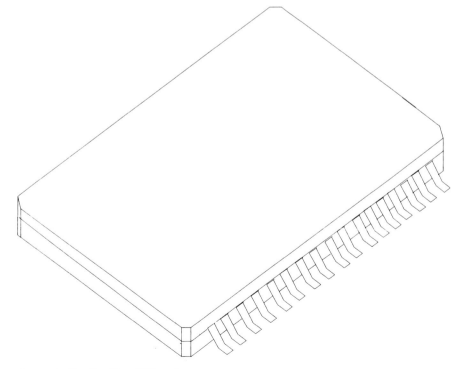

Figure 7.4 Small outline (SO) package.

Sockets exist for the standard sizes and make contacts to the leads without the necessity for solder.

Leadless chip carriers have bonding pads on the bottom of the substrate used to solder the LLCC to the substrate or printed circuit board (PCB). A common type of LLCC is the leadless ceramic chip carrier (LCCC). Ceramic chip carriers are of two basic types: single layer and multilayer. The term *multilayer chip carrier* can be somewhat confusing in that it has been used to describe chip carriers with several layers of alumina designed to mount a single device, and also chip carriers with several layers of metallization designed to mount and interconnect a number of devices. For this discussion, a multilayer chip carrier will be considered to be designed to package only a single chip. Figures 7.5 and 7.6 illustrate LLCC and LCCC packages, respectively.

Most of the multilayer chip carriers have three layers of ceramic material. The chip is mounted on the bottom layer, with wirebond pads on the second layer and the seal ring for the lid on the top layer. The wire bonding pads are routed to the outside between layers of ceramic and are connected to semicircular contacts called *castellations*. By far, the most common material is alumina, which is metallized with a refractory metal during fabrication and then successively plated with nickel and gold. Most multilayer chip carriers are designed to be sealed with solder, usually 80% gold (Au)–20% tin (Sn).

Configurations of chip carriers for military applications have been standardized by JEDEC in terms of size, lead count, lead spacing, and lead orientation, although nonstandard carriers can be used for specialized applications. The most common lead spacing is 0.050 in (1.3 mm), with high lead count packages having a spacing as low as 0.030 in (0.76 mm).

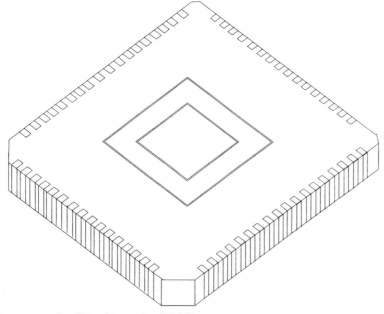

Figure 7.5   Leadless chip carrier (LLCC).

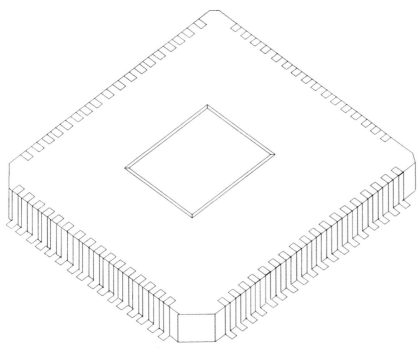

Figure 7.6 Leaded chip carrier (LCC).

The removal of heat from chip carriers in the standard cavity-up configuration has been a problem, since the only path for heat flow is along the bottom of the carrier out to the edge of the carrier, where it flows down to the substrate through the solder joints. This problem can be alleviated to a certain extent by printing pads on the bottom of the carrier, which are soldered directly to the substrate. This greatly lowers the thermal impedance. If this does not prove adequate, carriers with the cavity pointing down can be utilized. In this configuration, the chip is mounted on the top of the carrier in the upside-down position, the lid is mounted in a recess on the bottom of the carrier, and a heat sink is mounted to the top to enable the heat to be removed by convection. Heat transfer can be improved by manufacturing chip carriers from beryllia or aluminum nitride.

Devices mounted in chip carriers can be thoroughly tested and burned in before mounting on a substrate or printed circuit board. This process can be highly automated and is frequently used in military applications. Sockets exist for the standard sizes, which make contacts to the castellations without the necessity for solder.

Single-layer chip carriers (SLAMs) may either be purchased or fabricated by the user. SLAMs are typically manufactured by drilling holes in a substrate with a laser to form a number of SLAM patterns and scribing lines for separation at a later time. Contact is made through the holes by printing platinum–gold (Pt–Au) or palladium–silver (Pd–Ag) thick film paste over the holes, drawing it through with a vacuum, and firing it in the conventional manner. This step must generally be performed both on the top and on the bottom of the substrate. A mounting pad and wire bonding pads, which extend out to the holes, are printed on the top of the substrate and fired, which completes the process for SLAMs designed to be sealed with glass. For SLAMs designed to be sealed with solder, a ring of dielectric followed by a ring of Pt–Au or Pd–Ag must be printed and fired. In this manner, a number of SLAMs may be fabricated simultaneously.

In most cases, it is most economical to complete the assembly process while the substrate is in the array form, breaking it apart after the lids are attached. SLAMs may be printed in the compact form, where the contacts of adjacent devices touch each other, or in the expanded form, where the contacts are separated. The expanded form occupies more space on the substrate, but may be tested in the array form before breaking apart. The lids are dome-shaped and may be attached with solder or glass.

Both multilayer chip carriers and SLAMs are used in military applications, with the multilayer carriers predominating. The prime reasons for this are the high temperature required for glass seals and the difficulty in maintaining hermeticity of a glass seal during temperature cycling.

The use of chip carriers has proven to be a viable approach for packaging semiconductor devices in a minimum size compared to that of a ceramic DIP package. Although available with pin counts up to 128, chip carriers have proved to be a reliability risk when the pin count is greater than 84, since the net expansion of the carrier at temperature extremes, and, therefore, the stress on the solder joints, is proportional to carrier size. Furthermore, the temperature at which chip carriers may be used on PCBs is limited due to the difference in temperature coefficient of expansion (TCE) between the carrier and the board material.

As the solder joint is made higher, the difference in TCE becomes less significant. The highest solder column that can be made by ordinary means is about 0.007 in (0.18 mm). Above this height, a molten solder column begins collapsing due to its own weight. Various methods have been introduced to increase the height of a solder column, including the use of an organic preform containing solder in holes spaced to conform to the outline of a chip carrier. After soldering, the organic material is simply washed away, leaving the solder behind.

Power cycling, in which the device in the carrier is powered on and off at periodic intervals, has proven to be more of a serious reliability risk than temperature cycling when power devices are mounted. While the device in the carrier is being power-cycled, the carrier and the board are in a nonequilibrium state with respect to temperature. This causes considerable stress on the solder joints, ultimately resulting in failure due to metal fatigue.

7.3.4 Pin grid array (PGA)

For cases where chip carriers provide an inadequate pin count, pin grid arrays can be used. Pin grid arrays are through-hole packages which have rows of pins on the bottom, designed to plug into a printed circuit board. Since the pins may be placed inside the perimeter of the package, the lead density is substantially greater than with chip carriers, ranging up to several hundred pins.

PGAs have been designed with high and low temperature cofired ceramic tape systems as well as polyimides. Polyimide PGAs offer improved high frequency response due to polyimide's lower dielectric constant. These packages have been utilized at frequencies up to several gigahertz with excellent results. However, final selection of material for PGA fabrication may depend on the mechanical strength, the thermal conductivity, and the hermeticity property. A typical PGA package is depicted in Fig. 7.7.

7.3.5 Flat pack

Flat packs are used for higher I/O count than DIP packages can provide. Flat packs having leads on all four sides are generally known as *quad flat packs* (QFPs). QFPs have been fabricated in plastic, ceramic, and ceramic laminate technologies. Flat packs are generally surface mount packages, where the chip is mounted on the surface of the substrate. Maximum flat pack lead count is up to 208 for some plastic QFPs (PQFPs). Figure 7.8 illustrates a QFP.

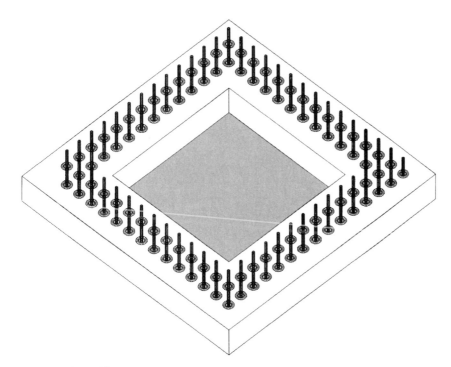

Figure 7.7 Pin grid array.

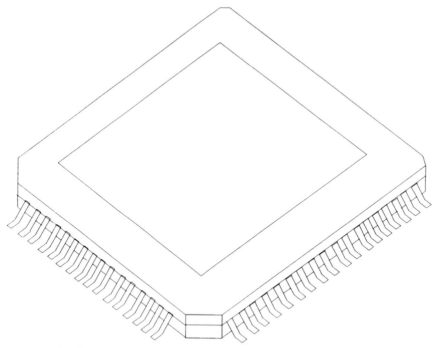

Figure 7.8   Quad flat pack (QFP).

7.3.6   Metal packages

Metal packages are fabricated primarily from American Society for Testing and Materials (ASTM) F-15 alloy, which consists of 52% iron (Fe)–29% nickel (Ni)–18% cobalt (Co) (known also as Kovar®). Kovar has many advantages in this application, including ease of welding and plating.[7] The tub-type package is fabricated by forming a sheet of the F-15 alloy over a set of successive dice. Holes for the leads are punched in the bottom for plug-in packages and in the side for flat packages. A layer of oxide is then grown over the package body. Beads of borosilicate glass, typically Corning 7052 glass, are placed over the leads and placed in the holes in the package body. A reactive glass to metal seal is formed by heating the structure above the melting point of the glass (approximately 500°C). The molten glass dissolves some of the oxides on the alloy (primarily iron oxide) which provide the adhesion mechanism on cooling. The glass to metal seal formed in this manner has four distinctive layers:

1. Metal
2. Metal oxide
3. Metal oxide dissolved in glass
4. Glass

After the glass to metal seals have been formed, the oxide not covered by the glass must be removed and the metal surface plated to allow the package to be sealed and to allow the package leads to be soldered to the next higher assembly. The prime plating material is electrolytic nickel, although gold is frequently plated over nickel to aid in sealing and to prevent corrosion. In either case, the package leads are plated with gold to allow wirebonding and to improve solderability. Although electroless nickel has better solderability, it tends to crack when the leads are flexed, and the latest revision of MIL-M-38510 requires that electrolytic nickel be used for improved lead integrity. The glass to metal seal formed in this manner provides an excellent hermetic seal and the close match in TCE between the glass and the F-15 alloy (approximately $5.0 \times 10^{-6}/°C$) maintains the hermeticity through temperature cycling and temperature storage.

7.4 Material Considerations for Electronic Packaging

The materials used for the fabrication of electronic packages can be classified as polymers and plastics, ceramics, and metals. In the following discussion the first two of these classes will be dealt with.

7.4.1 Polymers and plastics

Polymer and plastic packages are the most common types of electronic packages.[8] Plastic packaging material can be further classified as thermosetting or thermoplastic.

7.4.1.1 Thermosetting plastics. Thermosetting plastics are not fully polymerized in resin form. When these plastics are poured in their final forms, they are heated to a specific temperature to cause curing or polymerization. A copolymer material is often added to initiate the polymerization process. Copolymers remain in the plastic when it has cured to its final form. On the other hand, catalyst materials may be added to the polymer to initiate the polymerization process. The curing of thermosetting plastics is an exothermic process and the heat generated from the chemical reaction can sometimes cause cracking of the curing plastic, if heat dissipation techniques are not considered. When curing these types of plastics, shrinkage between 1 and 10% occurs.[3]

Thermosetting plastics are alkyds, allyls, epoxies, phenolics, unsaturated polyesters, polyimides, polyurethanes, and silicones. Alkyds are available in powder form. These powders can then be injected, compressed, or transferred into the mold and cured. They are formulated from polyester resins, fillers, monomers and catalyst. The fillers can be mineral or glass fibers, or a combination of both.

Allyls can also be injected, compressed, or transferred into the mold and cured. Allyls are used because they provide good dimensional control, better electrical properties, and high thermal and moisture resistance. The fillers can be mineral, glass, or synthetic fibers, or a mixture of these materials.

Epoxies are materials containing epoxide rings. Epoxies are available as either solids or liquids. They can be cast, compressed, or transferred to form a package. They have the advantages of low shrinkage, high rigidity, and good chemical resistance. Epoxies can be made with various viscosities and often contain some type of filler.

Phenolics are low cost materials. They are formed by a reaction between phenol and formaldehyde. They have high mechanical strength and can withstand high temperatures. Phenolics can be injected, compressed, or transferred to form a package.

Unsaturated polyesters are formed by a reaction of an organic acid, alcohol, vinyl monomer, and organic peroxide. They are low in cost, have low chemical resistance, and good electrical properties. Unsaturated polyesters of interest are made from polyester resin, fillers, and glass fiber to form laminates. The package can then be formed by either molding, transfer, or lamination.

Polyimides are available in powder or solution form. For use in electronic packages, polyimides are available as coatable materials or laminates. They are mostly used for multilayer circuits. They possess good heat and radiation resistance and are used in many space applications. They are, however, sensitive to moisture, organic acids, and alkalis. When used in conjunction with other materials, such as ceramics, the high melting point can create a large thermal stress at the interface.[10]

Polyurethanes are formed by reaction between a glycol and a diisocyanate and are sometimes used as conforming materials. These materials have high strength and, when polyether-based polymers are used, provide better moisture resistance.

Silicones are high temperature resistance materials produced as liquids and resins. The silicone molding compound is formed by combining silicone with some mineral filler, glass fiber, and one or more resins.

A potential advantage of plastic packages is their low cost, and some thermoset materials are emerging for microwave frequency applications.[9]

7.4.1.2 Thermoplastic plastics.

Thermoplastic plastics are fully polymerized in resin form. They can be melted and solidified repeatedly to form a package. However, thermal aging of the thermoplastic material, as a result of repeated heating and cooling, degrades the plastic properties.

Acrylic is a type of thermoplastic. It is made by polymerizing methyl methacrylate. It has high resistance to humidity and possesses good optical qualities.

Fluoropolymers are formed by polymerization of polytetrafluoroethylene. They possess superior properties, including low dielectric loss, high dielectric resistance, moisture resistance, and inertness to chemicals. They are produced as tapes, moldings, coatings, release films, and micro fibers.

Liquid crystal polymers retain their crystalline structure even in solid form. They can be molded into any shape. They possess low thermal expansion, high radiation resistance, resistance to chemicals, and high strength.

Nylons are polyamides with good mechanical strength. They are very sensitive to moisture; moisture absorption affects the electrical properties of the material. Nylons are available as reinforced polymers, where the reinforcement is provided through the addition of a glass film.

7.4.2 Ceramics

Ceramic packages are usually high performance electronic packages. Most of the ceramic materials used in packaging technologies can easily fulfill the packaging requirements of today's devices and hybrid circuits. They provide high dielectric strength, moderate to high thermal conductivity, thermal stability at the processing temperatures, ruggedness and high strength, and thermal expansion coefficients that are more closely matched to that of semiconductor devices. Alumina (Al_2O_3) is the most widely used ceramic with aluminum nitride (AlN) and beryllia (BeO) also used especially in high power applications.

7.4.2.1 Alumina. Alumina substrates in electronic packaging can be found in three different configurations:

1. High-temperature-fired alumina substrate

2. High-temperature-cofired ceramic (HTCC) alumina

3. Low-temperature-cofired ceramic (LTCC) alumina

The high-temperature-fired alumina, also called prefired alumina, substrates are available in Al_2O_3 concentrations ranging from 90 to 99.8 wt %, with 96% being the most commonly used composition. The remainder is glass frit binder. Alumina substrates are formed by either compression molding or tape casting.

In compression molding, the alumina slurry is poured into a mold and compressed under a uniaxial pressure of 8000 to 20,000 psi (55,000 to 138,000 kPa). The molded substrate is then taken out of the mold and preheated to remove the solvents, surfactants, and some binders. Preheat is usually done between 300 to 600°C. The substrate is sintered (fired) at a higher temperature, about 1700°C, and shrinks to its final size. Sometimes, an isostatic press may be used to compress the alumina slurry in the mold. Isostatic pressing has the advantage of delivering a uniform pressure over the surface of the substrate, thus reducing any chance of nonhomogeneity due to pressure variations.

In tape casting, the alumina slurry is poured onto a clear cellulose acetate material, such as Mylar™. To make the final cast tape flexible, a plasticizer is often added to the slurry. The Mylar is transported through the oven by a conveyer belt. The thickness of the poured slurry is determined by the height of a doctor blade. The cast tape is dried to make it easier to handle. The tape is then blanked to produce substrates of the desired shape. If the thickness of the cast tape is sufficient, monolayers may be fired; otherwise a number of layers are laminated together and cofired to obtain a fired alumina substrate. When cofired, the cast tape is also referred to as HTCC alumina tape.[11-14] Specially formulated pastes are used with the HTCC tapes so as to match the shrinkage of the substrate. Some variations of the process provide controlled shrinkage of the substrate when the substrate is fired under compression.

LTCC tape casting is identical to the HTCC tape casting process. The type of glass and material used in LTCC alumina tapes determine the firing temperature and profile. LTCC tapes are preheated at 350 to 400°C and fired at 850°C (peak firing temperature).

7.4.2.2 Beryllia. Beryllia use is primarily in high power circuits, where its high thermal conductivity offers a significant advantage. Beryllia in powder form is extremely toxic. However, when it has been fired to form a substrate, the chances of coming in contact with the BeO powder are very much reduced if necessary precautions are taken during processing. High thermal conductivity is achieved only when highly pure BeO is used. The extra precautions in handling, the need for high purity, and higher processing temperature make beryllia substrates more expensive.

7.4.2.3 Aluminum nitride. Aluminum nitride is becoming more widely used in packaging applications as metallization methods are improved. Because of its higher thermal conductivity at elevated temperature, it is replacing BeO in some applications. The thermal conductivity of AlN can be tailored to be in the range of 100 to 260 W/(m•K) by varying the ratio of AlN and the impurities, especially oxygen.[15] The presence of oxygen during substrate formation, even in quantities as small as 0.1%, can cause a considerable reduction in the thermal conductivity. By comparison, alumina has a thermal conductivity of 22 W/(m•K) and beryllia has a thermal conductivity of 260 W/(m•K) at room temperature. AlN substrates have a better thermal expansion match with silicon than either alumina or beryllia. Kovar® is also well-matched to AlN, and Kovar lids may be hermetically sealed to the AlN packages.

AlN substrates can be metallized with thick film pastes if preoxidation of the surface is performed. Preoxidation results in a layer of Al_2O_3 on the surface of the substrate, providing a means of adhesion for the thick film materials.

7.4.3 Metals

Most high frequency packages are made from metal, which is usually gold-plated to facilitate sealing and to minimize corrosion.[7] Kovar and aluminum are the two most widely used metals, with copper and molybdenum occasionally used in power applications. Metal matrix composite materials, such as aluminum silicon carbide (AlSiC) are also used to make packages for microwave applications.[16] Kovar[17] is usually preferred, however, because the coaxial feedthroughs form a hermetic seal when fired at 1000°C in conjunction with the metal. At this temperature, the borosilicate glass reacts with the metal oxide to form the seal. In aluminum packages, the feedthroughs must be soldered to the plating surface to achieve the desired hermeticity.

Aluminum has an advantage in that it has a high thermal conductivity, 167 W/(m•K) compared to 17 W/(m•K) for Kovar, and it is approximately one-tenth the cost. The major advantage of Kovar is its low TCE, approximately 20% that of aluminum, which is a much better match to silicon and ceramic components.

7.5. Electronic Package Fabrication

The fabrication processes for the various types of packages depend on the technology used. This section provides the fabrication processes for thick film low-temperature-cofired ceramic tape, thin film, and plastic packages.

7.5.1 Thick film LTCC fabrication

The low-temperature-cofired ceramic tape comes in the form of a rolled sheet of ceramic tape. The tape is usually cast on Mylar™, coated with a silicone release agent. Some tapes are fabricated with a glass frit as the binder. Due to the higher glass and solvent content of the tape, these tapes shrink more during the firing process. To compensate for the variability in shrinkage in the x and y directions, also referred to as the *machine* and *transverse* directions, successive layers of the tape are rotated 90° to each other. The amount of shrinkage depends on the thickness of the sample as well as the amount of conductive paste used. This shrinkage can be estimated quite precisely and can be accounted for in the design of the circuit or structure layout. A typical LTCC fabrication sequence is outlined below.[18]

1. The fabrication process starts with computer-aided design (CAD) of the interconnects, interconnect pads, power and ground lines and planes, and sometimes sealing metallization.

2. The required number of layers is blanked out from the ceramic tape roll. The blanking die can cut the tape into the specified dimensions. During blanking, the registration holes can also be punched into the blanked tape. These registration holes are very important as they help in alignment as well as in making sure that successive layers are rotated 90°.

3. To connect two different tape layers, vias are achieved in the different tape layers and filled with conducting pastes. Vias can be punched or drilled by a programmable machine tool or cut with a laser.

4. In order to fill vias, each layer is placed on a substrate holder which keeps the layer in place with the help of vacuum action. The vias are then filled with the appropriate paste through either a metal mask or a screen. The masks possess the same via pattern as the tape. Paste is forced through the holes in the mask or screen and pulled through the via in the LTCC tape layer by vacuum. After via filling, the layers with the via pastes are dried in an oven for a period of 5 min at 120°C.

5. The remainder of the metallization layers, containing the interconnects, pads, and power and ground planes, are now screen-printed and dried.

6. After the printing process is completed, the ceramic layers are cut to the desired shape and size by either a laser or a cutting blade.

7. All the layers comprising the package are now stacked together in the laminating die, with each successive layer rotated 90° with respect to its neighbors. Lamination may be done in either a uniaxial press or an isostatic press to support structures with cavities present. The lamination is usually performed under a temperature of 70°C and a pressure of 3000 psi (21,000 kPa).

8. The package is placed on a very flat and clean quartz substrate (a *setter*) and placed in the preheat furnace for burnout or ashing. The ramp-up rate is set to 12°C/min, while the ramp-down rate goes from 3°C/min to 1°C/min. The maximum temperature is set to 375°C and the holding time at this temperature is set to 60 min.

9. After the LTCC package is taken out of the preheat furnace, it is placed in the belt furnace for firing. A firing temperature, usually 850°C, and a belt speed, usually 2 in/min, corresponding to the requirements of the material are used. During the firing process, the package shrinks to its final dimensions.

7.5.2 Thin film fabrication

Thin film packages are generally formed by evaporation or sputtering. The substrates can be either a ceramic, a polyimide, or a semiconductor. The choice of substrate is sometimes dictated by thermal, electrical, or mechanical considerations. Typically, thin film packages consist of a combination of conducting films, interfacial adhesive films, resistive films, and dielectric films. Copper, gold, and aluminum are the typical conductors. The interfacial adhesive layer may be titanium, nickel, or chromium. Resistor films are usually nichrome, tantalum nitride, or tantalum.

In evaporation, a metal is heated in a vacuum chamber until it changes to a gaseous state. This metal gas then is deposited on the substrate, which is kept at a relatively cool temperature. The mechanism for production of the metal atoms is a thermal mechanism, and can deposit metal at rates as high as 750,000 Å/min. The deposition is dependent on the arrival rate and angle of the metal atoms. The heating mechanism for the metal can be resistive heating or electron beam heating. In resistive heating, a large current is passed through a metal container, usually made from tungsten, which contains the evaporation metal. This type of evaporation mechanism can sometimes cause contamination problems due to the evaporation of the container metal itself. Resistive wires coated with metals can also be used to evaporate metal. In electron beam heating, a focused, high current electron beam is directed toward the metal in the container. The high current causes the metal to evaporate and deposit on the substrate. The metal holder in this case is usually a water-cooled copper container. This method of heating greatly reduces the contamination of the metal film by container atoms. For the electron beam to be accurately controlled, no impurity may be present in its path. For this reason, the vacuum chamber is usually evacuated to a very low pressure to eliminate all those substrate materials which might outgas during the metal deposition process. To achieve a uniform coverage, sometimes multiple metal sources are used. Multiple sources may also be used to deposit a metal alloy. The evaporation rate of the different metal sources can be controlled to tailor the metal alloy to the desired combination.

High throughputs may be achieved by using a load-locked metallizing system. In this type of system, the substrates are placed in a substrate carrier, and the carrier is placed on a conveyer belt. The overall process is subdivided into a preheat chamber, a deposition chamber, and an unload chamber. Each chamber is isolated and has its own conveyer belt. The substrates are moved in batches from one chamber to another. The overall speed of the metallization is then a function of how many substrates can be metallized in one batch.

In sputtering deposition, the deposition mechanism is momentum transfer from an inert gas to the metal target. The sputtering chamber is usually filled with the inert gas and a high DC potential applied between the two electrodes inside the chamber. This DC potential breaks down the inert gas. The regions inside the chamber glow as a result of the creation of plasma, sometimes referred to as *glow discharge*. Charged particles are nonuniformly distributed inside the chamber, resulting in a nonuniform potential between the two electrodes. When an argon atom is attracted toward the negatively charged cathode, it accelerates and gains energy. As this atom strikes the cathode, it imparts most of its energy to the atoms of the material constituting the cathode. If enough energy is imparted to cathode atoms, they are ejected from the cathode. These atoms then deposit on the substrates as a thin film. The deposition rates for sputtering systems are significantly lower (about 24,000 Å/min) than for evaporation systems. Secondary emission of electrons from the target further ionizes the argon atoms and can result in a self-sustaining glow discharge.

Another variation of the sputtering system uses a magnetic field to divert the electrons and concentrate them in front of the target. This type of system is known as a *magnetron sputtering system*.

7.5.3 Plastic package fabrication

There are many different ways of fabricating plastic packages. The fabrication process is usually controlled by the type of plastic used. Thermosetting plastics take their final shape when the material solidifies after a chemical reaction. On the other hand, thermoplastic plastics solidify with a drop of temperature. The fabrication techniques for these plastic packages can be categorized as follows:

7.5.3.1 Compression molding. Compression molding is used with thermosetting plastics. The thermoset plastic material may be in free-flowing granular form or a viscous material. The material is placed in a hot mold, and the mold is closed by a hydraulic press. A typical compression molding process is performed at a mold temperature of 350°F and mold pressure of 100 psi (180°C and 700 kPa) with curing time of 3 min. After the material is cured, the mold is opened and the plastic package is pushed out. Compression molding is a low cost process and is capable of high volume production.

7.5.3.2 Transfer molding. Transfer molding is also used for thermosetting materials. Transfer molding is very similar to compression molding. In transfer molding, the thermoset material is placed in a hot crucible, and a ram pushes the material into the mold through channels known as *runners*. Transfer molding provides better dimensional control, and inserts in the package can be prepositioned in the mold prior to the ramming of the plastic material. Transfer molding equipment is more expensive than compression molding equipment.

7.5.3.3 Injection molding. Injection molding is used for thermoplastic materials. In injection molding, the thermoplastic material is initially in the form of small pellets. These pellets are forced through a screw, which pushes the material forward through some heated zones. During this transport, the heat from the zones and the pressure applied by the screw melt the plastic and compress it to about 20,000 psi (140,000 kPa). Once the proper conditions for the melt have been reached, the screw is pushed forward as a ram to force the molten plastic through the runner tubes into the molds. The mold is then immediately cooled to freeze the plastic. Injection molding processes are very low cost processes due to the simple nature of the equipment and the reusability of the residual thermoplastic material. The molds used in the injection molding process are very expensive and in some equipment may increase the cost.

An injection molding process, with certain modifications, may also be used for some thermosetting plastics. The screw shape may be changed to handle the thermosetting material and heating elements may be added to the mold.

7.5.3.4 Thermoforming. Thermoforming is also used for thermoplastic materials. In thermoforming, the thermoplastic material is heated enough to make it soft. A sheet of thermoplastic material is held above the mold. A heater above the sheet provides the heat to soften the thermoplastic sheet. The sheet is then forced onto the mold. A vacuum on the underside of the mold helps ensure that the thermoplastic sheet conforms to the mold. The heater is then removed and the material is left in the mold to harden. After the mold cools down, the molded part is removed. Scrap material left over from the molding process can be used again. Thermoforming is very inexpensive; however, it does not provide good dimensional control over the molded part and the process does not have high throughput.

7.5.3.5 Laminating. Lamination is a special type of compression molding. Lamination is usually used with thermosetting materials; however, some types of thermoplastic materials may also be laminated. Lamination presses a number of thin sheets together to form a part. The sheet materials are fiber-resin and polyimide. They are either cast with proper solvents or later permeated with solvents. The solvents are then dried off in an oven, and the sheets are stacked together in a mold and pressed in a hydraulic press. The laminates are then cured under pressure and temperature, with the usual laminating conditions of 500 psi (3500 kPa) at 350°C for 1 h. The laminating process itself is inexpensive and good dimensional control can be obtained, but the laminating equipment may be expensive.

7.5.3.6 Reaction injection molding. Reaction injection molding is similar to injection molding. It is used for thermosetting materials. The different constituents are mixed together and the mixture is forced into a heated mold. The mold may contain fiber reinforcements as required by the design. The part is cured in the mold under compression. Reaction injection molding is a low cost process capable of fabricating large parts. However, poor dimensional control, poor physical properties, and limited choice of available materials restrict its use.

7.6   Package Sealing

Packages can also be divided into hermetic and nonhermetic categories. A true hermetic seal would prevent intrusion of contaminants (liquid, solid, or gas) for an indefinite period of time. In practice, however, this is not realistic. Even in a perfectly sealed structure, diffusion phenomena will occur over time, allowing the smaller molecules, such as helium or water vapor, to penetrate the barrier medium and ultimately reach equilibrium within the package interior.

A true hermetic package must be made of metal, ceramic, or glass. Organic packages, or packages with an organic seal, may initially pass the leak rate test described above, but will eventually allow water vapor to pass in and out and are not truly hermetic. Interconnections through a metal package may be insulated by glass to metal seals utilizing glass that matches the thermal coefficient of expansion to the metal.

A hermetic package seals the circuit mounted inside in a safe environment, generally nitrogen, which is obtained from a liquid nitrogen source. Nitrogen of this type is extremely dry, with a moisture content of less than 10 ppm. As a further precaution, the open package with the enclosed circuit mounted inside is subjected to an elevated temperature, usually 150°C, in a vacuum, to remove absorbed and adsorbed water vapor and other gases prior to sealing. For added reliability, the moisture content inside a package should not exceed 5000 ppm. This figure is below the dew point at 0°C, ensuring that any water that precipitates out will be in the form of ice, which is not as damaging as water in the liquid form.

A hermetic package adds considerably to the reliability of a circuit by guarding against contamination, particularly of the active devices. An active device is susceptible to a number of possible failure mechanisms, such as corrosion and inversion, and may be attacked by something as clean as distilled, deionized water, which can leach phosphorus out of the passivating oxide to form phosphoric acid, which in turn can attack the aluminum bonding pads.

7.6.1   Methods of sealing metal packages

There are three types of lids commonly used on metal packages: domed, flat, and stepped. These are fabricated with ASTM F-15 alloy with the same plating requirements as the packages. The domed lid is designed for use with platform packages and may be projection-welded or soldered. The flat lid is designed for use with the tub package and is usually soldered to the package.

The stepped lid is fabricated by photoetching a groove in a solid sheet of F-15 alloy, resulting in a flange about 0.004 in (0.1 mm) thick. This lid is designed to be seam-welded to a tub package. When lids are intended for soldering, a preform of the desired solder material is generally attached to the outer perimeter of the bottom of the lid.

A flat lid or a stepped lid may be soldered to the package by hand, by a heated platen, or in a furnace. While the platen is somewhat faster, the metal package acts as a heat sink, simultaneously drawing heat away from the seal area and raising the temperature inside the package, unless the glass beads used for insulating the leads extend entirely around the periphery of the package, a clearly impractical requirement. In addition, leaks through the solder, or *blow holes,* caused by a difference in pressure between the inside and outside of the package, will result unless the ambient pressure outside the package is increased at the same rate as the pressure inside created by heating the package. Because of the temperature rise inside the package, it is risky to use epoxy to mount components, unless the glass beads extend around the periphery of the package as described earlier. Solder sealing may be accomplished in a belt furnace which has a nitrogen atmosphere. The nitrogen prevents the oxidation of the solder and also provides a benign environment for the enclosed circuit. Furnace sealing usually requires a certain degree of fixturing to provide sufficient pressure on the lid to eliminate leaks without squeezing the solder from under the lid when it becomes molten.

Parallel seam welding is accomplished through creation of a series of overlapping spot welds by passing a pair of electrodes along the edge of the lid. Alignment of the lid to the package is critical, and can be a lengthy and tedious job, since the package must be handled through bulky gloves inside the sealing chamber. As a result, the alignment process is frequently accomplished outside the sealing chamber, with the lid being tacked to the package in two places by small spot welds. A stepped lid greatly facilitates the process and improves the yield, since it requires considerably less power than a flat lid of greater thickness. While the sealing process is comparatively slow compared to one-shot welding, a package sealed by parallel seam welding can be easily delidded by grinding the edge of the lid away. Since the lid is only about 0.004 in (0.1 mm) thick in the seal area, this may be readily accomplished in a single pass of a grinding wheel. With minimal polishing of the seal area of the package, another lid may be reliably attached.

Certain classes of packages with a flange may be sealed by a process called *projection,* or *one-shot,* welding. In this process, an electrode is placed around the flange on the package and a large current pulse is passed through the lid and the package, creating a welded seam. It should be noted that heavy duty resistance welding equipment capable of supplying 500 psi (3500 kPa) pressure and 12,000 A per linear inch of weld is required for these packages. The major advantages of one-shot welding are shorter process time and a less expensive package. The major disadvantage is the difficulty of removing the lid to repair the circuit inside, which usually results in the destruction of the package, which must then be replaced.

7.6.2 Methods of sealing ceramic packages

Ceramic packages in this context are considered to be structures that permit a thick or thin film substrate inside in much the same manner as a metal package. Ceramic structures that have metallization patterns which allow direct mounting of components will be referred to as *multichip ceramic packages*. Ceramic packages for hybrid circuits generally consist of three layers of alumina. The bottom layer may be metallized, and this usually depends on how the substrate is to be mounted. A ring of alumina is attached to the bottom layer with glass and a lead frame is sandwiched between this ring and a top ring with a second glass seal. The top ring may be metallized to allow a solder seal of the lid or may be left bare to permit a glass seal.

The most common method of sealing ceramic packages is solder sealing. During the manufacturing process, a coating of a refractory metal or combination of metals, such as tungsten or an alloy of molybdenum and manganese, is fired onto the ceramic surface around the periphery of the seal area. Upon completion, the surface area is successively nickel-plated and gold-plated. A lid made from ASTM F-15 alloy is plated in the same manner and soldered onto the package, with an alloy of 80% Au–20% Sn, and usually in a furnace with a nitrogen atmosphere.

A less expensive, but also less reliable, method of sealing is to use a glass with a low melting point to seal a ceramic lid directly to a ceramic package. This avoids the use of gold altogether, lowering the material cost considerably. The glass requires a temperature of about 400°C for sealing, as opposed to about 300°C for the Au–Sn solder. The glass seal is somewhat susceptible to mechanical and thermal stress, particularly at the interface between the glass and the package.

These two techniques have a common problem; it is difficult to remove the lid for repair without rendering the package useless for further sealing. An alternative approach seeing increased use is to braze a ring of ASTM F-15 alloy, which has been nickel- and gold-plated as described above, onto the sealing surface of the ceramic package. It is then possible to use parallel seam welding with its inherent advantages for repair. This approach is also frequently used for ceramic multilayer packages designed for multichip packaging.

7.6.3 Nonhermetic packaging approaches

The term *nonhermetic package* encompasses a number of configurations and materials, all of which ultimately allow the penetration of moisture and/or other contaminants to the circuit elements. Most techniques involve encapsulation with one or more polymer materials, with the most common being the molding and fluidized-bed approaches.

Both injection and transfer molding techniques utilize thermoplastic polymers, such as acrylics or styrenes, to coat the circuit. In transfer molding, the material is heated and transferred under pressure into a closed mold in which the circuit has been placed. On the other hand, the material is heated in a reservoir and forced into the mold by piston action in injection molding.

The fluidized-bed technique uses an epoxy powder kept in a constant state of agitation by a stream of air. The circuit to be coated is heated to a temperature above that of the melting point of the epoxy and is placed in the epoxy powder. The epoxy melts and clings to the circuit, with the thickness controlled by the time and the preheat temperature.

Both methods are used to encapsulate hybrids and individual devices, and are amenable to mass production techniques. The overall process may be performed at a cost of only a few cents per circuit. The coatings are quite rugged mechanically. These coatings are also are resistant to many chemicals, and have a smooth, hard surface suitable for marking.

7.7   Evaluation of Electronic Packages

Electronic packages are frequently measured and tested to evaluate their electrical, thermal, and mechanical performance. To fully characterize the package, it is necessary to evaluate both the construction and the materials. The evaluations may be electrical, mechanical, or thermal in nature.

7.7.1   Electrical evaluation

Electrical evaluation can be performed to evaluate the electronic packages as well as to characterize the materials used in electronic packaging applications.[19–28] The materials of interest in this respect include dielectric systems as well as conductor systems. For dielectric systems, the electrical performance parameter of interest is the complex relative permittivity (complex dielectric constant). The electrical performance parameter of interest for conductor systems is the electrical conductivity.

7.7.1.1   Electrical evaluation of electronic packages.   Electronic package evaluation can be performed to obtain the DC, time domain, and frequency domain characteristics.[29–31]

DC evaluation of electronic packages.   The DC evaluation of electronic packages consists of measuring the DC ohmic loss of the conductors and measurements for open and short circuits. The DC resistance of the transmission lines must be small compared to the load resistance. In cases where the DC resistance is high, the signal may be attenuated at the output pin. A voltage source, an ammeter, and an ohmmeter are required to conduct these tests.

Time domain evaluation of electronic packages.   The time domain analysis setup consists of a time domain network analyzer. The test package is first installed in a test fixture, which may be a metal bracket with two SMA connectors at two ends. The SMA connectors are then connected to the signal strip of the package. The TDR pulse is sent in from one end of the signal strip while the other end is either open or terminated in a 50-Ω resistance load. Portions of this pulse will be reflected from any discontinuities that the pulse encounters as it propagates along the length of the signal strip.

The TDR waveform can show the discontinuities, and by comparing the waveform with the package dimensions, the higher reflections can be traced to the regions causing the discontinuities.

Frequency domain evaluation of electronic packages. The frequency domain analysis setup consists of a network analyzer with data acquisition capabilities. The network analyzer is first calibrated with a standard calibration kit. The package to be evaluated is placed in a test fixture and connected to the coaxial launchers of the network analyzer through coaxial connectors. The network analyzer may be used to measure the insertion loss (S_{12} or S_{21}) or the reflection loss (S_{11} or S_{22}) of the signal strip of the package. Cross-talk analysis may also be performed by launching a signal in the driver line and measuring the leakage signal in the driven line. For cross-talk analysis, the network analyzer is configured to measure S_{12} or S_{21}.

Similar to time domain measurements, frequency domain measurements are very sensitive to the measurement setup. The test fixture should be designed so that there is no electrical discontinuity in the path of the signal strip that can cause bandwidth limitations. The test fixture connectors must have a bandwidth larger than the expected bandwidth of the package. Special care must be taken to ensure that the ground and signal connections are tight enough for repeatable measurements.

7.7.1.2 Electrical evaluation of packaging materials. The measurement and characterization of dielectric and conductor materials can be performed over a wide range of frequencies. The ranges may include very low frequency (VLF), less than 1 kHz; low frequency (LF), 1 kHz to 10 MHz; radio frequency (RF), 1 MHz to 100 MHz; and wideband, 50 MHz to 26 GHz.

In the VLF frequency range, only the conductivities of the conductor and dielectric materials are of interest. Conductivity of conductors may be measured using the Van der Pauw and the four-point probe methods. The conductivity of dielectrics may be measured using the three-contacts method.

In both the LF and RF bands, samples of the conductor and dielectric materials formed into resistors or capacitors can be characterized by using LF and RF vector impedance meters. The measured complex impedances may then be translated into an electrical model for the material, describing its conductivity and relative permittivity as functions of frequency over these lower frequency ranges.

The wideband characterization utilizes both transient (time domain) and spectral (frequency domain) scattering parameter measurement techniques. Both measurement techniques complement each other and can be used jointly in developing the full material specifications. Both measurement techniques require a sample of the material under test to be inserted in a transmission line system in which the transient or spectral scattering S parameters are to be measured.

In time domain measurements, the structure is excited by a pulse. The response waveforms and excitation pulse are acquired with a wideband sampling oscilloscope and analyzed to determine the desired information regarding the properties of the material under test. In frequency domain measurements, the structure is excited by sinusoid swept in frequency through the frequency band of interest. The setup uses a vector network analyzer and gives direct readings of the scattering parameters vs. frequency. Again, the results have to be analyzed and related to the material properties of interest.

In order to make the material characterization meaningful, the material properties of the samples to be used in the measurements have to be as nearly identical as possible to those of the constructed structures in the material application. Consequently, the samples for the characterization tasks should be prepared in the forms of planar and multilayer structures. This implies testing of transmission structures in the forms of microstrip lines, striplines, coplanar lines, slotted lines, coupled lines, and many forms of discontinuities.

DC measurements. DC measurements include measurement of the breakdown voltage of the dielectric as well as the DC resistance of the conductors.

Breakdown voltage. An important characteristic of substrate materials is the breakdown voltage. This value provides an upper limit for the application of the material in question. Breakdown voltage can be measured using a high voltage supply capable of generating potentials up to 20 kV in compliance with ASTM standard D 3755-86. A thin slice of the sample material is mounted between two electrodes. The sample thickness is determined with a precision micrometer. The voltage is then linearly increased until breakdown of the sample occurs. This point is determined by monitoring the current through the sample. The breakdown voltage is then expressed in terms of voltage per unit thickness of the sample. Careful control over the sample temperature, sample thickness, and relative humidity of the sample environment must be maintained.

RF measurements. RF measurements of the sample can be in either the time domain or the frequency domain.

Time domain measurements of packaging materials. Time domain techniques can be used for the determination of the complex permittivity of a dielectric material in a planar line geometry. In one of the techniques, the planar line under test is modeled as a frequency-dependent lossy transmission line, based on time domain reflectometry and time domain transmission (TDT) data. The model is developed in the frequency domain to account for the frequency-dependent material parameters, conductor skin loss, and line discontinuities. The model is then used to simulate the TDR and TDT measurements. The model parameters are iteratively adjusted until the simulated TDR/TDT waveforms match those measured waveforms. The optimum model parameters are then used to compute the material's complex permittivity or dielectric constant value.

This method has been successfully used to estimate the dielectric constant, but not the loss tangent, of low loss dielectrics such as Duroid's 5870 and 5880 and DuPont's Pyralux™.

Frequency domain measurements of packaging materials. This approach uses the same principles as that of the TDR/TDT techniques above. The prime difference is that measurements are performed in the frequency domain. The vector network analyzer is used along with a planar calibration technique [such as with the HP TRL (Hewlett-Packard Thru-Reflect Line)]. To minimize errors due to transition regions around the launch ends of the lines, several lines of different lengths are required with identical launching regions. These lines are measured and the data are processed to reveal the required information about the substrate's dielectric properties.

7.7.1.3 Very low frequency electrical characterization

Conductivity of good conductors. The conductivity for good conductors, such as interconnection films, can be measured by two techniques: van der Pauw and four-point probe measurements. Both techniques rely on measuring a voltage drop across some portion of the sample induced by a constant current through the sample. The van der Pauw method has the advantage of being valid for arbitrary sample geometry whereas the four-point probe technique requires a disk shape or rectangular geometry.

For the van der Pauw measurement, contacts are placed on the edge of the sample. The precise location of the contacts depends on the geometry of the sample. However, the contacts are typically located at the corners or edge of the sample. Two measurements are required for the determination of the resistivity. Four contacts at each corner are labeled a through d. The measurements are then made as

$$R_{ab,cd} = V_{ab} \Big|_{I_{dc} = 1} \tag{7.4}$$

$$R_{bc,da} = V_{bc} \Big|_{I_{ad} = 1} \tag{7.5}$$

The resistivity is then given by

$$\rho = \left(\frac{\pi d}{\ln 2} \right) \left(\frac{R_{ab,cd} + R_{bc,da}}{2} \right) F\left(\frac{R_{ab,cd}}{R_{bc,da}} \right) \tag{7.6}$$

where F is a correction factor numerically determined from the expression

$$\frac{(R_{ab,cd} - R_{bc,da})}{(R_{ab,cd} + R_{bc,da})} = F \operatorname{arccosh} \left(0.5 \exp\left[\frac{\ln 2}{F} \right] \right) \tag{7.7}$$

The four-point probe measurement utilizes a linear array of four equally spaced contacts. The two outer contacts are driven with a constant current and the two inner contacts are connected to a voltmeter. The sheet resistance of the sample can then be determined by the relation

$$R_s = \text{CF}\left(\frac{V}{I}\right) \tag{7.8}$$

where CF is a geometric correction factor determined by the probe spacing and the sample thickness.

The resistivity of the sample is determined from the four-point probe measurement by the relationship

$$\rho_s = R_s t \tag{7.9}$$

where ρ_s = resistivity, $\Omega \cdot$cm
t = sample thickness, cm
R_s = sheet resistance, Ω/sq

This method can be applied to various electronic materials, from insulators to superconductors, with a high degree of accuracy. The contacts for this technique are typically four needle-sharp probes, which approximate point sources. A plot of resistivity vs. temperature yields great insight into the conduction mechanism involved in a material.

Conductivity of dielectrics. The conductivity of insulating materials, such as substrate or dielectric materials, may be measured using a three-contact method. Contacts can be formed on the sample by vacuum evaporation. The sample geometry for these measurements is shown below in Fig. 7.9.

The volume resistance is measured by connecting a voltage source and a current meter to the inner top contact and the bottom contact, with the outer top contact used as a guard. The surface resistance is determined by measuring the resistance from the inner top contact to the outer top contact. From these measurements, the volume resistivity can be found by

$$\rho_V = \left(\frac{A}{t}\right) R_V \tag{7.10}$$

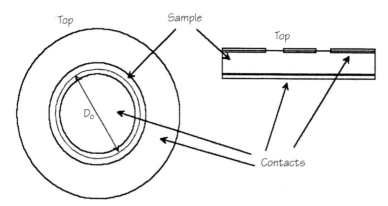

Figure 7.9 Sample geometry for conductivity measurement.

where R_V = volume resistance
  A = effective area of the inner top contact
  t = sample thickness

The surface resistance R_s is given by

$$R_s = \rho_s\left(\frac{g}{\pi D_0}\right) \tag{7.11}$$

where g = gap between the two top inner and outer electrodes
  R_s = surface resistance
  D_0 = diameter indicated

7.7.1.4 Low and radio frequency electrical characterization

Low and radio frequency conductor characterization. In these ranges, conductor material characterization may be performed through the measurement of the impedance of a test sample where the conductor has a rectangular form on the substrate. The ends of the sample may be designed to obtain a more uniform field distribution within the sample conductor. In order to compensate for the end effects, geometrical effects, and the substrate's contribution to the measured impedance, in particular in the RF range, two samples may be tested, one short and the other much longer. The measured electrical parameters of the short sample can be recovered from those of the long sample, thus yielding the compensated properties of the uniform middle portion of the long sample.

Low and radio frequency dielectric characterization. To fully characterize a dielectric material with respect to its electrical specifications, it is essential to measure its complex permittivity over a wide band of frequencies. The complex permittivity is a combination of the material's dielectric constant and its loss factor (conductivity). Knowing the complex permittivity as a function of frequency is sufficient to derive all the other electrical properties of the bulk material, such as the quality factor, as well as to predict electrical performance of fabricated structures. Knowing the complex permittivity and structure geometries, it becomes possible to estimate characteristic impedances, propagation constants, and cross-talk parameters. It also becomes possible to estimate the response of the structure to a known excitation. In the LF and RF bands, the dielectric characterization can be performed through the measurement of the impedance of a capacitor formed by the dielectric layered between two conductive deposits. It is fairly safe to assume that the conductor properties do not affect the dielectric measurement accuracy in this frequency range.

Capacitance measurement methods may be used for dielectric characterization over the low and radio frequency range. Depending on the dielectric material under consideration, capacitors of different sizes and geometries can be considered. Large capacitors resonate (due to the traveling wave nature of the structure at these frequencies) at the upper frequency limit, giving erroneous results, but they provide low fringing errors and thus improved results at lower frequencies. Smaller capacitors give better frequency characteristics but have a tendency to provide a higher than actual dielectric constant value due to fringing errors. Numerical methods, using Laplace's equation, are available to estimate and correct for the fringing component. This has resulted in the ability to estimate the error introduced by the fringing fields in capacitor structures. Calibration methods have also been developed and are used to improve the measurements on the vector impedance meter utilized for these studies. Another concern in the capacitance measurement is the determination of the thickness of the dielectric material under test. The thickness of the materials will be tested using a variety of means such as calipers, profilometer, and a scanning electron microscope (SEM).

Wideband electrical characterization. In order to measure the dielectric constant and loss tangent of the materials over the wideband frequency range, samples in both cylindrical and planar configurations are tested.

The cylindrical configuration offers the convenience of using the coaxial environment available with network analyzers and measurement equipment. The coaxial lines are convenient to calibrate using commercially available National Institute of Science and Technology (NIST) traceable standards. The drawback in using cylindrical configurations is that the field distribution, and hence polarization, in the tested samples is not typical of the material in planar packaging applications. Planar configurations yield material characterization results under field excitation and polarization similar to those experienced in planar applications and hence are more desirable. The complexity encountered in measurements of these configurations is related to the use of coaxial to planar adapters and to good calibration in the planar environment.

Cylindrical configuration—wideband dielectric-filled cavity. The wideband dielectric-filled cavity (WDFC) sample structure was designed in 1986 as an alternative to conventional methods that were known at that time.[32] Although the WDFC can be used as a resonant cavity, the idea adopted was to use it as a wideband sample holder. The scattering parameters measured either in the time domain or in the frequency domain were related to the geometrical and electrical parameters of the dielectric filling. A full field analysis was developed so as to evaluate the dielectric properties based on knowledge of the sample geometry and the S parameters of the filled cavity. The method has been used successfully to measure the dielectric constant of several ceramic and polymer substrate materials.

Transmission and reflection versions of the WDFC have been developed to provide estimates of the dielectric properties based on either S_{11}, or both S_{11} and S_{21}. The transmission cavity is called the *through* WDFC. Software packages have been developed for both the reflection and transmission WDFC configurations; they use numerical modeling methods to simulate the WDFC performance. Iterative methods to solve the inverse problem, utilizing numerical models, allow the determination of the dielectric properties. A point to note is that the errors can be serious without accurate determination of the sample thickness.

The separation of dielectric loss from the measured loss is achieved by a perturbation approach.[50] First, by assuming ideal conductor cavity walls, the surface current on these walls is estimated. The loss of the conductor walls is then evaluated from approximate knowledge of the conductor properties. Then the conductor loss is subtracted from the measured cavity loss to estimate the dielectric loss.

Planar configurations. Planar configurations include stripline, microstrip, coplanar, and other planar line formats. To characterize material in these configurations, measurements yield effective values for the material properties (e.g., effective dielectric constant) for all configurations except for the ideal stripline configuration. *Effective values* result from the fact that portions of the field excitation are distributed in the surrounding air as well as in the material under test; thus measured quantities average the material properties with those of the surrounding air. Effective values vary from one configuration to another and are highly dependent on geometrical parameters as well as the measurement frequency. A typical approach to this dilemma is to use empirical formulas that relate effective values to material properties, taking into consideration the geometry and the frequency.

To date, available empirical formulas for relating effective values to material properties are not highly reliable. Consequently, material samples in all planar configurations of interest should be constructed and measured. Databases for effective values of the different configurations as functions of geometrical parameters and frequency can then be established and maintained.

Planar resonator. The planar resonator method is capable of yielding accurate estimates for the dielectric constant as well as separate estimates for both the dielectric and conductor loss terms. Several resonator configurations are usable in this regard and typically yield similar accuracies. The common configurations are the linear, ring, and tee resonators. These configurations can be used with all forms of planar line structures.

The linear resonator structure is formed by providing two gaps in the central strip of the planar line. The section of the line between the gaps forms the resonator. The ring resonator is formed by replacing the central strip with a ring. In the tee resonator configuration, the resonator part is coupled at one end, via a small gap, to the measurement line while its other end remains open. The main line in this case is continuous, with no gaps.

The gap dimension is designed to provide a balance between two requirements:

1. To be large enough to minimize the source and termination, loading the Q of the resonator

2. To be small enough to allow for adequate signal to couple into the resonator (and to the termination on the other end in the linear and ring shapes)

Moreover, a small gap can minimize the uncertainty in determining the resonator effective length. It is relevant to state that several samples with either different strip widths or differing dielectric thicknesses are required to calculate the loss parameters.

7.7.2 Mechanical evaluation

Mechanical evaluation of the electronic package also may include evaluation of the mechanical properties of the packaging material. The topics of interest under mechanical evaluation are considered below.

7.7.2.1 Homogeneity and chemical composition.
Homogeneity of a material sample can be used as an indicator of the overall quality of the material. Homogeneity is a critical parameter for electronic materials, since strict quality control is an essential requirement. There are three principle techniques that can be employed to determine the homogeneity of a sample: scanning electron microscopy, energy-dispersive analysis of x-rays (EDAX), and auger electron spectroscopy (AES).

SEM allows magnification of very small features, less than a micrometer in size, on the sample's surface. This technique is very valuable for structural evaluation since it can be used to determine the grain size and structure of polycrystalline materials. The sample size is limited to approximately 1 in (2.5 cm) square. Additionally, materials with low electrical conductivity must be prepared in a special manner. The need for this step arises from the fact that the electron beam in an SEM causes the surface of an insulating sample to build up a charge; as a result, a crisp image is unattainable. Typically a thin layer of gold, of the order of 10 Å, is deposited on the surface of the sample. The thin gold layer provides an invisible conductive layer which carries away the charge.

Two variations of the SEM analysis are typically used, conventional and environmental SEM. Environmental SEM allows for heating of the sample, which would destroy the vacuum in a conventional system. Heating offers the advantage of allowing samples to be recrystallized or annealed while being viewed under the microscope.

Transmission electron microscopy (TEM) is another technique which affords even greater magnification than SEM. This equipment is capable of resolving details on the order of 100 Å. Sample size is very limited, since the samples must be thinner than 0.6 mm. TEM is most useful in situations where the structure of interest is beyond the range of SEM techniques.

EDAX is a technique closely linked to SEM, since most EDAX units are contained in an SEM. EDAX provides the chemical composition of any point on the surface of the sample. By performing EDAX measurement over the sample's surface in a systematic way, the chemical homogeneity of the surface can be determined. Sample preparation for this technique is similar to that required for SEM except for the gold deposition. An additional technique involves fracturing the sample and then viewing the cross section of the fractured segment with SEM/EDAX. This technique is especially valuable for film samples, since the exact film thickness can be determined, even for very thin films. Information concerning the diffusion or migration of layers into adjacent layers can also be obtained.

AES is a more precise way of measuring the chemical composition of a material. A scanning Auger microprobe can be used to precisely determine the chemical composition of the sample's surface. This technique also allows for depth profiling. Through a small hole milled in the sample, the Auger microprobe can gain access to the inner layers of material. A profile of the chemical composition of the sample, as a function of depth, can be generated. Combining this information with an Auger surface scan or an EDAX sampling of the surface provides a complete three-dimensional picture of the sample's chemical composition.

Samples used in AES must be handled very carefully, since any contact with finger oils or dirt can produce inaccurate results. Additionally, AES spectra often display a prominent carbon peak for scans performed on the surfaces. This is often caused by airborne carbon dioxide. In cases where very high precision measurements are required, much of this surface contamination can be removed by desiccating the sample under vacuum.

7.7.2.2 Porosity and density. The density, apparent porosity, and apparent specific gravity of a sample can all be measured by one experiment. The sample under study must first be dried; typically baking the sample at 110°C for a few hours is adequate. The sample is then carefully weighed to great precision. This represents the dry weight D. The sample is suspended in boiling water for 2 h and then left hanging in water for 12 h. A balance measures the sample's weight W while it is submersed in water. The sample is then removed and once again weighed; the saturated weight S is recorded. The apparent porosity P in percent, is then given by

$$P = \left(\frac{S - D}{S - W} \right) \times 100 \tag{7.12}$$

The bulk density D_b of the material in g/cm^3 is given by

$$D_b = \frac{D}{S - W} \tag{7.13}$$

Additionally, the apparent specific gravity G of the material can be determined by the relation

$$G = \frac{D}{D - W} \tag{7.14}$$

The key step in this experiment is the very precise measurement of the masses involved.

7.7.2.3 Shrinkage. Material shrinkage is of critical importance, since materials with a high degree of shrinkage can cause processing difficulties. The properties of interest are not only the magnitude of the shrinkage but also the variability of the shrinkage along the material axes.

In order to determine this information, all dimensions of the unfired material must be carefully measured with a precisely calibrated caliper. The sample in question is then fired. After firing, the dimensions are again measured, exactly as before. The difference between these values, expressed as a percentage, is the magnitude of the shrinkage. Uniformity can be determined by observing whether the sample shrinks by the same magnitude in all directions.

7.7.2.4 Solderability. The solderability of a material is a function of the adhesion of solder to the material as well as the degree to which the solder wets the surface of the material.

Adhesion is measured with a standard pull tester in accordance with ASTM standard F 692-80. The pull testing apparatus is composed of an outer housing and an inner rod. The bottom of the rod is coated with an adhesive. The apparatus is placed over the sample and the rod is lowered so that its end adheres to the sample. The adhesion of the solder is then determined by the amount of the solder film that is removed from the sample when the rod is returned to its original position.

The degree to which a solder wets a sample is determined by heating the sample to a temperature slightly higher than the solder's melting point. A small ball or bead of solder is then dropped onto the sample's surface. The temperature is maintained for 30 s and the sample is then cooled to room temperature. A microscope is used to inspect the bead to determine its diameter. Based on the size of the bead relative to its initial size, a solderability grade is assigned in accordance with ASTM standard F 357.

7.7.2.5 Stress and strain testing. For design and life prediction purposes, it is necessary to have the mechanical, viscoelastic, and fracture properties at the disposal of the designers. These properties are vital in determining the dimensions and geometries of the devices, their dimensional stability, and their durability.

The two types of testing used to determine the mechanical properties of materials are static testing and dynamic testing. Uniaxial, or tension/compression, tests are used to determine the static behavior of the material. Properties found by static testing are Young's modulus of elasticity, yield point, modulus of resilience, ultimate stress, and modulus of toughness. Dynamic testing determines a material's reactions under vibrating or oscillating forces. This is also known as *fatigue testing* and is performed by subjecting the specimen to cycles of loading and unloading. The number of cycles it takes the specimen to rupture determines its failure point.

Much of the mechanical property information can be obtained from static testing. The bulk specimen is subjected to a combination of normal and shear forces. The stress is obtained by finding the force exerted per unit area and the strain is obtained by finding the change in length of the specimen divided by the original length. The information acquired from this test is used to plot a stress-strain curve. This curve is then used to obtain the Young's modulus of elasticity, yield point, and modulus of resilience. The modulus of elasticity is defined as

$$E = \frac{\Delta \varepsilon}{\Delta \sigma} \qquad (7.15)$$

where $\Delta \varepsilon$ = change in strain and $\Delta \sigma$ = change in stress obtained anywhere along the stress-strain curve.

The yield point is that point on the stress strain curve where the relationship between stress and strain ceases to be linear. The area under this portion of the graph defines the modulus of resilience. The yield point defines where the material will begin to deform and the modulus of resilience defines how much energy the material can stand before it yields. The ultimate stress, or strength, of the material, σ_f, is the peak value on the stress-strain curve. The ductility is ε_f, the ultimate strain on the stress-strain curve. The modulus of toughness is defined as

$$U_f = \int_0^{\sigma_f} \sigma d\varepsilon \qquad (7.16)$$

which is the area under the entire stress-strain curve. The modulus of toughness is a measure of the energy the material can absorb before it reaches its fracture point (ruptures).

In certain applications, deflections in the material are undesirable. This would correspond to a low modulus of elasticity. Deflections in microelectronic circuit boards, for instance, could result in broken wire or conductor lines, which could cause the circuit to fail. Therefore, a high modulus of elasticity, high strength, and high ductility are preferred for most applications. Materials with high strength will be more able to resist load. High ductility will allow the material to bend under load. Unlike ductile materials, brittle materials fracture quickly under load.

Brittle materials, such as ceramics, follow Hooke's law,

$$E = \frac{\varepsilon}{\sigma} \qquad (7.17)$$

which makes the Young's modulus a linear relationship between stress and strain. There is no longer a yield point, only a fracture point, indicating that the material, due to its brittle nature, does not have plastic deformation.

A problem with many materials, especially ceramics, is that inherent flaws add to the brittleness. These flaws can be modeled as cracks in the material. To determine how materials with flaws will perform, the fracture toughness of the material can be studied. The fracture toughness K_{IC} is a material property that describes resistance to failure under static load for materials containing such cracks. The presence of cracks may result in lower strength due to fracture. The engineer must therefore consider the possibility of failure by fracture under load.

Fracture toughness testing involves a specimen of known geometry with a precrack of a given length. A specimen geometry is chosen whose characteristics are known either analytically or through numerical analysis. The fracture toughness can be calculated from test results as follows:

$$K_{IC} = F_a \frac{P_c}{tw^{0.5}}$$ (7.18)

where P_c = critical load when crack starts to propagate
  w = width of the specimen
  t = thickness of the specimen
  F_a = factor which depends on the ratio of crack length to the width of the specimen

Because temperature can also affect the material's brittleness, the specimens must be tested in a temperature-controlled environment. Most electronic materials will operate at two temperatures: ambient temperature, when the electronic system is powered down, and an operating temperature, when the electronic system is energized. Therefore, packaging materials should be tested at these two temperatures.

7.7.2.6 Platability. Plating is used with packaging materials for a variety of reasons: for conductors and ground planes with ceramic materials, for corrosion control and solderability with metallic material. Therefore platability can be an important consideration in the selection of packaging material.

ASTM standard B 571-91, "Standard Test Methods for Adhesion of Metallic Coatings," may be used to test electroless and electrolytic deposited copper, nickel, and gold. The burnishing test, the heat-quench test, and the peel test are particularly applicable to packaging materials. While quite subjective, this standard provides an excellent go, no go test to determine if a further quantitative test should be performed.

If the plating process passes the above tests, quantitative results may be obtained by soldering a small copper slug to the sample using controlled amounts of solder and flux. The sample is allowed to sit for a period of time to allow for stress relaxation of the solder. The sample is then clamped to a mechanical tester (such as the Instron 1331) and ramped in tension at a constant rate until either the solder, plating, or substrate fails. This yields bond strength data in the units of force per unit area which could prove useful to designers in evaluating plating systems or substrate materials. Consequently, this test also serves as an initial test for solderability of a plating material, since the failure mode of the test is recorded. This may be either failure of the solder, failure of the solder interface (and hence a measure of solderability), or failure of the plating.

In the event that the plating is not solderable, adhesives (such as epoxies) may be evaluated to substitute for solder in the above test. This type of test can work equally well for solder or adhesives.

7.7.2.7 Machinability. As outlined in the ASTM standard rating, the machinability of a metal is a relatively simple test and is performed on ordinary production equipment, such as a screw machine. Quite simply, a standard part design is used and the screw machine is set up with a standardized tool geometry to produce this part. The production rate is then increased until the dimensions of the finished parts begin to drop out of a predetermined tolerance range. This rate is then compared to the maximum rate with a reference material, typically type B1112 low carbon steel. The machinability is expressed as a percentage. For example, the machinability rating of A-2 tool steel (annealed) is 36%. Simply put, if the screw machine can produce 100 parts per hour with B1112 steel, it would have to be slowed down to 36 parts per hour with A-2 tool steel stock, a tougher and more difficult material to machine. Machinability ratings can also be greater than 100%, indicating a more freely machining material.

In the case of ceramics, no such standard rating system exists. However, a somewhat similar rating system may be used that includes the method of machining most appropriate to the material, such as conventional carbide or diamond tooling, or ultrasonic or laser machining. The machinability percentage is calculated by comparison to a common reference material such as 96% alumina.

This rating may be determined in a similar fashion as for metals, but using operations more consistent with the machining operations found in package production. A good indication of machinability is to determine the maximum feed rate possible while cutting a fixed-size bar of the material without unwanted side effects such as edge chipping and overcutting. Cutting conditions can be controlled as well, such as maintaining the same width of cut (all tests using a 0.5-mm kerf, for example). This maximum feed rate can be compared to the same operations performed on a commonly used reference material, such as 96% alumina, yielding a percentage as for metals. This information then provides the production engineer with an estimate of how difficult or easy it is to machine the material. It also provides an estimate of the final production rates to expect in manufacturing a product with this material.

This method may also be applied to polymers by using a slightly modified form of the ASTM standard where tool geometry would necessarily have to be different. Comparison could be made as before, with a commonly used material such as Delrin (acetal copolymer).

7.7.3 Thermal evaluation

7.7.3.1 Temperature coefficient of expansion. The temperature coefficient of expansion can be determined by direct measurement of the sample dimensions at room temperature and at elevated temperature. Two methods can be used: the direct measurement in accordance with ASTM standard C 372-88 and interferometric measurement in accordance with ASTM standard E 289-70.

Direct measurement. Initially, the sample is placed on a highly thermally conductive plate at room temperature. While the sample is in contact with the plate, the exact temperature and dimensions of the sample are measured. This initial temperature is defined as T_I whereas the initial length is defined as L_I. The plate and sample are then heated to an elevated temperature, determined by the type of material in the sample. Once this elevated temperature is reached, the whole assembly remains heated for several minutes. The sample temperature is then measured at a number of locations to ensure complete thermal equilibrium. Once complete equilibrium has been verified, the sample temperature and dimensions are again carefully measured in exactly the same manner as before. This second temperature is defined as T_F and the length of the sample at this temperature is L_F. From these measurements, the temperature coefficient of expansion can be determined by the relation

$$\text{TCE} = \frac{L_F}{L_I} \frac{1}{T_F - T_I} \tag{7.19}$$

Interferometric measurement. Samples are mounted in a vacuum chamber which is capable of cooling or heating the sample to the desired temperature. Temperature measurements are performed with a calibrated thermocouple or resistance temperature detector (RTD). A distance measuring interferometer is used to determine the linear thickness of the sample as a function of temperature. An interferometer with a resolution better than 1 μm is used. The instantaneous coefficient of linear expansion α_I can be determined from the relation

$$\alpha_I = \frac{\Delta L}{\Delta T} \frac{1}{L_0} \tag{7.20}$$

The mean coefficient of linear expansion α_M for a temperature range of T_1 to T_2 is given by

$$\alpha_M = \frac{\Delta L}{L_0 (T_2 - T_1)} \tag{7.21}$$

The instantaneous and mean values of the coefficient of expansion are determined over the temperature range of interest.

7.7.3.2 Specific heat. Differential calorimetry in accordance with ASTM standard E 1269-90 can be used to determine the specific heat of materials. This system has the advantages of very high accuracy and applicability to almost any material. Heat flow calibration is performed using a synthetic sapphire standard in accordance with ASTM standard E 968-83 before each measurement of the sample's specific heat.

A thermal curve is recorded for an empty sample holder, the sapphire standard, and the sample in question. The specific heat capacity of the sample $C_p(s)$ can then be determined from the relation

$$C_p(s) = C_p(st) \frac{D_s W_{st}}{D_{st} W_s} \tag{7.22}$$

where $C_p(st)$ = specific heat capacity of the sapphire standard
$\qquad D_s$ = difference in the response from the empty specimen holder and the sample
$\qquad D_{st}$ = difference in the response from the empty specimen holder and the sapphire standard
$\qquad W_s$ = mass of the sample
$\qquad W_{st}$ = mass of the sapphire standard

For these measurements the same sapphire sample holder is used for all three thermal runs, and is usually of purity greater than 99.9%.

7.7.3.3 Thermal conductivity.
Thermal conductivity can be measured by two methods: heat flow and thermal flash. The heat flow method has the advantage of providing a more direct measurement without the need for correction factors. The flash method provides a direct measurement of the thermal diffusivity, which allows for indirect determination of the thermal conductivity.

Heat flow technique. Thermal conductivity can be measured with a heat flow technique in accordance with ASTM standard C 408-88. This method has the advantages of being a direct measurement and suitable to almost any solid material. The apparatus for this measurement consists of a vacuum chamber which contains a long copper rod. The rod is split into two sections, and the sample is mounted between the two halves. One end of the rod is heated to a constant temperature, while the other end is cooled to a constant temperature. Four precisely calibrated thermocouples spaced along the length of the rod are used to determine the heat flow. From these temperature readings, the heat gradient and the thermal conductivity of the sample can be determined.

The cooling and heating sections of the unit are isolated from the center region, and the whole chamber is pumped down to a vacuum greater than 10^{-6} torr. This system is equipped with a rough vacuum pump and a turbomolecular vacuum pump.

Flash method. A thermal pulse method may be used to measure the thermal diffusivity of the sample material in accordance with ASTM standard E 1461-92. By combining the thermal diffusivity values with the density and specific heat capacity data, the thermal conductivity can be calculated as

$$\lambda = \alpha C_p \rho \tag{7.23}$$

where λ = thermal conductivity
      α = thermal diffusivity
     C_p = specific heat capacity
     ρ = density

For these measurements the samples must be thin to minimize heat loss corrections. Choice of the exact thickness of the sample is based on the estimated thermal diffusivity of the sample material. Typical samples are small disks 20 mm in diameter and 1 to 4 mm thick.

The samples are placed in vacuum which includes a flash thermal source and a thermal detector. The thermal source delivers a pulse of energy to the front face of the sample, while the detector measures the temperature of the back face of the sample. As the energy flows through the sample, the back surface temperature increases. A plot of time vs. temperature increase is obtained from the experimental apparatus. From this plot the time $t_{1/2}$ in seconds required for the back sample face to reach one-half its maximum temperature is determined. The thermal diffusivity, in m²/s, is then given by

$$\alpha = \frac{0.13879L^2}{t_{1/2}} \tag{7.24}$$

where L = sample thickness, mm. By heating or cooling the sample, these measurements can be performed over a wide range of temperature values.

Thermal evaluations of via structures have also been performed.[33] Via structures allow the designer to increase the thermal conductivity of the substrate by suitable placement of these vias.

7.8   Applications

Electronic packaging has applications ranging from passive components to optical devices[34–37] as well as microwave circuits.[38,39] In this chapter, the authors have elected to describe one such example, the multichip module (MCM). MCMs are emerging as the leading packaging technology for high speed, high density silicon CMOS circuits in computers.[40–43]

In order to improve circuit performance, reliability, and speed and to reduce circuit size and weight, the electronics industry has always upheld the idea of increasing circuit density. One way of achieving this is to integrate a maximum number of circuit elements on a single large semiconductor chip; however, chip cost increases greatly with surface area. It was demonstrated[44] that the normalized cost of all interconnects is equal for equal lengths. This finding gave birth to the idea of connecting multiple unpackaged dice, i.e., chips without individual packages, to each other as closely as possible in a single package, since the effect and cost would be almost the same as increasing the scale of integration on the individual die. This reduction in chip-to-chip interconnection lengths, by placing many unpackaged chips in a single package, not only is less expensive compared to using DIPs on a printed wiring board but has many performance advantages as well.

Better system performance, i.e., higher operating frequencies, are achieved as a result of shorter interconnections. Performance of high speed emitter-coupled logic (ECL) silicon and gallium arsenide chips having 2 to 10 GHz off-chip signal bandwidths can be improved by current multichip module technology. In addition, the multichip modules provide other classical benefits, such as mixing of several technologies. Of course, there is more risk in regard to packaging issues such as chip testing, replacement, and substrate rewiring.

A typical multichip module consists of a set of bare chips connected to the surface of a dielectric material. There are multiple conductor and power planes, necessitated by the higher number of interconnections, separated by layers of dielectric material. This multilayer structure is supported on a ceramic or metal substrate and enclosed in a hermetically sealed ceramic package. The input and output pins can be taken out of the package in many ways, depending on the type of application and the number of pins.

It seems simple enough to increase the interconnection density by placing more chips together and reducing the width and spacing of the interconnect conductors. However, this would result in increased cross-talk, unless the thickness of the dielectric material separating these interconnects from the ground plane is increased,[45] and/or a low dielectric constant material is used. Using low dielectric constant material between different interlayers of a multichip module ensures higher circuit density, which in turn affects the operating frequency, the cost, and the circuit size. Several low-loss dielectric materials possessing low dielectric constants have been produced, and among them polymers are considered the star materials for interconnect dielectrics. Conductor systems compatible with each of them have also been identified and are being used. Substrates of interest are the ones which can dissipate a large amount of power, support the structure mechanically without yield to stress or strain, and match the thermal coefficient of expansion of the semiconductor chips. Different processes and technologies have been developed to integrate these different elements.

LTCC tape systems have been used to generate high performance circuits.[46,47] Integrating passive components such as resistors into the package can increase the design flexibility and circuit density. LTCCs offer higher density, higher speed, and higher power density for microelectronics applications.[48,49] This technology has been extensively used by IBM, as well as a number of other large corporations like Hughes, TRW, and Westinghouse, for their high speed, high performance products. A typical MCM layout is shown in Fig. 7.10.

7.9 Conclusion

Electronic packaging of hybrid circuits has covered a significant distance in the last few years. Package interconnects are becoming denser while the lead pitch has dropped at the same time. The packaging community has done a tremendous job of opening new venues and exploring novel application, i.e., multichip module and optical packaging. For the electronic packages to keep pace with the ever shrinking monolithic devices, not only will the fabrication

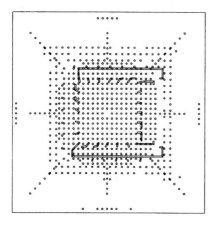
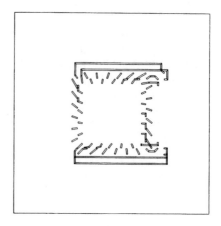

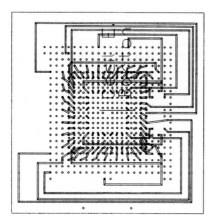
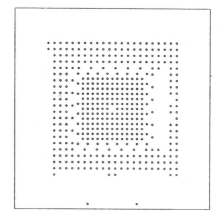

Figure 7.10 Multichip module (MCM) package.

process have to be improved, but new approaches will have to be developed and new areas will have to be explored.

References

1. N. Iwasaki, K. Katsura, and N. Kukutsu, "Wideband Package Using An Electromagnetic Absorber," *Electronics Letters,* vol. 29, no. 10, pp. 875–876, 1993.
2. Charles A. Harper, *Handbook of Thick Film Hybrid Microelectronics,* McGraw-Hill, New York, 1974.
3. Albert J. Blodgett, Jr., "Microelectronic Packaging," *Scientific American,* pp. 86–96, July 1983.
4. Yasuo Nakatsuka, T. Yamamoto, Y. Morita, M. Fujii, H. Ohtani, and E. Takahashi, "Fine Pitch and High Lead Count Multilayer Ceramic QFP," *IEEE Japan Intl. Electronics Manufacturing Technology Symp.,* pp. 176–180, 1993.

5. Herbert Reichl, "Packaging and Interconnection Techniques for Microsystems," *Proc. of IEEE Systems, Man, and Cybernetics Conf.,* pp. 423–428, 1993.
6. H. B. Bakoglu, "Circuits, Interconnections, and Packaging for VLSI," Addison-Wesley, Reading, Mass., 1990.
7. Eric F. Richardson, "Select Material to Balance Benefits for Packaged MICs," *Microwave & RF,* pp. 87–95, July 1989.
8. Philip Garrou, "Polymer Dielectrics for Multichip Module Packaging," *Proc. IEEE,* vol. 80, no. 12, pp. 1942–1953, 1992.
9. G. Robert Traut, "Thermoset Microwave Material Enhances Microwave Hybrids," *Applied Microwaves,* May 1989.
10. Hideo Arima, Hideo Sotokawa, Rohei Sato, Fusaji Shoji, Hidetaka Shigi, and Toshio Hatsuta, "Stress Analysis on Hybrid Multilayer Substrate Composed of Polyimide and Ceramics," *Proc. IEEE 1993 Japan Intl. Electronics Manufacturing Technology Symp.,* pp. 279–282, 1993.
11. John L. Sprague, "Multilayer Ceramic Packaging Alternatives," *IEEE Trans. CHMT,* vol. 13, no. 2, pp. 390–396, 1990.
12. Hidenobu Nishikawa, Manabu Tasaki, Seiichi Nakatani, Yasuhiko Hakotani, and Minehiro Itagaki, "Development of Zero X-Y Shrinkage Sintered Ceramic Substrate," *IEEE Japan Intl. Electronics Manufacturing Technology Symp.,* pp. 238–241, 1993.
13. Richard L. Wahlers, Sidney J. Stein, and Glenn P. Sykora, "Tapes and Thick Films for High Frequency Packaging," *Proc. of the 40th Electronic Components and Technology Conf.,* Las Vegas, Nevada, May 20–22, 1990, vol. 1, pp. 116–121.
14. Keiichiro Kata, Yuzo Shimada, and Hideo Takamizawa, "Low Dielectric Constant New Materials for Multilayer Ceramic Substrate," *IEEE Trans. CHMT,* vol. 13, no. 2, pp. 448–451, 1990.
15. Kevin J. Lodge, John A. Sparrow, Edward D. Perry, Elizabeth A. Logan, Martin T. Goosey, David J. Pedder, and Clive Montgomery, "Prototype Packages in Aluminum Nitride for High Performance Electronic Systems," *IEEE Trans. CHMT,* vol. 13, no. 4, pp. 633–638, 1990.
16. D. Wein, P. Anderson, J. Babiarz, J. Carter, N. Fulinara, and M. Goetz, "Microwave and Millimeter-Wave Packaging and Interconnection Methods for Single and Multiple Chip Modules," *IEEE GaAs IC Symp.,* pp. 333–336, 1993.
17. Margaret Hunt, "Electronics Packaging," *Material Eng.,* vol. 108, no. 2, pp. 25–28, February 1991.
18. Imran A. Bhutta, Tahir Qazi, Cecil B. Barton, Aicha Elshabini-Riad, Sedki M. Riad, "Electrical Characterization of a Multilayer Low Temperature Co-Fireable Ceramic Multichip Module," ASM International's 7th Electronic Materials and Processing Congress, Cambridge Mass., August 24–27, 1992.
19. Gerome Reeve, Roger Marks, and David Blackburn, "Microwave Monolithic Integrated Circuit-Related Metrology at the National Institute of Standards and Technology," *IEEE Trans. Instrum. Meas.,* vol. 39, no. 6, pp. 958–961, 1990.
20. Steven Lipa, Michael B. Steer, Arthur S. Morris, and Paul D. Franzon, "Comparison of Methods for Determining the Capacitance of Planar Transmission Lines with Application to Multichip Module Characterization," *IEEE Trans. CHMT,* vol. 16, no. 3, pp. 247–252, 1993.
21. Michael Gribbons, Andreas C. Cangellaris, and John L. Prince, "Finite-Difference Time-Domain Analysis of Pulse Propagation in Multichip Module Interconnects," *IEEE Trans. CHMT,* vol. 16, no. 5, pp. 490–497, 1993.
22. T. S. Horng, S. C. Wu, H. Y. Yang, and N. G. Alexopoulos, "A Generalized Method for the Distinction of Radiation and Surface-Wave Losses in Microstrip Discontinuities," *IEEE MTT-S Digest,* pp. 1055–1058, 1990.
23. V. I. Kazantsev and A. I. Kharitonov, "Choice of a Transmission Line for Operation in The Millimeter Band," *Izvestiya Vuz. Radioelektronika,* vol. 32, no. 10, pp. 67–72, 1989.
24. Hong You and Mani Soma, "Analysis and Simulation of Multiconductor Transmission Lines for High-Speed Interconnect and Package Design," *IEEE Trans. CHMT,* vol. 13, no. 4, pp. 839–846, 1990.
25. George E. Ponchak and Rainee N. Simons, "A New Rectangular Waveguide to Coplanar Waveguide Transition," *IEEE MTT-S Digest,* pp. 491–492, 1990.
26. Attilio J. Rainal, "Reflections from Bends in a Printed Conductor," *IEEE Trans. CHMT,* vol. 13, no. 2, pp. 407–413, 1990.
27. Harry K. Charles, Jr., and Guy V. Clatterbaugh, "Modeling for Electronic Packaging at APL," *Johns Hopkins APL Tech. Dig.,* vol. 11, nos. 1 and 2, pp. 137–144, 1990.
28. William P. Harokopus and Pisti B. Katehi, "Characterization of Microstrip Discontinuities on Multilayer Dielectric Substrates Including Radiation Losses," *IEEE Trans. Microwave Theory Tech.,* vol. 37, no. 12, pp. 2058–2066, 1989.

29. John R. Tyler and D. J. Gasper, "Evaluation and Characterization of Low-Temperature Cofired Ceramic Structures for Microwave Applications," *ISHM '89 Proc.,* pp. 390–399, 1989.

30. D. E. Carlton, K. Reed Gleason, Keith Jones, and Eric W. Strid, "Accurate Measurement of High-Speed Package and Interconnect Parasitics," *IEEE/CHMT '89 IEMT Symp.,* pp. 276–279, 1989.

31. Zhi-Yuan Shen, "New Time Domain Reflectometry Techniques Suitable for Testing Microwave and Millimeter Wave Circuits," *IEEE MTT-S Digest,* pp. 1045–1048, 1990.

32. Mohammad A. Saed, Sedki M. Riad, and Aicha Elshabini-Riad, "Wide-Band Measurement of the Complex Permittivity of Dielectric Materials Using a Wide-Band Cavity," *IEEE Trans. Instrumentation and Measurement,* vol. 38, no. 2, April 1989.

33. K. Azar, S. S. Pan, and R. E. Caron, "Effect of Via Density, Diameter and Arrangement on Thermal Performance of Single and Multichip Modules," *Proc. ASME Winter Annual Meeting,* pp. 1–15, 1993.

34. Eric M. Foster and Richard J. Wolff, "ESD Packaging Requirements for an Opto-Electronic Receiver Module," *IEEE Trans. CHMT,* vol. 13, no. 4, pp. 787–790, 1990.

35. D. Z. Tsang, D. L. Smythe, A. Chu, and J. J. Lambert, "A Technology for Optical Interconnections Based on Multichip Integration," *SPIE Integ. and Pack. Optoelectronic Devices,* vol. 703, pp. 122–127, 1986.

36. John Schlafer and Robert B. Lauer, "Microwave Packaging of Optoelectronic Components," *IEEE Trans. Microwave Theory and Techniques,* vol. 38, no. 5, pp. 518–523, 1990.

37. Scott Enochs, "A Packaging Technique to Achieve Stable-Mode Fiber to Laser Alignment," *SPIE Integ. and Pack. Optoelectronic Devices,* vol. 703, pp. 42–47, 1986.

38. Deborah S. Wein, "Advanced Ceramic Packaging for Microwave and Millimeter Wave Applications," *Proc. of the IEEE 1993 Intl. Symp. Dig. of Antennas and Propagation,* Ann Arbor, Mich., June 28–July 2, 1993, vol. 2, pp. 993–996.

39. Masao Ida and Toshio Nishikawa, "A Multilayered Package Technology for MMICs," *Proc. of the IEEE 1993 Intl. Symp. Dig. of Antennas and Propagation,* Ann Arbor, Mich., June 28–July 2, 1993, vol. 2, pp. 1013–1016.

40. Leonard W. Schaper, "Design of Multichip Modules," *Proc. of IEEE,* vol. 80, no. 12, pp. 1955–1964, 1992.

41. B. Gilbert, R. Thompson, G. Fokken, W. McNeff, J. Prentice, D. Rowlands, A. Staniszewski, W. Walters, S. Zahn, and G-W. Pan, "Advanced Multichip Module Packaging and Interconnect Issues for GaAs Signal Processors Operating Above 1 GHz Clock Rates," *Proc. SPIE Intl. Conf. Adv. Integ. and Pack.,* vol. 139, pp. 235–248, 1990.

42. Robert T. Crowley and E. Jan Vardaman, "Applications and Technology for Laminate Multichip Modules," *Proc. IEEE 1993 Japan Intl. Electronics Manufacturing Tech. Symp.,* pp. 151–154, 1993.

43. Rao R. Tummala, "Multichip Packaging—A Tutorial," *Proc. of IEEE,* vol. 80, no. 12, pp. 1924–1941, 1992.

44. W. H. Knausenberger and L. W. Schaper, "Interconnection Costs of Various Substrates—The Myth of Cheap Wire," *IEEE Trans. CHMT,* vol. 7, 261, 1984.

45. J. W. Balde, "Multichip Packaging and the Need for New Materials," *J. Electron. Mater.,* vol. 18, no. 2, pp. 221–227, February 1989.

46. A. L. Eustice et al., "Low Temperature Co-Fireable Ceramics—A New Approach for Electronic Packaging," *Proc. 36th Electron. Components Conf.,* pp. 37–47, 1986.

47. Imran A. Bhutta, Aicha Elshabini-Riad, and Sedki M. Riad, "Time Domain Analysis and Design of MIMIC Packages," *24th General Assembly of the International Union of Radio Science,* Kyoto, 1993.

48. Philip E. Rogren, "The Current State of Laminate Based, Molded MCM Technology," *Proc. IEEE/CHMT Intl. Electronics Manufacturing Technology Symp.,* pp. 485–489, 1993.

49. Joseph Reshey, "New Developments in Co-Fired Multilayer Ceramic for Custom Microelectronics," *Proc. IEEE Custom Integrated Circuits Conf.,* pp. 254–257, 1986.

50. M. Y. Andrawis, W. Davis, S. M. Davis, and A. Elshabini-Riad, "Dielectric Loss Determination Using Perturbation," *IEEE Trans. Instrumentation and Measurements,* vol. 42, no. 6, pp. 1032–1035, December 1992.

Discrete Passive Components for Hybrid Circuits

Jerry E. Sergent

8.1 Introduction

The hybrid technology has the capability of fabricating and integrating certain passive components into the manufacturing process, but there are practical constraints which require the use of discrete passive components to extend the value or range of a desired component or to provide a unique electrical performance function. The surface mount technology has also placed a greater demand on discrete component technology. The design of a hybrid or surface mount circuit requires an understanding of the properties of capacitors, resistors, and inductors.

8.2 Capacitors

Two conductors in proximity with a difference in potential have the ability to attract and store electric charge. This effect is enhanced by placing a material with *dielectric* properties between them. A dielectric material has the capability of forming electric *dipoles,* or displacements of electric charge, internally. At the surface of the dielectric, the dipoles attract more electric charge, thus enhancing the charge storage capability, or *capacitance,* of the system. The relative ability of a material to attract electric charge in this manner is called the *relative dielectric constant,* or *relative permittivity,* and is usually given the symbol K. The relative permittivity of free space is 1.0 by definition, and the absolute permittivity is

$$\varepsilon_0 = \text{permittivity of free space}$$

$$= \frac{1}{36\pi} \times 10^{-9} \quad \frac{\text{farads}}{\text{meter}}$$

8.2.1 Dielectric constant

A dielectric material is one in which the atoms or molecules have the ability to create electric dipoles, a displacement of charge which results in polarization when a potential is present. The relationship between the polarization and the electric field is

$$\overline{P} = \varepsilon_0 (K - 1)\overline{E} \quad Q/m^2 \tag{8.1}$$

where P = polarization, coulombs/m^2
$\quad\quad\;\, E$ = electric field, V/m

There are four basic mechanisms which create polarization: *electronic polarization,* in which the cloud of electrons is displaced relative to the nucleus in the presence of an electric field, *molecular polarization,* in which a permanent dipole moment exists as a result of the molecular structure, and *ionic polarization,* which exists in molecules having ionic bonds in which the positive ions are displaced relative to the negative ions in the presence of an electric field. In addition, a form of polarization, called *space charge polarization,* exists as a result of charges derived from contaminants or irregularities which exist within the dielectric. These charges exist in all crystal lattices to a greater or lesser degree, and are partly mobile. As a result, they will migrate in the presence of an applied potential. In a given material, more than one type of polarization can exist and the net polarization is given by

$$\overline{P}_t = \overline{P}_e + \overline{P}_m + \overline{P}_i + \overline{P}_s \tag{8.2}$$

where P_t = total polarization
$\quad\quad\;\, P_e$ = electronic polarization
$\quad\quad\;\, P_m$ = molecular polarization
$\quad\quad\;\, P_i$ = ionic polarization
$\quad\quad\;\, P_s$ = space charge polarization

Dielectric materials can also be classified as polar, which includes molecular polarization, water being a prime example, and nonpolar, which includes electronic and ionic polarization. Common table salt, NaCl, is an example of a material which exhibits ionic polarization. Numerous materials exhibit electronic polarization to a greater or lesser degree. A material is not generally considered to be a dielectric unless its resistivity is sufficiently high to eliminate or minimize rapid dissipation of the charge, as occurs in conductive material.

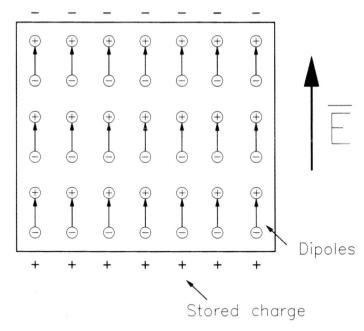

Figure 8.1 Orientation of dipoles in an electric field.

Normally, the dipoles are randomly oriented in the material and the resulting internal field is zero. In the presence of an electric field, the dipoles become oriented as shown in Fig. 8.1, and electric charge is attracted at the surface of the material. The dielectric constant is a measure of the relative ability of the material to store electric charge, and is a fundamental material property. Dielectric constants for various materials are given in Table 8.1.

Dielectric materials may be divided into two types: paraelectric and ferroelectric. When an electric field is applied to a paraelectric material and removed, the internal polarization immediately drops to zero. By contrast, when an electric field is removed from a ferroelectric material, a residual

TABLE 8.1 Dielectric Constants for Various Materials

Material	Dielectric constant
Air	1.004
Most polymers	2–6
Highest polymer (polyvinylidine difluoride, PVF_2)	16
Most ceramics	4–12
Al_2O_3	9
Ta_2O_5	25
TiO_2	90
$BaTiO_3$	1500
Ceramic formulation based	20–15,000

polarization remains. To reduce this remnant polarization to zero, an electric field of the opposite polarity must be applied. If the opposing field is increased beyond the point where the polarization is zero, a polarization in the opposite direction is created. This is illustrated in Fig. 8.2. Paraelectric materials tend to have a lower dielectric constant, be less sensitive to temperature, be more stable with time, and have a better characteristic at high frequencies.

The ceramic material based on the titanates (TiO_2, $BaTiO_3$, or the rare earth titanates) exhibits the highest usable dielectric constant with good electrical characteristics. The temperature dependence of $BaTiO_3$ is given in Fig. 8.3. The transformation between the paraelectric and ferroelectric phases is called the Curie point T_c. Pure barium titanate exhibits a high K at the Curie point and a large negative temperature coefficient on both sides of the peak.

To produce materials with a range of dielectric constants and useful temperature characteristics, additives are introduced that either shift the Curie temperature or depress the peak value of K at T_c. Depressors are not substitutionally soluble in the perovskite structure of the $BaTiO_3$ family of oxides and tend to act as grain boundary impurities. These additives (such as bismuth and magnesium) react with $BaTiO_3$ to increase sintering rates. Shifters are usually materials that can form solid solution with $BaTiO_3$ (such as titanates and zirconates of Ba, Ca, Sr, and the rare earth elements) and are used to shift the Curie point of the dielectric. A listing of shifters and depressors is given in Table 8.2.

In the presence of an electric field which is changing at a high frequency, such as a radio signal, the polarity of the dipoles must change at the same rate as the polarity of the signal in order to maintain the dielectric constant at the same level. Some materials are excellent dielectrics at low frequencies, but the dielectric qualities drop off rapidly as the frequency increases.

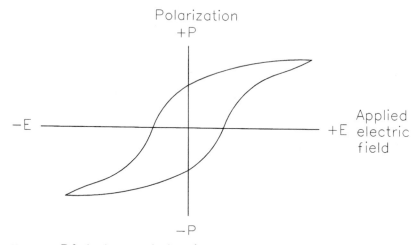

Figure 8.2 Polarization curve for ferroelectric dielectric.

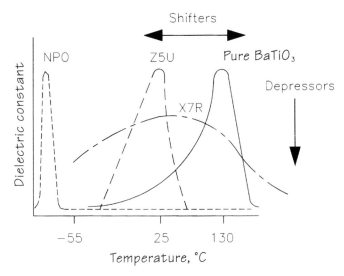

Figure 8.3 Temperature dependence of barium titanate.

TABLE 8.2   Additives for BaTiO₃ Formulation

Shifters		Depressors	
$BaSnO_3$	$SrTiO_3$	$CaTiO_3$	$MgSnO_3$
$BaZrO_3$	$PbTiO_3$	$MgTiO_3$	$Bi_2(SnO_3)$
$CaZrO_3$	La_2O_3	$MgZrO_3$	$Bi_2(TiO_3)_3$
$CaSnO_3$	$NaTiO_3$		

Electronic polarization, which involves only displacement of free charge and not ions, responds more rapidly to the changes in the direction of the electric field, and remains viable up to 10^{17} Hz. The polarization effect of ionic displacement begins to fall off at 10^{13} Hz, and molecular and space charge polarizations fall off at still lower frequencies. The frequency response of the different types is shown in Fig. 8.4, which illustrates that the dielectric constant decreases with frequency.

8.2.2 Capacitor structure

The efficiency of the system in storing charge is called the *capacitance* of the system and is defined by

$$C = \frac{Q}{V} \tag{8.3}$$

where C = capacitance of the system, farads
Q = charge on one conductor, coulombs
V = potential between the two conductors, V

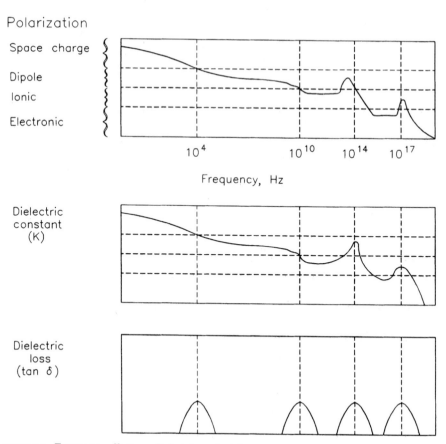

Figure 8.4 Frequency effects on dielectric materials.

The most fundamental capacitor structure is the parallel-plate capacitor shown in Fig. 8.5, where two conductive plates (electrodes) of area A are separated by a dielectric material of thickness t. The net charge on the plates is the sum of the charge due to polarization of the dielectric and the charge which would be present if the dielectric were removed.

$$\overline{D} = \varepsilon_0(K - 1)\overline{E} + \varepsilon_0\overline{E} = \varepsilon_0 K \overline{E} \qquad Q/m^2 \qquad (8.4)$$

where D = charge density, coulombs/m^2
$Q = DA$, coulombs
$E = V t$

Substituting these parameters in Eq. (8.3),

$$C = \frac{\varepsilon A}{t} \qquad \text{farads} \qquad (8.5)$$

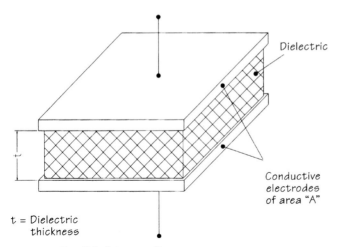

Figure 8.5 Parallel-plate capacitor.

where $\varepsilon = \varepsilon_0 K$, farads/m
A = area of plates, m^2
t = distance between plates, m

In English units, the capacitance of a parallel-plate system is given by Eq. (8.6).

$$C = 0.225 \, \frac{KA}{t} \tag{8.6}$$

where A = area of plates, in^2
t = distance between plates, in
K = dielectric constant of the material relative to air
C = capacitance in picofarads, pF

The three factors, dielectric constant K, area A, and dielectric thickness t, give rise to the electrical characteristics and performance limitations observed in the different types of capacitors.

Three types of capacitors are commonly used in hybrid microcircuits: ceramic capacitors, tantalum electrolytic capacitors, and MOS (metal-oxide-silicon) capacitors. Aluminum electrolytic capacitors are not widely used since they are not widely available in the surface mount configuration. Table 8.3 indicates, to a first order, if the property is relatively fixed or variable for the three types of capacitors.

8.2.3 Area and thickness

To aid in making the overall hybrid package as small as possible, it is desirable that the discrete capacitors have a high capacitance per unit area. The

TABLE 8.3   Capacitor Construction Variables

Type	Dielectric material	Dielectric constant	Area/ volume	Dielectric thickness	Capacitance value
MOS	Silicon dioxide	Fixed, small	Variable	Fixed	<50 pF
Ceramic	Various	Variable, 80–10,000	Variable	Fixed	1 pF–10 μF
Tantalum	Tantalum pentoxide	Fixed	Fixed	Variable	1–100 μF

single-layer structure depicted in Fig. 8.6 is not a very efficient method for achieving this goal, since the capacitance is directly proportional to the overall area of the plates. The density may be improved somewhat by adding multiple layers, which increase the effective electrode area without substantially increasing the volume. As shown in Fig. 8.7, the area is the effective overlap of the metal electrodes. By introducing common electrodes and interdigitating the plates, a multilayer capacitor with M number of layers has an effective area of $(M + 1) \times$ area of one layer. Multilayer structures are used to manufacture ceramic chip and thick film capacitors.

Another method used to increase area is to increase the effective area of the electrodes. This method is used with tantalum capacitors where the elec-

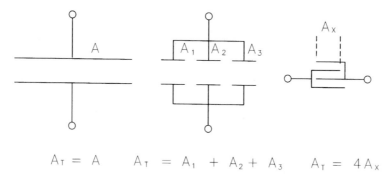

$$A_T = A \qquad A_T = A_1 + A_2 + A_3 \qquad A_T = 4A_x$$

Figure 8.6   Effective area of various capacitor configurations.

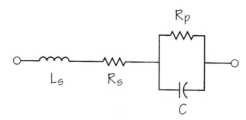

Figure 8.7   Equivalent circuit of a capacitor. L_s = equivalent series inductance. R_s = equivalent series resistance. R_p = equivalent parallel resistance. C = capacitance.

trode structure is fabricated by sintering tantalum powder into an open-pore solid structure. This spongelike arrangement results in an extremely high area/volume ratio with a corresponding increase in capacitance.

The ability to increase capacitance by reducing thickness required to fabricate capacitors is limited by two factors: (1) the dielectric breakdown voltage (in volts per mil) where the dielectric undergoes catastrophic current breakdown and (2) the manufacturing process utilized to fabricate the part. In MOS and tantalum capacitors, the dielectric thickness is submicron (1 μm = 10,000 Å) and the working voltage of the devices is rated only slightly above (10–30%) the dielectric breakdown strength of the dielectric material. In ceramic capacitors, the thickness is approximately 1 mil (0.001 in), established by the manufacturing process, and most devices can withstand an overload from 500–1000%.

8.2.4 Electrical properties and test criteria

The properties of many of the dielectric materials used to fabricate capacitors may vary somewhat with frequency, voltage, temperature, and time. The effect of these variables may be determined from the generalized equivalent circuit for a capacitor shown in Fig. 8.7.

8.2.4.1 Temperature coefficient of capacitance (TCC).

The TCC is a measure of the change in capacitance as a function of temperature and is defined by the following equation:

$$\text{TCC} = \frac{C_2 - C_1}{C_1(T_2 - T_1)} \times 10^6 \text{ ppm/°C} \tag{8.7}$$

where C_2 = capacitance at temperature T_2
C_1 = capacitance at temperature T_1 (reference temperature)
T_2 = temperature at which C_2 is measured
T_1 = reference temperature (usually 25°C)

When the TCC is high, it may be expressed in percent (10,000 ppm = 1%).

The TCC is a function of temperature but is usually expressed as a "hot" or "cold" TCC, both of which use room temperature as the reference temperature. The hot TCC is the average TCC between room temperature and an arbitrary elevated temperature, usually +125°C, and the cold TCC is the average TCC between room temperature and an arbitrary lowered temperature, usually −55°C.

8.2.4.2 Dissipation factor (DF).

Changing the polarity of the dipoles requires a finite amount of energy. This energy is dissipated as internal heat, which is quantified by a parameter called the "dissipation factor," or DF. In addition, dielectric materials are not perfect insulators. These phenomena may be modeled as a resistor in parallel with the capacitor. The DF, as expected, is a strong function of the applied frequency, increasing as the frequency increases.

In alternating current applications, the current and voltage across an ideal capacitor are nearly 90° out of phase. In actuality, the resistive component of the capacitor causes the current to lead the voltage. The DF is a measure of the real or resistive component of the capacitor and is given by

$$DF = \tan(X) = \tan(90° - Y) \tag{8.8}$$

where Y = angle by which the current leads the voltage

The DF is related to the power factor (PF) and the quality factor (Q). The PF is equal to cos Y, and for very small values of Y, PF = DF. The Q is equal to $1/\tan X$. These factors are significant only when the capacitor is excited by alternating current.

8.2.4.3 Insulation resistance (ir). The ir is the direct current analog of the DF. The ir is largely a bulk effect and is inversely proportional to the plate area. The resistance is usually expressed as the product of the ir and the capacitance with units of megohm-microfarads (MΩ-μF), a term that is constant for a given dielectric material. The ir has an inverse logarithmic relationship with the temperature and may vary by several orders of magnitude between +25 and +125°C. The ir is also inversely proportional to the leakage current.

8.2.4.4 Voltage coefficient of capacitance (VCC). The effective capacitance of a device changes under an applied voltage stress (volts applied per mil of dielectric). The VCC represents the percentage change for a given stress. The algebraic sign of the VCC depends on whether the applied potential is direct current (DC) or alternating current (AC). A DC potential decreases the capacitance of ferroelectric materials, and an AC potential increases it. The change for some materials may be considerable, and the test voltage must be carefully specified.

8.2.4.5 Self-resonance. In certain capacitor types, particularly the wound foil types often used in through-hole applications, the leads and the internal structure of the capacitor contribute a residual inductance component to the equivalent circuit of the capacitor. At low frequencies, the effect of this component is virtually negligible. As the frequency is increased, however, the inductive reactance begins to cancel out a portion of the capacitive reactance, which lowers the effective value of capacitance. At the frequency where the inductive and capacitive reactances are equal, the capacitor becomes a series self-resonant circuit. For the components commonly used in hybrid circuits, this effect is virtually negligible and will be a third or fourth order effect compared to other parameters, such as the DF.

8.2.4.6 Dielectric strength. The dielectric strength is a measure of the voltage required to cause breakdown for a given configuration. The dielectric strength is actually a field effect, and the term volts per unit length is constant for a given dielectric material. The dielectric strength is also a function of the temperature, but not to the extent that the insulation resistance is.

8.2.4.7 Aging. Ferroelectric dielectrics exhibit a decrease in capacitance with time, known as aging. The process results from the relation of the strain energy associated with the phase transformation and ferroelectric domain formation. The aging rate is linear per decade hour (1, 10, 10^2, 10^4 h, etc.) and is expressed as the percentage of change per decade hour (Fig. 8.8). Any time the dielectric is taken above the Curie point, the dielectric returns to its original state and begins at time zero upon cooling through T_c. Applying a DC voltage during the transition through T_c causes a decrease in capacitance and an "effective" time shift until aging occurs. For circuits where stability is required, a low aging dielectric or voltage conditioning may be required.

8.2.4.8 Quality factor. The quality factor, or Q, of a component is defined as the ratio of the amount of energy stored to the amount of energy dissipated per cycle. For a capacitor, this equates to

$$Q = \frac{1}{\omega R C} \tag{8.9}$$

where R = equivalent series resistance of the capacitor

8.2.5 Ceramic chip capacitors

8.2.5.1 Manufacturing process. Ceramic chip capacitors are manufactured in much the same manner as multilayer thick film or cofired ceramic substrates;

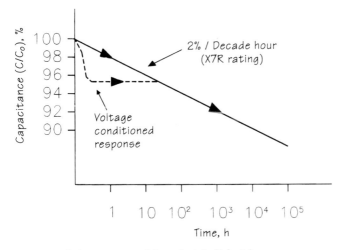

Figure 8.8 Aging response of ferroelectric dielectric.

alternating layers of conductor (electrodes) and dielectrics are formed. Two primary processes are used to fabricate ceramic chip capacitors: the dry process and the wet process.

In the dry process, the dielectric material is milled into a fine powder, mixed with a binding ingredient to form a viscous liquid, and formed into a thin film several inches wide on a mylar sheet using a doctor-blade technique to control the thickness. The ceramic films are dried, cut into lengths approximately 2 ft long, and peeled away from the mylar. The sheets are cut into shorter lengths and placed in a template with locating pins to allow the electrodes to be screen printed onto the film with a high degree of precision. The electrodes are typically a PdAg thick film composition similar to those used to fabricate thick film interconnections. The electrodes are screen printed, forming a number of capacitor electrodes simultaneously, and dried. A second ceramic film is placed over the electrodes, and the screen-and-dry process with the conductor is performed again. The cycle is repeated until the required number of layers are formed. The resulting structure is dried and laminated at an elevated temperature, pressed to force out any air bubbles, and separated into individual capacitors by cutting between the electrodes. The capacitors are fired in a two-step process, first at 800°C to remove the organic binders and then at 1200°C to sinter the PdAg electrodes and dielectric films together to form a monolithic structure. After firing, the capacitors are milled to remove any burrs or irregularities and to ensure that the edges of the electrodes are exposed at the ends. End terminations are applied by dipping the capacitors in a PdAg paste and firing at about 850°C. The structure of a multilayer ceramic chip capacitor is shown in Fig. 8.9.

In the wet process, the dielectric film is formed by spraying a slurry of the dielectric material onto a surface and drying. The electrodes are then screened onto the dielectric film and dried. The sequence is repeated until the

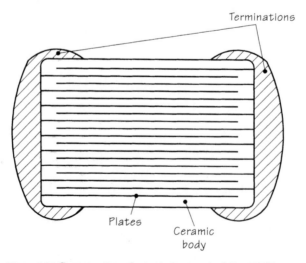

Figure 8.9 Cross section of a typical ceramic chip capacitor.

requisite number of layers are formed. The drying, pressing, firing, and electroding operations are identical to those used in the dry process.

The dielectric thickness of capacitors made by the wet process can be made thinner than those made by the dry process, and they will consequently have a higher capacitance per unit volume. However, they are more susceptible to dielectric breakdown and leakage for the same reason.

8.2.5.2 Dielectric class. Dielectric classes are categorized by the performance of the TCC of the materials. In ceramic chip capacitors, two types (Class I and Class II) are used. Class I dielectrics have low TCC and low dielectric constants and are fabricated from paraelectric materials, while Class II materials have high dielectric constants and higher TCC and are ferroelectric in nature. Code classifications have been developed by the Electronic Industries Association (EIA), defined in Standard RS-198-B, the DOD, and in MIL Specification C-55681, for both Class I and Class II dielectrics for capacitance change over a range of temperature performances. EIA designations for Class I and Class II dielectrics are shown in Tables 8.4 and 8.5.

Class I materials are made with low-K ($K<150$) dielectric materials and are paraelectric in nature. These materials have the following characteristics:

- Linear TCC but tending to be more negative

- High Q (DF<1%)

- High stability of both capacitance and DF with respect to time, voltage, and frequency

- No aging effects since they are paraelectric

- No VCC

Class I dielectrics can be made to have a TCC as low as ±30 ppm (NPO) with variations ranging to several thousand ppm with both positive and neg-

TABLE 8.4 EIA Designations for Class I Dielectrics (−55 to +125°C)

Significant figure of temp. coeff. of capacitance, ppm/°C (a)	Letter code for (a) (b)	Multiplier applicable to column (a) (c)	Numerical codes for (c) (d)	Tolerance of temp. coeff., ppm/°C (e)	Letter code for (e) (f)
0.0	C	−1.0	0	±30	G
1.0	M	−10	1	±60	H
1.5	P	−100	2	±120	J
2.2	R	−1000	3	±250	K
3.3	S	−10000	4	±500	L
4.7	T	+1	5	±1000	M
7.5	U	+10	6	±2500	N
		+100	7		
		+1000	8		
		+10000	9		

SOURCE: Courtesy of Novacap.

TABLE 8.5   EIA Designations for Class II Dielectrics (−55 to 125°C)*

Low temp. requirement, °C (a)	Letter symbol for (a) (b)	High temp. requirement, °C (c)	Numerical codes for (c) (d)	Max. %ΔC over temp. range (e)*	Letter symbol for (e) (f)
+10	Z	+ 45	2	± 1.0	A
−30	Y	+ 65	4	± 1.5	B
−55	X	+ 85	5	± 2.2	C
		+105	6	± 3.3	D
		+125	7	± 4.7	E
				± 7.5	F
				±10.0	P
				±15.0	R
				±22.0	S
				+22–33	T
				+22–56	U
				+22–82	V

*Column (e) is maximum change with 0 bias voltage.
SOURCE: Courtesy of Novacap.

ative values. Class II dielectrics are ferroelectric in nature and have the following characteristics:

- Nonlinear TCC expressed in percent (Fig. 8.10)
- Low Q (DF>2%)
- VCCs that decrease (DC) and increase (AC) with increase in effective dielectric constant (Fig. 8.11A,B)
- Aging effects, increasing from 1–2% per decade hour for BX/X7R to 5% per decade hour for high-K dielectrics

Class II dielectrics are divided into two subcategories based on the dielectric constant, the stability, and the temperature characteristics. The so-called

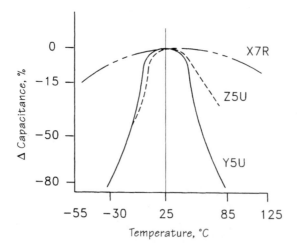

Figure 8.10 Temperature coefficient of Class II dielectrics.

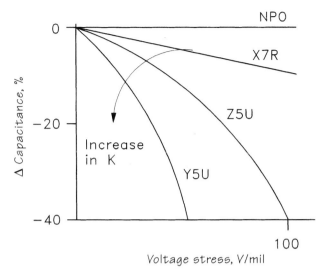

Figure 8.11A   DC voltage coefficient of capacitance.

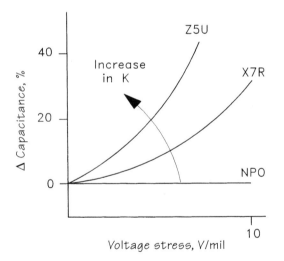

Figure 8.11B   AC voltage coefficient of capacitance.

mid-*K* materials (500<*K*<4000) include the BX class, which meets military standards, and the X7R class, which does not. The mid-*K* materials are more stable than the high-*K* materials (*K*>4000), which include the Z5U class. Class II materials have a higher capacitance per unit volume than Class I materials and are used primarily in coupling and decoupling applications. Class I materials are used in applications where stability and TCC are at a premium, such as tuned circuits and timing circuits. The electrical specifications for Class I and Class II dielectrics are summarized in Table 8.6. Typical characteristics of NPO, X7R, and Z5U capacitors are given in Figs. 8.12A, B, C, and D, 8.13A, B, C, and D, 8.14A, B, and C, and 8.15A, B, and C.

TABLE 8.6   Electrical Specifications for Class I and Class II Dielectrics—EIA RS 198 and MIL-C-55681

Properties	Dielectric characteristic			
	NPO or COG MIL-C-55681/ EIA RS 198	BX MIL-C-55681	X7R EIA RS 198	Z5U EIA RS 198
Operating temperature range, °C	−55 to + 125°C	−55 to + 125°C	−55 to + 125°C	+10 to + 85°C
Temperature coefficient, %ΔC Max 0 bias	0 ± 30 ppm/°C	±15%	±15%	+22%−56%
Temperature-voltage coefficient, %ΔC max at V_{DCW}	0 ± 30 ppm/°C	±15%−25%	Not applicable	Not applicable
Dissipation factor at 25°C	0.10% max	2.5% max	2.5% max	3.0% max
Insulation resistance at V_{DCW}, 25°C	Lesser of 100 GΩ or 1000 MΩ·μF	Lesser of 100 GΩ or 1000 MΩ·μF	Lesser of 100 GΩ or 1000 MΩ·μF	Lesser of 100 GΩ or 500 MΩ·μF
Insulation resistance at V_{DCW}, 125°C	Lesser of 1 GΩ or 10 MΩ·μF	Lesser of 10 GΩ or 100 MΩ·μF	Lesser of 10 GΩ or 100 MΩ·μF	Lesser of 10 GΩ or 50 MΩ·μF
Dielectric withstanding test voltage, 25°C*	$2.5 \times V_{DCW}$	$2.5 \times V_{DCW}$	$2.5 \times V_{DCW}$	$2.5 \times V_{DCW}$
Aging rate, %ΔC/decade hour, max %	0	−2.5	−2.5	−5.0
Test frequency, 25°C	<100 pF, 1 MHz >100 pF, 1 kHz	1 kHz	1 kHz	1 kHz
Test voltage, 25°C	$1.0 \pm 0.2 V_{rms}$	$1.0 \pm 0.2 V_{rms}$	$1.2 V_{rms}$ max $(1.0 \pm 0.2 V_{rms}$ typical)	$1.0 \pm 0.2 V_{rms}$

*$1.2 \times V_{DCW}$ for capacitors rated at or above 1000 V_{DC}.

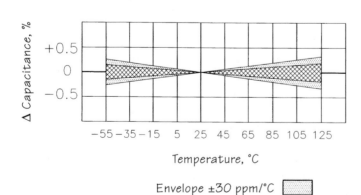

Figure 8.12A   Temperature coefficient of capacitance for an NPO dielectric. (*Courtesy of AVX Corp.*)

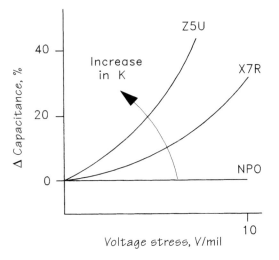

Figure 8.12B Aging rate of NPO capacitors. (*Courtesy of AVX Corp.*)

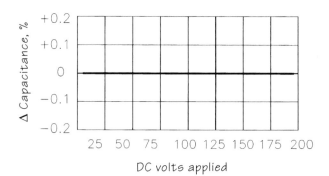

Figure 8.12C Voltage coefficient of capacitance for a typical NPO dielectric. (*Courtesy of AVX Corp.*)

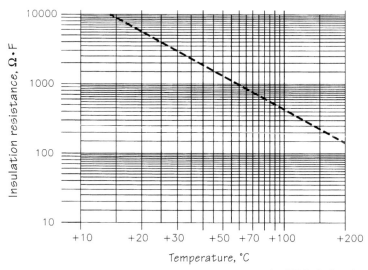

Figure 8.12D Insulation resistance vs. temperature for NPO dielectric. (*Courtesy of AVX Corp.*)

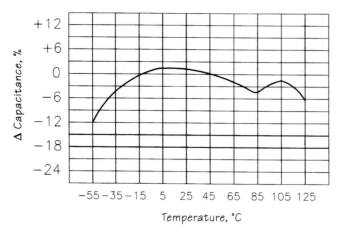

Figure 8.13A Temperature coefficient of capacitance for an X7R dielectric. (*Courtesy of AVX Corp.*)

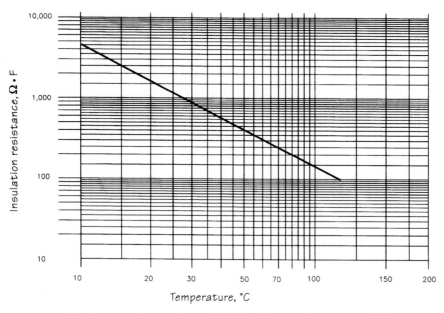

Figure 8.13B Insulation resistance of a typical X7R dielectric. (*Courtesy of AVX Corp.*)

Special purpose capacitors for higher voltage and higher frequency applications are also available. High voltage capacitors have a thicker dielectric material and are consequently physically larger than the more common 50 V rated versions. Capacitors designed for high frequency and microwave applications will generally be manufactured with a thicker inner conductor to lower the equivalent series resistance and improve the quality factor. Other dielectric materials, such as porcelain, may be substituted for one of the barium titanate structures to lower the loss tangent.

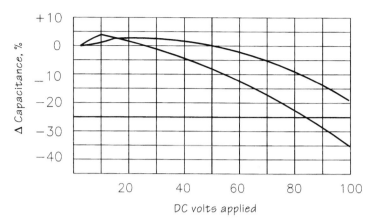

Figure 8.13C Voltage coefficient of capacitance for a typical X7R dielectric (note hysteresis). (*Courtesy of AVX Corp.*)

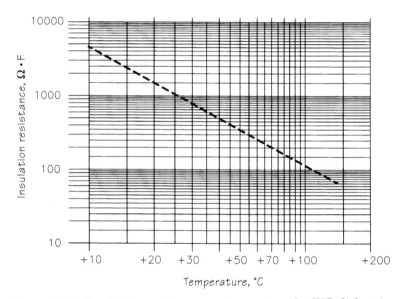

Figure 8.13D Insulation resistance vs. temperature for X7R dielectric. (*Courtesy of AVX Corp.*)

8.2.5.3 Configuration. The size of ceramic capacitors, as with many chip components, is characterized by a code which indicates the footprint requirement for the device. For chip capacitors, a four-digit code is utilized, where the first two digits indicate the length in terms of units and the last two digits represent width in terms of mils. As an example (see Fig. 8.16), the four-digit code 1206 represents a part 0.120 in long and 0.060 in wide. The thickness is nominally 50 mils and the termination width is specified by each manufacturer, starting at 10 mils for small sizes and averaging 20 to 25 mils for devices of 1210 size and larger. The typical range for three types of dielectrics for various sizes is given in Table 8.7.

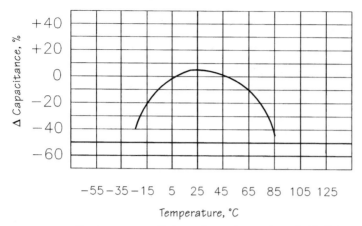

Figure 8.14A Temperature coefficient of capacitance for a Z5U dielectric. (*Courtesy of AVX Corp.*)

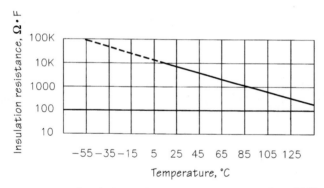

Figure 8.14B Insulation resistance vs. temperature for a Z5U dielectric. (*Courtesy of AVX Corp.*)

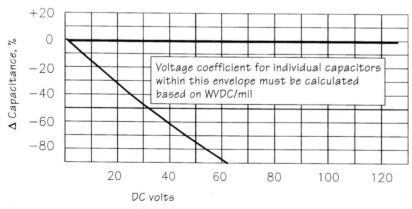

Figure 8.14C Voltage coefficient of capacitance for a typical Z5U dielectric. (*Courtesy of AVX Corp.*)

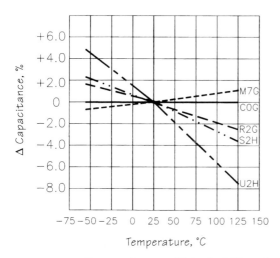

Figure 8.15A Temperature coefficient of Class I dielectrics. (*Courtesy of Novacap.*)

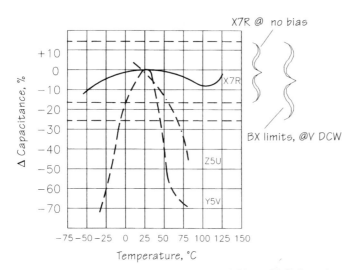

Figure 8.15B Temperature coefficient of Class II dielectrics. (*Courtesy of Novacap.*)

8.2.5.4 Termination. The selection of the end termination material is dependent on the expected assembly process, packaging requirements, and cost. The industry standard is palladium–silver for unsoldered components. To minimize solder leaching, a barrier layer, typically a plated nickel film, may be deposited over a silver termination for soldered parts.

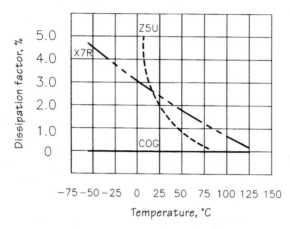

Figure 8.15C Temperature coefficient of dissipation factor. (*Courtesy of Novacap.*)

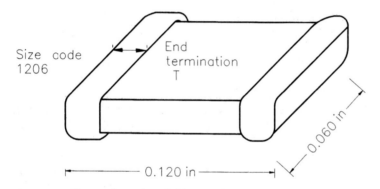

Figure 8.16 Size configuration of chip capacitor.

TABLE 8.7   Typical Maximum Capacitance vs. Size*

Dielectric	V (V_{DC})	Size					
		050	0805	1206	1210	1812	1825
NPO (pF)	50	470	1200	2700	4700	0.01 μF	0.027 μF
BX (μF)	50	0.0056	0.022	0.056	0.10	0.18	0.56
Z5U	50	N/A	0.047	0.12	0.22	0.47	1.0

*As new advances are made in the capacitor technology, this table will quickly become out of date. Manufacturer's catalogs should be checked for the latest information.

8.2.6 MOS capacitors

MOS capacitors are fabricated by thermally growing a thick layer of SiO_2 on a wafer of low resistivity silicon. The oxide is etched away over the capacitor areas and carefully regrown to obtain the proper thickness. To improve the stability, the breakdown voltage, and the capacitance per unit area, a thin layer of Si_3N_4 may be grown by chemical vapor deposition. Aluminum is deposited over the wafer by evaporation, and bonding pads are formed over the capacitor areas using a photoetch process. The bottom electrode is formed by evaporating gold on the bottom of the wafer and alloying. The device may be mounted onto the circuit by eutectic bonding or with conductive epoxy. The top electrode is wire bonded to the circuit metallization at the appropriate point, making the MOS capacitor compatible with active chip mounting. Capacitance values are limited by processing considerations to a few pF with a DF as low as 0.015%. Insulation resistance and stability are extremely high with respect to time, frequency, temperature, and voltage. The TCC is typically positive, varying from about +30 to +70 ppm/°C. MOS capacitors may be actively trimmed in two ways, both involving the disconnection of discrete sections of the capacitor. One method utilizes a laser trimmer to remove connections between capacitors of successively increasing values until the proper tolerance is achieved. A second method uses a capacitor with top electrodes of different areas that are wire bonded to a common point. Trimming is achieved by removing wire bonds from the different pads. The structure of an MOS capacitor with a range from 1 to 15 pF is depicted in Fig. 8.17.

8.2.7 Tantalum solid electrolytic capacitors

Tantalum capacitors have the highest capacitance × voltage (CV) rating of any type of capacitor for hybrid or surface mount applications. Whereas the ceramic capacitor obtains high capacitance value from the use of a high K dielectric material, tantalum capacitors use large effective surface area and a very thin dielectric to obtain high capacitance values.

The basic structure of a tantalum capacitor is a porous sintered tantalum anode body with an electrochemically formed tantalum oxide dielectric. The counterelectrode is formed by utilizing a conductive electrolyte and a conductor. Although liquid electrolytes are used in many commercial applications because of their high CV ratio, this discussion is limited to solid electrolyte

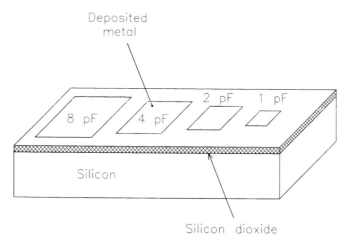

Figure 8.17 MOS capacitor with a range of 1 to 15 pF.

devices, since this type is primarily found in hybrid–surface mount applications. For a solid electrolyte device, a semiconducting film of manganese dioxide serves as the electrolyte and a graphite–silver coating is used to fabricate the counterelectrode (see Fig. 8.18).

The effective area of a tantalum slug is large due to the open structure of the sintered tantalum anode. For any given volume, the effective area is constant and area can only be considered a variable by change in component size. Since the effective area for any given volume of tantalum is the same, the capacitance is changed by varying the thickness of the dielectric. Dielectric films have typical thicknesses that range from 200 to 1000 Å (two orders of magnitude smaller than a ceramic capacitor). Since the thickness of dielectric determines the dielectric breakdown voltage and therefore the effective working voltage, tantalum capacitors are selected for both capacitance and voltage because the CV product is a constant for any given size (volume efficiency of 100,000 μFV/in^3).

Tantalum capacitors are found in various configurations. The high reliability solid tantalum devices are covered in MIL-C-55365. Table 8.8 represents the sizes and the minimum-maximum range of working voltage available, although eight voltage ranges exist.

Tantalum capacitors are polar, and catastrophic failure will occur if a reverse bias is applied to the device which is over 15% of the working voltage at room temperature. The polar nature of the device occurs because of the operation of the solid electrolyte, MnO_2. After manufacture, defect sites occur in the dielectric. Upon application of voltage, current flows through the defects and into the MnO_2. The increased localized temperature causes the MnO_2 to decompose into an insulating oxide phase, thus "healing" the defect site. Because the electron flow is highly localized at the MnO_2 layer, sufficient heating occurs. On a reverse bias, the current flow is not localized at the MnO_2 layer but in the tantalum layer, leading to heating and ultimately ignition of the tantalum without the "clearing" mechanism for protection.

Tantalum capacitors exhibit a high series inductance and are prone to leakage current failures if subjected to severe mechanical and environmental forces. The distributed inductance limits their frequency application to about 100 kHz.

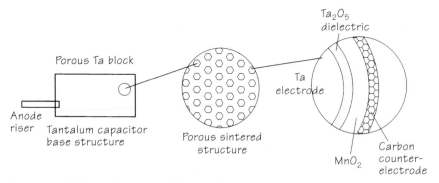

Figure 8.18 Solid tantalum electrolytic capacitor.

8.3 Inductors and Transformers

An electric current flowing in a conductor creates a magnetic field in the vicinity. The fundamental relationship which describes this phenomenon in a vacuum is

$$B = \mu_0 H \tag{8.10}$$

where B = magnetic flux density, webers/m^2
  μ_0 = permeability of free space
    = $4\pi \times 10^{-7}$ henrys/m
  H = magnetic field strength, ampere-turns/m

The total magnetic flux in webers in a given area is calculated by integrating the expression for the flux density as a function of position over the area of interest.

$$\Phi = \oiint \overline{B} \cdot \overline{dS} \tag{8.11}$$

where Φ = magnetic flux, webers
  dS = differential unit of area expressed as a vector

The amount of flux generated by a current in a conductor may be increased by winding the conductor in the form of a spiral coil as shown in Fig. 8.19. The efficiency of the coil in generating magnetic flux is called the *inductance* and is defined as

$$L = \frac{\Phi}{I} \tag{8.12}$$

where L = inductance, henrys
  Φ = flux, webers
  I = current, amperes

The inductance of the structure in Fig. 8.19 is given by Eq. (8.13).

TABLE 8.8   Sizes of Tantalum Capacitors per MIL-C-55365

Case	Size ($L \times W \times T$, ± 0.015 in)	Voltage rating	
		4 V_{DC}	50 V_{DC}
A	0.100 × 0.050 × 0.050	2.2	0.1
B	0.150 × 0.050 × 0.050	4.7	0.33
C	0.200 × 0.050 × 0.050	6.8	0.47
D	0.150 × 0.100 × 0.050	10.0	0.68
E	0.200 × 0.100 × 0.050	15.0	1.0
F	0.220 × 0.135 × 0.070	33.0	2.2
G	0.265 × 0.150 × 0.110	68.0	3.3
H	0.285 × 0.150 × 0.110	100.0	4.7

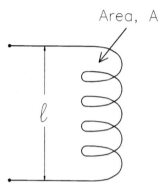

Area, A

ℓ

Figure 8.19 Basic structure of an inductor.

$$L = \frac{\mu_0 N^2 A}{l} \qquad (8.13)$$

where L = inductance of coil, henrys
A = area of core, m²
l = length of coil, m
N = number of turns

When a material which has the capability of forming magnetic dipoles is placed in the magnetic path, an additional magnetic field proportional to the field strength, as given in Eq. (8.14), is created in the material.

$$M = \chi_m H \qquad (8.14)$$

where M = magnetization of the material, webers/m²
χ_m = magnetic susceptibility

The magnetization is added to the field already present, making the total field equal to

$$B = \mu_0 H + \mu_0 M \qquad (8.15)$$

The overall relationship between B and H is therefore

$$B = \mu H \qquad (8.16)$$

where $\mu = \mu_0 \mu_r$
μ_r = permeability of the material relative to vacuum
$= 1 + \chi_m$

There are four basic types of magnetic materials; diamagnetic, paramagnetic, ferromagnetic, and ferrimagnetic, which are classified as to the nature of the B-H curve and the degree of permanent magnetism.

Diamagnetism is a weak, nonpermanent form of magnetism induced by changes in the orbits of the outer electrons of atoms of certain materials. The induced magnetic field in this instance is opposite that of the applied magnetic field, and the resultant magnetic field is actually lowered. Diamagnetic and paramagnetic materials have a linear relationship between B and H (Φ and l) as shown in Fig. 8.20, and do not exhibit residual magnetism when the applied field is removed. The permeability of certain diamagnetic and paramagnetic materials is listed in Table 8.9.

Ferromagnetic materials are metallic in nature and are characterized by a nonlinear relationship between B and H, a degree of hysteresis, and a residual permanent magnetism component, as shown in Fig. 8.21. As the field is increased, substantially all the magnetic dipoles eventually become aligned, resulting in no further increase in the magnetic field as the current is increased. This phenomenon is referred to as *saturation*. Since the magnetic field is nonlinear, the permeability is a function of the magnetic field strength and no single number can be assigned. At the maximum value, the relative permeability of certain ferromagnetic materials may range as high as 10^6 henrys/m. Typical ferromagnetic materials are iron, cobalt, and nickel, and certain of the rare earth metals.

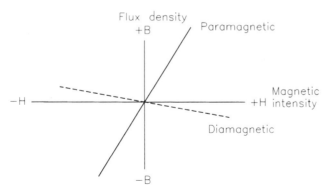

Figure 8.20 Magnetization curves for paramagnetic and diamagnetic materials.

TABLE 8.9 Relative Permeability of Diamagnetic and Paramagnetic Materials

Material	Type	Relative permeability
Al_2O_3	Diamagnetic	0.999982
Silicon	Diamagnetic	0.999996
Silver	Diamagnetic	0.999976
Gold	Diamagnetic	0.999966
Copper	Diamagnetic	0.999990
Aluminum	Paramagnetic	1.000002
Molybdenum	Paramagnetic	1.000012
Chromium	Paramagnetic	1.000031

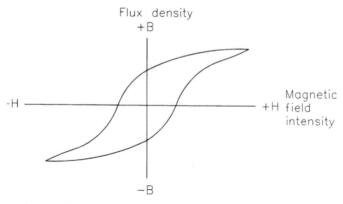

Figure 8.21 Magnetization curve for ferromagnetic and ferrimagnetic materials.

Ferrimagnetism is a form of permanent magnetism exhibited by some ceramic materials. The *B-H* curve of ferrimagnetic materials (*ferrites*) is similar to that exhibited by the ferromagnetic materials shown in Fig. 8.21, except that the saturation levels are somewhat lower. Another significant difference is that ferrites, being generally metallic oxides, are good electrical insulators, a desirable property for inductors or transformers used in high frequency applications. One of the most common ferrite materials is magnetite, which has the chemical formula Fe_3O_4, often referred to as "lodestone."

As a result of the hysteresis loop exhibited in ferromagnetic and ferrimagnetic materials, energy in the form of heat is absorbed by the core. This is an undesirable property, and considerable effort has been directed toward the development of so-called soft magnetic materials, or materials which have a relatively small hysteresis loop, as shown in Fig. 8.22. By contrast, "hard" magnetic materials have a larger hysteresis loop and are utilized in permanent magnets.

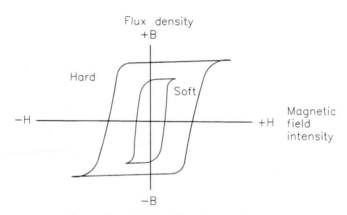

Figure 8.22 Comparison of soft and hard magnetization curves.

An increase in the temperature of a ferromagnetic or ferrimagnetic material causes the random motion of the magnetic dipoles to increase accordingly. This has the effect of randomizing the internal magnetic field, lowering the effective permeability and the saturation level. As the temperature is further increased, the saturation magnetism will gradually drop until the temperature reaches a critical temperature, called the "Curie temperature T_C," at which point it abruptly drops to zero and the material becomes paramagnetic in nature. For most applications, the Curie temperature is sufficiently high that it is not a factor. For example, the T_C for Fe_3O_4 is 585°C, and the permeability drops to approximately 90% of the room temperature level at 150°C.

8.3.1   Soft ferrite materials

Soft ferrites are the most common magnetic materials used to manufacture coils and transformers for hybrid microelectronic applications. The basic formula for a soft ferrite is $MO \cdot Fe_2O_3$, where MO represents a divalent metal oxide, such as manganese, manganese–zinc, or nickel–zinc. These are blended with 48 to 60 mole percent of iron oxide.[1] Ferrites typically have a low TCE (8 to 10 ppm/°C) and a high bulk resistivity.

Manganese ferrites are very stable materials with a very square hysteresis loop. They are used in such applications as switched-mode power supplies and high frequency power supplies. Manganese–zinc ferrites are highly vitrified and have the highest permeability. The volume resistivity may range as high as several thousand Ω cm. They are useful up to the low MHz range. Nickel–zinc ferrites may contain oxides of other materials, such as magnesium, manganese, copper, or cobalt. The volume resistivity may be >10 MΩ cm, and they may be used in higher frequency applications.

The electrical properties of soft ferrites are a function of temperature and frequency, although the temperature variations are generally small over the range of interest. The equivalent circuit of an inductor is shown for reference in Fig. 8.23. Owing to the distributed capacitance and resistance, the phase angle between the voltage and current will vary from the normal 90° exhibited in a perfect inductor. The actual relationship will be

$$\theta = 90° \pm \lambda \qquad (8.17)$$

where θ = phase angle between the voltage and current
λ = deviation from 90°

Note that λ may be positive or negative, depending on the frequency. The effective resistance of the inductor is dependent on the frequency due to a phenomenon known as the "skin effect." As the frequency increases, the cur-

---

[1]Fair-Rite Product Catalog, Fair-Rite Products Corp., P.O. Box J, One Commercial Row, Wallkill, NY 12589.

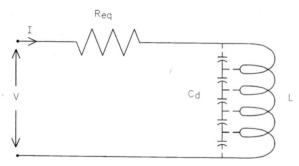

Figure 8.23   Equivalent circuit of an inductor. R_{eq} = equivalent series resistance, including skin effects. C_d = distributed interwiring capacitance. L = equivalent inductance.

rent density becomes nonuniform within the conductor, increasing toward the surface. This has the net effect of decreasing the effective cross section of the conductor, thereby increasing the resistance. The expression for the skin depth is

$$\delta = \frac{1}{\sqrt{\pi f \mu \sigma}} \qquad (8.18)$$

where δ = skin depth
   f = frequency, Hz
   μ = permeability, henrys/m
   σ = conductivity, siemens/m

A table of the skin depth for copper at various frequencies is shown in Table 8.10.

A resistive loss component exists in the core due to the hysteresis element and, to a lesser degree, eddy currents. Eddy currents are circulating currents which exist in the core of an inductor or transformer when alternating current is applied. Because of the high resistivity of ferrites, eddy current losses are substantially less than those in ferromagnetic materials. As a result of these losses, a resistive component in parallel with the inductance exists, which causes the magnetic flux to be out of phase with the current by an angle ϕ.

TABLE 8.10   Skin Depth of Copper vs. Frequency

Frequency	Skin depth, cm
60 Hz	0.00861
1 kHz	0.00211
10 kHz	6.67×10^{-4}
1 MHz	6.67×10^{-5}
100 MHz	6.67×10^{-6}

8.3.1.1 Loss factor.

The loss factor is defined as

$$\text{Loss factor} = \frac{\tan\phi}{\mu_i} \tag{8.19}$$

where μ_i = initial permeability, measured at Φ<10 gauss

8.3.1.2 Quality factor.

The quality factor Q is defined as the ratio of the energy stored per cycle to the energy dissipated, and is given by

$$Q = \frac{\omega L}{R_{\text{eff}}} \tag{8.20}$$

where $\omega = 2\pi f$, frequency, radians
R_{eff} = effective resistance at the frequency of interest

Referring to Eq. (8.20), R_{eff} varies inversely with the square root of the frequency and, therefore, the Q varies directly with the square root of the frequency. Consequently, the frequency must be specified when the Q is given. The method of attach can strongly affect the Q owing to the contact resistance. Solder attach or wire bonding is preferable to epoxy attach.

8.3.1.3 Self-resonant frequency.

The distributed capacitance in combination with the inductance will be resonant at a frequency f_0 which is defined by

$$f_0 = \frac{1}{\sqrt{2\pi LC_d}} \tag{8.21}$$

where C_d is the distributed capacitance of the windings, referred to in Fig. 8.23. Above the self-resonant frequency, the inductor actually becomes capacitive in nature and f_o may be considered to be the maximum usable frequency of the inductor regardless of the core characteristics. The distributed capacitance is also dependent on the surrounding environment and may be increased by potting or conformally coating the inductor, or by proximity to a ground plane.

The characteristics of typical soft ferrites are summarized in Table 8.11. There are many variations of these materials, including certain rare earth additives, which may be formulated for specific applications.

8.3.2 Inductors for hybrid and surface mount applications

Inductors for surface mount applications ("chip" inductors) are more troublesome to terminate than capacitors owing to the difficulty in attaching the wire leads to the circuit. Most frequently, these are wound as *toroids* or on square cores and attached to a lead frame which accommodates bonding to a substrate. A toroid is a doughnut-shaped structure which effectively contains

TABLE 8.11   Selected Properties of Typical Soft Ferrite Materials

Property	Unit	Symbol	NiZn[a]	Mn[b]	MnZn[c]	MnZn[d]
Initial permeability		μ_i	20	300	800	10,000
Flux density at field strength	mT A/m	B H	200 3200	390 800	250 800	400 160
Residual flux density	mT	B_r	100	345	135	125
Loss factor at frequency	10^{-6} MHz	$\tan \delta / \mu_i$	400 100	50 0.1	35 0.2	15 0.025
Curie temp.	°C	T_c	>500	>300	>150	>120
Resistivity	Ω cm	ρ	10^7	1.5×10^3	10^2	50
Recommended frequency range	MHz		<400[e]	<.15[f]	<3[e]	<0.5[e]

[a]Fair-Rite Material 68.
[b]Fair-Rite Material 83.
[c]Fair-Rite Material 33.
[d]Fair-Rite Material 76.
[e]For low flux density inductors and transformers.
[f]Special square-loop ferrite.
SOURCE: Courtesy of Fair-Rite Products Corp.

the magnetic flux within the structure as shown in Fig. 8.24. The toroid is especially convenient for transformers, as it couples the flux from winding to winding with a very high efficiency. The inductor may be potted or left open to reduce distributed capacitance. Properties of typical chip inductors are shown in Tables 8.12 and 8.13.

The following guidelines are beneficial for using chip inductors in hybrid circuits.

1. Attach with solder or wire bonding where possible. The higher ohmic resistance of conductive epoxy may be detrimental to the quality factor.

2. When breadboarding the circuit in the design phase, use the exact inductor or transformer intended for use in the actual hybrid, as characteristics of these components may vary widely between manufacturers.

8.4   Resistors

Chip resistors for hybrid circuit applications are generally of the thick film or thin film types described in earlier chapters. The characteristics of these technologies are summarized here for convenience.

Thick film resistors consist of particles of metal oxides interspersed in a glass matrix. Different sheet resistivities are made by combining the desired ratio of glass to metal oxide. Thick film resistors are fabricated by screen printing a resistor to the proper size and shape and firing at an elevated temperature. Thin film resistors are fabricated by depositing a layer of metal or

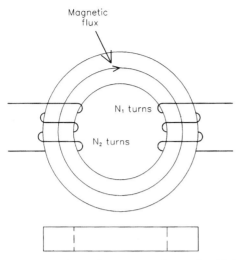

Figure 8.24　Transformer wound on a toroidal core.

TABLE 8.12　**Representative Parameters of TDK Multilayer Chip Inductors**

Part no.	Thickness, in	$L,$ μH	Q nominal	Test frequency, MHz	f_0 nominal	DC resistance, Ω	Rated current, mA
MLF3225C390	0.043	39	45	0.4	14.0	0.70	5
MLF3225C470	0.043	47	45	0.4	11.5	0.80	5
MLF3225C560	0.043	56	45	0.4	10.5	0.90	5
MLF3225C680	0.098	68	70	2.0	10.5	1.75	10
MLF3225C820	0.098	82	70	2.0	10.0	1.95	10
MLF3225C101	0.098	100	70	1.0	9.0	2.20	10
MLF3225C121	0.098	120	45	0.2	7.0	1.30	5
MLF3225C151	0.098	150	45	0.2	6.5	1.50	5
MLF3225C181	0.098	180	45	0.2	6.0	1.65	5
MLF3225C221	0.098	220	45	0.2	5.5	1.75	5

SOURCE: Courtesy of TDK Corporation.

TABLE 8.13　**Representative Parameters of TDK Ferrite Chip Inductors**

Part no.	Thickness, in	$L,$ μH	Q nominal	Test frequency, MHz	f_0 minimum	DC resistance, Ω	Rated current, mA
ACL3225S390	0.043	39	50	2.62	11.5	2.90	70
ACL3225S470	0.043	47	50	2.52	10.5	3.15	65
ACL3225S560	0.043	56	50	2.52	10.0	3.40	45
ACL3225S680	0.098	68	50	2.52	9.0	3.80	55
ACL3225S820	0.098	82	50	2.52	8.5	4.30	40
ACL3225S101	0.098	100	50	2.52	7.8	4.70	40
ACL3225S121	0.098	120	25	0.796	7.6	3.05	14
ACL3225S151	0.098	150	25	0.796	6.8	3.50	13
ACL3225S181	0.098	180	25	0.796	6.2	3.95	10
ACL3225S221	0.098	220	25	0.796	5.6	4.20	10

SOURCE: Courtesy of TDK Corporation.

metal alloy onto a substrate and photoetching the resistor to the desired geometry.

The conduction mechanisms present in thick film resistors have not been well defined in quantitative terms to date owing to the large number of variables involved and the difficulty in controlling these variables. The somewhat qualitative model described here explains many of the phenomena associated with thick film resistors and can be used to extrapolate further results. A discussion of basic conduction mechanisms is informative as an introduction.[2]

The conductivity of a material is proportional to the number of free carriers (holes and/or electrons) and to the mobility, or ease of movement, of those carriers as they move through the material. In a metal, the carriers are electrons that have sufficient thermal energy to escape from the positive attraction of the nucleus, leaving behind a fixed positive ion. Even at very low temperatures, the outer electrons of most metals possess enough thermal energy to escape, and at the temperature range of interest, the number of carriers is essentially independent of temperature. There are two velocity components of the carriers as they move through the metal: the thermal velocity, which is random, and the drift velocity, which is proportional to the strength of an applied electric field. When no field is applied, there is no net velocity of the carriers and therefore no net electric current. During the course of its travels through the metal, the carrier is deflected by the fixed positive ions, by adjacent electrons, and by such parameters as crystal boundaries. A carrier that undergoes such a deflection is said to have suffered a "collision." When a carrier suffers a collision, the net effect is to lower the drift velocity, thereby lowering the mobility and decreasing the net current for a given electric field. By definition, this increases the electrical resistance of the metal. As the temperature increases, the thermal velocity of the carriers is increased, causing the carriers to undergo more collisions and increasing the resistance. For most metals, the increase in resistance is virtually linear with respect to temperature in the temperature range of interest and may be described by a parameter called the temperature coefficient of resistance (TCR) as defined in Eq. (8.22).

$$\text{TCR} = \frac{R(T_2) - R(T_1)}{R(T_1) \times (T_2 - T_1)} \times 10^6 \text{ ppm/°C} \tag{8.22}$$

where
T_1 = base temperature (usually 25°C)
T_2 = temperature of interest (higher or lower than T_1)
$R(T_1)$ = resistance at T_1
$R(T_2)$ = resistance at T_2

Most metals have a positive TCR as high as several thousand ppm/°C. Alloys of metals such as NiCr, on the other hand, can have a TCR of only a

---

[2]Charles Schaffer and Jerry E. Sergent, "The Effect of Particle Size on the Properties of Thick Film Resistors," *Proceedings ISHM Symposium,* 1979.

few ppm/°C. By convention, a material that increases resistance with temperature has a positive TCR, and a material that decreases resistance with increasing temperature has a negative TCR.

When an electric field is applied to a metal, the number of carriers injected as a result is much smaller than the number of free carriers already generated by thermal energy and the number of collisions, and therefore the electrical resistance is essentially independent of the applied electric field when the magnitude of the field is relatively low. This is the definition of Ohm's law as depicted in Eq. (8.23).

$$I \propto V \text{ (conductor)} \tag{8.23}$$

where $\propto$ is read as "is proportional to."

In a semiconductor, the degree of ionization at moderate temperatures is somewhat less than for a conductor. As a consequence, the conductivity is somewhat lower and the number of free carriers (in this case both holes and electrons) is highly dependent on temperature. When an electric field is present in the semiconductor in the form of an applied potential, a current will flow. At low levels of current, the number of carriers injected from the applied potential will be substantially lower than the thermal carriers present, and the current will be proportional to the applied voltage as in a conductor. At moderate levels of current, defined by the point where the number of injected carriers is equal to the number of thermal carriers, the conduction mechanism becomes more complex and the current becomes proportional to the square of the voltage. At still higher currents, the carriers begin having more collisions as a result of the higher drift velocity and the current becomes proportional to the cube of the voltage. Equations (8.24), (8.25), and (8.26) describe the V-I relationships mathematically.

$$I \propto V \quad \text{(low current)} \tag{8.24}$$

$$I \propto V^2 \quad \text{(moderate current)} \tag{8.25}$$

$$I \propto V^3 \quad \text{(high current)} \tag{8.26}$$

In an insulator, there are virtually no thermal carriers and the only free carriers that exist are those injected by an outside potential. An insulator therefore behaves in much the same way as a semiconductor with moderate current flow.

$$I \propto V^3 \quad \text{(insulator)} \tag{8.27}$$

During firing, there are three types of contacts formed as shown in Fig. 8.25: direct sintered contacts between the active particles, tunneling contacts that form between the active particles as a result of a thin film of glass, and insulating contacts formed by a thick film of glass between the contacts. The sintered contacts act in the same way as conductors, the thin films of glass act in the same way as semiconductors, and the thick films of glass act as

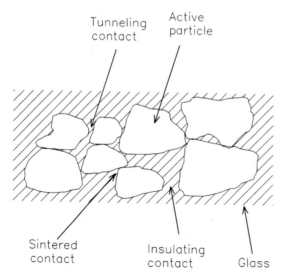

Figure 8.25 Contacts in thick film resistors.

insulators. Since thick film resistors contain all three types of contacts to a greater or lesser degree, the conduction mechanisms are quite complex.

1. High ohmic value resistors tend to have a preponderance of the thin films of glass and tend to have more of the properties associated with semiconductors, while low ohmic values tend to have more of the properties associated with conductors.

2. High ohmic value resistors tend to have a more negative TCR than low ohmic value resistors. This is not always the case in commercially available systems owing to the presence of TCR modifiers but always holds true in pure metal oxide–glass systems.

3. High ohmic value resistors exhibit substantially more current noise than low ohmic value resistors as defined in MIL-STD-202. Current noise is generated when a carrier makes an abrupt change in energy levels, as it must when it makes the transition from one metal oxide particle to another across the thin film of glass. When the metal oxide particles are directly sintered together, the transition is less abrupt and little or no noise is generated.

4. High ohmic value resistors are more susceptible to high voltage pulses and static discharge than low ohmic value resistors. The high voltage impulse breaks down the thin film of glass and forms a sintered contact, permanently lowering the value of the resistor. The effect of static discharge is highly dependent on the glass system used. Resistors from one manufacturer may drop by as much as half when exposed to static discharge, while others may be affected very little. This can be verified experimentally by heating a previously pulsed resistor at about 200°C. The value increases somewhat, indicating a regrowing of the glass oxide layer.

In general, metals exhibit positive TCRs and nonmetals exhibit negative TCRs. In metals, the electron cloud is more disordered with increasing thermal energy and resistance increases. Nonmetals (or semiconductors), which have electrons firmly bonded to crystal locations, become more mobile with energy, are better conductors as temperature is increased, and exhibit a negative TCR.

Thin film resistors, by contrast, are considerably more uniform and homogeneous. Electrons undergo collisions only with the grain boundaries or, in the case of extremely thin films, with the boundaries of the resistor itself, and the conduction process is more traditional.

8.4.1 Electrical properties of thick and thin film chip resistors

8.4.1.1 Temperature coefficient of resistance. The TCR of thick film resistors may vary widely in both sign and magnitude as the sheet resistivity is changed. Thick film resistors are readily available with TCRs in the $\pm$ 50 ppm/°C range. Thin film resistors made from tantalum oxide will generally have a TCR of about -70 ppm/°C, while those made from nichrome may be as low as 10 ppm/°C. Most frequently, the TCR of resistors is presented in terms of a hot and cold TCR, which measures the changes between room temperature and $+125$ and -55°C, respectively. This can be somewhat misleading, since the TCR of thick film resistors is somewhat nonlinear with temperature, particularly near the extremes of $+125$ and -55°C, while the TCR of thin film resistors is nearly linear.

8.4.1.2 Noise. Resistor noise is an effective method of measuring the quality of the resistor and its termination. On a practical level, noise is measured according to MIL-STD-202 on a Quantech noise meter. The resistor current noise is compared to the noise of a standard low noise resistor and reported as a noise index in dB.

On a fundamental level, noise occurs when an electron is moved to a higher and lower energy level. This change in energy of the electron is noise. The more difficult the transition between energy levels (the more energy required), the greater the noise. Metals with many available electrons in the "electron cloud" have low noise, while semiconductive materials have fewer free electrons and exhibit higher noise.

The noise index is expressed as microvolts per volt per frequency decade. The noise measured by MIL-STD-202 is this "1/f" noise, and the measurement assumes this is the only noise present. However, thermal or "white" noise is present in all materials that is not frequency dependent and adds to the 1/f noise. Measurements are taken at low frequencies to minimize the effects of this thermal noise. The noise index of low value resistors is lower than that of high value resistors because the low value resistors have more metal and more free electrons. Noise also decreases with increasing resistor area (actually resistor volume).

In thin film resistors, thermal or white noise, in which the noise level is independent of frequency, predominates, with the noise level measured in dB/Hz. The net rms noise level is therefore a function of the bandwidth. In thick film resistors, a second noise source called Johnson noise or current noise is present in which the noise level is inversely proportional to the frequency. The current noise predominates over the thermal noise at low frequencies (below a few kHz). The units of dB/Hz are not meaningful unless the actual frequency is specified. A more applicable set of units in this case is dB/decade, since it can be shown that the rms noise is the same for any range of frequencies in which the high frequency is related to the low frequency by a factor of 10 × or any other consistent ratio. It is difficult to directly compare thick film and thin film noise at low frequencies without specifying the actual frequency range. For most values, the noise level of thin film resistors is somewhat lower than that of thick film resistors, even at higher frequencies.

Noise information is particularly important for low signal applications as well as a quality check on processing. A shift in noise index, with constant resistor value, geometry, and termination indicates a process variation that must be investigated. For example, thin or underfired resistors generate higher than normal noise. The conductor–resistor interface can also be an important noise generator if it is glassy or otherwise imperfect. Finally, poor or incomplete resistor trimming also generates higher noise. A resistor noise test, such as Quantech or one of its variants, is an excellent method of measuring a resistor attribute not easily obtained by other methods.

The noise level of thick film resistors varies widely with the ohmic value, the specific composition, and the size. Thin film resistors exhibit a significantly lower noise figure. The manufacturers data sheets should be consulted for representative values.

8.4.1.3 Voltage coefficient of resistance.

Thick film resistors exhibit a change in resistance under the influence of an electric field as defined in Eq. (8.28).

$$\text{CR} = \frac{R(V_2) - R(V_1)}{R(V_1)(V_2 - V_1) \times 10^6} \qquad \text{ppm/V} \qquad (8.28)$$

where $R(V_1)$ = resistance at V_1
$R(V_2)$ = resistance at V_1
V_1 = voltage at which $R(V_1)$ is measured
V_2 = voltage at which $R(V_2)$ is measured

The VCR is most prevalent in high ohmic value resistors owing to the nature of the tunneling contacts and is invariably negative (as V_2 is increased, the resistance decreases). The term VCR is in some ways a misnomer, as the actual effect is due to the electric field on a resistor as opposed to the absolute value. Therefore, long resistors show less voltage shift than short resistors made from the same material when the same voltage is applied. The formula for the VCR is frequently presented to show the field or gradient effect:

$$CR = \frac{R(V_2) - R(V_1)}{R_1 \, dE} \qquad \text{ppm/V} \qquad\qquad (8.29)$$

where E = field applied, V/m
dE = gradient of the applied field

The VCR for a representative family of resistors is in the range of -30 to -50 ppm/V, and is significant only for high value resistors used in high voltage applications. In most applications, the voltage is not sufficiently high for the VCR to be observable.

8.4.1.4 High voltage discharge (ESD). High voltage discharge effects, more commonly known as electrostatic discharge or ESD, drive the resistance downward. However, unlike VCR effects, ESD is not reversible. After being subjected to an ESD pulse, the resistor takes a permanent lower value. To be more precise, the resistor takes a "semipermanent" lower value, since it is possible for the resistor to recover under some conditions.

The mechanism for this phenomenon seems to be breaking of metal oxide to metal contacts in the resistor body much the same way a hole is punched through a capacitor under high load. This rupturing of oxide layer is permanent except that, under conditions of heat and humidity, the oxide layer may partially regrow. It has been experimentally shown that resistors that have moved lower in value because of ESD have partially recovered under thermal aging.

Except for the fact that ESD causes a permanent shift in resistance, ESD and VCR have a great deal in common:

1. Both tend to drive the resistance value lower with increasing voltage.
2. Both have more pronounced effects on higher value resistors.
3. Both are gradients and must be expressed in volts per length of resistor.

As a rule, the voltage gradient should be kept below 1 V/mil to ensure minimum ESD effects on sensitive resistors. Unfortunately, there is no generally accepted test method for either ESD or VCR as there is with TCR testing or high temperature aging. However, MIL-STD-883C, method 3015.2, does specify an ESD test that is applicable for discrete resistors.

8.4.1.5 Stability. Because of the nature of the conduction mechanism and the internal structure, thin film resistors are inherently more stable than thick film resistors. With proper processing, thick film resistors can attain a stability of better than 0.5% after 1000 h at 150°C under load, and thin film about 0.1% under the same conditions. Values below 100 Ω/sq and above 100 kΩ/sq typically exhibit poorer stability than other values. Thin film resistors made from nichrome are susceptible to moisture and oxidation effects and must be passivated within silicon monoxide or silicon dioxide to achieve ultimate stability. Resistors made from tantalum may be self-passivated by baking in air for a period of time to form a layer of tantalum oxide, which is impervious to further oxygen diffusion and also stabilizes the resistor. Most chip resistors are sorted for value as opposed to trimmed, and the inherent small drift observed in laser-trimmed thick film resistors will not be observed.

8.4.1.6 Configuration. Thick film resistors typically have wrap-around terminations on the ends for reflow solder mounting and/or improved heat dissipation. Thin film resistors may be fabricated on a variety of substrate materials, such as alumina, silicon, or glass, and are usually designed to be interconnected to the circuit by wire bonding. Except for power resistors or resistor networks, chip resistors are usually of the 0805 or 1206 size.

Acknowledgments

The contributions of the AVX Corp, the Novacap Corp., The TDK Corp, the Fair Rite Products Corp, and Dr. Kinzy Jones, Florida International University, to this chapter are gratefully acknowledged

Cleanrooms for Hybrid Manufacturing

Roy Trowbridge

9.1 Introduction

9.1.1 Contamination and the product

Why worry about a little contamination? This is exactly what we need to worry about most—"a little contamination." As products become smaller and tolerances tighter, defects caused by contamination have a larger impact on our product. Many people are not aware of how the product can be affected by contamination. Contamination can be particulate (solid, liquid, or gaseous), vibrational influences, temperature, humidity, or anything unwanted at a particular time in the fabrication of the part, though particulate usually comes to mind first.

It is common to hear about integrated circuits (IC) fabrication being concerned with 0.5-μm-size particles and how particles in disk drives can cause problems. Many other areas of industry need to be concerned with contamination problems of some type. Contamination can cause problems in fluid flow through capillaries, cause unwanted localized reactions on surfaces of parts and tooling, scatter laser light, and cause pinholes in photolithography exposures and thin coating. The list can go on and on. Contamination buildup on tooling and equipment and its possible effect on the assembly process and product must be considered. The presence of a 1-mil (0.001-in) particle on tooling with tolerances of 2 mils will start making parts already close to being out of specification.

TABLE 9.1   Relative Size of Small Particles

Micrometers	Particle
450	Typical silicon wafer thickness
300	Razor blade
75–100	Human hair
65	Particulation from crumbling paper
50–500	Metal contamination from sliding surfaces (unlubricated)
40–50	Smallest particle visible to naked eye
30	Particulation from seating a screw
30	Particulation from belt drive movement
7–150	Pollen
3–8	Red blood cell
3–>1000	Particulation from skin flakes
0.3–15	Bacteria
0–0.5–10	Particulation from cosmetic
0.01–3	Tobacco smoke
0.01–>1000	Carbon black

$$1 \text{ micrometer} = 1 \times 10^{-6} \text{ m}$$
$$= 0.001 \text{ millimeters}$$
$$= 1000 \text{ angstroms}$$
$$= 0.00003937 \text{ in}$$
$$1 \text{ angstrom} = 1 \times 10^{-10} \text{ m}$$
$$1 \text{ mil} = 0.001 \text{ in} = 25.4 \text{ micrometers}$$

The unaided eye can detect particles greater than 50 μm. Table 9.1 shows the relative size of small particles for comparison purposes. Submicrometer particles can destroy IC and some hybrid assemblies. As the hybrid assembly is fabricated, it may become more tolerant of contamination. Once contamination-sensitive parts are sealed, it may be irrelevant if a dust particle gets onto the board or housing. The presence of dust may not interfere with the placement of devices. Many people do not grasp the relationship of such small particles that can impact their product. Posting of signs showing these small measurements (1 μm, 0.5 μm) in relation to some physical object people can relate to is often useful to help them grasp the truly small size of the particles of concern. Table 9.1 gives number values to small particles, but actually seeing a size relationship, as in Fig. 9.1, has more impact and understanding.

In the past, contamination has been considered as dust and dirt that could be seen. Considering the size of the particles that can adversely affect products, the hybrid industry must now be concerned with particles invisible to the naked eye. More and more hybrid products must be manufactured in a low contamination environment (Fig. 9.2).

The amount of energy and effort required to remove particulate increases as the particle size decreases. The smaller the particle the more the effect of interatomic attractions becomes involved in its removal.

From these considerations, it will be seen that contamination can adversely affect the product, and that it is easier to prevent contamination from settling on the part in the beginning than to attempt successful removal once it is already there. The product may also be affected by other factors in its assembly environment, including temperature, humidity, gases, vibration, radiation, pressure, lighting, sound, exhaust, and/or deviation from the desired set

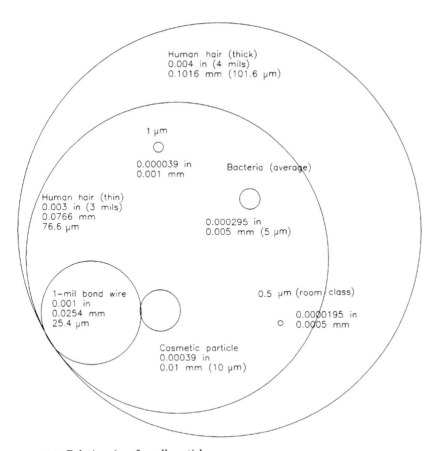

Figure 9.1 Relative size of small particles.

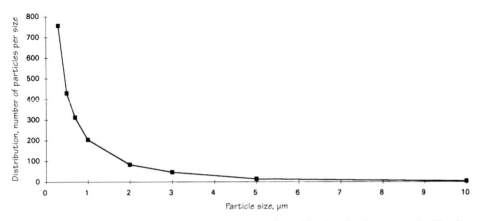

Figure 9.2 Particle size vs. distribution. Distribution of particle sizes in cleanroom air. The distribution of air particulate follows the above pattern. For a given size, there is more than double the number of particles half that size. The generation of particulate can also follow similar numbers, creating more small particles than large particles.

points of these variables. To overcome this, it is necessary to control the level of cleanliness in the hybrid assembly process area.

A successful hybrid facility must first look at three things: the product requirements, raw material and component characteristics and requirements, and the assembly process. Then look at the hybrid facility as it functions in relation to these items. Determine the requirements the product has at each stage and then tailor the assembly area to meet these requirements. The assembly process includes the process sequence, the layout logistics, the storage of both finished and in process items, and the packaging requirements.

Once the need and design for a cleanroom or clean manufacturing process controls have been established, the long extended process of building and maintaining a cleanroom begins. There are three major parts to a successful cleanroom facility. First and foremost is "management buy-in" of the cleanroom including the cleanroom concept of manufacturing and full support for its special requirements, not just support for rooms that cosmetically look clean. With management support for the cleanroom production facility, operators, technicians, engineers, facilities, and middle and low management will treat the facility as a cleanroom and abide by all the protocol required.

A cleanroom facility needs to be treated as such by everyone at all times. Management, sales, customers, vendors, contractors, and visitors are not to be exempted from cleanroom procedures. This can be accomplished through the second part of a successful facility program by educating everyone in how contamination control can affect the product. This would demonstrate the need for cleanroom procedures and what they actually are. The third part of the program is enforcement of the established cleanroom procedures. This is the area where management can help most. There must be an understanding of the financial impact on the product and what adequate procedures are for the product. This will help to allocate funds, time, and enforcement weight to lower management, facilities, and production so that a successful cleanroom facility can be maintained.

9.2   Cleanroom Design

9.2.1   Cleanroom classification and operation

Cleanrooms are typically classified by the quality of the air they are capable of maintaining. This quality is measured as the number of airborne particles 0.5 μm or larger found within a cubic foot of room air. This is expressed as multiples of 10, i.e., class 1, 10, 100, 1000, 10,000, 100,000. A class 1000 room will be capable of maintaining particulate contamination of the room air to 1000 or fewer particles (particle equivalents) of 0.5 μm and larger size per cubic foot. Table 9.2, taken from federal standard 209, shows the interrelationship of different particle size limits to given room classes. In a class 100 room, the number of particles $\geq$0.5 μm must be less than 100, the number of particles $\geq$0.3 μm must be less than 300, and the number of particles $\geq$0.2 μm must be less than 750. This is depicted in Fig. 9.3. The 0.5 μm count values are usually referenced for ease of convention.

TABLE 9.2   Class Limits of Particles*

	Measured particle size, μm				
Class	0.1	0.2	0.3	0.5	5.0
1	35	7.5	3	1	NA
10	350	75	30	10	NA
100	NA	750	300	100	NA
1,000	NA	NA	NA	1,000	7
10,000	NA	NA	NA	10,000	70
100,000	NA	NA	NA	100,000	700

*The class limit particle concentrations shown in Table 9.2 and Fig. 9.2 are defined for class purposes only and do not necessarily represent the size distribution to be found in any particular situation. Data obtained from federal standard 209.

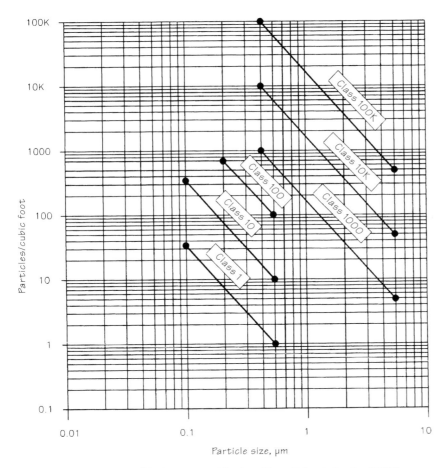

Figure 9.3   Particles per cubic foot vs. particle size. (*From federal standard 209.*)

It is impossible to measure all the air entering the room, so random and known samples of air are taken using air particle counters. Federal standard 209E[1] sets forth standardized methodology for classifying and testing cleanrooms. The American Society for Testing and Materials also prescribes testing procedures in its ASTM Standards for Cleanrooms.[2] An actual hands-on approach for testing is presented in Lutz.[3]

Unfortunately federal standard 209 and many other people in the past have been concerned with only the air quality and not surfaces that may be intimate with the product. It cannot be assumed that clean air will result in clean surfaces. Any particulate in the air can settle out onto surfaces and not be removed in the exhaust air. Such particulate could come from the process or equipment and never really enter the general room airflow. Thus we can develop an approach for room classification that will take into account the product level cleanliness of a process in a room.

The cleanroom air system is for the most part a closed, positively leaky system which conditions the air to desired temperature and relative humidity. This air is continually recycled back into the room through high efficient particulate absorbing filters (HEPA filters), also referred to as high efficient particle absolute filters. These filters are capable of filtering out 99.97% of particles >0.3 μm. Ultra low penetration air (ULPA) filters with greater efficiencies have become available for more critical applications. Their cost is higher and they present a greater pressure drop across the filter requiring HVAC capacity modifications.

The typical desired velocity out of a HEPA filter is approximately 70 to 100 fpm. There is a velocity loss as the air moves away from the filter and diffuses into the room. To achieve desired velocities at specified distances from the filter (product level) the face velocity must be increased. Caution must be observed since increasing airflow through a HEPA can increase particle penetration into the filter media, as shown in Fig. 9.4. The movement of air molecules and particulate in a filter can set up a vibration of the media. To reduce this impact onto the media a partial fixed damper is often placed across the

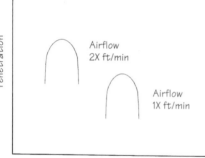

Figure 9.4 Particle penetration into HEPA filters. Increasing the airflow through a HEPA filter will result in an increased penetration of the filter by small particles. (*From Argot, "Contamination Control Technology Seminar Notes, 1993," Argot, Inc., P.O. Box 468, Honeoye Falls, N.Y. 14472, used with permission.*)

incoming air duct to diffuse the air and protect the filter media. Even with this plate the medium is subject to flexing and abrasion from airflow through it. Increasing the flow through the HEPA will reach a point where damage to the filter or shedding of material from the filter media can occur.[4] The resistance of filter media to airflow will cause a pressure drop across the HEPA. The pressure drop will increase with the filter efficiency (HEPA versus ULPA) and the age of the filter as it becomes loaded with filtered particulate. This pressure drop across a filter may also be dependent on the design, size, quality, and possibly manufacture. The amount of pressure drop across all filters in the system must be calculated into the overall HVAC (heating, ventilation, air conditioning) design to maintain the desired airflow and pressure within the cleanroom. Stabilization and knowledge of pressure and airflow are required to place equipment and operations within the cleanroom. Unbalanced air within the room can actually disrupt airflow and possibly move contaminates into sensitive areas even if they are under or near HEPAs.

A properly maintained HEPA filter should last for years (3 to 5 is common), but it does have a life span after which it will require replacement. Often, a HEPA filter becomes damaged from inside the room because of cleaning accidents, faulty equipment, or problems in installation of equipment. Once the medium is damaged it must be repaired or replaced as soon as possible. A hole or crack in a filter creates a high velocity dirt stream into the room. These leaks through the filters are undesirable since filters are usually placed over the most sensitive areas of the room or equipment. There are acceptable materials used to repair very small holes and cracks in a filter until replacement is feasible.[5] "The concern then becomes contamination (particulate and solvent) from the repair material and whether a good tight adhesive seal of filter media is attained.

The actual replacement of a filter must be performed so as not to release material into the room. A good HEPA change requires time, effort, and experience. Often this is done by isolating the affected area with a temporary plastic barrier. The unit is replaced and testing performed to audit the clean operation of the new filter unit and its installation. The area within the barrier and other affected areas are cleaned to a primary level before exposure to the rest of the cleanroom. A little extra care when working around HEPAs may avoid contaminating the room and the product.

Directionality and velocity may be maintained by extending a shield from the HEPA toward the sensitive area. This helps restrict diffusion and drag mixing with adjacent air. Air from the outside of the shield may creep up the inside of the shield, bringing with it particulate to mix with the clean HEPA filtered air. Velocity balancing should reduce creep. This is further explained by Karnick.[6]

A honeycomb structure placed in front of the HEPA to control the direction of the air is another way to overcome diffusive degradation of airflow without increasing flow through the HEPAs. There is a slight pressure drop across the honeycomb material that needs to be considered in each application. A plenum structure can produce enough static pressure to make the airflow across the honeycomb structure uniform. The internal surface finish of the ductwork and honeycomb material downstream from the HEPA must be a smooth, nonparticulating cleanroom finish. Directed airflow using honeycomb structures is ideal for complex workstations, conveyors, and localized assembly areas though possibly not for the general large room area. Large areas require more HEPA filters or higher flow.

Room design is usually driven by product and process, which will determine HEPA placement and room air changes per hour. The room is refined by establishing product and process layout, desired air quality at product level, and velocity required to give desired air quality. Extrapolated back, this will give face velocity at the HEPAs. HEPA size and number will give air volume needed to maintain air quality. HEPAs should be placed to optimize airflow patterns to the product and still maintain directed airflow quality.

The cleanroom is built on typical industrial ventilation designs as put forth in industry standard manuals.[7] The air characteristics of the cleanroom have to be maintained to tighter tolerances while being concerned about many other properties that normal industrial environments do not always have to consider: vibration, air patterns, invisible particulate, restrictive maintenance access, gowning procedures, and "clean" construction techniques. A full understanding of the cleanroom would require knowledge of HVAC practices and all other facilities practices. For most cleanroom users that is beyond their needs. Interface with the facilities personnel, presenting them with the requirements, and let them deliver the product. A basic understanding of HVAC items will help the interface with facilities. Disputes over the finished room can be tracked to misunderstanding of basic practices or not knowing the right questions to ask. Close work and communications with facilities people and contractors can help ensure that the cleanroom is built and maintained so as to provide an adequate clean environment for the product.

The HVAC system connected to the cleanroom is capable of tighter temperature and humidity control than normal building HVACs. Ranges of $68 \pm 2°F$ and $40 \pm 5\%$ relative humidity are easily obtainable for cleanroom air handling units. More exacting designs and equipment can give tighter tolerances but at increased cost. Cleanroom HVAC ductwork needs to be insulated and sealed for leaks more than normal HVAC systems because of the impact leaks may have on air quality and the cost to condition the air that may escape out of the system.

A class 100,000 cleanroom can have 20 to 30 air changes per hour and a class 10 as many as 500 changes per hour.[8] This amount of conditioned air relates to increased operational cost over noncleanroom areas, so conservation of conditioned air by recycling is very cost effective and desirable. Some fresh air has to be introduced into the cycle to maintain desired breathing levels and to make up for loss due to leakage and exhausts. This replacement air is brought in as "makeup" air which has to be conditioned to the same tight tolerances as dictated by the room.

Air is supplied to the room so as to establish the higher clean areas with an increased pressure level in relation to general building air pressure.[9] The cleanest area of a cleanroom complex is commonly 0.05 in of water column pressure higher than general building and 0.02 in higher than adjoining less clean areas. Set up the highest room pressurization in the most sensitive areas (cleanest room) and decrease the pressurization sequentially for each connecting room until reaching normal atmosphere pressure in the general building. This will ensure a flow of particulate away from sensitive areas and prevent the transfer of contamination from dirtier areas (Fig. 9.5). Cleanrooms are sometimes designed with airflow and pressure differentials within the room so as to provide pocket areas of higher pressure in sensitive areas. Such layouts are more complex to balance and maintain since they lack hard wall barriers to delineate pressure zones and movement of equipment within and near these zones can disrupt the air balance.

Pressurization should not be great enough to blow doors open. Doors should

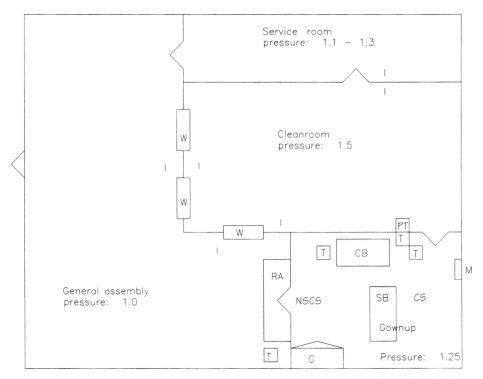

Figure 9.5 General room pressurization. A generalized layout of the rooms shows that the cleanroom should be at a higher pressure than the gownup area, which, in turn, is higher than the assembly area. Service room pressure should be less than that of any adjacent cleanroom. The net airflow from the cleanest areas toward the least clean areas reduces contamination migration into the cleanroom. Windows for tours reduce the number of nonessential personnel entering the cleanroom, but the thermal losses through the windows must be considered during cleanroom design. The gownup area layout segregates clean and less clean areas to protect the cleanroom. Double doors and movable furniture allow for the movement of large equipment into the cleanroom with minimal impact. CB = cleaning bench, CS = cleaner section, G = garment storage, I = intercom/phone, M = mirror, NSCA = not-so-clean area, P = passthru, RA = restricted area, SB = seating bench, T = trash/waste receptecle, and W = window.

have a closing mechanism that can overcome this pressurization differential. A door that does not close all the way is a potential area of back drafts for contamination and an imbalance to the designed room airflow. The room airflow and pressurization must be designed so that open doors do not present a problem for product cleanliness. Note that the design is for optimum operation which might deteriorate with time. Equipment and rooms do not get better with time, so the added up front cost for better design with longer useful life should be weighed against any cheaper way that may just look good.

The air volume lost due to leaks out of the room (seams, passthrus, doors, wall and floor penetrations) and captured exhaust air (solvent vapors and particulate) must be determined and replaced to maintain room pressurization. Efforts should be made to minimize room air loss since the cost to condition makeup air is so high. Excessive and unplanned loss of air may decrease room pressurization to the point that opening doors and passthrus may allow dirtier air to be sucked from less clean areas and disrupt desirable clean airflow patterns across product and equipment.

Air volumes captured by exhaust points such as solvent hoods and particle capture points (housekeeping vacuum) should be integrated into the overall room design. Many times the exhaust requirements of equipment and hoods added to a room are overlooked. A 3×4 ft hood opening could easily remove 1200 ft^3/min of air from the room. This is for maximum open area of the hood, i.e., sliding shields open. Machine ventilation, gas panel ventilation, and solvent vapor exhaust could easily take 100 to 200 ft^3/min each. Adding this up for all the equipment in the room can indicate a substantial volume of air lost from the cycle. Unfortunately, OSHA and good work environments set the minimum face capture velocities which can be implemented.[10] Reducing the face velocity at the hoods could lead to violations of regulations (a business problem) and inefficient removal of contamination from the operating area (a personnel and production problem).

There are several ways to reduce the face openings and still meet OSHA requirements. One is to extend rigid front shields down the front, forming an opening that just allows access to the hood for equipment and hands. This may not be a convenient method if the operator must move tall objects that cannot be tipped in and out of the hood. A second is to interlock moving shields to maintain a constant opening which is large enough for access. A third is to interlock the shield opening to a variable speed exhaust fan. As the shield is opened, the fan increases to maintain the appropriate velocity. Interlocks must be fail-safe! Test and document the interlocks on a periodic basis. This will help to keep the workers safe and the room balanced, and to maintain exhaust compliance records for reporting to the EPA.

A fourth is to utilize a workstation that cycles filtered air over a scrubbing mechanism. Such stations have been referred to as "biohoods" from their initial design for the pharmaceutical industry. A design consideration for this type of unit is to have enough capture velocity across the front bottom opening to prevent the escape of vapors from the front top of the station (Fig. 9.6). These units allow some of the exhausted air to be recycled through a scrubber

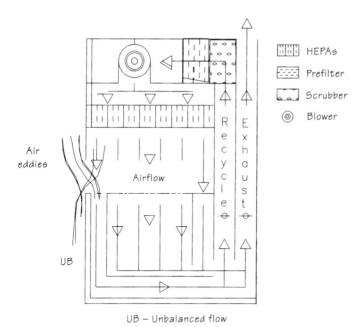

Figure 9.6 Hood elevation variable exhaust and recycling flow. Dampers in the recycle and exhaust ducts allow for adjustment of air intake through the hood opening. The use of solvents in the hood requires a net intake of air from the room. Placement of parts and assembly operations within the hood must take into consideration turbulent air eddies that can form at the front and back of the hood.

and prefilter system, then returned through a HEPA filter to the inside of the station. An adjustable plenum in the recycle air duct and a variable speed blower can adjust the hood's capture velocity to adequate levels. Further adjustments can be made by introducing makeup air into the recirculating loop. The directional, push-pull airflow at the front of the hood may allow a reduced face velocity and still meet exposure requirements and regulations. Placing honeycomb material across the front of the HEPA provides a focused laminar curtain of air which helps to reduce solvent vapor migration out the front of the hood. Such design modifications will require approval of the facilities' environmental, health, and safety representative. Installation of gauges and possible warning lights and alarms should be considered to monitor the loading of prefilter and scrubber so as not to overexpose the operator and to monitor any degradation of airflow. Another concern of this design is the ability to service the prefilter and scrubber without jeopardizing air quality of the cleanroom and hood.

The airflow for the type of hood used must be integrated with the cleanroom airflow patterns so as not to compromise product in the hood by the air taken from the room. A hood operating at any capture velocity can draw particulate dislodged from an operator's garments into the hood. Particulate can deposit onto product placed in the path of the hood opening and exhaust.

Vertical and horizontal air patterns may need to be designed into the placement and handling of product so as to avoid contamination falling onto the product.

9.2.2 Cleanroom types

Clean air production areas may be divided into three general types: bulk cleanroom, minienvironment, and combinations of the two. The typical cleanroom is a large scale, conditioned airflow which passes all air through HEPA filters as it enters the room. Minienvironments [also referred to as local process controllers (LPC) or filtered workbenches] take room air and direct it through HEPAs to localized areas of a room or equipment. Such local control typically does not condition the air it directs for temperature and humidity.

The bulk cleanroom has three basic designs depending on the placement of HEPA filters and return air ducts: vertical laminar flow, horizontal laminar flow, and mixed flow (Figs. 9.7 to 9.10).

Air entering the room must leave the room somehow. Cost dictates the recycling of most of this air though the cleanroom's HVAC system. The air system is designed to replace any air lost with makeup air. The less makeup air required the lower the cost of running the cleanroom. Cleanrooms are designed to minimize lost air. Typically 20% makeup air is added during recycling to help to keep the room air "fresh" to breathe. Air passed to areas outside the cleanroom through leaks and pressurization, air lost in fume extraction, and particulate capture (vacuum) must also be made up.

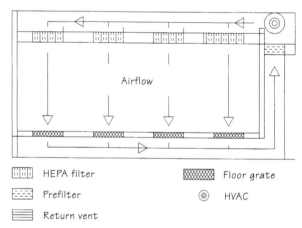

Airflow

	HEPA filter		Floor grate
	Prefilter	◎	HVAC
	Return vent		

Figure 9.7 Vertical flow cleanroom (also referred to as raised floor cleanroom). Air enters the room at the ceiling and is exhausted through grates in the floor. This design allows for tighter packing of equipment. Product can be contaminated by actions between the ceiling and the work surface. Particulation is quickly drawn to the floor and withdrawn from the room. The raised floor may not be adequate for vibration sensitive equipment, which will require isolation to the subfloor.

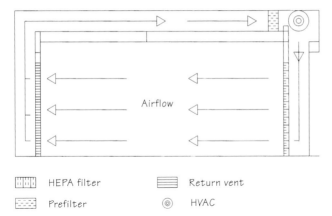

HEPA filter		Return vent	
Prefilter		HVAC	

Figure 9.8 Horizontal flow cleanroom—recycled airflow. Air enters the room from one wall and exits through vents in the opposite wall where it is directly carried to the HVAC unit via ductwork. All equipment after the one nearest the HEPA wall is downstream of potential contamination. Horizontal flow protects product from contamination generated above product level. Most sensitive product and assembly operations should be placed nearest the HEPA wall.

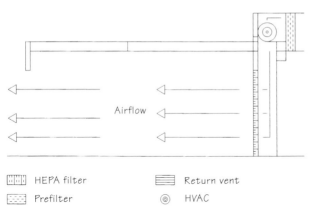

HEPA filter		Return vent	
Prefilter		HVAC	

Figure 9.9 Horizontal flow cleanroom—nonrecycled airflow. Air enters the room from one wall and is allowed to disperse itself into the general building. Replacement air is taken from the building proper rather than the cleanroom. All equipment after the one nearest the HEPA wall is downstream of potential contamination. Horizontal flow protects product from contamination generated above product level. Most sensitive product and assembly operations should be placed nearest the HEPA wall.

Airflow within the room should be predictable and for the most part laminar or unidirectional, which would make it predictable, which allows placement of the product into "clean air" zones. Laminar airflow helps prevent turbulent eddies that can trap contamination and release it in an uncontrolled manner *onto the product*. The laminarity of airflow can be affected by the placement

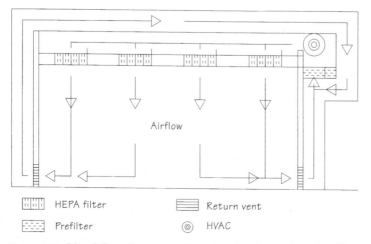

Airflow

HEPA filter		Return vent	
Prefilter		HVAC	

Figure 9.10 Mixed flow cleanroom. Air enters the room through ceiling HEPAs and is exhausted via wall vents, where it is immediately carried to the HVAC units via ductwork. More equipment can see first air than in horizontal rooms, but product and assembly is very susceptible to contamination from above. Equipment nearest the walls can also receive contamination generated by equipment in the center of the room. Turbulent eddies can form in the center of the room and between wall vents if air travel distances and vent spacings are not correctly planned.

and number of HEPAs and their relative position to return air ducts. Laminar flow allows the placement of product upstream of any potentially contaminating process or movement. Refer to Sec. 9.3 for further information.

By definition laminar flow is where the whole body of air moves with a uniform velocity in a direction parallel to its flow lines. Unidirectional flow is airflow parallel to its flow lines at a value of ±20% of its average airflow velocity. The airflow may not be truly laminar in all but the most exact designed cleanroom. Either full wall horizontal or full ceiling vertical raised floor rooms are preferred for providing true unidirectional laminar airflow.

A vertical laminar flow cleanroom, often called a raised floor cleanroom, is shown in Fig. 9.7. Air is forced into the room at a set velocity through HEPA filters set in the ceiling and exhausted through the floor. The floor is perforated or grated, allowing air to pass from the room to underneath. The true subfloor and perforated floor act as a return air duct. Adjustable slats in the floor provide a means to balance the laminar airflow and uniform capture velocity of the room air. Once a vertical flow cleanroom is balanced it is very important to maintain the location and slot openings of floor tiles. Floor tiles are picked up to clean under the floors, make utility connections, and move equipment. Changing locations or opening size of grates can affect the room balance such that the air is not drawn uniformly into the floor. This may set up undesirable airflow patterns that may result in contamination of the product and room. An advantage of this design is that contamination generated is quickly drawn toward the floor away from the sensitive product areas, unless contamination actions are performed upstream of or above the product.

HEPAs can be placed immediately over sensitive areas to ensure clean air-flow and adequate velocity and volume to sensitive areas.

Air enters the cleanroom via a wall in the horizontal laminar flow clean-room (Figs. 9.8 and 9.9). These are of two types: with and without return ducts, depending on whether the air is to be immediately recycled or not. The latter is more expensive to operate because of the noncapture of conditioned air. Return air ducts are placed on the wall opposite the HEPAs, creating a laminar flow across the room. A disadvantage of this type is that almost everything can be upstream of product. Only items closest to the HEPAs receive true clean air, as the airflow will pick up contamination from every object, machine, or person that it passes on its path to the return duct or out of the area.

Air enters through ceiling HEPAs and exits via wall ducts in mixed flow rooms (Fig. 9.10). The distance the air has to travel to ducts and the spacing of ducts around the room become critical in maintaining beneficial air pat-terns. Distances greater than 25 lin ft can result in turbulent eddy patterns in the room center that can trap contamination. Excessive duct spacing greater than 15 ft along the wall can result in turbulent eddies between the ducts. Taking air velocity, pressurization, HEPA, and duct placement into design consideration may give adequate airflow without such eddies.

A fourth type of room design combines the advantages of mixed flow rooms with vertical flow to better control air quality in selected areas of a room (Fig. 9.11). The placement of additional air return ducts in the center of a mixed flow room can overcome eddies caused by excessive distance to sidewall ducts. Further placement of grated raised floor air duct under product in the central areas of a room will ensure clean airflow to the product. Such combined designs lend themselves very well to conveyorized assembly and localized particle generating process steps. Particles generated are drawn toward the floor and exhausted rather than being swept across the room and possible other equipment.

The placement of HEPAs in relation to ducts, equipment, vents, hoods, and openings is very important in all these room types. The clean air from the HEPA should contact the sensitive product once and then exit the room. The filtered air may pass by sensitive product, then pass by less sensitive product, and then exit the room. Always try to keep the product upstream of any con-tamination generating operation or areas. The terms "first air" and "second air" have been used to describe airflow from HEPAs. Air that has not passed over any contaminating area or object is considered "first air." "Second air" has passed over a potentially contaminating area and may not be as clean as first air.

9.2.3 Local environments and local process control

Clean areas started as small areas and grew to encompass whole rooms. As the cost and need for very clean cleanrooms has increased, there has been a movement back to providing smaller, very clean environments within larger

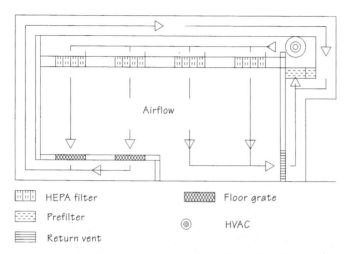

Airflow

▦ HEPA filter	▨ Floor grate
▧ Prefilter	◎ HVAC
▤ Return vent	

Figure 9.11 Combination flow cleanroom. Vertical airflow enters through ceiling HEPAs and is exhausted via both floor and wall venting. This can be used in larger rooms where the distance to the wall exceeds the design capability to maintain vertical unidirectional flow over the central assembly areas. Vents in the sidewalls of the raised floor area can also be used to exhaust air out of the room to maintain directionality of airflow. This combination concept is often applied to robotic conveyor assembly processes to keep the product under the cleanest air. Careful placement of HEPAs over conveyors and workstations can keep the product exposure to air classes 1 to 2 times cleaner than the bulk of the room.

cleanrooms of a higher class. This is being accomplished by minienvironments and the use of local process control.[6,11–18] HEPA filtered airflow strategically placed over or closely upstream of sensitive product and equipment can provide a very clean local environment. Aisleways accessing equipment that is always downstream from sensitive product do not have to be made as clean if the product is never brought into these areas unprotected, or if people do not routinely move through these less clean areas to access the higher clean areas. Thus these areas may have fewer HEPAs and the air quality can be less than in the sensitive areas. The general cleanroom can have a wide range of cleanliness levels throughout the day owing to operator movement, maintenance activities, equipment operation, loading and unloading, and product processing (Fig. 9.12). Because of these variations, the localized room quality can be one or two classes lower than the quality desired for the product or the established class of the room.[11]

Through the use of the minienvironments, isolated workstations, or local process control (LPC) the product can be kept in clean airflow, providing the desired cleanliness unaffected by fluctuations occurring throughout the room. The main effort in room layout is to keep the production area cleaner and pressurized with a slight net airflow toward the less clean areas. Local concentration of HEPA filtered air gives a more uniform and consistent airflow over the product and sensitive areas than may be provided by a general cleanroom. The optimized airflow can give a slight positive pressure to sensi-

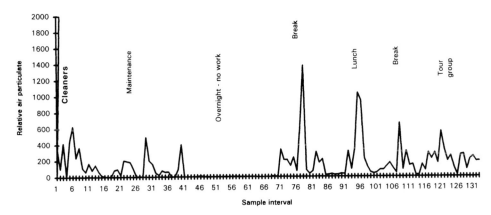

Figure 9.12 Clean area air cleanliness equilibrium. The relative contamination level within a cleanroom varies throughout the day as personnel enter or leave the room and perform various operations including cleaning. This demonstrates the need for limiting the number of people in a room and the number of entrance and exit times. The room requires some amount of time to recover from particulating operations. The room can easily meet the design classification during nonwork periods when most rooms are tested and certified. It is more important that the room be able to meet and maintain classification during operational work periods.

tive areas in respect to less sensitive neighboring areas in the same room. This will tend to move particulate out of the sensitive areas and provide a barrier to prevent particulate entering or coming back into the area. Minienvironments can be equipped with ionizers for ESD purposes. Having the ionizers closer to the sensitive area will provide better ESD protection than if they are located at ceiling level. Placement of ionizers should not be so close to the product as to place the product in the strong local ion fields of the emitter points. Emitter points must be maintained to remove particulate buildup, though studies have shown that the local minienvironment can recover from emitter particle bursts in less time than the general room.[18]

Understanding of the operation of the equipment in relation to particle generation and airflow patterns can help to determine the best type of local environment needed.[16] A gross ideal of airflow patterns and contamination generation can be determined by physical observation, extensive testing, and intuitive knowledge. Such initial tests may not reveal all the trouble areas since many particles generated are smaller than those visible to the naked eye. Current refinements of computer modeling are helping to design out contamination problems for these nonvisible particles in process equipment.[19] Such testing can be done in house, through a consultant or by a vendor.

The use of local environments can produce a 10- to 100-fold cleanliness differential of air quality within a room. Product can be kept under a class 100 cleanroom environment within a room built and maintained at class 1000 or even 10,000. Local environments can reduce the overall cost of constructing and maintaining a general cleanroom. Properly designed local environment rooms can allow use of the higher class gownup requirements if the product is always kept in the local environment, which is an additional cost saving. The lower costing class 10,000 frocks may be used to handle product in the class

100 area. This is easily achieved in robotic, module, and conveyorized applications. Rooms requiring the operators to carry product through different areas of cleanliness still require the higher class gownup. Remember, garments are protecting the product from the operator.

The load, unloading, and processing of product can generate particulate in bursts when the operation occurs and the material is handled.[11] Room scale handling of air quality is slower to attenuate control of such bursts. Minienvironments and LPCs tend to remove such challenges more quickly because of their closer proximity to the sensitive area. Thus local environments can reduce the duration of particle challenges to sensitive areas. The immediate environment is faster to recover from particle bursts than the overall room.[18]

The actual structure of minienvironments varies through different interpretations from a sealed glove-box type of enclosure, to an enclosed workstation, to a localized concentration of HEPA filters over equipment in a cleanroom. For discussion here the first two types are considered as minienvironments since there are some physical barriers demarcating the limits of the controlled environment. Minienvironments can be grouped as ceiling or equipment mounted depending on how they receive their conditioned air.[14]

Ceiling mounting employs shielding around the equipment connected to the ceiling. This may be just shields extending down 18 to 24 in from the ceiling to somewhat control the direction of the airflow to a full enclosure around the equipment. Such environments work best if HEPA placement, return air ducts, and utility placements are planned ahead. Retrofits for ceiling mounting need to consider the strength of the existing ceiling, whether the air handling system can accommodate the load, and if shielding will interfere with the room balancing. This mounting may make equipment maintenance a little more cumbersome. Shield design will have to address as many parameters as a vertical flow workstation, the more important being airflow differentials, creep from outside the shields, airflow eddies, and placement of product and tooling within the environment.[10] Ceiling mounting can be less flexible than other local environment alternatives and more expensive to alter as future equipment is added.

The equipment mounted minienvironment may prove to be more versatile and to provide a better local control. Cleanroom vendors have promoted the dropping of ducts from the ceiling which attach to the equipment. Thus the environment gets the same quality of air as provided to the rest of the room without adding equipment to the floor space of the cleanroom or increasing vibrational problems from the separate blower units. This approach is slightly flexible and still requires forward planning.

Another alternative is to utilize stand-alone air circulation systems to provide clean air circulation directly over the equipment. This is a very versatile and true local process control (LPC) approach to improving the local environment of a process. Tool mounting or LPC modules can be isolated from the equipment to reduce vibration transference from the blower fan. Connection of the blower to the equipment area can be via flexible or rigid ducting. Whatever ductwork is used, it should meet the same cleanroom requirements as any other piece of equipment. The air circulated by the blower must pass through a HEPA filter before it sees the local environment. Surfaces in front of the HEPA must have an acceptable cleanroom finish. This includes the inside of ductwork if the HEPA is on the blower unit end of the ductwork. The use of ducting will allow optimum placement of clean airflow over the product. Commonly such stand-alone units only circulate the air through HEPA filters and do not condition for temperature and humidity. Units can be made that would condition for temperature and humidity, but these would take up more space and not be as stand-alone. They would need to be larger, connected to water and drain, and isolate the conditioning part of their unit so as not to disrupt the air quality of the room due to thermal and humidity gradients which may be created, or from particulation of moving parts. Note that any blower and motor system placed inside the cleanroom must be contained so as not to release particulate into the room.

A unit placed in the room will need to be serviced sometime. Smaller units may present less potential contamination to work than larger sophisticated units. Removable duct connections at the equipment allow for ease and flexibility during maintenance. Another design plus is to have the LPC completely free-standing instead of hanging the blower, duct, and filters off a skeleton support frame. A maintenance person is more apt to reconnect a single unit than something that must be reassembled each time access to the tooling is required. An unassembled component type of unit takes up a lot of floor space when disassembled for maintenance of the equipment and can be a trip hazard.

Any connection of ductwork through the equipment or workstation should be a tight, gasketed seal. The movement of air may suck lower quality air through unsealed gaps interfering with designed clean air patterns potentially contaminating product. Such connections should be tight but easily detachable for maintenance access. Threaded bolts may be an acceptable fastener but the threads should be outside the desired minienvironment. The action of threading a nut onto a bolt can generate particles, so any maintenance assembly or disassembly operations should be assisted with a vacuum source to capture any particulate. Whenever possible, perform frequent maintenance actions remote to the sensitive areas, or provide some means of protecting sensitive areas when the work is being done. The choice of protection and tools used for maintenance should not be a source of particulate.

Blind pop rivets should be used for any ductwork or room shell applications. The center of regular pop rivets is plugged by the popped portion of the shank. This is not a tight seal and can loosen more owing to vibrational movements caused by the moving airstream. Leaky rivets can be a source of eddies and contamination, especially if the leak is upstream of the HEPA. Any gasketing, ductwork, or connectors should not outgas, particulate, or degrade the airstream that sees the sensitive area of the equipment. There is a need for a better finish on the inside of the equipment that sees the airstream than on the outside of the tooling which will not affect the product.

Optimum performance of tool mounted minienvironments requires additional design to integrate clean air supply to required areas of the equipment. This may be as easy as ducting directional air over a conveyor to redesigning the outer shell of a piece of equipment. A successful environment will reflect a joint effort between the equipment manufacturer, the LPC manufacturer, and the existing cleanroom.

9.2.4 Return air ducts

Air return duct openings must extend to the floor or else stagnant air eddies will be formed. Contamination can accumulate in such eddies and release to the room environment over time. The placing of materials and equipment should not block air return ducts unless the room airflow designs have specifically taken this into account. Interfering with designed airflow patterns can change room pressurization and decrease the efficiency of the room air handling system.

Return air ductwork should be sealed as tightly as supply ductwork. Leaks here can draw unconditioned air and particulate into the air cycle. Particulate will load up the prefilters faster, resulting in increased labor and materials for maintaining the air handling system. Prefilters are typically 90–99% efficient in removing medium to large particulate.[8] A full 1–10% of the particulate is passed through to challenge the HEPA filters, increasing the load on the HEPAs and shortening their usable life. HEPAs that are 99.97% efficient allow the passing of 0.03% of the particulate. The higher the challenge to the filters, the more particles that get through. This results in an increase of particulate entering the room. This will not be as noticeable with new filters and facilities, but as time goes on, the air from HEPAs will not be starting from the same "zero" baseline clean air.

Increased leaks will increase the amount of air needing conditioning, which will increase the operational cost. If the system is initially marginal in design, unplanned leaks can compromise the ability of the air handling unit to adequately provide conditioned air to the cleanroom.

The placement of return air ducts is integral to the design of a cleanroom. They establish unidirectional or laminar airflow patterns and the pressurization of the room. Excessive spacing between return vents results in dead space centered between the vents at the wall. Turbulent and possibly fluctuating air eddies are established in this space which can trap particulate, releasing it unpredictably back into the room. In vertical flow rooms excessive spaces between HEPA filters and return vents can cause air from subsequent filters to lose its downward directional thrust before reaching the floor area. Normally pressurization and airflow velocity from the filters will drive the air perpendicularly toward the floor. Close to the floor (hopefully below product level) the air will bend toward the return vents. Excessive distances and improper balancing of airflow will allow this bending to occur closer to the ceiling, giving a curved sweep from HEPA to vent. This again can allow dead air eddies to establish, which can trap and release particulate. Air patterns should be designed and tested to bend below product level and minimize such dead or turbulent areas. Whenever air velocity is modified, the altered cooling effect on operators working in the airstream and potential moving of tooling or parts must be considered, especially if the flow is increased.

Equipment should be placed in the airflow with the higher sensitive area upstream of lesser areas. The air will pick up contamination as it flows from the HEPAs to the ducts. Thus air in the immediate vicinity of return vents and ducts can be more contaminated than the air in the room proper. Sensitive parts should not be assembled or staged immediately in front of return vents. Once the room is characterized and the airflow balanced, careful thought should be given to any equipment movement within the room. Disrupting air patterns can cause local concentrations of particulate. *Do not block, restrict, narrow, or enlarge return duct openings without addressing any problems this may cause.* Their size has been calculated into the balanced operation of the cleanroom. Air patterns will need to be tested and possibly rebalanced when stand-alone local process control units are placed in the room, especially if their intakes are close to room return ducts.

9.2.5 Workstations

The layout of workstations, whether for manual, semiautomatic, or automatic operations, has the same concern: to present a clean environment to the product and perform the operation with the least amount of contamination generation. Layout of the assembly process should be with the sensitive components upstream of any particle or contamination generation. Ideally, provide smooth cleanable finishes on station areas between the filters, product, and any tooling that will touch or pass over the product.

A workstation may utilize multiple qualities of surface finishes depending on the relation of product to airstream. The best finish should be required for all surface areas the product will contact, pass through, or be downstream from. Surfaces below and downstream of the product path may be finished to a lesser quality. Surfaces isolated from the product zone by airflow barriers, physical partitions, or sealed enclosures may have an even lower quality finish. The latter finishes must still meet the overall cleanroom requirements. This zone concept works very well for robotic applications, conveyors, and tooling supports.[20] This is depicted in Fig. 9.13.

Zone A requires the best finish, 10 μin, and includes all surfaces from the product upstream to the HEPA. Zone B surfaces require a lesser finish, 35 to 60 μin finish, and includes the bases and undersides of tooling supports, channeling for wiring, and utilities. All of this zone is downstream of the product area. Zone C surfaces require a still lesser finish. Zone D, the lowest rated finish quality, requires a finish of 100 μin or greater. This can be remote surfaces and undersides of the station, utilities supplied to it from below, and control panels remote from the station. The actual finish numbers will vary according to the product requirements. Note that, if a utility feed or fixture passes through zone A, it will require a zone A finish. This type of lay-

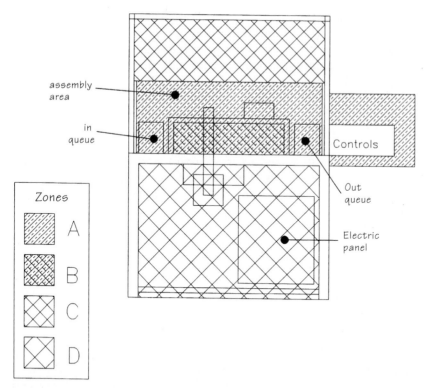

Figure 9.13 Workstation zone finish. Given unidirectional airflow from the back of a workstation, only areas intimate with or upstream from the product require very good surface finishes. Areas the product passes through or controls touched prior to touching the product will require good quality finishes. Areas isolated from the product by laminar flow or physical barriers (below assembly areas or station tables) can be finished to a lesser quality. Vertical airflow will require good finishes above the product but lesser quality behind. Horizontal flow can allow a lower quality above product but areas behind product must be of high quality finish. Airflow in relation to the product should be considered when designing assembly actions.

out works very well for conveyorized robotic systems and can present a cost savings when machining tooling supports that are not intimate with product versus tooling that is in contact with the sensitive product.

The zones A, B, C, and D must be related to the quality of air passing through them. Air is cleanest in zone A as it comes from the HEPA and can potentially pick up contaminates from tooling or the process as it travels to the room proper. Zone locations will vary slightly depending on airflow (horizontal and vertical) and number of sources. Further cost reductions may be realized with horizontal flow, which can present a more compressed primary clean zone A region.

Placement of sensitive material should be designed around the workstation or the workstation should be designed around any required placement and movement of product. The first 6 in of space from the HEPA represents a very turbulent airflow region. The front 6 in of a horizontal or vertical laminar flow workstation with solid side shielding is also an area of turbulence. The airflow from the station mixes with slower moving air outside the station, creating the possibility of creep of the outside air into the front section of the workstation. A restricted zone equal to 20% of the open height in the front of a workstation is suggested for vertical airflow systems because of this diffusion and mixing.[10] This percentage may be reduced if honeycomb material is used to directionalize air at the front of the workstation. Once a workstation is operational, the airflow patterns should be visibly tested. Some air particle counters have the capacity to measure airflow and perform lots of record keeping and statistical manipulations, besides the normal particle counting, temperature, and relative humidity measurements. This is useful in characterizing station and room air patterns. Process operations within the workstation often generate particulate, which should be captured and not allowed to settle on the product. Witness wafers and air particle counts are two ways to demonstrate this and help to optimize placement and correct deficiencies.

Plan for the product and sensitive items never to leave the primary zone of clean air and cleanest finish. All movement of product into fixtures, transport from station to station, or anything touching product and sensitive areas should remain in the same quality of clean air. Movements from clean airflow, through less clean airflow, and then back into clean area should be avoided to the extent possible. The bulk of the operator's arm or a robotic arm may be in a secondary clean zone, but the hand or actuator should never leave the clean zone, especially upstream of the product. Movement from a lesser quality area may deflect particulate into the cleaner area. When movement from secondary to primary zones is required, try to keep the most sensitive material upstream and move with slow, smooth motions. The faster the movements, the stronger the turbulent air eddies that will be established behind and beside the moving part which can carry particulate toward sensitive areas.

The distance the airflow will remain laminar can be increased by placing honeycomb or hexcell structured material in front of the HEPA. The hexcell material is a honeycomb structure which presents numerous parallel tubes for the air to pass through the structure. This straightens out the air's movement into a more directional parallel path which can travel a greater distance before diffusing away from perpendicular. This is demonstrated in Fig. 9.14. The honeycomb material may be made out of aluminum, steel, fiberglass, or a similar material, the main requirement being an adequately smooth, nonparticulating finish since it is immediately upstream of sensitive areas and of the product. The honeycomb structure should be finished so as not to generate particulate into the airstream. A polyurethane paint over a properly prepared base metal can be an adequate finish. Material should be degreased prior to priming and coating and placement as an assembly into the airstream. The honeycomb edges and frame should be sealed with as much care as that used to seal HEPA filter media to its frame. Mating of honeycomb pieces cut at various degree bevels can be used to bend air around corners with slight pressure drop and still maintain laminar and unidirectional airflow.

Ionization grids may be placed on either side of the honeycomb material. Upstream placement can see a reduction in the ionization's effective distance due to blocking by the honeycomb. Downstream placement will cause some

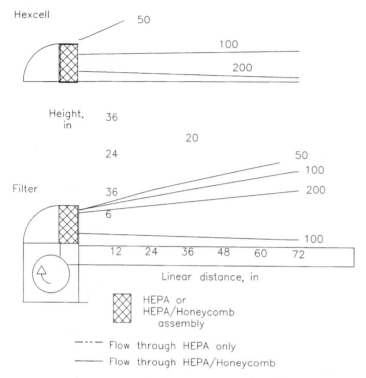

Figure 9.14 Airflow from a HEPA filter. Airflow forced through a honeycomb (hexcell) material will tend to maintain its laminarity longer and diffuse into the room less than from just a HEPA filter. Static pressure between the HEPA and the honeycomb must be adjusted to provide uniform velocity across the honeycomb face. Fewer turbulent eddies are formed around the outer edges of the honeycomb material than around the filter. The higher velocity of honeycomb-directed air can form a flow barrier to particulate falling into the directional airstream.

interference with laminar airflow having a very slight reduction of effective laminar distance of the airflow. Downstream placement is more convenient for maintenance of the ionizer emitter points. Testing or personal preference will determine the actual placement.

There is some energy loss as the air moves through the hexcell, evident as a small pressure drop through the hexcell. Design of the air driving system must account for pressure drops across the HEPA filter, the hexcell, and any ductwork between them, in order to maintain the airflow required by the product. The hexcell does not have to be right in front of the HEPA but can be separated by short distances of ductwork. This distance would be set by the maximum pressure drop that the air circulation system could overcome to still provide adequate face velocity at hexcell and product. To provide equal static pressure a plenum damper system may be used between the HEPA and ducting. This allows placing clean airflow inside a station so as not to blow past station and tooling supports. Such setups are very useful in robotic and conveyorized applications. Airflow to honeycomb discharge points may be supplied via ducting from the general room HVAC system or by stand-alone blower units as minienvironments or local process control units.

All airflow going into the workstation should be provided with a path of exit. This path should be tested to verify that desired airflow is provided. The air may be passed into the room proper or captured by the intake of minienvironment LPC units. This airflow should move down and away from any sensitive products. The airflow from known particulating processes (ultrasonic welding, screwing, scrapping, or sizing operations) should be removed immediately from the air and not be allowed to mix with the general room air. This air may be returned to the room or HVAC system once the particulate has been removed via prefilter and HEPA filtration. If this air is not returned to the system, its volume should be figured into the overall room loss and the HVAC system adjusted accordingly. Many overlooked points of pressure loss may add up to disrupted airflow patterns in the room and potential pressure drop in the area or room. The amount of room pressurization drops may be lessened by fitting stand-alone LPC units with ductwork to capture freed particulate in its intake airstream, needing then only to balance the room airflow around the LPC and workstation. One disadvantage is that the eventual maintenance of the LPC unit's filter system will be performed in the cleanroom. The more particulating the process the more often maintenance will be required. Any maintenance within a cleanroom will require good cleanroom technique so as not to contaminate the room or surrounding work areas.

9.2.6 Finish requirements

The design of operating stations and rooms should integrate airflow, product movement, and station operation. Ideally, the product should remain in "primary" or "first" clean air until it is not sensitive to contamination, via assembly or some type of packaging. The part should not pass downstream of any particulating action without cleaning or capturing of particulate. All station surface lines at product and upstream should be nonparticulating, smooth, visibly clean, and easily cleaned.[21] The quality of the finish will be dependent upon the sensitivity of the part and that area's location in respect to the part. The quality of the finish and how it will be verified should be put into the station's design specification given to a vendor. A single station may have several zones of finish quality, as explained in the workstation design section.

In the past, equipment has often been finished with a baked-on wrinkle finish, which is an industry standard. This is a good, durable finish and hides a multitude of surface defects that may occur during fabrication. Unfortunately, these wrinkles present places for the capture of particulate which is very hard to remove. Pieces of cleanroom wipes have been known to catch on the rough wrinkles, adding to the contamination load. The need for cleaner rooms requires cleaner and smoother surface finishes than in the past. The smooth finish requires a smoother base metal surface with fewer defects or a top coating which has very good leveling properties. All this means a more careful preparation and handling of material used to make a piece of equipment, which translates into increased cost. Surface finish quality may be a moot point if dealing only with experienced cleanroom equipment suppliers with proven product on the market. Many times the need for customized equipment requires working with vendors not familiar with cleanrooms or contamination control techniques. These vendors need exposure to cleanroom training programs and contamination control so they may understand cleanroom requirements and work to make equipment that will not put the product or the cleanroom at risk of contamination.

The life of the finishes used should also be considered.[8] Less durable coatings (baked enamel, latex, and acrylic) may slough off or crack with wear. Finishes may deteriorate with time through aging, solvent attack, usage, flexing, and rubbing actions of cleaning or process operation.

Epoxy and polyurethane paints are more durable coatings but may require more effort to prepare the surface correctly and apply. These coatings tend to absorb more impact without minor damage (small particle generation) and when they fail generate larger particles that are easier to detect and remove than those of other, softer paints.

The common anodized aluminum finishes have been found to wear off over time. Some anodizing (hard coat) particulate less than others (black) anodizing and may be an acceptable trade-off between cost and cleanliness. Electroplating (nickel on steel, nickel on aluminum) can provide good, hard finishes on parts. Slight disadvantages may be cost, finish quality, dimensional changes, and tempering of the metal, which inhibits reworking or tapping. Electropolishing of steel surfaces can give as smooth a surface as cost permits. Plating and electropolishing are very wear resistant coatings but have an added cost factor. Worked stainless steel surfaces should be passivated to ensure their corrosion resistance.

Once the station and tooling have a good finish, the interactions must be looked at between the surface and part, pallets, and trays. Minimize sharp or rough edges rubbing against any surface that could potentially scratch a surface and generate particulate. Any particulate that cannot be removed earlier must be kept downstream of the product and should be removed from the airstream as soon as possible. The worst approach to particulate generation is to do nothing. It is usually undesirable to let large amounts of particulate travel to the return air duct, leave the room in the return air, and be captured (hopefully) by the prefilters and HEPAs. There are several disadvantages to letting particulate just travel to the return ducts. The most obvious is that anything (product, raw material, storage devices, operators, workstations, or tooling) downstream of the point of generation can become contaminated and the contamination can be moved back into sensitive areas with subsequent actions. An often overlooked disadvantage is that this increased particulate in the HVAC will load up the prefilters and HEPAs, reducing their life span, which increases the room maintenance overhead. Changing of prefilters external to the cleanroom is relatively inexpensive and usually does not impact the quality of the cleanroom. The process of changing HEPAs inside the room without compromising room air quality is expensive and time consuming. The quality of a room would be more extensively harmed by any damage to HEPAs if its supply air carried this increased particulate load.

Also, remember that HEPAs are 0.03% inefficient in that 3 particles out of every 1000 can get through. If the supplied air is dirtier by a factor of 10, there can be 30 particles that get through in the same time that 3 particles would get through in a cleaner air supply. These are particles introduced in the most sensitive area of the room airflow, upstream of the product.

The finish choice for an item must also be able to withstand chemicals used in the process or environment and not add contamination to them. Solvent-based solutions can attack plastics while water-based solutions can attack some metals. Deionized water (DI) can attack some metals, dissolving out ions. These can be a contamination problem since the ions can be transported to sensitive areas of the product or of the laboratory. The loss of ions from the metal is manifested as corrosion, which causes pits in the metal surface. Particles can break off from these corroded pits and enter the supposedly pure water stream for transport to sensitive areas. Carried to extremes the metal may corrode through, releasing large flows of water into the equipment and cleanroom. Copper should never be used for deionized water transport. Stainless and tinned-stainless have an acceptable corrosion resistance for use with DI water. Glass is acceptable but requires gaskets, is fragile, and may leach out ions, such as sodium—a commonly unwanted impurity.

Plastics are usually inert to deionized water. Most plastics are formulated with modifiers such as plasticizers, fillers, stabilizers, and release agents to give them desired compounding and structural properties. DI water will leach out some of these modifiers, especially phthalate plasticizers. This will tend to make the plastic brittle and increase the water's total organic carbon (TOC) content. The former becomes a facility degradation problem over time. The latter becomes an immediate contaminate issue. Using components fabricated from pure plastics that contain little or no modifier reduces a lot of the above concerns. Such materials cost more and may not have as wide a range of physical properties as modified plastics. Certain grades of polypropylene, polyethylene, PVDF, and Teflon are available for use with deionized water and will still maintain ultra high purity water quality.

The realm of ultra high purity water is beyond the scope of this text. The reader is referenced to the local systems vendors, consultants, and various articles.[5,21–25]

It is interesting to look at the surfaces of equipment, tooling, cabinets, cleanroom, piping (inside and out), and exhaust ducting after they have been in use for a period of time. This should be part of scheduled cleanroom cleanings and audits. Examine them for visible wear, damage, ease of cleaning, and amount of particulation. If the finish is not lasting, it must be replaced or its effect engineered out of the product. Actual documentation of such findings can help reduce future problems and educate vendors, contractors, and successors.

9.3 Layout Considerations

9.3.1 Logistics

Most of the design effort for a cleanroom centers on fabricating a part without contamination. People are very aware of keeping the parts and equipment clean and protected while in the room. The logistics of moving raw material in, spent material out, and finished product is often given little thought. Pilot quantities of materials are easy to handle, to clean, or carry in and out of the room. The more parts at one time there are to move into the cleanroom, the greater the chance operators will skimp on cleaning the parts or the quality of cleaning given to each part. This suggests the need for machine cleaning of parts which will give consistent results when their volumes become large. Also there may not have been production time planning or space to clean items adequately. Needless to say, this slight oversight can become a major contamination problem that can impact the quality, yield, cost of the product, and integrity of the cleanroom.

All materials coming into the cleanroom must be cleaned sufficiently to remove any potential contamination. This includes raw material, parts, tools, and packaging material for finished and in process goods.

Vendors of cleanroom-specific materials often package their product in cleanroom-compatible containers, placed inside a clean poly wrap or bag which is placed in an overwrap or bag. These double-bagged items are shipped inside standard boxes or containers to customers. The double bags should not be opened by the receiving personnel but delivered intact to the gownup or prep room. Once in these rooms the outer bag is cleaned and removed, exposing the clean inner bag. Do not cut or penetrate the inner bag at this time. Inspect the inner bag for damage and contamination. If the inner bag is dirty or not intact or the part(s) inside show signs of contamination, the part(s) must be cleaned before they are passed into the cleanroom. A three-stage unwrapping operation is suggested. First, clean down the prep bench, open the outer bags on the dirtier side of the prep bench, place the inner bagged items on the cleaner side of the bench, dispose of packing material waste, and clean down the bench. Bags should be slice-cut, not torn, to open them since tearing generates more particulate and can throw it up into the airstream. Any equipment used to open bags or wrapping should be cleaned before and after use. Cleaning should be from the clean side toward the dirty side. Second, open and clean down any bagged parts suspect of being damaged or contaminated, rebag into appropriate containers if required, place the rebagged or cleaned parts on the clean side of the bench, dispose of wrapping material waste, and clean down the bench. Third, pass all cleaned items into the cleanroom proper. The inner bag can be opened inside the room when the material or item is ready to be used.

All materials received in protective wrapping should be audited to verify the quality of cleanliness provided by the overwrapping, compatibility of wrapping to the product, and condition received. If the package is received in suspect condition, it must be determined if damage was internal by receiving or material handlers, external by the transporting carrier or by the original packager. Damage caused by internal people can be addressed by educating them in the handling methods desired. Include them in cleanroom training programs so they may come to understand why so much effort has been put into special packaging and handling of these items. Damage resulting from external handling to the company should be addressed to the vendor. If the vendor is at fault, their handling and packaging of the items must be improved. If the carrier is at fault, the vendor may need to use a more durable package that will withstand typical transportation problems and still comply with federal Department of Transportation regulations.[26,27]

If a local environment workstation is used as the prep area, the airflow patterns generated during the actual cleaning operation must be considered. The net airflow should be away from the cleaner side of the prep and gownup room while also being away from any entrance or wall penetrations to the cleanroom. Air downstream from the actual cleaning area should flow toward the dirty side of the prep and gownup room and be removed from the room as soon as possible. Any directional air should be arranged to flow toward the dirty side of the bench. Remember, first air (the cleanest air) must be directed to the sensitive product first, and then to other areas.

The operators should be garbed while cleaning and unbagging parts to prevent any contamination from their hands, body, or clothes from falling onto the clean area or cleaned parts. Any items received that are grossly dirty should be given an initial cursory cleaning prior to introduction into the prep and gownup area. Once in the prep and gownup area parts will still need to be cleanroom cleaned for passage into the cleanroom. The location of cleaning supplies and appropriate procedures for handling packaging material and cleaned parts should be reviewed with all operators involved.

There are at least five items to note for the operators involved with cleaning parts. One, do not assume everyone naturally knows how items entering this particular cleanroom are to be cleaned and handled. Two, show the operators the techniques that this part and room require. The operator should do the actual cleaning during training to demonstrate that they understand correct procedures and techniques. Three, review the cleaning procedures with all people involved on a periodic basis. These sessions can be used as a refresher session for reviewing existing procedures, to introduce any changes that are required, to discuss special problems, and as a forum for concerns of the operators. Four, all materials including the materials used for cleaning need to be handled in a manner suitable to cleanroom requirements. Five, audit the operators' techniques from time to time. Over time, deviations from set procedures can cause a drift to decreased cleaning efficiency which will result in increased contamination introduction into the cleanroom. It is also suggested that management, sales, and purchasing realize packaging requirements for each cleanroom since it will require more time and space and will increase cost of the finished product.

Any local environment or LPC used should be positioned to provide a net flow of air away from desired cleaner areas, toward the dirty area and the exit from the gownup or prep area. Prep benches with horizontal HEPA airflow can be used, but the downstream effect of the airflow must be considered. The clean garment storage, in use garments, and the actual final gowning areas should not be downstream of the unbagging operation. Contamination generated by the unbagging operation can be picked up by the airflow and deposited downstream and carried back into the clean area on garments.

It is desirable for the airflow from LPCs or minienvironments to be left on continually. This will help to predict and maintain desired airflow patterns in the room. Also particulate may settle on the face of the filter media and any directing ducting when the airflow is turned off. The sudden surge of air through the filter and ducting can dislodge this particulate, introducing it into the room air and possibly onto sensitive product.

A cleaning area should have enough waste receptacle capacity to accommodate the normal amount of waste generated during cleaning operations. As with any cleanroom, a covered waste receptacle is desirable over open containers. Airflow over open containers can set up turbulent eddies which can pick up contamination from material inside the container and reintroduce it to the room airflow. These would be particles under 100 μm, most of which are very hard to detect. The unaided eye can only detect particles greater than 50 μm. All waste receptacles in gownup, prep, and inside the cleanroom should be lined with removable plastic bag liners whenever feasible. Removal of material from the receptacle should be downstream away from any sensitive areas, closing the bag tight at its top, and removing the entire bag. Taking trash out of a waste receptacle by hand to transfer to another bag should be discouraged. Such actions disrupt particulate on surfaces inside the container (materials are in there because they are dirty), possibly reintroducing particulate back into the room airflow. All surfaces of waste receptacles should be wiped down in appropriate cleanroom manner. Bags of waste should not be raised over product or sensitive areas since particulate can fall off the bags.

Electrostatic discharge (ESD) is a major concern when removing parts from containers or unwrapping parts (Table 9.3). Static charges can form by the proximity of two dissimilar materials, as seen in Table 9.4. Generated static charges may attract particulate to the parts or to the operator from which they can fall onto the parts or clean areas where parts are placed. Static charges can also damage ESD sensitive parts. All static sensitive parts should be labeled as such and adequate ESD precautions followed when handling them. Refer to Sec. 9.3.6 for more information. Ionizing the airflow over the unwrapping and cleaning area will help to reduce potential ESD damage and dirt clinging. ESD precautions should be tested on a periodic basis to ensure functional effectiveness.

Any contamination sensitive part, tooling, or equipment which must be removed from the cleanroom must be protected from contamination by some type of barrier (bag, wrap, or box). The bag or wrap should be durable enough to prevent rupture, not abrade or slough when used, and have the necessary ESD requirements for the appropriate area.[28] Some parts may require a dry air, inert nitrogen–argon or vacuum atmosphere inside the bag or container. Provide labels for nontransparent packaging so people do not have to open them to see what is in them. *Do not allow* sensitive parts to be opened outside clean environments.

TABLE 9.3   Typical Electrostatic Voltages

Event	Relative humidity, %		
	10	40	55
Walking across carpet	35,000	15,000	7,500
Walking across vinyl floor	12,000	5,000	3,000
Motions of bench worker	6,000	800	400
Remove DIPs from plastic tubes	2,000	700	400
Remove DIPs from vinyl trays	11,500	4,000	2,000
Remove DIPs from Styrofoam	14,500	5,000	3,500
Remove bubble pack from PWBs	26,000	20,000	7,000
Pack PWBs in foam-lined box	21,000	11,000	5,500

SOURCE: From Ref. 43.

TABLE 9.4   Triboelectric Series

Materials	Polarity (+ or −)
Asbestos	Acquires a more positive charge
Acetate	
Glass	
Human hair	
Nylon	
Wool	
Fur	
Lead	
Silk	
Aluminum	
Paper	
Polyurethane	
Cotton	
Wood	
Steel	
Sealing wax	
Hard rubber	
Acetate fiber	
Mylar	
Epoxy glass	
Nickel, copper, silver	
UV resist	
Brass, stainless steel	
Synthetic rubber	
Acrylic	
Polystyrene foam	
Polyurethane foam	
Saran	
Polyester	
Polyethylene	
Polypropylene	
PVC (polyvinyl chloride)	
Teflon	
Silicone rubber	Acquires a more negative charge

SOURCE: From Ref. 43.

Automatic packaging machines are available that are cleanroom compatible. Standard industrial packaging machines will probably not be cleanroom compatible owing to finish, location of moving parts, type of moving parts, lubricants used, and similar reasons. Modifications by the vendor may limit contamination generation during the operation of the machine, and application of local process control will improve contamination protection.

Small items can be bagged in polybags, which may also be used to protect from ESD if required. These bags should be sealed to prevent migration of contamination into the bag as the bag is handled outside the cleanroom. Bag closure rankings in providing barriers from least to most effective are: (1) fold over bag end, (2) folded, taped bag, (3) zip-lock bag, (4) heat seal bag, and (5) double bag. Larger tools and equipment can be wrapped, bagged, or containerized for protection.

Specifying parts provided by noncleanroom-specific suppliers to be delivered in cleanroom-suitable packaging will reduce the long term overhead in the cleanroom. This may require an initial investment of time and energy but will result in vendors providing clean packaged items. The vendor may need to manufacture in a cleanroom or install clean cleaning and clean packaging equipment. The cleanroom training program, room controller, and consultants will help vendors to understand and provide clean packaged parts.

9.3.2 Passthrus

Passthrus are very convenient to "pass through" material from one room to another, but they can also be a source of contamination into the cleanroom. All material should be cleaned prior to entry into the cleanroom. A passthru should be cleaned in and out as often, if not more, than the room and work areas are cleaned. Open-cell foam is not a desirable material for gasket material, since the cells can trap particulate and release it when compressed. Any gasket material should be replaced if it is noted to be deteriorating. Repairs should be made with appropriate low-particulating material. Poor gaskets on passthrus can shed particles into the room or their enclosure. Some of the most sensitive material (product) can pass through the immediate fallout area. Particles can also get on the product and clothing and become carried throughout the cleanroom. ESD properties should be considered in relation to the passthru and to the product.

Plan for some means of communication (intercom or phone) at the passthru. This will inhibit the opening of both sides of the passthru to talk, which may result in the direct passing of items from higher class areas without cleaning. Interlocking passthru doors so only one can be opened at a time is a good idea, but interlocks may be defeated. Set an example and *do not allow* others, especially managers and engineers, to break passthru rules. Enforce the concept that passthrus are only a means to pass material from one room to another. If the room receiving the material is a cleaner class room, the materials must be appropriately cleaned prior to placement into the passthru. Items should be removed from the passthru as soon as possible. Do not let passthrus become a short or long term storage area. Include passthrus in shift and daily cleanings and all primary cleanings when either room is cleaned.

Properly functioning rooms should provide positive airflow outward from the cleaner room even if both doors of the passthru are open. Marginal designs and modifications of a passthru may result in backflows if other doors are open at the same time as the passthru. The design and balancing of the cleanroom complex need to provide acceptable static and dynamic conditions. The location of a passthru should be planned for operator convenience but also needs to consider the airflow, contamination sources, and work patterns of both rooms. Avoid locating a passthru between a cleanroom and a dirty general production room, no matter how strongly production feels this would speed up their work flow. Their presence here could compromise the cleanroom. If passthrus must be located in a dirty area, it is advisable to install a soft clean area around the passthru. This would allow a relatively clean area for staging and cleaning of items before they are passed into the passthru. Do not locate passthrus where strong airflow sources (air vents or machine exhaust) can blow into the open passthru. There should be a slight net airflow out of the cleaner room if both doors are open. Note that both doors should be allowed to be open at the same time. Large or long items that will not conveniently fit into the passthru should be taken into the cleanroom via doors. Product flow and its various stages of sensitivity to contamination should also be considered when placing a passthru in a wall. Design passthrus to be large enough to handle the volume of material needing to be passed between the rooms. Passthrus are for small to medium-sized objects. It will be more convenient to use doors for large objects. Consider the height of passthrus in relation to product weight, and height above floors of the rooms served, especially if the wall separates a raised floor and nonraised floor cleanroom.

Passthrus are as varied in construction as there are variations in process flow. They may be simple boxes with gasketed doors, tightly sealed nitrogen-purged boxes with time delay on opening of doors, HEPA filtered air which is exhausted, or robotic interlocked stations. Their operation and condition should be put on a scheduled maintenance review with that of the cleanrooms. All these designs are dictated by product, logistics, convenience, cost, and governing cleanroom theory at time of installation. Overkill may be nice because rooms do not improve with age and use but may take money and effort away from areas which could benefit more from similar effort and money.

Providing a staging area and table with cleaning supplies at passthrus may facilitate the cleaning of materials entering the cleanroom. A clean area around the passthru in the dirtier room should be provided. Treat the passthru as a prep area for entrance into the cleanroom on an equal basis as a gownup room entering the cleanroom.

9.3.3 Gownup

Gownup rooms are part of the cleanroom and should be treated as such, restricting particulating actions, similar room finishes, and rules. Effort should be put into designing the gownup area so that it will not contaminate the cleanroom, can service the number of people having to enter the cleanroom, and provides clean access for equipment and materials required to go into the cleanroom. The air quality should be of similar class or no more than a factor of 10 less than the cleanroom it services. The gownup should be slightly lower in pressurization than the cleanroom but still higher than the room before it. This will provide a net flow of particulate away from the cleanroom. Refer to Fig. 9.5 for visualization of this concept.

The gownup room also requires as stringent enforcement of cleanroom procedures as inside the cleanroom. Consider the gownup room or area as two areas, a clean area around the cleanroom entrance and a dirty area a distance away. Making some type of physical distinction between clean and dirty areas will help the operators realize they are about to enter a cleanroom and cleanroom behavior must now prevail. This can be accomplished by training or by physically denoting the cleaner side with benches, floor stripes, or colored mats. The physical means will tend to accumulate particulate over time from the act of cleaning the room over and above the particulate which settles out of the air. All items in the gownup room should conform to the same finish requirements as set for the cleanroom. The smoother the finish the easier and better the cleaning can be.

A sign restricting entry to authorized persons properly garbed can be placed outside the entry of the gownup. Proper gowning procedures should be detailed in training sessions and a listing of such or photograph showing a properly gowned person can be posted in the gownup room. The use of the gownup area should be reserved for access into the cleanroom not as meeting rooms, storage rooms, or conversational areas. Entering the gownup room signifies someone is there to enter the cleanroom or to deliver something which will soon be taken into the cleanroom. Emergency and service exits should be labeled as such together with appropriate entry restrictions posted. It is noted that airflow through a door opened to a service area should be from the cleanroom, but the act of swinging the door open or closed can set up turbulent air currents which can trap particulates and transport them into the cleanroom. If people get in the habit of opening service doors to talk, they start disregarding the gownup and cleaning procedures and pass noncleaned items into the cleanroom or touch items inside the cleanroom without proper precautions.

A sign listing general and room-specific procedures should be posted as a reminder of desired behavior in the cleanroom and gownup area. Table 9.5 is a suggested beginning of cleanroom protocol and procedures listing. The location for posting is debatable. Two suggestions are outside the gownup room to prevent restricted items from entering the gownup room and by the gownup mirror as a final reminder before entering the cleanroom. Signs should be short and easily read and their number should be limited so their message is not lost among the clutter. All signs, notices, or holders should be cleanroom compatible. Smooth-surfaced vinyls or laminates are preferred over rough surface signs. Plastic sheet protectors may be acceptable in certain select

TABLE 9.5   Cleanroom and Gownup Protocol and Procedures

DO NOT ENTER CLEANROOM UNLESS PROPERLY GARBED
DO NOT USE DAMAGED GARMENTS
DO NOT REACH OR REMOVE ITEMS FROM UNDER CLEANROOM CLOTHES
NO COSMETICS
NO GUM CHEWING, FOOD, OR BEVERAGES
NO UNAUTHORIZED MATERIALS*
NO CARDBOARD, WOOD, TISSUE, OR REGULAR PAPER MATERIALS†
NO FAST WALKING OR RUNNING
NO SITTING OR STANDING ON WORK SURFACES OR BENCHES
DO NOT BLOCK AIR RETURNS OR ALTER AIRFLOWS
KEEP PRODUCT UPSTREAM OF ANY POTENTIAL PARTICULATE SOURCES
CLEAN ALL ITEMS ENTERING THE CLEANROOM

*Unauthorized materials list can be generated and posted separately outside the gownup area. Additional items restricted may be pens, pencils, markers, paper labels, sprays, abrasives, certain lubricants, solvents, and cleaners.

†Unless sealed from exposure to the room proper.

Protocol and procedures may be relaxed or made more specific depending on the actual room class and room and product requirements. Any modifications should be reviewed with respect to the overall impact on the product and room quality. The cleanroom controller may temporarily exempt certain items for use in the cleanroom, although this should be the exception and not the rule.

areas. Xerographic copies in sensitive areas should be sealed in plastic and left inside their covering since the toner can particulate off from wear and handling.

Place a full-length mirror in such a location that personnel can conveniently look into it before they actually enter the cleanroom. This will allow people to check their cleanroom appearance to verify adequate gownup protocol before entering. The person entering the cleanroom should check to ensure that garments are properly snapped and zipped, all hair contained, and that there is no damage to the garments. Stress that people are to look into the mirror to check if they are properly gowned with hair and skin covered. Once they have checked themselves they may enter the cleanroom. It is not advisable to allow this mirror to be used as a primping mirror for entrance back into the building proper. Combing and arranging hair and clothing generates particulate, and this is in a sensitive area of the room. Personal grooming items are not to be used or allowed uncovered in the gownup or cleanroom. Place a mirror outside the gownup in the building proper and encourage operators to use it to arrange themselves.

It may help to reinforce correct gownup procedures by posting gownup procedures and large photos of an adequately gowned person labeled with gowning areas to be checked. Table 9.6 presents an overall review of typical gownup requirements. The actual gowning requirements will depend on the classification of the room, the process, the work performed, and the airflow within the room.

Airflow into the gownup room should be away from the cleanroom, both the entrance and any passthru. All contamination-generating actions should be performed on the dirty side or such that contamination is directed toward the dirty side. Any particulates generated should be captured and taken out of the general room air as soon as possible since they may resettle on people and objects going into the cleanroom.

Some facilities make use of air showers at entrances to the cleanroom from the gownup room. Air showers are small rooms, with one door opening to the gownup and the other to the cleanroom. Once a person is properly gowned for the cleanroom, they enter the shower room and jets of air are blown over the operator for a preset cycle time. During the air shower process, the operator twists and turns, exposing all parts of the body and garment to the shearing action of the airstream. Ionized air is preferred to increase the removal of statically charged clinging particles. All this air is captured and exhausted out of the room. Once the cycle is complete, the door into the cleanroom is allowed to open and the operator can enter the cleanroom. Doors of the showers must seal tight, preventing any particles freed from the operator from entering the cleanroom. Normal pressurization of the shower should not be greater than the cleanroom unless the door is tightly sealed. Most showers are optimized

TABLE 9.6   Typical Gownup Requirements

Cleanroom class	Garment	
100,000	Head cover	ARP
	Foot cover	ARP
	Frock and lab coat	
	Gloves	ARP
10,000	Head cover	
	Facial cover	ARP
	Shoe cover	
	Frock	
	Gloves	
1,000	Head cover	
	Dressing gloves	ARP
	Facial mask or cover	ARP
	Coverall	
	Shoe cover	
	Gloves	
100	Dressing glove	
	Hood	
	Facial mask or cover	
	Coverall	
	Boot style shoe covers	
	Gloves	
10	Dressing gloves	
	Head cover	
	Hood (with full coverage style mask)	
	Goggles	
	Coverall	
	Boots	
	Gloves	

ARP: as required by product. The above may at times be modified to meet product, customer, and contamination requirements. Overall impact of any change must be reviewed in respect to product and room requirements.

for single use, and multiple people in the shower at a time should be discouraged. Operators should not be allowed to abbreviate or bypass the cycle or open both doors at the same time.

Some feel the cleaning efficiency of air showers is questionable. This is true, especially if they are not properly maintained. Possibly the best value of the showers is to identify a demarcation line from nonclean to clean attitudes, which is good only if air shower protocols are rigidly enforced.

Materials and a cleaning area must be provided in the gownup room to clean any items that need to be brought into the cleanroom, whether by the door or through a passthru. A bench large enough to segregate noncleaned items from cleaned items is desirable. There is little benefit to placing cleaned items back with dirty items while cleaning is continued. Clean parts such that any loosened contamination will not settle onto cleaned parts or items about to enter the cleanroom. The gownup room should also be provided with utilities (air, nitrogen, instrument vacuum, housecleaning vacuum, and electricity) conveniently located for access with minor contamination to the room. Such utilities will be useful for assembling and debugging equipment, testing for particle generation, and cleaning prior to taking materials and equipment into the cleanroom. All equipment must be cleaned immediately before it is taken into the cleanroom. The time span over which a cleaned surface will remain clean will depend on the air quality of the surrounding environment. Particles can settle out onto cleaned surfaces. Again remember, any particles generated must be captured and not allowed to enter the room proper.

Surface finishes and equipment in the gownup room should follow guidelines similar to those of the cleanroom it supports. Air ionization systems are beneficial in increasing the drop-out rate of small particulate from the airstream and off of an item. The low (40%) relative humidity of a cleanroom tends to increase static charges, which can hurt parts and attract particulate to surfaces (see Sec. 9.3.6). Ionization may be point of use via air guns, bench passive and active airflow, or by ionization grids covering the HEPAs. All will give some benefit if placed correctly but will require maintenance. High voltage point sources tend to accumulate particulate balls. These must be cleaned off before they break off and enter the airflow. Establishment of a preventive maintenance program that includes all ESD items will lessen their contamination. Point of use air guns should be discouraged because by their nature of operation they generate particles from surfaces, causing unplanned air eddies which disrupt normal airflow patterns. These particles are propelled away from the surface in turbulent air eddies and can be redeposited downstream. The high velocity that must be used to remove small particulate can impart enough energy to remove particles, causing them to travel great distances, possibly across and against the room airflow. These blow-offs can lead to upstream contamination of the work area and contamination of neighboring work areas and also garments. Once a garment is contaminated, the contamination can be spread throughout the room and to other sensitive areas as the operator moves within the room. If air or nitrogen blow-off must be used, try to capture all particulate generated. Ionization will increase the efficiency of blow-offs, increasing the need for particulate capture.

9.3.4 Communication

The layout of the cleanroom should include windows so people can view the clean process area and observe the practices. Sales, engineers, managers, and visitors can all look in without having to physically come into the room. The more people who come into the cleanroom the more contamination they can bring into the room or generate within the room. They can even cause contamination by not knowing and following cleanroom procedures. Placement of intercoms at windows will help those in the cleanroom answer questions without coming out of the room or extra people coming into the gownup areas and holding talk sessions in the gownup room. Refer again to Fig. 9.5 for suggested placement of intercoms. Remember, gownup areas are to prepare for entry into the cleanroom and not for use as a halfway point to have meetings.

Place phones and intercoms so they or their cords will not be upstream of sensitive areas even during use. A person generates particulate as he or she talks. Place phones near equipment and work zones, not right at them. Place phones and intercoms outside the gownup room in the assembly room to communicate with the cleanroom, not inside the gownup room. Placement inside the gownup room will lead to an increased flow into the room to ask questions. The gownup area should be restricted to those individuals who are preparing to enter the cleanroom.

Cleanroom phones and intercoms are available with smooth finishes, membrane surface keypads, flush mounting, and no-hands options. They cost more than regular phones but may be necessary in critical cleanrooms. The choice is dependent on the room and product criteria. Whatever type is used should be easily cleanable. Phones and intercoms should not be overlooked during cleanings. This includes their cords, which swing through the air during use and are a potential source of particulate. Avoid the use of paper labels, signs, and lists on or near phones. The use of cleanroom paper, vinyl, or polyester materials is more desirable.

A board or area should be provided for posting of operational notes, meetings, messages, and communication between shifts. Locating the board outside the gownup room in the assembly area will reduce contamination in the cleanroom and gownup areas. The ink used on white boards will, by nature, generate large particles when wiped. Reserve the clean areas for long term room-specific postings. Papers that must be taken into the cleanroom areas should be on cleanroom paper, laminated in plastic or contained inside plastic page protectors. Paper and cardstock sheets in protectors should not be removed inside the cleanroom. Remember, clean everything going into the cleanroom, including pages, notebooks, and pens. All surfaces that will be exposed in the cleanroom must be cleaned.

Cleanroom pens are available that have low sodium content ink for use where required. Some ball-point pens will glob, and when the ink dries it can send particulate into the room. Pencils and erasers are not to be allowed in the cleanroom or gownup area. Writing utensils should be provided inside the room, and people should not be allowed to bring or use their own in the room. People will rarely remember to clean them before entering the room, and have a tendency to put their pens in their pockets under their cleanroom garments. A person should not reach inside or under the cleanroom garment once gowned. Such actions release particulate. Pens inside the room should be included in the daily or shift cleaning. All pens, paper, notes, and supplies should be kept downstream of sensitive areas as much as possible.

9.3.5 Ergonomics

Repetitive motion syndrome has become an increasing problem in today's assembly areas, and the cleanroom is no exception. However, it is only the beginning of the ergonomic concerns for the assembly area. Companies should have an ergonomic awareness throughout the whole facility which is adapted to the cleanroom activities and requirements. Instruct all operators in the safe ergonomics of their job, and whenever possible incorporate friendly ergonomic designs into equipment and processes in the room.

Look into each operation as it is performed over time to see how the parts are handled, how tightly the muscles contract, and how many times per minute or hour the cycle is repeated. Can the parts be handled with little strain to the hand and wrist muscles? Avoid the continued use of a selected few muscles or full tension over a limited range and minimize overreaching by operators. Design work areas so the body can maintain a neutral position, not overstretched or twisted in any position but comfortable and relaxed.

Standing and walking on hard floors can cause strain to the feet, legs, and back. Providing a softer surface through antifatigue mats can reduce some of these problems. Mats are available as static-dissipative for ESD sensitive areas. Placement of mats over the entire work area is not feasible, so they should be placed in areas where operators most frequently stand. Unfortunately operators may tend to stand on the mat and twist or overreach for items outside the immediately matted area. This can result in pulled muscles. Mats will become dirty and compacted with use, requiring cleaning and eventual replacement. Placement of mats inside cleanrooms does have problems. They can block airflow in raised floors and generate particles. A workable alternative to mats in clean area is to provide workers with cushioned sole inserts for their shoes. These will ease strain to the legs and back wherever the operator walks or stands and do not present contamination problems to the clean areas. Good quality industrial inserts will retain cushioning effect longer than lightweight foam inserts. Insoles will need to be replaced over time as the material becomes compressed, losing its effectiveness. The cost for good insoles can easily be outweighed by the increase in comfort to workers, increased productivity, and lower rate of worker injury.

Provide seated workers with a footrest type of support to relieve strain weight of the upper legs and provide support to the back. Chairs should be adjustable for stool height, lower back support height, and lower back pressure. These adjustments will vary by operator owing to stature. It is very important to instruct operators how to adjust the chair to fit their individual stature. Sitting in improperly adjusted chairs can cause injury to or irritate backs. Any seating used in the gownup and cleanroom should be cleanroom and product compatible. Conductive or static dissipative items should be used in ESD areas depending on requirements.

A proper seating posture is with the knees bent 90° or less, feet firmly supported but not fully extended, back straight with slight but firm pressure on the small of the back, head tilted slightly forward (17°) of the back plane, arms bent 90° at the elbows, and wrists not strongly flexed. Forearm supports can be used if the arms must be maintained in a constant elevated position for a long time, such as keying in data or at inspection systems.

Video screen displays are most comfortable if placed 20 in from the operator. Tilt should be 0 to 10° if directly in front of the operator or 20° if elevated. Seated workplaces have suggested maximum display heights of 45 and 65 in (primary and secondary). Standing workplaces have suggested maximum heights of 62 and 69 in (primary and secondary). Unfortunately operators are not all the same height, so adjustable seating is required to bring individuals into comfortable work zones. Once systems are in place, review them for ease of access to controls and displays and ability to read all signs and displays. Labeling on displays sometimes becomes blocked by buttons when displays are raised to secondary positions.

All ergonomic modifications in the placement and design of equipment should be reviewed in respect to possible impact on the cleanroom. Displays should not be moved to secondary positions over sensitive areas unless the areas are protected by barriers, such as LPC type airflow or actual physical shields. The airflow from cooling fans should be captured or directed away from sensitive areas. Some manufacturers of standard equipment for cleanrooms have designed cleanroom and ergonomics into their product. The biggest concern is from custom equipment manufactured by noncleanroom suppliers. Suppliers will have to incorporate cleanroom requirements into a functional, ergonomically friendly piece of equipment. Again the cleanroom controller, expert, or consultant should interface early in the design of custom equipment to reduce its impact on the cleanroom.

The overlying design consideration for ergonomically friendly workstations is to allow the body to maintain a neutral position as much as possible. The reader is referred to consultants and various books as a start for ergonomic design considerations.[29–31]

9.3.6 Electrostatic discharge (ESD) concerns

Electrostatics in hybrid assembly manifests as two components of contamination. The more common is the potential damage from discharging of acquired electrical charge between surfaces of different electrical potential, also known as electrostatic discharge or ESD. Another component of electrostatics is the potential for static attraction of charged particulate to surfaces. This has two components: direct attachment of particulate to the part and attachment to surfaces in the airstream. The latter could act as nucleating points ultimately resulting in larger aggregates which can fall onto the product. An often over-looked component of ESD is the potential for its explosive interaction with flammable liquids used in the process or in cleaning. Solvents should be stored and used properly and the work area should be properly grounded to minimize vapor spark interactions.

The reader is referred to various authors for a more in-depth discussion of cleaning theory and application.[8,24,25,33–38] For discussion, particulate can be divided into three behavioral groups according to size: large, medium to small, and extra small. Large particles (>100 μm) tend to quickly fall out of all but the most vigorous of airstreams because of gravity. These particles may be seen with the naked eye and are easily removed from a surface. Extra small particles, less than 0.1 μm, will tend to remain in the airstream indefi-nitely and thus contribute little initially to contamination, but they can nucleate to form larger particles which will settle out and become a concern. Small particles that do attach to surfaces are very hard to remove. Medium to small particles move with the airflow and can attach to surfaces. Removing of such particles can be accomplished with minimal energy expenditure. Small particles move with the airflow, attach themselves to surfaces, and require a lot of energy to remove them once attached. Note that the human eye can dis-tinguish particles >50 μm in size. Therefore, many particles that can affect the product may not always be visible to the naked eye. For every particle that can be seen there are many more that cannot be seen and can still be detrimental to the product. It is easier to prevent attachment of particulates than to remove them from surfaces. The longer a particle is in contact with a surface, the harder it may be to remove it. When looking at cleaning methods, be concerned with two items: how small a particle the method can remove and how efficiently those particles are removed. Various methods do not always have a 100% cleaning efficiency, but a 95% efficiency may be accept-able in some instances balancing cost, energy requirements, and time and product requirements.

The operation of a cleanroom air system removes particulate but also removes most charged ions as the air passes through the HEPA filters (sub-ject to media and humidity). The room will tend to retain whatever local charges occur. The movement of two surfaces next to each other can produce electrical charges as shown in Table 9.3. The charge acquired and intensity is in relation to the two materials' relative position in the triboelectric series, listed in Table 9.4. The uppermost material will become positively charged and farther apart in the series the greater the charge.

If an item is conductive and properly grounded, a charge can rapidly drain away. Static dissipative materials drain charges more slowly. Materials with a surface resistivity of <10,000 Ω/sq are considered conductive. This is usual-ly a bulk property of a material but can also be accomplished by coating or impregnation with conductive material.

Conductive materials may drain the charge fast enough to cause electrical damage to ESD sensitive parts. Static dissipative materials will drain the charge more slowly and are usually preferable in these instances. Static dissi-pative materials have a surface resistivity of 100,000 to 10,000,000,000 Ω/sq.

This can be accomplished by impregnation of low concentrations of conductive materials or coating with conductive materials. The slight moisture film of water present on surfaces will act as a conduction path and will drain charges. The higher the humidity the greater the surface film, and thus the faster the charge will drain. Unfortunately, high humidity can be detrimental to some product and operation of some equipment. Consideration should be given to all materials that come in contact or come close to ESD sensitive parts.

Static charge may be neutralized by ionizing air. This can be accomplished by nuclear isotopes and intense electric discharge.[39] Alpha radiation from nuclear isotopes (polonium is commonly used) will ionize air particles without ozone generation and electromagnetic interference. Their effects are limited more than electric discharge and require airflow to distribute the ionized particles. They present no danger to the user under normal conditions, but such devices should not be dismantled. Nuclear devices are commonly used for local area protection (point-of-use blow-off, bench blowers, and bars) and not for full scale room protection.[40]

Isotopes need to be handled in accordance with state and federal regulations including but not limited to tracking to a particular location at a site, periodic certification for safety compliance, controlled transfer to different locations, and return of used devices to the manufacturer for disposal. Each site using radioisotopes should implement a radiation compliance program including training for affected employees and have a person knowledgeable in how to handle such devices. Nuclear ionizers have a limited useful life of 1 year. Periodic cleaning and testing of an ionizer's effectiveness should be included in the ESD program.

Air particles can also be ionized by electric discharges. The three types used are AC (alternating current) ionization, pulsed DC (direct current) and constant DC ionization.[39,40] AC high voltage ionization uses closely spaced emitter points. The alternating current cycling produces alternating positive-negative ions. High airflows are required to move ions away from electrodes preventing ion recombination. AC ionization is used in grid systems and bars.

Pulsed DC systems use separate emitter points powered by separate supplies for positive and negative ions. Cycling power between positive on and negative on generates waves of positive and negative ions. The ratio of ion types can be varied by varying the appropriate on times of each power supply. Such systems are used in laminar hoods, full room ionization, and areas of low airflow but not in high flow applications such as blow guns and blowers.

Constant DC ionization applies high voltage to pairs of emitter points. It is commonly used with high airflow, but proper emitter spacing can allow its use in low flow. Emitter point spacing is critical for even distribution of ions. Such systems are used in laminar hoods, full rooms, blowers, and blow-off guns.

It is significant that there is a potential for ozone generation with electric discharge ionization. This and the potential effects of the electromagnetic fields generated with high voltage may need to be considered in relation to the product and equipment.

As with any piece of equipment, ionization systems require periodic testing and maintenance. Emitter points, support bars, and grid structures all present surfaces that can collect particulate and need periodic cleaning. Remember to disconnect power prior to cleaning and follow any required lock out and tag out procedures.

Emitter points are sharp and the electrodes wear with time and should be replaced. The selection of low particulating, wear resistant materials for emitter points can reduce particle generation from erosion to acceptable levels.[41] The use of stainless steel points has all but been phased out for the lower particulating tungsten. Steinman showed particle generation for various point materials varies over time and can be subject to particulate bursts.[18] High airflow can reduce the effect to the airstream of these burst events. Point material can be ranked (least particulating to most) as (1) nonmetallic points, (2) ground titanium points, (3) tungsten, (4) thoriated tungsten points, and (5) thoriated tungsten wire.[18,42]

Emitter points will "plate out" material from the air, as evidenced by the formation of white, crystalline balls of ammonium nitrate onto the negative emitter points. These balls need to be removed cleanly before they become large and brittle enough for small particles to break off, falling into the airstream. This requires periodic inspection and cleaning of all ionizers. The formation of this precipitate can be reduced by excluding moist air from the immediate vicinity of the point or by alternating the charge on the points.[39,41] The latter works for point-point systems, not for point-grid systems.

Equipment consideration is an integral part of any ESD program. Moving parts on equipment can generate static charges. Machines are often provided with nonconductive plastic feet, which allows the machine to build up charges. Such equipment should be connected to common ground by supplemental means to either conductive floors or provided grounding paths. Vinyl floors must use conductive adhesives. ESD floors also need to be connected to ground and verified; otherwise a large floating potential can be generated which can damage the product. Coating ESD floors with non-ESD coatings (the common wax used in other parts of the building) can negate all of the ESD control efforts. All incoming material should be audited for particle contamination or charges they may acquire from packaging or transit through noncleanroom environments.

An effective ESD program can make work areas "ESD-safe" by applying the above ideas and by training operators and all people who will need to be in the area on safe ESD procedures.[43] Treat everything in an ESD sensitive area as ESD sensitive and always follow prescribed ESD procedures. Do not handle any ESD sensitive parts unless properly grounded. Always store sensitive parts in static safe environments. Eliminate the use of insulators as much as possible, utilizing static dissipative materials or additives as much as possible to prevent charge buildup. Utilize air ionizers to reduce space charges. Ground everything to a common ground. This reduces the development of charge differentials between adjacent areas which could result in a static discharge. Label parts and work areas as ESD sensitive and do so consistently. Specifying peelable adhesives on labels used to seal packages in the cleanroom can reduce paper fiber generation if the label is be torn to open the package. There are three disadvantages to this: one, peelable labels may not provide indication of tampering; two, such labels cost more; and three, the most important, the act of removing labels from their backers or other surfaces can generate triboelectric charges. Any such action (removing labels, tearing open packages and labels) should be performed downstream of any contamination-sensitive items using sufficient airflow and ionization to minimize potential contamination problems, both particulate and electrostatic.

Once an ESD sensitive room has been set up, everything should be tested to verify adequate ESD protection has been provided and is operational. Testing should be performed when the room is functioning, equipment is operating and operators are in the room, since their movements can generate static charges and move parts through potential statically charged zones. The cleanroom maintenance program should periodically monitor all ESD precautions in the room and correct any discrepancies. Topical ESD coatings wear off and need to be replaced, but when coatings wear off, they can generate particles which must be removed.

9.4   Cleanroom Management

9.4.1   Training

A functioning cleanroom is as good as the sum of its parts: the room (including air handling system), equipment, processes, raw materials, and personnel. Of these, only contamination from the raw material will decrease with time. This discussion will help realize and limit contamination from all of these. All efforts will be in vain if a cleanroom attitude is not consistently enforced. The easiest way to establish a cleanroom program at the facility is with the buy-in of management, which should also include production, facilities, maintenance, quality, and sales. Everyone needs to abide with the established cleanroom procedures. All people who come in contact or who may affect the cleanroom need to be trained in cleanroom protocol, some more extensively than others.

One person or group should coordinate this program to maintain standardization. The program may be developed using outside consultants or internal expertise. A cleanroom program should be approached with the goal of establishing a self-sufficient training staff who will be responsible for monitoring and continuing the program. There are several parts of cleanroom programs generic to all sites, but any program must also be customized to the particular site and product.

A training program should:

Explain what cleanrooms are, including the different types of airflows used to achieve cleanrooms.

Explain how the product is affected by contamination.

Explain and demonstrate particle monitoring with an air particle counter. A witness wafer set out on the table will demonstrate particle settling of visible particles.

Review cleanroom garments and proper gownup techniques.

Review cleanroom behavior, including an explanation of allowed and forbidden cleanroom actions, and how to handle product. The posting of a "Cleanroom Protocol and Procedures" listing in the gownup area will help to reinforce desired behavior. Some of the basic restrictions generic to most cleanrooms are given in Table 9.5. This list can be expanded for each specific room and area or split into separate prohibited actions and restricted materials listings.

Demonstrate appropriate cleaning technique for the cleanroom and the cleaning materials to be used. Explain why only certain materials are used and why everything going into the room must be cleaned.

Review which articles are restricted from the cleanroom. Provide cleanroom paper and pens inside the room and restrict people from bringing in their own.

Stress forbidden actions in the cleanroom. This is repeated because of the importance and impact they can have on the product and cleanroom. Specific to the site is to show the location and class of the cleanrooms, any special procedures required in each individual room, and whom they can go to for further information.

Review any room-specific safety measures they must follow.

Let people know who is in charge of the room and whom they can go to for further questions and clarification.

A brief quiz will indicate their understanding of the training. A signed attendance sheet showing an outline of material covered should be kept for the records. This step is required to become certified to good manufacturing practices (GMP), ISO 9000 and similar programs.

Most of the above can be covered using available videos. Videos and a written outline help to ensure consistency.

People need to be updated as conditions in the room change and retrained on a periodic basis as a refresher. The initial training program should be updated to incorporate any changes the room or products require.

One person should be designated as the room controller or supervisor. The room controller's responsibilities can include ensuring that everyone receives training and updates as needed, familiarizing people with the individual room's special procedures, working with facilities on cleaning and maintenance of the room, helping in coordinating contractor work that must be performed inside the room to ensure clean methods are used, and acting as an adviser on any cleaning of the room or equipment. The room controller acts as a focal point for anything that can affect the quality of the cleanroom and associated areas. The controller should be knowledgeable so as to start to answer cleanroom and contamination questions. The controller should have enough influence or access to proper channels to implement beneficial changes or stop anything which may adversely affect the quality of the cleanroom.

9.4.2 Garments

Garments used in the cleanroom should be as nonparticulating as possible consistent with cost, cleanliness, and operator comfort. Ideal garments for the cleanroom would not particulate, would prevent any particulation of the wearer from reaching the room environment, and would be comfortable to the wearer. Wrapping the operator in sheets of plastic would contain particulation but be very uncomfortable to the wearer. The garments worn are a compromise of comfort, contamination, and the ever-present cost factor.

Garments are of two types: disposable and reusable. Garment construction, application, and testing for particulation have been reviewed by various authors.[44–49] Disposable garments are usually made of Tyvek material. Their cost in bulk may be similar to that of reusable garments. They are clean, have a slight "plastic" odor when the package is first opened, and are readily available as class 100 garments. As a warning, Tyvek garments are available in noncleanroom quality for general industrial use at a lower cost. They may look the same and look clean enough for use, but they are not of the cleanliness quality required for cleanroom use. The main disadvantage of Tyvek garments is the discomfort to the wearer when worn for long periods of time or during strenuous work. The material does not breathe and contains body heat, which may cause sweating. Some people are sensitive to Tyvek in contact with their skin, which increases as the body perspires. Such people should wear clothes with full sleeves and legs and nylon undergloves to minimize this irritation. Tyvek garments can be worn more than once but start to break down with extended and abusive wearing. Plan on a small inventory of disposable garments for emergencies and visitors. Reserve the more comfortable, reusable garments for everyday use by operators.

Reusable garments are tightly woven synthetics resistant to thread breakage or "linting." Rooms of class 100,000 may allow Dacron polyester garments, but more sensitive areas use only polyester nylon material. The material is tightly woven, a herringbone type weave.

The gownup requirements may be generalized by room classification as shown in Table 9.6. These guidelines are commonly modified to meet different unique product and room requirements. The use of frocks, head covers, shoe covers, and gloves may be allowed in lower class cleanrooms if the product is protected enough by airflow and/or other barriers.

Rooms utilizing local process control minienvironments for product assembly may allow less stringent gowning procedures if the parts and equipment are always upstream of the operator, preventing any contamination generated by the operator from getting to the product or upstream work area of the workstation. Barriers may be laminar airflow, enclosed work environments, garments, and gloves.

A closed barrier between gloves and sleeves is a necessity in many contamination-sensitive areas, especially if the operator has to reach toward or handle sensitive parts. The movement of an arm in a sleeve can act as a piston to force air out of the garment. This air picks up particulate from inside the garment and from the operator's clothing and skin as it moves. The air takes the path of least resistance and can easily flow out from loose-fitting cuffs and open spaces between snaps on the front of a gown. The operator can in effect be propelling streams of particulate onto any part reached for if the cuffs are not adequately sealed.

Cuffs can be sealed by tightly snapping them, having elastic cuffs, or using gloves long enough to overlap the sleeve cuff. Often, glove cuffs over sleeves are used along with the first of the two means. Glove cuffs must be long enough to be pulled over the sleeve cuff without riding down with normal hand-arm actions. Short cuffed gloves or those not properly fitted can ride down the arm and bunch up on the wrist. This provides little protection from bellowing actions of arm movements, can be uncomfortable to the wearer, and often exposes bare skin, a further particulate source. Real critical areas may also utilize a tape seal around the glove cuff once it is in place over the garment sleeve.

All gloves used in sensitive areas should be powder-free. It is recommended that samples of gloves used be inspected to check for contamination generation and compatibility with the process and product. Some powder-free gloves are manufactured with mold release agents that can leave a residue on items they touch. Particulate can show up when the glove is looked at under ultraviolet light and incident visible light. Residue can be seen by inspecting witness plates of clean bare silicon surface, glass slides, or clean pieces of smooth plastic that have been touched by the glove. Notify suppliers of any problems found. Rotate gloves in stock (first in first out), as some gloves manufactured without modifiers can deteriorate with time. There can be a difference in the cleanliness quality of gloves from different manufacturers and suppliers. The quality and cleanliness of gloves should be checked on an audit basis, as should all cleanroom supplies. Gloves come in many varieties and packages, from bulk to individually wrapped. Bulk packed gloves are usually suitable for rooms greater than class 10,000 while class 100 require the cleanliness quality of individually packaged gloves, though product and room class can dictate different glove selection. The cleanliness of bulk packed gloves is less than that of individually packaged gloves. Repeated opening of a bulk glove package may leave particulate from the operator's hands on the remaining gloves.

Maintain the cleanliness of gloves by opening their inner wrappers only in a cleanroom or gownup area. Particulate in the air can settle on glove surfaces and be transported into the cleanroom, creating a contamination problem. Gloves may be reused in less sensitive areas if adequate precautions are taken. Do not put a used right glove on the left hand or a used left glove on the right hand. Once a glove is used, the inside becomes covered with skin particles. Using the glove inverted will present all these skin particles to any sensitive areas.

It is advisable to have a supply of extra gloves inside the cleanroom in case a glove becomes contaminated or damaged. An operator can then put on a clean pair of gloves and not be tempted to continue working with improper gloves. A new glove may be slipped over the first glove with little contamination to the room. If the glove must be removed, do so downstream of any sensitive areas and do not allow the used glove to be upstream of any sensitive area as it is placed in a waste receptacle. Limit contact of clean gloves with the inside of the used glove to avoid contaminating the new glove. Remember the action of removing and putting on a glove can release particulate from the skin and the inside of used gloves. Thus one should not remove gloves in sensitive areas.

All garments will break down with use and can become a source of particulate. Knit elastic cuffs can trap particles and break down owing to wear sooner than snap cuffs. This points to the fact that all garments must be inspected by the wearer before use. Garments that become damaged or noticeably soiled during use should be changed as soon as possible. Operators should limit their actions in damaged or soiled garments to minimize expelling particulate out of or off the garment as they move to the gownup area to change into a good outfit.

Body movements can force air out of the gaps between the snaps. Zippered front closures on frocks and coveralls will reduce this source of contamination. Open necks will allow air to escape but hopefully into an airstream that will be above and downstream from the product or sensitive areas. Hoods can reduce this contamination source by better sealing of the neck area.

Once these air leak areas are sealed, particles inside a frock can fall toward the floor. This should not be a problem, since airflow at the floor should be drawn to the return air ducts and removed from the room. Care should be taken not to place sensitive parts near the floor in this contaminated airflow. Also, if an operator has to work above product or a piece of equipment, any parts or sensitive areas of equipment should be protected from particulate falling from the operator.

Cleanroom garments are meant to contain particulate and the insides of used garments become dirty with particulate. Such garments should be handled as potential sources of contamination when being removed, stored, or used again. Used garments should not be passed over sensitive parts or clean garments. Garments intended for reuse should be stored in the gownup area. Allow enough space between garments to allow particulate to fall off and not rub off onto its neighbor. HEPA-filtered gown storage cabinets force a flow of clean air across the garments to knock off some of the particulate. Garments packed tightly together will not be cleaned as well by this airflow as loosely packed garments. The inside of such cabinets should be periodically cleaned to remove all particulate removed from the garments. The amount of dust, hair, or other particles that can accumulate over time from garments can become substantial. The airflow of such cabinets should be integrated with the airflow of the gownup room so as to reduce turbulent eddies and possible updrafts of contamination. If HEPA-filtered cabinets are not available, the garments can be hung inside out on hangers in the open area of the gownup room. The airflow from the room HEPAs will help to wash particles from the garments. Hanging garments inside out will prevent particulate from settling on the outer surfaces of the garment during storage. Hangers, hooks, and cabinets used for garment storage must be included in the periodic cleaning of the gownup area.

Garments that are not to be reused should be placed in laundry receptacles which are placed within the dirtier side of the gownup. Airflow in the room can pick up particulate from garment surfaces as it passes over open laundry. Receptacles can vary in design from laundry bags suspended on wire rod frames to stainless steel cupboards with swinging covers. The more critical the room area, the more precautions that must be taken to store the garments once they have been used so they cannot contribute contamination to the room. The receptacle must be included in the room cleaning schedule.

Cleanroom support should maintain a large enough inventory of garments, gloves, and other supplies to allow for a potential vendor–supplier delivery problem or a sudden surge of use due to customer visits, maintenance actions, or other unexpected traffic. Adequate noncontaminating storage for all these garments, gloves, cleaning materials, and supplies must be provided. The gownup area should be capable of holding short term quantities, but cardboard boxed items may not be stored in the room. The space for storing extra supplies is often overlooked in building planning. Having to relax gowning protocol due to short supplies will affect the product and prove to be a disciplinary problem for gowning over the long term.

The reader is referred to Dixon[45] and Zorich[25] for a detailed sequence of how to put on specific cleanroom garments. No matter what types of cleanroom garments are being used, a few governing ideas should be followed when gowning. Inspect all items prior to use and do not use any damaged or soiled items. Handle all items cleanly, minimizing any contact with any dirty or particulating surfaces. Dress in cleanroom garments from the top down. Particulate generated from putting on hair covers should fall onto street clothes, not the outside of a frock or coverall or gloves. Periodic audits and refresher training are advised for all persons who may have contact with the cleanroom. Reserve the area immediately in front of the cleanroom for fully garbed persons. The use of tape on the floor, a dressing bench dividing the room, or colored floor mats may be helpful to demarcate this area. Finally, *enforce gowning requirements for everyone entering the cleanroom.* Salespersons, customers, engineers, managers, vendors, and maintenance workers should not be exempt from any prescribed gownup procedures. They can contaminate the area and product if not properly dressed. A knowledgeable potential customer will notice any improprieties and this could adversely affect their decision to do business. Periodic audits and refresher training are advised for all persons who may have contact with or support any cleanroom areas.

9.4.3 Construction

Care must be taken so that construction or servicing of the HVAC system does not leave materials or large debris inside the ductwork. Such items can be moved by the airflow to impact the HEPA filter media. This can load up the media or damage it. Both reasons would require changing the HEPA filter, an involved process if performed correctly so as not to contaminate the room.

The construction of a cleanroom is another realm of expertise, especially if it is to be cost effective and functional to adequate standards over time. A full discussion of this subject is beyond the scope of this chapter; the reader is referred to vendors, training programs,[12] and various authors[9,17,50-55] for more in-depth information. To ensure that vendors and contractors build the cleanroom to the proper specifications, the cleanroom user should be aware of design, building, testing, qualification, and acceptance criteria. The build specifications for the cleanroom should include the above in addition to the testing and performance criteria to which the room will be qualified.

It is very useful for a cleanroom controller to meet with all people who will be involved in designing and building the cleanroom. There is a definite plus to presenting a training course to contractors and equipment vendors, especially if they do not commonly build cleanrooms or equipment used in cleanrooms. Such training should include basic cleanroom orientation, plus cleaning for clean construction, cleaning for the cleanroom, and clean workstation design requirements.

The preparation of an area and actual building of a cleanroom must follow clean construction techniques. Any dirt or contamination generated during construction or equipment installation must be removed. Try to limit the particulate generation as much as possible. The less contamination generated the less there will be to remove. The longer a particle is in contact with a surface the harder it will be to remove that particle.

An effective technique for building is to clean the area in accordance with the facility's standard industrial cleaning practice. Prepare a staging area to rough clean all incoming materials and equipment. Move materials and equipment into the cleanroom area and reclean with approved cleanroom cleaners. All mating surfaces should be cleaned prior to mating. Ideally all surfaces to be mated should be vacuumed, wiped, then vacuumed again immediately before mating. The longer a surface is exposed after cleaning the more chance there is for particles to settle out of the airflow onto it. Remember, any vacuuming inside a cleanroom area should utilize a house vacuum system which is exhausted outside the room or by a HEPA filtered vacuum. HEPA filtered vacuums require monitoring for efficiency. A damaged HEPA filter will propel high velocity streams of particulate across room airflows. Cleaning efficiency is greatly reduced if the HEPA filter is saturated. Contamination left between mated surfaces can break down into smaller pieces and eventually work its way into the room. This may be a very small amount of contamination but it is a continual source, like a virtual leak in a vacuum system—it's always there and hard to find. The filtering system will have a difficult enough time removing contaminate from ordinary sources during operation without having to worry about leaks from seams.

All particulate generated inside the work area must be captured and removed. A HEPA filtered vacuum cleaner is suggested for this purpose. When possible, it is advisable to cover all cleaned and installed equipment with plastic to prevent recontamination from other work in the room.

All edges of penetrations through the walls and floor material should be sealed to prevent particulation into the room. The seal can be a full gasket seam filling the void between hole and items passing through it or just lining for the hole, depending on the item penetrating and location. Sealing or lining the internal sides of the hole is preferred. Openings for conveyors need to be lined. There should be a tight seal between any surfaces and liners or sleeves. Penetrations which do not have the space filled will allow air to escape from the room. This quantity of lost air needs to be calculated and designed into the room HVAC system. Also, once the room is operational, all penetrations and openings should be tested to ensure pressurization has been maintained and that the flow is from the cleaner room through the opening.

Final floor and wall surfaces must be protected once installed if subject to any potential damage. The simultaneous installing of equipment while the cleanroom is being completed around them requires protection of not only the walls and floors but also any equipment. Any protective covering used should be selected so as not to damage or particulate in use or leave residues when removed. Construction of the cleanroom can generate metal and wood filings. Rigging can grind in any such particulate. Ground-in particulate is hard to remove from vinyl flooring once the room is in operation. Normal room operation, walking, and moving carts can continually abrade small particles from this ground-in material.

Plan to be able to move large equipment into and out of a cleanroom with minimal impact to the room. This can be as a set of large double doors as access points into a cleanroom. Even though today's modular cleanrooms have wall panels which are easily removed, the time, effort, and disruption of the cleanroom to move equipment is much less when only two doors are opened and closed as opposed to removing and reinstalling one or two panels. The integrity of the cleanroom is easier to maintain if the walls, ceilings, and floors are not disrupted once they are in place.

Doors, even emergency doors, should never open directly to the outside of a building. Provide some type of intermediate room to cushion the cleanroom if a fire exit has to open to the outside of the building. Operators in large hybrid facilities have been known to "step out the back door" for a few minutes. This should not be allowed. Entrance to and from a cleanroom under all but emergency circumstances should be by designated gownup rooms only. This needs to be enforced during construction of the cleanroom once the room is together.

Cleanrooms are not exempted from any required building or safety codes. They just have to meet these requirements using cleaner techniques. There is a whole new industry out there to provide these cleaner items. The cleanroom industry will constantly improve to meet the increased cleaner requirements that tomorrow's tighter tolerances will demand.

9.4.4 Cleaning

Ordinary industrial cleaning equipment and supplies should not be used in cleanrooms. Mop fibers break off, cleaning wipes shed, and soaps can leave residues or unwanted impurities. Take a good look at the finished product of standard industrial cleaning staff and their equipment. Not all dirt is removed, some is just moved around, and the quality may vary between times and operator. This is not acceptable results for sensitive cleanrooms and the product. Adequate cleaning of cleanrooms may be addressed by implementing a cleanroom cleaning program which covers four basic areas: supplies, training, staffing, and audits.

Cleanroom supplies include cleanroom wipes, cleanroom solutions, dispenser bottles, dedicated ladders, vacuum systems (HEPA filters, standalone units, or a central vacuum system with cleanroom compatible equipment), trash bags, deionized water, dedicated cleanroom mops, sponges, wringers, and buckets.[56] Most of these items are dedicated because every time something enters the cleanroom it must be cleaned adequately. Use of items outside the cleanroom can get them so dirty that it is very hard if not impossible to clean them back to acceptable levels for entrance into the cleanroom. Squeeze or pour bottles are preferred over aerosol spray bottles because aerosol drift may become a contaminate as it moves with the airflow. Aerosols may attach themselves to sensitive areas. Spray lubricants should be discouraged for the same reason. A small quantity of aerosol can contaminate a large area of sensitive material and be very hard to remove. Remember, *all* containers must be labeled with contents and hazard per OSHA regulations.[57] This includes the small squeeze bottles of cleaning solutions and any bottles of deionized water present in the room. By law, companies are required to have on hand a material safety data sheet of all chemicals, including cleaning solutions, used in the room.

Cleaning supplies should be tested and periodically audited for compatibility with the product, equipment, and cleanroom, and to determine the quality grade of supplies needed for both the product and the room prior to use. Using high end materials may not give any better cleaning results in some circumstances than a lower grade material. On the other hand, the higher grade material may be required to achieve desired results. Standardization of materials over different classes of room to the higher grade helps prevent the use of inappropriate lower grades in higher sensitive areas. Standardization may also help in purchasing by reducing vendor and parts base, possible quantity discounts, and also in storage and training.

Points to consider when evaluating materials include possible residues from use, any effect on the product, tooling, equipment, or finishes, the ability to stand up to solutions used for cleaning, the required purity of materials and the effect of any impurities, any necessary handling and disposal requirements of used materials, and the economic trade-offs of acceptable performance to the room and product. Cleanroom supplies cost more than normal industrial supplies so their use should be reserved for the cleanroom areas. Cleaning supplies should be handled so as not to introduce contamination in receiving, storage, or use.

A training program needs to be established of how to clean, what to clean, when to clean, and what to use. Training should include standard cleanroom orientation training, specific cleanroom cleaning techniques using the selected materials, and any room-specific precautions and requirements.[5,58,59] All personnel involved in the cleaning process must be trained. Training may be done in-house or by external consultants. During training give the reasons why these extensive procedures are necessary. Explain the cleaning scheme, how small particles and residues affect the room and product quality. Explain why only certain cleaning materials and supplies are allowed to be used in the room. Explain and demonstrate the cleaning techniques required in the room. Stress that *no* arbitrary changes are to be made; the room controller must always be consulted. Provide contacts for future questions or problems.

Cleaning may also be done in-house or by external contractors. Each approach has pros and cons. The room controller or someone on site should be knowledgeable in cleaning to coordinate the cleaning, supervise, and help in cases of immediate need. The type of cleaning involved determines who does it. Commonly, operators are responsible for their areas while some equipment is reserved for select technicians. Operators or specified cleaners may be used for the cleaning of the general cleanroom on a daily basis, though production priorities may leave only the select cleaning staff for this task. Operators, technicians, and engineers should always be required to clean up after themselves.

The largest cleaning task for the cleanroom is the primary or precision cleaning. This requires the meticulous cleaning of every surface, including equipment and storage cabinets (both inside and out). Raised floor cleanrooms require the cleaning of all surfaces of the raised floor, the support pedestals, all piping and materials under the floor, and the actual subfloor. Cleaning is performed in a top to bottom sequence: the ceilings, walls, equipment and furniture, and floors. It may be advisable to have technical staff clean delicate equipment in conjunction with the cleaning crew's work in the room. Cleaning should start in the most sensitive area, working toward less sensitive areas and then into the chases and gownup areas. Surfaces are vacuumed, wiped or mopped with surfactant cleaner and rinsed with deionized water to remove any residue from the cleaner, and then the surface is vacuumed again. Vacuuming and wiping should be performed in one direction only. Surfaces are then checked for the presence of contamination and recleaned if found unacceptable.

Primary cleandowns should be performed on a semiannual basis in most cleanrooms but may be more frequent in more sensitive cleanrooms (class 10, 100, and 1000). Their frequency can also depend on the established schedule, room history, mechanical malfunctions, and occurrence of contamination actions in the room. Primary cleaning is very time-consuming and thus is expensive. An improperly cleaned room can contaminate the equipment or product and will shortly require cleaning again. Cleaning needs to be coordinated with production, since the act of cleaning generates particulate. Primary cleaning is usually done with production shut down, which will adversely affect the manufacturing cost.

Never touch the HEPA filter media when cleaning HEPA filters in the ceiling or local process control areas. Extreme care should be taken so as not to damage HEPA filters with mop and vacuum handles. The room controller or supervisor should be notified immediately if a filter is damaged or even thought to be damaged. A damaged filter becomes a steady stream of contamination, and most HEPAs are placed over sensitive areas. Deactivate any ionization systems and verify by testing that they are at zero energy state. Establish and follow lock-out tag-out procedures while working on equipment.[57] Allow the use of only approved materials for cleaning. Surfaces sensitive to the moisture of cleaners or deionized water should be cleaned with isopropyl alcohol (IPA) or other tested material that is compatible with the equipment or surface. More detailed procedures are described by Lieberman,[8] Thompson,[56] and Tolliver.[24]

New construction will require gross and intermediate cleaning prior to a final primary cleaning. Gross cleaning can use typical industrial cleaning techniques while intermediate cleaning introduces the more precise techniques of the cleanroom. Materials used in these cleanings should not leave residues that are hard to remove and should not affect equipment, product, or the cleanroom. Construction areas for cleanrooms need to be kept clean as well. Construction waste such as saw and drill filings can damage the finishes of the floor and walls to such an extent that they will continue to particulate and contaminate the room once it is active. Care must be used in selecting temporary floor coverings that will not break down with wear while covering the floor. Various consultants and authors have put forth guidelines stressing some of the unique problems of building a cleanroom.[8,12,24,53]

Constructing a cleanroom in a clean manner will take more time and initially cost more, but it will result in a cleanroom that is clean and will stay cleaner longer than rooms in which compromises are made to minimize construction costs. Any new cleanroom should look clean and test good when it is first built. The true test of a room is how long it stays clean and whether it starts to deteriorate over a short time and jeopardize the product. Most contractors do not have experience in cleanly building a cleanroom. Many cleanroom contractors subcontract parts of the work out to subcontractors who may lack expertise in "clean building techniques." It is advisable to interface with each of the contractor's workers to ensure that they understand and abide by the specifications and construction practices. The cleanroom controller or outside consultant can help with such orientations.

The selected cleaning staff is usually responsible for the weekly and monthly cleaning of the room. Unless substantial cleaning resources are available internally, the primary cleaning is usually subcontracted to an external source. Requiring operators to be knowledgeable or perform one primary cleaning will help them appreciate the cleanroom and the cleaning required. One advantage of using outside contractors is a less frequent need for primary cleaning. This is their specialty, and they can provide an immediate large force to accomplish the task quickly. They are responsible for the training and quality of their people and for the finished product. Always audit the results of a cleaning, whether done by a subcontractor or by in-house personnel.

Review the items on the following list with the cleaning staff:

What is to be cleaned

Materials that can and cannot be used

The desired cleaning procedure for sensitive areas

The means for disposing of used cleaning materials

Any cleaning verification measures to be used

The contact person in case of problems or concerns

The time constraints

Cleaning can put particles into the airstream of the room and temporarily alter airflow patterns (Fig. 9.12). The actions of cleaners should be such as to minimize this particulation. Allow time after any major cleaning for the cleanroom's air handling system to clean the air before resuming production or exposing sensitive product. A few air changes may be a sufficient time frame, but actual monitoring of air particle counts should be taken. Verification of room air and surface quality should be performed prior to the start-up of production and exposure of sensitive parts.

Solvent vapors diffuse into the air and are not typically removed by the filtering mechanism of the air handling units. These vapors are continually recycled by the air units and can cause discomforts to operators and residue problems on equipment and product as they adsorb to surfaces in the room. Cleaning with such potentially contaminating solvents should be performed in an externally exhausted hood or with the utilization of local point-of-use vapor capture systems employing adequate carbon filtering or scrubbing to maintain safe levels of vapors in the room. The exhaust of stand-alone units should be HEPA filtered if exhausted into the room and monitored for continued correct operation. Maintenance of such units should be done in the dirty side of the gownup area so as not to recontaminate the cleaned room.

Cleaning should start at the most sensitive areas, moving outward to less sensitive areas and finally into the gownup areas. Gownup areas require cleaning as much if not more than the actual cleanroom, since it should not be a source of contamination to be taken into the cleanroom. Daily (or even once per shift) cleaning of floor and bench surfaces in gownup areas may be required to ensure a clean entrance to the cleanroom.

Once the area has been cleaned it should be tested for contamination. Unfortunately, most particle contamination in a cleanroom cannot be seen with the naked eye (Table 9.1). Since particles less than 50 μm are not visible to the naked eye, we have to rely on testing equipment to tell us if the room is really clean. Particulate in the air is measured by an air particle counter. Some units show the number of particles in different size groups. Air particle counters range from basic counters with user-selectable particle size to those that monitor particle size, temperature, relative humidity, differential pressure, and air velocity and perform statistical analysis of gathered information. Units can be connected to a computer or printer to download data. They can take timed interval samples of variable subsample sizes. Particle counters can also be hand held, cart mounted, or remotely controlled by a central computer. An example of readouts from two different counters is presented in Fig. 9.15. Air particle counters are sensitive electronic instruments and should be treated as such.

Readout from a basic unit:

```
     02/21          04:46
     UM    CTS  1MIN
     0.5            337
  TEMP = 69F RH = 60%
```

Readout from a more advanced unit with SPC capability:

Air Counter 94/01/27 07:39:57 S# 1

SIZE	DIFF	ACCUM		FINALDEL
μM	N	N	ANALOG SUMMARY	
0.3	682	1379	TEMP	AVERAGE
0.5	270	697	+ 073.73	DEG F
0.7	202	427	REL HUM	AVERAGE
1.0	141	225	+ 012.72	PERCNT
2.0	53	84	DIFPRESS	AVERAGE
3.0	20	31	NOT ENABLED	
5.0	9	11	AIR VEL	AVERAGE
10.0	2	2	+ 000.00 FT MIN	

FLOW: 1.0 CFM LREF: 8.9V 0K SI:00:01:00
DATA:94/01/27 07:39:57 VOL:0001.000CF

Figure 9.15 Air particle counter readouts. The more elaborate air particle monitor can give data on number of sized particles per volume of air, temperature, relative humidity, air pressure, air velocity, airflow, sample intervals, and sample size. Other screens of this unit have the capability of statistical analysis of data collected. Rooms are characterized at the 0.5 μm particulate level. The ability to determine other particle sizes is often very useful for varying product requirements. Units capable of detecting 0.1 μm are available. Air particle counters are very sensitive electronic machines and should be treated as such.

Impact collectors can capture the particles in different size groups, allowing for later analysis to determine composition. Particle counters give a more immediate result. Viewing surfaces under a black light (ultraviolet light) will show contamination as white fluorescent particles. This is another immediate test and is very good to demonstrate an operator's surface cleaning effectiveness. Final surface cleaning is often performed in conjunction with black light viewing for immediate correction of cleaning. High intensity incident light will expose particulate via shadow casting, though detection of particles less than 25 μm requires its use with a microscope. This is not as versatile as black light viewing but is very useful for examining witness plates set out to measure surface contamination from particles settling out of the room air.

If visible particles can be seen (particles greater than 50 μm), the area or surface is not clean. Testing has shown that for every particle of one size, there are many particles smaller than that which we cannot see (Fig. 9.2).

Cleaning should be scheduled in advance so as to minimize conflicts with production. The cleaning schedule should be posted so all affected employees are aware of cleaning downtimes. Extensive cleaning puts contamination into the room air and can be harmful to any production occurring or equipment and parts that may be open and unprotected during or immediately after cleaning. Provide feedback to the cleaners and operators as to the cleaning effectiveness. Solicit feedback from the people in the room as to cleaning effectiveness. Cleaners are very valuable to the quality of the product. Their work quality has a profound impact on product yield and quality.

9.4.5 Equipment cleaning

All equipment and the room should be cleaned on a periodic basis. This may be accomplished by a cleaning crew or by the operators themselves, depending on site requirements. The operators should act as if they own the room and are responsible for all functions inside the room including cleaning.

Cleaning should be performed if possible without product in the equipment or at the station. The act of cleaning will generate particulate into the airstream, so all downstream product should be protected. Cleaning of stations should start with the area closest to the airflow and work in a uniform process back toward the operator or toward the least sensitive area. Sprays should not be used in sensitive areas, since the aerosol may become a contaminate if it remains in the airstream and settles out on sensitive areas. Squeeze bottles or removable-top containers may be preferred types of dispensers. Note: OSHA requires all containers to be labeled with contents, hazard, and action to be taken. This includes cleaning solutions in the cleanroom. The use of printed-on labels or vinyl-type labels is preferred over paper labels because of their lower particulation. Material safety data sheets (MSDS) are required for all materials used in the cleanroom, including any cleaning materials, no matter how nonhazardous they appear to be. If the MSDS booklets are to be kept in the cleanroom or gownup area, have them copied onto cleanroom paper and placed in plastic notebooks, not cardboard. Also, any labels placed on the books should be non- or low particulating, such as vinyl or polyester.

Remember, static charges may be generated by the act of removing labels from their backers and residue may be left behind on surfaces where labels have been removed. Polymeric labels on polymeric backing which leave no residue are available but are more expensive than paper labels. The trade-off of cost versus low contamination must be considered in respect to the product. Also, consider that the movement of polymer materials in relation to countertops has the potential for generating static charges.

Wipe cleaning should be done with as low a particulating material as required in the room. Class 100,000 rooms may allow the Kem-wipe® type of material, but inside better class cleanrooms only cleanroom wipes should be used. Cleanroom wipes are made of synthetic low linting, low particulating materials. There are many types, brands, and grades of cleanroom wipes. The lower particulating wipes cost more but may not be detrimental to the room and product to the extent that the extra cost is justified. The room controller will have to make these trade-off considerations. The wipe should be wetted and then wiped across the surface. One surface of the wipe should contact an equipment surface once and be wiped in only one direction. This can be done by folding a wet wipe into quarters, wiping with one quarter surface, folding the wipe to expose a new quarter, wiping, refolding, and repeating the process. Each new wipe on the equipment should slightly overlap the previous wipe. The act of handling, folding, and unfolding wipes can shed material from the wipe or dislodge particulate that has wiped off cleaned surfaces. Care should be taken so as not to endanger sensitive areas or recently cleaned areas when cleaning. Used wipes should be placed in appropriate waste receptacles downstream of any sensitive parts. Do not shake or throw used wipes, as this may dislodge the particulate which has just been removed from the equipment. Continue until all surfaces have been cleaned. Cleaning can be time consuming and costly in materials if done correctly; but, if not performed correctly, the equipment may retain contamination which could get onto the product.

Cleaning frequency may vary depending on the room, equipment, and process. An individual workstation may need cleaning three times per shift (breaks and lunch) and the whole room less often. Again, the sensitivity of the product to contamination and the cleanliness of the work area and room will determine the cleaning frequency.

Any particulate produced during the operation of equipment during a process step should be captured. Examples are the residue from the blow-off of nest areas prior to placement, plastic debris from ultrasonic welding, and particles sloughed off as parts are handled. Particulate capture may be accomplished by local vacuum sources or housekeeping vacuum. Areas should be periodically monitored to test efficiency of removal and corrected as needed. Housekeeping vacuum is typically exhausted outside the room, so the only concern is how the room airflow patterns are being disrupted when the vacuum is in use. Local stand-alone vacuum sources exhaust into the room, so the interruption of airflow patterns by the vacuum, the airflow patterns set up by the vacuum's discharge, and the particulate contained in the discharge are all matters of concern. Discharged air should be filtered through HEPA filters before it is released into the room. Cleanroom vacuum cleaners are designed for noncontinuous use and may not stand up to continued use in equipment or processes requiring particulate capture. Particulate capture systems, on the other hand, have components designed for continuous use. Such units are available with variable motors to adjust the rate of airflow in the capture stream. Particulate captured units are quieter and should last longer but have a higher initial cost than vacuum cleaners. Cleanroom vacuum cleaners are better than nothing until more permanent equipment is available.

9.4.6 Testing

All cleanroom facilities should be capable of testing air particle counts, temperature, humidity, and airflow (velocity and patterns). Ideally, the necessary equipment should be dedicated to each cleanroom area. This reduces the contamination to the equipment as well as the potential for damage and loss of calibration as it is moved from area to area. Each time the equipment leaves the cleanroom proper, it must be cleaned down before being allowed back into the cleaner room. A class 10,000 room is dirtier than a class 1000 room, and there is the potential for equipment to become dirty from contamination settling out on it while in the 10,000 area. This settled dirt may contaminate a class 1000 area if not removed when the equipment is returned to the cleaner area. It is easier to enforce cleaning of all equipment entering a room if this requirement is not exempted for several minor instances.

The life and calibration of air particle counters may be prolonged by restricting its use to select trained individuals or training all area operators. People from other area or groups should be discouraged from borrowing air monitors for testing questionable and possibly corrosive atmospheres in generally dirty assembly areas. The monitor may be forgotten and remain in the questionable area for some time, potentially contaminating and damaging the unit. The few thousand dollars for replacement or repair of this equipment becomes a large unneeded expense. Also, nonknowledgeable people may return the dirty equipment into the cleanroom without proper cleaning.

Like all equipment in the assembly area, the air testing equipment should be put on a scheduled cleaning and preventive maintenance program. This should include periodic traceable calibration as per manufacturers' recommendations or per the internal quality schedule. Traceable calibration will be useful if reporting quality assurance and equipment performance on these units. Drift in the reading of a particle counter could lead to overlooking contamination problems or trying to solve problems that are not real.

All workstations should be tested for compliance to design for reducing contamination and periodically retested to look for changes or degradation of cleanliness. Testing should be performed upon initial installation and after final debug. Minor engineering changes are usually made with the objective of machine operation and often overlook the effects on clean operation.

Along with workstation testing, the entire room including the ceiling, the HEPAs, seams, doors, passthru, return air ducts, air patterns, and air velocity should be periodically tested. This is especially needed after moving major equipment or after room modifications.

The full testing of a cleanroom and equipment within it is time consuming and requires expertise in cleanrooms and contamination control. Shortcuts in testing often do not reveal flow patterns, contamination sources, and other potential problems.

Such testing should look at air and surface cleanliness, temperature, humidity, pressurization, lighting, ESD properties, radiation (ultraviolet, etc.), sound, vibrations, airflow patterns, personnel exposures (OSHA compliance), and exhaust and drain discharges (EPA compliance). Reference to federal standard 209,[1] ASTM standards,[2] and EOS/ESD standards[60] can help to determine what tests to perform. Reference to Lieberman[8] and Tolliver[24] can provide understanding about the methodology and purpose of these tests. Various training courses[5] and articles[3,61–63] show how testing may be performed.

Consistency between tests must be maintained. Establish the tests to be performed, the methodology, locations, and frequency of testing, and uniformly follow through. True correlations to previous tests must be established if different methods or conditions are employed. Basing the testing on established standards will help with vendor's discussions. Vendors and contractors may have to be educated in cleanroom requirements and testing procedures. Often this will help in the understanding of their product and testing in relation to cleanroom contamination control needs. Some testing requirements should be established with the vendors, facilities, and testers in the design phase of the cleanroom and equipment. Whether testing is performed in-house or subcontracted externally, it is necessary to have someone in-house, the cleanroom controller, who understands and can coordinate and determine tests and scheduling. Any work in the cleanroom (testing, cleaning, maintenance, or installation of equipment) can generate contamination and should be coordinated with production and the cleanroom controller.

The final test of a cleanroom is its effect on the product in production, and not how impressive the sales staff can make it appear. Building and maintaining a good production cleanroom can make good product and look clean, but building and maintaining a room that is only clean-looking may not guarantee the ability to continually produce good product.

References

1. Federal Standard 209, General Services Administration, Specification, Room 6654, 7th & D Streets SW, Washington, D.C. 20407.
2. ASTM, ASTM Standards for Cleanrooms, American Society for Testing and Materials, 1993.
3. D. Lutz, "Air Velocity in Cleanroom: Measurement and Analysis," in *Cleanrooms '92 Conference Proceedings,* Atlantic City, N.J., Witter Publishing Co., 1992, p. 80.
4. B. K. Bhola, "Designing and Constructing the Next Generation of HEPA Filters," *Microcontamination,* vol. 11, no. 11, p. 31, 1993.
5. Argot, "Contamination Control Technology Seminar Notes—1993," Argot, Inc., P.O. Box 468, Honeoye Falls, N.Y. 14472.
6. R. Karnick, "Mini Environment Design Methodology," in *Cleanrooms '92 Conference Proceedings,* Atlantic City, N.J., Witter Publishing Co., 1992, p. 104.
7. ACGIH, *Industrial Ventilation: A Manual of Recommended Practice,* 21st ed., American Conference of Government Industrial Hygienists, Inc., Cincinnati, Ohio, 1992.
8. A. Lieberman, *Contamination Control and Cleanrooms: Problems, Engineering Solutions, and Applications,* Van Nostrand Reinhold, 1992.
9. R. K. Schneider, "Specifying Cleanroom Design Parameters," in *Cleanrooms '92 Conference Proceedings,* Atlantic City, N.J., Witter Publishing Co., 1992, p. 377.
10. J. Greiner, "Workstation, History and Future," in *Cleanrooms '92 Conference Proceedings,* Atlantic City, N.J., Witter Publishing Co., 1992, p. 143.
11. S. Abuzeid, "Comparing Particle Contamination in Conventional and Mini-environment-based Cleanrooms," *Microcontamination,* vol. 17, no. 7, p. 33, 1993.
12. Argot, "Design Considerations for Cleanrooms—1994," Argot, Inc., P.O. Box 468, Honeoye Falls, N.Y. 14472.
13. R. Avery, "Directional Air Flow," in *Cleanrooms '93 Conference Proceedings,* Boston, Mass., Witter Publishing Co., 1993, p. 119.
14. T. DiNapoli, "Points to Consider When Applying Mini Environment Technology," *Cleanrooms,* vol. 8, no. 10, p. 12, 1993.
15. M. Lenich, P. Swift, and K. Golstein, "Integration of Facilities and Mini-Environments," in *Cleanrooms '93 Conference Proceedings,* Boston, Mass., Witter Publishing Co., 1993, p. 128.
16. G. Marvell, " Minienvironment Air Flow Dynamics," *Solid State Technology,* vol. 36, no. 8, p. 47, 1993.
17. W. Soules, "Cleanroom Design and Construction: Starting from Scratch," in *Cleanrooms '92 Conference Proceedings,* Atlantic City, N.J., Witter Publishing Co., 1992, p. 210.
18. A. Steinman, "Test Chamber Measures Particle Emissions from Ionizers," *Cleanrooms,* vol. 9, no. 11, p. 32, 1993.
19. A. Busnaina, "Solving Process Tool Contamination Problems," *Semiconductor International,* vol. 16, no. 10, p. 72, 1993.
20. T.R.S., "Generic Workstation Design: Zone Finish Requirements," 1992, T.R. Services, P.O. Box 93, Piffard, N.Y. 14533.
21. G. Devloo, "Effect of Surface Finish on Contamination Level," in *Cleanrooms '92 Conference Proceedings,* Atlantic City, N.J., Witter Publishing Co., 1992, p. 340.
22. M. A. Accomazzo, G. Ganzi, and R. Kaiser, "Deionized (DI) Water Filtration Technology," in D. L. Tolliver (ed.), *Handbook of Contamination Control in Microelectronics: Principles, Applications and Technology,* Noyes Publications, 1988, p. 210.

23. T. L. Faylor and J.J. Gorski, "Ultra High Purity Water—New Frontiers," in D. L. Tolliver (ed.), *Handbook of Contamination Control in Microelectronics: Principles, Applications and Technology,* Noyes Publications, 1988, p. 185.
24. D. L. Tolliver (ed.), *Handbook of Contamination Control in Microelectronics: Principles, Applications and Technology,* Noyes Publications, 1988.
25. R. Zorich, *Handbook of Quality Integrated Circuit Manufacturing,* Academic Press, 1991.
26. *Code of Federal Regulations (CFR) Title 49,* U.S. Department of Transportation, 400 Seventh Street, SW, Washington, D.C. 20590.
27. *ATA Hazardous Materials Tariff,* ATA 111-L, American Trucking Associations, Inc., 2200 Mill Road, Alexandria, Va. 22314, 1992.
28. J. Stuerzel and B. Phelan, "Utilization of Flexible Cleanroom Packaging," in *Cleanrooms '93 Conference Proceedings,* Boston, Mass., Witter Publishing Co., 1993, p. 392.
29. Kodak, S. M. Rogers (ed.), *Ergonomic Design for People at Work,* vol. 1, Eastman Kodak Company, Van Nostrand Reinhold, 1983.
30. Kodak, S. M. Rogers (ed.), *Ergonomic Design for People at Work,* vol. 2, Eastman Kodak Company, Van Nostrand Reinhold, 1986.
31. B. S. Matisoff, *Handbook of Electronics Packaging Design and Engineering,* Van Nostrand Reinhold, 1982, p. 23.
32. NIOSH, *The Industrial Environment—Its Evaluation and Control,* U.S. Department of Health and Human Services, National Institute for Occupational Safety and Health (NIOSH), 1973.
33. D. J. Elliott, *Integrated Circuit Fabrication Technology,* 2d ed., McGraw-Hill, 1989.
34. K. L. Mittal (ed.), *Treatise on Clean Surface Technology,* vol. 1, Plenum Press, 1987.
35. K. L. Mittal (ed.), *Particles of Surfaces 1: Detection, Adhesion, and Removal,* Plenum Press, 1988.
36. K. L. Mittal (ed.), *Particles of Surfaces 2: Detection, Adhesion, and Removal,* Plenum Press, 1989.
37. K. L. Mittal (ed.), *Particles of Surfaces 3: Detection, Adhesion, and Removal,* Plenum Press, 1991.
38. R. Trowbridge, "CO_2 and Selected Cleaning Techniques for Small Particle Removal," UNY-VAC American Vacuum Society, 1993 Annual Symposium, June 2, 1993, Rochester, N.Y.
39. S. Gehlke and A. Steinman, "Air Ionization: Theory and Use," in *Cleanrooms '92 Conference Proceedings,* Atlantic City, N.J., Witter Publishing Co., 1992, p. 176.
40. R. McCraty, "Electrostatics in Clean Rooms," in D. L. Tolliver (ed.), *Handbook of Contamination Control in Microelectronics: Principles, Applications and Technology,* Noyes Publications, 1988, p. 153.
41. P. C. D. Hobbs, J. S. Batchelder, V. P. Gross, and K. D. Murray, "Ultra-Clean Air Ionizers for Suppression of Particulate Surface Contamination," in K. L. Mittal (ed.), *Particles of Surfaces 3: Detection, Adhesion, and Removal,* Plenum Press, 1991, p. 249.
42. S. Gehlke and A. Steinman, "Particles Emission from Air Ionization Equipment: Analysis and Strategies," in *Cleanrooms '93 Conference Proceedings,* Boston, Mass., Witter Publishing Co., 1993, p. 182.
43. AT&T, *Electrostatic Discharge Handbook,* AT&T, 1989.
44. C. W. Berndt, E. S. Burnett, and R. Spector, "Selecting the Right Garment System," in *Cleanrooms '92 Conference Proceedings,* Atlantic City, N.J., Witter Publishing Co., 1992, p. 386.
45. A. M. Dixon, "Guidelines for Clean Room Management and Discipline," in D. L. Tolliver (ed.), *Handbook of Contamination Control in Microelectronics: Principles, Applications and Technology,* Noyes Publications, 1988, p. 136.
46. T. Finley, J. Nesbitt, and D. Wadkins, "A Comparison of Garment Cleanliness Using Helmke Drum and ASTM F-51 Methods," in *Cleanrooms '92 Conference Proceedings,* Atlantic City, N.J., Witter Publishing Co., 1992, p. 409.
47. B. W. Goodwin, "Clean Room Garments and Fabrics," in D. L. Tolliver (ed.), *Handbook of Contamination Control in Microelectronics: Principles, Applications and Technology,* Noyes Publications, 1988, p. 110.
48. W. Seemayer, "Cleanroom "Gowning and Behavior for Non-Sterile and Sterile Environments," in *Cleanrooms '93 Conference Proceedings,* Witter Publishing Co., 1993, p. 1.
49. R. Spector, C. Bernt, and E. Brunnett, "Reviewing Methods for Evaluating Cleanroom Garment Fabrics," *Microcontamination,* vol. 11, no. 3, p. 31, 1993.
50. G. Devloo, "Fundamentals of Cleanroom Mechanical Systems," in *Cleanrooms '93 Conference Proceedings,* Boston, Mass., Witter Publishing Co., 1993, p. 318.

51. F. Gerbig and M. Houge, "How to Write a Request for Quotation and Receive the Desired Cleanroom," in *Cleanrooms '93 Conference Proceedings,* Boston, Mass., Witter Publishing Co., 1993, p. 308.

52. J. Greiner, "Retrofits: Adventure into the Unknown," in *Cleanrooms '92 Conference Proceedings,* Atlantic City, N.J., Witter Publishing Co., 1992, p. 229.

53. R. Kraft, "A Sensible Quality Assurance Program for Cleanroom Construction," in *Cleanrooms '92 Conference Proceedings,* Atlantic City, N.J., Witter Publishing Co., 1992, p. 354.

54. D. C. Oberlies and T. S Wayer, "Programming Cleanroom Design," *Microelectronics Manufacturing Technology,* vol. 14, no. 4, p. 34, 1991.

55. K. B. Vincent, "Renew or New: Planning for Cleanroom Renovation of Construction," *Microcontamination,* vol. 11, no. 4, p. 41, 1993.

56. C. L. Thompson, "Cleaning During Construction, Janitorial Cleaning and the Cleanroom," in *Cleanrooms '92 Conference Proceedings,* Atlantic City, N.J., Witter Publishing Co., 1992, p. 9.

57. OSHA, Code of Federal Regulations (CFR) Title 29 CFR, 1993.

58. D. H. Shears and M. C. Shears, "Cleanroom Training," in *Cleanrooms '92 Conference Proceedings,* Atlantic City, N.J., Witter Publishing Co., 1992, p. 21.

59. T.R.S., "Cleanrooms: An Overview—1992," 1992, T.R. Services, P.O. Box 93, Piffard, N.Y. 14533.

60. EOS/ESD Association Air Ionization Standard, EOS/ESD-S3.1-1991. EOS/ESD Association, 200 Liberty Plaza, Rome, N.Y. 13440.

61. R. Vijayakumar, "Air Filtration and Filter Testing," in *Cleanrooms '93 Conference Proceedings,* Boston, Mass., Witter Publishing Co., 1993, p. 350.

62. T. A. Barber, "HEPA Filter Function and Integrity Testing: An Evaluation of Leak Detection Methodologies," in *Cleanrooms '94 Conference Proceedings,* Philadelphia, Pa., Witter Publishing Co., 1994, p. 147.

63. P. T. Quintin, "Cleanroom Qualification of Equipment and Supplies," in *Cleanrooms '94 Conference Proceedings,* Philadelphia, Pa., Witter Publishing Co., 1994, p. 82.

10

Failure Analysis of Hybrid Microelectronics

John A. Buono

10.1 Introduction

Failure analysis is an analytical process whereby the analyst applies a disciplined mindset and a defined set of procedures to determine the root cause of failure of a material or manufactured assembly. Failure analysis is usually performed to improve products, to improve manufacturing yields, and to minimize downtime and costly rework. Failure analysis can also help to minimize present and future financial and safety liability. In addition, failure analysis provides the benefit of reducing manufacturing and warranty service time and costs.

10.1.1 Goals of a failure analysis

The goals of a failure analysis are:

1. To determine the root cause of a failure
2. To predict if, how, and when similar units will fail in the future
3. To determine a corrective action to eliminate failures of already built parts
4. To determine a corrective action to eliminate the failure on subsequently built parts

While the "ideal" failure analysis will meet all of these goals, in practical situations they might not all be achievable owing to constraints on the number or type of failed samples, or the availability of sufficient time, talent, or money to perform the analysis. Realizing this, it is imperative for everyone concerned with the failure analysis to choose and agree upon which goals are of greatest importance before embarking on any failure analysis.

1. Root cause. The determination of the root cause of a failure is not always as clear-cut as it may appear to be. While a hybrid might have failed because of an internal lead bond failure, the root cause might have been contamination on the bond pad, thermal stress, improper bonding conditions, contaminated wire, or mechanical stress. Each of these causes would require a different corrective action to rework present lots, if possible, and to eliminate the failure in future lots.

2. Prediction whether similar (same lot) parts will fail. To predict if, when, and how similar failures will occur requires that the failure analyst, design engineers, manufacturing engineers, and often, the end user share all that they know about the failed part and its operating environment as quickly as possible before the start of the failure analysis. Similarly, at its conclusion the failure analyst should share the information developed in the failure analysis. The input of all these parties will increase the probability that predictions of future reliability will be accurate. Careful thought must go into not only what caused the original failure but also whether the environmental conditions under which the failure occurred are likely to be repeated in the future.

3. Determination of a corrective action to eliminate failures of already built parts. The failure analysis often provides information to support manufacturing and/or process engineering as they choose corrective actions. When there is an inventory of already built parts, the failure analysis must show whether the failure is isolated to an individual circuit or component part, if it is related to an entire lot of components or parts, or if it is independent of the actual hybrid circuit. On the one hand, failures caused by a one-time contaminant introduced into one hybrid or into an entire production lot in one process step will each lead to different corrective actions. On the other hand, a design problem that affects all subsequent production runs will require yet a different set of corrective actions than the contamination induced failures.

In the case of a contamination induced failure, the initial corrective action would be to inspect similar parts from the same lot to verify whether or not they are contaminated. If they are contaminated, the contaminant will have to be removed from the already built devices or the devices will have to be scrapped. In addition, the source of the contaminant will have to be located and a determination will have to be made as to whether the source of the contamination was a one-time occurrence, a random occurrence that might recur, or an ongoing problem.

4. Determination of a corrective action to eliminate the failure on subsequently built parts. In the example above, if the contamination was a result of a recurring process step, e.g., a faulty cleaning step, a different set of corrective actions, e.g., a design or process step alteration, or a change in cleaning materials or equipment might be necessary to ensure that the contamination does not recur in subsequent production runs. In addition, the failure analysis should show whether the cause of failure was internal or external to the hybrid. If the failure was caused by an internal problem which resulted from either the manufacturing process or a design defect, the corrective action might be to modify the design, the process, or the materials required to build the hybrid. If the failure was caused by an external influence, i.e., electrical overstress or thermal overload, the corrective action might be to change the circuit's specified operating environment or to design an improved hybrid which would be better able to withstand the external conditions.

In summary, failure analysis can be useful in solving immediate field problems, increasing yields, and engineering a better product. To maximize the benefits of failure analysis, there should be clear objectives for the analysis and maximum communication between the failure analyst and the designers, manufacturers, and (in the case of a field failure) end users of the failed part.

10.2   Who Should Perform a Failure Analysis?

Failure analysis should be performed by a trained analyst who usually will have a degree in electrical engineering, solid state physics, material science, material engineering, or chemistry. The actual college education or "degree" of a failure analyst is less important than the way they use their education and how much hands-on experience and training they have received after attaining their degree. A "good" failure analyst has most (if not all) of the following attributes:

1. An analytical mind

2. A natural curiosity

3. Ability to switch between linear and random thinking at will

4. Ability to make decisions on the fly

5. Relentless persistence and perseverance

6. Good interpersonal and communication skills

10.2.1   Attributes of a "good" failure analyst

Most of these traits are inherent, in varying degrees, in the person who becomes a "good" failure analyst. All these traits are inherent in the person who becomes an "excellent" failure analyst. In addition, these traits can be augmented by the education and training received in college and graduate school, and honed by the broad practical experience and mentoring that one can find only in industry.

1.   Analytical mind. A keen analytical mind is needed to view all possible options and diverse directions that a failure analysis might take, organize these options, evaluate their probability for success, and select the best approach available. Since each failure is usually unique, the analytical approach will also be unique.

2. Natural curiosity. A person with a natural curiosity about how things work or what makes things tick will find that performing a failure analysis is a challenging way to indulge that curiosity. This type of person is much less likely to be put off or worried about the enormity of the task or the potential frustration of completing the analysis without getting an answer, even though they did everything "right" according to established procedures. In addition, because of their natural curiosity, this type of person will have acquired a vast library of useful information from their experiences. They can draw on this information base to make faster, more informed decisions on their next failure analysis.

3. Ability to switch between linear and random thinking. Failure analyses are much like proceeding through a maze. To navigate through a maze, a person must choose a course of direction and proceed linearly until a branch in the maze is reached. The maze walker must then stop and evaluate the options before them, reevaluate the original plan, choose which way to turn, and then proceed linearly to the next branch. The process of a failure analysis is similar, but far more difficult. In a conventional maze, the walls of the maze are well defined, whereas in a failure analysis, they are not as well defined and are often not even seen until the analyst hits one. In a conventional maze, the locations of branches are easily discernible, while in a failure analysis, the opportunities to branch or change directions are not as discernible even though they are more numerous. Also, the person in the maze always has the option of retracing steps and starting over in a different direction. This is not always achievable in failure analysis, particularly if there is only one failed sample.

An analyst who can look at the "big picture," choose an appropriate starting approach, and doggedly pursue this approach is invaluable in a failure analysis. At the same time, the analyst must be able to recognize when the search is heading down a blind alley and change approaches accordingly.

4. Ability to make decisions on the fly. The ability to make decisions on the fly is critical in failure analysis because at each step of the analysis, the analyst can be faced with divergent, inconclusive, and sometimes contradictory data. A good failure analyst must be able to evaluate these data and select the next step to take. This is made more challenging by the fact that much of the testing done in failure analysis is destructive testing. If the failure analyst takes a wrong turn, it is not possible to go back with the same sample and start again. This can be a major problem with single unit, critical application failures.

5. Relentless persistence and perseverance. A good failure analyst must be persistent in gathering as much information as possible, both before and during the analysis, and must be capable of hitting dead ends as leads are followed and having the persistence to keep going. The analyst must be relentless in looking for all possible causes of failure, particularly when confronted with a potential cause that appears to be too obvious or when someone with a vested interest has already predicted the results of the failure analysis before any actual analysis has even been started.

6. Good interpersonal and communication skills. Lastly, a good failure analyst must be a good people person and a good communicator. In any failure analysis, it may be necessary to deal with several different engineers, scientists, and managers, each of whom might bring both valuable insights and preconceived biases. The failure analyst must receive all this information, use what is needed, perform the analysis, and then communicate findings and recommendations to all interested parties. This must be done even when delivering adverse information to people who might be inclined to "shoot the messenger."

In addition to the above traits, a "good" failure analyst has a working knowledge of the design, manufacturing processes, and intended application of the failed circuit or assembly. Ideally, this information will have been garnered from prior in-house experience as a processing or manufacturing engineer. Alternatively, the failure analyst will have accumulated this information from previous failure analyses or will pick up as much information about the present failure as needed from conversations with the manufacturing and processing engineers who designed and built the hybrid.

10.3 What You Should Expect to Get Out of a Failure Analysis

Ideally, a failure analysis should meet all four objectives:

1. Determine the cause of the failure.
2. Predict whether similar parts from the same lot will fail.
3. Determine how to correct the problem in similar parts from the same lot.
4. Help to develop a corrective action plan to avoid the recurrence of the problem in subsequent lots.

Unfortunately, not all failure analyses successfully find the root cause of failure. However, even when a failure analysis does not provide the root cause of the failure, at a minimum it should systematically rule out specific causes of failure and provide enough information for the analyst and the engineers to make realistic assumptions on the answers to objectives 2–4.

10.4 Why Hybrid Circuit Failure Analysis is Different from Other Microelectronic Failure Analysis

Hybrid circuit failure analysis is different from other microelectronic failure analysis because of the complexity of hybrids. The combination of many different discrete components and integrated circuits inside one package or potted assembly increases the probability of failure. In addition, there are many more processing steps in the production of hybrids versus discrete components or integrated circuits, any one of which may introduce contamination or push a marginal component toward failure. The complexity of hybrid packages and potting systems often subjects components to greater thermal and mechanical stresses than the individual parts might ordinarily be exposed to under manufacturing processes in a nonhybrid system.

The increased complexity of hybrids increases their cost, making the cost of performing a hybrid failure analysis reasonable when compared to the cost of scrapping the circuits. Also, the sooner a hybrid fails in the production cycle, the easier (and less costly) it will be to isolate the area of interest and perform the failure analysis. The sooner in the manufacturing process that a failure is found and analyzed, the easier (and cheaper) it will be to perform corrective actions and the less likely it will be that the corrective actions will damage or interfere with another component or part of the hybrid circuit.

Hybrid failures can be broken down into generic types according to where in the assembly they occur. The types are:

- Package (Sec. 10.4.1)
- Substrate (Sec. 10.4.2)
- Thick and thin film (Sec. 10.4.3)
- Component or die (Sec. 10.4.4)

While the same basic analytical approach will be used independent of type of failure, the tools used for the failure analysis might be different or their order of use might be different. For example, the failure analysis of package-related failures relies heavily on hermeticity testing, residual gas analysis, radiography, optical microscopy, and scanning electron microscopy. In contrast to package failures, substrate-related failures usually require a combination of optical microscopy, scanning electron microscopy, and substrate cross-sectioning. Failure analysis related to thick film problems also requires optical microscopy but usually only as a precursor to scanning electron microscopy. Thin film–related failures generally require surface analysis tools, while the analysis of component- or die-related failures generally requires more extensive sample handling and preparation procedures than other types of failures.

Because of their complexity, hybrid failure analyses usually require more analytical tools and greater technical experience and flexibility of approach by the analyst than other types of microelectronic failure analysis.

10.4.1 Package level failures

Typical package level failures include:

1. Loss of hermeticity
2. Particulate contamination

3. Gross contamination

4. Cosmetic staining

5. Package corrosion

6. Cracked glass-to-metal seals

7. Ionic contamination

While any one of the problems above can occur by itself, it is not unusual in hybrids to have more than one type of problem going on at the same time. The role of the failure analyst is to determine when this situation occurs, which problem is the root cause of failure, and what, if any, contributory effect the other problems had on the failure.

Failed glass-to-metal seals, loss of package hermeticity, cosmetic staining, and particulate contamination can usually be spotted during in-house testing, before product shipment. Package corrosion and gross contamination will often cause failure later in the product's life. The magnitude of the effect of gross contamination and corrosion on the hybrid will vary with the operating environment of the device. For example, external package corrosion will be dependent on the presence of ionic material, and the rate of the corrosion can be accelerated by the presence of moisture, current, or increased temperature.

A trace amount of ionic contamination within the package might manifest itself initially as a small leakage current. This leakage current might increase if the internal package conditions are suitable for metal migration. With time, this condition might build to a complete electrical short within the package, across the lead frame, or between metal runs or components on the substrate, manifesting itself as a catastrophic failure.

Commercial potted hybrid assemblies and plastic packages share some of the same failure mechanisms as the more typical metal or cermet packages. They also have their own failure mechanisms related to the potting materials used and their reduced resistance to moisture or contaminant intrusion. Potted or encapsulated hybrids are also much harder to failure analyze since extra care and effort must be taken to get to and expose the area where the failure occurred without destroying the evidence of the cause of the failure.

10.4.2 Substrate level failures

Substrate level failures usually occur during production or burn-in. Gross substrate failures are usually found in-house, making their failure analysis relatively easy. Typical substrate level failures include:

1. Substrate cracking

2. Delamination or lifting

3. Die or component detach (cleanly from the substrate)

4. Conductor or dielectric film delamination

Substrate level failures are usually fairly easy to analyze in that the areas to be analyzed are readily accessible. These types of failure usually occur early in the manufacturing cycle and therefore are more readily detected and handled.

Substrate cracking of a good substrate, i.e., one that has past numerous quality inspections, occurs as a result of mechanical stress due to thermal mismatch of the package and the substrate. Occasionally, substrate cracking of marginal substrates occurs when mechanical stress from temperature cycling, vibration, or particle impact noise detection testing causes the propagation of minor cracks or flaws already present in the substrate.

Substrate delamination can result from thermal mismatch of components, improper epoxy curing conditions, mechanical stress, or package, substrate, or bonding epoxy contamination. Often a number of analytical techniques are required to determine the actual cause of failure.

A component or die that cleanly detaches from the substrate is usually caused by contamination of the substrate, the component, or the attach material, and/or thermal stress. Conductor or dielectric film delamination from the substrate is usually caused by contamination of the substrate or a problem with the film deposition process or equipment.

10.4.3 Thick and thin film failures

Thick and thin film failures can occur at any time in the product life. These failures include:

1. Delamination (adhesive failure)

2. Delamination (cohesive failure)

3. Thick film cracking

4. Electrical opens

5. Corrosion

6. Leakage currents

7. Electrical shorts

Since there are many potential causes for failure, some of which are very subtle, film failures are more difficult to analyze than other hybrid failures.

When delamination of films occurs, it is imperative to analyze both sides of the separated layers to determine if the failure occurred between layers (an adhesive failure) or within a specific layer (a cohesive failure). This is particularly important with thin film systems, since a preliminary optical inspection of a thin film delamination can often be misleading. What looks like a clean separation between layers to the naked eye or in an optical microscope can often be a breakdown within one layer. In this case, the choice of analytical tool will be based on the actual thickness of the films. Thick films may crack as a result of improper firing or laser trimming. These cracks may propagate slowly during operation, causing shifts in resistivity or ultimately an open.

Contamination of or between thin films can cause films to change resistivity and/or delaminate over time. If the contamination is ionic in nature, it may cause corrosion of the film. The corrosion by-products may create a path for the formation of leakage currents. Over time, a short might form from the corrosion by-products, their reaction with other package materials, or metal migration under the presence of heat and power.

10.4.4   Die and component level failures

Die and component failures include the discrete components and integrated circuits, their attach mechanisms, and the electrical connections between them and the rest of the hybrid. Typical failures include:

1. Lifted bonds

2. Component burnout

3. Electrostatic discharge damage (ESD)

4. Leakage currents

5. Component detach

One of the most common causes of hybrid failures is the lifting of bonds to the discrete devices or integrated circuits. This can be more prevalent with hybrid circuits than with integrated circuits because of the many types of bonds and bonding parameters used within a hybrid. In addition, the problem is made worse by the increased number of mechanical and thermal process steps used in hybrid manufacturing, each of which can contribute to the failure of a marginal bond.

Component burnout usually is found in electrical screening but occasionally occurs after the hybrid is placed in field service.

Electrostatic discharge is usually caused by improper handling of the hybrid or the individual components. This type of failure occurs more frequently in drier (low humidity) weather or climates. It most often hits the weakest part of the integrated circuits on a hybrid. ESD damage usually manifests itself as a resistive leakage path or a short. Occasionally it manifests itself as an open circuit due to blown metallization. While not as common, ESD can also occur in other areas of the hybrid.

Leakage currents are generally the result of contamination on the die, between conductors, or around bond pads. Leakage currents also occur vertically through a dielectric as a result of pinholes or hairline cracks in the dielectric or electrical overstress conditions.

Component detach most often occurs with massive discrete components, e.g., ceramic capacitors. Detach failures can occur in any of the materials involved in the bond, the substrate, the attachment compound, or the component. Detach failures often result from interdependent problems in more than one of these areas. The most frequent causes of detach include:

1. Contamination of the substrate or component terminations

2. Contamination of the bonding epoxy

3. Thermal stresses from temperature cycling

4. Mechanical stress from vibrational and shock testing

5. Using the wrong epoxy

6. Incomplete curing of the attach epoxy

A different analytical approach will be required for each of these potential failure causes.

10.5   Analytical Approach to Failure Analysis

The probability of success of a failure analysis can be greatly increased by approaching the problem with the proper mindset. Unfortunately, during the "heat" of catastrophic failures, common sense and a disciplined approach usually are the first things discarded in the rush to find the "answer" or to "assess" damages. An investment of a few hours' time getting organized at the beginning of a failure analysis will save hours or even days of wasted time and effort. When everyone is screaming for "action" and instantaneous answers, good failure analysts will appear to be dragging their feet. In reality, the time will be wisely used to map out the strategy and analytical approach.

An effective failure analysis approach will consist of most or all of the following actions:

1. Select the failure analyst and the FA team.

2. Set objectives.

3. Gather information.

4. Organize information (and fill in gaps where possible).

5. Postulate failure mechanisms and possible causes.

6. Develop an analytical flowchart.

7. Start the analysis, evaluate the data versus postulated mechanisms, and redefine the postulated mechanisms or the analytical flowchart (when necessary).

8. Evaluate data with the FA team, and confirm the failure cause and/or mechanism.

9. Report findings with corrective actions.

10. Follow up and test the effectiveness of the corrective actions.

11. Document the FA process for future review.

While it might appear that the complexity of this approach would lengthen the amount of time it takes to achieve results, this approach will usually speed up the process. Not all steps will be performed in every failure analysis and a number of steps may be performed simultaneously.

However, when the timing of the failure analysis is critical, e.g., evaluation of a system failure which holds up a space launch, the failure analysis approach should be modified (but not abandoned) to accommodate the increased value of a timely resolution. This should be done without jeopardizing the ability to get as much information as is needed to meet the objectives of the analysis. In a time-sensitive failure, the analytical approach might be split into two parallel paths. The purpose of one path might be to find a quick fix for the immediate problem, while the purpose of the second path might be to provide the information necessary to make sure that the failure does not recur.

1.   Select the failure analyst and the FA team.   When it becomes apparent that a failure has occurred, the first order of business is to select a failure analyst and help assemble the team. The team might consist of design and process engineers, quality engineers, a product manager, and (in the case of a field failure) a service engineer and/or a customer representative. While not all of these people will be working full time on this one failure, their combined experience and their "on-call" availability will ensure that the analysis is effectively performed. The diversity of this type of team helps in the information gathering and ensures that the failure analysis will not bog down while someone reinvents the wheel.

2.   Set objectives.   The failure analysis team must set reasonable, reachable, measurable objectives. These objectives might be as simple as to find the root cause of the failure or as complex as to make certain that this type of failure will never occur again.

3.   Gather information.   A good failure analyst will gather as much information as possible about the failed part and its operating environment when it failed. Data will be included from previous similar failures to see if this is a repeat problem that has already been analyzed. The failure analyst will also gather information on the manufacturing and processing conditions relative to this lot of parts to determine whether there was something different in terms of material or processing that might make this device or lot of devices behave differently from others. The failed part's designed application and operating environment, versus its actual application and operating environment, will also be considered.

4.   Organize available information.   The information gathered above should be organized so that any piece of it is readily accessible to the team. Missing information can then be garnered from the team members, from the manufacturer, and in the case of a field failure, from the end user. While there are many different ways to organize the data, one commonly used organization scheme is to mirror manufacturing steps, i.e., keep all the design information together, the processing information together, etc. Regardless of which organizational style or system is chosen, the key is to have easy access to as much information about the circuit as possible, while the analysis is being performed. This will enable the failure analyst to make quicker and more effective decisions on the fly, even as the search accumulates additional data during the analysis.

5. Postulate failure mechanisms and possible causes.   Once all the available information is in hand, the failure analyst and the team should brainstorm possible causes of failure or failure mechanisms. This 15- to 30-minute exercise can save weeks of chasing the wrong problem. No matter how improbable, no cause or failure mechanism should be rejected at this time. When the team is convinced that it has listed every possible mechanism or cause, it can reject some as very unlikely when compared to the information available about the failed sample.

After the list of possible causes has been whittled down to the most likely candidates, the failure analyst and the team should organize the potential causes on the basis of their probability of occurrence. A different approach might be to organize possible causes into types of failure, organized along hybrid specific categories, e.g., package-related, component-related, etc., and then prioritize these candidates into the most likely causes of failure. As in the actual failure analysis, the approach used is less important than having an organized, consistent, and well-documented approach.

6. Develop the analytical flowchart.   At this point, the analyst has enough information to develop the analytical flowchart. The purpose of the chart is to serve as a guideline or road map to follow during the course of the analysis. A condensed flowchart might include only the major parts of the analysis and the key decision points (see Fig. 10.1). On the other hand, a complete, detailed flowchart might include the most likely steps of the analysis along with the purpose or objective for undertaking each step (see Fig. 10.2). The level of detail is a matter of preference for each failure analyst. The flowchart helps the failure analysis team and the analyst keep control of the analysis and minimizes the temptation to proceed randomly through the analysis, pursuing whatever turns up at each step.

The flowchart also serves as a model to keep the data organized as the failure analysis proceeds, enabling the analyst and others who might be only peripherally involved in the analysis to make more effective decisions on the fly. In cases where more than one sample is to be analyzed, the analytical flowchart ensures consistency of approach and data from sample to sample.

The analytical flowchart also serves to give secondary analysts, who might be called into the analysis to perform a specific analytical technique which is their specialty, an overview of how and where their work fits into the rest of the failure analysis.

7. Start the analysis, evaluate the data, and redefine the postulated mechanisms or the analytical flowchart.   Once the flowchart has been developed, the actual "testing" part of the analysis can begin. The failure analyst will perform or supervise the performance of the testing steps as specified in the flowchart. At the completion of each step, all the data generated should be organized

1. External microscopic examination

2. Radiography (x-ray) examination

3. Electrical failure verification; functional and pin-to-pin curve tracing

4. High temperature bake (biased and unbiased)

5. Delid and deencapsulate

6. Cross sectioning

7. Internal microscopic examination

8. Electrical probing

9. Voltage contrast imaging in SEM

10. Electron beam induced current (EBIC) imaging in the SEM

11. Liquid crystal fault detection

12. Bond strength pull test

13. Angle lapping and staining

14. Metallization removal

15. Oxide removal

16. Other analytical techniques

Figure 10.1 Condensed analytical flowchart.

and reviewed. Typical points of review include the following:

1. Do the data make sense?

2. Do they lend credence to or support one or more of the postulated failure mechanisms for this sample?

3. Should analysts proceed to the next step in the flowchart or should they modify the flowchart to take into account these new findings?

4. Should a second sample (for failure analyses with more than one sample) be started at this point and analyzed the same way as the first, to corroborate the data?

5. Has the root cause of failure been found?

6. Is enough information available to meet the remaining objectives of the failure analysis?

The focus of the review will be to keep everyone informed, to see that the maximum information is gathered from the sample, and to guide the failure analyst and analysis to a successful conclusion.

8. Evaluate data with the FA team and confirm the failure cause and/or mechanism. After methodically proceeding through the above exercise, enough data will have been generated to enable the failure analyst and/or failure

1. External microscopic examination, optical microscopy, and scanning electron microscopy
   a. Electrical leakage due to contamination between leads
   b. Fractures in package seam
   c. High depth of field imaging
   d. Identifying elemental constituents of contamination

2. Radiography (x-ray) examination
   a. Viewing device construction prior to deencapsulation
   b. Viewing wire bond integrity
   c. Viewing die placement

3. Electrical failure verification, functional and pin-to-pin curve tracing
   a. Verify operation of device
   b. Test device to specification
   c. Characterize a device
   d. Compare to known good device

4. High temperature bake (biased and unbiased)
   a. Identify the presence of mobile ions

5. Delid and deencapsulate (mechanical and/or chemical)
   a. Expose the working components

6. Cross sectioning (abrasive disk or low speed diamond saw)
   a. Package construction
   b. Bond wires and die attach
   c. Plating thickness and uniformity
   d. Junction depths

7. Internal microscopic examination, optical microscopy, and scanning electron microscopy
   a. Metallization and oxide defects
   b. Contamination and/or corrosion

8. Electrical probing
   a. Localize electrical faults at die level

9. Voltage contrast imaging in SEM
   a. Visualize voltage levels in a semiconductor
   b. Locate opens and shorts

10. Electron beam induced current (EBIC) imaging in the SEM
    a. Visual examination of current flow in a semiconductor
    b. Locate opens or shorts

11. Liquid crystal fault detection
    a. Thermal characteristic of a semiconductor surface
    b. Locate hot spots

Figure 10.2 Detailed analytical flowchart.

12. Bond strength pull test
    a. Verify bond integrity

13. Angle lapping and staining
    a. View semiconductor junctions
    b. View metallization, vias, etc.

14. Metallization removal
    a. View subsurface defects, EOS or ESD damage

15. Oxide removal (chemical and/or plasma etch)
    a. View subsurface defects

16. Other analytical techniques

    Scanning Auger microscopy (SAM)
    a. Contamination identification
    b. Corrosion analysis
    c. Stain identification
    d. Lifted lead bond evaluation
    e. Material delamination analysis
    f. Metal embrittlement evaluation

    Secondary ion mass spectrometry (SIMS)
    a. Location of low level ionic contamination
    b. Inversion studies
    c. Doping level investigation

    Fourier transform infrared spectroscopy (FTIR)
    a. Identification of contaminants on microelectronic packages and devices
    b. Identification of organic stains
    c. Inspection for the chemical degradation or decomposition of component or potting materials

Figure 10.2   (*Continued.*)

analysis team to draw conclusions about the cause of the failure, the probability of additional failures, and corrective actions to be undertaken. On sufficiently expensive or sensitive application failures, the analyst will corroborate these conclusions with subsequent analyses of additional samples.

9. Report findings with corrective actions.   The failure analyst will put together a report which documents the findings. Generally a failure analysis report will consist of the following sections:

1. Summary report (optional)

2. Purpose (objectives) of the analysis

3. Description of samples

4. Analyses performed

5. Conclusions

6. Analytical results

7. Recommendations

Segmenting the report into the above categories allows a variety of readers to find the information they need with a minimum of effort. A typical failure analysis report is described in detail in Sec. 10.6.

10. Follow-up and test corrective actions. As part of the failure analysis, the analyst should, where possible, list corrective actions to be undertaken and the results which are expected from the implementation of these actions. In addition, the failure analyst should work with design or processing engineers to develop a follow-up testing plan. The purpose of this testing plan will be to verify that the corrective actions have been performed as specified and that the corrective actions keep similar failures from recurring.

11. Document the FA process for future review. The failure analysis process should be reviewed and documented for each failure analysis performed. Proper documentation will meet a variety of objectives that might seem to have little bearing on this failure analysis but which will have significant ramifications for future failure analyses performed either by the same team or on similar hybrids. The main purposes for process documentation are:

1. Evaluation of where the process might be streamlined

2. Documentation of the relevant points of this failure analysis for use in future failure analyses

3. Estimation of the time required for future, similar failure analyses

4. Determination of the actual cost of performing this analysis

When the failure analyst takes the time to document the FA process, management is helped to make intelligent decisions as to the value of this and future failure analyses.

The best time to evaluate how well a particular failure analysis went is immediately after its completion. All parties involved should have input into what worked well in the process and how it might be streamlined or improved for the future. This can save significant time and resources for future failure analyses.

In addition the documentation process should include a key word synopsis of each failure analysis. Typical categories include part type, manufacturer, failure mode, failure manifestation or symptoms, failure analysis report number, date, and location, and the name of the lead failure analyst. While the lead failure analyst has this information, it is important to get it into a readily accessible database for future use by other failure analysts, engineers, and managers.

The estimation of the time required for future similar failure analyses helps management decide if the analysis of the next failure can be completed within a time frame and at a cost that is relevant to the actual failure. If a manager needs to make a marketing or manufacturing decision in two days, it will be very helpful to know if the failure analysis results can be available prior to making the decision.

Similarly, the estimation and the tracking of the time and other directly related costs of a failure analysis enable management to determine the value of performing the failure analysis and the true costs of maintaining a failure analysis department.

10.6 Failure Analysis Documentation and Reports

The failure analyst will put together a report which documents the findings. Generally a failure analysis report will consist of the following sections:

1. Summary report (optional)

2. Purpose (objectives) of the analysis

3. Description of samples

4. Methods of analysis

5. Conclusions

6. Analytical results

7. Recommendations

The report should be clear, concise, readable, and understandable by a variety of engineers, scientists, and managers.

1. Summary report. For a very lengthy or complex failure analysis, the report might also contain a one- to two-page summary report. This summary report serves the purpose of getting the important details of the analysis across to a variety of managers, engineers, sales personnel, and customers, etc., who need to know the end result or recommendations for corrective actions or lot disposal but do not need all of the detail of the analysis.

2. Purpose (objectives) of the analysis. The purpose of the analysis should be clearly stated so that subsequent readers of the report can understand the reasoning behind the analytical approach and the conclusions. This section should consist of short, clear sentences or a bullet list of objectives.

3. Description of samples. Similarly, a description of the samples should be concise. It should include relevant information about each sample that was analyzed. This will help to put this analysis into perspective and will help future failure analysts locate data on similar types of failures.

4. Methods of analysis. The methods of analysis section should list each analytical procedure that was performed in the failure analysis to help people interested in this failure understand how the data were generated and to guide future failure analyses of similar circuits. This section should be just a brief listing of the procedures used. A more detailed account of procedures should be addressed in the analytical results section. If desired, a brief description of the basics of each analytical technique can be included as an appendix to the report.

5. Conclusions. The conclusions should clearly state what was found to be the cause of failure. These should not be a listing of the results or analytical data. The conclusions should be the logical inferences drawn from the data by the failure analyst. Wherever possible, conclusions should be substantiated by the data in the results section. It should be clearly stated which conclusions are fully substantiated by the data and which represent the experienced opinions of the failure analyst or the team.

6. Analytical results. The analytical results section should be a complete compilation of all the results and data generated in the analysis. This might be compiled by analytical technique used or by sample analyzed. Where possible, large groups of spectra should be summarized in qualitative and quantitative tables to facilitate comparisons between multiple samples. In an alternative report format, the results section is often combined with the analysis section and written as a running narrative of the failure analysis.

7. Recommendations. Recommendations encompass the opinions of the failure analyst and the failure analysis team. They include corrective actions for similar parts and, where necessary, substitution of materials and changes to hybrid designs and/or their manufacturing processes. Recommendations might also include suggestions for follow-on testing to verify that any corrective actions have indeed fixed the problem or removed the cause of the failure. Lastly, this section of the report might present recommendations on the most expeditious way to perform subsequent failure analyses of similar parts for the use of other failure analysis teams.

Appendix 10.1 is a typical report for a failure analysis of a multichip module. Since this report is fairly concise, there is no need for a summary report.

10.7 Failure Analysis Tools

The failure analyst has a variety of tools which can be used to perform the failure analysis. These tools can be divided into two categories: "basic," which are readily available in-house; and "advanced," which are either available in-house or readily available from outside laboratories. The choice of which tools to use and in which order to use them is usually specified in the analytical flowchart. Selection and timing of the use of analytical tools will be dependent on whether or not the information they provide is needed, their availability, whether or not they are fully or partially destructive, their cost, and the amount of time required for the application of each tool. Usually, the failure analyst will select tools from the basic group and reserve the selection of tools from the advanced group for when a specific piece of information or data is required to confirm the cause of failure.

10.7.1 Basic failure analysis tools

The more common or readily available failure analysis tools are:

1. Electrical or parametric testing
2. Radiography
3. Curve tracing
4. Optical microscopy
5. Preparation techniques
6. Liquid crystal imaging
7. Submicrometer probing
8. Basic scanning electron microscopy
9. Bond pull
10. Die shear

When a company has an in-house failure analysis laboratory, these tools are usually readily available there. When the company does not have an in-house failure analysis laboratory, most of these tools are available within other departments, e.g., quality control or manufacturing. While the failure analyst is free to use all these tools or as few of them as deemed necessary, they will typically be used in the order that they are listed above. This order enables the failure analyst to get the most information about the circuit and the failure site with a minimum of time and effort. This analysis order also maximizes the probability of finding the cause of the failure while minimizing the possibility of destroying the evidence before the analyst has documented it.

1.   Electrical or parametric testing. Electrical or parametric testing, is usually reserved for three types of failures:

1. Failures of very complex circuits or assemblies
2. Field failures which have very little documentation
3. Failures which are intermittent

In the case of failures of very complex circuits or assemblies, or field failures which have very little documentation, electrical testing helps the failure analyst determine the extent of the failure. The electrical testing results will enable the failure analyst to narrow the area of interest in the circuit, allowing the analysis to proceed at a faster pace. On the basis of electrical testing results, the analyst can often shorten the list of potential failure mechanisms and refine the analytical flowchart, thereby saving time and money.

   In the case of failures which are intermittent, extensive electrical and parametric testing will often point to the area of the circuit or the individual component which is behaving intermittently. The most useful analytical data will often be in the form of a varying resistance or intermittent opens or shorts. In rare cases, the electrical testing will have to be performed in the same (or simulated) environment as the intermittent failure. This can be the case when mechanical or thermal stresses cause a leakage path or a marginal connection to open or close intermittently.

Figure 10.3 Positive print of an overview radiograph of a hybrid.

2. Radiography. Radiography enables the failure analyst to nondestructively obtain an internal view of the mechanical layout of the hybrid circuit or assembly. The quality of workmanship can readily be ascertained and the more obvious mechanical causes of failure in a high quality x-ray radiograph can be seen. Crushed or lifted lead bonds, die or component detach, and die cracking can usually be readily seen in the radiograph. In addition, the original radiographs (and subsequent localized area radiographs) will serve as road maps during the preparation stages of the analysis. The analyst will use these to guide the mechanical or chemical depackaging and depotting steps which are necessary to expose the failure site. By referring to the radiographs and taking additional radiographs as needed, the analyst will ensure that a minimum amount of damage is done to the areas or components of interest.

Figure 10.3 is a typical overview radiograph of a hybrid circuit. Conventional radiographs are produced at 1× magnification and examined either with an eye loupe or under an optical microscope to view very small details about the circuit and its construction. As an alternative during the analysis, the failure analyst will often make one or more photographic enlargements of the original radiographs. The enlargement, seen in Fig. 10.4, of the original

Figure 10.4 Photographic enlargement of the radiograph in Fig. 10.3.

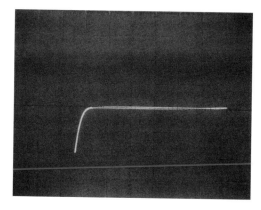

Figure 10.5 Current versus voltage curve trace of a good circuit junction.

radiograph can be used to examine the failure site more closely. It may also be used as a map of component location which will serve as a guide for subsequent mechanical or chemical preparation techniques. A third use of the radiographic enlargement is as a reference document in the final report.

3. Curve tracing. Curve tracing is used on specific components or parts of the circuit to verify that each part of the circuit or component is operating up to specification. It is particularly valuable for locating leakage paths, opens, shorts, and blown components. Figure 10.5 is a curve trace from a "good" circuit junction, and Fig. 10.6 is a curve trace from an identical circuit junction which shows a leakage in parallel with the junction due to corrosion of the substrate metallization.

4. Optical microscopy. Optical microscopy is used to perform a quick, nondestructive inspection of the failed circuit looking for signs of poor workmanship, failed or leaking seals, rough handling, thermal and mechanical damage, external contamination, or corrosion. A skilled analyst can gather a lot of information about the circuit and its environment from a thorough optical inspection performed at 20× to 100× magnification. The information gath-

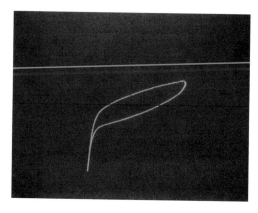

Figure 10.6 Current versus voltage curve trace of a failed circuit junction.

ered at this step will be used either to verify the logic of the original analytical flowchart or to modify it.

Optical microscopy will be used again and again, during and after sample depackaging and depotting, to determine the presence of gross contamination, complete or partial component detach, lifted lead bonds, gross die or component cracking, or metallization migration. Extensive photographic documentation is usually performed to keep a record of everything seen in the optical inspection.

Figure 10.7 is a typical overview optical micrograph of a hybrid circuit which shows circuit construction, component layout, and gross anomalies (if present). Figure 10.8 is a typical higher magnification optical micrograph detailing the construction of a transistor on the same hybrid circuit.

5. Preparation techniques. After the external condition of the failed circuit has been photographically documented, the analyst reviews the flowchart and decides a plan of attack for opening or exposing areas of interest. The analyst will use a variety of techniques including but not limited to:

1. Chemical etching
2. Mechanical milling and grinding
3. Plasma etching
4. Sawing, cross sectioning, and polishing

The analyst will switch between techniques as necessary to expose the area or component of interest. The preparation techniques require a lot of skill, patience, and manual dexterity, particularly when applied to encapsulated or

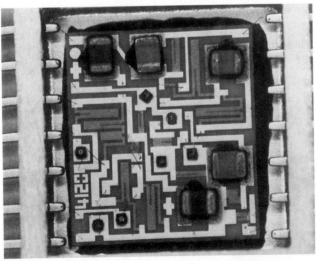

Figure 10.7   Optical photograph of a failed hybrid.

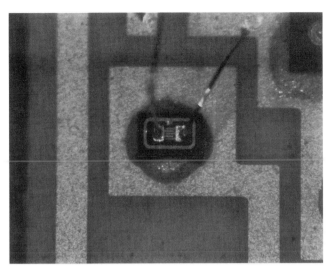

Figure 10.8 Optical photograph of a transistor on the hybrid from Fig. 10.7.

potted circuits. It is important that adequate photographic documentation be performed during all the steps of this process since the sample, the failure site, or the evidence of the cause of failure may be destroyed at any step.

Chemical etching is usually performed on encapsulated hybrid assemblies and on encapsulated components within a hybrid. It is the one step in the entire failure analysis that has the most uncertainty and the highest risk of destroying the sample or the evidence of failure without learning the cause of failure. The best approach to chemical etching entails evaluating the chemicals and the processing conditions on a control sample (when one is available) prior to working on a one-of-a-kind failure. Lacking a control sample, the analyst will usually nondestructively analyze the package or potting material by Fourier transform infrared spectroscopy (FTIR). This will enable the determination of the best combination of acids and solvents which will remove the potting material while minimizing damage to the areas of interest. Prior experience at chemical etching is very useful and often spells the difference between successfully exposing the area of interest without adding artifacts and destroying the failed sample along with the evidence of the cause of the failure.

Mechanical milling and grinding encompass a variety of tools. It might be as simple as using a precision modeler's file or a sanding block to open metal packages. Or it could be as complicated as using a milling machine, a model maker's Dremel tool, or a lathe to remove potting or encapsulating material in a precise, controlled manner. Often the failure analyst will take additional x-rays and optical photographs during the procedure to verify the proximity to critical components or to sites which are either suspected to be the source of the failure or which might contain evidence of the source of the failure.

Plasma etching can be used to remove the last layers of polymeric encapsulants while leaving the circuit undisturbed. It can also be used to remove passivation at the die level to facilitate subsequent optical and scanning electron microscopic evaluation of circuit metallization.

Low speed diamond sawing and cross sectioning are useful to analyze the internal structure of components and to assess the quality of die and component attach, plating materials, thick film materials, circuit interconnections, and solder joints. This procedure entails potting or embedding the circuit in epoxy, diamond sawing through the area of interest, and subsequently grinding and polishing the sample to highlight the structure. Since this process subjects the sample to a maximum of physical abuse and potential chemical contamination, it is usually performed toward the end of the failure analysis.

6. Liquid crystal imaging. Liquid crystal imaging can be used to detect electrical leakage, opens, shorts, and in some cases, electrostatic static discharge (ESD) damage. In liquid crystal imaging, a series of liquid crystals, which change from translucent to opaque at different temperatures, is applied to the circuit. Wherever there is a hot spot or change in localized temperature on the device, the liquid crystal shows an "apparent" color shift when viewed under a microscope with a polarizing filter. The location of the hot spot can be documented photographically for subsequent higher magnification (optical or scanning electron microscopic) examination. Figure 10.9 is an optical, liquid crystal photograph of a defect hot spot on a failed capacitor.

7. Submicrometer probing. Submicrometer probing uses a micromanipulator stage in conjunction with a curve tracer to electrically probe and test localized areas on a circuit or device. It is particularly useful for locating small opens, shorts, or leakage paths. It is also useful for verifying that the failure is still present and intact, on or within a circuit, after each depackaging or deprocessing step. A typical probe station is capable of probing devices or areas smaller than 1 μm^2.

Figure 10.9 Optical photograph of a "hot spot" as revealed by liquid crystal imaging.

8. Basic scanning electron microscopy. Basic scanning electron microscopy (SEM) is useful for examining the failed hybrid at higher magnifications than is possible with an optical microscope. Typical failure causes and mechanisms that may be seen in a scanning electron microscope include:

1. Contamination

2. Metal migration

3. Microcracking of components or substrates

4. Film delamination

5. Solder homogeneity

6. ESD damage

All these potential causes of failure can be easily spotted in an SEM if they are on the surface of the hybrid or its components, since SEM has a range of magnifications from 5× to 300,000× magnification and a depth of field which is 500 times greater than that of an optical microscope. The great depth of field is particularly helpful when performing a failure analysis on a large hybrid with components of varying heights and geometries. See Fig. 10.10 for a comparison of an optical micrograph and a scanning electron micrograph of the same hybrid at the same magnifications. Figure 10.10a is an optical micrograph of a hybrid and Fig. 10.10b is a secondary electron micrograph of the same hybrid. The increased depth of field and the "3-D" perspective of the SEM image make it easier for the failure analyst to spot anomalies.

A scanning electron microscope operates by scanning a finely focused beam of electrons across the surface of a sample. The electron beam interacts with the atoms on the surface of the sample, generating secondary electrons, backscattered electrons, and x-rays. These signals are collected by specific detectors and displayed on a viewing cathode ray tube. The raster on the cathode ray tube corresponds to the raster of the electron beam on the sample surface, and the brightness on the cathode ray tube corresponds to the amount of signal generated at each point on the sample.

The brightness of the secondary electron signal varies with surface topography or roughness. The secondary electron signal is used to image what the sample looks like at higher magnifications than an optical microscope. The image is documented photographically on film as a secondary electron micrograph.

The backscattered electron signal is dependent on the atomic number of the material of the sample. The higher the atomic number of the material, the greater the probability that the incoming electron will be backscattered out of the sample and the brighter the backscattered signal will be. Hence backscattered electron microscopy is an excellent tool for imaging changes in composition on a sample. Such changes in composition can be related to plating or thick film thickness, contamination, metal migration, corrosion products, or material diffusion. Figure 10.11 shows the differences between a secondary electron image (Fig. 10.11a) and a backscattered electron image (Fig. 10.11b)

(a)

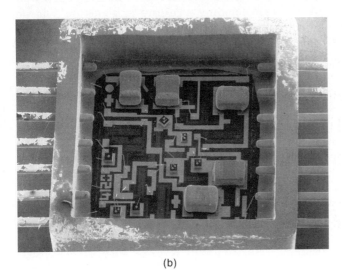

(b)

Figure 10.10 Comparison of an optical photograph (a) with a secondary electron micrograph (b).

taken at the same magnification on a hybrid circuit. The two images complement each other in that the secondary electron image shows topography while the backscattered electron image shows changes in composition.

9. Bond pull. Wire bond pull is used when lifted lead bonds are found in a hybrid. Usually the failure analyst will pull a representative number of bonds similar to the one that failed, to determine if other bonds will fail on the same or on subsequently produced circuits. In addition, bond pull is often used to

(a)

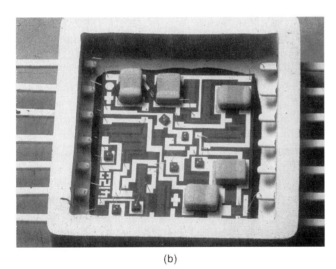

(b)

Figure 10.11 Comparison of a secondary electron micrograph (*a*)
with a backscattered electron micrograph (*b*).

expose a fresh interface at the bond site. This fresh surface is useful for subsequent elemental analysis to determine the presence of the proper intermetallics and the absence of contamination on either the wire or the bond pad.

10. Die shear.   Die shear is used in failure analysis when the failure appears to be caused by delamination of the die or components from the substrate. The measurement of a low die shear force will verify and localize the presence of the failure site. As in wire bond pull, die shear is occasionally used to expose a fresh area under the component or die for further elemental or chemical analysis. This helps the analyst determine whether contamination of the component, the substrate, or the die attach epoxy, or incomplete epoxy curing, was the cause of delamination.

10.7.2   Advanced failure analysis tools

The advanced failure analysis tools are:

1. X-ray microradiography

2. Acoustic imaging

3. Thermal imaging

4. Advanced scanning electron microscopy

5. Energy dispersive x-ray spectroscopy

6. Micro–Fourier transform infrared spectroscopy

7. Scanning Auger microanalysis

While these tools are not as readily available to every failure analyst and require more experience to use and interpret, they can be more specific in determining the root causes of failure than the common tools.

1.   X-ray microradiography.   X-ray microradiography (XRM) utilizes a small spot, tightly collimated x-ray source and an x-ray sensitive video camera to perform real-time x-ray imaging of microelectronic devices. Presently available XRM systems can image through most hybrid packages. The circuit can be rotated and moved in the x, y, and z directions via a joy stick controlled, movable stage. The use of the small-sized x-ray source enables direct x-ray imaging in video mode at magnifications from 1 to 50 times. The entire examination may be videotaped and selected fields of view may be captured on conventional x-ray negatives or on Polaroid™ positive or positive-negative film.

2.   Acoustic imaging.   Scanning laser acoustic microscope (SLAM) imaging produces an image of the internal structure of a hybrid by imaging the propagation of sound waves through a material. Since the sound wave propagation stops whenever there is a break in the material, this is a very good, nondestructive technique for imaging voids in die attach, cracks in or under substrates, and localized areas of film delamination. While the technology is best suited for routinely screening devices before they are sealed, it can be valuable for the nondestructive inspection of parts suspected of failing because of attach problems.

3.   Thermal imaging.   Thermal imaging utilizes an infrared microscope and camera system to image and map thermal gradients across a circuit. It can have greater spatial resolution than liquid crystal imaging and it has the benefit of mapping thermal dissipation over a wide range of temperatures simultaneously. This technology can be helpful in locating very subtle hot spots which might cause failures in the future.

4. Advanced scanning electron microscopy. Advanced scanning electron microscopy combines the benefits of basic SEM with submicrometer probing. Electron beam induced current (EBIC) imaging is used to map current flow through a circuit or device. The scanning electron beam from the SEM will induce a current within semiconductor and microelectronic devices. A contact is made to the leads or connectors of the circuit, and the EBIC current, generated within the sample by the beam, is amplified and superimposed over the secondary electron image. This produces a map of current flow through the circuit. It is an excellent tool for locating surface and subsurface shorts, leakage paths, and opens. Figure 10.12 is a secondary electron micrograph of a circuit. Figure 10.13 is the EBIC micrograph of the same circuit at the same magnification. The dark contrast shows where current is generated and how it flows as the electron beam is scanned across the circuit. Using EBIC to compare a "good" or known circuit to a "failed" circuit, the failure analyst can quickly locate electrical problems such as opens, shorts, or leakage paths.

5. Energy dispersive x-ray spectroscopy. Energy dispersive x-ray spectroscopy is an accessory technique to the scanning electron microscope. It enables the analyst to qualitatively and quantitatively analyze materials and contaminants imaged in the SEM. In addition to SEI and BEI imaging signals, the scanning electron beam of the SEM also produces x-rays. If the SEM is equipped with an x-ray detector and a multichannel analyzer, the failure analyst can verify that the correct materials are present on or in the hybrid. In addition, the EDS can identify the elements in any anomalies imaged in the SEM. This is particularly useful for identifying gross contamination, corrosion products, or metals which have migrated to form leakage paths. Figure 10.14 is an EDS spectrum from some particulate contamination found on aluminum metallization inside a hybrid package. The EDS system identified the particulate as a flake of silver. EDS element mapping can be used to image the distribution of elements across a circuit at a depth of 0.1 to 1.0 μm. In addition, EDS can provide a quantitative analysis of particles or features imaged in the SEM.

6. Micro–Fourier transform infrared spectroscopy. Micro–Fourier transform infrared spectroscopy (FTIR) is an analytical tool that identifies organics and polymers and corroborates the presence of inorganic salts. In the FTIR technique, the hybrid is subjected to a beam of infrared light. This light is absorbed at specific frequencies as it excites the molecular bonds in the atoms of the material. The FTIR plots out spectra which can be used to identify whatever was in the path length of the beam. With the addition of an

Figure 10.12 Secondary electron image of an integrated circuit.

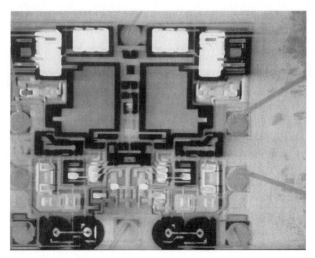

Figure 10.13 Electron beam induced current image of an integrated circuit.

infrared microscope, particulates or sample areas as small as 10 μm in diameter can be analyzed, often directly on the circuit with a minimum of sample handling or preparation. Micro-FTIR is also very useful for determining whether or not adhesives, encapsulants, or potting compounds are completely cured. Figure 10.15 is an FTIR spectrum of organic contamination found inside a hybrid package. Interpretation of the spectrum showed that the contamination was a plasticizer which had leached out of the plastic tubing in a cleaning system.

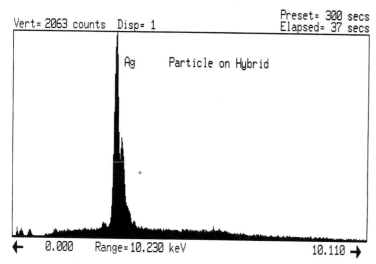

Figure 10.14 Energy dispersive x-ray spectroscopy spectrum of contamination on a hybrid package.

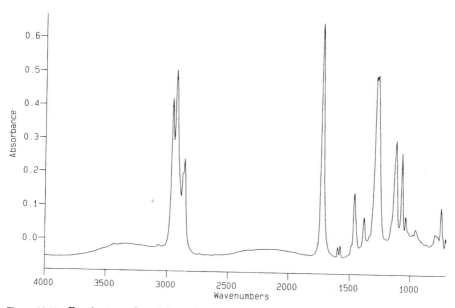

Figure 10.15 Fourier transform infrared spectrum of contamination.

7. Scanning Auger microanalysis. Scanning Auger microanalysis (SAM) is an analytical tool that identifies elements on surfaces. Like an SEM, a SAM focuses an electron beam on the surface to be analyzed. Secondary electrons are collected to provide an image of what the surface of the sample looks like, enabling the analyst to focus on features of interest. Auger electrons, which are also produced by the interaction of the electron beam with the surface atoms of the sample, are also ejected from the surface, collected, and sorted by energy. A spectrum consisting of a plot of Auger electrons versus their energy qualitatively identifies the elements on the surface. In addition to the qualitative analysis, most SAM systems have the capability of providing quantitative analyses at concentration ranges from 0.1 to 100%. Figure 10.16 is an optical photograph of a stain which was found on a hybrid package. A SAM spectrum of the stain can be seen in Fig. 10.17. The concentrations of the elements found in the stain can be found in the box in the lower right-hand corner of the SAM spectrum. In addition, the distribution of these elements can be "mapped" across the surface of the package by setting windows for Auger electrons of specific energies. Figure 10.18 shows a combination of SAM element maps for nickel, calcium, potassium, and sulfur taken on the surface of the package. SAM element maps complement EDS element maps in that the SAM maps are monolayers deep while EDS maps are from an area with an average depth of about 1 μm, which is approximately 3000 monolayers deep.

The SAM technique is inherently a shallow analysis tool allowing the analyst to identify contaminants a few monolayers thick. The SAM also has the capability of using an ion sputter gun to remove material as it is being analyzed, allowing the analyst to produce a "depth profile" for 1 to 16 elements. Figure 10.19 is a representative depth profile showing the change in concentrations of nickel, calcium, oxygen, sodium, and phosphorus as a function of depth in the stain. The profile shows that the stain in Fig. 10.16 is 200 to 400 Å (20 to 40 nm) thick.

10.7.3   Summary of analytical techniques

This is a compilation of the most often used failure analysis tools. While it presents an overview of the use and usefulness of each tool, it is not intended

Figure 10.16 Optical photograph of a stain on a hybrid package.

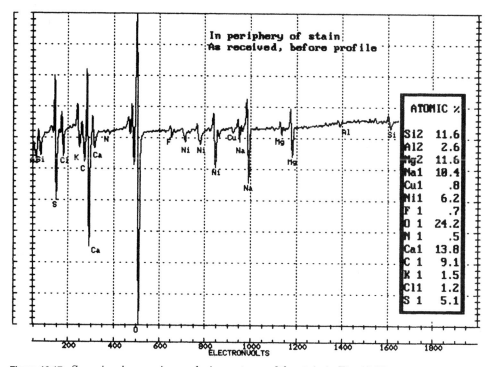

Figure 10.17 Scanning Auger microanalysis spectrum of the stain in Fig. 10.16.

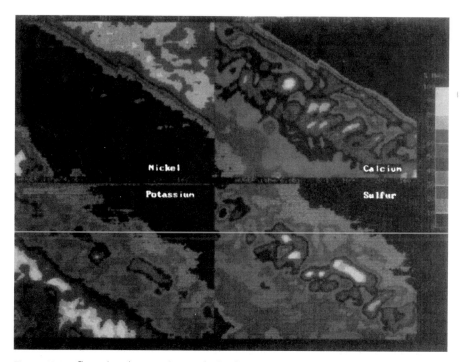

Figure 10.18 Scanning Auger microanalysis element maps of nickel, calcium, potassium, and sulfur.

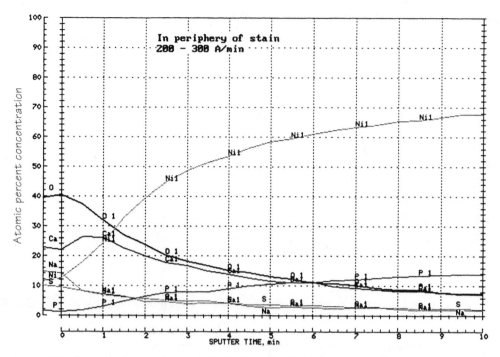

Figure 10.19 Scanning Auger microanalysis depth profile of a stain on a hybrid package.

to be an all-encompassing survey of every available analytical tool. When performing a failure analysis, one should constantly bear in mind that the most sophisticated tools available are useless unless the analyst understands how the tool works, where it should be applied, and how to interpret the results obtained from that application. To help in the use of these analytical tools and in their proper application, Fig. 10.20 presents a list of acronyms versus analytical tool names.

Acronym	Technique name
EBIC	Electron beam induced current imaging in the SEM
EDS	Energy dispersive x-ray spectroscopy
EOS	Electrical overstress
ESD	Electrostatic discharge damage
ET	Electrical and curve tracing
FTIR	Fourier transform infrared spectroscopy
Micro-FTIR	Fourier transform infrared spectroscopy with an infrared microscope
SAM	Scanning Auger microanalysis
SEM	Scanning electron microscopy
SIMS	Secondary ion mass spectrometry
VC	Voltage contrast imaging in the SEM
XPS	Electron spectroscopy for chemical analysis
XRM	X-ray microradiography

Figure 10.20 List of acronyms.

10.8   Typical Hybrid Failure Modes

While the term "typical" hybrid failure is almost an oxymoron, there are several types of failures that are common to a variety of hybrids. It is easiest to think of these failures in terms of the area of the circuit in which they occur or the hybrid manufacturing process that might have caused them. Figure 10.21 lists failure symptoms and the root causes for failures that are commonly found at the package level. A large number of package related failure mechanisms and failure symptoms can be traced to contamination of the package, usually during manufacture. Thin film related failures are also often linked to some form of contamination of the thin film, the deposition system, or occasionally, raw materials. Figure 10.22 lists failure symptoms

Package failures and problems	Common root caused
Leaking seals	Mismatched materials Contamination of package or seal Improper sealing parameters
Corrosion	Ionic contamination (with or without moisture)
Contamination	Cleaning solvent residue Epoxy residue or bleed Human Flux, resist, masking materials
Lifting lead bonds	Organic contamination Improper bonding parameters Oxidation of pad or wire
Electrical shorts	Gross contamination Ionic contamination Overstress Misplaced conductive paste Metal migration
Leakage currents	Ionic contamination Metal migration

Figure 10.21   Package level failures.

Thin film failures and problems	Common root causes
Contamination	Improper cleaning Contaminated sputtering systems Solvent residue Vacuum leak
Delamination	Organic contamination Oxide formation
Oxidation	Air leak
Corrosion	Ionic contamination with or without moisture

Figure 10.22   Thin film related failures.

and the root causes for failures that are commonly found at the thin film level of hybrids.

Thick film related failures are often linked equally to contamination or to processing problems. The increased thermal, mechanical, and chemical processing steps that thick film hybrids commonly are subjected to yield a greater opportunity for failure. Figure 10.23 lists failure symptoms and the root causes for failures found in thick film systems.

Lastly, die and component level failure symptoms and their root causes are listed in Fig. 10.24. These failures are more likely to be found in in-house quality control screening or burn-in. However, while less probable than package or thin or thick film failures, these failures can occur in the field.

10.9   Typical Failure Analysis Examples

10.9.1   Particulate contamination

A hybrid was subjected to a final internal visual inspection just prior to sealing the package. Figure 10.25 is an optical photograph of a particle found on

Thick film failures and problems	Common root causes
Contamination	Raw materials Solvent or cleaning residue
Diffusion	Thermal parameters Material incompatibility
Delamination	Gross contamination Improper drying or firing
Resistor cracking	Overpower Contamination Incorrect firing parameters

Figure 10.23   Thick film related failures.

Die and component failures and problems	Common root causes
Lifted lead bonds	Organic contamination of bond pad Oxidation of the pad or the wire
Delamination	Organic contamination Oxide formation Epoxy contamination Incomplete epoxy curing Thermal mismatch of components, epoxy, or substrates
Leakage paths	Ionic contamination Metal migration
Corrosion	Ionic contamination with or without moisture
Opens or shorts	Electrostatic discharge damage Transient damage

Figure 10.24   Die and component related failures.

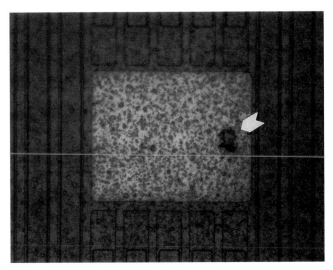

Figure 10.25 Optical photograph of hybrid contamination.

metallization. The failure analysis team had the following questions to answer:

1. What is the particle?
2. Is it conductive?
3. Is it corrosive?
4. Is it attached or could it become mobile over time?

Figure 10.26 is a scanning electron micrograph of the particle. It does not appear to be firmly attached to the metallization and does not appear to be from the metallization or a contaminant trapped in the metallization. Subsequent energy dispersive x-ray spectroscopy produced the spectrum seen in Fig. 10.27. The particle was a bit of silver flake from a conductive epoxy. While the particle was not corrosive, it was conductive. If it became detached over time, it could have shorted out metallization or discrete components.

10.9.2   Electrical transient damage

A hybrid power supply failed when its output dropped by 35%. Subsequent electrical testing isolated the failure location to an IC in an op amp. While optical examination did not show any gross anomalies on the IC, SEM inspection showed an area of damage to the metallization. The passivation was removed from the metallization, and a scanning electron micrograph (see Fig. 10.28) was taken of the suspect area. The photograph shows nodules of silicon fused with aluminum, and voiding damage to the underlying insulator, characteristic of transient damage. This type of failure usually manifests itself as a leakage current between conductors or between conductor and ground.

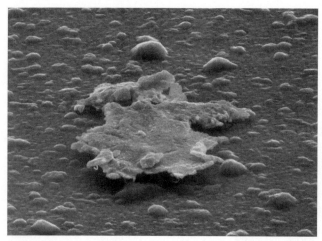

Figure 10.26 Scanning electron micrograph of the same contamination as in Fig. 10.25.

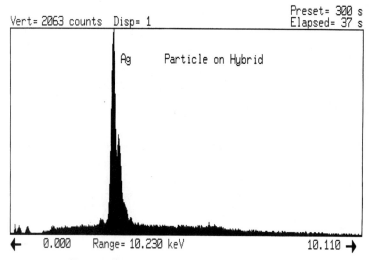

Figure 10.27 Energy dispersive x-ray spectroscopy spectrum of the same contamination as in Fig. 10.25.

10.9.3 Process contamination

A small lot of hybrids failed internal visual inspection when each displayed one or more "blobs" of a contaminant. Figure 10.29 is an optical photograph of a typical "blob." Scanning electron microscopy (see Fig. 10.30) showed that the "blob" was relatively uniform and had probably been introduced into the package as a liquid. Energy dispersive x-ray spectroscopy produced a spectrum (see Fig. 10.31) that contained carbon, oxygen, silicon, chlorine, and sulfur. The presence of the carbon and oxygen implied that at least part of the contamination was organic. Micro–Fourier transform infrared spectroscopy

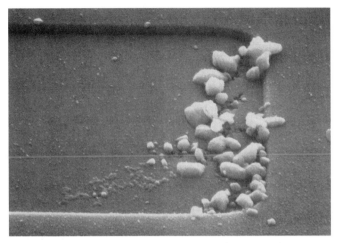

Figure 10.28   Scanning electron micrograph of transient damage.

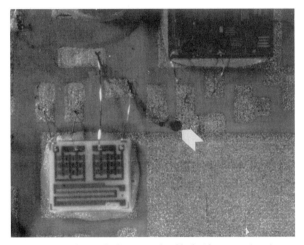

Figure 10.29   Optical photograph of hybrid contamination.

was then used to identify the contaminant as flux residue (see the FTIR spectrum in Fig. 10.32).

10.9.4 Process contamination

After the plasma cleaning of circuits and assemblies, it is not uncommon to find traces of fluorine on the surface by scanning Auger microanalysis. However, when this fluorine is in concentrations on the order of several percent and in depths greater than a few monolayers [10 to 20 Å (1 to 2 nm)] it can cause subsequent leakage current or corrosion problems within the hybrid. Figure 10.33 is a SAM spectrum taken at a depth of 20 Å on a power hybrid showing the presence of 2.5% fluorine. This was found to be the cause

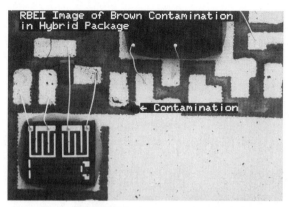

Figure 10.30 Scanning electron micrograph of the same contamination as in Fig. 10.29.

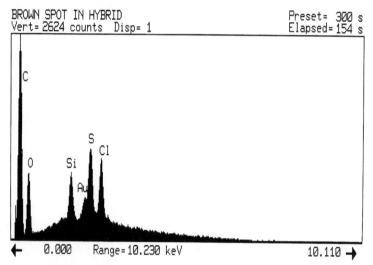

Figure 10.31 Energy dispersive x-ray spectroscopy spectrum of the same contamination as in Fig. 10.29.

of corrosion induced leakage currents and metal migration within the power hybrids.

10.9.5 Human contamination

Another source of hybrid failures is human contamination. This can be identified by a combination of energy dispersive x-ray spectroscopy which identifies the salts prevalent in biological fluids and oils, with Fourier transform infrared spectroscopy which corroborates the presence of the salts and identifies trace amounts of human contamination, e.g., skin oils, hand creams, saliva, etc. When the failure analyst needs to know the root source of an organic

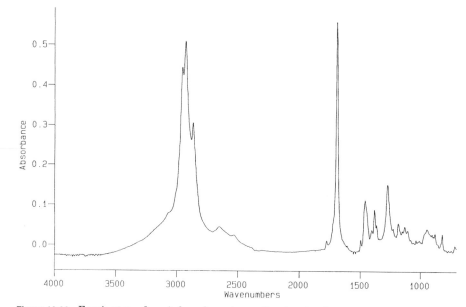

Figure 10.32 Fourier transform infrared spectrum of contamination.

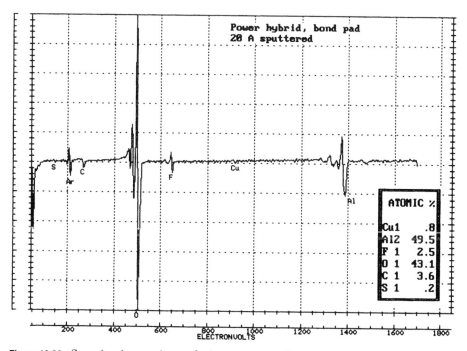

Figure 10.33 Scanning Auger microanalysis spectra on a plasma cleaned bond pad.

contaminant, FTIR is specific enough to be able to tell not only that a contaminant is a hand cream but also the brand of hand cream.

10.9.6 Improper processing

The laser trimming of resistors often leaves significant microcracks in the resistor material. Figure 10.34 is a scanning electron microscope backscattered electron image of a cross section of a laser trimmed resistor. Arrow 6 indicates the area of the trim in cross section. Arrow 1 indicates the crack as it has grown with thermal stressing from operation. The ruthenium dioxide resistor, dielectric, silver conductor, and alumina substrate are labeled with arrows 2 through 5, respectively.

10.9.7 Cleaning process contamination

With the advent of aqueous based cleaning systems, hybrids are failing from the presence of salts left behind by either contaminated rinses or incomplete rinsing, or by biological (microbial) induced corrosion. As cleaning baths and filtration systems sit around, particularly at elevated temperatures, they become a breeding ground for microbial growth. When this residue is left behind in spots on hybrids, the microbes will concentrate free ions, which later, upon dehydration of the microbes, can attack metals and produce leakage current carrying ionic by-products. A combination of surface and microanalysis techniques can find and identify these types of residues both before and after a failure has occurred.

Figure 10.35 is an optical micrograph of a typical bioresidue. The energy dispersive x-ray spectroscopy spectrum seen in Fig. 10.36 shows the presence of the corrosive elements chlorine, sulfur, sodium, potassium, and calcium.

Figure 10.34 Scanning electron micrograph of the cross section of a laser trimmed resistor.

Figure 10.35 Optical photograph of hybrid biocontamination.

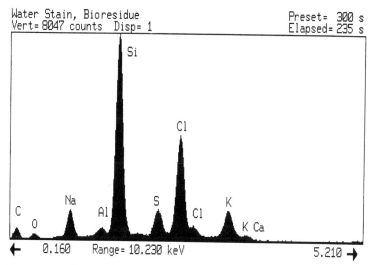

Figure 10.36 Energy dispersive x-ray spectroscopy spectrum of the same contamination as in Fig. 10.35.

Since these elements will also carry a charge, they could produce low level leakage currents even before or while causing severe corrosion. Figure 10.37 is the Fourier transform infrared spectrum of this contamination which indicates that the dark residue is a bioresidue.

10.10   Failure Analysis in the Future

10.10.1   Failure mode effects analysis (FMEA)

Failure mode effects analysis (FMEA) is an analytical tool that enables design engineers to predict and eliminate potential cause of failures and fail-

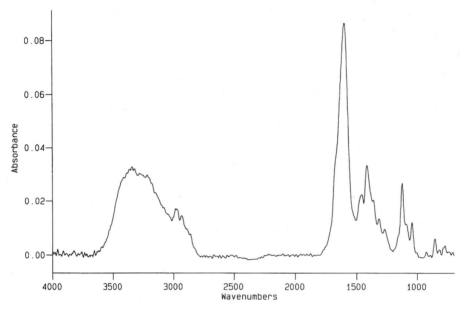

Figure 10.37   Fourier transform infrared spectrum of biocontamination.

ure sites before the circuit or assembly is built. Under FMEA, an engineer will critically examine the design layout and all the components specified in the design and procurement documents. The engineer will single out a component, consult failure probability guidelines to see how many different ways the component can fail, and determine the statistical probability of each failure. It will then be determined how each failure would affect the next component or piece of the circuit, and ultimately the end result of each component failure on the hybrid circuit performance will be predicted.

Repeating the above process for each component, the engineer methodically creates a massive spreadsheet of data which can be sorted to indicate the weakest areas and/or components within the circuit. Then a cost analysis is performed to determine which of these components should be changed or replaced. The circuit is then redesigned to eliminate or minimize the problem areas, and either a prototype is built and tested or the second generation design is resubmitted for another FMEA.

While this process may take as much as 1 to 4 work weeks of an engineer's time, it can save costly prototype development, testing, and rework time. The more complex and expensive the assembly or circuit, the greater the potential savings in time and manufacturing cost. As FMEAs become more widespread, there exists the potential for significant cost savings by applying FMEAs to "blocks" or modules of a circuit, enabling the FMEA analyst to use data from previously analyzed, similar modules. Ultimately, software is being developed that will perform a lot of the lookup and tabulate part of this job, speeding up the entire process and considerably reducing the amount of engineering time involved.

10.11 Summary

Failure analysis can be a very cost effective part of hybrid circuit manufacturing. However, failure analysis will only be as effective as the resources, workforce, analytical tools, planning, and organization that are allocated to it. The combination of a good failure analysis team with a well-planned approach and the right tools can greatly increase the yield and future reliability of hybrids. This chapter outlines some of the steps, resources, and skills needed to perform good failure analyses. Appendix 10.2 is a reference list of additional sources of information on failure analysis of hybrids.

Appendix 10.1   Sample: Basic Failure Analysis Report

January 12, 1993

Ms. Ima Engineer
Hybrids To Go

Report No: 77734   P.O. No: 123456

PURPOSE OF ANALYSIS
To determine the cause for failure of a D/A converter module.

SAMPLES
One (1) 14 Bit D/A converter multichip module assembly.

METHOD OF ANALYSIS
Scanning electron microscopy (SEM)
X-ray microradiography (XRM)

CONCLUSION
The module failed owing to an electrical transient which damaged the input circuitry of one of its integrated circuits. Inspection of the system schematic indicates that the source of the transient was a board supply-level switch directly connected to the damaged input pin. In addition, the IC was found to have manufacturing-caused bond wire defects. However, they are not believed to have contributed to the failure.

ANALYTICAL PROCEDURE AND RESULTS
Functional testing of the module indicated erratic operation at the B4 input. Subsequent x-ray inspection did not reveal any gross defects or damage. Only a gross inspection was possible by x-ray owing to the limited resolution imposed by the metallic housing. The case was opened and the defective component identified by electrically probing the internal circuitry. The component, a 74LS00 plastic DIP integrated circuit (IC), was removed from the module and subjected to a component-level failure analysis.

   X-ray inspection of the IC revealed deformed bond wires and two die-level bond anomalies (see micrograph 10.4). Electrical characterization of the IC indicated a parasitic 1 kΩ parallel resistance at the pin 5 input.

   Internal optical and SEM inspection following encapsulant removal revealed crushed bond wires and two (2) double bond sites (see micrographs 10.6 to 10.8). No defects or damage to the die circuitry were apparent at this stage in the analysis. After deprocessing (passivation and metallization removal), optical and SEM inspection revealed the area of electrical transient damage (see micrographs 10.10 to 10.12).

RECOMMENDATIONS
1. Verify the operation of the switching mechanism connected to the inputs to rule out the possibility of contact bounce or corrosion.
2. Use a protection network at the IC inputs to protect similar assemblies.

The enclosed data present the results of the analysis.
The enclosed data sheet further describes the SEM analytical technique.

Keith L. Hybrid
Senior Failure Analysis
Engineer

KLH:ks

Enclosures: Samples:
             Micrographs:
             Spectra:
             Tables:
             Data Sheets:

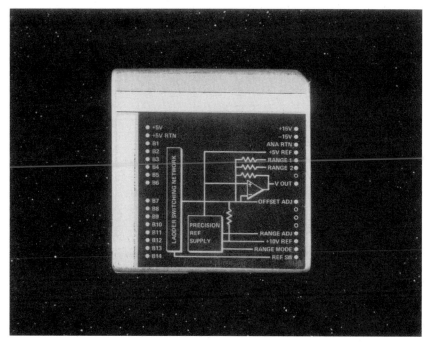

Micrograph 10.1   External optical view of the failed module "as received."

Micrograph 10.2   X-ray image of the failed module.

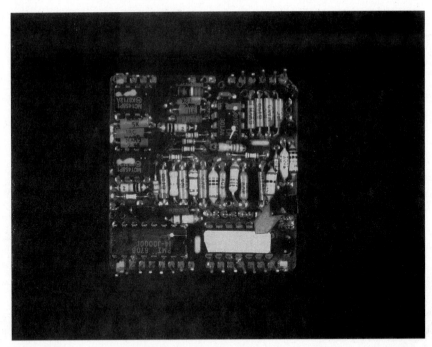

Micrograph 10.3 Internal optical view of the module assembly following package opening. The arrow indicates the defective integrated circuit.

Micrograph 10.4 X-ray image of the defective IC. Note that the bond wires are deformed (1) and there are two die-level bond anomalies (2).

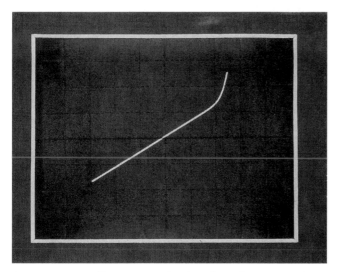

Micrograph 10.5 Curve trace showing the voltage vs. current characteristics of the IC from pin-5 to ground. Note that the junction has a parasitic, parallel 1 kΩ resistance, as indicated by the sloped trace. Scale: horizontal = 0.2 V/div, vertical = 0.2 mA/div.

Micrograph 10.6 Internal optical image of the die region of the IC following encapsulant removal. Note that the bond wires are crushed, confirming the findings in the x-ray.

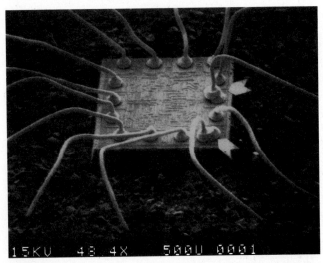

Micrograph 10.7 Scanning electron microscopic image of the die area. Note that in addition to the crushed wires, there are also two double bonds. All these defects were produced during the manufacture of the IC.

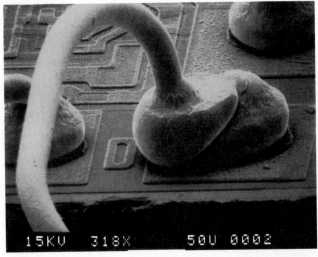

Micrograph 10.8 Higher magnification SEM image showing one of the double bond sites

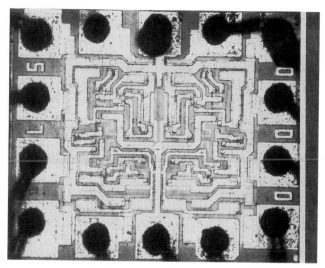

Micrograph 10.9   Optical microscopic image of the die surface before deprocessing. No defects or damage are apparent.

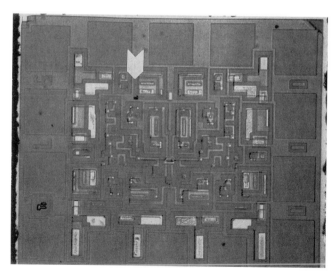

Micrograph 10.10   Optical microscopic image of the die surface following passivation and metallization removal. The arrow indicates an area of damage near pin 5.

Micrograph 10.11 Higher magnification optical microscopic image of the damaged area.

Micrograph 10.12 SEM image of the damaged area. Damage of this type is typically due to an electrical transient.

Appendix 10.2   Sources of Additional Reading on Failure Analysis

Thomas Lee (ed.), *Microelectronic Failure Analysis Desk Reference,* ASM International, Materials Park, Ohio 44073.

E. I. Cole, Jr., "Advanced Scanning Electron Microscopy Methods and Applications to Integrated Circuit Failure Analysis," *Scanning Microscopy,* vol. 2, no. 1, pp. 133–150, 1988.

Proceedings of the International Symposium for Testing and Failure Analysis, Annual Symposia, ASM International, Materials Park, Ohio 44073.

Proceedings of the International Society for Hybrid Microelectronics, Annual Symposia, International Society for Hybrid Microelectronics, Reston, Va. 22090.

International Journal of Microelectronics & Electronic Packaging, International Society for Hybrid Microelectronics, Reston, Va. 22090.

Semiconductor Reliability News, DM Data, Inc., Scottsdale, Ariz. 85260.

Physics of Semiconductor Failures, Dicken, Howard, DM Data, Inc., Scottsdale, Ariz. 85260.

Microelectronics Digest, Saratoga, Calif. 95070.

Devaney, Hill, and Seippel, *Failure Analysis Mechanisms, Techniques and Photoatlas,* Failure Recognition & Training Services, Inc., Tacoma, Wash., 1987.

Chandler, Denson, Rossi, and Wanner, *Failure Mode / Mechanism Distributions,* Reliability Analysis Center, Rome, N.Y. 13440, 1991.

Design Methods for Hybrid Circuits

Jerry E. Sergent

11.1 Introduction

The successful design of a hybrid microcircuit requires not only that the design meets the technical requirements but also that it meets the cost, quality, and schedule requirements. The designer must provide the customer (internal or external) a complete drawing package which documents the materials, components, assembly instructions, manufacturing processes, and test procedures required to make the finished product conform to the specifications and requirements. The design package includes as a minimum the items listed in Table 11.1.

The role of the hybrid design engineer is quite varied. The design process is a series of tradeoffs of performance vs. cost, and the designer must be capable

TABLE 11.1 The Design Package

Circuit schematic
Complete parts and materials lists
Complete process sequence
Process specifications
Layout with individual layer drawings
Assembly diagram and instructions
Electrical test procedure
Environmental test procedure
Troubleshooting procedure
Qualification procedure
Handling and packaging instructions
Special instructions

of making these decisions quickly and accurately. The hybrid engineer must therefore be part electrical engineer, part mechanical engineer, part materials engineer, part chemical engineer, part manufacturing engineer, and part cost analyst.

The hybrid design engineer may enter into the design process at any of the following stages:

1. *Black box design.* The hybrid designer is presented with a set of electrical, mechanical, and environmental specifications and must design both the circuit and the packaging scheme. This approach generally results in the optimum and most economical hybrid design, since the designer is able to consider both the advantages and disadvantages of the hybrid technology during the initial stages. The design of electronic circuits is beyond the scope of this book, and it will be assumed that a schematic exists. When an entire system is to be packaged, partitioning of the system into individual circuits will also be necessary. The hybrid designer should also play an active role in this procedure as well.

2. *Packaging design.* The hybrid designer is presented with a schematic, the mechanical requirements, and the environmental requirements, and must design a packaging scheme to meet the specifications. Although the hybrid technology permits a one-to-one conversion from the schematic to the finished circuit in most cases, there are usually some minor changes necessary to fully utilize the attributes of the hybrid technology.

3. *Producibility design.* The hybrid designer is presented with an existing hybrid design for the purpose of redesigning it to lower the cost. This effort may range all the way from a minor layout modification to a complete redesign of the circuit.

4. *Build-to-print.* This phase, in theory, requires no input from the hybrid designer, since a design package exists. However, even with a mature design, there are enough nuances between hybrid facilities to warrant a design review to effect any changes to accommodate the peculiarities of a particular facility.

5. *Production support.* After the design has been turned over to production, some producibility redesigns may be necessary. In addition, one or more of the discrete components, such as an integrated circuit chip, may become unavailable. In either instance, the hybrid design engineer must provide support.

11.2 The Design Process

The design process begins with the formation of a design team consisting, as a minimum, of representatives from production, design, and quality (referred to as the PDQ team[1]). At times, it may be appropriate to include representatives from finance as well. The purpose of the PDQ team is to assure that the technical, manufacturing, quality, and cost issues are considered at each decision point, and it is critical that the PDQ team be involved in the design process from the beginning. The cost and schedule impacts of a design change increase dramatically as milestones are passed, and early involvement of the PDQ team will minimize the consequences when a change is required.

Typically, the production function will include production engineering, production control, planning, cost control, and assembly. The role of the production engineer is to develop and monitor the manufacturing processes and to plan the manufacturing flow. The production control coordinator is responsible for interfacing between the various groups, and is responsible for scheduling and material control. The responsibility of the planning function is to develop the overall program schedule and to assure that all elements, such as capital equipment, are in place when needed. The cost control group is responsible for analyzing and reporting cost information. The assembly group is responsible for fabricating the product.

The design group, in addition to performing all the necessary analyses and designing the package, is responsible for component engineering and manufacturing support. The design group must also provide test and inspection criteria to the quality group.

The function of the quality group is to assure that the product meets the specification. To this end, quality monitors the product from the time of order to the time the finished product is shipped. Quality is responsible for assuring that the incoming parts and materials meet the requirements, for monitoring the manufacturing processes, and for performing the final acceptance tests. The quality group may also act as the customer advocate within the manufacturing facility.

A series of reviews at appropriate points in the design process helps to ensure that the design is proceeding satisfactorily. In these reviews, the members of the team review the design to date and identify potential problems as risks, which are weighted from *advisory* to *unacceptable,* depending on the degree of risk. The design team must address each of these risks to the satisfaction of the team before the design can proceed.

11.3   The Design Sequence

The design sequence, in most instances, depends on the nature of the product and the point at which the design engineer enters the design process, but generally follows a logical pattern as depicted in Table 11.2, assuming that the basic electronic circuit design has been completed.

11.3.1   Partitioning

The first step in the hybrid design process is to partition the system into individual circuits. This is a critical step in achieving a good hybrid design and considerable effort must go into this phase. The following items must be considered to obtain a good design, but individual circumstances will dictate which items must be prioritized and where compromises must be made. Some of these items relate to the technical performance of the circuits, some to cost, and some to both.

TABLE 11.2   Hybrid Circuit Design Sequence

Operation	Output
Partitioning	Division of system into individual circuits
Initial concept review	Preliminary packaging concepts 　Circuit schematics 　Risk analysis
Circuit analysis	Verification of electrical design 　Design parameters centered 　Sensitivity analysis 　Voltage, current, and power levels in each component
Breadboard tests	Verification of circuit analysis
Component selection	Preliminary parts list
Preliminary thermal analysis	Indication of potential thermal problems
Sizing analysis	Approximate size of circuit
Technology selection	Determination of substrate technology
Process sequence	Selection of manufacturing process
Material selection	Selection of attachment materials
Circuit layout	Location of components and interconnection traces 　Layer drawings 　Assembly drawings
Detailed thermal analysis	Temperature profile of circuit
Preliminary design review	Review prior to prototype build 　Exceptions to design guidelines 　Risk analysis
Prototype build	Verification of performance 　Electrical performance 　Conformance to quality standards
Documentation release	Assembly drawings Process instructions Travelers
Detailed design review	Review of prototype build and documentation
Preproduction build	Verification of design and documentation
Production release review	Review of preproduction build prior to production
Production	Manufacturing of hybrid circuit

1. *Group like technologies together.*   This applies when more than one substrate metallization technology is to be used in the system. While multiple technologies can be used on the same substrate when required, it is more cost effective to use them separately.

2. *Manage tolerance distribution.*   To achieve high yields on all hybrid circuits in a system, it is essential that electrical, mechanical, and environmental tolerances not be distributed such that one circuit has impossibly tight tolerances. Better overall system yields are attained when tolerances are apportioned appropriately throughout.

3. *Minimize system input/output interfaces.* Minimizing the lead count simplifies the interconnection scheme between the hybrid circuits, decreases the cost, and increases the reliability. In addition, the cost of the individual packages decreases with lead count.

4. *Manage power distribution.* Spreading the power dissipation as evenly as possible among the individual hybrid circuits adds considerably to the reliability of the system, since the rate of growth of many failure mechanisms increases exponentially with temperature. The degree of thermal management is also minimized with a consequent reduction of cost and space requirements. This step is particularly important when power hybrid circuits are involved.

5. *Partition functional blocks.* Partitioning the system into functional blocks greatly simplifies the testing process and minimizes the peripheral test equipment required. Frequently, intelligent partitioning can make the difference between testing automatically and testing manually, with tremendous cost savings.

6. *Consider size distribution.* Every effort should be made to use standard packages for as many circuits as possible. This will minimize tooling costs, lower the cost of the package, and minimize changeover costs in production. Obviously, the available area and form factor will influence the choice of package sizes and shapes.

7. *Analyze electrical considerations.* Circuits in which cross-coupling of signals may be detrimental to performance should be isolated from each other. Analogously, circuits in which lead length may be a critical factor should be located in close proximity.

8. *Ensure component compatibility.* The system must be examined for components that are not compatible with the hybrid technology or that would create extreme difficulties in packaging due to size or other constraints. Examples are large value capacitors or components that require extensive heat sinking. Consideration must be given to including these components in the next highest level of packaging to minimize cost.

11.3.2 Initial concept review

The purpose of the initial concept review is to review the partitioning process and to identify any potential risks that may influence the design. The outcome of this review should be acceptance of the schematics of individual circuits and the preliminary packaging concepts, and addressing the risk analyses to the satisfaction of the PDQ team.

11.3.3 Circuit analysis

The hybrid designer will typically not be involved in the initial circuit design, but must be skilled in both circuit design and analysis. A complete circuit analysis is the first critical step in the hybrid design process, and provides the designer with the information required to select the components, materials, and assembly processes required to manufacture the circuit. Specifically, the objectives of the circuit analysis are:

1. To establish whether the circuit as designed will meet the performance specifications with nominal parameters for the components. A Monte Carlo type analysis reveals whether the circuit parameters are centered within the allowable range as determined by the circuit specifications.

2. To establish the specifications for the active and passive components. It is not sufficient to simply establish the inductance required of an inductor, for example. Such parameters as the core material, the quality factor Q, and the self-resonant frequency f_0, may also be necessary to fully characterize the component.

3. To establish the sensitivity of the circuit parameters to changes in the values of circuit elements and, at the same time, to establish the tolerances of the circuit components. Monolithic integrated circuits in the chip form are difficult to test thoroughly and, consequently, the variations in parameters are greater than for packaged devices. The circuit analysis must determine whether the circuit can tolerate the variance. The sensitivity analysis is especially crucial when the circuit is to be actively trimmed or when it is to be exposed to temperature extremes. Given the ease with which high precision resistors can be fabricated, there is a temptation to specify resistor tolerances very tightly, even when not required for circuit performance. Overtolerancing of film resistors can cause a considerable yield loss at both the substrate level and the laser trim level and must be avoided. Tolerances must be as loose as possible for all components, both active and passive, consistent with circuit performance.

4. To establish the applied voltage, current, and power dissipation in each component. This information determines whether the component is overstressed and provides data for the thermal analysis.

11.3.4   Breadboard tests

A discrete breadboard, while not exactly matching the hybrid version due to such factors as lead lengths and temperature characteristics, can be used to verify the circuit analysis. It may be advisable to install variable components at critical locations identified by the circuit analysis to confirm the sensitivity and tolerance analyses. If inductors are required, the breadboard should use the same version as will be used in the hybrid. The parameters of inductors, such as the distributed capacitance or the quality factor may vary widely among inductors of the same inductance, and the circuit performance may yield misleading results.

11.3.5   Component selection

Unlike integrated circuits, most hybrid circuits can be translated from the schematic to the finished product on a 1:1 basis in that each circuit element will correspond to a discrete or printed component. The design engineer must select the discrete components on the basis of the functional requirements, cost, and availability. It is important not to *over*specify the components, either in terms of functionality or in terms of the quality requirements. This will only add to the cost without improving the performance.

11.3.6 Preliminary thermal analysis

The preliminary thermal analysis is performed using approximate methods to determine whether a potential problem exists, and considers only one device at a time. The effect of adjacent heat-generating sources is not considered until the detailed thermal analysis is performed. While not precise, approximate methods can usually indicate whether the junction temperature will be excessive, and may determine whether a detailed thermal analysis is necessary. The methods of device attachment, substrate metallization, and substrate attachment are largely dictated by the results of this analysis.

11.3.7 Sizing

There are a number of empirical methods used to approximate the size of the finished circuit, all of which are based on assigning a weighting figure to each component to allow for the interconnection pattern. The weighting factors are, of course, different for thin film and thick film and are based on the standard line widths of 0.002 in for thin film and 0.010 in for thick film. For wider lines, the weighting factors must be adjusted proportionately. The weighting factors for different components are shown in Table 11.3.

It should be noted that the size is often limited by the pin count requirements. The maximum pin count is limited by the length around the periphery of the package and, if the size as determined by the empirical formula is smaller than that required to accommodate the required pin count, a larger size must be used. For example, a package which is 1.0×1.0 in with 0.1-in pin spacing has 36 leads.

TABLE 11.3 Weighting Factors for Components for Substrate Size

Component	Weighting factor	
	Thick film	Thin film
IC (chip-and-wire)	5	4
Transistor (small signal)	4	3
Diode (small signal)	4	3
Transistor (power)	5	5
Diode (power)	4	4
Resistor	3	3
Capacitor	2	2
Chip resistor	3	2
Inductor	3	2

Note: Do not load substrate more than 80% of available area.

11.3.8 Selection of substrate technology

It is frequently possible for more than one substrate metallization technology to meet the technical requirements of the circuit. Once these are satisfied, cost becomes the prime consideration. The technology selection process always begins by assuming that the circuit will be fabricated with the least expensive process, or with the standard processes that exist within a given company. This assumption remains in effect until it is proven that the standard processes cannot meet the technical requirements. In this way, the most economical design will result.

11.3.9 Selection of materials

Once the substrate technology has been selected, the materials used to metallize the substrate and to assemble the circuit must be selected. Many metals, when metallurgically joined, form intermetallic compounds which may have electrical and mechanical characteristics radically different from the pure materials. The formation of most of these compounds is accelerated by exposure to elevated temperatures, which can lead to a severe reliability risk under conditions of high temperature storage. Further, certain materials in simple contact with each other can create a battery effect due to a difference in Fermi potential, which can initiate corrosion. Where a device generates heat during operation, the thermal conductivity of the attachment materials is a critical consideration. The materials selection process must therefore take into account the characteristics of the components to be attached, the expected environmental conditions, the thermal management requirements, and the temperature at which each process is to be performed.

11.3.10 Layout

Most of the information required to create the layout is contained in the design guidelines or in the software of the computer-aided design (CAD) system. The layout is a graphical representation of each of the individual layers on the substrate, the location of the individual components on the substrate, and the location of the substrate with respect to the input/output pins. An important part of the layout process is to perform a worst-case analysis on the dimensional tolerances to determine whether a potential interference fit exists.

No CAD system to date can perform the complete layout of a chip and wire hybrid circuit. Although many of them come close, none can handle such unique aspects of hybrid layouts such as using wirebonds from the silicon device to the substrate metallization to jump over a conductor line. Most CAD systems want to use a separate metallization layer to handle this problem, which adds to the overall complexity and cost.

In addition to performing the drafting function, some CAD systems have the following features:

- *Automatic checking.* With the schematic entered into the system, the CAD system can perform a point-to-point continuity check of the layout.

- *Component boundaries.* The computer prevents the designer from placing a component within a predetermined distance from another component to prevent interference.

- *Parts list.* A complete parts list, or bill of materials, can be created. When thick film circuits are being laid out, the area of each print can be calculated and the amount of paste required can be determined from an algorithm.

- *Visual aids.* The visual aids needed for fabricating the circuit can be created by the CAD system directly from the layout.

- *Resistor design.* The data from the resistor characterizations can be fed into the computer, which will then calculate empirical design equations allowing for power dissipation, tolerance, and termination effects.

11.3.11 Preliminary design review

The preliminary design review (PDR) should be performed by representatives from design, production, and quality after the completion of the layout phase. The design, including the electrical design, the thermal analysis, the mechanical design, and the layout should be checked for any errors. Any deviations from the design guidelines should have a sound technical basis and should be carefully documented in terms of potential technical risk and cost. Any new processes should be identified along with the skill levels required.

11.3.12 Prototype build

The purpose of the prototype build is to verify the design and to prepare the process documentation needed to transfer the design to the production phase. The quantity of prototypes to be fabricated varies with the particular situation, but should be sufficient to obtain a reasonable statistical distribution of parameters. Among the specific items that should be addressed in this phase are:

- *Distribution of electrical parameters.* The output variables should be in the center of the spread of allowed values for maximum producibility. If one or more of the parameters are skewed toward one extreme, the normal variation of component tolerances will result in a reduced yield in the production phase.

- *Optimization of component tolerances.* It is expensive both to overtolerance and to undertolerance resistors and other components. Overtolerancing results in unnecessarily low yields at laser trim and in paying premium costs for other components, while undertolerancing results in low yields due to the failure of the circuit to meet the electrical specifications.

- *Operation at temperature extremes.* Many circuits are required to operate at temperature extremes, particularly in military and automotive applications. While the individual components may work well, cumulative effects may cause circuit performance to deteriorate to the point where it no longer meets specifications.

- *Proximity effects.* While the short leads associated with the hybrid technology tend to minimize the effects of parasitics, the compact nature enhances the probability that cross-coupling will occur to a greater degree than on the breadboard.

- *Thermal effects.* The thermal design can be verified by measuring the junction temperatures with an infrared (ir) microscope, which detects hot spots on the surface of the chip, substrate, and surrounding areas to determine the degree of spreading.

- *Process sequence compatibility.* Any process that disturbs a previous process can be detected during the prototype build, in particular a process that is at a lower temperature than a subsequent process.

- *Verification of resistor designs.* The pretrim values of the resistors should be measured and the percentage below the nominal calculated. The percentage should be nearly the same for each resistor or problems with resistor yield will occur during production due to the narrow process window.

- *Reliability.* Enough prototypes should be fabricated to enable a sample to be subjected to the qualification tests required in the specification, especially accelerated life tests. Potential reliability problems detected at this point can save a substantial amount of money and time.

- *Sensitivity of active trim parameters.* If applicable, the sensitivity of the output variables to the laser trim rate and/or the bite size should be determined in the prototype phase. In addition, the sensitivity of any components to the laser light should be ascertained.

- *Documentation.* During the prototype phase, the complete documentation package needed to fabricate the circuit in production, including process specifications, is prepared. Photographs of the prototype units may be used for assembly drawings, if desired.

11.3.13 Detailed design review

The detailed design review is scheduled immediately after the results of the prototype phase are available. The results of the electrical, thermal, and reliability tests are compared with those predicted from earlier analyses. Any deviations from the predictions should be thoroughly analyzed and any problems should be corrected before proceeding. The documentation is examined for obvious errors and clarity.

11.3.14 The preproduction phase

During the preproduction phase, a limited quantity of circuits are fabricated on the production line using the documentation prepared during the prototype phase and using parts and materials ordered according to the specifications. The preproduction phase is utilized to check the following items:

- *Producibility.* The design, materials, and processes are examined for any problems with fabrication and for process compatibility.

- *Documentation.* The documentation is checked for clarity, accuracy, and completeness.

- *Time standards.* The time standards for each operation are measured and compared with those used for quoting. At the same time, the efficiency factor for each operation is determined. If any significant differences exist, the reasons must be established before full production can begin.

- *Test results.* The results of the electrical test are examined to determine whether the parameters are centered or skewed toward one side of the allowable spread of values.

11.3.15 Production release review

The production release review is the most stringent and detailed of all. Every aspect of the design, documentation, and results of the preproduction build are examined for accuracy, conformance to specifications, and cost.

11.3.16 Summary

The design cycle and cost obviously depend on the complexity of the circuit and the extent to which the state of the art is pushed. To be economically successful, the hybrid facility must continually compromise between using established processes and components that might not be optimum for a given product and continually changing the design and documentation to implement changes. The decision is rarely clear-cut, and many times must be made on incomplete data. The use of an established design procedure with design reviews at appropriate intervals and the establishment of accurate time standards, realistic efficiency factors, and real-time cost analysis can greatly enhance the chances for success.

11.4 Design of Thick Film Resistors

The spread in a group of resistor values after printing and firing may vary by as much as $\pm 20\%$. While laser trimming can adjust the value of resistors upward, those resistors which are greater than the maximum acceptable value (design value + tolerance) may not be used and may result in the entire substrate being lost.

The resistance for a rectangular solid of resistive materials is given by the equation:

$$R = \frac{\rho_B L}{WT} \tag{11.1}$$

where ρ_b = bulk resistivity, $\Omega \cdot$cm
L = length of resistor, cm
W = width of resistor, cm
T = thickness of resistor, cm

Bulk parameters are characteristic of a particular homogeneous material and are not a function of the dimensions of the sample.

For materials which are utilized in the form of a film $(T \ll W)$ it is useful to define a parameter called the sheet resistivity ρ_S where

$$\rho_S = \frac{\rho_B}{T} \qquad (11.2)$$

The quantity ρ_S is considered to be a *sample* parameter, in which the properties are a function of the dimensions of the sample. A film made from the same material with twice the thickness will have the same bulk resistivity, but the sheet resistivity will be halved.

With this definition,

$$R = \rho_S \frac{L}{W} \qquad (11.3)$$

For the case where $L = W$ (i.e., the sample has the shape of a square), the value of the resistor is equal to the sheet resistivity. The units of sheet resistivity are thus expressed as ohms per square (Ω/sq). In common use, the standard unit of thickness used for this measurement is 25 μm in the dried state. The dried state has been adopted as the standard for measuring thickness as opposed to the fired state, since it is more readily measured during processing. To measure a resistor in the fired state, it is necessary to wait for the resistor to be fired, which may take more than an hour. The precise set of units for the sheet resistivity of thick film resistors is, therefore, ohms/square per 25 μm of dried thickness, which is usually expressed as simply Ω/sq. Resistor paste, as furnished by a manufacturer, is specified by its sheet resistivity in units of Ω/sq of Ω/sq/mil. Typically, resistor pastes are furnished in decade values with sheet resistivities ranging from 1 Ω/sq to as high as 100 MΩ/sq in decade steps. Resistor pastes of the same chemistry which are blendable are referred to as a *family*. Normally, a single family will not encompass the entire range of values as described.

For a film of uniform thickness with no termination effects, the resistance is simply equal to the sheet resistivity times the ratio of the length to the width (the so-called aspect ratio)

$$R = \rho_s A_R \qquad (11.4)$$

where A_R = aspect ratio = L/W.

According to Eq. (11.4), a resistor with dimensions of 1 mil or 1 mile would have the same ohmic value if the thickness is uniform. While there are physical limitations that tend to make larger square resistors higher in value than smaller ones, the basic formula is valid, and resistors may be designed by manipulating squares and fractions of squares in series and parallel.

11.4.1 The effect of thickness

According to Eq. (11.1), the thickness of a resistor has a direct effect on value. If the thickness is doubled, the value is halved. If the thickness is halved, the value doubles. Experimentally, it has been found that very thin resistors have higher resistance than a simple application of the formula would indicate. This is due in part to the porosity that is introduced into very thin resistors during the firing process. The relationship is generally true for fired resistor films between 0.5 and 1.5 mils, as shown in Fig. 11.1.

11.4.2 Aspect ratio

From Eq. (11.4), for a given material of a given sheet resistivity, the resistance can be modified by changes in the aspect ratio. Thus, if the resistor is made twice as long as it is wide, so that $L = 2W$, the resistance of the resistor will be twice as great; if the resistor is made four times longer than it is wide, the resistance is four times as great. This is shown in Fig. 11.2. As can also be seen from Fig. 11.2, if the length is held constant and the width is reduced by half, four squares also occur. The number of squares is independent of the absolute dimension, but only in the ratio of length to width.

11.4.3 Effect of dimensions

Previously, it was illustrated that a large and a small square resistor should theoretically have the same resistance value. However, there are practical limitations that make small square resistors lower in value than large resistors. One limitation is the diffusion of conductor metal into the resistor that occurs at the firing temperature. Both silver and gold are highly mobile species that migrate quickly into the resistor body. A large resistor, with greater distance between conductor pads, will have less volumetric metal than a small resistor, as shown in Fig. 11.3. Note that, although the length of diffusion is the same, the volumetric fraction is greater for the small resistor.

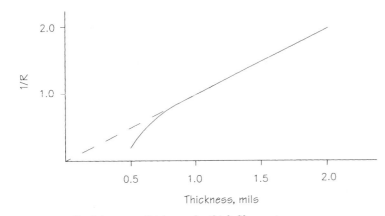

Figure 11.1 Resistance vs. thickness for thick film paste.

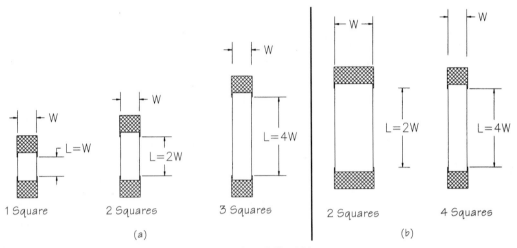

Figure 11.2   Variation of aspect ratio with (*a*) length and (*b*) width.

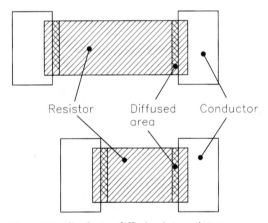

Figure 11.3   Conductor diffusion into resistor.

Since both silver and gold are more conductive than the ruthenium dioxide (RuO_2) resistor material, the effect is to make small square resistors lower in value than large square resistors. This effect also tends to make smaller resistors have a more positive TCR than larger resistors.

Another limitation results from the nature of the screen printing process. The printer squeegee, when moving over closely spaced conductors found in smaller resistors, tends to deposit a thicker film than when moving over a flat surface. This effect, somewhat exaggerated, is illustrated in Fig. 11.4.

11.4.4   Resistor design curves

There are two basic constraints on the thick film resistor design process:

1. The values of the resistors can only be increased.

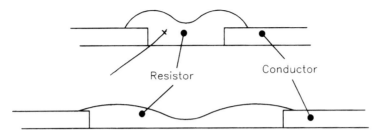

Figure 11.4 Thickness difference with size.

2. The values of rectangular resistors cannot be increased by more than a factor of 2. If a range of trim greater than 2 is required, a nonrectangular design is necessary.

The resistor design process, therefore, must produce a distribution of resistors, all of which are at or below the nominal values of the resistor plus the tolerance, and none of which are more than 50% below the desired value minus the tolerance. Obviously, good process controls are an integral part of obtaining high yields. When a sample of resistors is fired, the distribution of values is typically as shown in Fig. 11.5. It can be shown that process control must be sufficient to control the standard deviation of a group of resistors within 10% of the mean if all of the resistors in the group are to be trimmable.

Because of the variation in values caused by geometric effects, thick film resistor design is much more readily accomplished by creating a set of design curves based on the established processes to be used in manufacturing for a given paste. This is accomplished by the following procedure[5]:

1. For each sheet resistance, print and fire a number of samples (at least 100) of varying lengths and widths to ensure a valid statistical sample. This will usually be done on a test pattern, which has a standard width and a number of lengths. Several test patterns with different widths must be used. To be comprehensive, this step should include not only the decade

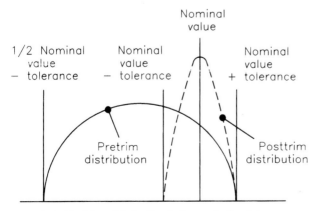

Figure 11.5 Resistor distribution before and after trim.

values but also blends which are to be used. If more than one conductor material or brand of substrate is to be used, this step must be duplicated.

2. For each size, calculate the mean value and standard deviation.

3. Calculate the design value as follows:

$$R_D = R_M + 3\sigma \tag{11.5}$$

where R_D = design value of resistor
R_M = mean value as calculated in step 2
σ = standard deviation

4. For each sheet resistance, plot R_D vs. length with width as a parameter.

An example of a set of resistor design curves is shown in Fig. 11.6.

11.4.5   Power considerations

Thick film resistors are limited in the amount of power they can dissipate. If the power is excessive, the temperature of the resistor may increase to the point where the value changes. Resistors must be designed such that the power density in watts per square inch is below the level which will create resistor instability. Resistors may be increased in physical size without changing the value if the aspect ratio is maintained, allowing more power to be dissipated. For example, a 40- × 40-mil resistor can have the same resistance as a 100- × 100-mil resistor. However, the latter will dissipate almost 6 times the power. A rule of thumb for modern thick film resistors is that they can handle about 100 W/in² of active resistor area. This approximation becomes less valid

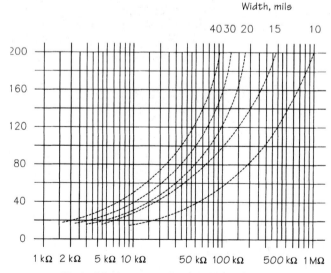

Figure 11.6 Typical design curves for 10 kΩ resistor paste.

as the size of the resistor is increased. The actual power handling capability depends on many other factors such as substrate heat spreading capability, other hot devices in the package, operating temperature of the package, and allowable resistor drift at end of life. While using power density may not be completely accurate to describe the power handling capability of thick film circuits, a figure of 50 W/in^2 is adequate in the design stage.

There is considerable evidence that high power stress causes drift by heating of the resistive element. Unlike thermal aging, power aging heats the point-to-point metal contacts and stresses the glass–metal interfaces. If resistors are to be used at power densities greater than 100 W/in^2, a more detailed thermal analysis of the complete package should be performed.

The power requirement of the resistor of determines the minimum resistor dimension. For the normal case, where $L>W$, the minimum width in terms of the power requirements is given by

$$W_{MIN} = \sqrt{\frac{2\rho_S P_D}{P_R R}} \tag{11.6}$$

where P_D = power dissipated by the resistor
ρ_S = sheet resistivity
P_R = rated power of the resistor (50 W/in^2 for alumina at 25°C; 100 W/in^2 for beryllia at 25°C)
R = nominal value of the resistor

The factor 2 is introduced to allow for trimming, since as much as half the material may be effectively removed by the trimming process. The minimum width for a given tolerance is given in Table 11.4.

To determine the resistor dimensions, select the paste to be used, select the minimum width based on the most restrictive of the power and tolerance requirements, and read the length from the resistor design curves.

The selection of the paste requires some elaboration. It is critical for yield purposes that the number of resistor screenings be held to three or less when possible. A secondary consideration, but still important, is to hold the aspect ratio within the limits of $0.2 < A < 5$. This process is made much easier by blending high and low values of thick film paste to produce an intermediate value.

TABLE 11.4 Minimum Width vs. Tolerance for Thick Film Resistors

Tolerance, %	Minimum width	
	Preferred, %	Absolute, %
20	20	20
10	25	20
5	30	25
2	35	30
1	40	35

11.4.6 Blending of resistor pastes

A group of thick film materials with identical chemistries that are blendable—a family—will generally have a range of values from 10 Ω/sq to 1 MΩ/sq in decade values, although intermediate values are available as well. There are both high and low limits to the amount of material that may be added. As more and more material is added, a point is reached where there is not enough glass to maintain the structural integrity of the film. A practical lower limit of sheet resistivity of resistors formed in this manner is about 10 Ω/sq. Resistors with a sheet resistivity below this value must have a different chemistry and often are not blendable with the regular family of materials. Sometimes at 1 to 10 kΩ/sq, there is a change in chemistry that necessitates a *transition* paste for some resistor systems. At the other extreme, as less and less material is added, a point is reached where there are not enough particles to form continuous chains, and the sheet resistance rises very abruptly. Within most resistor families, the practical upper limit is about 2 MΩ/sq. Resistor materials are available to about 20 MΩ/sq, but the chemical composition of these materials is not amenable to blending with lower value resistors.

While many circuits can be fabricated with the standard decade values, the use of intermediate values can be utilized to help meet these criteria. Intermediate values can be obtained directly from the paste manufacturer, usually at no extra cost, or by blending. The use of intermediate values can also aid in minimizing the area required for resistors. To fabricate a 5-kW resistor with a 5-kΩ/sq paste requires only one square; to fabricate it with 10-kΩ/sq paste requires two square, and with 1-kΩ/sq paste, five squares. A blending curve may be developed by the following procedure:

1. Print, dry, and fire a sample of the lower value paste (low end member), calculate the mean value, and plot it on a sheet of semilog graph paper as indicated by the point R_L in Fig. 11.7.

2. Print, dry, and fire a sample of the higher value paste (high end member), calculate the mean value, and plot it on the same sheet of semilog graph paper as indicated by the point R_H in Fig. 11.7.

3. Thoroughly mix the high and low value pastes in equal amounts in a clean jar. Print, dry, and fire a sample of the blend, calculate the mean value, and plot it on the same sheet of semilog graph paper as indicated by the point R_T in Fig. 11.7.

4. Connect point R_L to point R_T and then on to point R_H.

Intermediate values can be obtained by matching up the desired value on the y axis with the ratio on the x axis. Although any resistor paste may be blended with any other paste in the same family, it is preferable to blend adjacent decades for additional control. For two or more decades of separation, the slope of the curve is very steep, and the blend may change dramatically with a relatively small change in one of the end members.

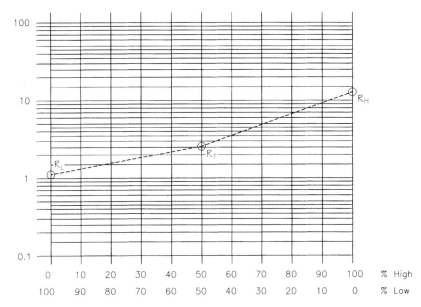

Figure 11.7 Blending curve for thick film pastes.

11.4.7 Special resistor configurations

When the aspect ratio exceeds 5, or the trim range requirement exceeds 2 times, two special resistor designs are incorporated: the serpentine and top hat, shown in Fig. 11.8. The serpentine pattern allows for a large number of squares to be executed without consuming an inordinate length of substrate. The number of squares is calculated as in a standard linear resistor except that the corners are counted as only half a square.

The top hat resistor allows for a far greater range in resistor trimming. Trimming allows the selective removal of material to reduce the resistor width and change the effective number of squares. For a standard resistor, a safe design margin is a 50% reduction, allowing an increase of 2 times the resistor value. The top hat resistor is somewhat wider than the basic resistor, but the extra width contributes little to determining the effective contribution to the resistance. The number of squares is approximately the length of the top hat divided by the width of the non–top hat section. However, on trim-

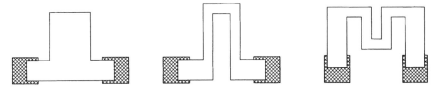

Figure 11.8 Serpentine and top hat resistors.

ming, the top hat approaches a serpentine pattern with an appropriate increase in the number of squares, without compromising the minimum resistor width or trimming.

11.5   Design of Thin Film Resistors

The design methods for thin film resistors are substantially different from those for thick film resistors for a variety of reasons:

- The resistivity of thin films can be held to a close tolerance by controlling the deposition parameters, by the precision of the etching process, and by controlling the stabilization time.
- The designer is not limited to simple straight-line patterns. Right angle and meandering patterns are also available.
- Linewidths down to 0.001 in are possible, although 0.002 in is more common. This requires special design techniques if laser trimming is required.
- Only a single sheet resistivity is available unless special deposition techniques are used. This necessitates resistors with extremely high or low aspect ratios if high or low values are required.
- No termination effects are observed, and the sheet resistivity is not a function of the resistor size.

To design a thin film resistor with a given sheet resistivity, three pieces of information are required: the nominal value of the resistor, the tolerance, and the power dissipation. The design is accomplished by the sequence of steps outlined below[6]:

1. Determine the minimum width from

$$W_{MIN} = \sqrt{\frac{\rho_S P_D}{P_D R}}$$

$$(11.7)$$

where P_D = power dissipated by the resistor
ρ_S = sheet resistivity
P_R = rated power of the resistor (20 W/in² for alumina at 25°C; 40 W/in² for beryllia at 25°C)
R = nominal value of the resistor

The value obtained from Eq. (11.7) should be rounded up to the nearest standard width.

2. Calculate the number of squares from

$$N = \frac{R}{\rho_S}$$

$$(11.8)$$

The total effective length of the resistor is

$$L_{EFF} = N \times W \qquad (11.9)$$

Resistors deviating from a straight line path must have allowances made for the corners since the current density tends to increase in the corners. This effect is shown in Fig. 11.9 for $L_1 > 4W$.

3. Make allowances for tighter tolerances by adding trim pads. The resistance increase capacity necessary for precision trimming is shown in Table 11.5. Note that Table 11.5 takes the tolerance of the sheet resistivity into account.

The geometry of the trim pad depends on the form of the resistor:

- For the meander resistor, the trim pad is fabricated as shown in Fig. 11.10 by filling in a portion of the bend. The effective number of squares and the amount of increase which can be expected is shown in Fig. 11.11.

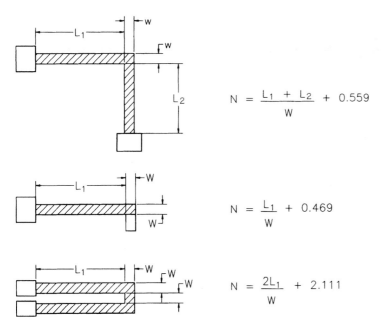

$$N = \frac{L_1 + L_2}{W} + 0.559$$

$$N = \frac{L_1}{W} + 0.469$$

$$N = \frac{2L_1}{W} + 2.111$$

Figure 11.9   Effective number of squares for turns.

TABLE 11.5   Tolerance of Thin Film Resistors vs. Increase Capacity

Tolerance, %	Increase capacity, %
1	25
5	20
10	15
15	8
20	0

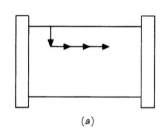

(a)

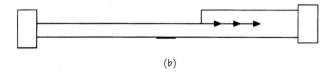

(b)

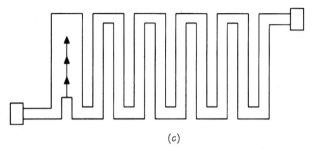

(c)

Figure 11.10 Trim pads for thin film resistors. (a) Straight line resistor ($L/W < 5$); (b) straight line resistor ($L/W > 5$); (c) meander resistor.

- For the straight line resistors with $L/W > 5$ and right angle resistors, the trim pad is made by extending the width of a portion of the resistor with the overall width of this portion equal to 2 × the width of the resistor. Half a square is added to the trim pad to compensate for current crowding at the pad. This is illustrated in Fig. 11.10.

- For straight line resistors with $L/W < 5$, the width is increased and the resistor is trimmed by conventional trim techniques as shown in Fig. 11.10.

These principles are illustrated in the following examples.

Example 11.1   10-kΩ, 10% tolerance, 50-mW dissipation with 100-Ω/sq material.

1. $$W_{MIN} = \sqrt{\frac{100 \times 0.05}{20 \times 1000}} = 0.016 \text{ in}$$

2. $N = 10,000/100 = 100$ squares

3. $L_{EFF} = 100 \times 0.005 \text{ in} = 0.5 \text{ in}$

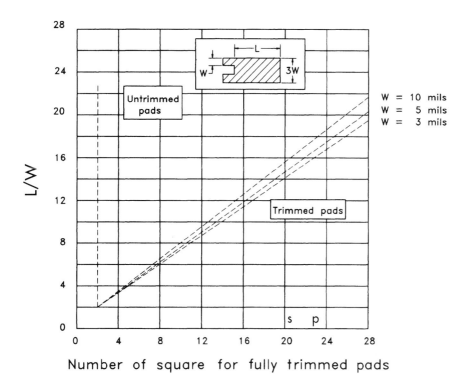

Figure 11.11   Trim pad dimensions for thin film resistors.

4. From Table 11.5, for a 10% resistor, a pad must be designed which will allow an increase of 15% in the value of the resistor:

$$0.15 \times 100 = 100 \text{ squares}$$

5. From Fig. 11.10, an untrimmed pad for $L/W > 2$ adds 1.8 squares. Therefore, the trim pad after trimming should have $15 + 1.8 = 16.8$ squares. To find the L/W ratio from Fig. 11.10, find 16.8 squares on the abscissa and use the $W = 0.005$ in curve. This corresponds to $L/W = 13$ when rounded up to the next highest integer. Since $W = 0.005$ in, the trim pad length would be 0.065 in. Also, since the untrimmed pad contributes 1.8 squares, an additional 98.2 squares are needed to obtain the 100 squares total. The completed resistor is shown in Fig. 11.12.

Example 11.2   1-kΩ, 5% tolerance, 50-mW dissipation with 100-Ω/sq material.

1. $W_{\text{MIN}} = 100 \times 0.05/20 \times 1000 = 0.016$ in

   This equals 0.01 in when rounded up to the next highest integer.

2. $N = 1000/100 = 10$ squares

3. $L_{\text{EFF}} = 10 \times 0.016$ in $= 0.160$ in. With this relatively small length, it is possible to use a straight line resistor.

4. From Table 11.5, for a 5% resistor, a pad allowing 20% increase is required. This corresponds to $0.20 \times 10 = 2$ squares.

5. The two-square trim capability is added by increasing a portion of the width as shown in Fig. 11.11. The trim pad is twice as wide as the resistor, or 0.032 in. The 2-square

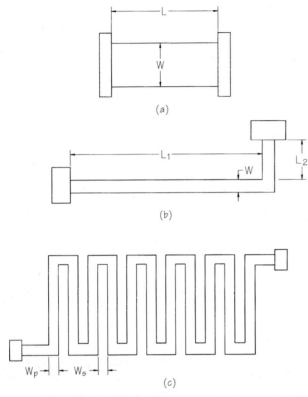

Figure 11.12 Thin film layout patterns. (*a*) Straight line resistor; (*b*) right angle resistor; (*c*) meander resistor.

trim pad is then 0.064 in long. An additional 8 squares are required at the calculated width. To compensate for the nonlinear current flow at the width transition, the pad is extended by half the calculated width, or 0.008 in, without adjusting the total resistor length.

Example 11.3   200-Ω, 10% tolerance, 100-mW dissipation with 50-Ω/sq material.

1. $W_{MIN} = \sqrt{\dfrac{100 \times 0.05}{20 \times 200}} = 0.036$ in

This equals 0.036 in when rounded up to the next highest integer.

2. $N = 200/50 = 4$ squares

3. $L_{EFF} = 4 \times 0.036$ in $= 0.144$ in. With this relatively small length, it is possible to use a straight line resistor.

4. From Table 11.5, for a 10% resistor, a pad allowing 15% increase is required. This corresponds to

$$0.015 \times 0.036 \text{ in} = 0.0054 \text{ in}$$

This equals 0.006 in when rounded up to the next highest integer.

5. The total resistor width is rounded up to

$$W = 0.036 \text{ in} + 0.006 \text{ in} = 0.042 \text{ in}$$

6. The total resistor length is

$$L = 0.042 \text{ in} \times 4 \text{ squares} = 0.168 \text{ in}$$

The finished resistor is shown in Fig. 11.12.

11.6 Electrical Considerations in Hybrid Design

Hybrid microcircuit technology provides the packaging engineer with many opportunities to improve the performance of an electronic circuit. The ability to miniaturize electronic circuits minimizes stray capacitance and inductance, and the high performance characteristics of thick and thin film resistors offer potential enhancements not available with other packaging technologies. On the other hand, the close proximity of signal paths increases the probability of cross-talk and other detrimental effects associated with adjacent circuitry. This section describes examples of how the properties of film resistors can be used to the circuit designer's advantage, and how stray capacitance, inductance, and mutual inductance can be modeled.

11.6.1 Temperature compensation of thick film hybrid circuits

Thick film resistors with a temperature coefficient of resistance (TCR) of ± 50 ppm are readily available in the range of a few hundred to several hundred thousand ohms per square and with a TCR of ± 100 ppm in ranges beyond that. High ohmic value resistors tend to have a more negative TCR than those of low ohmic value, although the addition of TCR modifiers during the paste fabrication process may alter this somewhat by producing resistors of the same ohmic value but with TCRs of the opposite sign. This characteristic may be used to provide resistor combinations with a substantially lower TCR than obtainable with a single paste. By combining resistors in series or parallel with opposite TCRs, a cancellation effect is observed. While the TCR of thick film resistors varies with temperature, the relative shape of the resistance vs. temperature curve will usually be similar from one resistor value to another, and the cancellation effect remains viable across the temperature range.

Consider two resistors in series.[2] The formula for the total resistance is given by

$$R_T = R_1 + R_2 \tag{11.10}$$

When the temperature changes, the resistors also change by an amount

$$R_T + \Delta R_T = R_1 + \Delta R_1 + R_2 + \Delta R_2 \tag{11.11}$$

Subtracting Eq. (11.10) from Eq. (11.11),

$$\Delta R_T = \Delta R_1 + \Delta R_2 \tag{11.12}$$

For complete compensation of the combination, $\Delta R_T = 0$, from which

$$\Delta R_1 = - \Delta R_2 \tag{11.13}$$

The actual TCR of the compensation resistor, therefore, depends on the ohmic value.

For resistors in parallel, the formula for the total resistance is

$$R_T = \frac{R_1 R_2}{R_1 + R_2} \tag{11.14}$$

Assuming again that a change in temperature produces a change in each of the resistors and assuming that ΔR_1 and ΔR_2 are much smaller than R_1 and R_2, for complete compensation ($\Delta R_T = 0$),

$$\frac{\Delta R_1}{R_1} = - \frac{\Delta R_2}{R_2} \tag{11.15}$$

indicating that the TCRs of the respective resistors must be equal in magnitude and of opposite sign. Note that the actual value of the compensating resistor does not enter into the compensating equation.

In using these principles, it is important to note certain of the properties of thick film resistors.

- Short resistors have a more positive TCR than long resistors due to the diffusion effects of the conductor metal termination.

- Resistors terminated on silver-bearing conductor materials tend to have a more positive TCR than resistors terminated on gold, since silver diffuses more into the resistors during firing.

- The TCR of thick resistors is more positive than that of thin resistors.

- The TCR of thick film resistors as given by the manufacturers is an average over a range of temperature; the hot TCR, measured between room temperature and +125°C, and the cold TCR, measured between room temperature and −40°C. In actuality, the TCR varies with the temperature in a nonlinear fashion.

- The ratio of resistors printed with the same material will track over temperature to a higher degree than will an individual resistor. Where possible, design circuits to depend on resistor ratios than on individual values.

11.6.2 Functional trimming of hybrid circuits

A desirable feature unique to hybrid microcircuit technology is the ability to trim resistors with the circuit under power while monitoring a circuit function, such as voltage, current, frequency, or phase. Where the desired output can be controlled by one or more resistors, other components, active and passive, can have wider tolerances with a corresponding savings in cost. It is possible to eliminate potentiometers altogether, while also saving on the labor necessary to adjust them.

In order to fully utilize the advantages of functional trimming,[3] it is necessary to properly design the circuit electrically to incorporate certain features.

1. Use resistor ratios as opposed to individual resistors. If one of the resistors is overtrimmed during adjustment, the other resistor may be trimmed to adjust the output in the other direction. This results in much higher yields during the trimming process.

2. Design circuits so that the variables of interest are independent and controlled by only one resistor. Otherwise, it will be necessary to alternate trimming of resistors, or to use a predict-and-trim routine, both of which require considerable laser trim.

Example 11.4 As an example of these two principles, consider the active filter in Fig. 11.13a. The transfer function of this filter is given by

$$\frac{E_o(s)}{E_i(s)} = \frac{-As}{s^2 + (BW)s + \omega_0^2}$$

(11.16)

where

$$A = \frac{1}{R_1 C}$$

(11.17)

$$BW = \frac{2}{R_2 C} = \text{bandwidth of filter}$$

(11.18)

$$\omega_0 = \sqrt{\frac{1}{R_1 R_2 C^2}} = \text{center frequency of filter}$$

(11.19)

The optimum method of trimming is to set the center frequency first and the bandwidth last. This is impossible in this circuit since both the bandwidth and the center frequencies of the filter are a function of R_2. If the center frequency is adjusted first, it will change when R_2 is changed to adjust the bandwidth. In order for the center frequency and the bandwidth to both converge at the desired values, it is necessary for an algorithm to be developed which will measure the initial frequency and bandwidth and trim to intermediate values based on these data. The result is a lengthy and somewhat inaccurate trim program.

By comparison, consider the design shown in Fig. 11.13b. The center frequency of this filter is given by

$$\frac{E_o(s)}{E_i(s)} = \frac{\dfrac{-s}{R_1 C}\left(1 + \dfrac{R_4}{R_3}\right)}{s^2 + s\left(\dfrac{2}{R_2 C} - \dfrac{R_4}{R_1 R_3 C}\right) + \dfrac{1}{R_1 R_2 C^2}}$$

(11.20)

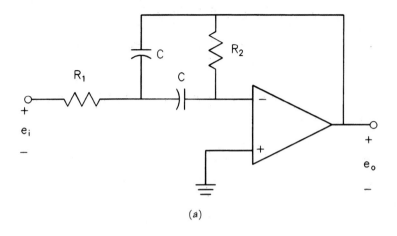

(a)

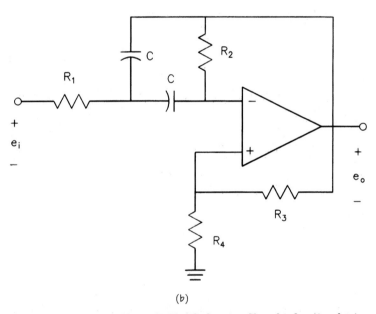

(b)

Figure 11.13 (a) Active filter. (b) Modified active filter for functional trimming.

The center frequency is

$$\omega_o = \sqrt{\frac{1}{R_1 R_2 C^2}} \qquad (11.21)$$

and the bandwidth is

$$BW = \frac{2}{R_2 C} - \frac{R_4}{R_1 R_3 C} \qquad (11.22)$$

This circuit has several advantages over the previous circuit. The center frequency may be trimmed to the desired value by trimming R_1 or R_2, and the bandwidth may be adjusted by trimming R_3 or R_4, which do not affect the center frequency. Note that it is possible to adjust the bandwidth higher or lower by adjusting R_3 or R_4, respectively.

11.6.3 Stray inductance and capacitance

Many hybrid circuits can be translated directly from the schematic to the circuit on a 1:1 correspondence. As the frequency or speed of operation increases, however, the small size of hybrid circuits may create stray inductance or capacitance due to the close proximity of the components and the traces which must be accounted for in analyzing and predicting circuit performance. There are three effects which must be considered; self-inductance, mutual inductance, and coupling capacitance.

The self-inductance of a conductor or a wire bond will act to reduce the frequency response of an electronic circuit by adding inductive reactance in series. The inductance of a nonmagnetic wire in free space is given by[4]

$$L \ (\text{nH}) = 5.08 \times 10^{-3} \times Z\left[\ln\left(\frac{Z}{D}\right) + 0.386\right] \qquad (11.23)$$

where Z = length of wire, mils, and D = diameter of wire, mils.

This expression holds true for both gold and aluminum wire since neither is magnetic. The inductance of a wire 0.001 in in diameter as a function of length is shown in Fig. 11.14.

Equation (11.23) may also be used to approximate the inductance of a thick or thin film conductor trace. A conductor trace of width W and thickness T may be considered to be made up of N wires of diameter T in parallel, where

$$N = \frac{W}{T} \qquad (11.24)$$

At low frequencies, the inductance of the trace is therefore equal to the inductance of one of the wires divided by N. A typical thick film is about 0.5 mils thick, and a typical gold thin film conductor is about 50 μin thick. A typical thick film conductor which is 0.010 in wide with a length of 0.300 in will have an inductance of 0.05 nH, while a thin film conductor of the same length which is 0.002 in wide will have an inductance of 0.015 nH. For higher frequencies, the trace must be considered a strip or microstrip transmission line, and the inductance must be calculated accordingly.

The mutual inductance is important in applications where the coupling between two adjacent conductor films is critical. Consider the structure in Fig. 11.15 where two conductor traces of widths W_1 and W_2 and thickness, respectively, are separated by a distance D. Assuming that the current density is uniform, the current, dI, in the incremental trace dx_1 is

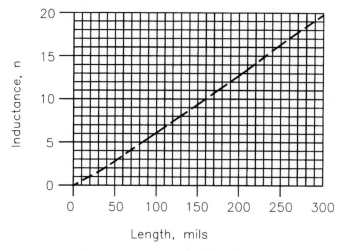

Figure 11.14   Inductance vs. length for 0.001-in-diameter wire.

$$dI = I \, \frac{dx}{W_1} \qquad (11.25)$$

This creates a magnetic field in the increment, dx_2 given by

$$dB(x_2) = \frac{\mu_o I \, dx_1}{4\pi \, W_1(x_2 - x_1)} \qquad (11.26)$$

In a unit length Z of conductor trace, the magnetic field creates an increment of flux in the incremental area bounded by Z and dx_2:

$$d\Phi_2 = \frac{\mu_o I Z dx_1 \, dx_2}{4\pi \, W_1 \, (x_2 - x_1)} \qquad (11.27)$$

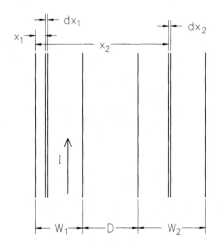

Figure 11.15   Mutual inductance between two conductors.

The mutual inductance per unit length between the two traces is defined by

$$M = \frac{\Phi_2}{ZI} \quad \text{H/m} \tag{11.28}$$

which is obtained by integrating Eq. (11.27) twice.

$$M = \int_0^{W_1} \int_{W_1 + D}^{W_1 + W_2 + D} \frac{\mu_o \, dx_1 \, dx_2}{4\pi W_1(x_2 - x_1)} \quad \text{H/m} \tag{11.29}$$

The result is a rather lengthy, cumbersome equation,

$$M = \frac{\mu_0}{4\pi} \frac{1}{W_1} \left[-D \ln\left(\frac{W_2}{D} + 1\right) + W_2 \ln\left(\frac{W_2}{D}\right) - W_2 \ln\left(\frac{W_2}{D} + 1\right) \right. \tag{11.30}$$

$$\left. + W_1 \ln\left(\frac{W_1 + W_2 + D}{W_1 + D}\right) + W_2 \ln\left(\frac{W_1 + W_2 + D}{W_1 + D}\right) + D \ln\left(\frac{W_1 + W_2 + D}{W_1 + D}\right) - W_2 \ln\left(\frac{W_2}{W_1 + D}\right) \right]$$

For two conductors 0.010 in wide and separated by 0.010 in, there is about 1.33 nH/in of mutual inductance at low frequencies.

Stray capacitance from one conducting surface to another may also occur in hybrid circuits. This is particularly true in thick film multilayer circuits where traces may be in close proximity with only a thin layer of dielectric between them. Stray capacitance may also occur from component to trace, from component to component, and from trace or component to nearby conductors at ground potential. The capacitance between conductors in different layers of a multilayer thick film circuit may be calculated from the classic formula,

$$C = \frac{\varepsilon_0 \varepsilon_r A}{T} \tag{11.31}$$

where C = coupling capacitance between conductors
  ε_0 = dielectric constant of free space
  ε_r = relative dielectric constant of thick film dielectric = 4–7 (typical, depends on filler)
  A = area of overlap
  T = thickness of dielectric

Stray capacitance can be difficult to predict. The geometry effects for irregular or nonsymmetrical surfaces are not easy to model, and it may be easier and more accurate to simply measure the capacitance.

A thick film or thin film resistor may exhibit considerable distributed capacitance to ground if the substrate is mounted to a ground plane, and may be modeled as a distributed RC network at high frequencies. At low frequencies, the capacitance may be calculated from Eq. (11.31) and modeled as a simple tee or pi network.

11.7   Thermal Considerations

The failure rate of hybrid microcircuits is directly related to time at temperature. At elevated temperatures, the rate of intermetallic compound formation increases and the rate of corrosion increases, both of which can be detrimental to reliability. In order to accurately calculate the reliability of a hybrid circuit, it is essential that the temperature at each point in the circuit be known, particularly the temperature of the semiconductor junctions.

11.7.1   Background theory

Heat is generated in electronic components as a natural consequence of circuit operation and may be generated by both active and passive components. For active devices, such as transistors, diodes, and integrated circuits, heat is generated at the junctions and may be considered to be generated at the surface inside the area indicated by the device boundary. In resistors, the heat source is the occupied surface area and is a direct function of the voltage and current. In trimmed resistors, the heat distribution may be highly nonlinear due to the nonlinear distribution of the current density. Where fine resolution of the temperature within the body of the resistor is critical, it may be necessary to divide the resistor into cells by the use of curvilinear squares. The heat generated in a capacitor is a function of the dissipation factor and the leakage current, and is influenced by whether the dielectric is NPO, BX, or tantalum. Ceramic dielectric materials classified as NPO have a low dielectric constant and a low leakage current. BX materials have a higher dielectric constant and a higher leakage current. Tantalum capacitors have a higher leakage current than either NPO or BX materials. See Chap. 8 for further details. Capacitors are not generally a heat source unless operated above the rated frequency and/or voltage. The entire body of the capacitor must be considered as the heat source, since the plates are interspersed throughout the body of the capacitor. The heat generated in inductors is a function of the current in the windings, which creates I^2R losses, and of the frequency (high frequency creates losses in the core). Inductors are not generally a heat source unless operated above the rated current and/or frequency. Conductor runs are not generally considered a heat source and may even assist in heat spreading. The application must be carefully considered here, since the most common problem with conductors is insufficient width to handle electrical transient conditions.

The actual temperature that the component attains is a function of many variables, including the configuration, the basic material properties, and the power. The material properties of most interest are the thermal conductivity and the thermal capacity. The thermal capacity is defined as the amount of energy required to raise the temperature by 1°C, with units of W•s/°C. A more common expression is the specific heat, which is the thermal capacity divided by the mass, with units of W•s/(g•°C).

Heat is transferred from one medium to another by one or more of three phenomena: conduction, convection, and radiation. Conduction requires direct physical contact between the media. Convection requires a fluid, which may be in the gaseous form. Radiation may be transmitted even through a total vacuum.

The phenomenon of conduction is described by Poisson's equation in three dimensions,

$$\frac{\partial T^2}{\partial X^2} + \frac{\partial T^2}{\partial Y^2} + \frac{\partial T^2}{\partial Z^2} + \frac{q_i}{k} = 0 \tag{11.32}$$

where q_i = internal heat generation in the system, W, and k = thermal conductivity, W/(m•°C).

Convection is the process by which heat is transferred to a surrounding fluid by conduction. Convection takes two forms. Free, or natural, convection occurs when movement of the fluid away from the heated surface is caused by the density differences between the hot fluid near the body and the cooler differences between the hot fluid near the body and the cooler fluid. Forced convection occurs when the fluid is forced past the body by a fan or a pump.

Heat flow due to convection is proportional to the surface area and the temperature differential between the surface and the surrounding fluid is as shown in Eq. (11.33)

$$q = hS\,(T_s - T_f) \tag{11.33}$$

where S = surface area, m^2
T_s = surface temperature, °C
T_f = fluid temperature, °C
h = heat transfer coefficient, W/(m^2•°C)

In practice, the heat coefficient transfer coefficient, h, is quite difficult to obtain except for the most elementary cases.

Energy may be transmitted between two media not in direct contact by radiation. While conduction and convection require matter to accomplish heat transfer, radiation is electromagnetic in nature and may transmit heat through a vacuum. The radiation equation is

$$q = Cf_af_eS(T_s^{\,4} - T_r^{\,4}) \tag{11.34}$$

where C = constant
f_a = arrangement factor
f_e = emissivity factor
S = surface area, m^2
T_s = source temperature, °C
T_r = receiver temperature, °C

The radiation equation is easily analyzed only for a point source, as the emissivity and arrangement factors are nonlinear and functions of temperature.

11.7.2 Thermal properties of materials

Three basic types of materials are used in the fabrication of hybrid microcircuits; metals, ceramics, and plastics. These materials are frequently used in conjunction, and a critical aspect of hybrid design is to ensure that the variations in properties of these materials are properly considered. The most important properties are the thermal conductivity, the heat capacity, and the temperature coefficient of expansion.

11.7.2.1 Thermal conductivity.

The *thermal conductivity* of a material is defined as

$$q = -k \frac{dT}{dx} \tag{11.35}$$

where k = thermal conductivity, W/(m•°C)
   q = heat flux, W/cm^2
   dT/dx = temperature gradient, °C/m in steady state

The negative sign denotes that heat flows from areas of higher temperature to areas of lower temperature.

There are two mechanisms which contribute to thermal conductivity; the movement of free electrons and lattice vibrations, or phonons. When a material is locally heated, the kinetic energy of the free electrons in the vicinity of the heat source increases, causing the electrons to migrate to cooler areas. These electrons undergo collisions with other atoms, losing their kinetic energy in the process. The net result is that heat is drawn away from the source toward cooler areas. In a similar fashion, the increased lattice vibrations as a result of a higher temperature transmit phonons, which also carry energy away from the source. The thermal conductivity of a material is the sum of the contributions of these two mechanisms.

$$k = k_p + k_e \tag{11.36}$$

where k_p = contribution due to phonons and k_e = contribution due to electrons.

Metals have a high concentration of free electrons, $k_e \gg k_p$, and the thermal conduction is primarily due to electron movement. The electrical and thermal conductivities are related by the Wiedemann-Franz law,

$$L = \frac{k}{\sigma T} \tag{11.37}$$

where L = Lorentz number=2.44×10^{-8} Ω•W/K^2 (theoretical)
   k = thermal conductivity, W/(mm•K)
   σ = conductivity, S/mm
   T = temperature, K

The Lorentz number is independent of temperature if it is assumed that the heat energy is transported entirely by electrons. Table 11.6 gives the

TABLE 11.6 Lorentz Number for Certain Metals Used in Hybrid Microelectronics Applications

Metal	Lorentz number at 0°C, 10^{-8} W•Ω/K²	Lorentz number at 100°C, 10^{-8} W•Ω/K²
Silver	2.31	2.37
Aluminum	2.24	2.28
Gold	2.35	2.40
Copper	2.23	2.33
Molybdenum	2.61	2.79
Lead	2.47	2.56
Platinum	2.51	2.60
Tin	2.52	2.49
Tungsten	3.04	3.20

Lorentz numbers for some of the more common metals used in the fabrication of hybrid circuits.

In ceramics, the heat flow is primarily due to phonon generation, and the thermal conductivity is generally lower than that of metals. Crystalline structures, such as alumina and beryllia, are more efficient heat conductors than amorphous structures such as glass. Organic materials used to fabricate printed circuit boards or epoxy attachment materials are electrical insulators and highly amorphous, and tend to be very poor thermal conductors.

Impurities or other structural defects in metals and ceramics tend to lower the thermal conductivity by causing the electrons to undergo more collisions, which lowers the mobility and lessens the ability of the electrons to transport heat away from the source. For example, alloys of metals have a lower thermal conductivity than the pure metal constituents. By the same token, as the ambient temperature increases, the number of collisions increases, and the thermal conductivity of most materials drops. In most practical cases only the temperature dependence of ceramics is critical, as these materials are frequently the limiting factor in the amount of heat flow. The thermal conductivity as a function of temperature for several common ceramics is shown in Fig. 11.16.

11.7.2.2 Heat capacity. Heat capacity is defined as

$$C = \frac{dQ}{dT} \tag{11.38}$$

where C = heat capacity, W•s/(mol•K)
$\quad\quad\ Q$ = energy, in W•s
$\quad\quad\ T$ = temperature, K

This is the amount of energy required to raise one mole of material by one kelvin. The specific heat c is defined in a similar manner and is the amount of heat required to raise the temperature of one gram of material by one degree celsius, with units of W•s/(g•°C). The quantity *specific heat* in this context

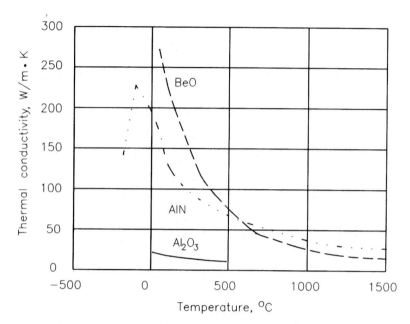

Figure 11.16 Thermal conductivity vs. temperature for ceramic packaging materials. (*Courtesy Tegmen–Brush Wellman Corp.*)

refers to the quantity c_V, which is the specific heat measured with the volume constant, as opposed to c_p, which is measured with the pressure constant. At the temperatures of interest, these numbers are nearly the same for most solid materials. Thermal capacity is primarily the result of an increase in the vibrational energy of the atoms when heated, and the specific heat of most materials increases with temperature up to a temperature called the *Debye temperature,* at which point the specific heat becomes essentially independent of temperature. The specific heat of several common ceramic materials as a function of temperature is shown in Fig. 11.17.

11.7.2.3 Temperature coefficient of expansion. The temperature coefficient of expansion (TCE) arises from the asymmetrical increase in the interatomic spacing of atoms as a result of increased heat. Most metals and ceramics exhibit a linear, isotropic relationship in the temperature range of interest, while certain plastics may be anisotropic in nature. The TCE is defined as

$$\alpha = \frac{l(T_2) - l(T_1)}{l(T_1)(T_2 - T_1)}$$ (11.39)

where α = temperature coefficient of expansion, 1/°C
T_1 =   initial temperature
T_2 = final temperature
$l(T_1)$ = length at initial temperature
$l(T_2)$ = length at final temperature

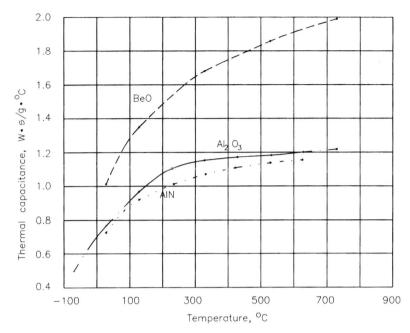

Figure 11.17 Specific heat vs. temperature for ceramic packaging materials. (*Courtesy Tegmen–Brush Wellman Corp.*)

The thermal properties of common materials are listed in Table 11.7.

11.7.3 Thermal resistance

The exact solution of the heat transfer equation for a given set of boundary conditions is quite complex even when convection and radiation are not considered, and, in general, a numerical solution must be performed. For many cases of practical interest, however, an approximate solution will suffice.

The simplest case is that where the heat flows in only one dimension. Consider a wall of thickness L in the x direction and with infinite dimensions in the y and z dimensions. The wall has a thermal conductivity k which is uniform and homogeneous, and the temperature is uniform along each wall, which restricts the heat flow to only the x direction. For this case, the temperature distribution is given by

$$T(x) = \frac{(T_2 - T_1)x}{L} + T_1 \qquad (11.40)$$

Note that the temperature distribution is independent of the thermal conductivity.

For the more realistic case where heat is applied to a finite area and diffuses through to the other side, the temperature differential is given by

TABLE 11.7a   Thermal Properties of Selected Materials at Room Temperature

Material	Thermal conductivity, W/(°C•in)	Density, g/in³	Specific heat, W•s/g•°C
Air	0.000665	0.0193	1.006
Alumina (94%)	0.70	58.99	0.88
Alumina (96%)	0.89	60.63	0.88
Alumina (99.5%)	0.93	63.91	0.88
Aluminum	5.5	44.25	0.92
ASTM-F15 Alloy*	0.49	134.37	0.46
Beryllia	3.9	47.19	1.04
Borosilicate glass	0.026	36.1	
Copper (Cu)	9.6	146.83	0.39
Cu-clad molybdenum (1–1–1)	6.27†		
	7.82‡		
Cu-clad molybdenum (1–6–1)	5.23†		
	5.86‡		
Cu-clad Invar (1–1–1)	0.79†		
	6.70‡		
Cu-clad Invar (1–6–1)	0.44†		
	4.30‡		
Epoxy–glass laminate	0.007		
Epoxy resin, unfilled	0.004	36.1	
Epoxy, silver-filled	0.0515§	§	
Epoxy, nonconductive	0.0057	36.1	
Epoxy, nonconductive film¶	0.0080	36.1	
Gold	7.5	316.6	0.13
94% gold–6% silicon	7.5	299.9	0.146
80% gold–20% tin solder	6.22	277.24	0.154
Lead	0.83	185.7	
Molybdenum	3.7	167.48	0.25
Nickel	2.29	145.84	0.44
Palladium	1.79	196.33	0.243
Phosphor bronze	5.256		
Platinum	1.75	351.5	0.13
Plastics	0.0438	34.4	
Quartz	0.05		
Silicon	2.13	38.33	0.75
Silicon dioxide	0.035		
Silicone (room temperature vulcanizing)	0.004		
Silver	10.6	171.19	0.234
Thick film dielectric	0.038		
Tin	1.63	119.79	0.23
63% tin–37% lead solder	0.091	144.18	0.154

*54% Fe–29% Ni–17% Co.
†Normal direction.
‡Tangential direction.
§Depends on silver concentration.
¶Fiberglass mesh filler.

TABLE 11.7b   Conversion Factors for Thermal Conductivity and Specific Heat

To convert from	Thermal Conductivity			
	To			
	$\dfrac{C_a l}{s \cdot {}^\circ C}$	$\dfrac{W}{cm \cdot {}^\circ C}$	$\dfrac{W}{in \cdot {}^\circ C}$	$\dfrac{t}{h \cdot ft \cdot {}^\circ F}$
	Multiply by			
$\dfrac{Cal}{s \cdot {}^\circ C}$	1	4.18	10.62	241.9
$\dfrac{W}{cm \cdot {}^\circ C}$	0.239	1	2.54	57.8
$\dfrac{W}{in \cdot {}^\circ C}$	0.0943	0.393	1	22.83
$\dfrac{Btu}{h \cdot ft \cdot {}^\circ F}$	0.00413	0.0173	0.0438	1

	Specific heat				
To convert	$\dfrac{cal}{g \cdot {}^\circ C}$	to	$\dfrac{W \cdot s}{g \cdot {}^\circ C}$	multiply by	4.17
To convert	$\dfrac{W \cdot s}{g \cdot {}^\circ C}$	to	$\dfrac{cal}{g \cdot {}^\circ C}$	multiply by	0.24

$$T_2 - T_1 = \frac{PL}{kA} \tag{11.41}$$

where A is the area of the path and P is the power in watts.

There is a direct analog between this expression and Ohm's Law, which relates the potential difference between two points to the current and the electrical resistance between the points. The expression L/kA is an analog to the electrical resistance, and is referred to as the *thermal resistance*. The power is an analog to the current. The electrical-thermal analog is summarized in Table 11.8.

In the general case, heat flows in three dimensions, which makes a closed-form solution of Poisson's equation very difficult for all but a few configurations, which are not normally found in practice. Therefore, either a numerical solution or an approximate solution must suffice. Most numerical solutions utilize the method of finite element analysis (FEA) to obtain the solution, while most approximate solutions assume that the heat spreads at a constant angle, usually 45°, as shown in Fig. 11.18. The so-called spreading angle approach must be used with some discretion, but is sufficiently accurate to indicate whether a potential thermal problem exists. For materials with a high thermal conductivity, the spreading angle is usually less than 45°; conversely, for materials with a low thermal conductivity, the angle is greater

TABLE 11.8 Electrical-Thermal Analog

Electrical	Thermal
Voltage (V)	Temperature (°C)
Current (A)	Heat flux (W)
Resistance (Ω)	Thermal resistance (°C/W)
Capacitance (F)	Thermal capacitance (W·s/°C)

than 45°. When heat flows from a material with a high thermal conductivity to one with a low thermal conductivity, heat will flow laterally out to the boundaries of the high thermal conductivity material and the spreading angle will be quite shallow.

The diagram in Fig. 11.19 can be used to calculate the thermal resistance of a rectangular solid with a spreading angle of 45°. Consider a heat generating element with lateral dimensions L and W on an infinite surface of thickness T and thermal conductivity k. The thermal resistance of the differential element dz is given by

$$dR_{TH} = \frac{dz}{kxy} \tag{11.42}$$

In terms of the variable z

$$x = W + 2z \tag{11.43}$$

$$y = L + 2z \tag{11.44}$$

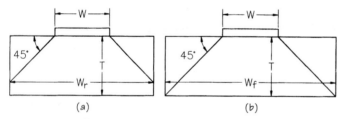

Figure 11.18 Heat spreading angle. (*a*) Restricted spreading ($W_r < W + 2T$). (*b*) Full spreading ($W_f \geq W + 2T$).

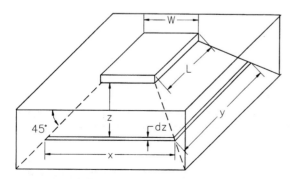

Figure 11.19 Calculation of thermal resistance.

Substituting Eqs. (11.43) and (11.44) into Eq. (11.42).

$$dR_{TH} = \frac{dz}{k[4z^2 + 2z(L + W) + LW]} \tag{11.45}$$

and

$$R_{TH} = \frac{1}{k} \int_0^T \frac{dz}{4z^2 + 2z(L + W) + LW} \tag{11.46}$$

The solution to this integral has two forms. For the case where $L = W$, the thermal resistance is given by

$$R_{TH} = \frac{T}{kL(L + 2T)} \tag{11.47}$$

The expression

$$L(L + 2T) \tag{11.48}$$

may be considered to be the *effective area* of the sector under the heat generating element, and the expression

$$L + 2T \tag{11.49}$$

may be considered to be the dimension of the heat generating element for subsequent layers of material below.

For the case where L is not equal to W, the expression for thermal resistance is

$$R_{TH} = \frac{1}{2k(L - W)} \ln \frac{L(2T + W)}{W(2T + L)} \tag{11.50}$$

For the cylindrical configuration, as shown in Fig. 11.20, the thermal resistance from one face to the other is given by

$$R_{TH} = \frac{H}{k\pi R(R + H)} \tag{11.51}$$

where R = radius of the circle and H = length of the cylinder.

The rules for combining thermal resistances in series and parallel are the same as for electrical resistances. For thermal resistances in series,

$$R_{TH} = R_1 + R_2 + R_3 + \ldots \tag{11.52}$$

and in parallel,

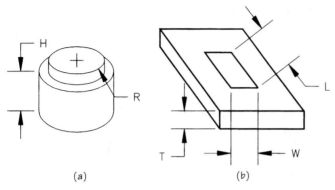

(a)   (b)

Figure 11.20 Models for thermal resistance of solids. (a) Cylindrical. (b) Rectangular.

$$\frac{1}{R_{\text{TH}}} = \frac{1}{R_1} + \frac{1}{R_2} + \frac{1}{R_3} + \dots \tag{11.53}$$

Example 11.5 Consider the case of an integrated circuit chip dissipating 0.2 W mounted with nonconductive epoxy onto an alumina substrate, as shown in Fig. 11.21. The IC is $0.100 \times 0.100 \times 0.010$ with an active heat generating area of 0.080×0.080, and dissipates 0.2 W. The parameters of the materials are summarized in Table 11.9.

Integrated circuit. The thermal resistance of the integrated circuit is calculated from Eq. (11.47).

$$R_{\text{TH}}(\text{IC}) = \frac{0.010}{2.13\{0.080[0.080 + 2(0.010)]\}} = 0.587°\text{C/W}$$

Nonconductive epoxy. From Eq. (11.40), the effective length of the heat source at the interface between the integrated circuit and the nonconductive epoxy is

$$L_{\text{eff}} (\text{epoxy}) = L + 2T = 0.080 + 2 \times 0.010 = 0.1 \text{ in}$$

The thermal resistance of the epoxy layer is

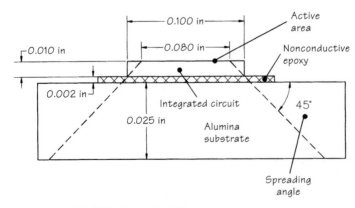

Figure 11.21 Model for Example 11.5.

TABLE 11.9   Parameters of Materials Used in Examples

Parameter	Integrated circuit	Nonconductive epoxy	Substrate
W_T, in	0.080	0.100	0.104
W_B, in	0.100	0.104	0.154
T, in	0.010	0.002	0.025
A_T, in^2	0.0064	0.010	0.0103
A_B, in^2	0.010	0.0103	0.0237
V, in^3	8.13×10^{-5}	2.08×10^{-5}	4.212×10^{-4}
d, g/in^3	39.33	36.05	60.63
m,g	0.0032	7.502×10^{-4}	0.0255
S, W•s/(gm•°C)	0.751	0.942	0.876
C_{TH} W•s/°C	0.0024	7.067×10^{-4}	0.0224
k, W/(°C•in)	2.13	0.0057	0.89

$$R_{TH}(\text{epoxy}) = \frac{0.002}{0.0057[0.100(0.100 + 2 \times 0.002)]} = 33.738°\text{C/W}$$

Substrate.   The effective length of the heat source at the interface between the epoxy and the substrate is

$$l_{\text{eff}}(\text{substrate}) = 0.100 + 2 \times 0.002 = 0.104 \text{ in}$$

and the thermal resistance is

$$R_{TH}(\text{substrate}) = \frac{0.104}{0.89[0.104(0.104 + 2 \times 0.025)]} = 1.754°\text{C/W}$$

Summary.   The total thermal resistance is

$$R_{TH}(\text{total})=R_{TH}(\text{IC}) + R_{TH}(\text{epoxy}) + R_{TH}(\text{substrate}) = 0.587 + 33.378 + 1.754 = 36.079°\text{C/W}$$

For a power dissipation of 0.2 W, the temperature rise of the integrated circuit over the bottom of the substrate is

$$\Delta T = P_D \times R_{TH}(\text{total}) = 0.2 \times 36.079 = 7.216°\text{C}$$

The results of Example 11.5 are summarized in Table 11.10. Note that most of the thermal resistance is due to the nonconductive epoxy. Considerable improvement can be attained by using a conductive epoxy or solder to attach the integrated circuit die. Table 11.11 summarizes the temperature rise when these materials are used.

11.7.4   Thermal capacitance

The thermal capacitance of a material is related to the specific heat and the mass by the relationship

$$C = mS \tag{11.54}$$

TABLE 11.10   Results of Example 11.5

Element	L, in	W, in	T, in	R_{TH}, °C/W
Integrated circuit	0.080	0.080	0.010	0.587
Nonconductive epoxy			0.002	33.738
Alumina substrate			0.025	1.754
Total				36.079
Power dissipated				0.2 W
Temperature rise				7.216°C

Assumptions:
1. Complete spreading takes place in all elements.
2. No voiding is present at any interface.
3. The active area of the integrated circuit is 0.080 × 0.080 in.

TABLE 11.11   Results of Example 11.5 with Alternative Die Attach Materials

Die attach material	Thermal resistance of die attach layer, °C/W	Total thermal resistance, °C/W	Temperature rise, °C
Nonconductive epoxy	33.738	36.079	7.216
Conductive epoxy	3.734	6.075	1.215
60% tin–40% lead solder	0.139	2.48	0.496

where C = thermal capacitance, W•s/°C
m = mass of the solid = volume × density
S = specific heat of the solid, W•s/(g•°C)

The thermal capacitance provides a convenient means to calculate the approximate time required for the temperature to come to equilibrium. The product $R_{TH}C_{TH}$ has the units of time and may be interpreted as the thermal time constant, which determines the rate of temperature rise. The thermal capacitance may also play an important role in thermal management. When the duty cycle of the input power is small, the temperature rise may be limited by simply adding more mass or by using materials with a high thermal capacitance.

Example 11.6   Calculate the thermal capacitance of each of the components in Example 11.5 and construct an equivalent circuit. The parameters of interest are given in Table 11.9. This example uses the value of specific heat at 300 K to illustrate the principle. It must be noted that, to obtain a more accurate solution, the variation of the specific heat with temperature must be taken into consideration. A cursory analysis shows that an iterative solution is required, as the value of the specific heat will change from the time the power is applied at $t = 0$ until steady state is reached. This situation requires the solution of a complex algorithm, and is beyond the scope of this book.

Integrated circuit.   When the heat spreading angle is considered, the components in Example 11.5 have the shape of a truncated pyramid which has a volume given by Eq. (11.55).

$$V = \tfrac{1}{3}h(A_T + A_B + \sqrt{A_T A_B})$$

(11.55)

where V = volume of section
h = height of section
A_T = area of section top
A_B = area of section bottom

For the integrated circuit, the top dimension is 0.080 in and the bottom dimension after spreading is 0.100 in. The volume is

$$V(\text{si}) = \frac{0.01(0.0064 + 0.010\sqrt{0.0064 \times 0.01}\ )}{3} = 8.133 \times 10^{-5}\ \text{in}^3$$

The mass is the product of the volume and the density.

$$M(\text{si}) = V(\text{si}) \times d(\text{si}) = 8.133 \times 10^{-5} \times 39.33 = 0.0032\ \text{g}$$

The thermal capacitance is

$$C_{\text{TH}}(\text{si}) = 0.0032 \times 0.751 = 0.0024\ \text{W}\cdot\text{s/}^\circ\text{C}$$

The calculation of the thermal capacitance of the nonconductive epoxy and the substrate proceeds in an identical manner. The equivalent circuit is shown in Fig. 11.22. For a constant power level, the temperature will quickly come to equilibrium with the temperature rise as shown in Example 11.5. For the case where the input power has a duty cycle or is a more complex function of time, the circuit in Fig. 11.22 may be analyzed by conventional means to determine the actual temperature rise at any point in the integrated circuit–epoxy–substrate structure as a function of time.

11.7.5 Cooling by convection

Convection cooling takes place by transferring the heat from a surface to a fluid, which may be air or a liquid, as described in Eq. (11.33). The heat increases the kinetic energy of the fluid, which transports heat away due to the increased motion. There are two basic types of convection: free convection, in which the motion is generated by the heat source, and forced convection, in which the fluid is forced across the heated surface by an external source. As with thermal resistance, a closed-form solution for the convection coefficient is very difficult to obtain except for a few geometries.

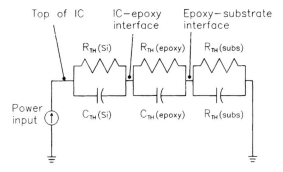

Figure 11.22 Thermal equivalent circuit for Example 11.5.

To understand the mechanics of convection, it is necessary to understand the basics of the relationship between fluid flow and heat transport. There are two fundamental types of fluid flow: laminar flow, which is due primarily to density differences, and turbulent flow, which is created primarily by external means, such as a pump or fan. The distinction between and nature of these types of fluid flow are determined empirically by the values of a series of dimensionless numbers. The properties of air and water relating to heat transport are tabulated in Tables 11.12 and Table 11.13, respectively.

The Reynolds number Re is the ratio of the inertial forces to the viscous forces in a coolant flow and is defined as

$$Re = \frac{\rho v d}{\mu} \tag{11.56}$$

where ρ = density of the fluid, g/cm^3
v = velocity of the fluid, cm/s
d = a critical dimension, such as the edge of the circuit under consideration, in
μ = viscosity of the fluid, g/(s•cm)

TABLE 11.12   Thermal Transport Properties of Air

Parameter	Temperature, °K	Temperature, °C	Value	
			W/cm•°C	W/in•°C
k	200	−73	0.000181	0.0000712
	300	27	0.000262	0.0001030
	400	127	0.000336	0.0001321
	500	227	0.000404	0.0001590
			W•s/g•°C	
c_p	200	−73	1.006	
	300	27	1.006	
	400	127	1.014	
	500	227	1.030	
			kg/m•s	g/in•s
μ	200	−73	13.28×10^{-6}	0.000337
	300	27	18.46×10^{-6}	0.000469
	400	127	22.86×10^{-6}	0.000581
	500	227	26.70×10^{-6}	0.000678
			kg/m^3	g/in^3
ρ	200	−73	1.769	0.02899
	300	27	1.177	0.01929
	400	127	0.882	0.01445
	500	227	0.706	0.01157
			1/°C	
α			0.00367	

TABLE 11.13   Thermal Transport Properties of Water

Temperature, °F	Temperature, °C	c_p, Btu/lb•ft	k, Btu/h•ft•°F	μ, lb/ft•h	ρ, lb/ft³	α, 1/°F	Pr
32	0	1.029	0.337	4.320	62.54	1.10×10^{-4}	13.20
100	37.8	0.999	0.365	1.620	62.20		4.43
200	93.3	1.004	0.393	0.738	60.20		1.88
		W•s/(g•°C)	W/(m•°C)	kg/(m•s)	kg/m³	1/°C	
32	0	4.305	0.583	0.00866	1001.9	1.98×10^{-4}	13.20
100	37.8	4.180	0.631	0.00325	996.4		4.43
200	93.3	4.201	0.680	0.00148	964.4		1.88

The Reynolds number is a function of temperature since both ρ and μ are functions of temperature. For external flow, as opposed to confined flow as in a pipe, the value of Re classifies turbulent and laminar flow.

Laminar flow: $\text{Re} < 5 \times 10^5$

Turbulent flow: $\text{Re} > 5 \times 10^5$

The Prandtl number Pr is the ratio of fluid and thermal boundary layer thicknesses, and is equal to 0.7 for air over a flat plane. The value of the Prandtl number for air is relatively insensitive to temperature. The Prandtl number for water as a function of temperature is given in Table 11.13.

The Nusselt number Nu is defined as

$$\text{Nu} = \frac{hd}{k} \tag{11.57}$$

where h = convection constant, W/m²•°C
d = critical dimension, m
k = thermal conductivity, W/m•°C

The Grashof number Gr is the ratio of buoyancy forces to viscous forces and is defined as

$$\text{Gr} = \frac{\rho^2 g \alpha d^3 (T_S - T_F)}{\mu^2} \tag{11.58}$$

where g = acceleration of gravity = 9.78 m²/s
α = coefficient of expansion of the fluid
T_S = temperature of surface
T_F = temperature of fluid

The Rayleigh number Ra is the product of the Grashof and Prandtl numbers. The dimensionless numbers defined here are related by a series of empirical equations. For free convection,

$$\text{Nu} = c(\text{Gr Pr})^n = c(\text{Ra})^n \tag{11.59}$$

where c and n are constants which depend on the value of the Rayleigh number as shown in Table 11.14.

For forced convection,

$$\text{Nu} = c\text{Re}^n \text{ Pr}^m \tag{11.60}$$

where c, n, and m are given in Table 11.14.

For zero to moderate external airflow, heat is transported away from the source by a combination of free and forced convection. The contribution due to free convection is by far the most difficult to calculate analytically, since a trial-and-error solution is required. The Grashof number is a function of the difference between the fluid temperature and the plate temperature. In most cases, the fluid, or ambient, temperature is known and the plate temperature is unknown. To obtain a solution for equations of this type, a value must be estimated for the plate temperature and the system of equations solved for the actual plate temperature on the basis of the estimate. If the estimate is different from the solved value, another estimate is made and the process is repeated until the two converge. The complexity of this process requires a computer routine to obtain an acceptable degree of accuracy. By contrast, the effect of forced convection can be calculated from Eq. (11.60) by determining whether the flow is laminar or turbulent from the Reynolds number, and using the appropriate form from Table 11.14.

Example 11.7 Consider the configuration of Example 11.5 mounted such that the bottom of the substrate is pointed upward. A stream of air at ambient temperature with a velocity of 1000 ft/min (200 in/s) is directed across the surface. Calculate the temperature at the surface directly underneath the die if it can be assumed that free convection is negligible.

The critical dimension here is the length of the bottom of the truncated pyramid which bounds the heat emanating from the top of the integrated circuit, previously calculated as 0.154 in.

TABLE 11.14   Rayleigh and Nusselt Numbers for Free and Forced Convection

	Free Convection		
Orientation	Rayleigh number	c	n
Vertical plate	10^4–10^9	0.59	0.25
	10^9–10^{12}	0.13	0.33
Horizontal plate (heated surface up)	10^4–10^7	0.54	0.25
	10^7–10^{11}	0.15	0.33

	Forced Convection	
Flow type	Orientation	Nusselt number
Laminar	Flat plane	$0.664 \text{ Re}^{0.5} \text{ Pr}^{0.33}$
Turbulent	Flat plane	$0.037 \text{ Re}^{0.5} \text{ Pr}^{0.33}$

For this case, the Reynolds number is

$$\mathrm{Re} = \frac{0.0193 \; \frac{g}{in^3} \times 20000 \; in/s \times 0.154 \; in}{0.000469 \; g/(in \cdot s)} = 1.267 \times 10^5$$

indicating laminar flow. From Table 11.14, the Nusselt number is given by

$$\mathrm{Nu} = 0.664 \; (\mathrm{Re})^{0.5} \; (\mathrm{Pr})^{0.33}$$

$$= 0.664 \; (1.267 \times 10^5)^{0.5} \; (0.7)^{0.33} = 210.14$$

The convection constant is

$$h = \frac{\mathrm{Nu} \; k}{d}$$

$$= \frac{210.44 \times 0.000103}{0.154} = 0.141$$

The temperature rise from the airflow to the substrate is

$$T_S - T_F = \frac{q}{ha}$$

$$= \frac{0.2}{0.141 \times 0.154 \times 0.154} = 60.0°C$$

This result is higher than the actual temperature since free convection was not considered.

Based on the author's experience, the correlation with experimental data can be expected to be within $\pm 20\%$, and it must be emphasized that the methods described here are only approximations designed to provide the hybrid engineer with a means to estimate the temperature rise. However, this approach will enable the hybrid engineer to determine the points in the thermal path which are the most potentially troublesome and will serve to compare various techniques of thermal management.

11.7.6 Thermal management

Thermal management is the technology of directing heat away from circuit elements, especially semiconductor devices, and toward the outside world, where it can be safely dissipated. There are three basic approaches to thermal management which derive directly from the expression for thermal resistance noted in Eq. (11.47):

1. *Minimizing path length.* The thermal resistance of a system is directly proportional to the path length. It is sometimes possible to dramatically reduce the thermal resistance by simply using a thinner substrate material.

2. *Maximizing thermal conductivity.* This applies to both substrates and die attach material. Table 11.11 compares the thermal resistance of the system in Example 11.5 when different die attach materials are used. Substrate materials such as beryllium oxide or aluminum nitride offer improved thermal conductivity over aluminum oxide.

3. *Maximizing area.*   By the innovative use of heat spreading techniques, it is possible to increase the effective area for heat flow by a considerable amount. For example, mounting the die on a molybdenum tab spreads the heat out before it begins to flow through the substrate, which has a lower thermal conductivity. Example 11.8 further illustrates the effect of heat spreading techniques.

Example 11.8   Consider the structure in Example 11.7. Assume that the integrated circuit is mounted on a substrate which is 1×1 in and which has a layer of direct bond copper (DBC) on the bottom side. The DBC spreads the heat out laterally so that the critical dimension is now 1 in and the effective area is now 1 in². Calculate the temperature rise of the bottom of the substrate above that of the fluid. For this case,

$$\text{Re} = 8.23 \times 10^5 \text{ (turbulent)}$$

$$\text{Nu} = 1.776 \times 10^3$$

$$h = 0.183$$

$$T_S - T_F = 1.093°\text{C}$$

which represents a substantial improvement in junction temperature.

11.7.7   Thermal measurement

The measurement of the temperature at a given point on a hybrid circuit is not an exact science. There are three fundamental methods by which it may be accomplished.

1. *Noncontact.*   It is known that objects emit infrared radiation when heated. The amount of heat is proportional to a number of variables, including the temperature and the emissivity. The emissivity is a function of the surface characteristics; smooth, dark surfaces are more efficient radiators than rough, light surfaces. Infrared temperature measurement systems which scan the surface must be calibrated for each material in the circuit to give an accurate reading, which can be a formidable task. A further complication is that the surface may change during the reading as a result of surface oxidation or other reaction. The two major advantages are that the temperature of the entire surface may be scanned in a short time, and there is no direct contact with the surface to disturb the reading.

2. *Contact.*   Contact methods use a thermocouple or thermistor probe to measure the temperature at a given point on the hybrid circuit. While there are no surface irregularities to be concerned about, the probe can act as a heat sink to draw heat away from the point in question and affect the reading.

3. *Junction voltage.*   The current in a perfect forward-biased diode at low current is given by

$$I = Ae^{qV/kT} \tag{11.61}$$

where I = current, A
A = constant

q = charge on one electron
V = voltage across the diode
k = Boltzmann's constant
T = temperature, K

Taking the logarithm of both sides and solving for T,

$$T = \frac{qV}{k \ln(I/A)} \qquad (11.62)$$

Note that T is directly proportional to V. This characteristic can be used to obtain accurate temperature measurements by passing a small constant current through the diode and measuring the voltage. Bipolar devices can be measured directly, and field-effect transistors (FETs) can be measured by using the built-in diode. Diodes can also be used as temperature-monitoring devices at various points in a circuit, serving as protection mechanisms for higher powered devices. Each diode must be calibrated individually when used for this application, as the constant A may vary from device to device.

11.7.8 Thermal stress analysis

The temperature coefficient of expansion phenomenon described in Eq. (11.39) has serious implications in the design of hybrid microcircuits. When a sample of material has one end fixed, which may be considered to be a result of bonding to another material that has a much smaller TCE, the net elongation of the hotter end per unit length, or *strain E,* of the material is calculated by

$$E = TCE \times \Delta T \qquad (11.63)$$

where E = strain in length/length and ΔT = temperature differential across the sample.

Elongation develops a stress in the sample as given by Hooke's law.

$$S = EY \qquad (11.64)$$

where S = stress, psi (Pa), and Y = Young's modulus, psi (Pa).

When the total stress exceeds the tensile strength of the material, mechanical cracks will form in the sample and may even propagate to the point of separation. Certain of the parameters of interest are shown in Table 11.15. In using Table 11.15, it is important to note that there is a wide variation of certain of these parameters in the literature, and values for some of the alloys and composite materials may not be available at all. Material parameters also vary widely with temperature, and more detailed information should be utilized where applicable.

The above analysis is simplistic, but should provide a basic understanding of the mechanical considerations in joining two materials together.

TABLE 11.15   Mechanical Properties of Selected Materials at Room Temperature

Material	TCE	Young's modulus, 10^6 psi	Tensile strength, 10^3 psi
Alumina (96%)	6.4	50	25
Alumina (99.5%)	6.8	50	25
Aluminum	25.2	10	8
ASTM-15 Alloy	5.0	20	80
Beryllia	9.0	50	20
Beryllium copper	16.74	19	
Copper	16.56	17	32
Cu-clad molybdenum (1–1–1)	9.66		
Cu-clad molybdenum (1–6–1)	6.40		
Cu-clad Invar (1–1–1)	10.60		
Cu-clad Invar (1–6–1)	6.50		
Epoxy resin (unfilled)	45–90	0.55	4–13
Epoxy, silver-filled	0.4*		
Epoxy, nonconductive	0.5		
Epoxy, film	0.6†		
Gold	14.22	10.8	17
94% gold–6% silicon eutectic	13.67		
Lead	28.8	2.0	
Magnesium	25.2	6.4	
Molybdenum	5.4	40.0	80
Nickel	13.32	31.0	46
Platinum	9.0	21.3	22
Silicon	5.04	16.0	‡
Silicon (room temperature vulcanizing)	2–4	3–4	
Silver	19.80	10.5	18
Steel, high carbon	12.06	30.0	89
Steel, stainless	17.28	28.0	80
Tin	19.80	6.0	
63% tin–37% lead solder	24.6	4.5	7.5
Tungsten	4.50	50	430
Zinc	34.20	12	

*Varies widely—depends on filler.
†Fiberglass mesh reinforcement.
‡Dependent on the number and type of defects.

11.7.8.1 Considerations in component mounting.   When an active device is mounted to a ceramic substrate, a three-layer structure composed of the substrate, the die attach material, and the active die is formed, as shown in Fig. 11.23. For a structure of this type, excessive stress can occur when either of two conditions is present:

1. When the TCE of the die is sufficiently lower than that of the substrate, such that the resultant stress is greater than that of the tensile strength of the die material when the structure is in thermal equilibrium. This condition may occur when no power is applied to the die and the structure is subjected to temperature excursions, or temperature cycling.

2. When a nonequilibrium condition exists such that the temperature differential between the die and the substrate is sufficiently large to create a stress condition greater than the tensile strength of the die. This condition may be brought about when the thermal resistance of the die bond is high

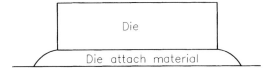

Figure 11.23 Die–die attach–substrate structure.

due to improper selection of materials, when voiding occurs at the die bond interface during transient power conditions, or during power cycling, when the die is operating.

In many practical situations, both conditions exist to a greater or lesser degree. In this case, the strain is given by

$$E = [TCE(D) - TCE(S)][T(D) - T(A)] + TCE(S)[T(D) - T(S)] \quad (11.65)$$

where TCE(D) = temperature coefficient of expansion of the die
TCE(S) = temperature coefficient of expansion of the substrate
T(D) = temperature of the die
T(S) = temperature of the substrate
T(A) = ambient temperature with the power off

Two situations are of further interest.

1. T(D) = T(S). In this case, the power dissipated in the die is negligible or nonexistent, and Eq. (11.65) reduces to

$$E = [T(D) - T(S)][T(D) - T(A)] \quad (11.66)$$

   This is the situation that prevails in temperature cycling.

2. TCE(D) = TCE(S). In this case, the TCEs of the die and the substrate have been made equal by proper material selection during the design process. Eq. (11.65) reduces to

$$E = TCE(S)[T(D) - T(S)] \quad (11.67)$$

   This is the condition that prevails in power cycling or where thermal resistance is high.

The analysis presented here may be considered a "best-case" analysis, since most experimental results will result in lower figures than calculated. There are many other considerations that will determine whether or not a failure occurs, such as a fault in the material, which concentrates the stress at a given point, or failures that occur as a result of excessive mechanical fatigue when a structure is subjected to a number of temperature excursions.

Frequently, the maximum stress is observed when the structure is cooled down from the die attach process, since the process temperature is greater than the operating temperature. Bolger and Mooney[7] developed the relationship for calculating the stress under this condition.

$$E_m = K[\text{TCE(S)} - \text{TCE(D)}][T(\text{P}) - T(\text{A})] \sqrt{\frac{E_A E_S L}{x}} \qquad (11.68)$$

where E_m = maximum stress at the die corners, psi
   K = geometric constant, dimensionless; function of die shape and amount of filleting; must generally be determined experimentally
   E_A = adhesive tensile modulus, psi at 25°C
   E_S = substrate tensile modulus, psi at 25°C
   L = maximum substrate dimension, mils
   x = adhesive bond thickness, mils
   $T(\text{P})$ = processing temperature, °C
       = glass transition temperature for epoxies
       = solidification temperature for solders
   $T(\text{A})$ = ambient temperature or temperature of interest

Note:

   1 psi = 6.90×10^{-3} N/mm^2

   1 N/mm^2 = 1.45×10^2 psi

   1 N/mm$^{3/2}$ = $28.77 \times$ lb/mil$^{3/2}$

   1 lb/mil$^{3/2}$ = 3.48×10^{-2} N/mm$^{3/2}$

Van Kessel et al.[8] proposed a criterion to predict whether or not a given die will crack based on the depth of scratches, chipouts or other flaws in the die as shown in Eq. (11.69).

$$E = m \frac{K_{1c}}{\sqrt{\pi a}} \qquad (11.69)$$

where m = 1.20 for a surface flaw with a semi-elliptical shape
   K_{1c} = fracture toughness of silicon = approximately 0.90 lb/mil$^{3/2}$
   a = depth of flaw/2 defined in Fig. 11.24

For brittle, crystalline materials (such as silicon), the tensile strength is not a good criterion to use for predicting fracture, since stress will tend to concentrate at any flaws that are present. In this case, Eq. (11.69) provides a more accurate result.

Van Kessel et al. also present data obtained by mounting strain gauges with epoxy and by eutectic means. The data indicates that such factors as bond thickness and bond voids may reduce the applied stress by more than an order of magnitude from that calculated by Hooke's law.

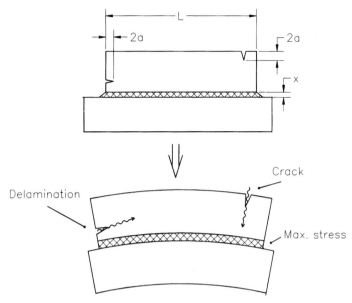

Figure 11.24 Cracks and chip-outs in semiconductor devices.[8]

Figures 11.25 and 11.26 use Eq. (11.69) to plot the predicted stresses required to cause cracking. Note that in Fig. 11.25 the maximum stress is observed at cold temperature. This is due to the fact that, as the temperature drops, the strain as described in Eq. (11.65) increases.

Figure 11.27 illustrates the types of failures that can be observed in an active die.

Example 11.9 Consider a silicon die soldered to a copper header.

Size:	0.250×0.250 in
Junction temperature:	130°C
Header temperature:	80°C
Ambient temperature:	25°C
Surface flaws:	0.0001 in maximum

Calculate the stress and determine if a potential for cracking exists. From Eq. (11.65), the strain per unit length is

$$E = [\text{TCE(D)} - \text{TCE(S)}][T(D) - T(A)] + \text{TCE(S)}[T(D) - T(S)]$$

$$= (5.04 - 16.56)(130 - 25) + (16.56)(130 - 80)$$

$$= -1209.6 + 828.0$$

$$= -381.6 \times 10^{-6} \text{ in/in}$$

The negative sign indicates compressive stress. Note that the power dissipated in the die actually relieves the stress by causing the die to expand and more nearly match the copper.

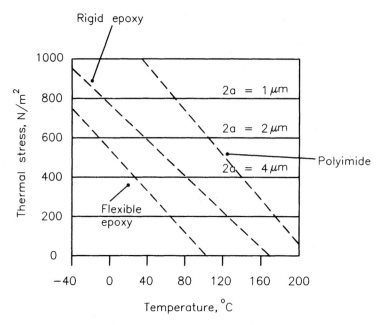

Figure 11.25 Thermal stress vs. temperature (predicted) for flexible epoxy, rigid epoxy, and polyimide.[7]

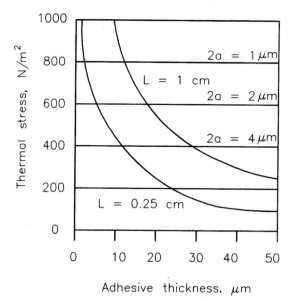

Figure 11.26 Effect of die length and adhesive thickness on thermal stress.[7]

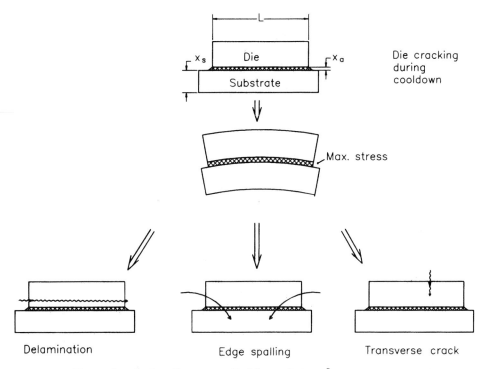

Figure 11.27 Types of cracks in a die as a result of thermal stress.[7]

The maximum dimension of the die is $0.25 \times 1.414 = 0.354$ in. The net strain is

$$E = -381 \times 10^{-6} \times 0.354$$

$$= -134.7 \times 10^{-6} \text{ in}$$

From Eq. (11.64), the stress is

$$S = EY$$

$$= -134.7 \times 10^{-6} \times 16 \times 10^{6}$$

$$= -2154.9 \text{ psi}$$

The predicted fracture strength from Eq. (11.69) is

$$E = 1.20 \, \frac{0.9}{\sqrt{0.05 \, \pi}} = 2.725 \text{ psi}$$

This analysis predicts that cracking will almost certainly occur, which is consistent with experimental results.

Example 11.10 Repeat the analysis if an alumina substrate with 96% purity is used. From Eq. (11.65), the strain/unit length is

$$E = [TCE(D) - TCE(S)][T(D) - T(A)] + TCE(S)[(T(D) - T(S)]$$

$$= (5.04 - 6.4)(130 - 25) + (6.4)(130 - 80)$$

$$= -142.8 + 320.0$$

$$= 177.2 \times 10^{-6} \text{ in/in}$$

The positive sign is indicative of tensile stress, which tends to pull the die apart. The maximum dimension of the die is $0.25 \times 1.414 = 0.354$ in. The net strain is

$$E = (177.2 \times 10^{-6})(0.354)$$

$$= 62.73 \times 10^{-6} \text{ in}$$

From Eq. (11.64), the stress is

$$S = EY$$

$$= (62.73 \times 10^{-6})(16 \times 10^6)$$

$$= 1004 \text{ psi}$$

This figure is still substantially higher than the predicted fracture stress, indicating that cracking is still very likely. This indicates that, although the TCEs of the die and the substrate are more closely matched, a problem still exists.

Example 11.11   It is apparent from Examples 11.9 and 11.10 that the temperature differential between the die, the substrate, and the ambient contributes greatly to the stress in the die. When temperature differentials of these magnitudes are observed, it is indicative of poor thermal design. For a more realistic example, consider the die soldered to an alumina substrate of 96% purity under the following conditions:

Junction temperature: 33°C
Header temperature: 31°C
Ambient temperature: 25°C

From Eq. (11.65), the strain/unit length is

$$E = [TCE(D) - TCE(S)][T(D) - T(A)] + TCE(S)[(T(D) - T(S)]$$

$$= (5.04 - 6.4)(33 - 25) + (6.4)(33 - 31)$$

$$= -10.88 + 12.8$$

$$= 1.92 \times 10^{-6} \text{ in/in}$$

The maximum dimension of the die is $0.25 \times 1.414 = 0.354$ in. The net strain is

$$E = (1.92 \times 10^{-6})(0.354) = 0.68 \times 10^{-6} \text{ in}$$

From Eq. (11.64), the stress is

$$S = EY = (0.68 \times 10^{-6})(16 \times 10^6) = 10.87 \text{ psi}$$

While this figure is still higher than that calculated from Eq. (11.69), it is within the margin of error as described by Van Kessel. If this approach is used, it can be assumed with reasonable certainty that minimal cracking will occur.

11.7.8.2 Factors that minimize stress. Hooke's law, as demonstrated above, typically results in values of stress that are much greater than those observed experimentally. A variety of factors contribute to the reduction in stress.

Bond thickness. An increase in the bond thickness contributes to a reduction in stress on the die by giving it a greater ability to flex when a force is applied. This principle is commonly employed by increasing the thickness of solder joints when chip carriers, chip capacitors, and other ceramic components are soldered to a printed circuit board.

Bonding voids. It has been experimentally observed that small voids in the bond distributed over the area of the die reduce the stress. However, voids in the bond area increase the thermal resistance and, consequently, the temperature of the die, which counters the positive effect somewhat. Large voids tend to concentrate the stress at the point of bonding and increase the probability of cracking.

Compliant bonding materials. The use of a compliant bonding material, such as a "soft" solder or epoxy, enables the bond to absorb much of the stress, minimizing the stress on the die. Examples of soft solders include lead–indium solders and high tin content solders, such as 96% tin–4% silver. It must be noted, however, that even filled epoxies are relatively poor thermal conductors and tend to increase the junction temperature of the die.

Intermediate layers. The use of an intermediate layer of material with a TCE somewhere between the die and the substrate is a design approach commonly used to reduce stress. One very common material used for this purpose is molybdenum, which has a TCE of 5.4 ppm/°C as compared to 5.04 ppm/°C for silicon and 6.4 ppm/°C for alumina. Typically, the molybdenum tab, or *molytab,* is plated with gold and the die is eutectically mounted to it. The molytab with the die attached may be mounted to the substrate with solder or epoxy. The molytab also has the effect of a heat spreader, increasing the effective area of the bond and reducing the junction temperature of the die.

11.8 Determination of Conductor Width

The ability of a film conductor to carry large currents is of critical interest to the designer in both steady-state and the transient conditions. In the steady state, the conductor may increase in temperature because of internal resistive heating and contribute significantly to the overall temperature of the circuit, and the voltage drop across the conductor may be detrimental to circuit performance. In the transient case, the conductor may be required to handle currents substantially larger than normal for short intervals of time, and the conductor may reach a temperature high enough to evaporate, creating an open circuit. The method described here[9] determines the minimum width based on a predetermined allowable temperature rise in the steady-state condition, and based on the melting point in the transient case.

11.8.1 Theoretical analysis—steady state

In the steady-state condition, the current passing through a conductor will generate heat because of resistive losses in the conductor. The temperature rise of the conductor depends on the thermal path of the system from the conductor through the intervening layers to the heat sink. The *heat sink* in this context is the boundary layer, which could be simply air, a liquid-cooled plate, or an actual heat sink. For this case, the substrate will be considered to be mounted on a heat sink of constant temperature and will be the only thermal barrier. This is a common configuration for power hybrids, since it is highly desirable to minimize the thermal path. The principle may be extended to other configurations by simply expanding the thermal analysis to include the entire system.

For this analysis, it is assumed that $K_C >> K_S$, where K_C is the thermal conductivity of the conductor and K_S is the thermal conductivity of the substrate. The implication of this assumption is that the substrate represents the only thermal path.

$$R_{TH} = \frac{T_S}{K_S(W_C)\,(W_C + 2T_S)} \tag{11.70}$$

The power dissipated in one square of the conductor is given by

$$P = I^2R = I^2\frac{R_B}{T_C} = I^2R_S \tag{11.71}$$

where R_B = bulk resistivity of the conductor
$\quad R_S$ = sheet resistance of the conductor
$\quad R$ = electrical resistance of the conductor
$\quad I$ = steady-state direct or root mean square (rms) alternating current in the conductor

The net temperature rise of the conductor is therefore

$$T_{RISE} = PR_{TH} = I^2\,R_S\,\frac{T_S}{K_S W_C(W_C + 2T_S)} \tag{11.72}$$

From Eq. (11.72), the maximum current which the conductor can handle for a given width is

$$I_{MAX} = \sqrt{\frac{T_{RISE}\,K_S\,W_C\,(W_C + 2T_S)}{R_S T_S}} \tag{11.73}$$

Solving Eq. (11.73) for the width gives the minimum width for a given current.

$$W_{C(MIN)} = \sqrt{\frac{T_S I^2 R_S}{K_S T_{RISE}} + T_S^2} - T_S \tag{11.74}$$

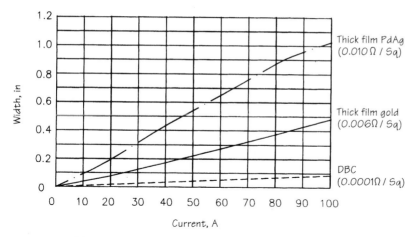

Figure 11.28 Minimum width vs. current for several conductor materials on a 0.025-in alumina substrate.

The results of Eq. (11.74) are depicted in Fig. 11.28.

11.8.2 Theoretical analysis—transient

In the transient case, the designer must consider the effect of very high currents for a short period of time. This condition can occur when an output device shorts, for example, or when the circuit is connected to the AC line and a surge occurs on the line. A common specification for relay and motor control circuits is that they be able to withstand a surge current of up to 20 times the rated current for an interval of one cycle, or 0.0167 s.

Assume that the conductor is subjected to an impulse of current of magnitude I which continues for the interval $0 \le t \le t_D$, where t_D is the time duration of the pulse. For simplicity, alternating-current signals can be modeled as a pulse having the same magnitude as the rms value of the current. Assume also that the conductor is at a steady-state temperature T_0. The power dissipated in the conductor is given by Eq. (11.71).

The thermal analog of Kirchhoff's current law is

$$P = I^2 R_S = P_{R(\text{TH})} + P_{C(\text{TH})} \tag{11.75}$$

which translates to

$$P = \frac{T_{\text{RISE}}}{R_{\text{TH}}} + C_{\text{TH}} \frac{d}{dt} T_{\text{RISE}}(t) \tag{11.76}$$

Solving Eq. (11.76) for T_{RISE} and adding T_0 gives

$$T(t) = T_{\text{RISE}}(t) + T_0 = I^2 R_S R_{\text{TH}}[1 - e^{1/R_{\text{TH}}C_{\text{TH}}}] \qquad (0 \le t \le t_D) \tag{11.77}$$

The thermal time constant is the product of the thermal resistance and the thermal capacitance. It is noteworthy that the only significant dimension is the thickness of the substrate.

$$R_{TH}C_{TH} = \frac{T_S}{K_S A_{EFF}} A_{EFF} T_S D\, S = \frac{T_S^2 DS}{K_S} \qquad (11.78)$$

Inserting Eq. (11.78) into Eq. (11.77) gives

$$T(t) = T_{RISE}(t) + T_0 = I^2 R_S R_{TH}[1 - e^{-tK_S \backslash T_S^2 DS}] + T_0 \qquad (0 \le t \le t_D) \qquad (11.79)$$

At time $t = t_D$, when the power source is removed, the temperature reaches the maximum, T_{MAX}, and will begin to decay toward T_0 at the same rate at which it rose. The expression for the temperature decay is

$$T(t) = T_{MAX} e^{(-t + t_D)/R_{TH}C_{TH}} + T_0 \qquad (t > t_D) \qquad (11.80)$$

In the transient case, the designer is generally concerned about conductor failure, which may occur either where the adhesion mechanism fails or where the conductor opens as a result of melting or vaporizing. The susceptibility of the adhesion mechanism to heat for a given conductor may vary from lot to lot depending on the quality of the metallizing process, and it is difficult to establish a maximum temperature for this variable. Often, the failure mode is more dramatic; the conductor vaporizes, accompanied by a massive flash of light, and opens, resulting in permanent failure of the circuit.

When a metal simply melts, which requires somewhat less energy than evaporation, the surface tension of the molten metal will tend to form it into small, isolated spheroids, thereby creating an open circuit and causing a failure. If the melting point T_M of the conductor metal is adopted as the temperature at which failure occurs, Eq. (11.77) may be used to establish criteria for determining the conductor width. With $t = t_D$ and $T(t_D) = T_M$, Eq. (11.77) becomes

$$T_M = I^2 R_S R_{TH}(1 - e^{-t_D K_S/T_S^2 DS}) + T_0 \qquad (0 \le t \le t_D) \qquad (11.81)$$

Solving Eq. (11.81) for W_C gives

$$W_C = T_S \left[\left\{ \sqrt{\frac{I^2 R_S}{K_S T_S (T_M - T_0)}[1 - e^{-t_D K_S/T_S^2 DS}] + 1} \right\} - 1 \right] \qquad (11.82)$$

Equation (11.81) can also be solved for the maximum current for a particular conductor width and time duration.

$$I_{MAX} = \sqrt{\frac{T_{RISE} K_S W_C (W_c + 2T_S)}{R_S T_S}} \qquad (11.83)$$

$$t_{D(MAX)} = -\frac{T_S^2 DS}{K_S} \ln\left[1 - \frac{[(T_S + W_C)/T_S]^2 - 1}{I^2 R_S}(K_S T_S)(T_M - T_0)\right] \quad (11.84)$$

The methods described here for determining the minimum conductor width for steady-state and transient conditions are based on the approximate methods presented in the section on thermal analysis. They have been verified experimentally and should be applicable in most situations.

By comparison, MIL-H-38534B requires that a conductor dissipate no more than 4 W/in^2. For a gold thick film conductor with a sheet resistance of 0.006 Ω/sq on alumina, this corresponds to a current of 25.82 A. Equation (11.72) predicts a temperature rise of 0.64°C for this structure.

11.9   Determination of Wirebond Size

MIL-H-38534B establishes the maximum current density for materials in contact with the substrate as shown in Table 11.16. The current density here may be expressed either as direct current or as rms alternating current. This number does not apply to wirebonds, however, since they are in a free space environment, and lose heat only by radiation, convection, or by conduction through plastic encapsulation, which is a relatively poor thermal conductor.

A wirebond must be able to carry the necessary current with an adequate safety factor such that the wire is not stressed during operation. MIL-H-38534B also establishes criteria for the maximum current for different configurations:

$$I_{MAX} = Kd^{3/2} \quad (11.85)$$

where I_{MAX} = maximum current in DC amperes or rms AC amperes
d = diameter of wire, in, or equivalent diameter for ribbon or other conductor which corresponds to the same cross-sectional area
K = applicable constant from Table 11.17

TABLE 11.16   Maximum Current Density in Bonding Wires from MIL-H-38534B

Conductor material	Maximum allowable current density	
	A/cm^2	A/in^2
Aluminum (99.99% pure or doped) without glassivation or without glassivation layer integrity test	2×10^5	1.29×10^6
Aluminum (99.99% pure or doped), glassivated	5×10^5	3.23×10^6
Gold	6×10^5	3.87×10^6
All other (unless otherwise specified)	2×10^5	1.29×10^6

TABLE 11.17   Current Constants for Wirebonds from MIL-H-38534B

Composition	Value of K	
	Length ≤ 0.040 in (0.1 cm)	Length > 0.040 in (0.1 cm)
Aluminum	22,000	15,200
Gold	30,000	20,500
Copper	30,000	20,500
Silver	15,000	10,500
All other	9,000	6,300

The requirements from Eq. (11.85) are summarized in Table 11.18. Many prefer to derate the currents calculated from this formula by a factor of 0.5.

11.10   Selection of Substrate Technology

The hybrid engineer may choose from one of at least seven options when packaging an electronics circuit:

1. Thick film hybrid circuits

2. Thin film hybrid circuits

3. Monolithic circuits

4. Printed circuit boards (PCBs)

5. Direct bond copper (DBC) circuits

6. Plated-copper circuits

7. Multichip modules

The fact that many circuits may be manufactured by all of the above technologies makes the decision process even more complex. Each of the listed technologies has advantages and disadvantages which may make it suitable or unsuitable for a given application.

TABLE 11.18   Maximum Current for Wire Size

Material	Diameter, in	Maximum current, A	
		$L < 0.040$ in	$L > 0.040$ in
Gold	0.001	0.949	0.648
	0.002	2.683	1.834
Aluminum	0.001	0.696	0.481
	0.002	1.968	1.360
	0.005	7.778	5.374
	0.008	15.742	10.876
	0.012	28.920	19.981
	0.015	40.417	27.924
	0.022	71.789	49.600

The basic rule for selection, however, is simple: determine the technologies that meet the technical requirements for the circuit and choose the least expensive. The primary scope of this discussion is hybrid technology, which, in this context, includes thick film circuits, thin film circuits, direct bond copper circuits, and plated-copper circuits. The selection of one or a combination of these approaches must be based on the criteria for a particular application, because force fitting a technology into an application for which it is not suited will invariably lead to disaster.

11.10.1 Line resolution

The screen printing process used to manufacture thick film circuits is capable of resolving lines and spaces down to 0.005 in or less. However, in high volume production, it is advisable to restrict the resolution to 0.010 in. The thin film process is manufactured by a photoetching process which is capable of resolving lines and spaces less than 0.001 in in production. DBC and plated-copper circuits also utilize a photoetching process, but the additional thickness limits the resolution which can be attained because of undercutting of the lines during the etching process. A rule of thumb is that the line width should be no less than the thickness. DBC has a lower limit on thickness due to the nature of the fabrication process, and, consequently, has the worst resolution. Plated copper can be deposited thinner, and lines down to 0.002 in can readily be etched.

11.10.2 Resistor technology

Thick film technology permits the fabrication of a wide range of resistors on a single substrate, since there are a variety of pastes with different sheet resistivities available. Thin film technology is limited to a single sheet resistivity without extra processing, and very low value and very high value resistors are difficult to attain without using excessive space. DBC and plated copper are not directly compatible with nitrogen-firable resistor materials, but copper thick film terminations may be printed where needed as an interface.

The stability of thin film resistors which have been properly passivated is excellent, with less than 0.1% drift readily attainable for the life of the resistor without added processing. Thick film resistors are also quite stable, with less than 0.5% drift expected except at the extreme high and low ranges. If thick film resistors are trimmed 2% low, baked at 200°C for 48 h, and retrimmed in an area of low sensitivity, stabilities approaching those of thin film resistors may be attained. The nitrogen-firable resistors used with DBC and plated copper are not as stable because of their inherent properties.

11.10.3 High frequency properties

High frequency circuits require precise, smooth conductors to avoid excessive losses. The thin film technology is clearly superior here, followed by plated copper. While DBC conducts well, the thicker conductors cannot be etched to the precision that thin film and plated copper can. The surface and edges of thick film conductors are too irregular for operation at very high frequencies.

11.10.4 Multilayer structures

Multilayer structures are readily attained with thick film conductors and dielectrics. The availability of dielectric materials which can resolve small vias makes this technology the most economical in this area. Nitrogen-firable dielectric materials are compatible with DBC and plated copper, and the latter can be used in conjunction with low-temperature-cofired dielectric materials. Multilayer thin film structures can be fabricated with organic dielectric materials or sputtered dielectric materials.

11.10.5 Power and current handling capability

DBC is exceptional for handling high power or high current applications. The thick layers of copper attainable with this technology allow the heat to be spread over a wide area, with dramatic improvements in the thermal resistance. Plated copper can carry high currents, but cannot be made thick enough to achieve the same degree of spreading as DBC. The thick and thin film technologies provide little heat spreading and are limited in current carrying capability by the relatively high sheet resistivity of the conductors.

Table 11.19 summarizes the properties of the substrate metallization technologies.

TABLE 11.19   Summary of Substrate Technology Characteristics

Technology	Advantages	Disadvantages

Thick film	Resistor capability
	  Wide range
	  Low TCR (±50 ppm)
	  Stability (<0.5%)
	Multilayer capability
		Relatively high resistor noise

Thin film	Resistor capability
	  Low noise
	  High stability (<0.1%)
	Fine line capability (0.001 in)
	High frequency capability
		Limited resistor range
		Limited multilayer capability

Direct bond copper	High power handling capability High current capability	Limited resolution due to high thickness of copper Limited multilayer capability Limited resistor technology
Plated copper	Fine line capability Through-hole plating capability High frequency capability	Limited thickness Limited multilayer capability Limited resistor technology

11.11 Design Guidelines

Design guidelines for hybrid microcircuits are unique to every company in that they must mirror the capabilities of that company. The design guidelines should reflect the ability to perform each of the indicated processes to the extent indicated, and serve as the foundation for the risk analyses completed as part of the design reviews. The guidelines presented here[10] reflect the consensus of the hybrid industry as a whole and should be considered in that context.

11.11.1 Thick film conductor guidelines—general considerations

1. Do not screen curved lines.
2. Straight lines screened at 45° angles to the edges of the substrate are permissible.
3. Component mounting pads on conductor lines must be so indicated by making the line slightly wider as shown in Fig. 11.29. The purpose of this rule is to indicate the mounting position of the chip.
4. All wirebonding pads and mounting pads should be on the top conductor layers.
5. Both terminations for a given resistor shall be on the same conductor print.
6. The power density shall not exceed that determined from Sec. 11.8.
7. Do not place wirebond pads on vias.

11.11.2 Thick and thin film conductor guidelines

See Figs. 11.30 to 11.38.

11.11.3 Thick film dielectric guidelines—general considerations

See Figs. 11.39 to 11.43.

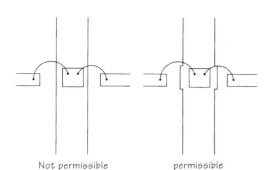

Not permissible   permissible

Figure 11.29 Die mounted on conductor path.

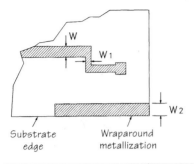

Substrate   Wraparound
edge   metallization

Dimension	Thick film, mils		Thin film, mils		Remarks
	Preferred	Minimum	Preferred	Minimum	
W	As required for current-carrying				*
	capability or electrical resistance				
W_1	10	5	5	2	
W_2	20	15	15	10	†

*I_{MAX} may be calculated from Eq. (11.74).
†Wraparound metallization for thick film printing requires special tooling. Do NOT use laser-scribed substrates for either thick film or thin film fabrication with this process unless the substrates have been properly annealed, as the metallization will not adhere well to the initial laser-scribed surface.

Figure 11.30 Conductor linewidths.

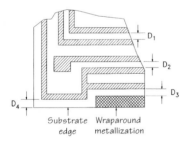

Substrate   Wraparound
edge   metallization

Dimension	Thick film, mils		Thin film, mils		Remarks
	Preferred	Minimum	Preferred	Minimum	
D_1	10	7.5	5	2	*
D_2	15	10	7.5	5	†
D_3	15	10	7.5	5	
D_4	10	10	7.5	7.5	‡

*Conductor lines ≤ mils in length.
†Conductor lines > 15 mils in length.
‡From minimum substrate size.

Figure 11.31 Conductor line spacing.

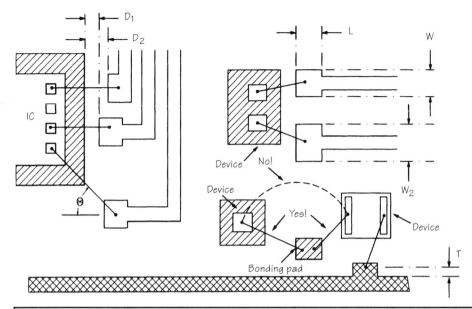

Dimension	Thick film, mils		Thin film, mils		Remarks
	Preferred	Minimum	Preferred	Minimum	
L	10	7.5	7.5	5	(1)
W	10	7.5	7.5	5	(1)
W_2	15	12.5	12.5	10	(2)
Θ	90	45	90	45	(3)
D_1	75	100 max.	75	100 max.	(4)
D_2	30	15	30	15	(5)
W_2	15	10	NA	NA	(6)
T	25	20	12.5	10	(7)

1. Use 10×10 minimum for automatic wire bonding.
2. Width for double wire.
3. Locating angle, die edge to center of wirebond site.
4. *Maximum* distance from edge of die to edge of pad.
5. Distance from mounting pad to wirebonding pad.
6. Distance from center of wirebond site to dielectric.
7. Wraparound metallization distance to wirebond site.

Figure 11.32 Bonding pad sizes.

1. Two layers of high temperature dielectric are required in multilayer and crossover applications.

2. Do not place wire bond pads on vias.

3. Via fills must be used when printing three conductor layers or more. The top via fill may be eliminated. The only exception to this rule is when vias are used as heat sinks under components.

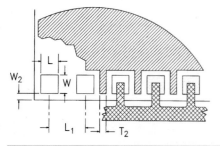

| Dimension | Thick film, mils | | Thin film, mils | | Remarks |
	Preferred	Minimum	Preferred	Minimum	
L	60	30	60	30	
W	60	50	60	50	
L_1	100	50	100	50	*
W_2	10	5	10	5	†
T_2	10	7.5	NA	NA	‡

*Pad spacing, center to center.
†Distance from edge of substrate.
‡Screened dielectric between pads.

Figure 11.33 Pads for lead frame attachment.

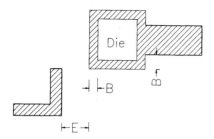

| Dimension | Thick film, mils | | Thin film, mils | | Remarks |
	Preferred	Minimum	Preferred	Minimum	
B	10	5	10	5	*,‡
B	10	10	10	10	†,‡
E	15	10	10	2	§

*Distance from edge of chip to edge of mounting pad; chip size≤100 mils.
†Chip size > 100 mils.
‡Distance=10 mils *maximum*.
§Pad distance to other metallization.

Figure 11.34 Mounting pads for semiconductor die.

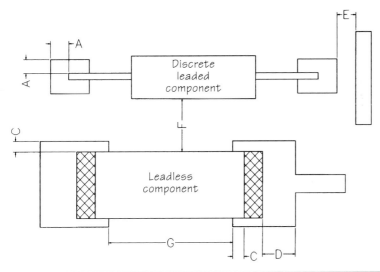

| Dimension | Thick film, mils | | Thin film, mils | | Remarks |
	Preferred	Minimum	Preferred	Minimum	
A	10	10	10	10	(1)
C	10	5	10	5	(2)
D	20	10	20	10	(3)
E	15	15	15	15	(4)
F	40	40	40	40	(5)
G	AR	30	AR	30	(6)

1. Metallized pad overlap for leaded component.
2. Distance from edge of pad to chip metallization on leadless chip side and underneath.
3. Distance from edge of pad to chip metallization end.
4. Pad distance to other metallization.
5. Distance from leaded component.
6. Spacing between termination for leadless chip components.

Figure 11.35 Pads for discrete components.

11.11.4 Thick film resistor guidelines—general considerations

See Figs. 11.44 to 11.46.

1. It is preferred that no more than three resistor values be used.

2. Do not use curved or right-angle resistors.

3. Do not penetrate more than halfway into the resistor when trimming.

4. When trimming to less than 5% tolerance, use the double-plunge cut or L-cut mode. Trim to −5% or until $W/2$ is reached with a single-plunge cut, then switch to double-plunge cut or L cut.

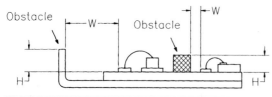

H, mils	W,* mils
20	40
40	45
60	50
80	55
100	65

*The W dimension may be smaller in thermosonic bonding if a tip with a smaller angle is used.

Figure 11.36 Wirebond distance from obstacle.

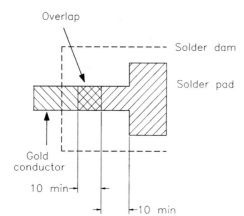

Figure 11.37 Gold conductor extension under solder pad.

11.12 Assembly Materials and Process Selection

The selection of the materials and processes to be used in the assembly of a hybrid circuit is an integral part of the design process. The electric and thermal design criteria and the materials and processes selected to fabricate the circuit must ultimately converge into a design which meets the technical requirements and is economically manufacturable. To this end, the designer must always consider the manner in which the circuit is to be assembled during the entire design procedure.

From an economic standpoint, it is desirable to assemble all the components to the substrate with a single material and a single process, and, unless there are exacting technical criteria, this can often be accomplished. In surface mount technology, for instance, it is common to screen-print solder paste onto a ceramic substrate or PC board, place all the components, and attach them by heating the assembly above the melting point of the solder.

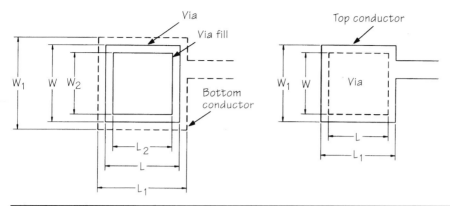

| Dimension | Thick film, mils | | Thin film, mils | | Remarks |
	Preferred	Minimum	Preferred	Minimum	
L	12.5	10	NA	NA	(1), (7), (8)
L_1	$L + 5$	L	NA	NA	(2)
L_2	L	L	NA	NA	(3), (9)
W	12.5	10	NA	NA	(4), (7), (8)
W_1	$L + 5$	L	NA	NA	(5)
W_2	W	W	NA	NA	(6), (9)

1. Via length.
2. Insert top and bottom.
3. Length of via fill.
4. Via width.
5. Conductor overlap, top and bottom.
6. Width of via fill.
7. It is preferred for inspection purposes that $L=W$.
8. It is preferred for inspection purposes that all vias on a substrate be the same size.
9. Via fills are required with more than two metallization layers. For more than two layers, the top via fill may be omitted.

Figure 11.38 Via and conductor pad sizes.

While the matrix for material selection is quite complex, ultimately it can be narrowed down to two choices: metallic attachment (solder or eutectic) or organic attachment (epoxy or polyimide). Often it is the most exacting requirement which dictates the choice of materials. For example, power devices may require solder for improved thermal conductivity, while other components may be attached by either means. In this case, it may be more practical to solder all the components rather than use two processes. Table 11.20 compares some of the features of organic and metallic attachment.

For complete assembly, including the packaging process, the vast majority of hybrid circuits will require at least two processes. Where this is the case, the materials and processes must be selected such that subsequent operations are lower in temperature than prior operations in order not to disturb processes already performed. Table 11.21 lists the process temperatures of certain selected materials, while Table 11.22 gives some of the criteria for selecting both materials and processes.

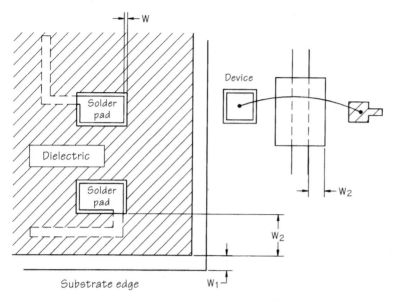

Dimension	Thick film, mils		Remarks
	Preferred	Minimum	
W	10	5	*, †
W_1	10	5	‡
W_2	10	5	§

*Distance to metallized pad for solder or epoxy attach.
†Low temperature dielectric preferred for solder attach, high temperature preferred for epoxy attach.
‡Distance to edge of substrate.
§Dielectric overlap when covering conductor.

Figure 11.39 Thick film dielectric spacing and size.

11.13 Reliability Considerations

11.13.1 Reliability and failure mechanisms

The *reliability* of a system is the ability of that system to meet the required specifications for a given period of time. The reliability is measured in a number of ways, including the failure rate, expressed as a probability distribution function $f(t)$, and the mean time to failure (MTTF), which represents the length of time the system is expected to operate until the first failure. The MTTF is related to the failure rate by the expression

$$\text{MTTF} = \bar{t} \int_0^\infty tf(t)\,dt \tag{11.86}$$

The failure rate of a hybrid circuit typically varies with time as shown in Figure 11.47. During the initial phase, referred to as the *infant mortality* phase, the failure rate is relatively high and is decreasing, as inherent defects in the devices and the manufacturing processes surface. This is followed by a phase in which the failure rate is somewhat lower and is relatively constant,

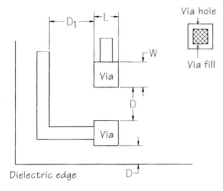

	Thick film, mils		
Dimension	Preferred	Minimum	Remarks
L	20	10	*
W	20	10	*
D	20	10	†
D_1	15	10	‡
D_2	10	10	§

*It is preferred for inspection purposes that all vias on a substrate be the same size.
†This also applies to vias on adjacent layers. Vias must be staggered unless used for heat dissipation purposes.
‡Via distance to metallization.
§Conductor distance to metallization.

Figure 11.40 Dielectric via sizes and locations.

referred to as the *operating life* of the product, which in turn is followed by a phase characterized by an increasing failure rate, referred to as the *end of life* phase.

When $f(t)$ in Eq. (11.86) is a constant, t_f, as in the operating life phase, it can be shown that the MTTF is simply the reciprocal of the failure rate.

$$\text{MTTF} = \frac{1}{t_f} \tag{11.87}$$

This is not the case, however, where the failure varies with time, as in the infant mortality and end of life phases.

There are a number of mechanisms which can cause failure in a hybrid circuit. The majority of these can be divided into four categories:

1. *Failures due to mechanical interconnections.*   This includes failures due to wirebonds, die bonds, lead frame bonds, tape augmented bonding (TAB) bonds, and other bonding schemes. These may be caused by inadequate process controls, leading to bonds which are not optimally formed, or to selection of materials which are not compatible with each other or with the process.

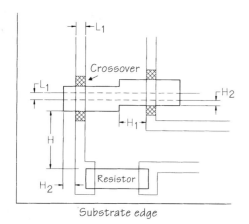

| Dimension | Thick film, mils | | Remarks |
	Preferred	Minimum	
H	50	30	*
H_1	20	10	†
H_2	20	10	‡
L_1	15	10	§

*Distance to resistor.
†Conductor distance to dielectric step. Use staggered crossovers or H patterns when two or more adjacent lines must cross in the same direction. (See Fig. 11.40.)
‡Overlap distance on crossover.
§Conductor lines overlap.

Figure 11.41   Dielectric distance on crossovers.

2. *Failures due to chemical reactions.*   This includes failures such as corrosion or the formation of intermetallic compounds, which can be manifested as a mechanical failure.

3. *Failures due to inherent manufacturing defects in active devices.*   This includes defects due to pinholes in the insulating oxide, defects or impurities in the body of the semiconductor, or mask defects.

4. *Failures due to electrical overstress.*   Failures in this category can be created either by overstress during operation or test, or due to exposure to electrostatic discharge (ESD).

This section describes the methods by which reliability can be predicted, methods of accelerating the failure rate (screening), and methods of designing for reliability.

11.13.2   Failure rate prediction

Many of the chemical reactions involved in the processes used to fabricate hybrid microcircuits continue to occur throughout the life of the circuit, and

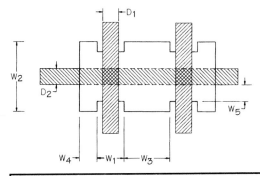

| Dimension | Thick film, mils | | Remarks |
	Preferred	Minimum	
D_1	10	5	(1)
D_2	10	5	(2)
W_1	$D_1 + 10$	$D_1 + 10$	(3)
W_2	$D_1 + 10$	$D_1 + 10$	(4)
W_3	10	5	(5)
W_4	15	10	(6)
W_5	10	5	(7)

1. Top conductor widths.
2. Bottom crossover width.
3. Width of crossover.
4. Overlap of bottom conductor.
5. Width of dielectric print between crossovers.
6. Width of crossover arm.
7. Extension of crossover past conductor.

Figure 11.42 H pattern crossovers.

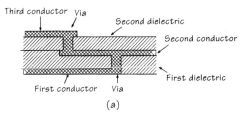

(a)

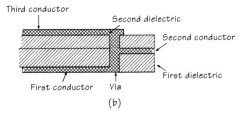

(b)

Figure 11.43 Via location on adjacent layers. (a) Permitted; (b) not permitted. The only exception to this rule is when vias are used as heat sinks under components.

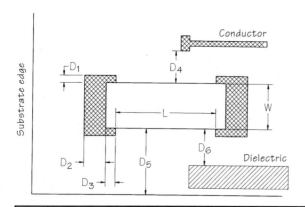

Figure 11.44   Thick film resistor dimensions.

| | Thick film, mils | | |
Dimension	Preferred	Minimum	Remarks
L	40	20	(1)
W	See Table 11.4		(2)
D_1	10	5	(3)
D_2	10	5	(4)
D_3	10	7.5	(5)
D_4	20	15	(6)
D_5	30	20	(7)
D_6	20	20	(8)

1. Aspect ratio $0.5 \leq L/W \leq 5$ preferred, $0.3 \leq L/W \leq 10$ maximum.
2. Depends on the more restrictive of tolerance and power.
3. Minimum excess conductor width.
4. Minimum excess conductor length.
5. Resistor overlap onto conductor.
6. Conductor distance from resistance.
7. Distance from edge of substrate.
8. Distance to multilayer or crossover dielectric.

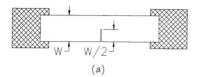

(a)

(b)

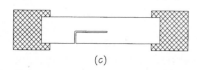

(c)

Figure 11.45 Laser trim cut modes. Do not penetrate more than halfway into the resistor. When trimming to less than 50% tolerance, use the double-plunge-cut or L-cut mode. Trim to -5% or until $W/2$ is reached with a single-plunge cut, then switch to double-plunge or L cut.

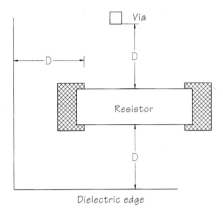

Dielectric edge

Dimension	Thick film, mils		Remarks
	Preferred	Minimum	
D	20	20	*, †, ‡

*Distance to edge of dielectric or via when resistors are printed on top of dielectric.
†The same dimensions for resistors in Fig. 11.43 also hold.
‡The use of resistors on dielectric requires an alternative set of resistor curves.

Figure 11.46 Dimensions of resistors on dielectric.

TABLE 11.20   Comparison of Metallic and Organic Attachment

Metallic	Organic
Better electrical and thermal contact	More pliable
Stronger mechanically	No cleaning
May require cleaning	Lower temperature process
Higher temperature process	Possible smearing*
Higher surface tension*	

*In the molten state, metallic attachment materials tend to wick back onto the mounting pads. This tends to center the components on the pads, a highly desirable feature, and to minimize the tendency for short circuits between pads, which permits the use of finer pitch devices. Organic materials, on the other hand, have a low surface tension, and components tend to remain exactly where they are placed. This also increases the probability of short circuits due to smearing, since organic materials will not wick onto the pads during curing.

will eventually lead to failure if the circuit operates long enough. One example of this is the gold–aluminum intermetallic compounds that form when a gold wire is thermosonically bonded to an aluminum pad on an active device. The aluminum continues to diffuse into the gold throughout the life of the device and will eventually cause failure. However, under normal operating conditions, the diffusion rate is so low that the circuit will probably become obsolete before failure occurs! The thermodynamic theory of chemical reactions states that many reactions that are dependent on time, current, and

TABLE 11.21 Processing Temperatures of Selected Organic and Metallic Attachment Materials

Organic Attachment Materials	
Material	Temperature, °C
Polyimide	250–350°C*
Epoxy	150°C

Metallic Attachment Materials		
Alloy†	Liquidus, °C§	Solidus, °C
In 52–Sn 48‡	118	118
Sn 62.5–Pb 36.1–Ag 1.4‡	179	179
Sn 63–Pb 37‡	183	183
In 60–Pb 40	185	174
Sn 60–Pb 40	188	183
Sn 96.5–Ag 3.5‡	221	221
Pb 60–Sn 40	238	183
Pb 70–Sn 27–Ag 3	253	179
Pb 92.5–Sn 5–Ag 2.5	280	179
Sn 90–Ag 10	295	221
Pb 90–Sn 10	302	275
Au 88–Ge 12‡	356	356
Au 96.4–Si 3.6‡	370	370
Ag 72–Cu 28‡	780	780

*Polyimide materials may also require a precure step at 70°C to remove solvents.
†Numerical values are percentages. In = indium, Sn = tin, Pb = lead, Ag = silver, Au = gold, Ge = germanium, Si = silicon, Cu = copper.
‡Eutectic composition.
§The processing temperature of most alloys is ≥20°C above the liquidus.

TABLE 11.22 Process Considerations

Criterion	Comments
Thermal management	A device which dissipates a great deal of power will require soldering to help transfer the heat to the outside. The thermal conductivity of organic materials is too low to allow effective thermal management.
Temperature sensitivity	High temperatures are detrimental to the reliability of certain components and metallic alloys (Al–Au compounds, for example).
Interference with previous processes	For illustration, it is most economical to apply epoxy for component mounting by screen printing. If a device has been previously mounted, however, this will interfere with the screen printing process and necessitate applying the epoxy by other means.

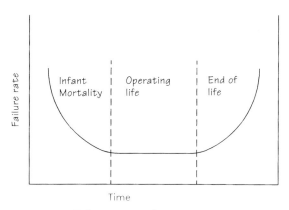

Figure 11.47 Failure rate vs. time.

temperature follow the relationship given in Eq. (11.88), the so-called Arrhenius equation:

$$R(i,t,T) = R(i)\,R(t)\,e^{-E_a/kT} \tag{11.88}$$

where $R(i,t,T)$ = reaction rate of the chemical reaction
$\qquad\quad i$ = electrical current, A
$\qquad\quad t$ = time, s
$\qquad\quad T$ = temperature, K
$\qquad\quad k$ = Boltzmann's constant = 8.625×10^{-5} eV/K
$\qquad\quad E_a$ = activation energy, eV

Some reactions, particularly those involving corrosion and/or migration, are also accelerated by the presence of electric current. The electric current can transport active ions to the vicinity of other oppositely charged ions, causing a reaction to occur.

The rate of the reaction with respect to time depends on the amount of material present and its ease of movement. In a liquid, which can be kept constantly agitated and where there is a vast amount of material present, the reaction rate with respect to time is essentially constant. In a solid where a limited amount of material is present and where agitation is impossible, the rate will decrease with respect to time. A rather simplistic explanation for these phenomena is that the probability of a reaction occurring between molecules is proportional to the probability that the molecules will be in close enough proximity for the reaction to occur. For a fixed amount of material, as a reaction progresses, there is less material left to react and, therefore, the probability of the molecules being in proximity decreases. Analogously, at an increased temperature, the molecules are more active and are more likely to come in proximity to other molecules and react with them.

For most cases, the dependence on temperature is substantially greater than either the dependence on current or time, and Eq. (11.88) may be reduced to the following relationship:

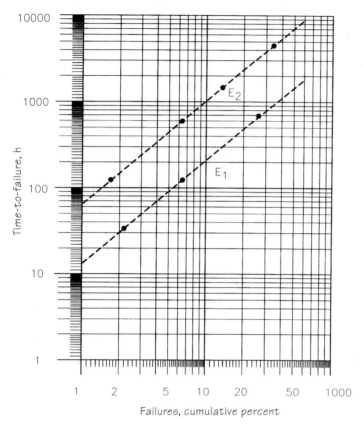

Figure 11.48 Cumulative percent failures vs. time to failure for two different activation energies. $(E_2 > E_1.)$[12]

$$R(T) = R_0\, e^{-E_a/kT} \qquad (11.89)$$

where R_0 = a constant, which depends on a number of variables, including the amount of material available and the physical configuration. R_0 is a difficult parameter to define analytically for a number of practical cases, and Eq. (11.89) is generally used to extrapolate data points obtained for a specific situation at a specific temperature to another temperature.

The temperature in Eq. (11.89) is the temperature at the point of reaction. In a hybrid circuit, the temperature may vary widely from one point to another during normal operation, and a thermal analysis must be performed to determine the temperature at a given point.

If the cumulative percentage of failures is plotted against time on a log–log scale, most failure mechanisms common to hybrid circuits plot as a straight line as shown in Fig. 11.48. This is indicative of the so-called normal, or bell-shaped, distribution of failures. An S-shaped curve indicates that the failure rate is initially higher and that the sample is prone to infant mortality. If a plot is made at a second temperature, and if the slope is the same, the

Arrhenius equation, Eq. (11.88), applies and the following relationships also apply.

The acceleration equation compares the rate of reaction at different temperatures:

$$\frac{r_1}{r_2} = e^{E_a\,(1/T_2\,-\,1/T_1)/k} \qquad (11.90)$$

Once the acceleration rate is known, the activation energy may be calculated.

$$E_a = k \ln \frac{r_1/r_2}{1/T_2 - 1/T_1} \qquad (11.91)$$

The most detailed and comprehensive information on activation energies and failure rates of active devices is found in MIL-HDBK-217. Failure rates for specific active devices at different temperatures can be obtained in Secs. 5.1.2 and 5.1.3 of this handbook.

If the time-to-failure t_1 is known for a temperature T_1, the time-to-failure t_2 for a temperature T_2 is given by

$$t_2 = t_1\,e^{-E_a(1/T_2\,-\,1/T_1)/k} \qquad (11.92)$$

The same basic form applies for the failure rate.

$$F_2 = F_1 e^{-E_a(1/T_2\,-\,1/T_1)/k} \qquad (11.93)$$

where F is the failure rate at a given temperature.

Typical values of the activation energy for common failure modes of semiconductor devices in hybrid circuits are shown in Table 11.23.[11]

Failure modes with activation energies in the range of 0.30 eV are characteristic for failures of the infant mortality type and are only slightly affected by temperature, while those in the 1.50-eV range are characteristic of end-of-life phenomena and are greatly accelerated by temperature. As a rule of thumb, the lower the activation energy, the higher will be the failure rate. If the Arrhenius equation applies, a plot of time to failure on a log scale against temperature on a reciprocal scale is a straight line, as shown in Fig. 11.49. From observation of Table 11.23, it is apparent that the range of activation energies is quite broad. This factor makes it difficult to extrapolate Eq. (11.93) accurately across a range of temperatures. Referring to Fig. 11.49, each of the three different failure mechanisms dominates at a given temperature. To accurately predict the failure rate at a different temperature, the nature of the failures must be known. Each failure mode can then be individually predicted and the net failure rate is the sum of the individual failure rates.

For example, assume that a sample of 500 active devices is subjected to a powered life test for 1000 h at 125°C and that 32 fail. The net failure rate is, therefore, 6.4%/1000 h.

TABLE 11.23   Activation Energies for Hybrid Circuit Failure Modes[12]

Device type or test condition	Activation energy, eV
Transistors	
Germanium	0.88–1.24
Silicon	
Surface inversion failures	1.02
Au–Al bond failures	1.02–1.04
Metal penetration into silicon	1.65
$pnpn$	1.65
Diodes	
Four-layer	1.41
Varactors	2.31–2.38
Others	1.13–2.77
Integrated circuits	
Oxide defects	0.30
Silicon defects	0.30
Mask defects	0.50
Ball bond lifts*	0.35–0.44
Ball Bond lifts*	0.80–1.10
Electromigration	1.00–1.10
Contamination	1.00–1.40
Surface charge	0.50–1.00
Charge injection	1.30
Electrolytic corrosion	0.30–0.70
MIL-STD-883	
Method 1015.3 burn-in	0.44
Method 1008.1 high temperature storage	1.00
Method 1005.3 steady-state life	1.00

*Dependent on cause of failure mode.

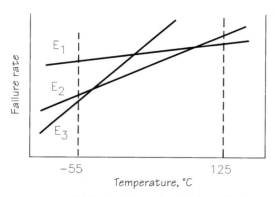

Figure 11.49   Failure rate vs. temperature for three different activation energies. ($E_3 > E_2 > E_1$.)

Referring to Table 11.23, the activation energy assumed in Method 1005.3 of MIL-STD-883 is 1.0 eV. This value may be used in Eq. (11.93) to estimate the failure rate at 25°C.

$$F_2 = 6.4\% \; e^{-1.0(1/298 \; - \; 1/398)/(8.625 \times 10^{-3})}$$

$$= 0.000364\%/1000 \text{ h at } 25°C$$

Further assume that 20 of the failures were caused by oxide defects, 8 by contamination (assume $E_a = 1.00$ eV for this failure), and 4 by wire bond lifts (assume $E_a = 0.80$ eV for this failure). In this case, the failure rates for each of the failure modes are:

$$\text{Oxide defects} \quad = 20/500 \; = 4.0\%$$

$$\text{Contamination} = 8/500 \quad = 1.6\%$$

$$\text{Wirebonds} \quad\;\; = 4/500 \quad = 0.8\%$$

The failure rate for oxide defects at 25°C is

$$F_2 = 4.0\% \; e^{-0.30(1/298 \; - \; 1/398)/(8.625 \times 10^{-3})}$$

$$= 0.21\%/1000 \text{ h}$$

The failure rate for contamination at 25°C is

$$F_2 = 1.6\% \; e^{-1.0(1/298 \; - \; 1/398)/(8.625 \times 10^3)}$$

$$= 0.00009\%/1000 \text{ h}$$

The failure rate for wire bonds at 25° is

$$F_2 = 0.8\% \; e^{-0.8(1/298 \; - \; 1/398)/(8.625 \times 10^{-3})}$$

$$= 0.000321\%/1000 \text{ h}$$

The net failure rate at 25°C is

$$0.21 + 0.00009 + 0.000321 = 0.210411\%/1000 \text{ h}$$

This failure rate is substantially higher than that predicted by MIL-STD-883! This is due to the preponderance of failures from oxide defects that have a low activation energy and that dominate at lower temperatures.

A further consideration is the position of the devices on the failure rate vs. time curve shown in Fig. 11.48. A larger sample run for a shorter time may accumulate the same number of device hours as a smaller sample run for a longer time, but will show a substantially higher failure rate since it appears on the infant mortality portion of the curve.

In summary, to obtain accurate failure rates of active devices, it is necessary to obtain actual failure mode data at the actual time of occurrence.

Given the extremely high reliability of most semiconductor devices made today, this can be a lengthy and expensive proposition. To accomplish this task, it is necessary to operate either a small number of devices for a very long time or a large number of devices for a shorter time. Translation of failure rate data from one temperature to another must take the activation energy of the actual failure mode into account to avoid serious errors. Failure rates, to be usable, must be expressed in terms of infant mortality, normal operation, and wearout.

The failure rate of film resistors is difficult to define and is highly application and process dependent. In one application, a drift of 0.1% away from the initial value may result in a failure of the circuit to operate, while in another, a drift of 50% may not be discernible. In addition, a change in furnace profile or in the laser trim parameters can influence the drift rate. The failure rates of other components can also be found in MIL-HDBK-217. Discrete semiconductors are covered in Sec. 5.1.3, packaged resistors in Sec. 5.1.6, and capacitors in Sec. 5.1.7 of the handbook.

11.13.3   Failure rate prediction of hybrid microcircuits

The failure rates of hybrid microcircuits may be predicted by the use of MIL-HDBK-217, which contains the established failure rates of a number of components and instructions on how to use them. It should be noted, however, that the failure rate of relatively new devices is difficult to establish and must frequently be estimated.

The hybrid failure rate model as described in MIL-HDBK-217 is given by the following equation:

$$\lambda_p = [\Sigma N_C \lambda_C \pi_G + (N_R \lambda_R + \Sigma N_I \lambda_I + \lambda_S) \pi_F \pi_E] \pi_Q \pi_D \qquad (11.94)$$

where
λ_p = failure rate of the hybrid circuit in failures/1,000,000 h

$\Sigma N_C \lambda_C P_G$ = sum of the adjusted failure rates for the active components, packaged resistors, and capacitors in the hybrid circuit (see Tables 11.29 to 11.36 for active devices and Tables 11.37 and 11.38 for capacitors)

N_C = number of each particular component in the circuit

λ_C = component failure rate

π_G = die correction factor (see Table 11.24).

$N_R \lambda_R$ = number and failure rate contributions of the chip or substrate resistors (see Table 11.25)

$\Sigma N_I \lambda_I$ = sum of the failure rate contribution λ_I of each interconnection (see Table 11.26).

λ_S = failure rate contribution of the hybrid package [see Eq. (11.95)]

π_E = environmental factor for the film resistors (see Table 11.27)

π_Q = quality factor (see Table 11.28)

π_D = density factor (see Table 11.29)

TABLE 11.24   Die and Capacitor Correction Factors

Component	π_G
Integrated circuits	1.0
Transistors	0.4
Diodes	0.2
Capacitor chips	0.8

TABLE 11.25   Failure Rates for Chip or Substrate Resistors

Failure rate	Package temperature T
0.00010	$T \leq 50°C$
0.00015	$50°C \leq T \leq 80°C$
0.00020	$80°C \leq T \leq 100°C$
0.00025	$100°C \leq T \leq 125°C$
0.0003	$125°C \leq T \leq 150°C$

π_F = circuit function factor (1.0 for digital hybrids; 1.25 for linear or linear–digital combinations)

$$\lambda_S = 0.11S\,(1 - e^{S^2/50})(e^{-5203(1/T + 273) - 1/298}) \tag{11.95}$$

where T = package temperature, °C, and S = seal perimeter, in.

11.13.4   Failure rate prediction models for active devices

The failure rates for active devices depend largely on the complexity of the device. For digital circuits, this can be quantified by the number of logic gates present. Digital devices with less than 100 gates are considered to be small scale integration (SSI) or medium seal integration (MSI). As an approximation, transistors and diodes may be considered a special case of integrated circuits with the number of gates or transistors as appropriate equal to one. Devices with more than 100 gates are considered to be large scale integration (LSI).

The device categories are

1. Monolithic bipolar and metal–oxide–semiconductor (MOS) digital

2. Monolithic bipolar and MOS linear

3. Monolithic bipolar and MOS random logic LSI and microprocessor

4. Monolithic bipolar and MOS memories

The failure rate in failures/1,000,000 h for all categories except read-only memory (ROM) is given by:

$$\lambda_P = \pi_Q[C_1\pi_T\pi_V + (C_2 + C_3)\pi_E]\pi_L \tag{11.96}$$

TABLE 11.26   Failure Rates for Interconnections

Temperature, °C	F_{11}	F_{12}
25	0.000174	0.000174
30	0.000230	0.000218
35	0.000302	0.000271
40	0.000394	0.000334
45	0.000508	0.000410
50	0.000650	0.000499
55	0.000826	0.000604
60	0.00104	0.000727
65	0.00130	0.000871
70	0.00162	0.00103
75	0.00201	0.00123
80	0.00247	0.00145
85	0.00302	0.00170
90	0.00367	0.00199
95	0.00444	0.00231
100	0.00534	0.00268
105	0.00639	0.00310
110	0.00762	0.00356
115	0.00904	0.00409
120	0.0106	0.00467
125	0.0125	0.00531
130	0.0147	0.00603
135	0.0171	0.00682
140	0.0199	0.00770
145	0.0231	0.00866
150	0.0266	0.00971

1. The indicated temperature is the package temperature.
2. F_{11} is for bimetal bonds (gold–aluminum).
3. F_{12} is for single metal bonds (aluminum–aluminum or gold–gold).
4. One wire, one solder joint, or one epoxy bond counts as one interconnection.
5. A bond is considered bimetallic if any one of the bond interfaces involves more than one type of metal.
6. Active die attach bonds (die to substrate bonds) are not counted as interconnections.
7. Redundant interconnections shall be counted as only one interconnection.
8. Only active current-carrying interconnections are counted.
9. Approximations of the number of interconnections can be made as follows:

Component	Number of interconnections
Each IC chip bonding pad	1
Each transistor	2
Each diode	1
Each capacitor	2
Each external lead	1
Each chip resistor	2

For ROM, the expression is

$$\lambda_P = \pi_Q[C_1\pi_T\pi_V\pi_{PT} + (C_2 + C_3)\pi_E]\pi_L \qquad (11.97)$$

where λ_P = device failure rate, failures/1,000,000 h
π_Q = quality factor (see Table 11.29)

TABLE 11.27   Environmental Factor for Resistors, Packages, and Interconnections

Environment*	π_E
G_B	0.20
G_F	0.78
G_M	2.2
M_P	2.0
N_{SB}	0.99
N_S	1.7
N_U	3.2
N_H	3.1
N_{UU}	3.4
A_{RW}	4.5
A_{IC}	1.5
A_{IT}	1.5
A_{IB}	2.5
A_{IA}	2.0
A_{IF}	3.0
A_{UC}	2.5
A_{UT}	2.5
A_{UB}	4.0
A_{UA}	3.0
A_{UF}	4.0
S_F	0.32
M_{FF}	2.1
M_{FA}	2.9
U_{SL}	6.1
M_L	7.0
C_L	120.0

*See Table 11.41 for definitions of abbreviations.

C_1, C_2 = circuit complexity failure rates based on gate count (see Table 11.30)

π_T = temperature acceleration factor (see Table 11.31)

π_V = voltage derating stress factor (see Table 11.32)

C_3 = package complexity factor (see Table 11.33)

π_E = application environment factor (see Table 11.34)

π_L = device learning factor (see Table 11.35)

π_{PT} = programming technique factor (see Table 11.36)

$$\pi_T = 0.1 \, e^x \tag{11.98}$$

$$x = -A\left(\frac{1}{T_J + 273} - \frac{1}{298}\right) \tag{11.99}$$

where T_J = worst case junction temperature, °C, and A = value from Table 11.31.

TABLE 11.28 Quality Factor π_Q

Quality level	Description	π_Q
S	Procured in full accordance with MIL-M-38510E, Class S requirements	0.5
B	Procured in full accordance with MIL-M-38510 Class B requirements	1.0
B-0	Procured in full accordance with MIL-M-38510, Class B requirements, except that device is not listed on QPL.	2.0
B-1	Procured to all the screening requirements of MIL-STD-883, Method 5004, Class B, and in accordance with electrical requirements of MIL-M-38510, DESC drawings, or vendor/contractor electrical requirements. The device shall be tested to all the quality conformance requirements of MIL-STD-883, Method 5005, Class B.	3.0
B-2	Procured to vendor's equivalent of the vendor's screening requirements of MIL-STD-883, Method 5004, Class B.	6.5
C	Procured in full accordance with MIL-M-38510 Class C requirements.	8.0
C-1	Procured to screening requirements of MIL-STD-883, Method 5004, Class C, and the qualification requirements of Method 5005, Class C.	13.0
D	Hermetically sealed part with no screening beyond the manufacturer's quality assurance practices.	17.5
D-1	Commercial part, encapsulated or sealed with organic materials	35.0

TABLE 11.29 Density Factor π_D

Density	π_D
20	0.87
40	1.15
60	1.36
80	1.54
100	1.70
120	1.84
140	1.97
160	2.10
180	2.21
200	2.32
220	2.42
240	2.52
260	2.62
280	2.71
300	2.80

$$\text{Density} = \frac{\text{number of interconnections}}{A_S + 0.10}$$

A_S = area of substrate, in^2

$\pi_D = 0.2 + 0.15 \sqrt{\text{density}}$

TABLE 11.30   Device Complexity Factors

Technology	Factor
Bipolar SSI/MSI	$C_1 = 7.48 \times 10^{-4} N_G^{0.654}$ $C_2 = 2.19 \times 10^{-4} N_G^{0.364}$
MOS SSI/MSI	$C_1 = 2.17 \times 10^{-3} N_G^{0.357}$ $C_2 = 3.11 \times 10^{-4} N_G^{0.178}$
Linear	$C_1 = 1.57 \times 10^{-3} N_T^{0.780}$ $C_2 = 8.00 \times 10^{-4} N_T^{0.535}$
Bipolar LSI	$C_1 = 1.48 \times 10^{-3} N_G^{0.506}$ $C_2 = 3.20 \times 10^{-4} N_G^{0.279}$
MOS LSI	$C_1 = 1.75 \times 10^{-3} N_G^{0.400}$ $C_2 = 2.52 \times 10^{-4} N_G^{0.226}$
Bipolar RAM*	$C_1 = 2.20 \times 10^{-3} N^{0.576}$ $C_2 = 4.00 \times 10^{-5} N^{0.364}$
MOS RAM†	$C_1 = 5.00 \times 10^{-4} B^{0.610}$ $C_2 = 3.00 \times 10^{-5} B^{0.585}$
Bipolar ROM, programmable ROM (PROM)†	$C_1 = 8.80 \times 10^{-4} B^{0.388}$ $C_2 = 4.50 \times 10^{-5} B^{0.378}$
MOS ROM, PROM†	$C_1 = 1.20 \times 10^{-3} N^{0.425}$ $C_2 = 6.60 \times 10^{-5} N^{0.399}$

N_G = number of gates $\leq$ 20,000.
N_T = number of transistors.
*B = number of bits $\leq$ 16,384.
†B = number of bits $\leq$ 65,536.

TABLE 11.31   Temperature Factors

Technology	Package type	A
TTL, HTTL, DCL, ECL	Hermetic	4,635
	Nonhermetic	5,214
LTTL, STTL	Hermetic	5,214
	Nonhermetic	5,794
LSTTL	Hermetic	5,794
	Nonhermetic	6,373
ITL, MNOS	Hermetic	6,952
	Nonhermetic	9,270
PMOS	Hermetic	5,794
	Nonhermetic	8,111
NMOS, CCD	Hermetic	6,373
	Nonhermetic	9,270
CMOS, CMOS/SOS, LINEAR	Hermetic	7,532
	Nonhermetic	10,429

TABLE 11.32   Voltage Derating Stress Factor π_V

Technology	π_V
CMOS, $V_{DD} = 5$ V	1.0
CMOS, 12 V $\leq V_{DD} \leq 15.5$ V	Eq. (11.100)
CMOS, 18 V $\leq V_{DD} \leq 20$ V	Eq. (11.102)
All other technologies	1.0

V_{DD} is the supply voltage.

TABLE 11.33   Package Complexity Factor

Package type	C_3
Hermetic DIPs with solder or weld seal	$2.8 \times 10^{-4} N_P^{1.08}$
Hermetic DIPs with glass seal	$9.0 \times 10^{-5} N_P^{1.51}$
Nonhermetic DIPs	$2.0 \times 10^{-4} N_P^{1.23}$
Hermetic flat packs	$3.0 \times 10^{-4} N_P^{1.82}$
Hermetic cans	$3.0 \times 10^{-5} N_P^{2.01}$

N_P = number of pins on device package.

TABLE 11.34   Environmental Factors

Environment	P_E
G_B	0.38
G_F	2.50
G_M	4.2
M_P	3.8
N_{BB}	4.0
N_B	4.0
N_U	5.7
N_H	5.9
N_{UU}	6.3
A_{RW}	8.5
A_{IC}	2.5
A_{IT}	3.0
A_{IB}	5.0
A_{IA}	4.0
A_{IF}	6.0
A_{UC}	3.0
A_{UT}	4.0
A_{UB}	7.5
A_{UA}	6.0
A_{UF}	9.0
S_F	0.90
M_{FF}	3.9
M_{FA}	5.4
U_{BL}	11.0
M_L	13.0
C_L	220

Note: For environmental codes, see Table 11.41.

TABLE 11.35   Learning Curve Factors

P_L = 10 under any of the following conditions:

1. New device in initial production

2. Where major changes in design or process have occurred

3. Where there has been an extended interruption or a change in line personnel, such as a radical expansion

4. For all new and unproven technologies

The factor of 10 can be expected to apply until conditions and controls have stabilized. This period extends for 4 months of continuous production.

P_L = 1.0 under all production conditions not stated above

TABLE 11.36   Programming Technique Factors

Device type	Technology	Programming	p_{PT}
ROM	Bipolar	Mask	1.0
	MOS	Mask	1.0
PROM	Bipolar	Nickel–chromium or titanium– tungsten links	*
		Polysilicon links	*
		Shorted junction (AIM)	*
	MOS	Ultraviolet and electrically erasable	†

$*p_{PT} = 0.985 + 9.5 \times 10^{-5}\,B$
$\dagger p_{PT} = 0.950 + 7.5 \times 10^{-5}\,B$
where B = number of bits.

For supply voltages between 12 and 15.5 V,

$$\pi_V = 0.110e^x \tag{11.100}$$

where

$$X = \frac{0.168V_S(T_J + 273)}{298} \tag{11.101}$$

For supply voltages between 18 and 20 V,

$$\pi_V = 0.068e^x \tag{11.102}$$

where

$$x = \frac{0.135V_S(T_J + 273)}{298} \tag{11.103}$$

and V_S = actual supply voltage and T_J = worst case junction temperature.

11.13.5 Sample reliability calculations for a hybrid microcircuit

A work sheet summarizing the information required is given in Table 11.39, with the appropriate information provided for this example. From Eq. (11.94), the hybrid circuit failure rate is

$$\lambda_p = [\Sigma N_C \lambda_C \pi_G + (N_R \lambda_R + \Sigma N_I \lambda_I + \lambda_S) \pi_F \pi_E] \pi_Q \pi_D \qquad (11.104)$$

11.13.5.1 Component failure rates

Integrated circuits

Digital integrated circuits.   The failure rate for integrated circuits is given by Eq. (11.96).

$$\lambda_p = \pi_Q [C_1 \pi_T \pi_V + (C_2 + C_3) \pi_E] \pi_L \qquad (11.105)$$

From Table 11.29, $\pi_Q = 1.0$. From Table 11.30,

$$C_1 = (7.48 \times 10^{-4})(180)^{0.654} = 2.23 \times 10^{-2}$$

$$C_2 = (2.19 \times 10^{-4})(180)^{0.364} = 1.45 \times 10^{-3}$$

From Table 11.31, π_T is calculated as follows:

$$A = 4635$$

$$x = -4635\left(\frac{1}{90 + 273} - \frac{1}{298}\right)$$

$$= 2.79$$

$$\pi_T = 0.1e^{2.79} = 1.62$$

From Table 11.32, $\pi_V = 1.0$. From Table 11.33, the package may be considered a hermetic dual in-line package (DIP) with 26 pins, and C_3, the package derating factor, is

$$C_3 = (2.80 \times 10^{-4})(26)^{1.08} = 9.45 \times 10^{-3}$$

From Table 11.34, the environmental factor $\pi_E(A_{IB}) = 5.0$. From Table 11.35, the learning factor $\pi_L = 1.0$.

The net failure rate of each digital integrated circuit is

$$\lambda_p = 1.0 [2.23 \times 10^{-2} \times 1.62 \times 1.0 + (1.45 \times 10^{-3} + 9.45 \times 1 0^{-3})5.0] 1.0$$

$$= 9.06 \times 10^{-2} \text{ failures}/1{,}000{,}000 \text{ h}$$

From Table 11.24, the die correction factor is 1.0.

Analog integrated circuits.   From Table 11.29, $\pi_Q = 1.0$. From Table 11.30,

$$C_1 = (1.57 \times 10^{-3})(120)^{0.780} = 6.57 \times 10^{-2}$$

$$C_2 = (8.00 \times 10^{-4})(120)^{0.535} = 1.04 \times 10^{-2}$$

From Table 11.31 π_T is calculated as follows:

$$A = 7532$$

$$x = -7532 \left(\frac{1}{120 + 273} - \frac{1}{298} \right)$$

$$= 6.11$$

$$\pi_T = 0.1 \, e^{6.11} = 45.0$$

From Table 11.32, $\pi_V = 1.0$. From Table 11.33, the package may be considered a hermetic DIP with 26 pins and C_3, the package derating factor, is

$$C_3 = (2.80 \times 10^{-4})(26)^{1.08} = 9.45 \times 10^{-3}$$

From Table 11.34, the environmental factor is $\pi_E(A_{IB}) = 5.0$. From Table 11.35, the learning factor $\pi_L = 1.0$.

The net failure rate of each analog integrated circuit is

$$\lambda_P = 1.0 \, [6.57 \times 10^{-2} \times 45.0 \times 1.0 + (1.04 \times 10^{-3} + 9.45 \times 10^{-3})5.0] \, 1.0$$

$$= 3.05 \text{ failures/1,000,000 h}$$

From Table 11.24, the die correction factor is 1.0.

Passive components

Capacitors. From Table 11.37, the failure rate of the capacitors is

$$\lambda_C = 0.20 \text{ failures/1,000,000 h}$$

assuming that the capacitor temperature is the same as the package temperature. (Note that Table 11.38 gives failure rates for a higher temperature.) From Table 11.24, the die correction factor for the capacitors is 0.8.

Resistors. From Table 11.25, the failure rate of the thick film resistors is

$$\lambda_R = 0.00015 \text{ failures/1,000,000 h}$$

and the total contribution is

$$N_R \lambda_R = 22 \times 0.00015 = 3.3 \times 10^{-3} \text{ failures/1,000,000 h}$$

The net sum of the failure rates of the integrated circuits and capacitors is

$$\Sigma N_C \lambda_C \pi_B = (10)(9.06 \times 10^{-2})(1.0) + (6)(3.05)(1.0) + (4.0)(0.20)(0.8)$$

$$= 19.85 \text{ failures/1,000,000 h}$$

Interconnections. For the digital integrated circuits, there are 180 wire bonds, or 360 interconnections, with 180 of these bonded to the aluminum contacts on the devices and 180 bonded to the thick film gold. The bonds on the integrated

TABLE 11.37   Failure Rates for Chip Capacitors Rated at 125°C

Temperature, °C	S = ratio of operating to rated voltage									
	0.1	0.2	0.3	0.4	0.5	0.6	0.7	0.8	0.9	1.0
0	0.00062	0.00077	0.0012	0.0020	0.0034	0.0054	0.0 082	0.012	0.017	0.023
5	0.00063	0.00078	0.0012	0.0020	0.0034	0.0054	0.0 083	0.012	0.017	0.023
10	0.00063	0.00079	0.0012	0.0021	0.0034	0.0055	0. 0084	0.012	0.017	0.023
15	0.00064	0.00080	0.0012	0.0021	0.0035	0.0056	0. 0085	0.012	0.017	0.024
20	0.00065	0.00081	0.0013	0.0021	0.0035	0.0056	0. 0086	0.013	0.018	0.024
25	0.00066	0.00082	0.0013	0.0021	0.0036	0.0057	0. 0087	0.013	0.018	0.024
30	0.00067	0.00083	0.0013	0.0022	0.0036	0.0058	0. 0088	0.013	0.018	0.024
35	0.00067	0.00084	0.0013	0.0022	0.0037	0.0059	0. 0089	0.013	0.018	0.025
40	0.00068	0.00085	0.0013	0.0022	0.0037	0.0059	0. 0090	0.013	0.018	0.025
45	0.00069	0.00086	0.0013	0.0022	0.0038	0.0060	0. 0091	0.013	0.019	0.025
50	0.00070	0.00088	0.0014	0.0023	0.0038	0.0061	0. 0093	0.013	0.019	0.026
55	0.00071	0.00089	0.0014	0.0023	0.0039	0.0062	0. 0094	0.014	0.019	0.026
60	0.00072	0.00090	0.0014	0.0023	0.0039	0.0062	0. 0095	0.014	0.019	0.026
65	0.00073	0.00091	0.0014	0.0024	0.0039	0.0063	0. 0096	0.014	0.020	0.027
70	0.00074	0.00092	0.0014	0.0024	0.0040	0.0064	0. 0097	0.014	0.020	0.027
75	0.00075	0.00093	0.0014	0.0024	0.0040	0.0065	0. 0099	0.014	0.020	0.027
80	0.00076	0.00094	0.0015	0.0025	0.0041	0.0066	0. 010	0.015	0.020	0.028
85	0.00076	0.00096	0.0015	0.0025	0.0042	0.0066	0. 010	0.015	0.021	0.028
90	0.00077	0.00097	0.0015	0.0025	0.0042	0.0067	0. 010	0.015	0.021	0.028
95	0.00078	0.00098	0.0015	0.0025	0.0043	0.0068	0. 010	0.015	0.021	0.029
100	0.00079	0.00099	0.0015	0.0026	0.0043	0.0069	0 .010	0.015	0.021	0.029
105	0.00080	0.0010	0.0016	0.0026	0.0044	0.0070	0. 011	0.015	0.022	0.029
110	0.00081	0.0010	0.0016	0.0026	0.0044	0.0071	0. 011	0.016	0.022	0.030
115	0.00082	0.0010	0.0016	0.0027	0.0045	0.0072	0. 011	0.016	0.022	0.030
120	0.00084	0.0010	0.0016	0.0027	0.0045	0.0072	0. 011	0.016	0.023	0.031
125	0.00085	0.0011	0.0016	0.0027	0.0046	0.0073	0. 011	0.016	0.023	0.031

TABLE 11.38   Failure Rates for Chip Capacitors Rated at 150°C

Temperature, °C	S, ratio of operating to rated voltage				
	0.2	0.4	0.6	0.8	1.0
0	0.00074	0.0019	0.0051	0.011	0.022
25	0.00079	0.0020	0.0055	0.012	0.023
50	0.00083	0.0022	0.0058	0.013	0.024
75	0.00089	0.0023	0.0061	0.014	0.026
100	0.00094	0.0024	0.0065	0.014	0.028
125	0.0010	0.0026	0.0069	0.015	0.029
150	0.0011	0.0027	0.0073	0.016	0.031

circuit are at a temperature of 90°C, and those on the substrate are assumed to be at the package temperature of 70°C. From Table 11.26,

$$\Sigma \, N_I \lambda_I \, (\text{digital}) = 180 \, (0.00103) + 180 \, (0.00367)$$

$$= 0.846 \text{ failures}/1{,}000{,}000 \text{ h}$$

For the analog integrated circuits, there are 104 gold wire bonds, or 208 interconnections, with 104 of these bonded to the aluminum contacts on the devices and 104 bonded to the thick film gold. The bonds on the integrated circuits are at 120°C and those on the substrate are assumed to be at the package temperature, 70°C. From Table 11.26,

$$\Sigma\, N_I\lambda_I \,(\text{analog})\, = 104\,(0.00103) + 104\,(0.0106)$$

$$= 1.21\ \text{failures}/1{,}000{,}000\ \text{h}$$

For the package pinouts, there are 26 wires, or 52 interconnections, which are connected from the gold thick film on the substrate to the gold plating on the package pins. The temperature of all interconnections may be assumed to be at the package temperature of 70°C.

$$\Sigma\, N_I\lambda_I \,(\text{package}) = 52\,(0.00103)$$

$$= 0.0536\ \text{failures}/1{,}000{,}000\ \text{h}$$

For the capacitors, there are 8 interconnections of unlike metal (silver to gold), and they may be assumed to be at the package temperature of 70°C. The net failure rate for the capacitor interconnections is

$$\Sigma\, N_I\lambda_I \,(\text{capacitors}) = 8\,(0.00103)$$

$$= 0.00824\ \text{failures}/1{,}000{,}000\ \text{h}$$

The net failure rate due to interconnections is

$$\Sigma\, N_I\lambda_I \,= 0.846 + 1.21 + 0.0536 + 0.00824$$

$$= 2.12\ \text{failures}/1{,}000{,}000\ \text{h}$$

Package. From Eq. (11.95), the failure rate of the package may be calculated.

$$S = 4 \qquad\qquad T = 70°C$$

$$\lambda_S = (0.11)(4.8)(1 - e^{4.8^2/50})(e^{-5203[1/(70\,+\,273)\,-\,1/298]})$$

$$= 1.93\ \text{failures}/1{,}000{,}000\ \text{h}$$

General. Since the circuit contains both analog and digital devices, the circuit function factor is $\pi_F = 1.25$. From Table 11.27, the environmental factor is $\pi_E(A_{IB}) = 2.5$. From Table 11.28, the quality factor is $\pi_Q = 1.0$. The density factor is calculated from the equations in Table 11.29 or read from the chart.

$$A_S = 1.3 \times 0.7 = 0.91\ \text{in}^2$$

$$\text{Density} = \frac{16 \times 18 \times 2 + 26 \times 2 + 4 \times 2}{0.91 + 0.10}$$

$$= 629.7$$

$$\pi_C = 0.2 + 0.15 \, (629.7)^{0.5}$$

$$= 3.96$$

Summary. The net failure rate of the hybrid circuit is

$$\lambda_H = 19.85 + (0.0033 + 2.12 + 1.93)(1.25)(1.0)(1.0)(3.96)$$

$$= 98.76 \text{ failures/1,000,000 h}$$

This equates to a mean circuit lifetime of

$$\text{MTBF} = \frac{1}{\lambda_H} = \frac{1,000,000}{98.76}$$

$$= 10,135 \text{ h}$$

$$= 1.16 \text{ years}$$

Table 11.39 is the worksheet used to record information for this example of reliability calculations.

11.13.6 Accelerated testing

In certain applications, such as medical and military, which require hybrid circuits with very high reliability, it is very desirable to remove those units that are prone to infant mortality prior to releasing the hybrid circuits to the field. This is usually accomplished by subjecting the circuits to some form of accelerated testing. Accelerated testing is designed to increase the failure rate of circuits by subjecting them to one or more extremes of temperature, humidity, bias, or mechanical stress. In this manner, units with inherent defects will fail earlier and may be removed from further processing.

Before accelerated testing can be effective, the acceleration factor must be known in order to ensure that the units are just far enough into the useful life region that all units prone to infant mortality have been removed without stressing the remaining units so hard that a considerable part of their useful life is also removed. It must also be assured that the stresses placed on the units are not so severe that failure mechanisms are introduced which would not be observed in normal use.

High temperature storage. High temperature storage is effective in removing those failures which are caused by chemical reaction. These failures include intermetallic formation, diffusion, and corrosion. It can be shown from Eq. (11.90) that the rate of a chemical reaction approximately doubles every time the temperature is increased by 10°C. By this rule of thumb, one hour at 125°C is equivalent to 2^{10} h, or 1024 h at 25°C. Many failures are further accelerated by the presence of bias or moisture. Elevated temperature in combination with an applied potential will accelerate inversion phenomena, particularly in the vicinity of device contacts. Temperature storage in the presence of moisture will accelerate corrosion, and, with the further addition of electric potential, electromigration.

Temperature cycling. Temperature cycling, or alternating the unit between hot and cold extremes at a predetermined rate, is an effective method for stressing wire bonds, solder joints, die bonds, and hermetic seals.[12] The alternating hot and cold temperatures act to flex the junctions, promoting the propagation of microcracks or voids which occur as a result of intermetallic compound formation, mismatched TCEs, improper wirebonding parameters, and similar phenomena. Temperature cycling under bias is becoming more prevalent, particularly in the automotive industry. At least two failure mechanisms are exhibited in this procedure which are reflective of actual performance conditions. As a circuit cools, any moisture present tends to condense on the surface of the hybrid. In ordinary temperature cycling with no bias, this would likely have no effect. In the presence of bias, however, the condi-

TABLE 11.39   Worksheet for Reliability Calculations

Package type	Plug-in
Seal length	4.8 in
Number of pins	26
Style	Seam welded
Hermetic?	Yes
Temperature	70°C
Substrate	Alumina
Size	1.3 × 0.7 in
Metallization	Thick film gold
Active devices	16
Digital (number)	10
Pinouts for each	18
Junction temperature	90°C
Analog (number)	6
Pinouts for each	18
Junction temperature	120°C
Learning factor	Well-established
Capacitors	4
Temperature rating	125°C
% voltage rating	90%
Interconnections	
Wirebonds	Gold wirebonds
Like metal	18 × 16 + 2 × 26 = 340
Junction temperature	180 @ 90°C, 52 @ 70°C, 108 @ 120°C
Unlike metal	18 × 16 = 288
Junction temperature	70°C
Solder or epoxy	8
Reliability level	Screened to MIL-STD-883, Method 5008, in accordance with App. G of MIL-M-38534B
Environment	Airborne inhabited bomber (A_{IB})

tions exist for metal migration. The most stressful condition which a die bond, particularly a power die bond, can be subjected to is to have power applied at cold temperature. The resulting heat generated creates larger temperature gradients and higher stresses between the die and the substrate than exhibited under any other conditions. From observation, a significant percentage of circuit failures occur at the point when power is applied. Typical temperature cycling schedules are 10 to 50 cycles from −55 to +150°C for use as a screen, and up to 1000 cycles at the same extremes to simulate end of life.

Temperature shock.   Temperature shock is similar to temperature cycling, except that the circuit is transferred abruptly from one temperature extreme to another. This may be performed air to air or liquid to liquid. The latter is a more severe shock, as heat is transferred much more efficiently in a liquid than in air.

Centrifuge.   The centrifuge test subjects the circuits to high g forces by spinning them rapidly about an axis. This process applies additional stress to bonds which have been subjected to previous temperature cycling.

Burn-in.   Burn-in simulates operational conditions by applying potential to a circuit under conditions of elevated temperature. This process may extend from several hours up to 168 h (1 week). End of life is considered to be 1000 h.

11.13.7   Derating factors

A common practice to improve the reliability of electronic circuits is to use devices at lower than the rated limits. MIL-HDBK-217 uses an acceleration factor defined by

$$A = e^{m(p_1 - p_0)} \tag{11.106}$$

where p_1 = percent of maximum rated electrical stress
p_0 = reference percent of rated stress (usually 25%)
m = component factor (see expressions in MIL-HDBK-217)

The term p refers to the stress variable of primary interest to the component in question. For capacitors, p is the applied voltage, and for thick film resistors, p may refer to either the applied voltage or dissipated power, depending on which parameter is of most significance in a given application.

11.13.8   Designing for reliability

While the acceleration tests described earlier are effective in preventing most potentially bad units from being placed in service, they are no substitute for good design practices, which prevent them from being manufactured in the first place. The hybrid design engineer must design high reliability into the circuits by using design practices which promote high performance. Some of these are described in Table 11.40. Table 11.41 lists symbols and descriptions of environments affecting design for reliability.

TABLE 11.40 Considerations in Designing for Reliability

Design parameter	Comments
TCE matching	Matching of TCEs at die bond interfaces, substrate to package interfaces, and other attachment points will minimize the number of interconnection failures by lowering the mechanical stress on the joint.
Use of derating curves	Using components at lower than the maximum ratings decreases the number of failures due to component breakdown.
Compatible metals	Certain metals in combination form intermetallic compounds which increase the electrical resistance and lower the mechanical strength of the bond.
Conformal coatings	Water in the liquid form is the greatest detriment to reliability. Even pure distilled water can cause corrosion by leaching phosphorus from the protective oxide of semiconductor devices, forming phosphoric acid which attacks the aluminum device metallization. Hybrid circuits designed for use in a moist environment should be protected by applying a compatible conformal coat which has a low mobile ion (LMI) content, by applying a suitable encapsulation, or by hermetically sealing the circuit in a metal or ceramic package.

TABLE 11.41 Symbols and Descriptions

Environment	π_E symbol	Description
Ground, benign	G_B	Nonmobile, laboratory environment readily accessible to maintenance; includes laboratory instruments and test equipment, medical electronic equipment, business and scientific computer complexes.
Ground, fixed	G_F	Conditions less than ideal such as installation in permanent racks with adequate cooling air and possible installation in unheated buildings. Includes permanent installation of air traffic control, radar, and communications facilities.
Ground, mobile	G_M	Equipment installed on wheeled or tracked vehicles; includes tactical missile ground support equipment, mobile communication equipment, and tactical fire direction systems.
Space flight	S_F	Earth orbital. Approaches benign ground conditions. Vehicle neither under powered flight nor in atmospheric reentry; includes satellites and shuttles.
Manpack	M_P	Portable electronic equipment being manually transported while in operation; includes portable field communications equipment and laser designators and rangefinders.
Naval, sheltered	N_S	Sheltered or belowdeck conditions, protected from weather; includes surface ship communication, computer, and sonar equipment.
Naval, unsheltered	N_U	Nonprotected surface shipborne equipment exposed to weather conditions; includes most mounted equipment and missile/projectile fire control equipment.
Naval, undersea unsheltered	N_{UU}	Equipment immersed in saltwater; includes sonar sensors and special-purpose antisubmarine warfare equipment.
Naval, submarine	N_{SB}	Equipment installed in submarines; includes navigation and launch control systems.

(Continued)

TABLE 11.41   **Symbols and Descriptions (*Continued*)**

Environment	π_E symbol	Description
Naval, hydrofoil	N_H	Equipment installed in a hydrofoil vessel.
Airborne, inhabited cargo	A_{IC}	Typical conditions in cargo compartments occupied by aircrew without environmental extremes of pressure, temperature, shock, and vibration, and installed on long-mission transport aircraft.
Airborne, inhabited trainer	A_{IT}	Same as A_{IC}, but installed on high performance aircraft, such as trainer aircraft.
Airborne, inhabited bomber	A_{IB}	Typical conditions in bomber compartments occupied by aircrew without environmental extremes of pressure, temperature, shock, and vibration and installed on long-mission bomber aircraft.
Airborne, inhabited attack	A_{IA}	Same as A_{IC}, but installed on high performance aircraft such as used for ground support
Airborne, inhabited, fighter	A_{IF}	Same as A_{IC}, but installed on high performance aircraft such as fighters and interceptors.
Airborne, uninhabited cargo	A_{UC}	Bomb bay, equipment bay, tail, or where extreme pressure, vibration, and temperature cycling may be aggravated by contamination from oil, hydraulic fluid, and engine exhaust. Installed on long-mission transport aircraft.
Airborne, uninhabited trainer	A_{UT}	Same as A_{UC}, but installed on high performance aircraft such as trainer aircraft.
Airborne, uninhabited bomber	A_{UB}	Bomb bay, equipment bay, tail, or where extreme pressure, vibration, and temperature cycling may be aggravated by contamination from oil, hydraulic fluid, and engine exhaust. Installed on long-mission bomber aircraft.
Airborne, uninhabited attack	A_{UA}	Same as A_{UC}, but installed on high performance aircraft such as used for ground support.
Airborne, uninhabited fighter	A_{UF}	Same as A_{UC}, but installed on high performance aircraft such as fighters and interceptors.
Airborne, rotary wing	A_{RW}	Equipment installed on helicopters; includes laser designators and fire control systems.
Missile, launch	M_L	Severe conditions related to missile launch (air and ground), and space vehicle boost into orbit, vehicle reentry, and landing by parachute. Conditions may also apply to rocket propulsion powered flight.
Cannon, launch	C_L	Extremely severe conditions related to cannon launching of 155-mm and 5-in guided projectiles. Conditions apply from launch to target impact.
Undersea, launch	U_{SL}	Conditions related to undersea torpedo mission and missile launch.
Missile, free flight	M_{FF}	Missiles in nonpowered free flight.
Airbreathing missile, flight	M_{FA}	Conditions related to powered flight of air breathing missile; includes cruise missiles.

11.14 Summary

The design of a hybrid microcircuit requires a set of diverse skills in order to accomplish the ultimate goal: a design which meets the technical, cost, and schedule requirements. Given the wide variety of decisions which must be made during the design process, it is virtually impossible to develop a suitable design by trial and error in a reasonable length of time. Instead, a strong analytical approach incorporating the scientific method must be part of the repertoire of skills of the hybrid design engineer.

Computer-aided design systems are an invaluable asset. Systems developed for use in the hybrid design process range from workstation-based systems as supplied by companies such as Intergraph, Mentor, and Harris, to PC-based systems supplied by companies such as ViewLogic. These systems all have the ability to perform circuit analysis, thermal analysis, and layout with a single schematic entry point, which dramatically lowers the design time and the probability of error. These systems can perform much of the design rule checking (DRC) as well. Most layout systems, however, have difficulty in handling such features as jumping traces with wirebonds, and require a degree of manual intervention by the operator.

References

1. Peggy A. Sergent and Jerry E. Sergent, "The Hybrid Microcircuit: From Design to Production," *Proc. ISHM Symp.,* 1981.
2. Elissa S. Klingman, Thomas Teel, and Jerry E. Sergent, "Design Methods for the Temperature Compensation of Thick Film Hybrid Microcircuits," *Proc. ISHM Symp.,* 1978.
3. Jerry E. Sergent and Robert Van Hooser, "Active Laser Trimming of Hybrid Circuits," *Proc. NEPCON West,* 1980.
4. Richard Brown, *Materials and Processes for Microwave Hybrids,* ISHM Monograph, 1991.
5. Jerry E. Sergent and George Lane, "A Design Method for Thick Film Resistors," *Electronics Packaging and Production,* October 1978.
6. Don Mattox, "Thin Film Technology," Sandia Labs technical report, 1972.
7. J. C. Bolger and C. Mooney, "Die attach in Hi-Rel P-Dips: Polyimides or Low-Chlorine Epoxies?," *Proc. IEEE Components Conf.* 1984.
8. C. G. M. Van Kessel, S. A. Gee, and J. J. Murphy, "The Quality of Die Attachment and Its Relationship to Stresses and Vertical Die-cracking," *Proc. IEEE Components Conf.* 1983.
9. Jerry E. Sergent and Gary Shawan, "Determination of the Current-Carrying Capability of Film Conductors", *J. Hybrid Microelectronics,* ISHM, September 1991.
10. Jerry E. Sergent, *The Hybrid Microelectronics Technology,* ISHM Monograph.
11. Dan Epstein, "Application and Use of Acceleration Factors in Microelectronics Testing," *Solid State Technology,* November 1982.
12. Erwin A. Herr and Alfred Poe, "Transistor High-Reliability Program," *Proc. Industry Application Soc.,* IEEE, pp. 833–839, 1977.

Index

Index

Page numbers followed by an *f* or *t* indicate the term is from a figure or table.

Active metal braze (AMB), **5**-2, **5**-21 to **5**-27
  adhesion mechanisms, **5**-21 to **5**-23
  alloys for, **5**-23*t*
  processes for, **5**-24 to **5**-27, **5**-25*f*
    blushing, **5**-26

Capacitors, **8**-1 to **8**-24
  ceramic, **8**-11 to **8**-22
    configuration of, **8**-19, **8**-22*f*
    dielectrics, classes of, **8**-13 to **8**-19, **8**-13*t*,
      8-14*t*, **8**-16*t*
    manufacturing process of, **8**-11 to **8**-13
    properties of, **8**-13 to **8**-19
    terminations, **8**-21
  dielectric constant of, **8**-2 to **8**-5, **8**-3*t*
  electrical properties of, **8**-9 to **8**-11
    aging, **8**-11, **8**-11*f*
    dielectric strength, **8**-11
    dissipation factor (DF), **8**-9 to **8**-10,
      8-22*f*
    frequency response of, **8**-6*t*
    insulation resistance (IR), **8**-10, **8**-17*f*,
      8-18*f*, **8**-19*f*, **8**-20*f*
    quality factor, **8**-11
    self-resonance, **8**-10
    temperature coefficient of capacitance
      (TCC), **8**-9, **8**-16*f*, **8**-18*f*, **8**-20*f*
    voltage coefficient of capacitance (VCC),
      8-10, **8**-15*f*, **8**-19*f* to **8**-20*f*
  MOS (metal-oxide-silicon), **8**-22 to **8**-23
  structure of, **8**-5 to **8**-9, **8**-7*f*
  tantalum, **8**-23 to **8**-24, **8**-24*f*, **8**-25*t*
Cleanroom, design of, **9**-4 to **9**-27
  air flow in, **9**-4 to **9**-12
  classification, **9**-4, **9**-5*f*, **9**-5*t*
  finish requirements, **9**-25 to **9**-27
  HEPA filters, **9**-6 to **9**-8
  local environments in, **9**-15, **9**-20
  return air ducts, **9**-20 to **9**-21
  types of, **9**-12 to **9**-15
  workstations, **9**-21 to **9**-25

Cleanroom, layout of, **9**-27 to **9**-44
  communication areas, **9**-38 to **9**-39
  electrostatic discharge (ESD) concerns,
    9-40 to **9**-44
  ergonomics, **9**-39 to **9**-40
  gownup areas, **9**-33 to **9**-38
  logistics of, **9**-27 to **9**-32
  passthrus, **9**-32 to **9**-33
Cleanroom, operation and management of,
  9-44 to **9**-60
  cleaning of, **9**-51 to **9**-58
  construction of, **9**-49 to **9**-51
  garments for, **9**-44 to **9**-49
  testing of, **9**-58 to **9**-60
  training of personnel, **9**-44 to **9**-45
Cofired ceramics, **1**-11 to **1**-12, **1**-11*f*
Component attachment, **6**-46 to **6**-52
  adhesive attach, **6**-46 to **6**-50
    application, **6**-48 to **6**-49
  metallurgical attach, **6**-50 to **6**-52
Conductors, thick film, **3**-1 to **3**-38
  adhesion mechanisms, **3**-19
  aluminum, **3**-14
  applications, **3**-1, **3**-2
  constituents, **3**-3, **3**-4*t*
  copper, **3**-12 to **3**-14
    tin, intermetallic formation with, **3**-14
  cost, **3**-5
  gold, gold alloys, **3**-6 to **3**-7
    bondability, **3**-7
    solderability, **3**-7
  inorganic binders, **3**-15 to **3**-19
    effect on bondability and solderability,
      3-18
    glass compositions, **3**-15 to **3**-17
    oxide additives, **3**-17 to **3**-19
  leach resistance, **3**-7, **3**-11
  melting point, **3**-4
  metallurgy, **3**-3, **3**-5, **3**-6*t*
  metals, base, **3**-4
  metals, noble, **3**-4, **3**-6
  nickel, **3**-14

Conductors, thick film (*Cont.*):
  organic vehicles, **3**-19
  polymeric, **3**-14 to **3**-15
  properties, desired, **3**-2
  properties, related to structure, **3**-34 to **3**-38
  properties, typical, **3**-7*t*
  refractory (W, Mo, and Mo–Mn), **3**-14
  rheology, **3**-19
  sheet resistivity, **3**-4, **3**-14*t*
  silver, silver alloys, **3**-8 to **3**-12
    migration, **3**-8
    multilayer applications, **3**-9
    palladium–silver alloy, **3**-9 to **3**-12
  sintering, **3**-19 to **3**-23, **3**-20*f*, **3**-21*t*, **3**-22*f*
  test methods, **3**-23, to **3**-34
    adhesion, **3**-27, **3**-29*f*, **3**-31*f*
    aging characteristics, **3**-32
    conductivity, **3**-24
    leach resistance, **3**-32 to **3**-33
    microstructure, **3**-33 to **3**-34
    printing characteristics, **3**-24
    solderability, wettability, **3**-25, **3**-26*f*
    test pattern, **3**-24
Contamination, effect on product, **9**-1 to **9**-4
  relative size of particles, **9**-2, **9**-2*t*, **9**-3*f*
Copper metallization, **5**-1*t*, **5**-3 to **5**-7, **5**-27 to **5**-30
  applications of, **5**-48 to **5**-49
  assembly methods for, **5**-30 to **5**-36
  comparisons between technologies, **5**-27 to **5**-30, **5**-31*t*
  electrical performance of, **5**-39 to **5**-41
    current handling, **5**-39 to **5**-40
    high frequencies, **5**-40 to **5**-41
  market for, **5**-3
  reliability of, **5**-41 to **5**-48
    adhesion, **5**-43 to **5**-46
    corrosion and migration, **5**-47 to **5**-48
    effect of TCE mismatch on, **5**-42 to **5**-43
    solderability, **5**-47
  substrates for, **5**-4 to **5**-7, **5**-5*t*
  thermal performance of, **5**-36, **5**-39
  (*See also* Active metal braze; Direct bond copper; Plated copper)

Decision support system (DSS), **6**-5 to **6**-10
  building of, **6**-5 to **6**-6
  decision models for, **6**-5
  DSS for assembly technology selection, **6**-9 to **6**-10
  model design process, **6**-6 to **6**-7
  support requirements for, **6**-7 to **6**-9
Design, hybrid microcircuit:
  assembly, materials and processes, **11**-72 to **11**-73, **11**-80*t*

Design, hybrid microcircuit (*Cont.*):
  conductor width, determination of, **11**-59 to **11**-63
  design guidelines, **11**-67 to **11**-72, **11**-68*f* to **11**-79*f*
  electrical considerations, **11**-25 to **11**-31
    capacitance, stray, **11**-31
    functional trimming, **11**-26 to **11**-29
    inductance, stray, **11**-29 to **11**-31
    wire bond, **11**-29
    temperature compensation, thick film, **11**-25 to **11**-26
  reliability considerations, **11**-74 to **11**-102
    accelerated testing, **11**-98 to **11**-100
      burn-in, **11**-100
      centrifuge, **11**-100
      temperature cycling, **11**-99 to **11**-100
      temperature shock, **11**-100
      temperature storage, **11**-98 to **11**-99
    derating factors, **11**-100
    design considerations, **11**-100 to **11**-101, **11**-101*t*
    failure mechanisms, **11**-74 to **11**-76
    failure rate prediction, **11**-76 to **11**-98
      activation energy, **11**-84*t*
      Arrhenius equation, **11**-81, **11**-82*f*
      MIL-HDBK-217, **11**-86 to **11**-98, **11**-99*t*
  sequence, **11**-3 to **11**-11
    circuit analysis, **11**-4 to **11**-5
    component selection, **11**-6 to **11**-7
    design reviews, **11**-4, **11**-9, **11**-11
    flow chart, **11**-4*f*
    layout, **11**-8 to **11**-9
    material selection, **11**-8
    partitioning, **11**-3 to **11**-5
    preproduction, **11**-10 to **11**-11
    prototype build, **11**-9 to **11**-10
    sizing, **11**-7
      weighting factors for, **11**-7*t*
    substrate technology, **11**-8
    thermal analysis, preliminary, **11**-7
  stages of, **11**-2
  substrate technology, selection of, **11**-64 to **11**-66
  team, **11**-2 to **11**-3
  thermal considerations, **11**-32 to **11**-59
    (*See also* Thermal analysis)
  thick film resistor design, **11**-11 to **11**-20
    (*See also* Resistor design, thick film)
  thin film resistor design, **11**-20 to **11**-25
    (*See also* Resistor design, thin film)
  wire bond size, determination of, **11**-63 to **11**-64, **11**-64*t*
Design rules for, **5**-14*t*

Dielectrics, thick film:
  capacitor dielectrics, **3**-39 to **3**-40, **3**-46 to
    3-48
    constituents, **3**-46, **3**-47
    densification of, **3**-50
    desirable properties of, **3**-39 to **3**-40
    properties of, **3**-55 to **3**-57
  composition of, **3**-40 to **3**-46, **3**-43*t*
    ceramic-filled glasses, **3**-40 to **3**-41, **3**-40*t*
    crystallizing dielectrics, **3**-41 to **3**-42
    high *K* dielectrics, **3**-47*t*
    insulator dielectrics, **3**-40
    low *K* dielectrics, **3**-42
    nitrogen-firable, **3**-42 to **3**-44
    substrates other than alumina, **3**-44 to
      3-46, **3**-46*t*
  crossover dielectrics, **3**-39
    (*See also* Insulator dielectrics)
  functions, **3**-38
  insulator dielectrics, **3**-48 to **3**-50
    crystallization of, **3**-50
    organic removal from, **3**-48
    properties of, **3**-55
    sintering of, **3**-49
  multilayer dielectrics, **3**-29
    (*See also* Insulator dielectrics)
  properties of, **3**-51 to **3**-64
    coefficient of thermal expansion (CTE),
      3-61 to **3**-63
    dielectric constant, **3**-51 to **3**-57
      measurement of, **3**-64
    dielectric loss, **3**-57 to **3**-59
      measurement of, **3**-64
    dielectric strength, **3**-60 to **3**-61
    effects of processing on, **3**-67 to **3**-69
      drying, **3**-68
      firing, **3**-69
      printing, **3**-67
    hermeticity, **3**-63 to **3**-64
    insulation resistance, **3**-59 to **3**-60
    measurement of, **3**-64 to **3**-67
      dielectric constant, **3**-64
      dielectric loss, **3**-64
      dielectric strength, **3**-64
      hermeticity, **3**-64 to **3**-65
      life test, **3**-65 to **3**-67
      warpage, **3**-64, **3**-65*f*
    stability in humid environment, **3**-63 to
      3-64
  sealing glasses, desirable properties of,
    3-38
Direct bond copper (DBC), **5**-2, **5**-7 to **5**-16
  alumina, bonding with, **5**-7 to **5**-12, **5**-8*f*
    copper-alumina eutectic, **5**-8 to **5**-9
    interface characteristics, **5**-9 to **5**-10
    process sequence, **5**-10 to **5**-12, **5**-9*f*

Direct bond copper (DBC) (*Cont.*):
  aluminum nitride, bonding with, **5**-15 to
    5-16
  etching of, **5**-12 to **5**-13, **5**-12*f*
    effects of thickness on, **5**-13
    etch factor, **5**-12
  multilayer structures, **5**-15

Factory automation, **6**-90 to **6**-99
Failure analysis:
  analytical approach to, **10**-10 to **10**-17
    flowchart for, **10**-13*f*, **10**-14*f*
  documentation of, **10**-17 to **10**-18
  goals of, **10**-1 to **10**-3
  hybrid microelectronics, **10**-5 to **10**-10
    die and component failures, **10**-9 to
      10-10
    metallization failures, **10**-8 to **10**-9
    package level failures, **10**-6 to **10**-7
    substrate level failures, **10**-7 to **10**-8
  personnel qualifications for, **10**-3 to
    10-5
  tools for, **10**-19 to **10**-34
    acoustic imaging, **10**-28
    bond pull, **10**-26 to **10**-27
    curve tracing, **10**-21
    die shear, **10**-27 to **10**-28
    electrical or parametric testing, **10**-19
    liquid crystal imaging, **10**-24
    list of, **10**-19
    optical microscopy, **10**-21 to **10**-22
    radiography, **10**-20 to **10**-21
    sample preparation, **10**-22 to **10**-24
    scanning electron microscopy (SEM),
      10-25 to **10**-26
    submicrometer probing, **10**-24
    x-ray microradiography, **10**-28
Flip chip bonding, **6**-73, **6**-78 to **6**-80

Hybrid microelectronics:
  advantages of, **1**-20, **1**-21*t*
  applications of, **1**-20 to **1**-28
  assembly of, **1**-2*t* to **1**-5*t*, **1**-16 to **1**-20
    epoxy, **1**-16
    eutectic, **1**-16
    flip chip, **1**-17 to **1**-18, **1**-18*f*
    solder, **1**-17
    tape automated bonding (TAB), **1**-18 to
      1-19, **1**-19*f*
    wire bonding, **1**-17
  definition of, **1**-1
  density of, **6**-8*t*
  package types for, **1**-19 to **1**-20
  rework and repair of, **6**-41 to **6**-46

Hybrid microelectronics (*Cont.*):
  substrates for, **1**-2
  thick film, **1**-2 to **1**-7
  thin film, **1**-7 to **1**-11, **1**-9*t*, **1**-10*t*
    delidding of metal packages, **6**-44
    epoxy bonded devices, **6**-42
    eutectically bonded devices, **6**-44
    fine-pitched devices, **6**-42
    flip chip devices, **6**-42 to **6**-43
Hybrid microelectronics, failure modes in,
      10-5 to **10**-10
  die and component failures, **10**-9 to **10**-10
  metallization failures, **10**-8 to **10**-9
  package level failures, **10**-6 to **10**-7, **10**-35*f*
  substrate level failures, **10**-7 to **10**-8

Inductors and transformers, **8**-25 to **8**-33
  hybrid applications, **8**-31 to **8**-32
  materials for, **8**-26 to **8**-31, **8**-27*f*, **8**-27*t*
    diamagnetic, **8**-27
    ferrimagnetic, **8**-27 to **8**-29
      Curie temperature of, **8**-29
    ferromagnetic, **8**-27
    paramagnetic, **8**-27
    soft ferrite, **8**-29 to **8**-31, **8**-28*f*, **8**-32*t*
      skin effect, **8**-29 to **8**-30
  permeability, definition of, **8**-25 to **8**-26
  properties of, **8**-31
  surface mount applications, **8**-31 to **8**-32,
      8-33*t*
Interconnect technologies, comparison of,
      6-77 to **6**-78, **6**-78*t*
Intermetallic formation, gold–aluminum,
      6-52 to **6**-62, **6**-56*f* to **6**-61*f*
  bond interface degradation due to, **6**-60 to
      6-62
  Kirkendall voids, **6**-54 to **6**-56
  purple plague, **6**-53
  time-temperature effects on, **6**-56 to **6**-60

Multichip modules MCM, **6**-2

Packaging and packages:
  applications of, **7**-42 to **7**-43
  electrical evaluation of, **7**-26 to **7**-34
    DC, **7**-26
    frequency domain, **7**-27
    low frequency (LF) and radio frequency
        (RF), **7**-31 to **7**-32
      conductor characterization, **7**-31
      dielectric characterization, **7**-31 to **7**-32
      wideband characterization, **7**-32 to
          7-34

Packaging and packages, electrical evalua-
    tion of (*Cont.*):
    packaging materials, **7**-27 to **7**-29
      DC, **7**-28
      RF, **7**-28 to **7**-29
    time domain, **7**-6*t*, **7**-26 to **7**-27
    very low frequency (VLF), **7**-29 to **7**-31
      conductivity of dielectrics, **7**-30 to **7**-31
      conductivity of metals, **7**-29 to **7**-30
  fabrication of, **7**-19 to **7**-23
    low temperature co-fired (LTCC), **7**-19 to
        7-20
    plastic package formation, **7**-21 to **7**-23
    thin film methods, **7**-20 to **7**-21
  interconnections, electrical properties of,
      7-2 to **7**-7
    crosstalk, **7**-3 to **7**-4
    design issues, **7**-5 to **7**-7
    modeling of, **7**-5
    propagation delay, **7**-2
    propagation velocity, **7**-2
    reflection coefficient, **7**-3
    simulation of, **7**-4 to **7**-5
      coplanar line, **7**-7
      microstrip, **7**-7
      striplines, **7**-5 to **7**-7
  mechanical evaluation of, **7**-34 to **7**-39
    homogeneity and chemical composition,
        7-34 to **7**-35
    machinability, **7**-39
    platability, **7**-38 to **7**-39
    porosity and density, **7**-35
    shrinkage, **7**-36
    solderability, **7**-36
    stress and strain, **7**-36 to **7**-38
  nonhermetic, **7**-25 to **7**-26
  requirements of, **7**-2
  sealing of, **7**-23 to **7**-26
    ceramic packages, **7**-25
    metal packages, **7**-23 to **7**-24
  thermal evaluation of, **7**-39 to **7**-42
    specific heat, **7**-40 to **7**-41
    thermal coefficient of expansion (TCE),
        7-39 to **7**-40
    thermal conductivity, **7**-41 to **7**-42
  types of, **7**-7 to **7**-18
    ceramic packages, **7**-17 to **7**-18
      alumina, **7**-18
      aluminum nitride, **7**-18
      beryllia, **7**-18
    chip carriers, leadless and leaded, **7**-9 to
        7-12, **7**-10*f*, **7**-11*f*
    dual-in-line (DIP), **7**-8, **7**-8*f*
    flat pack, **7**-13 to **7**-14, **7**-14*f*
    metal packages, **7**-14 to **7**-15, **7**-18
    pin grid array (PGA), **7**-7*f*, **7**-12 to **7**-13

Packaging and packages, types of (*Cont.*):
  plastic packages, **7**-15 to **7**-16
    thermoplastic, **7**-16
    thermosetting, **7**-15 to **7**-16
    small outline (SO), **7**-8 to **7**-9, **7**-9*f*
Plated copper, **5**-2, **5**-17 to **5**-21
  adhesion mechanisms, **5**-17 to **5**-19
  via metallization, **5**-20 to **5**-21
  processes for, **5**-17 to **5**-19, **5**-20*f*
Printed wiring boards, **1**-12 to **1**-15, **1**-12*t*,
    1-13*t*, **1**-14*f*, **1**-15*f*

Resistor design, thick film, **11**-11 to **11**-20
  aspect ratio, **11**-13
  blending of resistor paste, **11**-18 to **11**-19,
    11-19*f*
  design curves, **11**-14 to **11**-16, **11**-16*f*
  dimensional effects, **11**-13 to **11**-14
  effect of power density on, **11**-16 to **11**-17
  special resistor configurations, **11**-19 to
    11-20, **11**-20*f*
  thickness effects, **11**-13
Resistor design, thin film, **4**-23 to **4**-31, **11**-20
    to **11**-25
Resistors, discrete, **8**-32 to **8**-40
  conduction mechanisms, thick film, **8**-34 to
    8-37
  properties of, thick and thin film, **8**-37 to
    8-40
    high voltage discharge (ESD), **8**-39
    noise, **8**-37 to **8**-38
    stability, **8**-38 to **8**-40
    temperature coefficient of resistance
      (TCR), **8**-37
    voltage coefficient of resistance (VCR),
      8-38 to **8**-39
Resistors, thick film:
  conduction mechanisms, **3**-98 to **3**-99
  constituents, general, **3**-69 to **3**-71, **3**-72*f*
    cermet resistors, **3**-75*t*, **3**-76
    low ohmic value resistors, **3**-76, **3**-76*t*,
      3-78*t*
    mid-high range resistors, **3**-77 to **3**-80
    TCR modifiers, **3**-77 to **3**-79
  laser timming of, **1**-15 to **1**-16, **3**-99 to
    3-103
    drift due to, **3**-102 to **3**-103
    laser parameters, **3**-99 to **3**-101, **3**-101*f*
    types of cuts, **3**-101 to **3**-102
      double plunge cuts, **3**-102
      L-cuts, **3**-101 to **3**-102
      plunge cuts, **3**-101, **3**-102*t*
      serpentine cuts, **3**-102
  overglazing of, **3**-103
  properties of, **3**-85 to **3**-91, **3**-80*t*

Resistors, thick film, properties of (*Cont.*):
  effects of variables on, **3**-91 to **3**-98
    firing time at peak temperature, **3**-92
    particle size, **3**-92, **3**-94*f*
    peak firing temperature, **3**-91 to **3**-92
    termination effects, **3**-95 to **3**-98
  microstructure development, **3**-85
  noise, **3**-85 to **3**-86
    noise index, **3**-74
  sheet resistance, **3**-86 to **3**-87, **3**-87*f*
    sheet resistivity, **3**-71 to **3**-74
  surge properties, **3**-89
  temperature coefficient of resistance
    (TCR), **3**-87 to **3**-89
  thermal stability, **3**-89 to **3**-91
  voltage stability, **3**-89
resinates, **3**-83 to **3**-84
strain gauge resistors, **3**-82 to **3**-83, **3**-84*f*
  gain factor of, **3**-82
thermistors, **3**-80 to **3**-81, **3**-82*f*, **3**-83*f*
voltage coefficient of resistance (VCR),
  3-74
Rheology of thick film paste, **3**-103 to **3**-139
  classifications, **3**-106 to **3**-109, **3**-105*f*,
    3-108*f*
    Newtonian, **3**-106
    non-Newtonian, **3**-106 to **3**-109
      Bingham plastic, **3**-107
      shear thickening, **3**-107
      shear thinning flow, **3**-107
      thixotropy, **3**-107 to **3**-109, **3**-109*f*
  definition, **3**-103 to **3**-104
  formulation, effects on, **3**-119 to **3**-125,
    3-120*t*
    dispersion and other additives, **3**-123 to
      3-125, **3**-126*f*
    solid concentration, **3**-122*f*
    vehicle, **3**-123
  interparticle forces affecting, **3**-113 to
    3-118
  suspension structures of fluids, **3**-116 to
    3-119
  viscosity, **3**-104 to **3**-112
    definition of, **3**-104 to **3**-106
    measurement of, **3**-109 to **3**-113, **3**-110*t*
      Deborah number, **3**-112
      methods of, **3**-111 to **3**-113

Solder assembly, **6**-10 to **6**-16
  alloys for, **6**-15*t*
  failure modes of, **6**-11 to **6**-16, **6**-12*f*, **6**-13*f*,
    6-15*f*
  processes, **6**-23*t*, **6**-26 to **6**-28
Statistical process control (SPC), **6**-80 to **6**-90,
    6-84*f* to **6**-88*f*

Statistical process control (SPC) (*Cont.*):
  arachnoid tool for, **6**-87 to **6**-90, **6**-89*f*
  control charts, **6**-81 to **6**-82
  reasons for, **6**-82 to **6**-87
Substrates, ceramic, **2**-1 to **2**-34, **2**-3*t*, **2**-4*t*
  chemical properties of, **2**-11 to **2**-12
  cleaning of, **2**-25 to **2**-26
  composition of, **2**-17 to **2**-20, **2**-20*t*
    alumina, **2**-18 to **2**-19
    aluminum nitride, **2**-19, **2**-20*t*
    beryllia, **2**-19 to **2**-20
  cost of, **2**-33
  electrical properties of, **2**-12 to **2**-15
    dielectric constant, **2**-12, **2**-13*f*
    dielectric strength, **2**-14, **2**-15*f*, **2**-15*t*
    dissipation factor, **2**-12
    permeability, **2**-14 to **2**-15
    volume resistivity, **2**-12, **2**-13*f*
  manufacturing methods of, **2**-20 to **2**-24,
      2-25*t*
    aluminum nitride, **2**-24
    extrusion, **2**-23
    isostatic powder pressing, **2**-22 to **2**-23
    powder pressing, **2**-22
    roll compaction, **2**-20 to **2**-21, **2**-21*f*
    sintering, **2**-23 to **2**-24
    tape casting, **2**-21 to **2**-22, **2**-22*f*
  mechanical properties of, **2**-2 to **2**-8
    compression strength, **2**-4 to **2**-5
    dimensional stability, **2**-5
    tensile strength, **2**-5
    thermal coefficient of expansion (TCE),
        2-5, **2**-6*f*
    thermal shock, failures due to, **2**-6 to **2**-8
    Young's modulus, **2**-3 to **2**-4
      coefficient of endurance, **2**-7, **2**-7*t*
  physical properties of, **2**-8 to **2**-11
    camber (warpage), **2**-10 to **2**-11
    specific gravity, **2**-11
    surface finish, **2**-8 to **2**-10
      measurement of, **2**-8 to **2**-10, **2**-9*f*,
          2-10*f*
    water absorption, **2**-11
  postfire processing of, **2**-26 to **2**-32
    annealing of, **2**-31
    laser drilling, machining, and scribing,
        2-30 to **2**-31
    laser drilling, with boring head, **2**-29 to
        2-30
    laser processing, **2**-27 to **2**-29
    ultrasonic impact grinding (UIG), **2**-32
  sources of, **2**-32 to **2**-33
  thermal properties of, **2**-15 to **2**-16
    specific heat, **2**-16
    thermal conductivity, **2**-15 to **2**-16, **2**-16*f*,
        2-17*f*, **2**-17*t*, **2**-18*f*

Surface mount technology (SMT), **6**-2, **6**-17 to
    6-29
  assembly equipment for, **6**-17 to **6**-29
    component placement, **6**-17 to **6**-21
    reflow solder, **6**-21 to **6**-28
  cleaning of, **6**-29 to **6**-40
    aqueous cleaning, **6**-38 to **6**-39
    equipment for, **6**-39 to **6**-41
    measurement of, **6**-36
    Montreal protocol, effect of, **6**-36 to **6**-37,
        6-38*f*
    solvents for, **6**-31 to **6**-35, **6**-32*t*, **6**-34*t*,
        6-35*t*
      chlorinated, **6**-33
      chlorofluorocarbons (CFCs), **6**-33 to
          6-34
  rework and repair of, **6**-41 to **6**-46

Tape automated bonding, **6**-73 to **6**-76
Thermal analysis:
  conduction, **11**-33
  convection, **11**-33
  cooling, convection, **11**-45 to **11**-49
  heat capacity, **11**-35 to **11**-36, **11**-37*f*, **11**-38*t*
  radiation, **11**-33
  temperature coefficient of expansion
      (TCE), **11**-36
  thermal capacitance, **11**-43 to **11**-45
  thermal conductivity, **11**-34 to **11**-35,
      11-36*f*, **11**-38*t*
  thermal management, **11**-49 to **11**-50
  thermal measurement, **11**-50 to **11**-51
  thermal resistance, **11**-37 to **11**-43
  thermal stress analysis, **11**-51 to **11**-59,
      11-52*t*
Thick film, general description of, **1**-2 to **1**-7
  (*See also* Conductors, thick film; Dielec-
      trics, thick film, Resistors, thick film;
      Rheology, thick film)
Thick film, nonhybrid applications of:
  flat panel displays, **3**-141 to **3**-142, **3**-142*f*
  solar cells, **3**-141, **3**-142*f*
Thick film paste, quality control of, **3**-137 to
    3-141
  paste production, **3**-139
  raw material, **3**-137 to **3**-139
Thick film processes, **3**-125 to **3**-141
  drying, **3**-140
  firing, **3**-140 to **3**-141
  screen printing, **3**-125 to **3**-137, **3**-140,
      3-129*f*, **3**-135*f*
    leveling, **3**-131 to **3**-132
    performance examples, **3**-134 to **3**-137
    resolution, vias and lines, **3**-132 to **3**-134
    rheology, effects of, **3**-125 to **3**-130

Thick film processes, screen printing (*Cont.*):
  wetting, **3**-130 to **3**-131
Thin film:
  capacitor materials, **4**-15 to **4**-18, **4**-19*t*
    manganese oxide–tantalum oxide, **4**-17
    silicon monoxide, **4**-16 to **4**-17
    silicon monoxide–tantalum oxide, **4**-17 to
      4-18
    structure of, **4**-15, **4**-16*f*
    tantalum oxide, **4**-17
  characterization of, **4**-34 to **4**-37
    adhesion, **4**-35
    chemical methods, **4**-36 to **4**-37
    film thickness, **4**-34 to **4**-35
    structure, **4**-35 to **4**-36
  conductors, **4**-10 to **4**-11, **4**-11*t*
  definition of, **4**-1
  deposition methods, **4**-1 to **4**-8
    anodization, **4**-7
    electroplating, **4**-7
    evaporation, **4**-2 to **4**-5, **4**-3*f*, **4**-4*f*
    laser ablation, **4**-7
    sol-gel coatings, **4**-8
    sputtering, **4**-5 to **4**-6, **4**-5*f*
  design guidelines for, **4**-23 to **4**-31
    capacitors, **4**-24, **4**-28 to **4**-30
    inductors, **4**-30
    resistors, **4**-23, **4**-24 to **4**-28, **4**-25*f*, **4**-26*f*,
      4-27*t*
  diamond films, **4**-20 to **4**-22
  fabrication process, **4**-31 to **4**-34
    pattern generation and etching, **4**-33
    photolithography, **4**-31 to **4**-32
    protection of, **4**-34
    resistor trimming methods, **4**-32 to **4**-34
    stabilization, **4**-34

Thin film (*Cont.*):
  inductor materials, **4**-18
  modifications of, **4**-37
  multichip modules (MCM-D), applications
    in, **4**-37 to **4**-41
    definition of, **4**-37 to **4**-38
    dielectrics for, **4**-30 to **4**-41
    metallization for, **4**-41
    process sequence, **4**-38
    substrates for, **4**-38 to **4**-39
  optical materials, **4**-22 to **4**-23
  resistors, **4**-11 to **4**-15, **4**-15*t*
    cermet films, **4**-15
    chromium films, **4**-14
    nickel—chromium films (nichrome), **4**-13
      to **4**-14
    ruthenium films, **4**-14 to **4**-15
    tantalum films, **4**-12 to **4**-13
    tantalum nitride, **4**-13
    tantalum-oxynitride films, **4**-13
    TCR of, **4**-11
  substrates for, **4**-8 to **4**-10
  superconducting materials, **4**-18 to **4**-20,
    4-20*t*

Wire bonding, **6**-62 to **6**-73
  automated, **6**-70 to **6**-73
  ball bonding, **6**-63 to **6**-65, **6**-64*f*
  contamination of surfaces, **6**-62 to **6**-63
  intermetallic growth in, **6**-52 to **6**-62
  thermocompression bonding, **6**-65 to
    6-66
  thermosonic bonding, **6**-68 to **6**-69
  ultrasonic bonding, **6**-66 to **6**-67
  wedge bonding, **6**-63 to **6**-65

ABOUT THE EDITORS

JERRY E. SERGENT is president and CEO of BBS PowerMod, where he oversees the design, development, engineering, and marketing of a variety of electronic products. He is currently Chairman of the Publications Committee for the International Society for Hybrid Microelectronics (ISHM), a past president of this Society, and a recipient of the Daniel C. Hughes Award.

CHARLES A. HARPER is president of Technology Seminars, Inc. He is the coeditor of *Electronic Materials and Processes Handbook* and the author or editor of numerous other technical books.